Anaerobes in Human Disease

Anaerobes in Human Disease

Edited by

Brian I. Duerden

Professor of Medical Microbiology, University of Wales
College of Medicine, Cardiff

and

B. S. Drasar

Professor of Bacteriology, London School of Hygiene and
Tropical Medicine, London

WILEY-LISS
A JOHN WILEY & SONS, INC., PUBLICATION
New York • Chichester • Brisbane • Toronto • Singapore

© 1991 Brian I Duerden and B S Drasar

First published in Great Britain 1991

Published in the United States of America 1991 by Wiley-Liss, a division of John Wiley & Sons, Inc., 41 East 11th Street, New York, NY 10003-4672.

Library of Congress Cataloging-in-Publication Data
is also available

ISBN 0-471-56091-X

All rights reserved. No part of this publication may be reproduced or transmitted in any form or by any means, electronically or mechanically, including photocopying, recording or any information storage or retrieval system, without either prior permission in writing from the publisher or a licence permitting restricted copying.

Printed in Great Britain.

Contents

Preface		vii
Contributors		ix
1	Introduction *B S Drasar and B I Duerden*	1
2	Classification of anaerobes *B S Drasar*	4
3	Characteristics of anaerobic metabolism *J G Morris*	16
4	Genetics of anaerobes *N P Minton and D E Thompson*	38
5	Bacteroides and Fusobacterium: classification and relationships to other bacteria *H N Shah and S E Gharbia*	62
6	The clostridia *G Hobbs*	85
7	Anaerobic cocci *B Watt*	100
8	The Spirochaetes *M J Hudson*	108
9	Actinomyces and Arachnia *G H Bowden*	133
10	Bifidobacteria *A K Roberts*	151
11	Anaerobes in the normal flora of man *B S Drasar and B I Duerden*	162
12	Laboratory diagnosis of anaerobic infection *M W D Wren*	180
13	Abdominal sepsis *A Trevor Willis*	197
14	Anaerobes in genitourinary infections *B I Duerden*	224
15	Dental and oral infection *J M Hardie*	245
16	Respiratory, ENT and CNS infections *H R Ingham and P R Sisson*	268
17	Superficial ulceration and necrosis *B Adriaans, B S Drasar and B I Duerden*	287
18	Gas gangrene and clostridial cellulitis *A Trevor Willis*	299
19	Anaerobic bacteraemia *R C Spencer*	324
20	Anaerobes in intestinal homeostasis *S P Borriello and F E Barclay*	343
21	Anaerobic bacteria in diarrhoea and colitis *S P Borriello*	351
22	*Clostridium perfringens* type A food poisoning *P R Berry and R J Gilbert*	362
23	Systemic toxigenic diseases (tetanus, botulism) *J G Collee and S van Heyningen*	372
24	Metabolic and carcinogenic effects of anaerobes *M J Hill*	395
25	Antibiotics and anaerobes *D Greenwood, B Watt and B I Duerden*	415
Index		**431**

Preface

This book sets infection and other pathological activities due to anaerobes in the context of our knowledge of the biology of anaerobiosis and the fundamental aspects of the physiology, biochemistry and genetics of anaerobic bacteria. The modern history of anaerobic microbiology as a science practised by man began with the father figures of microbiology—Antonie van Leeuwenhoek and Louis Pasteur—but as a biological phenomenon it pre-dates history. Anaerobes are primary life forms and the earliest forms of life on earth were the anaerobic microflora of the pre-oxygen era. Nevertheless, in medical microbiology the study of anaerobes remained a cinderella subject until the technological advances in anaerobiosis of the 1960s led to a great expansion of interest in the medical importance of anaerobes in the 1970s and 1980s.

The expansion of interest in anaerobic bacteria has resulted in an extensive literature—articles, reviews, supplements and books, most of which focus on particular aspects of the subject, eg, biochemistry, the normal flora, clinical infections. In this book we have attempted to provide an integrated view of all these varied aspects of the world of anaerobiosis as they relate to the maintenance of health and to the wide range of human disease to which anaerobic bacteria contribute. We hope that it relates the work of scientists working to reveal the mysteries of anaerobic metabolism to the problems of anaerobic infection as seen by medical microbiologists and clinicians, and that through such integrated understanding, scientists, doctors *and their patients* will benefit. However, although concentrating on anaerobes, we do not ignore or eschew the more recently evolved facultative and aerobic organisms; anaerobes cannot be separated from the rest of the microbial world with which they form an interactive whole, and moreover, an understanding of the ancient anaerobic microbial forms contributes enormously to the greater understanding of the aerobic and facultative organisms that evolved from them.

This book has been brought together through the varied interests and expertise of members of the Society of Anaerobic Microbiology (formerly the Anaerobe Discussion Group). The ADG was founded in 1975 when 16 people working in various aspects of anaerobic microbiology met at the Central Public Health Laboratory, London, to exchange information and experience. They included people interested in human infection, gut metabolism, dental and veterinary bacteriology, and toxicology. The aims of the group were to raise the general awareness of anaerobic bacteria in health and disease and to improve technical standards of anaerobic microbiology. Through three or four half-day meetings each year and a major international symposium held at Churchill College, Cambridge every second year since 1979, the ADG continued to further these aims. Membership expanded over the years to some 250 and in 1990 its widening scientific activities and its high reputation made a change of name to the Society for Anaerobic Microbiology a welcome inevitability. The contributors to the book have all been active members of the Society or have contributed to its meetings, and the style and content of the book reflect the aims and interests of the Society. We are grateful for the support of the contributors and of the Committee of the ADG/SAM in the preparation of this book, and for the efforts of our publishers for bringing it into print. Finally, we are grateful to our wives for their help and support during this extended period of anaerobiosis.

BRIAN I. DUERDEN,
B. S. DRASAR.
1990

Contributors

B Adriaans MB, ChB, MRCP
London School of Hygiene and
 Tropical Medicine
Keppel Street
London
UK

F E Barclay FIMLS
Clinical Research Centre
Watford Road
Harrow
Middlesex
UK

P R Berry BSc, PhD
Central Public Health Laboratory
61 Colindale Avenue
London
UK

S P Borriello BSc, PhD, MRCPath
Clinical Research Centre
Watford Road
Harrow
Middlesex
UK

G H Bowden PhD
Department of Oral Biology
Faculty of Dentistry
University of Manitoba
Canada

J G Collee MD, FRCPath, FRCP(Edin)
Department of Bacteriology
University of Edinburgh Medical
 School
Teviot Place
Edinburgh
UK

B S Drasar PhD, DSc, MRCPath
London School of Hygiene and
 Tropical Medicine
Keppel Street
London
UK

B I Duerden BSc, MD, FRCPath
Department of Medical Microbiology
University of Wales College of
 Medicine
Cardiff
UK

S E Gharbia BSc, PhD
Department of Oral Microbiology
The Royal London Hospital Medical
 College
Turner Street
London
UK

R J Gilbert M.Pharm, PhD, Dip.Bact, FRCPath
Central Public Health Laboratory
61 Colindale Avenue
London
UK

D Greenwood Dsc, MRCPath
Department of Microbiology
Queen's Medical Centre
University of Nottingham
Nottingham
UK

J M Hardie PhD, Dip.Bact, MRCPath
Department of Oral Microbiology
The Royal London Hospital Medical
 College
Turner Street
London
UK

S van Heyningen MA, PhD, FRSC
Department of Biochemistry
University of Edinburgh Medical
  School
Teviot Place
Edinburgh
UK

M J Hill DSc, FRCPath
PHLS Centre for Applied
  Microbiology and Research
Porton Down
Salisbury
Wiltshire
UK

G Hobbs BSc, PhD, FIFST, FRSH
Torry Research Station
PO Box 31
135 Abbey Road
Aberdeen
Scotland

H N Shah BSc, PhD
The Royal London Hospital Medical
  College
Turner Street
London
UK

M J Hudson BSc
PHLS Centre for Applied
  Microbiology and Research
Porton Down
Salisbury
Wiltshire
UK

H R Ingham MB, ChB, Dip.Bact, FRCPath
Newcastle General Hospital
Westgate Road
Newcastle upon Tyne
UK

N P Minton BSc, PhD
PHLS Centre for Applied
  Microbiology and Research
Porton Down
Salisbury
Wiltshire
UK

J G Morris FRS
Department of Botany and Microbiology
School of Biological Sciences
The University College of Wales
Aberystwyth
UK

A K Roberts BSc(Tech), PhD
PHLS Centre for Applied
  Microbiology and Research
Porton Down
Salisbury
Wiltshire
UK

P R Sisson BSc
Department of Microbiology
Newcastle General Hospital
Westgate Road
Newcastle upon Tyne
UK

R C Spencer MB, BS, MSc, FRCPath
Royal Hallamshire Hospital
Glossop Road
Sheffield
UK

D E Thompson BSc, PhD
AFRC
Institute of Food Research
Reading
Berkshire
UK

B Watt MD, FRCPath, C.Biol, FI.Biol
City Hospital
Greenbank Drive
Edinburgh
UK

A T Willis PhD, DSc, MD, FRCPa, FRACP, FRCPath
Public Health Laboratory
Luton
Bedfordshire
UK

M W D Wren FIMLS
University College Hospital
Clinical Microbiology
Grafton Way
London
UK

1

Introduction

B S Drasar and B I Duerden

Anaerobic bacteria may be described as those bacteria that will grow in conditions of low redox potential (Eh) and in the absence of significant amounts of free oxygen. Generally, they have a fermentative type of energy-producing metabolism in which organic molecules, or some inorganic molecules (but not oxygen) are the final electron acceptors. Their vegetative cell forms die when exposed to the free molecular oxygen in the normal atmosphere, although there is wide variation between species in the degree of oxygen susceptibility or tolerance, but this does not apply to the metabolically inert spores of the clostridia.

The major physiochemical factor in the development of the advanced and highly varied life forms on earth has been the presence of an atmosphere containing about 20 per cent of free molecular oxygen. This has allowed the development of highly efficient forms of aerobic metabolism and the evolution of the higher forms of life. It might be thought that this environment would be inimical to the survival of anaerobes, or at least would curtail their influence on an aerobic biological system, but anaerobic bacteria and their close relatives have probably been the dominant life forms on this planet for much of the time since life developed. Anaerobic bacteria outnumber all other forms of aerobic life by several logarithmic factors. Indeed, evolution could readily be described in terms of a steady progression of anaerobic bacteria towards achieving more comfortable ecological niches. The warm-blooded 'advanced' mammals succeeded the cold-blooded 'ancient' dinosaurs, but both groups of animals, which contain the largest forms of animal life that have evolved, depend upon the cellulolytic and degradative activity of anaerobic bacteria for the acquisition of their nutrients and for the recycling of organic matter, without which life would cease.

Man, in common with all other animals, birds, fish and even insects, plays host to a vast number of anaerobic bacteria that make up the predominant normal microbial flora of the mucosal surfaces and hollow viscera of the gastrointestinal, genitourinary and respiratory tracts. These commensal bacteria, like most of the aerobic and facultative species with which they are associated, are beneficial to their host, or are at least harmless, and they are essential to the normal function of the body. However, the anaerobic group also includes a number of important pathogens responsible for serious diseases in man and other animals. These pathogens include *Clostridium botulinum* and *C. tetani*, whose toxins are the most potent poisons known, and a variety of other clostridia and non-sporing anaerobes that can cause severe morbidity in necrotizing and gangrenous infections, with significant mortality.

Despite their dominance in nature, anaerobic bacteria were neglected for much of the twentieth century. They were the 'poor relations', particularly in *medical* microbiology during the decades when bacteriologists concentrated on aerobic pathogens such as staphylococci and streptococci, neisseriae and the enterobacteria. This situation changed quite dramatically in the 1960s and 1970s. Advances in anaerobic technology reawakened interest in the neglected non-clostridial anaerobes and resulted in a series of wide ranging and major investigations of the many roles played by anaerobic bacteria in health and disease. This was justifiably termed the 'anaerobic renaissance', and answers were provided in terms of diagnosis and specific therapy for many previously undiagnosed 'culture-negative' infections. However, we must not presume that all the contributions to the 'decade of anaerobes' represented new scientific knowledge. The history of anaerobic microbiology is as old as the history of microbiology itself, but much of the knowledge of anaerobes established by workers in the nineteenth century was either forgotten or never reached general medical microbiological perception, and anaerobic microbiology was given a very low priority in many laboratories.

The existence of anaerobes has been known since the earliest years of bacteriology. Antonie van Leeuwenhoek, in his 39th letter to the Royal Society dated 17 September 1683, described bacteria from the scrapings of his teeth that must have included anaerobic spirochaetes. The first description of the cultivation of a recognizable anaerobe was written by Pasteur and Joubert (1877) and their description of *C. septicum* (*Vibrion septique*) is still quoted. Indeed, Pasteur, the founding father of modern microbiology, is credited with the first use of the term 'anaerobe' for such organisms. The description of *Sarcina ventriculi* (Goodsir, 1842) is the earliest recognizable description of a bacterial species and, although it was not cultivated at that time, it is also an anaerobe.

During the final years of the nineteenth century and the early years of the twentieth, many species of anaerobic bacteria, both sporing and non-sporing, were isolated and their role in infection was recognized. Three major clostridial pathogens were known before 1900—*C. tetani*, described by Nicolaier in 1884 and isolated by Kitasato in 1889, *C. perfringens*, described by Welch and Nuttall in 1892, and *C. botulinum*, isolated by van Ermengen (1896) from a ham that had caused botulinum poisoning. Amongst the non-sporing anaerobes, Miller described fusiform bacteria in patients with ulcerative stomatitis in 1883 and Loeffler in 1884 first associated '*Bacillus necrophorus*' (now *Fusobacterium necrophorum*) with calf diphtheria, a form of necrobacillosis. See Leiner 1907. Plaut (1894) and Vincent (1896) independently described the association of a fusiform organism and a spirochaete with necrotic, ulcerative lesions of the mouth and throat that resembled diphtheria. Vincent also reported finding similar bacteria in gangrenous lesions and Veillon and Zuber (1897, 1898) isolated non-sporing anaerobic bacilli, that are now clearly recognizable as *Bacteroides* spp., from suppurative appendicitis, pyogenic arthritis and abscesses of the lung, brain and pelvis. Subsequently, the group of workers at the Institut Pasteur that included Veillon, Guillemot and Rist showed that anaerobic bacteria played a significant role in septicaemia, empyema, otitis media and genital infections, including puerperal infections. Thus, by the outbreak of World War I the role of clostridia in disease was well established and many other anaerobes, including fusobacteria, anaerobic cocci and bacteroides, had been recognized and associated with clinical infections. The period of discovery culminated during World War I with the dramatic results of the use of tetanus antitoxin for the treatment and prevention of tetanus. Subsequently, although a few workers kept interest alive, anaerobic bacteriology dropped from the mainstream of scientific activity until the revival of the past two decades.

It is now widely recognized that anaerobes are of major importance both as objects for biological investigation and as the cause of much morbidity and mortality. Their primitive origins in a world devoid of free oxygen and their unique evolutionary diversity means that their taxonomy, and most particularly the chemotaxonomy based upon biochemical analyses and the technology of molecular biology, can illuminate our theories about evolution. Some anaerobes may represent direct lineages from primeval forms, whereas others represent more recent convergent evolution.

Anaerobes are a vital area of study for microbiologists with a wide range of fields of interest: in the energy industry—the benefits of methanogenesis and the problems of oil pipeline corrosion; in agriculture—silage production and the role of rumen anaerobes in the conversion of vegetation into animal protein; in the food industry—the problems of spoilage and toxin production contrast with the use of microbial fermentation in food production; in veterinary medicine—the cause of major economic loss through anaerobic infections in farm animals. In the present context this book is particularly concerned with anaerobes as causes of disease in man. However, it is important that the basic biology of the organisms is understood if we are to understand the pathogenesis and the natural history of the diseases. We have, therefore, striven to set knowledge of disease in the context of the physiology, genetics, ecology and taxonomy of the bacteria. We have emphasized their contribution to the normal microbial flora of man and have contrasted their roles in disease with their importance in the microecological ecosystems associated with the human body.

With these concepts in mind, initial chapters devoted to the basic physiology and genetics of anaerobes are followed by individual considerations of the major groups of anaerobic bacteria associated with man. These are linked by chapters on the normal anaerobic flora of the human body and laboratory approaches to diagnosis to a series of chapters on various groups of anaerobic infections. The importance of synergic ecosystems and the possible effects of the metabolic activities of anaerobes in non-infective disease are included in sections devoted to colonization resistance and carcinogenesis. This discussion of anaerobic infections concludes with a chapter on the activity of antibiotics against anaerobes and the treatment of anaerobic infections.

We do not intend to denigrate the importance of aerobic life; higher organisms, including man, depend upon it and many important infective diseases are caused by aerobic or facultative bacteria. However, it could be argued that, because of their key role in vital biological activities, the whole complex pattern of life depends upon the activity of the anaerobic bacteria.

References

van Ermengen E. (1896). Untersuchungen über Fälle von Fleischvergiftung mit Symptomen von Botulismus. *Centralblatt für Bakteriologie Parasitenkunde und Infektionskrankheiten I. Abt.* **19**: 442–444.

Goodsir J. (1842). History of a case in which a fluid periodically ejected from the stomach contained vegetable organisms of an undescribed form. *Edinburgh Medical School and Surgical Journal* **57**: 430–443.

Kitasato S. (1889). Ueber den Rauschbrandbacillus und sein Culturverfahren. *Zeitschrift für Hygiene* **6**: 105–116.

Leiner, K. (1907). Ueber anaerobe backterien bei diphterie. *Zentralblatt für Bakteriologie, Parasitenkunde und Infektionskrankheiten* 1 Abt. **29**: 4–22.

Nicolaier A. (1884). Ueber infectiosen Tetanus. *Deutsche Medizinische Wochenschrift* **10**: 842–844.

Pasteur L, Joubert M. (1877). Etude sur la maladie charbonneuse. *Comptes Rendus Hebdomadaires des Seances de l'Academie des Sciences* **85**: 900–906.

Plaut H C. (1894). Studien zur bakteriellen Diagnostik der Diphtherie und Angienen. *Deutsche Medizinische Wochenschrift* **20**: 920–923.

Veillon A, Zuber A. (1897). Sur quelques microbes strictement anaerobies et leur role dans la pathologie humaine. *Comptes Rendus des Seances de la Societe de Biologie et de ses filiales* **49**: 253–255.

Veillon A, Zuber A. (1898). Recherches sur quelques microbes strictement anaerobies et leur role en pathologie. *Archives de Medecin Experimentale, Anatomie et Pathologie* **10**: 517–545.

Vincent H. (1896). Sur l'etiologie et sur le lesions anatomo-pathologique de la pourriture d'hopital. *Annales de l'Institut Pasteur* **10**: 488–510.

Welch W H, Nuttall G H F. (1892). A gas-producing bacillus (*Bacillus aerogenes capsulatum* Nov. Spec.) capable of rapid development in the body after death. *Johns Hopkins Hospital Bulletin* **3**: 81–91.

2
Classification of anaerobes

B S Drasar

Principles of classification
Classification of anaerobic bacteria
Culture of anaerobic bacteria
   Anaerobic jar
   Prereduced anaerobically sterilized media (PRAS)
   Anaerobic chamber
Identification of medically important anaerobes
   Anaerobic habit
   Gram's stain
   Cell morphology
   Spore formation
References

Principles of classification

The development of systems for the classification and identification of bacteria has long been a major activity of bacteriologists. However, with a few honourable exceptions most bacteriologists have concentrated their efforts on aerobic and facultative organisms. The classification of anaerobic bacteria seems to have been regarded as an adjunct to the classification of other bacteria. There may have been good empirical reasons for this in the past, but these can no longer be maintained and, both on theoretical and practical grounds, the taxonomy of anaerobic bacteria is among the central activities of major significance in biology. Indeed 'the anaerobic bacteria may be, in one way or another the lineal descendants of the first forms of life to appear on the earth' (Smith, 1967).

Taxonomy may be regarded as an amalgam of separate but related activities:
classification—the arrangement of units in categories (taxa) and the definition of these categories;
nomenclature—the naming of the categories;
identification—the assignment of units to previously described categories.
None of these activities is well developed with respect to anaerobic bacteria and the dependence of anaerobic groupings upon concepts derived from the study of aerobic and facultative organisms is a particular impediment to, and constraint upon, the development of an anaerobe classification system and the allocation of names to anaerobes.

Identification of bacterial strains isolated from

infections is of major concern to medical microbiologists. The systems available for the identification of the major groups of anaerobes will be described later, in the present context we are concerned with natural classifications that reflect the biological properties of the organism. Such approaches to classification include:

phenetic classification based upon the phenotype of organisms;

patristic classification reflecting genetic relationships or similarity;

cladistic classification based upon evolutionary consideration.

It must be remembered that while phenetic and patristic classifications result directly from analysis of observations and experimental data, cladistic classifications depend upon evolutionary theories.

Historically, bacterial classification has been based on a study of phenotypic characters, but advances in molecular biology, e.g., determination of DNA base-pair sequence similarity (DNA homology) and G + C content, have made the study of genetic relationships possible and have, perhaps, cast light upon evolutionary linkages.

Not all bacteriologists believe in the possibility of cladistic classifications. However, examination of the concept of the evolutionary clock has established the value of some biologically important macromolecules, in particular ribosomal RNA, as molecular chronometers (Woese, 1987). These studies are based upon the demonstration that some essential molecular structures are highly conserved as organisms differentiate and develop different phenotypic characters. The studies are not yet complete, but so far the data collected suggest that the groupings delineated by such analyses do not correspond to the currently accepted groupings (Table 2.1). As is immediately apparent, not all important groups of anaerobes have been examined and this system must therefore be regarded as a provisional sketch, but in the long term its importance cannot be exaggerated. In this scheme, methanobacteria and other archaebacteria are placed in a separate kingdom. Indeed, the number of kingdoms needed to accommodate all living organisms in this system is a major problem (Last, 1988).

Bacteriologists accept the need for a single kingdom of *Procaryotae* to separate bacteria and their relatives from plants and animals. However, the group of micro-organisms that we regard as bacteria

Table 2.1 The major groups of anaerobic Eubacteria*

Eubacterial phyla: 16S rRNA phylogeny
I Gram-positive eubacteria
   A. High G + C (>55 mol %)
      Bifidobacterium
      Propionibacterium
      Actinomyces
      Megasphaera
      Selenomonas

   B. Low G + C (<50 mol %) sub-division
      Anaeroplasma
      Clostridium
      Sarcina
      Butyrivibrio
      Eubacterium
      Acetobacterium
      Peptococcus
      Ruminococcus

II Spirochaetes
     Treponema
     Borrelia

III *Bacteroides–Flavobacterium* phylum
     Bacteroides
     Fusobacterium

*After Woese (1987)

may include representatives of several kingdoms; this is particularly true of the anaerobic organisms. Acceptance of subdivisions must depend on the accumulation of adequate data about the large numbers of bacterial strains representing all the diverse groups. In practical terms, most microbiologists deal with bacteria at the genus and species level. The problems with the higher taxa were such that they were excluded from the 8th edition of *Bergey's Manual of Determinative Bacteriology* (Buchanan and Gibbons, 1974). The discussion as to their importance has been reopened as a result of advances in molecular biology and they are included in *Bergey's Manual of Systematic Bacteriology* (Holt and Krieg, 1984; Holt *et al.*, 1986). For most practical microbiologists, these considerations may not be of particular day-to-day importance, but it should be remembered that they form the theoretical basis of the subject.

Advances in chemistry and molecular biology have greatly increased the range of techniques available for the acquisition of data on which bacterial classifications are based. Whole-cell or cell-envelope biochemical analysis and genetic analysis of bacteria have been used in attempts to examine the fundamental biological basis of taxonomic groups, as described on the basis of classical bacteriological criteria, and to develop further groupings. This process has been aided by advances in computerized data handling procedures and development of the conceptual basis of numerical taxonomy.

Techniques that have been used widely include examination of DNA-DNA and DNA-RNA homology, analysis of nucleic acids (particularly rRNA), analysis of cell walls and membranes, examination of cytochromes and other respiratory pigments, analysis of lipids (including quinones) and analysis and sequencing of proteins. Analysis of the end products of metabolism, particularly the acid end products from glucose metabolism, and 'fingerprinting' by means of solubilized cellular proteins separated in polyacrylamide gels (PAGE) have proved to be particularly valuable techniques when applied to anaerobic bacteria.

The classification of bacteria remains in a state of flux but it is clear that major advances have been, and will continue to be made, although the techniques that may answer many of our problems are too demanding for routine use. Medical microbiologists deal with bacterial strains which they assign to species and consider to be harmless commensals or pathogens. Historically, the definitions of these species, and the identification of the bacterial strains isolated in the routine diagnostic laboratory, were based on biochemical tests and bacterial morphology. The range of the tests used in phenotypic characterization has greatly expanded and has been further supplemented by routine application of advances in molecular biology. Genetic and molecular biological analyses have been used to confirm the identity of groups, which were defined by other means. Less frequently, bacterial groups have been delineated on the basis of DNA-DNA homology. The net result of these advances is that bacterial species and genera are more firmly based and better understood than ever before. The relationships of the genera to each other and the nature of the higher taxa is only now beginning to be understood.

Classification of anaerobic bacteria

The general classification of anaerobes has received relatively little attention in comparison with the efforts devoted to aerobic and facultative bacteria; the reasons for this include the technical difficulties associated with the isolation, purification and study of anaerobes. However, the lack of interest reflects the comparatively recent general realization of their ecological, medical and evolutionary significance. For a long time interest was limited to a few species of medical importance, mainly clostridia (Willis, 1977). The most accessible account of the historical developments in the classification of bacteria is that of Breed (1948).

Prévot and his co-workers at the Pasteur Institute made the first major historical attempt to produce a general classification of anaerobes; their system is most readily accessible to English readers in the translation by Fredette (Prévot, 1966) (Table 2.2). In this system many genera commonly regarded as consisting of aerobic and facultative organisms, e.g., *Neisseria*, include anaerobic species. The importance of this classification is in part historical, reflecting changes in the perception of anaerobes by the French school over a long period (Weinberg *et al.*, 1937), and in part scientific. A feature of this system is the use of a separate kingdom—*Schizomycetes*—for bacteria, and the use of a wide range of characters including DNA-DNA homology. Although the classification has not been widely used, it has served as an inspiration for other developments. The descriptions of the bacterial species, collected together form a valuable source of detailed information from workers on all aspects of anaerobic bacteriology, and the monograph by Prévot *et al.* (1967) remains the most complete compilation of anaerobic bacteria.

The study of anaerobic bacteria has been influenced not only by the needs of medical microbiology but also by advances in microbial ecology. Studies of the rumen (Hungate, 1966) and the intestine have been of particular importance. Studies of the pathogenic anaerobes by Smith and colleagues (Smith, 1955, 1975; Smith and Holdeman, 1968; Smith and Williams, 1984) highlight the advances that have been made and document the rise in awareness of the clinical importance of non-sporing anaerobes. Studies at the Anaerobe Laboratory of Virginia Polytechnic Institute and State University have been of particular importance. These studies have resulted in the characterization and identification of many strains of bacteria, representing both new and previously described species (Cato *et al.*,

Table 2.2 The anaerobic members of the Kingdom Schizomycetes*

Class I: Non-sporing Eubacteria or Asporulales	Class II: Sporeforming Eubacteria or Sporulales	Class III: Actinomycetales
Order I: *Micrococcales*	Order II: *Clostridiales* (gram-negative)	Order I: *Actinobacteriaes* (Gram-positive)
Family I: *Neisseriaceae* (gram-negative)	Family I: *Endosporaceae*	Family: *Actinomycetaceae*
Genus I: *Neisseria*	Genus I: *Endosporus*	Genus I: *Corynebacterium*
Genus II: *Veillonella*†	Genus II: *Paraplectrum*	Sub-genus: *Propionibacterium*†
Family II: *Micrococceae* (gram-positive)	Family II: *Clostridiaceae* (gram-positive)	Genus II: *Actinobacterium*
Tribe I: *Streptococceae*	Genus I: *Infabilis*	Sub-genus: *Bifidobacterium*†
Genus I: *Diplococcus*	Genus II: *Welchia*	Genus III: *Micromonospora*
Genus II: *Streptococcus*	Genus III: *Clostridium*†	
Subgenus: *Ruminococcus*†	Order III: *Plectridiales* (gram-positive)	
Tribe II: *Staphylococceae*	Family I: *Terminosporaceae*	
Genus I: *Gaffkya*	Genus I: *Terminosporus*	
Genus II: *Staphylococcus*	Genus II: *Caduceus*	
Tribe III: *Micrococceae*	Family II: *Plectridiaceae*	
Genus I: *Sarcina*†	Genus I: *Plectridium*	
Genus II: *Micrococcus*	Genus II: *Acuformis*	
Order II: *Bacteriales*	Order IV: *Sporovibrionales*	
Family I: *Parvobacteriaceae* (gram-negative)	Family: *Sporovibrionaceae*	
Tribe I: *Pasteurelleae*		
Genus I: *Pasteurella*		
Tribe III: *Hemophileae*		
Genus III: *Dialister*		
Family II: *Ristellaceae* (gram-negative)		
Genus I: *Ristella*		
Genus II: *Zuberella*		
Genus III: *Capsularis*		
Family III: *Sphaerophoraceae* (gram-negative)		
Genus: *Sphaerophorus*		
sub-genus: *Fusiformis*		
Genus: *Leptotrichia*†		
Family III: *Bacteriaceae* (gram-positive)		
Genus I: *Eubacterium*†		
sub-genus: *Zymobacterium*		
sub-genus: *Butyribacterium*		
Genus II: *Catenabacterium*		
Genus III: *Ramibacterium*		
Genus IV: *Cilliobacterium*		
Order III: *Spirillales*		
Family I: *Vibrionaceae*		
Genus I: *Vibrio*		
sub-genus: *Butyrivibrio*†		
sub-genus: *Succinovibrio*†		
sub-genus: *Selenomonas*†		

*Based on Prévot (1966);
† Extant genera

Table 2.3 The anaerobic members of the Kingdom Prokaryotae*

Kingdom: *Prokaryotae*

Division I: *Gracilicutes*
  Class I: *Scotobacteria*
    Order: *Spirochaetales*
      Family: *Spirochaetaceae*
        Genus: *Spirochaeta*
        Genus I: *Treponema*
    Order: *Bacteriales*
      Family: *Bacteroidaceae*
        Genus: *Bacteroides*
        Genus: *Fusobacterium*
        Genus: *Leptotrichia*
        Genus: *Butyrivibrio*
        Genus: *Succinimonas*
        Genus: *Succinivibrio*
        Genus: *Anaerobiospirillum*
        Genus: *Wolinella*
        Genus: *Selenomonas*
        Genus: *Anaerovibrio*
        Genus: *Acetivibrio*
        Genus: *Lachnospira*
    Order: *Veillonellales*
      Family: *Veillonellaceae*
        Genus: *Veillonella*
        Genus: *Acidaminococcus*
        Genus: *Megasphaera*

Division II: *Firmicutes*
  Class I *Firmibacteria*
    Order: *Micrococcales*
      Family: *Streptococcaceae*
        Genus: *Streptococcus*
      Family: *Peptococcaceae*
        Genus: *Peptococcus*
        Genus: *Peptostreptococcus*
        Genus: *Ruminococcus*
        Genus: *Coprococcus*
        Genus: *Sarcina*
    Order: *Actinobacteriales*
      Genus: *Arachnia*
      Genus: *Propionibacterium*
      Genus: *Eubacterium*
      Genus: *Acetobacterium*
      Genus: *Lachnospira*
      Genus: *Butyrivibrio*
      Genus: *Actinomyces*
      Genus: *Bifidobacterium*

Division III: *Tenericutes*
  Genus: *Anaeroplasma*

Division IV: *Mendosicutes*
  Class I: *Archaebacteria*
    Order: *Methanobacteriales*
      Family: *Methanobacteriaceae*
        Genus: *Methanobacterium*
        Genus: *Methanobrevibacter*
    Order I: *Methanococcales*
      Family: *Methanococcaceae*
        Genus: *Methanococcus*
    Order: *Methanomicrobiales*
      Family: *Methanomicrobiaceae*
        Genus: *Methanomicrobium*
      Family: *Methanosarcinaceae*
        Genus: *Methanosarcina*

* Based on Holt and Krieg (1984) and Holt *et al.* (1986) with various additions and modifications by Balch *et al.* (1979)

1969; Holdeman and Moore, 1972, 1975; Holdeman et al., 1977). Strains derived from these and other studies (e.g., Balch et al., 1979) have been used to examine the basis of the taxonomy of anaerobes by using molecular biological techniques. Such studies, drawing on a wide range of techniques, have provided the underpinning to our present system for the classification of anaerobes (Table 2.3).

Culture of anaerobic bacteria

In practical terms anaerobic bacteria are defined by the means used for their cultivation. Indeed, isolation procedures have probably been the dominant influence in delineating taxonomic groups. The use of heat shock to isolate clostridia, and the other specialized techniques for the isolation of spirochaetes and methanobacteria, undoubtedly affect our perception of these groups. The important advances in anaerobic bacteriology that have occurred during the last 20 years stem from the refinement and adoption of reliable anaerobic techniques. The effect of these advances has been to ensure that anaerobes are isolated sufficiently often in routine clinical laboratories to attract attention. Physiological testing of anaerobes can now be undertaken routinely and the existence of reliable anaerobic methods has also facilitated attempts to identify these clinical isolates. Similarly, the ability to grow anaerobes reliably has led to the use of chemotaxonomic techniques and molecular biology in an attempt to resolve their taxonomy.

Anaerobic jars are used in the isolation and study of anaerobes. Although first described many years ago, it is only since the advent of more complex techniques, which have provided a reference standard, that such jars have been used routinely and satisfactorily. Those interested in the development of anaerobic methods are referred to the reviews of Hall (1929) and Sonnenwirth (1972).

Oxygen dissolved in culture media which have been prepared and stored aerobically may prevent the growth of anaerobic bacteria by causing the oxidation of organic substances in the medium, by raising the oxidation–reduction potential of the medium, by inhibiting oxygen-sensitive enzymes, by the destruction of oxygen-sensitive compounds containing disulphide bridges and by the diversion of metabolism to produce toxic metabolites such as hydrogen peroxide. Individual bacteria separated from one another by the plating out of specimens on media containing dissolved oxygen are maximally exposed to these toxic influences. Dissolved oxygen can be removed from media by boiling or by allowing time for equilibrium with an oxygen-free atmosphere, and some of the effects of dissolved oxygen can be overcome by the addition of reducing agents to the culture media. Anaerobic systems utilize one or more of these techniques to obtain an anaerobic environment.

Anaerobic jar

The anaerobic jar is the device most often used to maintain an oxygen-free atmosphere around bacterial cultures. Several types of jar are available commercially, but any gas-tight container that will withstand the stress of evacuation, such as a milking machine pail or a pressure barrel, can be used. Anaerobiosis is achieved by partially evacuating (-650 mmHg) the jar and refilling with an oxygen-free mixture incorporating at least 10 per cent hydrogen. Any residual oxygen is combined catalytically with the hydrogen; catalyst active at room temperature has largely replaced the electrically heated catalyst used in earlier models.

As an alternative to this type of evacuation–replacement system a disposable gas generator may be used. This system relies on the generation of excess hydrogen and the catalytic removal of all the oxygen in the jar. The activated steel wool system of Parker (1955) may be similarly employed. Specialized jars have been devised for particular applications, e.g., jars containing hydrogen at high pressure have been used in studies of methanobacteria.

Prereduced anaerobically sterilized media (PRAS)

The use of tightly stoppered, individually gassed tubes of anaerobically sterilized media was described by Hungate (1950) for the culture of cellulolytic bacteria from the bovine rumen. In this system each culture tube is an individual anaerobic container. Any bacteriological culture medium may be prepared in a prereduced form and sterilized anaerobically. The dry ingredients are placed in a flask with water, any necessary mineral salt solutions and a redox potential (Eh) indicator; resazurin is the most commonly used Eh indicator but others are available and have specific applications. The mixture is boiled until the Eh indicator shows that the dissolved oxygen has been removed. The medium may be gassed with oxygen-free gas during this process but for the routine study of medically important bacteria it is usually sufficient to fit the flask with a chimney and allow the steam to displace the air.

The oxygen-free medium must now be dispensed and sterilized; reducing agents and buffer solutions may be added at this stage. During dispensing both the medium and the tubes are flushed with oxygen-free gas. As the tube is stoppered the gas tube is withdrawn, taking care to exclude air; butyl rubber stoppers are impermeable to oxygen. Some investigators prefer to dispense media aseptically after sterilization, this enables heat-sensitive materials to be added. Before autoclaving, tubes or flasks containing medium are closed and the stoppers wired in place or held in clamps. Media and cultures are gassed during all manipulations involving the prereduced anaerobically sterilized media (PRAS) technique; hydrogen, nitrogen and carbon dioxide have all been used. Oxygen-free carbon dioxide is readily available and has the advantage that it can be used with a bicarbonate buffer. These techniques are best adapted to bulk growth of bacteria and for the use of roll tube techniques. The plate-in-bottle system of Mitsuoka *et al.* (1976) can be adapted in these systems for standard plating procedures. Details of these techniques and their application to the characterization of anaerobes from clinical samples and human microecologies can be found in the publications of the Virginia Polytechnic Institute Anaerobe Laboratory (Cato *et al.*, 1969; Holdeman and Moore, 1972, 1975; Holdeman *et al.*, 1977).

For bulk culture it is also possible to make use of the deoxygenated state of media immediately after autoclaving. Media are prepared and autoclaved without special precautions and removed from the autoclave immediately the pressure falls. If allowed to cool under a stream of oxygen-free gas such media remain anaerobic.

Anaerobic chamber

When an anaerobic jar is used for the culture of anaerobic bacteria all manipulations are performed on the laboratory bench and the organisms are exposed to atmospheric oxygen. To protect specimens and cultures from the atmosphere, various glove boxes with an anaerobic atmosphere have been devised. The chambers make it possible to perform conventional bacteriological procedures under anaerobic conditions.

Studies on the indigenous bacteria of man and animals became a major area for serious study between 1960 and 1970. Non-sporing anaerobic bacteria were shown to predominate in the large intestine and to be important in the mouth. The realization of the need for a laboratory system able to deal with many isolates and a variety of procedures resulted in the development of the anaerobic cabinet (Socransky *et al.*, 1959–1960; Drasar, 1967). The first anaerobic chambers were made specifically to apply the lessons already learned to clinical problems, most notably to the culture of spirochaetes (Rosebury and Reynolds, 1964). The early cabinets were individually built and several attempts were made to simplify the design (Leach *et al.*, 1971) and to improve performance (Gordon and Dubos, 1970). The first routinely practicable cabinet was the flexible isolator system designed by Aranki *et al.* (1969). The combination of a flexible isolator (with an airlock that could be evacuated when the system was operated) with a gas mixture containing hydrogen and a palladium catalyst was widely adopted and many versions were developed. Indeed, this type of cabinet is still widely used for research; of particular interest is the double chamber cabinet for handling very fastidious anaerobes (Edwards and McBride, 1975).

Manufacturers of cabinets for the clinical laboratory have favoured the rigid format rather than the flexible enclosures described by some early investigators; several designs are available. It is remarkable how, over the last 20 years, the anaerobic cabinet has developed from an exotic research tool to be an accepted part of hospital microbiology laboratories.

Identification of medically important anaerobes

In the clinical laboratory there is seldom the need, or the opportunity, for using the full range of tests that a taxonomist might consider valuable. As mentioned previously, use of the techniques described in the manuals produced at the VPI Anaerobe Laboratory (Cato *et al.*, 1969; Holdeman and Moore, 1972, 1975; Holdeman *et al.*, 1977) has made a wide range of anaerobic bacteria accessible to clinical investigators. The adaptation of these methodologies for clinical laboratory use and the integration of these advances with more usual systems and with methods derived from their own research is an underlying theme of the laboratory manuals produced by the group at the Veterans Administration Wadsworth Medical Center (Sutter *et al.*, 1972, 1975, 1980, 1985).

Identification of anaerobic bacteria to genus depends on a small number of tests, most noticeably Gram's stain, and on the production of spores. To these should be added the analysis, usually by gas–liquid chromatography (GLC), of the acid end products produced by metabolic reactions. Identification to species level is a more complex procedure and will be dealt with in the individual chapters; the principles involved are the same as those used for other (aerobic) organisms. The four primary criteria for the allocation to groups of anaerobes important in human health and disease are:
proof of anaerobic habit;
reaction in Gram's staining method;
cell morphology;
the ability to form spores.

Anaerobic habit

If bacteria are to be identified as anaerobes it is important that their ability to grow anaerobically and their failure to grow aerobically is verified. This may prove more difficult than would at first appear. CO_2-dependence may result in failure of aerobic or anaerobic growth. Metronidazole sensitivity may be useful but some anaerobes, e.g., *Bifidobacterium* spp. are resistant. Some workers advocate the use of shake cultures to verify growth habit. The use of plates of suitable media incubated in different atmospheric conditions usually proves adequate for clinical isolates.

Gram's stain

Judgement of the reaction of a bacterial strain to Gram's stain influences the choice of the tests used for species identification. A major problem is the tendency of old cultures to become gram-negative. The reaction with Gram's stain may be supplemented by determination of vancomyin sensitivity in a disc diffusion test with a 5 µg disc. A refinement of Gram's staining method is the use of buffered crystal violet to control the environment in which staining occurs. These strategies reduce error, but if a strain remains unidentified as a gram-negative organism consideration should be given to performing the identification appropriate to gram-positive organisms.

Cell morphology

Cell morphology is usually judged on the basis of the examination of gram-stained preparations. However, these are often subject to vigorous heat fixation and cell morphology can be better perceived by the examination of unfixed material by phase-contrast microscopy. If fixed material is examined simple stains, e.g., methylene blue, can be used with advantage on air-dried preparations.

Spore formation

The ability to produce spores is crucial for the definition of the medically important anaerobic genus *Clostridium*. The definitive demonstration that an organism does not produce spores is seldom achieved. It is important that all avenues be explored. Cultures should be stained with a spore stain and tested for heat and alcohol resistance (Dowell and Hawkins, 1974).

Initial classification of bacteria based on these four criteria is shown in Table 2.4. This list of bacteria comprises the relatively narrow range of bacteria that are of medical importance. Even among the endospore-forming anaerobes where clostridia predominate (see Chapter 6) other genera, including *Desulfotomaculum* and *Sarcina*, may be isolated from intestinal contents. Furthermore, it must be stressed that it is essential to test *at least* the four basic criteria of classification and, preferably, to examine the end products of metabolism before assigning a strain to a genus.

Many isolates obtained during studies of clinical samples or examination of the normal flora will be gram-negative anaerobes (see Chapter 5). Such bacteria, and in particular some bacteroides and fusobacteria, are comparatively easy to isolate using standard anaerobic jar techniques. However, other members of this group are more difficult to handle

Table 2.4 Genera of anaerobic bacteria represented in clinical material

Gram-reaction; Spore detection	Morphology	Genus	Major fermentation products of glucose
Gram-positive; Spores detected	Rods	*Clostridium*	Various; may be very complex
Gram-positive; Spores not detected	Rods	*Propionibacterium*	Propionic
		Eubacterium	Various; may be very complex
	Branched irregular rods	*Actinomyces*	Acetic, lactic and succinc acid
		Bifidobacterium	Acetic and lactic acid
	Cocci	*Peptococcus*	None
		Peptostreptococcus	Various; may be very complex
Gram-negative; Spores not detected	Rods	*Fusobacterium*	Butyric acid
		Bacteroides	Various
	Cocci	*Veillonella*	(Propionic and acetic acid from lactate)

Table 2.5 The gram-negative non-sporing rods

Genus	Morphology			Flagella					Volatile acids			Non-volatile acids			Gas
	Straight	Curved	Spiral	Motile	Polar	Sub-polar	Lateral	Peritrichous	Acetic	Butyric	Propionic	Other acids and volatile products	Lactic	Succinic	Hydrogen
Bacteroides	+	−	−	±	−	−	−	+	+	+	±	+	+	+	±
Fusobacterium	+	−	−	−	−	−	−	−	±	±	++	±	±	−	+
Leptorichia	+	−	−	−	−	−	−	−	±	−	−	±	++	±	−
Butyrivibrio	−	+	−	+	+	+	−	−	−	±	++	−	+	±	±
Succinimonas	+	−	−	+	+	−	−	−	++	−	−	−	−	++	−
Succinivibrio	−	−	+	+	+	−	−	−	++	−	−	±	±	++	−
Anaerobiospirillum	−	−	+	+	+*	−	−	−	++	−	−	±	±	++	−
Wolinella†	+	+	+	+	+	−	−	−	−	−	−	−	−	−	−
Selenomonas	−	+	−	+	−	+	+‡	−	++	++	−	−	++	±	±
Anaerovibrio	−	+	−	+	+	−	−	−	+	++	−	−	+	±	−
Acetivibrio	+	+	−	+	−	−	+	−	+	−	−	+	−	−	+
Lachnospira	+	+	−	+	−	+	+	−	+	−	−	−	+	−	+

*Biopolar tufts;
† non-fermentative;
‡ tufts on concave side of cell

or require a special search to be made; the range of gram-negative anaerobes is shown in Table 2.5. However, few of these bacteria will be isolated by standard techniques in the clinical laboratory and it is as well to be aware of the limitations imposed by the methods used. It may be that, if they were looked for, members of other genera of gram-negative anaerobes would be encountered more frequently. Motile anaerobes that seem to be gram-negative often turn out to be clostridia that are easily decolourized. The examination of the type and location of flagella, essential for identification of many anaerobes, is a difficult procedure.

Among the anaerobic cocci (see Chapter 7) members of the genus *Peptostreptococcus* are most important as causes of sepsis, whereas *Ruminococcus* and *Coprococcus* spp. are prominent members of the normal flora (Tables 2.6 and 2.7). The isolation of anaerobic cocci from the normal flora is best accomplished by use of PRAS media or an anaerobic chamber. Strains from infection can usually be isolated in an anaerobic jar. Unless a special search is made, anaerobic spirochaetes (Chapter 8) will usually be missed.

The status of gram-positive non-sporing rods in clinical microbiology is equivocal (Chapters 9 and 10). While *Actinomyces* spp. are well recognized as specialist pathogens and *Bifidobacterium* spp. are considered to be significantly associated in the faeces with breast-feeding, other members of this group (Table 2.8) are often dismissed as anaerobic diphtheroids. Propionibacteria are important members of the skin and gut flora; eubacteria are common in the gut flora and may also be a significant cause of sepsis, but definitive studies are lacking. They have been isolated frequently from clinical samples in the USA, but are seldom reported in the UK, this may reflect differences in laboratory practise or differences inherent in the patient groups studied. Eubacteria are an extremely heterogeneous group of organisms. The genus includes all those gram-positive, anaerobic, rod-shaped bacteria that cannot be assigned to other groups. All species are

Table 2.6 The gram-negative cocci

Genus	Substrates fermented			Volatile acids produced			
	Carbohydrates	Amino acids	Lactate	Acetatic	Propionic	Butyric	Other
Veillonella	−	−	+	+	+	−	−
Acidaminococcus	−	+	−	+	−	+	−
Megasphaera	+	−	+	+	+	+	+

Table 2.7 The gram-positive cocci

Genus	Arrangement of cells								Volatile acids				Non-volatile acids	
	Chains	Clumps	Cubical packets	Spores formed	Sugars essential for growth	Cellulose fermented	Carbohydrate fermented	Peptone fermented	Acetic	Propionic	Butyric	Other acids and volatile products	Lactic	Succinic
Peptococcus	−	+	−	−	−	−	−	+	+	±	+	+	±	±
Peptostreptococcus	+	+	−	−	−	−	D	D	+	−	+	+	+	+
Ruminococcus	+	+	−	−	+	+	+	−	+	−	−	+	+	+
Coprococcus	+	+	−	−	+	−		−	±	±	+	±	±	−
Sarcina	−	−	+	+	−	−	+		+	−	−	+	−	−

D, different species give positive and negative results

Table 2.8 The gram-positive non-sporing anaerobic rods

Genus	Morphology				Volatile acids				Non-volatile acids	
	Straight	Curved	Irregular	Motile	Acetic	Propionic	Butyric	Other acids and volatile products	Lactic	Succinic
*Arachnia**	−	−	+	−	+	++	−	−	+	+
*Propionibacterium**	−	−	+	−	±	++	−	±	+	±
Eubacterium	+	−	+	±	+	+	+	+	+	+
Acetobacterium	+	−	−	+	+	−	−	−	−	−
Lachnospira†	−	+	−	+	+	−	−	−	+	−
Butyrivibrio†	−	+	−	+	−	±	++	−	+	±
Actinomyces	−	−	+	−	+	−	−	−	+	+
Bifidobacterium	−	−	+	−	+	−	−	−	+	+

*Aerobic growth;
†gram-reaction variable

strict anaerobes; some strains and species are extremely sensitive to oxygen and require PRAS media for growth. They are separated from clostridia solely on the basis of spore formation and both genera include groups that are widely separated on the basis of the G + C content of their DNAs. It might be more sensible if the species making up the genera *Clostridium* and *Eubacterium* were reclassified without reference to spore formation. Eubacteria are widely distributed in nature; at present 34 species are recognized in the genus, of these, 17 have been isolated from clinical material.

Finally, it must be remembered that the resurgence of interest in infections due to anaerobes is comparatively recent and it is unlikely that all the significant pathogens are at present recognized. Do not expect to identify all the anaerobes that you isolate!

References

Aranki A, Syed S A, Kenney E B, Freter R. (1969). Isolation of anaerobic bacteria from human gingiva and mouse caecum by means of a simplified glove box procedure. *Applied Microbiology* 17, 568–576.

Balch W E, Fox G E, Magram L J, Woese C R, Wolfe R S. (1979). Methanogens: re-evaluation of a unique biological group. *Microbiological Reviews* 43, 260–296.

Breed, R S. (1948). Historical survey of classifications of bacteria, with emphasis of outlines proposed since 1923. In *Bergey's Manual of Determinative Bacteriology*, 6th edn, pp. 5–38. Edited by Breed, R S et al. Williams and Wilkins, Baltimore.

Buchanan, R E, Gibbons, N E. (1974). *Bergey's Manual of Determinative Bacteriology*, 8th edn, Williams and Wilkins, Baltimore, pp. XV–XVII; 4–9.

Cato E P, Cummins C S, Holdeman L V, Johnson J L, Moore W E C, Smibert R M, Smith L D S. (1969). *Outline of clinical methods in anaerobic bacteriology*. Virginia Polytechnic Institute and State University, Blacksburg.

Dowell V R, Hawkins T M. (1974). *Laboratory methods in anaerobic bacteriology*. Centers for Disease Control, Atlanta.

Drasar B S. (1967). Cultivation of anaerobic intestinal bacteria. *Journal of Pathology and Bacteriology* 94: 417–427.

Edwards T, McBride B C. (1975). New method for the isolation and identification of methanogenic bacteria. *Applied Microbiology* 29: 540–545.

Gordon J H, Dubos R. (1970). The anaerobic bacterial flora of the mouse caecum. *Journal of Experimental Medicine* 132: 251–60.

Hall I C. (1929). A Review of the development and application of physical and chemical principles to the cultivation of obligately anaerobic bacteria. *Journal of Bacteriology* 17: 225–301.

Holdeman L V, Cato E P, Moore W E C. (1977). *Anaerobe laboratory manual*, 4th edn, Virginia Polytechnic Institute and State University, Blacksburg.

Holdeman L V, Moore W E C. (1972). *Anaerobe laboratory manual*, 2nd edn, Virginia Polytechnic Institute and State University, Blacksburg.

(1975). *Anaerobe laboratory manual*, 3rd edn, Virginia Polytechnic Institute and State University, Blacksburg.

Holt J G, Krieg N R. (1984). *Bergey's Manual of Systematic Bacteriology*, vol 1, pp. 1–964, Williams and Wilkins, Baltimore.

Holt J G, Sneath P H A, Mair N S, Thorpe M E. (1986). *Bergey's Manual of Systematic Bacteriology*, vol 2, pp. 965–1599. Williams and Wilkins, Baltimore.

Hungate R E. (1950). The anaerobic mesophilic cellulolytic bacteria. *Bacteriological Reviews* 14: 1–49.

(1966). *The rumen and its microbes*. Academic Press, London.

Last G A. (1988). Musings on bacterial systematics: how many kingdoms of life? *A.S.M. News* 54: 22–27.

Leach P A, Bullen J J, Grant, I D. (1971). Anaerobic CO_2 cabinet for the cultivation of strict anaerobes. *Applied Microbiology* 22: 824–827.

Mitsouka T, Ohno K, Benno Y, Suzuki K, Namba K. (1976). Comparison of the newly developed method with the old conventional method for the analysis of intestinal flora. *Zentralblatt für Bakteriologie, Parasitenkunde, Infectionskrankheiten und Hygiene. I. Abteilung Originale. Rehie A. Medizinishe Mikrobiologie und Parasitologie* 234: 219–233.

Parker C A. (1955). Anaerobiosis with iron wool. *Australian Journal of Experimental Biology* 33: 33–38.

Prévot A R. (1966). *Manual for the classification and determination of the anaerobic bacteria*. Translated by V. Fredette (1965) Lea and Febiger, Philadelphia.

Prévot A R, Tarpin A, Kaur P. (1967). *Les bactéries anaerobies*. Dunod, Paris.

Rosebury I, Reynolds J B. (1964). Continuous anaerobiosis for the cultivation of spirochetes. *Proceedings of the Society for Experimental Biology and Medicine* 117: 813–815.

Smith L DS. (1955). *Introduction to the pathogenic anaerobes*. University of Chicago Press, Chicago, (1967). Anaerobes and oxygen. In *The Anaerobic Bacteria: proceedings of an international workshop*. pp. 13–24. Edited by Fredette V. The Institute of Microbiology and Hygiene of Montreal University, Laval des Rapides.

(1975). *The pathogenic anaerobic bacteria*, 2nd edn. Thomas, Springfield.

Smith L DS, Holdeman L V. (1968). *The pathogenic anaerobic bacteria*. Thomas, Springfield.

Smith L DS, Williams B L. (1984). *The pathogenic anaerobic bacteria*. Thomas, Springfield.

Socransky S, MacDonald J B, Sawyer S. (1959–1960). The cultivation of *Treponema microdentium* as surface colonies. *Archives of Oral Biology* 1: 171–172.

Sonnenwirth A C. (1972). Evolution of anaerobic methodology. *American Journal of Clinical Nutrition* 25: 1295–1298.

Sutter V L, Attebery H R, Rosenblatt J E, Bricknell K S, Finegold S M. (1972). *Anaerobic bacteriology manual*. University of California, Los Angeles.

Sutter V L, Citron D M, Edelstein M A C, Finegold S M. (1985). *Wadsworth anaerobic bacteriology manual*, 4th edn. Star Publishing Co, Belmont.

Sutter V L, Citron D M, Finegold S M. (1980). *Wadsworth anaerobic bacteriology manual*, 3rd edn. Mosby, St Louis.

Sutter V L, Vargo V LO, Finegold S M. (1975). *Wadsworth anaerobic bacteriology manual*, 2nd edn. University of California, Los Angeles.

Weinberg M, Nativelle R, Prévot A R. (1937). *Les microbes anaerobies*. Masson, Paris.

Willis A T. (1977). *Anaerobic bacteriology—clinical and laboratory practice*. Butterworths, London.

Woese C R. (1987). Bacterial evolution. *Microbiological Reviews* 51: 221–271.

3

Characteristics of anaerobic metabolism

J G Morris

Oxygen sensitivity of anaerobes
    Causes of oxygen toxicity
    Protection against oxygen toxicity
Redox potential
Anaerobic conservation of free energy
    Fermentation
Metabolism in the absence of oxygen
    Anaerobic bioconversions
Metabolic relationships in communities of anaerobes
References

The existence of obligately anaerobic organisms testifies to the fact that the fundamental metabolic processes of living cells are essentially anaerobic in character. The obligate anaerobe grows and reproduces itself in the total absence of molecular oxygen. It can evidently obtain free energy by processes other than aerobic respiration and cannot employ O_2 as a reactant in any of its very diverse catabolic and biosynthetic processes. Thus it must synthesize all of its genetic, enzymatic and structural components (DNA, RNA, proteins, phospholipids, polysaccharides) without recourse to molecular oxygen. Yet this is not so remarkable as might at first appear, for if we examine aerobic organisms (i.e., oxybiontic cells) we find the same to be true of their primary biosynthetic pathways. Most obligate aerobes which have evolved an absolute dependence on oxygen will be found to utilize the oxygen for a relatively limited range of purposes—primarily as the terminal electron sink in their respiration, but possibly also as an essential participant in a restricted number of somewhat specialized biochemical processes, e.g., catabolic oxygenation reactions or the biosynthesis of sterols. Thus it appears that oxygen-utilizing organisms have evolved by taking the basic stock of anaerobic metabolism and grafting upon it the means of additionally exploiting a more potent oxidant (O_2) than any previously available to them. But as well as the benefits that accrued from this enlargement of their metabolic repertoire, in utilizing molecular oxygen, living organisms also incurred several penalties. Certain of the enzymes and cofactors of the obligate anaerobe, particularly some of those that operated at the lowest redox potentials, were not compatible with an aerobic existence and there were side effects of oxygen utilization that were harmful to cells which had not acquired an appropriate defensive capability. Indeed, phylogenetic evidence suggests that obligate anaerobes were probably the earliest living organisms to emerge on our planet. A particularly persuasive case can be made for the evolutionary development of the necessary biochemical apparatus for aerobic respiration from that earlier elaborated for anaerobic respiration and/or photophosphorylation, with these in turn being predated by fermentative processes (Gest, 1980).

Something of the biochemical challenge and opportunities posed by exploitation of molecular oxygen, and of the elegant metabolic economy whereby this was accomplished, can be glimpsed in the physiological changes that attend the shift of a facultatively anaerobic bacterium from growth in anaerobic conditions to continued growth in aerobic conditions. In the case of *Escherichia coli*, Smith and Neidhardt (1983a) found that the anaerobic-to-aerobic transition provoked preferential synthesis of some 19 proteins, products specified by a global regulatory network (Gottesman, 1984) designated as the aerobic *E. coli* stimulon. In the same organism the converse aerobic-to-anaerobic shift elicited the synthesis of 18 proteins specified by its anaerobic stimulon (Smith and Neidhardt, 1983b). It is remarkable that such a dramatic change in the 'lifestyle' of the organism can be sustained with such an apparently modest alteration in its metabolic equipment. Genetic studies with *Salmonella typhimurium* and *E. coli* have suggested that these organisms possess about fifty anaerobically-regulated genes (Strauch *et al.*, 1985; Winkelman and Clark, 1986), again a rather modest proportion of a total chromosomal complement of about 1500 genes.

The facility to exploit O_2 productively, yet harmlessly, evidently favoured the emergence of the versatile facultative anaerobe and, in certain niches, microaerophiles (Stouthamer *et al.*, 1979) and obligate aerobes (Morris, 1984). As we shall see, however, it would be very wrong to consider contemporary species of obligate anaerobes as primitive organisms left behind by the forward march of evolution. These organisms are evidently well suited to the oxygen-free locations which they now so successfully inhabit, and where they consort with facultative anaerobes and microaerophiles, whose O_2-scavenging activities secure for them the conditions they most favour.

Why are obligate anaerobes aerointolerant (though not uniformly oxygen-sensitive) and not merely indifferent to the presence of oxygen? Why are microaerophiles facultatively anaerobic or obligately oxygen-requiring, but not aerotolerant? Such questions have puzzled microbiologists ever since Pasteur (1861) reported the existence of 'animalcules infusoires vivant sans gaz oxygene libre', or possibly for even longer; our Dutch colleagues would cite Antonie van Leeuwenhoek's much earlier microscopic observations of microbes which we would now recognize as anaerobes. In a review written some 20 years ago (Morris and O'Brien, 1971) we rather plaintively remarked that 'the literature on this topic is replete with studies which started with an idiosyncratic experimental system and lapsed into anecdotal reportage of a variety of disconnected "effects" of O_2'. Happily, in the interim, substantial additions have been made to our knowledge of the metabolic bases of oxygen toxicity which can form an appropriate starting point to our consideration of anaerobic metabolism.

Oxygen sensitivity of anaerobes

Molecular oxygen (i.e., O_2) is potentially toxic to all living organisms; it appears that no organism exists which is wholly indifferent to the presence of oxygen in its environment. Aerotolerant organisms have evidently acquired the means to counter the threat posed by exposure to oxygen at the concentration present in air at 1 atmosphere (which in equilibrium with a normal bacterial growth medium at 37°C would sustain a dissolved oxygen concentration of about 0.18 mmol l^{-1}). Obligate anaerobes, although manifestly aerointolerant, are not uniformly intolerant of oxygen. Indeed, they differ so markedly in their responses to oxygen that they can be considered to display a spectrum of oxygen tolerance ranging from the extremely oxygen sensitive (EOS) anaerobes to those 'moderate' anaerobes that can withstand exposure to quite high concentrations of oxygen for brief periods.

It has proved less easy than might be supposed to allocate individual species of micro-organisms to an invariable rating on any scale of oxygen tolerance. Although several attempts have been made to devise reproducible quantitative assay procedures, none has been wholly successful (for a review of early studies on clostridia see Morris and O'Brien, 1971). Such attempts have often consisted of various methods of determining the minimum concentration of atmospheric oxygen which just prevents dispersed growth in liquid culture or colonial growth from single cell inocula on agar plates. Loesche (1969) adopted this procedure in a pioneering study of the comparative oxygen sensitivities displayed by various oral anaerobes. Streaked plates were incubated in jars under gas mixtures of CO_2 10 per cent with N_2, H_2 and air, constituted so as to give O_2 concentrations ranging from 0 to 12 per cent. In a somewhat more elaborate procedure organisms were grown in deep agar tube culture (Schwartz, 1973) and an attempt was made to correct for the need to use different media for different organisms and for variations in growth rates and cell concentrations by defining and assaying values for 'the constant of oxygen sensitivity' and the 'modification factor' for each organism. Recently Kikuchi and Suzuki (1986) have, for the same reason, measured growth rates in a rich liquid broth medium in tubes maintained (a) aerobically and (b) anaerobically, and shaken for 24 h at 37°C. The 'aerotolerance' of the test organisms was recorded as the ratio of the culture densities achieved under the two atmospheres (termed the 'relative bacterial growth ratio'). An alternative approach (van Winkelhoff et al., 1986) seeks to monitor survival of single cells of the anaerobic bacterium during a period of exposure to air on the surface of a suitable (blood agar) medium. The findings are generally expressed as D values, being the time in hours required to kill 90 per cent of the initially viable cell population. Death rates on aeration of pH-buffered suspensions of obligate anaerobes have also been measured (Patel et al., 1984).

These and similar procedures are useful in that they enable bacterial species or isolates to be assigned to certain broad bands within the spectrum of oxygen tolerance. Yet it will be found that there are considerable strain differences within a single species and the apparent oxygen sensitivity will depend not only on the composition of the growth medium but even on the physiological state of the culture from which the test inoculum was taken (Tally et al., 1975). Furthermore, the behaviour of single cells when exposed to a given concentration of oxygen is not always a good indicator of how an already dense population of such cells will react to the presence of the same concentration of oxygen. It is frequently surprising to discover that many moderate anaerobes can consume relatively large quantities of oxygen when growing in culture without suffering apparent harm. Such tests are best performed in continuous (chemostat) culture and should take account of the fact that the culture may differ in its response to a given concentration of oxygen, depending on whether it is abruptly exposed to this concentration or is gradually adapted to it by prior exposure to gradually increased lower oxygen tensions (Pritchard et al., 1977). The organisms in such cultures are in contact with the oxygen 'in solution' and it is wise to report both the tension and the concentration of dissolved oxygen (see Morris, 1984). Unfortunately, the usual types of commercially available submerged oxygen electrodes (voltammetric or galvanic) are unable to measure oxygen tensions corresponding to dissolved oxygen concentrations smaller than about 1 μM, yet it is at such low concentrations that the effects of oxygen on cultures of obligate anaerobes may be most informative. Extraordinarily oxygen-sensitive probes have been devised, for example a membrane-covered *Photobacterium* probe whose photoemission varies linearly with oxygen concentration over the range 35–840 nM (Lloyd et al., 1981), but the technique of membrane inlet mass spectrometry has a more general use because it allows simultaneous measurement of other dissolved gases in the culture medium (Lloyd et al., 1983).

Causes of oxygen toxicity

There is no single cause of oxygen toxicity, nor any comprehensive unitary explanation for the supersensitivity of obligate anaerobes to challenge with oxygen. It has become evident that there is a multiplicity of routes whereby oxygen can harm bacterial cells, which in turn seek to protect themselves by diverse means.

Molecular oxygen is an avid electron acceptor (i.e., a potent oxidant) and the presence of free oxygen in solution in a culture is incompatible with that culture being able to sustain a low reduction–oxidation potential (see page 22). Since the metabolic electron flow in obligately anaerobic bacteria must be channelled through specific routes that serve the dual function of supplying the 'reducing power' for biosynthetic purposes and sustaining ATP generation, any non-productive diversion to an alien electron acceptor could prove lethal if unchecked. Short-circuiting of the electron flow by oxygen serving as an inexhaustible electron sink must, therefore, pose a significant threat to the viability of those anaerobes which can consume oxygen non-productively as a preferred oxidant. But oxygen can also interact directly with key enzymes and cofactors that play a central role in the metabolism of many anaerobes, causing reversible or irreversible inactivation. Although those intracellular components that operate at low redox potentials (including iron–sulphur proteins, folate coenzymes, cobamide coenzymes or the unique enzymes and cofactors that are implicated in the energy-conserving metabolism of methanogenic bacteria) come to mind immediately, other examples exist which illustrate the often unexpected vulnerability of anaerobic metabolic pathways to oxygen. The anaerobe literature is rich in examples of enzymes which rapidly lose activity when exposed to air and have, therefore, to be isolated and purified in its absence. Furthermore, it is rarely possible to predict in advance whether a particular enzyme is likely to be oxygen-labile. For example, the hydrogenase from the obligately aerobic bacterium *Alcaligenes eutrophus* is subject to autogenous inactivation in air (Schneider and Schlegel, 1981), those from the fermentative obligate anaerobes *Clostridium pasteurianum* and *Megasphaera elsdenii* are irreversibly inactivated by air, whereas the hydrogenases from the obligately anaerobic sulphate-reducing bacteria are reversibly inactivated by oxygen (Odom and Peck, 1984).

It would appear, however, that the major cause of oxygen toxicity is not attributable to molecular oxygen *per se* but to the peculiarly harmful nature of certain products of its 'partial' reduction. The complete (four electron) reduction of O_2 which is the normal terminal reaction in aerobic respiration poses no threat because it yields the harmless and ubiquitous water molecule as the end product. However, partial reductions of the O_2 molecule also occur which yield potentially lethal products.

$$O_2 + 4H^+ + 4e^- \rightarrow 2H_2O$$
$$O_2 + 2H^+ + 2e^- \rightarrow H_2O_2$$
$$O_2 + e^- \rightarrow O_2 \cdot ^-$$

Thus hydrogen peroxide (H_2O_2) formation in biological systems is the consequence of a two-electron reduction of molecular oxygen that can be catalysed by various agents including a number of enzymes (generally flavoproteins). Again, a one-electron reduction of the oxygen molecule yields the superoxide anion ($O_2 \cdot ^-$), a free radical whose potential to cause harm to living cells has been fully documented by Fridovich and his colleagues (Fridovich, 1986). An additional, particularly lethal, derivative of molecular oxygen is singlet oxygen, and production of this activated oxygen species is particularly associated with those circumstances where exposure to oxygen is combined with intense illumination (photodynamic cell death).

The production of H_2O_2 by various bacteria, resulting in some instances in its accumulation in their growth media, and the lethality of H_2O_2 to other organisms, are well documented (Morris, 1975). The toxicity acquired by rich growth media following exposure to air, which prevents the growth of obligate anaerobes, has similarly been attributed to the production of hydrogen peroxide or various organic peroxides, or both. Although over the years there have been many attempts to identify the prime cause of H_2O_2 toxicity, it is likely that more than one mechanism is involved and that living cells contain numerous vulnerable targets. However, DNA damage (single strand breaks in chromosomal DNA) in susceptible bacteria exposed to lethal concentrations of H_2O_2 have been particularly noted.

The cause(s) of superoxide anion toxicity have proved to be rather more contentious. It has been suggested that $O_2 \cdot ^-$ itself is not a very reactive radical, and that its toxicity should be attributed to its participation in the generation of the hydroxyl free radical $HO \cdot$, an indisputably highly dangerous reagent. The latter can be produced from H_2O_2 whenever $Cu(I)$ or $Fe(II)$ ions are available (via the Fenton reaction) but it is also formed by the interaction of the superoxide anion with hydrogen peroxide, and is catalysed by $Fe(III)$ (the so-called Haber–Weiss reaction). Reduction of $Fe(III)$ to $Fe(II)$ by $O_2 \cdot ^-$ is followed by the oxidation of $Fe(II)$ to $Fe(III)$ by H_2O_2, which is simultaneously reduced to yield OH^- and $HO \cdot$. The hydroxyl free radical is an exceedingly powerful reactant which is also produced during lethal irradiation of living cells by ionizing radiation. However, it is also a very short lived species, so that to contribute to oxygen toxicity it would have to be produced in the immediate vicinity of its susceptible target. This paradox may be solved by assuming site-specific localization of the Fenton reaction, e.g., polyanionic macromolecular structures such as chromosomal DNA and cytoplasmic membranes binding the necessary catalytic $Fe(II)$. While accepting that indirect effects of superoxide anion are indeed important in the aetiology of its toxicity, Fridovich (1986) contends that the anion is additionally toxic in its own right.

Protection against oxygen toxicity

The first 'unitary' theory of anaerobiosis propounded by McLeod and Gordon (1923) identified H_2O_2 accumulation as the prime cause of oxygen toxicity and hence sought to attribute the phenomenon of aerointolerance to non-possession of the usual agents of H_2O_2 destruction, viz., catalase and/or peroxidases.

$$H_2O_2 \rightarrow H_2O + \tfrac{1}{2}O_2$$

However, although a convincing case could be made for the beneficial effects of such enzymes in organisms routinely exposed to oxygen, the distribution of catalase amongst bacteria proved to be rather erratic. The presence of catalase in some aerointolerant bacteria (e.g., *Propionibacterium shermanii*) and its absence from some obligate aerobes (e.g., *Bacillus popilliae*) suggested that its presence or absence was of itself insufficient to explain aerotolerance/aerointolerance. Following the recognition that superoxide anion production would pose a threat to those organisms with no defence against this radical, a superoxide dismutase theory of anaerobiosis was promulgated (McCord *et al.*, 1971) which proposed that those organisms which possessed no means of scavenging $O_2\cdot^-$ ions would necessarily be aerointolerant. The superoxide dismutases (SOD) found in bacteria proved to be iron or manganese proteins (Fe-SOD and Mn-SOD) and one or other, or both, of these enzymes were present in all organisms which normally consumed or produced oxygen. The role of the enzyme was to catalyse the dismutation of $O_2\cdot^-$ to yield H_2O_2 and molecular oxygen,

$$O_2\cdot^- + O_2\cdot^- + 2H^+ \rightarrow H_2O_2 + O_2$$

so that concurrent possession of a H_2O_2-utilizing enzyme should prove additionally desirable. Indeed, it has even been proposed that molecular oxygen could act as a scavenger of free radicals, being transformed to $O_2\cdot^-$, and that these superoxide anions are then harmlessly transformed to H_2O and O_2 by the joint action of SOD plus catalase (Imlay and Linn, 1988). Although initially it appeared that SODs were restricted to aerotolerant bacteria, low levels of the enzyme were soon reported in a wide variety of anaerobes (Hewitt and Morris, 1975; Gregory *et al.*, 1978; Fulghum and Worthington, 1984). Some obligate anaerobes, e.g., *Eubacterium limosum* and *C. oroticum*, actually contain quite high concentrations of the enzyme. Virtually all of the moderate anaerobes commonly encountered in human disease were found to contain some SOD activity and, in some instances, exposure to oxygen enhanced the cell content of the enzyme (Carlsson *et al.* 1977; Rolfe *et al.*, 1978). Indeed, Tally *et al.* (1977) postulated that SOD may be a virulence factor in pathogenic anaerobes which allows the organisms to survive in oxygenated tissues until the oxygen is depleted and conditions of reduced oxygen concentration are established which will support their growth. Although, in general, strains of a given species that contained higher levels of SOD were more tolerant of oxygen than those that contained smaller quantities of the enzyme (Gregory *et al.*, 1978) total absence of SOD did not necessarily relegate an organism to EOS status. Thus SOD-minus strains of *Bacteroides* spp. and of *C. sporogenes* still displayed the characteristics of moderate anaerobes, while *C. tertium*, usually regarded as one of the most aerotolerant of clostridia, was also devoid of the enzyme (Gregory *et al.*, 1978). A more extreme example is supplied by *Neisseria gonorrhoeae*, which is an obligate aerobe and yet contains no SOD (Archibald and Duong, 1986).

With both catalase and the superoxide dismutases having thus been identified as being important components of the defences raised by aerotolerant organisms against oxygen toxicity, is it possible to determine which is the more important, or indeed is it sensible even to ask this question? Some indication of their respective roles can be gained from the manner in which their intracellular levels respond to changes in imposed O_2 tensions. Thus SOD is one of the enzymes specified by the *E. coli* aerobic stimulon and its specific activity increased some seventy-fold when an anaerobic culture was made aerobic (Smith and Neidhart, 1983a). *B. distasonis* must be supplied with haemin for it to be able to synthesize catalase and Gregory and Fanning (1983) exploited this dependence to produce cells with different catalase levels following growth in media supplemented with increasing concentrations of haemin. The resistance to H_2O_2 toxicity displayed by these cells paralleled their intracellular catalase contents. However, an even more convincing test is the behaviour of strains of bacteria that have lost their former ability to produce one or more of the putative protective enzymes. Hassan and Fridovich

(1979) isolated five aerointolerant mutants of *E. coli*. All contained the normal complement of Fe-SOD, but three were incapable of inducing catalase and peroxidase synthesis when exposed to air, while the remaining two strains were additionally unable to form the (inducible) Mn-SOD. A catalase-negative strain of *E. coli* isolated from a clinical source on the basis of its gentamicin resistance proved to be aerointolerant (Funada *et al.*, 1978) but selection of catalase-negative mutants of *E. coli* by other methods has not confirmed this direct linkage between possession of catalase and aerotolerance. Thus Loewen (1984) reported that although a catalase-deficient mutant strain of *E. coli* was some fifty times more sensitive to the lethal action of H_2O_2 than was its parent strain, it was still capable of aerobic growth. Meir and Yagil (1984) found that *E. coli* contained a constitutively synthesized catalase (pH optimum 10.5) which predominated in cells from stationary phase cultures, and a second catalase (pH optimum 6.8) which was the more active enzyme in exponential phase cultures and whose induced synthesis was provoked by H_2O_2. Mutants lacking the latter enzyme grew aerobically but would not decompose added H_2O_2, while the one mutant strain isolated by virtue of its hypersensitivity to H_2O_2 lacked the pH 10.5 enzyme, although it contained normal levels of peroxidase and SOD.

In an elegant series of experiments, Carlioz and Touati (1986) substituted ineffective mutant alleles for the *sod*A and *sod*B genes in the *E. coli* chromosome (which respectively specify the Mn-SOD and Fe-SOD enzymes). Absence of either enzyme alone had no major effect on aerobic growth of the organism in nutrient-rich or minimal media. The double mutant strain lacking both Fe-SOD and Mn-SOD remained able to grow aerobically in the rich medium but only anaerobically in the minimal medium. Aerobic growth could be achieved in the minimal medium by introducing a SOD-overproducing plasmid into the double mutant or by supplementing the medium with a mixture of some twenty amino acids. They concluded that total deprivation of SOD created in *E. coli* a conditional sensitivity to oxygen, due presumably to some inhibitory action of superoxide anion on amino acid biosynthesis. It was also relevant that although the double mutant had a normal content of catalase it was highly sensitive to exogenous H_2O_2.

The picture that emerges is one of independent but mutually supportive roles for catalase and superoxide dismutase so that the organism which encounters oxygen is the better equipped to combat the deleterious effects of O_2 consumption if it contains high levels of both enzymes, particularly if it utilizes O_2 at a high rate via mechanisms which are prone to generate H_2O_2 and $O_2 \cdot {}^-$. On the other hand, possession of these enzymes does not necessarily guarantee aerotolerance, e.g., some strains of *Campylobacter fetus* ss. *fetus* contain very high levels of SOD and contain catalase but are still microaerophilic (Kikuchi and Suzuki, 1984). Yet in *E. coli*, as in many other bacteria, still more enzymes have been identified as contributing to oxygen protection. Thus Carlsson and Carpenter (1980) found that the *rec*-A-dependent DNA repair system of *E. coli* was possibly even more important than its catalase in affording protection against exogenous H_2O_2. Intracellular peroxidases have also received attention. Zavadova *et al.* (1974) reported the isolation of an aerotolerant mutant of *C. perfringens* which had apparently acquired a non-flavin NADH peroxidase. An aerotolerant mutant of the normally microaerophilic bacterium *Spirillum volutans* lacked catalase activity, had no greater SOD activity than its parent, but possessed three-fold greater peroxidase activity (Padgett and Krieg, 1986).

Glutathione and glutathione dehydrogenase have also been singled out for mention as potential contributors to O_2 tolerance. From studies with glutathione-negative mutants of *E. coli* Morse and Dahl (1978) concluded that this thiol was the key protective against oxygen enhancement of radiation damage, while among mutants of *E. coli* super-sensitive to H_2O_2, some were catalase-negative, others lacked glutathione reductase and still others possessed normal levels of both these enzymes (Barbado *et al.*, 1983).

Taken together, these protective agencies serve as the 'last ditch' defence of an organism against oxygen toxicity. However, the need for these protectives would be lessened if the oxygen was unable to gain access to those intracellular sites where H_2O_2 and oxygen radicals are generated. Oxygen scavenging by some mechanism which did not give rise to harmful products would, therefore, constitute a valid primary defence. Several moderate anaerobes apparently have this ability. Exposure of cultures of such anaerobes to concentrations of oxygen greater than that at which they can still continue to grow may, in the short term, prove bacteriostatic rather than bactericidal, with the ill-effects being autogenously remedied as soon as anaerobic conditions are re-established. Quastel and Stephenson (1926) found that bubbling air through cultures of *C. sporogenes* for up to 24 h halted growth but was not lethal to the organism; when the O_2 was ultimately removed growth resumed, although after a pronounced lag. In *C. acetobutylicum* (O'Brien and Morris, 1971a) oxygen consumption in this first bacteriostatic phase of response to oxygen was effected by a soluble NADH oxidase that produced water, and not H_2O_2, as its product. A similar enzyme was reported in *Peptostreptococcus anaerobius* which behaves as a moderate anaerobe despite its total lack of catalase, peroxidase and SOD (Hoshino *et al.*, 1978). Only when such a reductive first phase defence is overwhelmed by excessive exposure to oxygen would the bactericidal second phase of oxygen toxicity supervene (Morris, 1976).

Thus it appears that the aetiology of oxygen toxicity in anaerobes is multifactorial. The degree of oxygen tolerance that is displayed by any anaerobe under any given circumstances will be the result of two opposing sets of tendencies:
1. Those that make for oxygen sensitivity (Table 3.1);
2. Those that predispose to oxygen tolerance.

Lessons learned from the ways in which living cells themselves combat oxygen toxicity can be applied to devising means of detoxifying the culture media used to support growth of obligate anaerobes. Addition of catalase or peroxidase to culture media can aid the recovery of obligate anaerobes plated on such media (e.g., Frölander and Carlsson, 1977; Harmon and Kautter, 1977). The beneficial effects are probably predominantly due to destruction of peroxides that have accumulated in the medium; exogenous catalase and/or SOD cannot, of course, afford protection against intracellular generation of oxygen radicals (e.g., Mateles and Zuber, 1963). The first phase protection by oxygen scavenging has also been reproduced in the procedure introduced by Adler and Crow (1981), which enables aerointolerant organisms to be grown in the presence of air in liquid culture or on the surface of an agar plate. A filter-sterilized suspension of comminuted cell membrane 'particles', derived from aerobically-grown *E. coli*, is added to the growth medium. The aerobic respiratory rate of this membrane preparation is sufficiently great to remove O_2 as quickly as it diffuses into the medium, without build-up of harmful concentrations of H_2O_2.

Redox potential

It has long been recognized that cultures of anaerobic bacteria are notable for their highly reducing character. Indeed many obligate anaerobes will grow only if inoculated into prereduced media, while other anaerobes will grow in liquid culture with shorter lag periods if a compact inoculum is used and the medium is not disturbed until growth is well established. In such circumstances it would seem that the organisms generate the reducing conditions they require in their immediate vicinity and progressively invade the remainder of the medium as the zone of reduction is extended. The major advances in isolating, purifying and handling obligate anaerobes have, therefore, not only consisted of improved means of incubation and transfer of cultures in oxygen-free atmospheres but also of the incorporation of appropriate reducing agents into culture media.

The reduction–oxidation potential (redox potential, Eh) of any redox couple is a measure of its spontaneous tendency to donate electrons to (or accept electrons from) a standard redox couple (the hydrogen electrode) which is decreed to possess a zero redox potential and thus serves as the null point of a scale of redox potentials. By convention, redox couples which are more oxidizing than the standard hydrogen electrode have positive values on this

Table 3.1 Multiple factors that predispose to oxygen sensitivity

1. Little excess 'reducing power' available for O_2 scavenging and for reduction of medium constituents
2. O_2 serves as an avid 'alien' electron acceptor, thus diverting normal fermentative electron flow
3. O_2 consumed at high rate by mechanisms yielding potentially toxic products ($O_2 \cdot^-$, H_2O_2 etc.)
4. Inadequate cellular levels of protective agents (especially catalase, SOD)
5. Catalysts of key metabolic (especially free energy conserving) pathways are especially sensitive to oxidative inhibition
6. Constituents of growth medium liable to autoxidation yielding toxic products

redox scale, and couples that are more reducing than the standard couple have negative values. As in all cases of electron flow, the potential (impulsive tendency to flow from donor to acceptor couple) is measured in volts (more practically, mV). The redox couple itself possesses a reduced form (reductant) and an oxidized form (oxidant) and their interconversion serves as the source or sink of electrons:

$$\text{reductant} \rightleftharpoons \text{oxidant} + \text{electron(s)}$$

The standard redox potential of the couple is displayed when reductant and oxidant are present at equal activities (ideal concentrations). For this reason it is also termed the mid-point potential (E_m) and is registered when the couple is 50 per cent oxidized. In many redox couples protons are additionally involved in the reduction–oxidation (which thus appears as hydrogen transfer), e.g.,

$$AH_2 \rightleftharpoons A + 2H^+ + 2e^-$$

Because of the involvement of H^+ ions (and, similarly, whenever either the reductant or oxidant is a weak acid or base) the prevailing pH will affect the redox potential of the couple considerably. For this reason the redox potentials of couples of biological interest are generally reported as modified standard values (E_0' values) at a given pH and stated temperature. Incidentally, this pH effect must never be overlooked in studies with anaerobes whose cultures can become very acidic. Thus, for example, the E_0' value of the NAD/NADH$_2$ couple, which is -320 mV at pH 7, is -230 mV at pH 4, while for the H^+/H_2 couple the E_0' at pH 7 is -420 mV, but at pH 4 it is -240 mV. When the reduced and oxidized components are not present in equal concentrations the measured redox potential (Eh) will in part be determined by the ratio of their concentrations. The relevant (Nernst) relationship predicts that at 30°C

$$\text{Eh} = E_0' + \frac{0.06}{n} \log \frac{[\text{oxidant}]}{[\text{reductant}]}$$

where n = number of mol of electrons available for donation per mol reductant converted to oxidant.

A redox couple whose oxidized form is strikingly different in colour from its reduced form can serve as a visual indicator of the Eh of a culture medium, the two forms adopting the concentration ratio determined by the Eh of the environment and the E_0' value of the couple. The Nernst relationship predicts that with such a redox indicator, if n = 2, the useful visual 'span' from 99 per cent fully oxidized to 99 per cent totally reduced, would represent an Eh change from about 60 mV below the indicator's E_0' value to 60 mV above it. However, by using suitable combinations of several redox indicators a much wider range of Eh values can be monitored.

However, difficulties are encountered (a) when the Eh indicator also has the properties of a pH indicator; (b) when its oxidation or reduction is not freely reversible, and (c) in anaerobic bacterial cultures when the indicator is a substrate for direct (preferential) reduction (Ruseler–van Embden and Both-Patoir, 1985). Direct measurement of Eh values of aqueous media or microbial cultures is very simply accomplished via an immersed, polished platinum electrode connected via a millivoltmeter to a reference electrode of known Eh (e.g., a normal calomel electrode of Eh +280 mV). However, one must remember to take account of the potential of the reference electrode when converting the ΔEh value registered by the millivoltmeter to the required Eh value of the test system. For example, if a given bacterial culture registered -350mV versus a normal calomel electrode its Eh value would, in fact, be -70 mV. For further information on the meaning and measurement of redox potential, see Hewitt (1950), Jacob (1970), Morris (1974).

If the measurement of culture Eh values is easy, their interpretation is by no means straightforward (Morris, 1975). The normal uninoculated culture medium, and quite certainly a growing bacterial culture, will contain a host of interactive redox couples. If all were freely reversible and in perfect equilibrium then one could argue that what is being reported to the sensing platinum electrode is the Eh value of the dominant (terminal) couple. Otherwise, the hope is that by monitoring the measurable changes in Eh, one is at least following shifts in the relative magnitudes of the contending reducing and oxidizing agencies that are the prime determinants of Eh. The redox potential of an aerated nutrient medium changes in proportion to the logarithm of the dissolved oxygen tension, and it has proved possible, by following changes in Eh', to reproduce required degrees of oxygenation in pure bacterial cultures growing in a simple defined medium (e.g., Wimpenny, 1969; Harrison, 1972). Just as a pH buffer possesses a quantifiable capacity to moderate pH change in a system to which it is added so, by the incorporation of suitable redox couple(s), can the redox buffering of an aqueous medium be enhanced. The use of such poising agents in the formulation of anaerobic culture media is now more or less routine, and a suitable redox indicator may also be incorporated to allow visual verification that the prepared, sterilized medium is deaerated and at a suitably low Eh. Resazurin is particularly useful in this role, since at pH 7 the reduction of resorufin (pink) to dihydroresorufin (colourless) is readily reversible and has a E_0' of -51 mV, so that it becomes colourless at an Eh of about -110 mV. Quastel and Stephenson (1926) highlighted the vulnerability to oxidation of intracellular $-$SH groups and simple thiols (e.g., cysteine, thioglycollate, dithiothreitol) are commonly employed as poising agents in anaerobic media. Thus it is the more disturbing to find that exogenous thiols (especially thioglycollate) can sometimes enhance the bactericidal action of oxygen on some moderate anaerobes (Griffiths and Shoesmith, 1977). Other reducing agents can be employed; for example, titanium (III) citrate (E_0' -480 mV), which has the additional advantage of acting as its own indicator, being blue–violet when reduced and colourless when oxidized (Moench and Zeikus, 1983).

The fact that aeration of a culture medium will invariably lead to an increase in its Eh has been a complication in the debate concerning the prime causes of oxygen toxicity to anaerobes. As a result of this, there have been microbiologists who have contended that nothing is of importance in anaerobiosis except maintenance of a low Eh value; to quote Louis DS Smith (1967), 'this is an interesting point of view; it would be more readily acceptable if there were direct evidence to support it'. In fact, with moderate anaerobes at least, the evidence tends to indicate that growth in the absence of oxygen is possible in media maintained at what might be considered to be unpropitiously high Eh values. Thus Knaysi and Dutky (1936) found that a butyric clostridium, which in anaerobic culture would normally develop an Eh of -265 mV, would not grow under aerobic conditions at Eh $+335$ mV, but *would* grow at this positive Eh value when it was maintained with ferricyanide instead of oxygen. O'Brien and Morris (1971a) made similar findings with *C. acetobutylicum* which, while ferricyanide was the oxidant, would grow at an Eh of $+370$ mV, but with oxygen as oxidant could tolerate an increase of culture Eh to only -50 mV (with dissolved oxygen < 1 μM). Although growth was halted at higher oxygen concentrations the culture maintained its viability for at least six hours, during which period there was a high rate of oxygen consumption due to NADH oxidase activity. Onderdonk et al. (1976) obtained similar results with *B. fragilis* ss. *fragilis* growing in chemostat culture. A ferricyanide-provoked increase in Eh from -56 mV to $+300$ mV (both at pH 5.6) showed no significant effect but exposure to oxygen, which raised the Eh to $+250$ mV, was bacteriostatic.

Such findings illustrate the truism that culture Eh values should not be considered independently of the agents that have conspired to produce them. Thus, although many attempts have been made to define upper and lower limits of media Eh values compatible with growth of various anaerobes, most such attempts are ill-advised, or subject to misinterpretation. It is evident that the upper Eh limit will be determined by several factors, e.g., inoculum size, nutritional status of the medium, whether or not the Eh is externally controlled, and if so by what agency. In a survey of the then published values of upper limits of Eh for growth of some clostridia, Morris and O'Brien (1971) found that these varied between species (from $+60$ mV to $+230$ mV) and even within one species, thus Eh values of $+90$ mV to $+238$ mV were recorded for *C. perfringens* growing in various rich media. The range of minimum Eh values developed in growing cultures of clostridia was equally broad (-138 mV to -450 mV).

The low Eh maintained by anaerobes is sensibly viewed merely as evidence that they grow best in reducing conditions whose establishment and maintenance calls for a reducing contribution from the organisms themselves. But anaerobes differ in their capacity to generate the required disposable reducing power. Possibly those EOS bacteria which are the most reliant on external reducing agents (chemical electron donors or companion microorganisms) channel their endogenous electron flow through minimally flexible, tightly coupled routes that cannot concurrently accommodate a requirement for reduction of additional (exogenous) electron acceptor.

Anaerobic conservation of free energy

Three main mechanisms of free energy conservation are available to suitably equipped anaerobic organisms:

> fermentation;
> anoxygenic photophosphorylation;
> anaerobic respiration.

In the second and third of these processes electron transport via a series of membrane-integrated redox carriers causes protons to be vectorially 'pumped' across the cell membrane. Since this membrane is proton impermeable and electrically insulating, the result is the creation of an electrochemical gradient of protons across it (the interior being electronegative and alkaline with respect to the exterior). By channelling the return flow of protons into the cell via a proton-translocating ATP synthetase, the transmembrane protic current is harnessed to generate ATP. The proton-motive force ($\Delta\bar{\mu}H^+$) is additionally exploited to accomplish selective, active ion and substrate transport via membrane-integrated 'porters' (Nicholls, 1982). Bacteria reliant on photophosphorylation are of little concern in a clinical context, nor indeed are those major groups of anaerobically respiring organisms that can utilize nitrate (denitrifiers), sulphate (sulphate reducers) or carbon dioxide (methanogens) as terminal electron acceptors. We shall, therefore, grant them unjustifiably scant attention. However, it should be noted that anaerobic respiration is not wholly restricted to the utilization of inorganic electron acceptors. The capacity to utilize fumarate as terminal oxidant in an anaerobic respiratory process (with succinate formation) is surprisingly widespread and can serve as the sole source of energy for growth of *Wolinella succinogenes* and as a supplementary source of ATP in other obligate or facultative anaerobes, including *B. fragilis* and *E. coli* (Kroger, 1977).

The anaerobic bacteria associated with man are predominantly fermentative organisms. It used to be possible to define fermentation as 'a metabolic process whereby an exogenous organic compound is utilized by a series of biochemical reactions to generate ATP solely via substrate level phosphorylation (SLP)'. In the light of recent findings this definition has now to be amended to comprehend the possibility of some additional ATP production arising from the simultaneous generation of a $\Delta\bar{\mu}H^+$.

Fermentation

Possibly the most striking feature of anaerobic bacterial fermentation is the enormous diversity of compounds that may be used as substrates and the biochemical ingenuity that this requires. The molar yield of ATP (i.e., mol ATP formed per mol of substrate consumed) is much less in a fermentation than in any other form of respiration. Thus, to achieve comparable rates of growth and of biomass yields, the fermentative organism has to consume substrate(s) proportionately more rapidly and in much larger quantities. It is, therefore, understandable that in the course of anaerobic evolution the metabolic machinery has been developed to ferment not only carbohydrates, amino acids, purines and pyrimidines but various other organic compounds besides. The voracious appetite of the fermentative organism could also explain why so many of them are able to create additional supplies of fermentable substrates by degrading the polymeric molecules in which these are present. They do this by secreting a range of degradative, hydrolytic enzymes. Thus, among the clostridia alone we find species that can degrade polysaccharides (starch, cellulose, pectin, chitin), others that secrete proteases and still others that produce lipases or nucleases. In a clinical context, several of the products recognized as invasive agents, virulence factors or toxins have been identified as exoenzymes, e.g., collagenases, hyaluronidases, phospholipases, neuraminidases. The gram-positive species are at an advantage in such enzyme secretion as they lack an outer cell membrane. However, some gram-negative anaerobes also attack macromolecules, although the responsible enzymic activities usually tend to remain more closely cell-associated. Furthermore, although some bacteria are specialized fermenters and are able to utilize only a very restricted range of substrates, others are able to secrete a battery of hydrolytic enzymes and ferment a wide selection of different monomers. Again, turning to the genus *Clostridium* for examples, we can identify saccharolytic species (e.g., *C. difficile*), proteolytic species (e.g., *C. histolyticum*), species that can ferment sugars and/or mixtures of amino acids (e.g., *C. perfringens*), and species that are neither saccharolytic nor proteolytic but display a much more limited fermentative capacity, for example utilizing purines (*C. acidiurici*) or single amino acids (*C.*

cochlearium, which degrades only glutamate or glutamine or histidine).

Whereas in aerobic respiration the organic substrate is generally totally oxidized (to yield CO_2 and water) in fermentative processes organic end products are accumulated in proportion to the quantities of substrate(s) consumed. Furthermore, although a few fermentations are lytic in character, e.g., the fermentation of arginine by *Streptococcus faecalis*, the vast majority are balanced oxidation–reduction processes and the end products are very often small chain length fatty acids and alcohols together with CO_2 and possibly H_2. The pH of a fermentative culture, therefore, very frequently falls as the fermentation proceeds, except in those instances when a nitrogenous substrate is being used and liberation of ammonia or various amines tends to moderate the drop in pH or, occasionally, even causes it to increase. The metabolic routes followed in the various fermentations are so diverse that we cannot consider them in detail here. Such information is readily available in many textbooks, e.g., Gottschalk (1979), Schlegel (1986), and review articles: general (Gottschalk and Andreesen, 1979), carbohydrates (Wood, 1961; Morris, 1985), amino acids (Barker, 1981), purines (Vogels and van der Drift, 1976; Schiefer-Ullrich *et al.*, 1984).

However elaborate the fermentation pathway, it can be rendered more comprehensible if the fact that it operates under two constraints is borne in mind. First, the process must produce a net yield of ATP, so that any initial ATP-consuming activation of the substrate must be more than compensated for in the pathway by ATP generation from SLP reactions. Second, perfect redox balance is sustained, so that accumulation of any product that is more oxidized than the substrate must be compensated for by production of an equivalent quantity of a more highly reduced end product. This balance is enforced by the need to recycle the electron carriers that, as essential coenzymes, participate in the fermentative process.

All of these elements are apparent in the simplest of all fermentations, the homofermentative production of lactic acid from glucose (Fig. 3.1). In this fermentation, two ATP molecules are directly or indirectly utilized in the preliminary conversion of glucose to fructose diphosphate. This molecule is cleaved to give two molecules of triose phosphate whose oxidation and concurrent phosphorylation

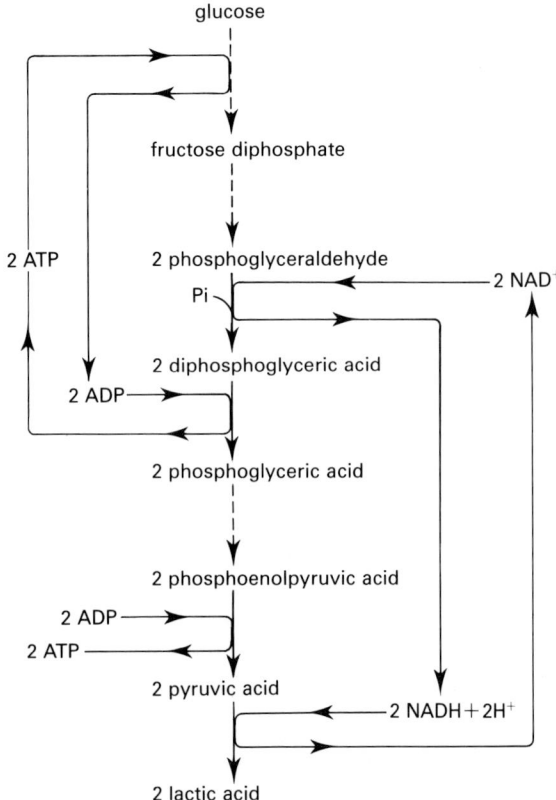

Fig. 3.1 Outline pathway for the fermentation of glucose by a homolactic fermentative bacterium

yields two molecules of 1,3-diphosphoglycerate. This compound possesses a sufficiently high phosphate transfer potential to phosphorylate ADP and, as a result of this SLP reaction, two molecules of 3-phosphoglycerate and two molecules of ATP are produced (thus making good the ATP used to produce fructose bisphosphate). Conversion of the two molecules of 3-phosphoglycerate to phosphoenol pyruvate (PEP) allows a further two ATP molecules to be produced via the SLP reaction wherein PEP yields pyruvate. Thus, each molecule of glucose gives rise to four molecules of SLP substrate(s) and the reaction sequence which yields pyruvate has achieved the required objective of net synthesis of (two) ATP molecules. But so far, glucose has been oxidized to two molecules of pyruvate at the expense of the reduction of two molecules of NAD at the triose phosphate dehydrogenation step. To ensure continuance of 'throughput', the $NADH_2$ must be reoxidized to supply the electron acceptor for more triose phosphate. This is accomplished by electron transfer from NADH$_2$ to pyruvate, with the resulting accumulation of two molecules of the lactate, which is the end product of the fermentation. Reviewing the pathway in the manner appropriate to any redox fermentation we note:

1. There is a net yield of ATP, viz., two ATP/mol of glucose consumed. (The SLP substrates are identifiable as 1,3-diphosphoglycerate and phosphoenolpyruvate.)
2. The necessary redox balance is maintained—in this instance there is a single end product (lactate) which is at the same reduction–oxidation level as the substrate (glucose).

The electron donating reaction (triose phosphate dehydrogenation) and the electron accepting reaction (pyruvate reduction) are identifiable, as is the intermediary electron carrier (the NAD/NADH$_2$ couple).

It is a surprising fact that despite the very varied chemical structures of the organic compounds fermented by anaerobic bacteria, all of these fermentations rely upon the operation of a very limited number of SLP reactions. Aside from the two implicated in glycolysis (Fig. 3.1) the other most common 'energy-rich' SLP substrates are (fatty) acid phosphates (e.g., acetylphosphate, propionylphosphate, butyrylphosphate) while carbamonyl phosphate and N^{10}-formyl-tetrahydrofolate are implicated in a few specialist fermentations (Thauer et al., 1977).

The 'simple' linear fermentation pathway of homolactic fermentation is a comparative rarity. More commonly the pathway is branched, leading to the production of the several end products of a 'mixed fermentation'. Thus the fermentation of glucose by E. coli, although based on the fructose diphosphate route to pyruvate, yields a mixture of end products—lactate, acetate, ethanol, succinate, formate, CO$_2$ and H$_2$. This occurs because, in addition to reduction to yield lactate, the pyruvate is further cleaved to yield acetyl CoA and formate, the latter in turn being degraded to CO$_2$ and H$_2$ (Fig. 3.2). Conversion of acetyl CoA to acetyl phosphate enables another molecule of ATP to be formed when the acetyl phosphate, an SLP substrate, yields acetate. Ethanol is produced by reduction of acetyl CoA by NADH$_2$. Here we have an example of the sort of compromise that many fermentative organisms have to strike between the opportunity to generate more ATP (acetyl CoA → acetate) and the need to sustain electron flow (in this instance by employing acetyl CoA as an ancillary electron acceptor). The succinate in figure 3.2 probably arises as the result of fumarate

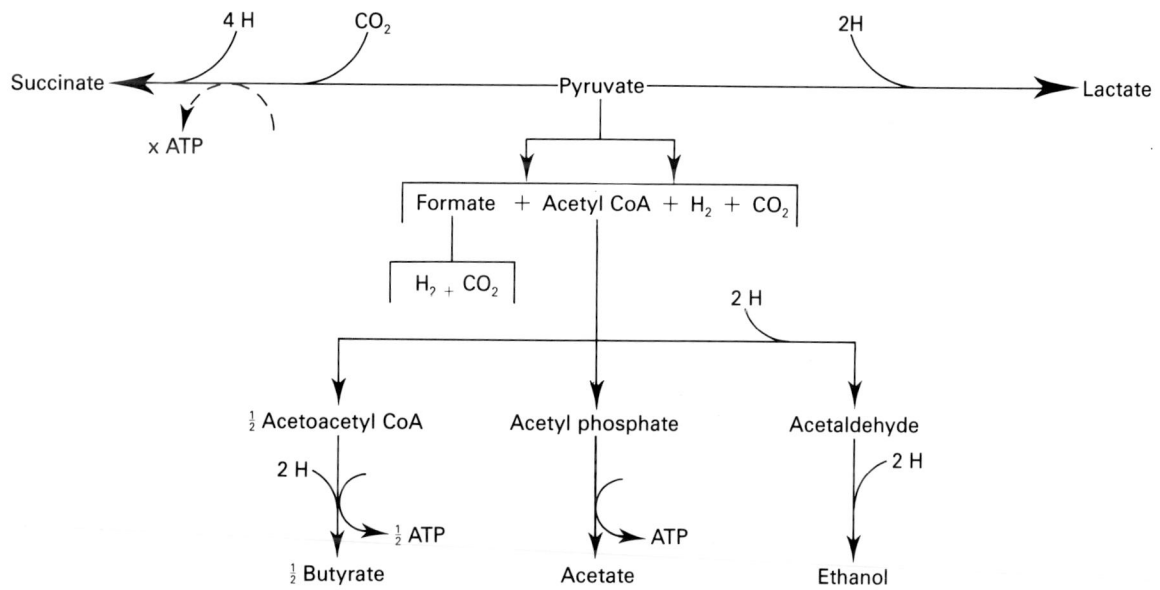

Fig. 3.2 Some of the fermentative fates of pyruvate

reduction accomplished by a membrane-integrated fumarate reductase. Succinate production thus fulfils a dual function—electron acceptance and indirect ATP production by the anaerobic respiratory device of $\Delta\bar{\mu}H^+$ generation. Other organisms utilize pyruvate by other means, yielding such products as butanediol, butyrate, butanol and isopropyl alcohol; all these have either associated ATP generation or electron acceptance in common (Morris, 1986). Such are the complexities that result from the different routes of utilization of a single key metabolite by a variety of organisms.

It is worth nothing that when any branched fermentation pathway is employed, the proportional yields of the various possible products (and the ATP gain) can vary depending upon growth conditions. These can include the composition of the culture medium, the growth rate of the culture, the prevailing pH and temperature and whether the liberated gases are allowed to accumulate (and so to increase in pressure) or are removed. It is for this reason that procedures of anaerobic bacterial identification based on patterns of end product accumulation (obtained, for example, by GLC analysis) should always be undertaken under the precise culture conditions stipulated.

By the homolactic pathway (see Fig. 3.1), the oxidation–reduction fermentation of a single substrate (i.e., glucose) is accomplished. In this process the substrate must be metabolized by a route which generates the electron donor(s) and the terminal electron acceptor(s). However, certain anaerobic bacteria require these to be supplied separately, as joint substrates of the fermentation. Thus *C. kluyveri* ferments a mixture of ethanol (as electron donor) and acetate (as electron acceptor) to yield butyrate and caproate (note that not all fermentations are catabolic processes). A more general example of a dual requirement for a pair of fermentable substrates is given by those proteolytic anaerobes that ferment pairs of amino acids via the so-called Stickland reaction (Seto, 1980) (Fig. 3.3). Thus *C. sporogenes* will ferment valine (as electron donor) only when simultaneously supplied with a second amino acid capable of serving as electron acceptor (e.g., proline). The fermentation, like all other Stickland-type fermentations (Bader *et al.*, 1982), generally results in deamination of both the amino acids, and the oxidative and reductive branches are associated with ATP generation. Even so, certain bacteria can ferment single amino acids, e.g., *C. tetanomorphum* and *Fusobacterium fusiformis* both ferment glutamate to yield acetate, butyrate, NH_3, H_2 and CO_2. However, despite the fact that they ferment the same substrate to yield identical products, these organisms employ quite different fermentative pathways. *C. tetanomorphum* ferments glutamate via β-methyl aspartate in a vitamin B_{12}-dependent pathway, while *F. fusiformis* degrades the glutamate via a route in which glutaconyl CoA and crotonyl CoA are intermediates. Thus it is not possible to deduce, from knowledge of the fermentative end products alone, by what metabolic route these have been formed.

At the outset we adopted a view of fermentation which did not exclude the possibility of ancillary ATP production by mechanisms other than SLP. We

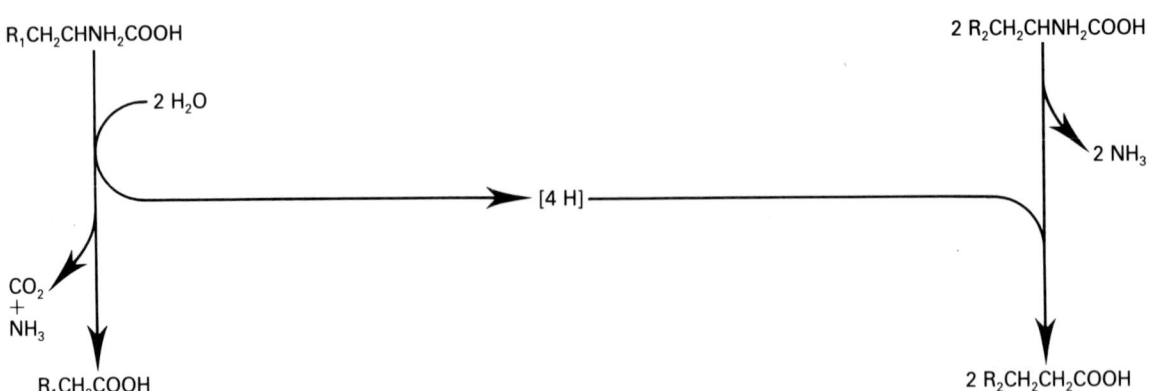

Fig. 3.3 Outline summary of the interaction of electron donor amino acid ($R_1CH_2CHNH_2COOH$) and electron acceptor amino acid ($R_2CH_2CHNH_2COOH$) in the classic Stickland reaction

have encountered one such example of $\Delta\tilde{\mu}H^+$-generation coupled to formation of succinate via fumarate reductase activity at the cell membrane, a process which helps to explain the unusually high molar growth yields of *B. fragilis* on carbohydrates (which it ferments to yield succinate and propionate as well as acetate and CO_2). A transmembrane $\Delta\tilde{\mu}H^+$ can also be generated as a result of:
1. the electrogenic excretion of certain fermentation end products in symport with protons, e.g., lactate excretion by *Str. cremoris* (Otto et al., 1980);
2. the activity of certain Na^+ ion dependent membrane-associated decarboxylases which act as Na^+ ion pumps, e.g., methylmalonyl CoA decarboxylase of *Veillonella alcalescens* and glutaconyl CoA decarboxylase of *Acidaminococcus fermentans* (Hilpert and Dimroth, 1982; Buckel and Semmler, 1983);
3. reduction of certain enoates (Bader and Simon, 1983);
4. reduction of proline by *C. sporogenes* (Lovitt et al., 1986).

For these, and other reasons (Thauer and Morris, 1984), the bioenergetics of anaerobic bacterial growth remains a fruitful field of research.

Mention has been made of the release of H_2 gas along with CO_2 in many anaerobic bacterial fermentations. In the *E. coli* mixed acid fermentation the H_2 has its origin in formate, but in many H_2-evolving obligate anaerobes, e.g., clostridia, H_2 production serves as an 'escape valve' for excess reducing power. The hydrogenases which catalyse this redox reaction are widespread in anaerobes (Adams et al., 1981). Some bacteria can even contain two forms of the enzyme, one preferentially catalysing proton reduction (the H_2-evolving hydrogenase) and another utilizing H_2 as an electron donor in various reductions (an H_2-uptake hydrogenase). The $H^+/\frac{1}{2}H_2$ redox couple at pH 7 has an E_0'' value of about -420 mV, so that a low-potential electron donor (generally reduced ferredoxin) serves as reductant in the hydrogenase catalysed production of H_2 gas. In the clostridia, reduced ferredoxin is generated directly by the oxidation of pyruvate to acetyl CoA plus CO_2 but additional ferredoxin reduction, and hence H_2 production, can have its origin in NAD(P)H-oxidation. These organisms contain soluble NAD(P)H-ferredoxine oxidoreductases, whose operation is vital for maintaining the redox balance during metabolism. In an organism such as *C. pasteurianum* the proportions of acetate and butyrate that will be produced must, of course, depend on the reducing power (i.e., NADH) available for reduction of acetyl CoA to butyryl CoA. This in turn will be determined by the proportion of NADH, generated during glycolysis, which serves as substrate for the NADH–ferredoxin reductase and hence is diverted (via hydrogenase) to the production of H_2. The coupling between the fates of acetyl CoA and NADH is effected via allosteric control of the NADH–ferredoxin reductase by the prevailing ratio of acetyl CoA/CoA (Thauer et al., 1977).

This provides a glimpse of the relatively simple, but effective, mechanisms of metabolic self-control that anaerobes must employ to ensure the most efficient operation of their fermentative processes. Little is currently known of the regulatory controls that must determine the choices anaerobes make when presented with a mixture of different fermentable substrates. Preferential use of certain carbohydrates could be explained by competition for shared transport mechanisms such as the PEP-dependent phosphotransferase system, but not all sugars will in fact enter the cell by a single common route (Booth and Morris, 1975). Even more interesting is the problem of substrate selection faced by an organism that is both saccharolytic and proteolytic when it is introduced into a glucose–peptone medium. We are aware that the concentration of glucose in the culture medium can greatly alter the fermentation products of organisms such as *C. bifermentans, C. sporogenes* and *Peptostr. anaerobius* (Turton et al., 1983), but perhaps not so expected is the finding that amino acid fermentation can sometimes take precedence over glucose utilization, as in *C. tetanomorphum* (Anthony and Guest, 1968). Furthermore, evidence is accumulating to suggest the comparative widespread use that may be made of mixed Stickland-type reactions in which a carbohydrate may supply the electron donor and an amino acid serve as the electron acceptor, e.g., fermentation of glucose plus proline by *C. sporogenes* (Lovitt et al., 1986).

Metabolism in the absence of oxygen

The absence of O_2, which enables cultures of anaerobic bacteria to accomplish many novel reductive processes, also limits their ability to accomplish those oxidative reactions which in aerobes are undertaken via an oxygenative mechanism. Thus it is notable that paraffins and aromatic compounds are not rapidly mineralized in anaerobic environments, while aerobically they are catabolized by a wide variety of microbes which possess appropriate mono- or dioxygenases. However, anaerobic cleavage of the aromatic ring, for example of substituted benzene compounds, can be accomplished by the device of initial ring reduction (Sleat and Robinson, 1984). Thus, in a mixed anaerobic population, compounds such as benzoate, phenylacetate and cinnamate can be completely degraded to yield methane and CO_2. The oxygen atom(s) that are incorporated into the intermediates of anaerobic aromatic degradation have their origin in water (Vogel and Grbić-Galić, 1986) and prehydroxylated aromatic rings are rather more liable to anoxic attack (Thauer and Morris, 1984). Oxygenases and other O_2-dependent enzymes may also be employed for some biosynthetic purposes in aerobes, but often they serve as alternative means to accomplish a synthesis that is achieved by an equivalent O_2-independent route in anaerobes. Aerobic routes for the synthesis of nicotinic acid, unsaturated fatty acids and ubiquinones are examples of such alternative pathways (Morris, 1985). On the other hand, since O_2 is necessary for the cyclization reaction whereby squalene is converted into lanosterol, anaerobes, although able to synthesize squalene, cannot form sterols. Incidentally, the capacity to synthesize haem (and hence cytochromes) is not limited to aerobic organisms; the presence of cytochromes in propionibacteria, *Bacteroides* spp., some *Clostridium* spp. and even certain methanogens, as well as in more blatant anaerobic respirers such as sulphate-reducing bacteria, is well established (Gottschalk, 1981). In those instances when cytochrome and catalase synthesis depends upon the provision of haemin in the growth medium (e.g., *B. fragilis*; Macy *et al.*, 1975) the presence or absence of O_2 is irrelevant to the biosynthetic lesion that prevents the organism from synthesizing its own haem. Interestingly, a blue bile pigment (bactobilin) has been identified in *C. tetanomorphum* (Brumm *et al.*, 1983).

Anaerobic bioconversions

As is to be expected, the most novel and spectacular biotransformations that are particularly associated with obligately anaerobic bacteria are those which implicate low redox potential reductions exploiting what Benemann and Valentine (1971) have loosely termed 'high energy electrons'. Low-potential electron donors (ferredoxins, flavodoxins) play a particularly prominent role in such processes, and obligate anaerobes in general are rich in metalloproteins (molybdenum, selenium, tungsten, vanadium) and also in novel coenzymes (some also contain transition metals such as cobalt or nickel) whose structures, although equipping them for their anaerobic roles, often cause them to be oxygen-sensitive. Besides hydrogenase, the most celebrated of the essentially anaerobic metallo–protein complexes is undoubtedly nitrogenase, whose possession confers N_2-fixing ability on a range of organisms. The methanogens, which are among the most O_2-sensitive of anaerobes, are particularly dependent on a number of unique cofactors and enzymes, which together accomplish the exergonic reduction of CO_2 by H_2. Though in some anaerobes, e.g., *C. butyricum*, production of formate, and thence C_1 units employed in biosynthesis, is accomplished by the activity of a pyruvate formate lyase, in many others formate is produced by the direct reduction of CO_2 with reduced ferredoxin, or sometimes NADPH, as the electron donor. Reduction of CO_2 via CO to acetate is another peculiarly anaerobic process accomplished by organisms such as *C. thermoaceticum* and *Acetobacterium woodii* (Pezacka and Wood, 1984) and is of considerable significance because it can form the basis of a novel route for anaerobic autotrophic growth with CO_2 (or CO) as the sole source of carbon (Fuchs, 1986). Other reductive carboxylations dependent on reduced ferredoxin as the low-potential electron donor include formation of pyruvate from acetyl CoA, 2-oxoglutarate from succinyl CoA and the formation, from fatty acyl CoA esters, of those oxo-acids that are the precursors of branched chain amino acids (valine, isoleucine, leucine) (Gottschalk, 1981; Monticello *et al.*, 1984).

$$R-CH_2COSCoA + CO_2 \xrightarrow{Fer_{red} \quad Fer_{ox}} RCH_2COCOSCoA$$

Studies of the reductive biotransformations accomplished by anaerobes have lagged well behind studies of the oxidative processes undertaken by aerobic micro-organisms. Our current knowledge is largely limited to the sporadic findings of those who have primarily been interested in the fate of some xenobiotic in an anaerobic environment (e.g., in the gut of man and animals, in sewage sludge, marine or lake sediments etc.) or who have been interested, for industrial purposes, in achieving a chiral reduction and/or substitution of a key intermediate in some chemical synthesis. Into the first category fall accounts of reductive dehalogenation of pesticides and related molecules (Jagnow et al., 1977; Tsuchiya and Yamaha, 1984) and anaerobic biotransformation of bile acids (e.g., MacDonald et al., 1983; Owen, 1985; Edenharder and Schneider, 1985). For obvious reasons, connected with the possibility of effecting highly specific chemical changes in clinically important steroid molecules, ketosteroid reductase and hydroxysteroid dehydrogenase activities have been studied in anaerobes such as C. innocuum, C. absonum, C. limosum, and C. paraputrificum. A new species of Clostridium, designated C. scindens, has been identified as a human intestinal bacterium with a desmolytic activity on corticoids (Morris et al., 1985). A cholesterol-reducing Eubacterium with a growth requirement for plasmenylethanolamine has been described (Mott and Brinkley, 1979) and inactivation of contraceptive steroid hormones by clostridia in the human gut has been reported (Bokkenhauser et al., 1983). Interest in anaerobes as agents of chiral syntheses has been stimulated by a series of reports, by Simon et al. (1985), on the specificity of the hydrogenation of various enoates, ketones and 2-oxocarboxylates.

Specific biotransformations undergone by individual species of anaerobes have also formed the basis for qualitative biochemical tests helpful in the identification of these species, e.g., formation of caproic acid and p-cresol during the growth of C. difficile on a nor-leucine–tyrosine broth (Nunez-Montcil et al., 1983), production of p-hydroxyhydrocinnamic acid from tyrosine by Peptostr. anaerobius (Lambert and Moss, 1980), production of indole 3-propanoic acid plus 3(p-hydroxyphenyl)propanoic acid in cultures of C. sporogenes (Jellet et al., 1980).

There is also considerable interest in the anaerobic reduction of nitrate and of aliphatic and aryl nitro compounds. This is not only because of the perceived dangers of nitrosoamine production in the human gut, but also because metronidazole is a nitroimidazole whose reduction is responsible for its 'antianaerobe' activity; while chloramphenicol can be inactivated by reduction of its aryl nitro group. When addition of nitrate to an energy-limited chemostat culture of an anaerobe provokes an increase in specific growth yield, the natural inclination is to infer the ability of the organism to effect additional ATP synthesis by anaerobic nitrate respiration (i.e., nitrate dissimilation via a membrane-associated nitrate reductase). This was concluded for C. perfringens (Hasan and Hall, 1975) although it was long ago reported that nitrate reduction by clostridia frequently yielded ammonia (Woods, 1938) and it was indeed subsequently shown that both the nitrate reductase and nitrite reductase of C. perfringens are soluble enzymes which employ reduced ferredoxin as electron donor (Seki et al., 1982; Sekiguchi et al., 1983). The extra ATP, whose production is reflected in an enhanced growth yield, doubtless arises from nitrate acting as an alien electron sink and causing a diversion of the electron flow in the mixed acid fermentation of C. perfringens, resulting in a lower yield of butyrate and a higher yield of acetate. However, nitrate reduction proceeds via nitrite, which may not only interact with amines to yield nitrosoamines but may also be toxic in its own right. Some anaerobes are more susceptible to nitrite inhibition than others; fortunately C. botulinum is nitrite-sensitive, possibly due to its direct interaction with iron–sulphur proteins such as ferredoxin and the clostridial pyruvate dehydrogenase (Reddy et al., 1983). Ferredoxin also plays a direct role in the reduction of aliphatic and aromatic nitro compounds by clostridia. Thus the anaerobic clostridium C. acetobutylicum was found to inactivate chloramphenicol by ferredoxin-mediated reduction of its nitro group (O'Brien and Morris, 1971b), although pyridine nucleotide-dependent enzymes can, in some systems, serve as aryl nitro reductases (Angermaier and Simon, 1983; Hill and Cook, 1986)

Reduction of metronidazole and other nitroimidazoles supplies us with a particularly important example of the capacity of obligate anaerobes (and certain microaerophiles or facultative anaerobes) to accomplish low-potential reduction. Again, with C. acetobutylicum (O'Brien and Morris, 1972) it was found that metronidazole acted as a preferential acceptor of electrons from reduced ferredoxin, being itself wholly reduced in a six electron-consuming process, to yield the corresponding (unstable) 5-amino derivative. As with any other preferred exogenous electron acceptor, such diversion of reducing power could prove injurious, but much more devastating is the fact that in the course of its reduction a highly toxic intermediate is produced, probably the 5-NO$_2\cdot$ (nitro free radical) derivative, which is formed as the product of the first one-electron reduction step. It has been reported (Edwards et al., 1982) that this toxic radical interacts with the DNA of the target organism causing fission of phosphodiester bonds adjacent to thymine residues, thus releasing thymidine nucleotides as a result of strand breakage. This could explain the particular susceptibility of anaerobes whose DNA is particularly rich in AT (e.g., *Bacteroides* and *Clostridium* spp. and *Trichomonas vaginalis*). Yet Sigeti et al. (1983) found no metronidazole-induced degradation of chromosomal DNA in *B. fragilis*, nor did they find any nicking or strand breakage in plasmid DNA molecules carried by the organism. They concluded that the primary effect of metronidazole was on DNA replication. Oxygen can interact with the free radical derivative to reform metronidazole (although simultaneously generating superoxide anion) and aminothiols such as glutathione and cysteine have some protective effect. Disproportionation of the nitro free radical derivative would yield nitrite ions.

In summary, therefore, the specificity of action of metronidazole against anaerobes resides in the ability of anaerobes to reduce it, as has now been demonstrated with many different species (Reynolds, 1981; Lockerby et al., 1985). It is intriguing that, while microaerophilic *Campylobacter* spp. are susceptible to metronidazole (Freyiere et al., 1984), certain obligately anaerobic organisms, such as propionibacteria and actinomycetes, are metronidazole-resistant (Beerens et al., 1986).

Metabolic relationships in communities of anaerobes

It would be wrong to conclude this chapter without acknowledging the fact that, although for investigative purposes individual species of anaerobes are grown in monoculture, in nature they are predominantly components of highly complex, mixed microbial communities. It is their intimate association with oxygen-scavenging facultative anaerobes, microaerophiles, and even aerobes, that helps to explain their widespread distribution and occupancy of niches which, superficially, would appear to be unpropitiously aerated. For example, the oral microflora, which has many obligately anaerobic components, inhabits an obviously complex and inconstant ecosystem within which it is possible to distinguish a mosaic of ecological niches, each supporting a bacterial community whose composition may be subject to temporal change (by autogenic and/or allogenic succession). Even in more stable anoxic environments anaerobes generally live in communities (consortia) sustained by complex syntrophic relations between component species. Indeed, by their concerted metabolic activities the component organisms in such a consortium will often effect biochemical transformations which none alone would be able to accomplish. A prime example of the potential benefits of such co-operativity between anaerobes is provided by the phenomenon of interspecies hydrogen transfer (Wolin, 1982). A hydrogen-producing fermentative anaerobe (hydrogenogen) can live in close association with a hydrogen-consuming partner organism (hydrogenotroph), which generally utilizes the H$_2$ as electron donor in its anaerobic respiration. Thus methanogens and sulphate-reducing bacteria are common hydrogenotrophs whose activities lower the prevailing hydrogen tension in the environment to such an extent that the hydrogen-producer benefits from the increased exergony of the reaction on which it relies for ATP synthesis. Thus, species of obligate anaerobes exist which have only been isolated in dual culture with a companion hydrogenotrophic bacterium. The phenomenon also extends the range of fermentative organisms, by making available to them substrates whose anaerobic oxidation would, under other circumstances, not be able to sustain their growth. For example, the acetogens, which play an essential role in methanogenic digestion of organic materials, are obligate hydrogenogens which convert short chain fatty acids or alcohols to acetate; both the acetate and the H$_2$ they produce are then avidly consumed by methane-generating bacteria. Yet H$_2$ transfer is not the only basis for metabolic co-operation in anaerobic consortia. It is easy to imagine how a compound produced as the end product of fermentation of one organism can serve as the substrate for the fermentation or respiration of a companion organism. Thus, among oral bacteria, the lactate produced by a streptococcus can be utilized in the acetate- and propionate-yielding fermentation of a *Veillonella* (Delwiche et al., 1985). Such relationships amongst the components of the anaerobic rumen microflora have been extensively investigated (Hobson and Wallace, 1982). There are surely lessons here that can be applied to mixed polymicrobial infections (with their component pathogens, commensals and symbionts) including some mixed infections in which both aerobes and anaerobes are essential contributors to the disease syndrome (Gorbach, 1982). Elucidation of the roles of the component organisms in such mixed infections, as has been pointed out by Willis (1984), 'will dictate not only the proper clinical approach to their prevention and treatment but also the most appropriate use of broad- to narrow-spectrum antibiotics'.

References

Adams M W W, Mortenson L E, Chen J S. (1981). Hydrogenases. *Biochimica et Biophysica Acta* 594: 105–176.

Adler H I, Crow W D. (1981). A novel approach to the growth of anaerobic microorganisms. *Biotechnology and Bioengineering Symposia 1981* 11: 533–540.

Angermaier L, Simon H. (1983). On the reduction of aliphatic and aromatic nitro compounds by clostridia; the role of ferredoxin and its stabilization. *Hoppe-Seyler's Zeitschrift für Physiologische Chemie* 364: 961–976.

Anthony C, Guest J R. (1968). Deferred metabolism of glucose by *Clostridium tetanomorphum*. *Journal of General Microbiology* 54: 277–286.

Archibald F S, Duong M N. (1986). Superoxide dismutase and oxygen toxicity defenses in the genus *Neisseria*. *Infection and Immunity* 51: 631–641.

Bader J, Rauschenbach P, Simon H. (1982). On a hitherto unknown fermentation path of several amino acids by proteolytic clostridia. *FEBS Letters* 140: 67–72.

Bader J, Simon H. (1983). ATP formation is coupled to the hydrogenation of 2-enoates in *Clostridium sporogenes*. *FEMS Microbiology Letters* 20: 171–175.

Barbado C, Ramirez M, Blanco M A, Lopez-Barea J, Pueyo C. (1983). Mutants of *Escherichia coli* sensitive to hydrogen peroxide. *Current Microbiology* 8: 251–253.

Barker H A. (1981). Amino acid degradation by anaerobic bacteria. *Annual Review of Biochemistry* 50: 23–40.

Beerens H, Neut C, Romond C. (1986). Important properties in the differentiation of gram-positive non-sporing rods in the genera *Propionibacterium, Eubacterium, Actinomyces* and *Bifidobacterium*. In *Anaerobic bacteria in habitats other than man*, Symposium No. 13, Society for Applied Bacteriology, pp. 37–59. Edited by Barnes E M, Mead G C. Blackwell Scientific, Oxford.

Benemann J R, Valentine R C. (1971). High-energy electrons in bacteria. *Advances in Microbial Physiology* 5: 135–172.

Bokkenhauser V D, Winter J, Cohen B I, O'Rourke S, Mosbach E H. (1983). Inactivation of contraceptive steroid hormones by human intestinal clostridia. *Journal of Clinical Microbiology* 18: 500–504.

Booth I R, Morris J G. (1975). Proton motive force in the obligately anaerobic bacterium *Clostridium pasteurianum*: a role in galactose and gluconate uptake. *FEBS Letters* 59: 153–157.

Brumm P J, Fried J, Friedmann C. (1983). Bactobilin: blue bile pigment from *Clostridium tetanomorphum*. *Proceedings of the National Academy of Sciences of the USA* 80: 3943–3947.

Buckel W, Semmler. (1983). Purification, characterization and reconstitution of glutaconylCoA decarboxylase, a biotin-dependent sodium pump from anaerobic bacteria. *European Journal of Biochemistry* 136: 427–434.

Carlioz A, Touati D. (1986). Isolation of superoxide dismutase mutants in *Escherichia coli*: is superoxide dismutase necessary for aerobic life? *The EMBO Journal* 5: 623–630.

Carlsson J, Carpenter V S. (1980). The $recA^+$ gene product is more important than catalase and superoxide dismutase in protecting *Escherichia coli* against hydrogen peroxide toxicity. *Journal of Bacteriology* 142: 319–321.

Carlsson J, Wrethen J, Beckman G. (1977). Superoxide dismutase in *Bacteroides fragilis* and related *Bacteroides* species. *Journal of Clinical Microbiology* 6: 280–284.

Delwiche E A, Pestka J J, Tortorello M L. (1985). The Veillonellae: gram-negative cocci with a unique physiology. *Annual Review of Microbiology* 39: 175–193.

Edenharder R, Schneider J. (1985). 12 β dehydrogenation of bile acids by *Clostridium paraputrificum, Clostridium tertium* and *Clostridium difficile* and epimerization of C-12 of deoxycholic acid by co-cultivation with 12 α-dehydrogenating *Eubacterium lentum*. *Applied and Environmental Microbiology* 49: 964–968.

Edwards D I, Knox R J, Skolimowski I M, Knight R C. (1982). Mode of action of nitroimidazoles. *European Journal of Chemotherapy and Antibiotics* 2: 65–72.

Freyiere A M, Gille Y, Tigaud D, Vincent P. (1984). In vitro susceptibilities of forty *Campylobacter fetus* subsp. *jejuni* strains to niridazole and metronidazole. *Antimicrobial Agents and Chemotherapy* 35: 145–146.

Fridovich I. (1986). Biological effects of the superoxide radical. *Archives of Biochemistry and Biophysics* 247: 1–11.

Frölander F, Carlsson J. (1977). Bactericidal effect of anaerobic broth exposed to atmospheric oxygen tested on *Peptostreptococcus anaerobius*. *Journal of Clinical Microbiology* **6**: 117–123.

Fuchs G. (1986). CO_2 fixation in acetogenic bacteria: variations on a theme. *FEMS Microbiology Letters* **39**: 181–213.

Fulghum R S, Worthington J M. (1984). Superoxide dismutase in ruminal bacteria. *Applied and Environmental Microbiology* **48**: 675–677.

Funada H, Hattori K-I, Kosakai N. (1978). Catalase-negative *Escherichia coli* isolated from blood. *Journal of Clinical Microbiology* **7**: 474–478.

Gest H. (1980). The evolution of biological energy-transducing systems. *FEMS Microbiology Letters* **7**: 73–77.

Gorbach S L. (1982). The pre-eminent role of anaerobes in mixed infections. *Journal of Antimicrobial Chemotherapy* **10** Suppl. A: 1–6.

Gottesman S. (1984). Bacterial regulation: global regulatory networks. *Annual Review of Genetics* **18**: 415–442.

Gottschalk G. (1979). *Bacterial metabolism*. Springer, New York.

(1981). The anaerobic way of life of prokaryotes. In *The Prokaryotes*, pp. 1415–1423. Edited by Starr M P et al. Springer, Heidelberg.

Gottschalk G, Andreesen J R. (1979). Energy metabolism in anaerobes. In *Microbial biochemistry*, vol 21. International Review of Biochemistry. pp. 85–115. Edited by Quayle J R. University Park Press, Baltimore.

Gregory E M, Fanning D D. (1983). Effect of heme on *Bacteroides distasonis* catalase and aerotolerance. *Journal of Bacteriology* **156**: 1012–1018.

Gregory E M, Moore W E C, Holdeman L V. (1978). Superoxide dismutase in anaerobes: survey. *Applied and Environmental Microbiology* **35**: 988–991.

Griffiths J L, Shoesmith J G. (1977). The enhancement of the bactericidal effect of oxygen on obligate anaerobes by sodium thioglycollate. *Proceedings of the Society for General Microbiology* **4**: 145.

Harmon S M, Kautter D A. (1977). Recovery of clostridia on catalase-treated plating media. *Applied and Environmental Microbiology* **33**: 762–770.

Harrison D E F. (1972). Physiological effects of dissolved oxygen tension and redox potential on growing populations of micro-organisms. *Journal of Applied Chemistry and Biotechnology* **22**: 417–440.

Hasan S M, Hall J B. (1975). The physiological function of nitrate reduction in *Clostridium perfringens*. *Journal of General Microbiology* **87**: 120–128.

Hassan H M, Fridovich I. (1979). Superoxide dismutase and its role for survival in the presence of oxygen. In *Strategies of Microbial Life in Extreme Environments*, pp. 179–193. Edited by Shilo M. Dahlem Konferenzen, Berlin.

Hewitt J, Morris J G. (1975). Superoxide dismutase in some obligately anaerobic bacteria. *FEBS Letters* **50**: 315–318.

Hewitt L F. (1950). *Oxidation–reduction potentials in bacteriology and biochemistry*, 6th edn. Churchill Livingstone, Edinburgh.

Hill M J, Cook A R. (1986). Nitrogen metabolism in the animal gut. In *Anaerobic bacteria in habitats other than man*, Symposium No. 13, Society for Applied Bacteriology, pp. 287–301. Edited by Barnes E M, Mead G C. Blackwell Scientific, Oxford.

Hilpert W, Dimroth P. (1982). Conversion of the chemical energy of methylmalonylCoA decarboxylation into a Na^+ gradient. *Nature* **296**: 584–585.

Hobson P N, Wallace R J. (1982). Microbial ecology and activities in the rumen. *CRC Critical Reviews in Microbiology* **9**: 165–225 and **11**: 253–319.

Hoshino E, Frölander F, Carlsson J. (1978). Oxygen and the metabolism of *Peptostreptococcus anaerobius* VPI 4330-1 *Journal of General Microbiology* **107**: 235–248.

Imlay J A, Linn S. (1988). DNA damage and oxygen radical toxicity. *Science* **240**: 1302–1309.

Jacob H-E. (1970). Redox potential. In *Methods in microbiology*, vol 2. pp. 91–123. Edited by Norris J R, Ribbons D W. Academic Press, London.

Jagnow G, Haider K, Ellwardt P C. (1977). Anaerobic dechlorination and degradation of hexachlorocyclohexane isomers by anaerobic and facultatively anaerobic bacteria. *Archives of Microbiology* **115**: 285–292.

Jellet J J, Forrest T P, MacDonald I A, Maric T J, Holdeman L V. (1980). Production of indole 3-propanoic acid and 3(*p*-hydroxyphenyl) propanoic acid by *Clostridium sporogenes*: a convenient thin-layer chromatography detection system. *Canadian Journal of Microbiology* **26**: 448–453.

Kikuchi H E, Suzuki T. (1984). An electrophoretic analysis of superoxide dismutase in *Campylobacter* sp. *Journal of General Microbiology* **130**: 2791–2796.

Kikuchi H E, Suzuki T. (1986). Quantitative method for measurement of aerotolerance of bacteria and its application to oral indigenous anaerobes. *Applied and Environmental Microbiology* **52**: 971–973.

Knaysi G, Dutky S R. (1936). The growth of a butanol *Clostridium* in relation to the oxidation–reduction potential and oxygen content of the medium. *Journal of Bacteriology* **31**: 137–143.

Kröger A. (1977). Phosphorylative electron transport with fumarate and nitrate as terminal hydrogen acceptors. In *Microbial energetics*, Symposium 27, Society for General Microbiology, pp. 61–93. Edited by Haddock B A, Hamilton W A. Cambridge University Press, Cambridge.

Lambert M A, Moss C W. (1980). Production of *p*-hydroxyhydrocinnamic acid from tyrosine by *Peptostreptococcus anaerobius*. *Journal of Clinical Microbiology* **12**: 291–293.

Lloyd D, James K, Williams J, Williams N. (1981). A membrane-covered probe for oxygen measurements in the nanomolar range. *Analytical Biochemistry* **116**: 17–21.

Lloyd D, Scott R I, Williams T N. (1983). Membrane inlet mass spectrometry-measurement of dissolved gases in fermentation liquids. *Trends in Biotechnology* 1: 60–63.

Lockerby D L, Rabin H R, Laishley E J. (1985). Role of the phosphoroclastic reaction of *Clostridium pasteurianum* in the reduction of metronidazole. *Antimicrobial Agents and Chemotherapy* 27: 863–867.

Loesche W J. (1969). Oxygen sensitivity of various anaerobic bacteria. *Applied Microbiology* 18: 723–727.

Loewen P C. (1984). Isolation of catalase-deficient *Escherichia coli* mutants and genetic mapping of *kat*E a locus that affects catalase activity. *Journal of Bacteriology* 157: 622–626.

Lovitt R W, Kell D B, Morris J G. (1986). Proline reduction by *Clostridium sporogenes* is coupled to vectorial proton ejection. *FEMS Microbiology Letters* 36: 269–273.

MacDonald I A, Hutchison D M, Forrest T P, Bokkenhauser V D, Winter J, Holdeman L V. (1983). Metabolism of primary bile acids by *Clostridium perfringens*. *Journal of Steroid Biochemistry* 18: 97–104.

Macy J, Probst I, Gottschalk G. (1975). Evidence for cytochrome involvement in fumarate reduction and adenosine 5'-triphosphate synthesis by *Bacteroides fragilis* grown in the presence of hemin. *Journal of Bacteriology* 123: 436–442.

Mateles R I, Zuber B L. (1963). The effect of exogenous catalase on the aerobic growth of clostridia. *Antonie van Leeuwenhoek* 29: 249–255.

McCord J M, Keele B B, Fridovich I. (1971). An enzyme-based theory of obligate anaerobiosis: the physiological function of superoxide dismutase. *Proceedings of the National Academy of Sciences of the USA* 68: 1024–1027.

McLeod J W, Gordon J. (1923). The problem of intolerance of oxygen by anaerobic bacteria. *Journal of Pathology and Bacteriology* 26: 332–343.

Meir E, Yagil E. (1984). Catalase-negative mutants of *Escherichia coli*. *Current Microbiology* 11: 13–18.

Moench T T, Zeikus J G. (1983). An improved preparation method for a titanium media reductant. *Journal of Microbiological Methods* 1: 199–202.

Monticello D J, Hadioetomo R S, Costilow R N. (1984). Isoleucine synthesis by *Clostridium sporogenes* from propionate or α-methylbutyrate. *Journal of General Microbiology* 130: 309–318.

Morris J G. (1974). *A Biologist's physical chemistry*, 2nd edn. Edward Arnold, London.
(1975). The physiology of obligate anaerobiosis. *Advances in Microbial Physiology* 12: 169–246.
(1976). Oxygen and the obligate anaerobe. *Journal of Applied Bacteriology* 40: 229–244.
(1984). Changes in oxygen tension and the microbial metabolism of organic carbon. In *Aspects of microbial metabolism and ecology*, pp. 59–96. Edited by Codd G A. Academic Press, London.
(1985). Anaerobic metabolism of glucose. In *Comprehensive biotechnology* vol 1, pp. 357–378. Edited by Cooney C L, Humphrey A E. Pergamon Press, Oxford.
(1986). Anaerobiosis and energy-yielding metabolism. In *Anaerobic bacteria in habitats other than man*, pp. 1–21. Edited by Barnes E M, Mead G C. Blackwell Scientific, Oxford.

Morris J G, O'Brien R W. (1971). Oxygen and clostridia: a review. In *Spore research 1971*, pp. 1–37. Edited by Barker A N, *et al*. Academic Press, London.

Morris G N, Winter J, Cato E P, Ritchie A E, Bokkenhauser V D. (1985). *Clostridium scindens* sp. nov., a human intestinal bacterium with a desmolytic activity on corticoids. *International Journal of Systematic Bacteriology* 35: 478–481.

Morse M L, Dahl R H. (1978). Cellular glutathione is a key to the oxygen effect in radiation damage. *Nature* 271: 660–661.

Mott G E, Brinkley A W. (1979). Plasmenylethanolamine: growth factor for cholesterol-reducing *Eubacterium*. *Journal of Bacteriology* 139: 755–760.

Nicholls D G. (1982). *Bioenergetics. An introduction to the chemiosmotic theory*. Academic Press, London.

Nunez-Monteil O L, Thompson F S, Dowell V R. (1983). Norleucine-tyrosine broth for rapid identification of *Clostridium difficile* by gas–liquid chromatography. *Journal of Clinical Microbiology* 17: 382–385.

O'Brien R W, Morris J G. (1971a). Oxygen and the growth and metabolism of *Clostridium acetobutylicum*. *Journal of General Microbiology* 68: 307–318.
(1971b). The ferredoxin-dependent reduction of chloramphenicol by *Clostridium acetobutylicum*. *Journal of General Microbiology* 67: 265–271.
(1972). Effect of metronidazole on hydrogen production by *Clostridium acetobutylicum*. *Archiv für Mikrobiologie* 84: 225–233.

Odom J M, Peck H D. (1984). Hydrogenase, electron-transfer proteins, and energy coupling in the sulfate-reducing bacteria *Desulfovibrio*. *Annual Review of Microbiology* 38: 551–592.

Onderdonk A B, Johnston J, Mayhew J W, Gorbach S L. (1976). Effect of dissolved oxygen and Eh on *Bacteroides fragilis* during continuous culture. *Applied and Environmental Microbiology* 31: 168–172.

Otto R, Sonnenberg A S M, Veldkamp H, Konings W N. (1980). Generation of an electrochemical proton gradient in *Streptococcus cremoris* by lactate efflux. *Proceedings of the National Academy of Sciences of the USA* 77: 5502–5506.

Owen R W. (1985). Biotransformation of bile acids by clostridia. *Journal of Medical Microbiology* 20: 233–238.

Padgett P J, Krieg N R. (1986). Factors relating to the aerotolerance of *Spirillum volutans*. *Canadian Journal of Microbiology* 32: 548–552.

Pasteur L. (1861). Animalcules infusoires vivant sans gaz oxygene libre et determinant les fermentations.

Comptes Rendus Hebdomadaies des Seances de l'Academie des Sciences **52**: 344–350.

Patel G B, Roth L A, Agnew B J. (1984). Death rates of obligate anaerobes exposed to oxygen and the effect of media prereduction on cell viability. *Canadian Journal of Microbiology* **30**: 228–235.

Pezacka E, Wood H G. (1984). The synthesis of acetylCoA by *Clostridium thermoaceticum* from CO_2, H_2, CoA and methyltetrahydrofolate. *Archives of Microbiology* **137**: 63–69.

Pritchard G G, Wimpenny J W T, Morris H A, Lewis M W A, Hughes D E. (1977). Effects of oxygen on *Propionibacterium shermanii* grown in continuous culture. *Journal of General Microbiology* **102**: 223–233.

Quastel J H, Stephenson M. (1926). Experiments on 'strict' anaerobes. 1. The relationship of *B. sporogenes* to oxygen. *Biochemical Journal* **20**: 1125–1132.

Reddy D, Lancaster J R, Cornforth D P. (1983). Nitrite inhibition of *Clostridium botulinum*: electron spin resonance detection of iron-nitric oxide complexes. *Science* **221**: 769–770.

Reynolds A V. (1981). The activity of nitro-compounds against *Bacteroides fragilis* is related to their electron affinity. *Journal of Antimicrobial Chemotherapy* **8**: 91–99.

Rolfe R D, Hentges D J, Campbell B J, Barrett J T. (1978). Factors related to the oxygen tolerance of anaerobic bacteria. *Applied and Environmental Microbiology* **36**: 306–313.

Ruseler-van Embden J G H, Both-Patoir H C. (1985). The applicability of redox-indicator dyes in strongly reduced media: their effect on the human fecal flora. *FEMS Microbiology Letters* **28**: 341–345.

Schiefer-Ullrich H, Wagner R, Dürre P, Andreesen J R. (1984). Comparative studies on the physiology and taxonomy of obligately purinolytic clostridia. *Archives of Microbiology* **138**: 345–353.

Schlegel H G. (1986). *General microbiology*. Cambridge University Press, Cambridge.

Schneider K, Schlegel H G. (1981). Production of superoxide radicals by soluble hydrogenase from *Alcaligenes eutrophus* H16. *Biochemical Journal* **193**: 99–107.

Schwartz A C. (1973). Anaerobiosis and oxygen consumption of some strains of *Propionibacterium* and a modified method for comparing the oxygen sensitivity of various anaerobes. *Zeitschrift für Allgemeine Mikrobiologie* **13**: 681–691.

Seki S, Higo K, Ishimoto M. (1982). Studies on nitrate reductase of *Clostridium perfringens*. III. Comparison of nitrate reductase from five strains of *Clostridium perfringens*. *Journal of General and Applied Microbiology* **28**: 541–550.

Sekiguchi S, Seki S, Ishimoto M. (1983). Purification and some properties of nitrite reductase from *Clostridium perfringens*. *Journal of Biochemistry, Tokyo* **94**: 1053–1059.

Seto B. (1980). The Stickland reaction. In *Diversity of bacterial respiratory systems*, vol 2. pp. 50–64. Edited by Knowles C J. CRC Press, Boca Raton.

Sigeti J S, Guiney D G, Davis C E. (1983). Mechanism of action of metronidazole on *Bacteroides fragilis*. *Journal of Infectious Disease* **148**: 1083–1089.

Simon H, Bader J, Günther H, Neumann S, Thanos J. (1985). Chiral compounds synthesised by biocatalytic reductions. *Angewandte Chemie International Edition, English* **24**: 539–553.

Sleat R, Robinson J P. (1984). The bacteriology of anaerobic degradation of aromatic compounds. *Journal of Applied Bacteriology* **57**: 381–394.

Smith L DS. (1967). Anaerobes and oxygen. In *The anaerobic bacteria*, pp. 14–37. Edited by Fredette V. Institute of Microbiology and Hygiene of Montreal University, Montreal.

Smith M W, Neidhardt F C. (1983a). Proteins induced by aerobiosis in *Escherichia coli*. *Journal of Bacteriology* **154**: 344–350.

Smith M W, Neidhardt F C. (1983b). Proteins induced by anaerobiosis in *Escherichia coli*. *Journal of Bacteriology* **154**: 336–343.

Stouthamer A H, de Vries W, Niekus H G D. (1979). Microaerophily. *Antonie van Leeuwenhoek* **45**: 5–12.

Strauch K L, Lenk J B, Gamble B L, Miller C G. (1985). Oxygen regulation in *Salmonella typhimurium*. *Journal of Bacteriology* **161**: 673–680.

Tally, F P, Stewart P R, Sutter V L, Rosenblatt J E. (1975). Oxygen tolerance of fresh clinical anaerobic bacteria. *Journal of Clinical Microbiology* **1**: 161–164.

Tally F P, Goldin B R, Jacobus N V, Gorbach S L. (1977). Superoxide dismutase in anaerobic bacteria of clinical significance. *Infection and Immunity* **16**: 20–25.

Thauer R K, Morris J G. (1984). Metabolism of chemotrophic anaerobes: old views and new aspects. In *The microbe 1984*, Part II, 36th Symposium of the Society for General Microbiology, pp. 123–168. Edited by Kelly D P, Carr N G. Cambridge University Press, Cambridge.

Thauer R K, Jungermann K, Decker K. (1977). Energy conservation in chemotrophic anaerobic bacteria. *Bacteriological Reviews* **41**: 100–180.

Tsuchiya T, Yamaha T. (1984). Reductive dechlorination of 1,2,4-trichlorobenzene by *Staphylococcus epidermidis* isolated from intestinal contents of rats. *Agricultural and Biological Chemistry* **48**: 1545–1550.

Turton L J, Drucker D B, Ganguli L A. (1983). Effect of glucose concentration in the growth medium upon neutral and acidic fermentation end products of *Clostridium bifermentans*, *Clostridium sporogenes* and *Peptostreptococcus anaerobius*. *Journal of Medical Microbiology* **16**: 61–67.

van Winkelhoff A J, van Steenbergen T J M, de Graaff J. (1986). Oxygen tolerance of oral and non-oral black pigmented *Bacteroides* species. *FEMS Microbiology Letters* **33**: 215–218.

Vogel T M, Grbić-Galić D. (1986). Incorporation of oxygen from water into toluene and benzene during anaerobic fermentative transformation. *Applied and Environmental Microbiology* **52**: 200–202.

Vogels G D, van der Drift C. (1976). Degradation of purines and pyrimidines by microorganisms. *Bacteriological Reviews* **40**: 403–468.

Willis A T. (1984). Anaerobic bacterial diseases now and then; where do we go from here? *Reviews of Infectious Diseases* **6** Suppl 1: S293–S299.

Wimpenny J W T. (1969). Oxygen and carbon dioxide as regulators of microbial growth and metabolism. In *Microbial growth*, 19th Symposium, Society for General Microbiology, pp. 161–197. Edited by Meadow P, Pirt S J. Cambridge University Press, Cambridge.

Winkelman J W, Clark D P. (1986). Anaerobically induced genes of *Escherichia coli*. *Journal of Bacteriology* **167**: 362–367.

Wolin M J. (1982). Hydrogen transfer in microbial communities. In *Microbial interactions and communities in biotechnology*, vol 1. pp. 323–356. Edited by Bull A T, Slater J H. Academic Press, London.

Wood W A. (1961). Fermentation of carbohydrates and related compounds. In *The bacteria*, vol 2. pp. 59–149. Edited by Gunsalus I C, Stanier R Y. Academic Press, New York.

Woods D D. (1938). The reduction of nitrate to ammonia by *Clostridium welchii*. *Biochemical Journal* **32**: 2000–2008.

Závadová M, Mikulik K, Sebald M. (1974). Aerotolerant mutant of *Clostridium perfringens* A. *Csl. Epidemiologie, Mikrobiologie Immunologie* **23**: 249–256.

4

Genetics of anaerobes

N P Minton and D E Thompson

Extrachromosomal genetic elements
    Cryptic plasmids
    Bacteriocinogenic plasmids
    Toxigenic plasmids
    Antibiotic resistance plasmids
    Bacteriophages
Genetic exchange in anaerobes
    Transformation
    Conjugation
    Shuttle vector development
Gene cloning
Conclusion
References

The attainment of our present level of understanding of the molecular complexities of the bacterial cell has relied heavily on genetic analysis. Although the genetic techniques available for studying *Escherichia coli* have reached an unparallelled level of sophistication, considerable progress has also been made in the genetic analysis of other aerobic and facultatively anaerobic bacterial species, i.e., *Bacillus*, *Pseudomonas*, *Streptomyces* and Enterobacteriaceae generally. In contrast, the genetics of strictly anaerobic bacteria have received scant attention. During the last decade, however, progress has been made towards remedying this situation with a number of anaerobic species. This is particularly true of members of the genera *Clostridium* and *Bacteroides*, and the developments made with these two genera form the basis of this review. However, it should also be noted that a rudimentary genetic analysis of other anaerobic bacteria has also taken place during the same period. For example, plasmids have been isolated from *Bifidobacterium* (Sgorbati et al., 1982), *Pediococcus* (Daeschel and Klaenhammer, 1985) *Fusobacterium* (Vandenbergh et al., 1982), *Actinomyces* (Vandenbergh et al., 1982). Bacteriophages have been identified in *Actinomyces* (Tylenda et al., 1985a,b), *Veillonella* (Totsuka, 1976) and *Propionibacterium* (Jong et al., 1975) and various genes have been cloned in *E. coli* from *Actinomyces* (Donkersloot et al., 1985), methanogenic archaebacteria (Rauhut et al., 1984; Wich et al., 1984, Cue et al., 1985; Muller et al. 1985), *Ruminococcus* and *Butyrivibrio* (Romaniec et al., 1986) and *Propionibacterium* (Yongsmith and Cole, 1986).

Members of the genus *Bacteroides* are some of the most predominant species in the human colon, accounting for approximately 30 per cent of all faecal isolates. They undoubtedly play an important role in the colonic ecosystem, and certain *Bacteroides* spp. are opportunist pathogens. *Bacteroides fragilis* and, to a lesser extent, *B. distasonis, B. thetaiotaomicron* and *B. ovatus*, can cause a variety of infections including bloodstream infections as well as abscesses of the brain, lung and abdominal cavity (Salyers, 1984). Certain species are also important in a veterinary context, e.g., *B. nodosus* causes ovine footrot. From taxonomic considerations, the genus *Bacteroides* appears to represent a fairly homologous assemblage of micro-organisms (but see Chapter 5).

In contrast, the genus *Clostridium* consists of a heterologous collection of micro-organisms whose principle characteristics may be defined as anaerobic, gram-positive, endospore-forming rods. Their importance to human and veterinary medicine resides in the severity of the toxic clostridial diseases (as in botulism and tetanus), the destructive nature of clostridial wound infections (as in gas gangerene) or the extent of outbreaks of disease (as in food poisoning by *C. perfringens* and antibiotic-associated diarrhoea and colitis due to *C. difficile*).

The elucidation of the molecular basis of bacterial pathogenicity and invasiveness would benefit considerably from genetic analysis of these two important anaerobic genera, while the application of recombinant DNA methodology could lead to the production of more effective vaccines. Similarly, genetic analysis can serve to define the molecular mechanisms responsible for the widely reported rise in antibiotic-resistant isolates. The development of genetic systems for *Clostridium* has a wider purpose outside medical confines; many of the saccharolytic clostridia may be used as biocatalysts in the production of chemical feedstocks from renewable biomass. These processes are now receiving renewed attention in the light of the finite nature of fossil fuel supplies (Rogers, 1986).

Extrachromosomal genetic elements

The procedures employed in the genetic analysis of bacteria depend almost totally on extrachromosomal elements (ECEs). These ECEs may take the form of either plasmid or bacteriophage DNA, both of which may reside within the cell either as autonomously replicating units or integrated into the host genome. In the following section naturally occurring ECEs isolated from *Clostridium* and *Bacteroides* spp. will be considered, while in a later section the construction of artificial elements will be described.

Cryptic plasmids

Plasmids are widespread in *Clostridium* and *Bacteroides* spp. and although functions have not been assigned to many of them (i.e., they are cryptic) those for which functions are known are listed in Table 4.1. The majority of the plasmids found in *Bacteroides* spp. are functionally cryptic. Stiffler *et al.* (1974) first reported the presence of plasmids with mol.wt(10^6) of 2.7, 4.0, and 16 in *B. fragilis* strains. Since then cryptic plasmids ranging in size from 2.5×10^6 to 100×10^6 mol.wt have been reported in *B.fragilis* (Tinnel and Macrina, 1976; Salaki *et al.*, 1976; Mays and Johnson, 1979; Welch *et al.*, 1979; Macrina *et al.*, 1981; Wallace *et al.*, 1981; Mays *et al.*, 1982; Callihan *et al*, 1983; Guiney *et al.*, 1983; Beul *et al.*, 1985), *B. thetaiotaomicron* (Tinnel and Macrina, 1976; Rashtchian *et al.*, 1982; Callihan *et al.*, 1983; Guiney *et al.*, 1983), *B. uniformis* (Mays and Johnson, 1979), *B. distasonis* (Wallace *et al.*, 1981; Callihan *et al.*, 1983; Beul *et al.*, 1985), *B. ovatus* (Callihan *et al.*, 1983; Smith and Macrina, 1984; Smith, 1985a), *B. vulgatus* (Callihan *et al.*, 1983) and *B. eggerthii* (Shoemaker *et al.*, 1985). Some effort has been made to classify such plasmids on the basis of size, DNA homology and restriction digest patterns. Mays and Johnson (1979) demonstrated that a 4.4-kb plasmid from *B. uniformis* was homologous with plasmids from 15 of 23 *Bacteroides* strains tested. Callihan *et al.* (1983) assigned the plasmids of 15 different *Bacteroides* isolates to three homology classes by Southern hybridization and restriction endonuclease mapping. A similar study by Beul *et al.* (1985) defined the plasmids of 11 different isolates into five size classes. Plasmids of equal size exhibited homology and gave identical restriction patterns with 17 endonucleases. The function or usefulness of these plasmids awaits further studies; however, a number of them are now being used as the basis of *E. coli–Bacteroides* shuttle vectors, e.g., pB8-51 from *B. eggerthii* (Shoemaker *et al.*, 1985) and pBI143 from *B.ovatus* (Smith, 1985a).

Screening of *Clostridium* spp. has also revealed an abundance of naturally occurring ECEs, the majority of which have no known function. Cryptic plasmids

Table 4.1 Plasmids isolated from *Clostridium* spp. and *Bacteroides* spp. to which a function has been assigned

Species	Plasmid	Size	Function	Reference
C. perfringens	pIP404	5.7 Md	bacteriocin	Ionesco and Bouanchaud (1973); Ionesco *et al.* (1976)
C. perfringens	pCW4	5.6 Md	bacteriocin	Mihelc *et al.* (1978)
C. perfringens	?	5.5 Md	bacteriocin	Li *et al.* (1980)
C. perfringens	?	75 Md	beta-toxin (?)	Duncan *et al.* (1978)
C. perfringens	pCW3	47 kb	Tc^r, Tra$^+$	Rood *et al.* (1978b)
C. perfringens	PJIR5/6	47 kb	Tc^r, Tra$^+$	Abraham and Rood (1985a)
C. perfringens	pJIR2	67 kb	Tc^r, Tra$^+$	Abraham and Rood (1985a)
C. perfringens	pJIR4	50 kb	Tc^r, Tra$^+$	Abraham and Rood (1985a)
C. perfringens	pJU124	38.8 kb	Tc^r, Tra$^+$	Heefner *et al.* (1984)
C. perfringens	pIP401	54 kb	Tc^r Cm^r, Tra$^+$	Brefort *et al.* (1977; 1978)
C. perfringens	pIP402	63 kb	Em^r Cc^r, Tra$^+$	Brefort *et al.* (1977; 1978)
C. perfringens	pHB101	2.1 Md	caseinase	Blaschek and Solberg (1981)
C. tetani	pCL1	75 Md	tetanus toxin	Finn *et al.* (1984)
C. cochlearium	?	?	methylmercury degradation, Tra$^+$	Pan-Hou *et al.* (1980)
B. ovatus	pBI136	82 kb	Cc^r Em^r, Tra$^+$	Smith and Macrina (1984)
B. fragilis	pBFTM10	14.6 kb	Cc^r Em^r, Tra$^-$	Tally *et al.* (1979)
B. fragilis	pBF4	41 kb	Cc^r Em^r, Tra$^+$	Welch *et al.* (1979)
B. thetaiotaomicron	pCP1	14.6 kb	Cc^r Em^r, Tra$^-$	Guiney *et al.* (1984a)
B. uniformis	pRYC3373	39.5 kb	Cm^r, Tra$^+$	Martinez-Suarez *et al.* (1985)

have been observed in *C. perfringens* (Ionesco *et al.*, 1976; Duncan *et al.*, 1978; Rokos *et al.*, 1978; Rood *et al.*, 1978a; Li *et al.*, 1980; Blaschek and Solberg, 1981; Heefner *et al.*, 1984), *C. cochlearium* (Pan-Hou *et al.*, 1980) *C. difficile* (Smith *et al.*, 1981; Muldrow *et al.*, 1982; Arai *et al.*, 1984; Hayter and Dale, 1984), *C. botulinum* and *C. sporogenes* (Scott and Duncan, 1978; Strom *et al.*, 1984; Weickert *et al.*, 1986), *C. absonum* (Roberts *et al.*, 1986), *C. acetobutylicum* (Truffaut and Sebald, 1983), *C. butyricum* (Minton and Morris, 1981; Urano *et al.*, 1983; Popoff and Truffaut, 1985), *C. beijerinckii* (Popoff and Trufaut, 1985) and *C. tetani* (Laird *et al.*, 1980). The plasmids isolated ranged in size from 1.9×10^6 to 81×10^6 mol.wt and, in a recent survey of *C. perfringens*, occurred in 69 per cent of the strains examined (Mahony *et al.*, 1986). In common with those species of *Bacteroides*, certain of these cryptic plasmids are being used as the basis of cloning vectors (Heefner *et al.*, 1984; Collins *et al.*, 1985; Luczak *et al.*, 1985; Brehm *et al.*, 1986).

Bacteriocinogenic plasmids

The production of inhibitory substances with bacteriocin-like properties by clostridia has been reported by numerous authors (see Nieves *et al.*, 1981) and schemes for the typing of *C. perfringens*, *C. difficile*, and *C. acetobutylicum* have been reported (Hongo *et al.*, 1968; Mahony, 1974, Mahony and Li, 1978; Satija and Narayan, 1980; Sell *et al.*, 1983). The production of bacteriocins by various strains of *Bacteroides* has also been reported (Salyers, 1984). Certain authors (Rowbury, 1977), believe that such inhibitory substances may only be classified as bacteriocins if they are plasmid encoded. The first evidence that the production of a clostridial bacteriocin was plasmid-mediated was presented by Ionesco and Bouanchaud (1973). Their studies showed that the production of bacteriocin BC5 by *C. perfringens* CPN50, and immunity to this inhibitory substance, could be cured by treating growing cells with acriflavin (0.5 per cent curing frequency). Cured derivatives (Per$^-$ mutants) lacked the satellite DNA observed in the parent strain after dye-buoyant centrifugation of cell lysates. The plasmid carried by strain CPN50 was subsequently called pIP404, and was shown to have a mol.wt of 5.7×10^6 (Ionesco *et al.*, 1976). In the same study pIP404 was shown to be mobilized to other strains of *C. perfringens* by a large conjugal cryptic plasmid, pIP405 (32.4×10^6 mol.wt), a co-resident in strain CPN50. Transconjugants inheriting pIP404 became Per$^+$. In a separate study Mihelc et al. (1978) obtained similar evidence that the bacteriocin of *C. perfringens* strain CW55 was encoded by a plasmid of 5.6×10^6 mol.wt, pCW4. In this case mobilization of pCW4 was not achieved with the conjugal tetracycline resistance (Tc^r) plasmid pCW3. Interestingly, two further bacteriocinogenic strains of *C. perfringens* have been shown to carry plasmids of remarkably similar size to pIP404 and pCW4. These are the plasmid pIP405, carried by strain CP590 (Brefort et al., 1977; 1978), and a plasmid of 5.6×10^6 mol.wt, carried by strain 28 (Li et al., 1980). In the latter case a Per$^-$ mutant was isolated which no longer contained the plasmid. However, the similar sizes of these plasmids are probably coincidental, as the properties of bacteriocin 28 (Li et al., 1980) differ from those of BC5 (Ionesco et al., 1976), while plasmids pIP404 and pCW4 give dissimilar restriction endonuclease patterns (Garnier and Cole, 1986).

C. butyricum NCIB 7423 produces a bacteriocin (butyricin 7423, Clarke et al., 1975) whose mode of action resembles that of the colicin E1 and K group of bacteriocins (Konisky, 1978), but in addition appears to inhibit the cell membrane ATPase of vegetatively grown *C. pasteurianum* (Clarke and Morris, 1976). This particular strain was shown to carry two plasmids, pCB101 (6.05 kb) and pCB102 (7.8 kb) which did not occur in other non-bacteriocinogenic strains of *C. butyricum* (Minton and Morris, 1981). However, whether either of these plasmids encode butyricin production remains to be determined, as the ability to produce this bacteriocin could not be cured with a range of curing agents. We have recently determined the entire nucleotide sequence of both plasmids (Minton et al., unpublished data) and a number of open reading frames (ORFs) capable of coding for a protein of mol.wt 32 000 have been identified. However, none of the encoded polypeptides contain the hydrophobic regions at their C-terminal ends which are characteristic of ionophores (Pugsley, 1984), nor do any of the predicted proteins exhibit significant primary sequence homology to known colicins (see Varley and Boulnois, 1984) or the perfringocin of pIP404 (Garnier and Cole, 1986).

Toxigenic plasmids

Plasmids have been found to be widespread in different toxigenic types of *C. perfringens* (Duncan et al., 1978; Rokos et al., 1978). Examination of the extrachromosomal DNA of five different toxigenic types, A–E, by electron microscopy revealed multiple (up to nine) size classes of plasmids, ranging in mol.wt (10^6) from 1.9 to 74.9. Out of a total of 22 strains examined, only four lacked plasmid DNA. In one particular strain, a toxigenic type C, curing of the ability to produce β-toxin correlated with the loss of a 75×10^6 mol.wt plasmid (Duncan et al., 1978), and a similar plasmid was also present in a second strain that produced β-toxin. However, no evidence was found to support the view that the production of α- or θ-toxin, or the possession of heat-resistant spores, was plasmid mediated (Rokos et al., 1978).

Two studies (Scott and Duncan, 1978; Strom et al., 1984) have shown that plasmids are widespread in both toxigenic and non-toxigenic *C. botulinum* strains and related organisms. In one study 56 per cent of the 68 toxigenic strains tested harboured one or more plasmids (Strom et al., 1984). However, although certain strain types appeared to carry seemingly identical plasmids (e.g., a common plasmid of mol.wt 81×10^6 in type G strains and an identical plasmid of mol.wt 11.5×10^6 in proteolytic type F isolates) no correlation between toxigenicity and a particular plasmid was established. Similar findings were obtained in the more recent study of Weckert et al. (1986).

Plasmid-associated toxigenicity in *C. tetani* was first suggested by Hara et al. (1977), and supportive evidence for this claim was obtained by Laird et al. (1980). In this latter study, 21 toxigenic strains were shown to carry a single large plasmid, while all non-toxigenic derivatives isolated from four of these strains had lost this plasmid DNA. Subsequently, a tetanus toxin-specific DNA probe (eight octadecamers, synthetic oligonucleotides complementary to the 5' end of the structural gene) was shown to hybridize to a 75×10^6 mol wt plasmid (pCL1) resident in a toxigenic strain, but not to the plasmid DNA of a non-toxigenic strain (Finn et al., 1984). A deletion variant, pCL2, of pCL1 was isolated from a non-toxigenic derivative of one of these strains. However, the DNA probe mixture still hybridized to pCL2 DNA, indicating that the DNA encoding the toxin N-terminus had not been deleted. Restriction endonuclease maps of both pCL1 and pCL2 were constructed, enabling the approximate localization and orientation of the structural gene on pCL1 to be determined. In more recent studies the structural gene has been cloned in *E. coli* cloning vectors as specific gene subfragments. However, in one study the source of the tetanus gene was a 75×10^6 mol.wt plasmid, presumably analogous to pCL1 (Eisel *et al.*, 1986), while in another, earlier study total chromosomal DNA was utilized (Fairweather *et al.*, 1986). Although in this latter study a large plasmid was not detected, the method employed in the purification of chromosomal DNA would not exclude high mol.wt plasmid DNA. Nucleotide sequence analysis (Eisel *et al.*, 1986; Fairweather and Lyness, 1986) has shown that the amino acid sequence of the tetanus toxin exhibits striking homologies with the partial amino acid sequence of botulinum toxins A, B, and E, indicating that the neurotoxins of *C. tetani* and *C. botulinum* may be derived from a common ancestral gene.

Antibiotic resistance plasmids

The selection of antibiotics for treatment of anaerobic infections in man has been limited to a few drugs, but until recently physicians caring for patients with such infections were assured that the activity of these drugs against anaerobes were stable. Little alteration in the susceptibility patterns of anaerobes has been seen since the mid-1950s. In the early 1970s, however, striking increases in resistance to tetracycline and erythromycin (Em) among isolates of *Bacteroides*, *Clostridium* and anaerobic cocci were reported (See Bawdon *et al.* 1982). By far the most widespread and best-documented system for the transfer of resistance determinants in aerobic bacteria involves conjugation mediated by plasmids, usually referred to as resistance factors or R-factors.

A class of R-factors that have been particularly well characterized in *Bacteroides* are those coding for lincosamide–macrolide resistance. Four conjugative plasmids encoding resistance to clindamycin (Cc) and Em have been reported. The largest, pBI136 (82 kb) was isolated from *B. ovatus* (Smith and Macrina, 1984), while the smallest, pBFTM10 and pCP1 (14.6 kb) were isolated independently from *B. fragilis* (Tally *et al.*, 1979; Malamy and Tally, 1981) and *B. thetaiotaomicron* (Guiney *et al.*, 1984a); they are probably identical. An intermediate plasmid of 41 kb in size was independently isolated by two groups, from *B. fragilis*, and called pBF4 (Welch *et al.*, 1979) and pIP410 (Magot *et al.*, 1981). A combination of restriction endonuclease mapping and DNA-DNA hybridization studies have shown that pBF4 (pIP410), pBFTM10 (pCP1) and pBI136 carry a common Cc–Em element but have dissimilar replication and transfer functions (Shimmel *et al.*, 1982; Marsh *et al.*, 1983; Guiney *et al.*, 1984a). More detailed analysis has indicated that the Cc–Em determinants of pBF4 and pBFTM10 share 90 per cent homology, while the equivalent determinant of pBI136 has diverged significantly (Smith and Gonda, 1985). Several observations led to the suggestion that this Cc–Em determinant was part of a transposable genetic element. Firstly, functionally cryptic Cc^s plasmids were isolated; these exhibited extensive or complete homology to the replicative and transfer functions of the above plasmids, but not to the Cc–Em determinant, e.g., pBI106 homologous to pBI136 (Smith and Macrina, 1984); pCP2 homologous to pCP1 (Guiney *et al.*, 1984a); pBF4Δ1 and pBF4Δ2 homologous to pBF4 (Welch and Macrina, 1981); pBFTM10Δ1 homologous to pBFTM10 (Malamy and Tally, 1981) pIP401Δ1 homologous to pIP410 (Magot *et al.*, 1981). Secondly, the Cc–Em determinant has been found to be integrated into the chromosome of a wide variety of 'plasmid-free', donor proficient, clindamycin resistant (Cc^r) *Bacteroides* spp. isolated from around the world (Guiney *et al.*, 1983; Marsh *et al.*, 1983; Smith and Macrina, 1984). Lastly, the Cc–Em element of both pBF4 and pBFTM10 was shown to be flanked by at least 550 base-pairs of directly repeated DNA (Shimmel *et al.*, 1982).

Direct evidence that the Cc–Em determinant could undergo transposition was first obtained from *E. coli*. These particular experiments were made possible by an unexpected discovery. Recombinant plasmids, consisting of an *E. coli* cloning vector carrying the Cc–Em determinants of pBF4 or pCP1, did not confer resistance to either Cc or Em on *E. coli* transformants (Guiney *et al.* 1984b). However, under aerobic conditions, these same transformants became tetracycline resistant (Tc^r). This Tc^r phenotype was later designated as $*Tc^r$ to distinguish it from other Tc^r genes (Shoemaker *et al.*, 1985) and the encoding gene itself was shown to be distinct from the Cc^r gene (Matthew and Guiney, 1986). Expression of $*Tc^r$ was not observed in *E. coli* under anaerobic conditions, presumably explaining why *Bacteroides* cells harbouring the Cc–Em determinant are not Tc^r. The availability of this $*Tc^r$

phenotype provided a convenient means for the selection of the Cc–Em element during manipulations in *E. coli*.

In the first series of experiments the Cc–Em element was shown to transpose from a shuttle vector pSS-2 (composed of RSF1010, a 33 kb *Pst*1 subfragment of pBF4, and the *trpE* gene of *B. fragilis*) into both the *E. coli* chromosome and the conjugal mobilizer, R751 (Shoemaker *et al.*, 1985). Once in the chromosome, the *Tc^r element could re-transfer to R751. The transfer of the *Tc^r determinant between pSS-2, R751 and the chromosome was *recA* independent. The transposon responsible was later designated Tn43451 (Shoemaker *et al.*, 1986a). In a similar series of experiments Robillard *et al.* (1985) demonstrated that the Cc–Em gene of pBFTM10 also resided on a compound transposon, Tn*4400*, which contained active insertion sequence (IS) elements directly repeated at its ends. Thus either Tn*4400* or IS*4400* was capable of mediating co-integrate formation between a *Bacteroides–E. coli* shuttle vector (pGAT500) carrying Tn*4400*, and the conjugal mobilizer pOX38. Tn*4400* was also shown to mediate inverse transposition. Subsequently, Tn*4351* was shown to transpose in *Bacteroides* spp. from several different R751::Tn*4351* suicide plasmids (R751 cannot replicate in *Bacteroides* spp.) and from pSS-2 into the *Bacteroides* chromosome (Shoemaker *et al.*, 1986b). Chromosomal insertion of Tn*4351* was shown to generate auxotrophic mutants, and to mediate integration of R751 into the *Bacteroides* genome. These experiments suggest that the Cc–Em transposons of the *Bacteroides* plasmids pBF4 and pBFTM10 are composite, class 1 transposons similar to Tn*9* (Kleckner, 1981).

The presence of other classes of conjugal R-factors in *Bacteroides* spp. has been demonstrated. The conjugal transfer of chloramphenicol resistance (Cm^r) among strains of *B. fragilis* (Rotimi *et al.*, 1981) and from *Bacteroides* spp. to *E. coli* (Mancini and Behm, 1977) has been demonstrated. However, the inferred involvement of plasmid DNA was not fully investigated. In contrast, the involvement of plasmid DNA in the conjugal transfer of Cm^r from *B. uniformis* to *B. fragilis* was clearly established by Martinez-Suarez *et al.* (1985). The plasmid involved, pRYC3373, was 39.5 kb in size and encoded a constitutive cytoplasmic chloramphenicol transacetylase (CAT) similar to an *E. coli* Type II enzyme. The same strain was also able to mediate the independent transfer of cefoxitin (Cf), Cc, Em, and Tc resistance, although the genetic element responsible was not investigated.

The presence of conjugal R-factors in *Clostridium* spp. has also been demonstrated. One of the earliest studies was concerned with an isolate of *C. perfringens* (strain CP590) which exhibited resistance to the antibiotics Tc, Em, Cc and Cm. Curing experiments indicated the segregational linkage of Em^r and Cc^r, and Tc^r and Cm^r (Sebald *et al.*, 1975). Further studies demonstrated that Tc–Cm resistance was transferable to suitably sensitive recipient strains by a conjugation-like process, using a plate mating technique. (Sebald and Brefort, 1975). Frequencies of transfer were in the range $(0.2–1.2) \times 10^{-4}$ per donor cell. Subsequently, CP590 was found to carry two plasmids, pIP401 and pIP402, of mol. wt 54 and 63 kb respectively (Brefort *et al.*, 1977, 1978), and pIP401 was shown to be associated with transferable resistance to Cm–Tc. Interestingly, in matings involving selection for Tc^r, a high percentage of the transconjugants were Tc^r Cm^s, heteroduplex analysis revealing that a 6-kb segment had been deleted from pIP401 concomitant with the loss of Cm^r. It is tempting to speculate that the Cm^r determinant may be part of a transposable genetic element, especially in the light of recent studies which have shown that pIP401 carries two inverted repeat regions (Magot, 1984).

Although conjugal transfer of pIP402 did not occur in plate matings, transfer was demonstrated by an *in vivo* mating technique using axenic mice. From this rather unwieldy mating procedure a Tc^r, Cc^r, Em^r and Cm^s strain was isolated (CP720), carrying two plasmids indistinguishable in contour length from pIP401 and pIP402. CP720 was capable of acting as a donor for both Tc and Em–Cc resistance in the more standard plate mating technique. However, as transfer of Em–Cc was always linked to Tc, the possibility remains that pIP402 was mobilized by pIP401. More surprising was the finding that CP720 could effect the mobilization of chromosomal markers. As CP720 was shown to contain both pIP401 and pIP402 in the autonomous state, Brefort *et al.* (1977, 1978) suggested that the mechanism of mobilization could not have been by F-like integration (Hfr) of one of these plasmids, and therefore represented the type of temporary interaction between plasmid and chromosome discussed by Holloway (1978). It is apparent that the exact relationship of the plasmids to both antibiotic resistance transfer and chromosomal mobilization would benefit from further studies.

In a separate study two plasmids were isolated from a *C. perfringens* isolate (CW92) resistant to Tc and Cm. Of the two plasmids, pCW2 (36.4 × 10^6 mol.wt) and pCW3 (30.6 × 10^6 mol.wt), the smaller was shown to mediate the conjugal transfer of Tc^r to suitable recipient strains at a frequency of 2.8 × 10^{-5} per donor cell (Rood *et al.*, 1978b). Subsequent work has indicated a widespread distribution of pCW3-like plasmids. These particular studies were initiated in 1978 with the isolation of antibiotic multi-resistant strains of *C. perfringens* from porcine faeces (Rood *et al.*, 1978a). The strains were composed of isolates resistant to Tc alone or Tc together with macrolide–lincosamide antibiotics. Although the majority carried plasmids, attempts to cure or transfer resistance determinants were unsuccessful. In a later, more detailed study (Rood, 1983) 15 of these 89 strains were shown to be capable of the conjugal transfer of Tc^r to antibiotic-sensitive strains in mixed-plate matings, at frequencies of between 1.3 × 10^{-3} and 1.9 × 10^{-6} per donor cell. Significantly, when the donor and recipient were isogenic, transfer frequencies increased dramatically to values of between 3.7 × 10^{-1} and 4.6 × 10^{-2} per donor cell. These levels of transfer represent the most efficient transfer frequencies reported for *C. perfringens*, and indeed anaerobes in general. Subsequently, the 15 donor-proficient strains were shown to carry large plasmids. Of these, eight carried a 47-kb plasmid with a restriction profile identical to that of pCW3, while the remaining seven harboured plasmids that had at least 29 kb of DNA in common with pCW3 (Abrahams and Rood, 1985a). These findings prompted the authors to suggest that a pCW3-core plasmid may form the basis of all conjugative Tc^r plasmids in *C. perfringens*, and indeed the restriction map of pIP401 shows striking similarities to that of pCW3 (Magot, 1984). To facilitate further analysis of this hypothesis Abrahams and Rood (1985b) have constructed a detailed restriction map of pCW3. In addition they have cloned the Tc^r gene in *E. coli* and confirmed its localization (Squires *et al.*, 1984) on two contiguous 2-kb *Eco*R1 restriction fragments. The ubiquity of this element within *Clostridium* spp. exhibiting Tc resistance can now be determined by DNA—DNA homology studies. The results with those strains in which conjugal transfer of Tc^r occurs in the absence of plasmid DNA will be particularly instructive.

Two further plasmids are worthy of note. Firstly, a plasmid has been isolated from *C. perfringens* ATCC 3626B (pHB101, 2.1 × 10^6 mol.wt) which, on the basis of curing experiments, appears to code for caseinase activity (Blaschek and Solberg, 1981). Secondly, a strain of *C. cochlearium* was shown to carry a conjugal plasmid of undetermined size which encoded methylmercury-decomposing ability (Pan-Hou *et al.*, 1980). Neither plasmid appears to have been further characterized. Plasmids isolated from *Clostridium* and *Bacteroides*, to which a function has been assigned, are listed in Table 4.1, page 40.

Bacteriophages

Although bacteriophages occur widely throughout the genus *Clostridium* (Ogata and Hongo, 1979), there are very few reports of *Bacteroides* phages (see Salyers, 1984). Both lytic and lysogenic clostridial phages have been isolated, and their appearance has caused disruption of commercial scale solvent production in *C. acetobutylicum* and *C. saccharoperbutylacetonium*. The involvement of phages in the toxigenicity of some clostridia (*C. perfringens*, *C. botulinum* and *C. novyi*) and in the intra- and interspecies conversion within the genus has been demonstrated (Eklund and Poysky, 1974; Eklund *et al.*, 1974; Schallehn and Eklund, 1980), while certain phages have proved useful in the typing of clostridial isolates (Sell *et al.*, 1983). However, most reports have been limited to morphological studies of the phage particles and infected host cell (Nieves *et al.*, 1981). The apparent lack of a reported example of a transducing phage is somewhat disappointing.

Genetic exchange in anaerobes

Gene transfer between micro-organisms occurs by three principle means—transformation, conjugation and transduction. The availability of such systems is essential to enable the genetic analysis of any group of bacteria. At present only the first two processes have been demonstrated in *Bacteroides* and *Clostridium*.

Transformation

Genetic transformation of bacteria is the process whereby naked DNA becomes translocated from the external environment to the cell's interior, where it may survive either autonomously or in an integrated state within the genome. The importance of such processes in the natural environment is debatable, but the availability of such a mechanism is crucial in the application of modern genetic engineering techniques to any micro-organism. In considering the development of transformation procedures for *Bacteroides* and *Clostridium* spp. it is apparent that the avenues explored reflect the fundamental difference between the two genera with respect to the presence or absence of an outer membrane. Thus the development of a transformation procedure for the gram-negative *Bacteroides* has employed whole cells, while procedures in the gram-positive *Clostridium* have concentrated on the transformation of 'wall-less' cells.

Clostridium

The technique of protoplast transformation, pioneered in *Bacillus subtilis* (Chang and Cohen, 1979) and *Streptomyces* (Bibb *et al.*, 1978), relies on the ability to form stable 'wall-less' cells and to subsequently induce their reversion to a normal 'walled' bacterium. The regeneration of three types of clostridial 'wall-less' cells have been studied. L-Phase variants of *C. perfringens* may be generated by growth in liquid medium containing penicillin or D-cycloserine (Mahony and Moore, 1976); however, such L-forms appear incapable of reversion to the vegetative state (Heefner *et al.*, 1984). Cells of certain *Clostridium* spp. undergo autolysis in media or when suspended in buffer. In the presence of an osmotic stabilizer, such as sucrose, the cells of *C. perfringens*, *C. botulinum*, *C. acetobutylicum* and *C. saccharoperbutylacetonium* do not lyse, but form autoplasts (Rogers, 1986). Certain of these autoplasts have been successfully regenerated into walled cells (Kawata *et al.*, 1968; Ogata *et al.*, 1981; Heefner *et al.*, 1984). The artificial induction of protoplast formation, by the treatment of exponential cells with lysozyme, has been obtained in a number of clostridial species. By modification of both the protoplast buffer and the isotonic solidified medium employed, a number of groups have reported the successful regeneration of protoplasts to bacillary rods. Regeneration frequencies of 1 to 10 per cent were reported for *C. pasteurianum* (Minton and Morris, 1983) and *C. saccharoperbutylacetonicum* (Yoshino *et al.*, 1984). Allcock *et al.* (1982) estimated the regeneration frequency of *C. acetobutylicum* protoplasts as being in excess of 80 per cent. However, the method of calculating this frequency was unconventional. Using this published procedure, Rogers (1986) obtained 1–5 per cent regeneration of *C. acetobutylicum* protoplasts. In *C. tertium*, although the majority of protoplasts gave rise to L-form colonies on regeneration medium, a derivative was isolated which exhibited regeneration frequencies approaching 100 per cent (Knowlton *et al.*, 1984).

The first report to appear of the successful transformation of clostridial 'wall-less' cells was by Reid *et al.* (1983). In these experiments protoplasts of *C. acetobutylicum* appeared to be transfected with DNA from the clostridial bacteriophage CA1, but in the absence of polyethylene glycol (PEG). No quantitative measure of the frequency of transformation could be obtained, however, as plaques were not visible on the regeneration medium employed. Successful transfection of protoplasts was demonstrated by scraping regenerated protoplasts off the regeneration medium and assaying for plaque-forming units on clostridial basal medium. The same group of workers later demonstrated that mixtures of protoplasts from two double auxotrophic strains of *C. acetobutylicum* could be induced to fuse in the presence of PEG (Jones *et al.*, 1985). Regenerated colonies gave recombinants and biparental diploid cells at frequencies of 0.3 to 2.0 per cent and 1.4 to 8.3 per cent, respectively. Lin and Blaschek (1984) subsequently showed that *C. acetobutylicum* protoplasts could be transformed with the *Staphylococcus* plasmid pUB110. Successful transformation required both the presence of PEG and preincubation of the protoplasts at 55°C for 15 min, the latter inactivated protoplast-associated DNAase activity. The frequency of transformation was estimated as being 6.6×10^{-4} transformants per μg of DNA, and Km^r transformants were shown to carry pUB110 plasmid DNA. Although this work represented a substantial breakthrough in the development of clostridial host:vector systems, it should be noted that other groups known to this laboratory have been unable to repeat these experiments.

In the same year, Heefner et al. (1984) demonstrated that both L-phase variants and autoplasts of *C. perfringens* could be transformed with the large conjugal plasmids pCW3 and pJU124. Transformation required the presence of PEG and, in the case of L-phase variants, Tc^r colonies arose at a frequency of 1–4 per viable cell. Plasmid DNA was prepared from these transformants and shown by restriction analysis to be identical to pJU124. As these L-forms were incapable of reverting to bacillary rods, further experiments were conducted with autoplasts of *C. perfringens*, generated by growth of strain VP1 1268 in medium containing sucrose as an osmotic stabilizer. Autoplasts were shown to transform at similar frequencies to L-forms, although the Tc^r colonies obtained consisted entirely of 'wall-less' cells. Reversion of these cells to normal bacillary rods could be elicited, however, by subsequent growth in liquid medium containing 25 per cent gelatin, in addition to sucrose.

Although all the above studies have been concerned with the transformation of 'wall-less' clostridial cells, the successful transformation of permeabilized whole cells of *C. thermohydrosulfuricum* has been reported (Soutschek-Bauer et al., 1985). This particular clostridium has a paracrystalline proteinaceous surface layer (S layer) in the cell wall and, in common with other bacteria which share this characteristic, is amenable to transformation by the alkaline-Tris procedure (Fornari and Kaplan, 1982). Transformation was demonstrated with pUB110 and a derivative carrying the Cm^r gene of pC194. As many clostridia are tolerant of choramphenicol (O'Brien and Morris, 1971), selection for Cm^r was achieved with thiamphenicol.

Bacteroides

The first successful report of transformation of a *Bacteroides* spp. was published in 1977, when it was shown that a strain of *B. thetaiotaomicron* could be transfected with DNA from the bacteriophage B1 employing a calcium shock method similar to that used in *E. coli* (Burt and Woods, 1977). This work was not pursued by this group, while attempts by others to transform *Bacteroides* strains using similar procedures proved unsuccessful (Salyers, 1984). Commonly used in the transformation of gram-positive bacteria, PEG has also been shown to facilitate the uptake of plasmid DNA in gram-negative bacteria (Klebe et al., 1983). The observation was subsequently used by Smith (1985b) in the development of a transformation procedure for *B. fragilis* 638. The procedure used PEG 1000 at a final concentration of 25 per cent (w/v) and yielded an average number of transformants of 4.2×10^3 per µg of pBFTM10 plasmid DNA.

Investigation of the parameters affecting transformation indicated that temperature was of primary importance. No transformants were observed when transformation reactions were incubated on ice, and at temperatures >30 or <15°C, a 60 per cent reduction in frequency was observed. Growth of the cells in brain–heart infusion broth containing MgCl$_2$ 20mmol/l was shown to have a stimulatory effect, while optimal transformation occurred when mid-logarithmic phase cells (8×10^{-8} ml^{-1}) were concentrated 10-fold. When the final concentration of PEG was <15 or >35 per cent, transformation was significantly inhibited. The relationship between DNA concentration and the number of transformants was linear through the range tested, while their appearance could only be elicited by postincubation (routinely for 3 h at 37°C) of the transformed cells in prewarmed broth. The optima for a number of other parameters, such as Mg^{2+} concentration and pH of the standard transformation buffer, were also determined, and the results were similar to those found in *E. coli*. A procedure was also developed for storing frozen competent cells. In this case PEG 6000 was found to be superior to PEG 1000.

Conjugation

Conjugation is a process by which genetic elements ensure their propagation within a bacterial population. Such elements induce the formation of a close association between their host (the donor) and neighbouring cells (the recipient) thereby allowing their self-transfer from donor to recipient. Evidence for the involvement of a conjugal process requires the demonstration of a requirement for cell-to-cell contact, and an insensitivity to the presence of DNAase. The classic genetic elements involved are large (>20 kb) plasmids capable of coding for the various conjugation and transfer functions required, which may exist within the cell either as an autonomous unit or integrated into the chromosome. Such conjugal plasmids may mediate solely their own transfer, or may elicit the transfer of other genetic material (i.e., coresident plasmids or chromosomally located genes). All such processes appear to operate in anaerobes. In addition it is apparent that conjugation occurs both inter- and intragenerically.

Intrageneric

The conjugal transfer of antibiotic resistance determinants between strains of *Bacteroides* spp. and between members of the genus *Clostridium* is widespread. As indicated above, many of the described examples of such conjugal transfer have been shown to be mediated by plasmids, and will not be discussed further. There are, however, numerous reports of the conjugal transfer of antibiotic resistance determinants which appear to be 'plasmid-free'.

The transfer of antibiotic resistance between clostridial species in the absence of detectable extrachromosomal DNA was first reported in *C. difficile*. Ionesco (1980) demonstrated that Tc^r could be transferred from a resistant to a sensitive strain, and although the involvement of plasmid DNA was not established, Tc^r was not curable. In a more detailed study Smith *et al.* (1981) confirmed the intraspecific transfer of Tc^r between strains of *C. difficile*. Transfer occurred by a conjugation-like event at frequencies of $(2-8) \times 10^{-7}$ per donor cell. The Tc^r transconjugants displayed a resistance phenotype identical to that of the donor and carried no detectable plasmid DNA. Although the donor strain carried plasmids of mol.wt 5.1 and 22, a cured Tc^s derivative was still shown to harbour both these plasmids. Similarly, a recent report (Wust and Hardegger, 1983) of mixed-culture filter matings of *C. difficile* showed transfer of Cc^r, Em^r and streptogramin resistance, in addition to Tc^r, to sensitive strains at low frequencies (0.1×10^{-7} to 5×10^{-7} per donor cell). Again, the mechanism of transfer seems to be by a conjugation-like process, requiring cell-to-cell contact, but plasmid transfer was not detected. Transfer of Cc^r and Em^r from *C. innocuum* to *C. difficile* in the absence of detectable plasmid transfer has also been reported (Magot, 1983) at a frequency of 4×10^{-6} to 7×10^{-6} per donor cell. In this case transfer of donor Tc^r was not demonstrated.

Malamy and Tally (1981) demonstrated that both their strains of *B. fragilis* (TMP10 and TM2000) were capable of transferring Tc^r to a Tc^s recipient strain in the absence of detectable plasmid DNA. Transfer of antibiotic resistance required cell-to-cell contact and was DNAase insensitive. In agreement with the findings of Privitera *et al.* (1979a), transfer could only be demonstrated after preincubation of the donor cells with tetracycline when 1×10^{-3} to 1×10^{-4} recipients per donor cell became Tc^r. Furthermore this unidentifiable Tc element (Tc^r ERL) was also shown to mobilize resident small cryptic plasmids and an Ap^r element, which also did not appear to be associated with plasmid DNA.

The transfer of Tc^r between strains of *Bacteroides* in the absence of detectable DNA was subsequently demonstrated by other laboratories. Smith *et al.* (1982) re-examined *B. fragilis* V479-1 (strain 92). In contrast to the results obtained by Privitera *et al.* (1979b) with strain 92, transfer of Tc^r was not linked to any extrachromosomal element, and was shown to be totally independent of the resident plasmid pBF4. *B. fragilis* strain V503 has also been shown to mediate the conjugal transfer of Tc^r to *B. uniformis* and *B. ovatus* at frequencies of between 10^{-5} and 10^{-6} per donor cell, in the absence of detectable plasmid DNA (Mays *et al.*, 1982). In this case, however, transfer was not inducible by preincubation with tetracycline. Inducible plasmid-free transfer of Tc^r was also demonstrated in three different strains by both Guiney *et al.* (1983) and Marsh *et al.* (1983).

Malamy and Tally (1981) proposed that the Tc^r ERL of *B. fragilis* exists within the chromosome, and that pretreatment of the cell with tetracycline induces synthesis of the transfer apparatus, as well as activating the excision or transposition of the Tc^r ERL. The autonomous form of the element transfers to the recipient where it integrates. Other resistant determinants adjacent to the ERL can also be co-transferred. This model resembles the 'conjugal transposon' model proposed by Clewell to explain an analogous Tc transfer system in *Streptococcus faecalis* (Franke and Clewell, 1981). Indeed, it would appear that, among the streptococci, most drug resistance determinants are parts of large nonhomologous insertions into the host chromosome (Horodiceanu *et al.*, 1981; Inamine and Burdett, 1985; Shoemaker *et al.*, 1979) several of which exhibit conjugal properties. Two such conjugal elements are Tn917 (Gawron-Burke and Clewell, 1982) and Tn3951 (Vijayakumar *et al.*, 1986). Similar mechanisms are presumably operating in both *Clostridium* and *Bacteroides* spp.

All of the Tc^r donor proficient strains of *Bacter-*

oides used in the above studies were also shown to be Cc^r, and the data obtained suggests that transfer of Cc^r in the majority of these strains is reliant on the resident Tc^r ERL. The first piece of evidence supporting this view comes from the observation that Tc^r and Cc^r appear to transfer *en bloc*. Thus all Tc^r transconjugants obtained in matings employing the *B. fragilis* strains TAL2300 (Malamy and Tally, 1981) and V503 (Mays *et al.*, 1982) as donors were shown to have also become Cc^r. Although the question of linkage was not addressed by Guiney *et al.* (1983) and Marsh *et al.* (1983), in the former study the frequency of Tc^r and Cc^r transfer was identical (2 × 10^{-7} per donor cell) while in the latter study the two frequencies were of a similar order of magnitude. The second piece of evidence is that induction of Tc^r ERL transfer by preincubation with Tc also induces transfer of Cc^r (Malamy and Tally, 1981; Marsh *et al.*, 1983). Third, the Tc^r ERL has been shown to effect the transfer of other elements, including autonomous plasmid DNA (Privitera *et al.*, 1979b; Malamy and Tally, 1981; Mays *et al.*, 1982; Smith *et al.*, 1982; Tally *et al.*, 1982). Studies have established that these donor-proficient strains, and their progeny, contain chromosomally located sequences homologous to the Cc–Em determinants of pBFTM10 (Malamy and Tally, 1981; Marsh *et al.*, 1983) and pBF4 (Mays *et al.*, 1982; Guiney *et al.*, 1983), i.e., transposons Tn*4400* and Tn*4351*. In at least one case there is evidence that the Cc–Em Tn element has integrated adjacent to a Tc^r ERL (Malamy and Tally, 1981). Interestingly, it is now evident that the conjugal transfer of pBFTM10 is reliant on the Tc^r ERL (Malamy and Tally, 1981; Shoemaker *et al.*, 1986b). Thus, in a strain carrying both these elements (TMP10), induction of the Tc^r ERL by preincubation with tetracycline also increases the transfer of pBFTM10 100-fold, an enhancement not seen in a strain carrying only pBFTM10. This is in contrast to a strain carrying the Tc^r ERL and pBF4, where transfer of the two elements is totally independent and the frequency of Cc^r transconjugants is not increased by preinduction of the donor (*B. fragilis* V479-1) with tetracycline (Smith *et al.*, 1982). Convincing evidence was obtained by Shoemaker *et al.* (1986b), who demonstrated that shuttle vectors containing pBFTM10, or pB8–51, were not transferred out of *B. uniformis* donors unless the conjugative Tc^r ERL element was coresident.

A variation on 'plasmid free' antibiotic resistance transfer was described by Rashtchian *et al.* (1982). Cefoxitin (Cf) resistance was shown to be transferable by a conjugation-like process from *B. thetaiotaomicron* to *B. fragilis* and *B. uniformis* in the absence of detectable plasmid DNA. The progeny could not be cured of Cf^r and could no longer act as donors in further matings. The mechanism was therefore thought unlikely to represent a conjugal transposon but may be analogous to transconjugants obtained after Hfr crosses, where only partial transfer of the integrated element occurs. On a note of caution, Smith and Macrina (1984) found the conjugal plasmid pIP136 was very difficult to isolate in most strain backgrounds, emphasizing the need for restraint in designating a strain 'plasmid free'.

Intergeneric

Although the transfer of R-factor determinants from *E. coli* to *Bacteroides* spp. was reported as early as 1976 (Burt and Woods, 1976), the identity of the recipient strains has recently been called into question (Salyers, 1984). Mancini and Behme (1977) reported on the successful transfer of multiple antibiotic resistance to *E. coli*, though the transconjugants obtained were unstable for all the acquired resistances except Tc^r. The involvement of plasmid DNA, though inferred, was not demonstrated. The conjugal transfer of pGD10 from *B. ochraceus* to *E. coli* (Guiney and Davis, 1978) would now appear not to be an example of gene transfer from an obligate anaerobe, as this strain has been reclassified as a microaerophile, *Capnocytophaga ochracea* (Guiney and Davis, 1982). Conjugal transfer of Cc^r, Em^r, Cm^r and Tc^r from *B. fragilis* and *B. theaiotaomicron* to *E. coli* was also demonstrated by Rotimi *et al.* (1981), while the conjugal transfer of ampicillin resistance (Ap^r) and resistance to mercury from *B. fragilis* to *E. coli* HB101 has also been reported (Soong *et al.*, 1984). Unfortunately, there have been no further reports of work with the strains described by the above workers. More recent studies have indicated that neither *Bacteroides* replicons nor resistance genes function in *E. coli*, and vice versa (see below). The definitive demonstration of conjugal transfer between *Bacteroides* spp. and *E. coli* has been made possible by the construction of cointegrate *Bacteroides*–*E. coli* vectors (see below).

Until very recently there were no examples of the

conjugal transfer of genetic information between *Clostridium* spp. and other genera. This situation was remedied by Oultram and Young (1985), when they investigated the possibility of transferring the broad host-range, Gram-positive plasmid pAMβ1 to saccharolytic clostridia. Originally isolated in *Str. faecalis*, pAMβ1 has been shown to undergo conjugal transfer to other *Streptococcus* spp. (including *Str. lactis*), *Staphylococcus aureus*, and various *Lactobacillus* and *Bacillus* spp. (Oultram and Young, 1985). Using a plate mating technique, pAMβ1 was transferred from *Str. lactis* to *C. acetobutylicum* at high efficiency (ranging from 1.4×10^{-3} to 4.1×10^{-5} per donor cell). The plasmid appeared to be stably maintained in *C. acetobutylicum* and could be retransferred at lower frequencies between strains of *C. acetobutylicum* (1.3×10^{-5} to 9.1×10^{-6}), transferred back to *Str. lactis* (1.4×10^{-7}) and conjugated to and from *Bacillus subtilis* (4×10^{-7}). Similar work, published shortly afterwards, demonstrated that, in addition to pAMβ1, the closely related plasmids pIP501 and pJH4 could also be transferred to *C. acetobutylicum* from *Str. faecalis* (Reysset and Sebald, 1985). The suggestion by Oultram and Young (1985) that this system might be usefully employed to introduce commonly used *B. subtilis* cloning vectors into *Clostridium* spp. appears to have been successfully tested by Yu and Pearce (1986). These authors claim to have used both pAMβ1 and the closely related conjugal plasmid pVA797 to mobilize the small non-conjugative plasmid pAM610 from *Str. lactis*, *Str. sanguis* and *Str. faecalis* donors to *C. acetobutylicum*. However, transfer of pAM610 was not conclusively demonstrated, and similar experiments undertaken by Reysset and Sebald (1985) and Oultram and Young (personal communication) were unsuccessful. The potential ability of these streptococcal conjugative plasmids to introduce small vectors into *Clostridium* spp. awaits further experimentation.

Shuttle vector development

In both *Bacteroides* and *Clostridium* the development of two types of cloning vector are actively being pursued. These two types are:

vectors which rely on their introduction into the cell by transformation;
vectors which may be transferred by a conjugal process.

In *Bacteroides* the development of a transformation procedure enabled Smith (1985b) to undertake the construction of classical cloning vectors. These were based on the replicon of pBI143, as the wide dissemination of cryptic plasmids of this class (Class I; see above) suggested that vectors derived from pIB143 may be able to undergo replication in many different *Bacteroides* spp. Two types of vector were constructed. *E. coli–Bacteroides* shuttle vectors (i.e., pFD173 and pFD176) which carried the *E. coli* ColE1 replicon and Ap^r determinant (pUC19 derived) and the pIB143 replicon and pBF4 Cc–Em determinant. A *Bacteroides* cloning vector (pIB191 carried only the replicon of pBI143, and the Cc–Em determinant of pBF4, but retained the multilinker cloning sites of pUC19. When prepared from the *B. fragilis* strain background, the frequency of transformation of all three cloning vectors was greater than 10^5 Cc^r transformants per μg of plasmid DNA. This represents a 25-fold increase in frequency over that typically obtained with pBFTM10. When the two shuttle vectors were prepared from an *E. coli* strain background, there was a 1000-fold decrease in transformation frequency indicative of a difference in the restriction/modification specificity of the two organisms. The two plasmids pFD173 and pFD176 exhibited marked differences in their stability during cultivation of *Bacteroides* cells, indicating that the *Xba*1 site of pIB143 lay close to the maintenance functions of this plasmid. Smith (1985a) indicated that the vector pIB191 had been used in preliminary shotgun cloning experiments in *Bacteroides*, demonstrating the utility of direct cloning in *Bacteroides* spp.

The development of a transformation system for autoplasts of *C. perfringens* allowed the construction of *E. coli–Clostridium* shuttle vectors (Squires *et al.*, 1984). Three small cryptic plasmids, pJU121 (3.3 kb), pJU122 (3.9 kb) and pJU281 (1.6 kb), were isolated from two different strains of *C. perfringens* and cloned as single fragments into the *Bam*H1 site of pBR322. To allow the selection of these plasmids in *C. perfringens*, the *tet* genes from pCW3 and the related pJU124 were introduced into them. The resultant plasmids (pJU7, pJU10, pJU12, pJU13, pJU16) all conferred Tc^r on *E. coli*, indicating that the clostridial *tet* gene expresses in *E. coli*. With the exception of the two plasmids derived from pJU281 (pJU7 and pJU10), all of these plasmids were shown to transform both L-phase variants and autoplasts of

C. perfringens, although the transformation of the autoplasts was between 10- and 100-fold more effective than that of the L-phase variants. The failure of pJU7 and pJU10 to transform clostridial 'wall-less' cells was most probably due to disruption of the replication region of pJU281 during the construction of the appropriate shuttle vector. Evidence that a restriction barrier exists between *E. coli* and *C. perfringens* was obtained when it was shown that heterologous DNA transformed either organism at significantly reduced frequencies compared to homologous DNA. Surprisingly pCW3 (47 kb) was equally efficient (3×10^{-6} to 10×10^{-6} transformants per viable cell per µg DNA) at transforming autoplasts as pJU12 (11.6 kb), auguring well for the future cloning of large DNA fragments (up to at least 30 kb) into *C. perfringens*.

The lack of reproducible transformation methods for saccharolytic clostridia has led other laboratories to adopt an alternative approach to the construction of shuttle vectors (Collins *et al.*, 1985; Luczak *et al.*, 1985). Various restriction fragments of the *C. butyricum* plasmids pCB101 and pCB102 (called pCBU2 and pCBU3 by Luczak *et al.*, 1985) have been cloned into the gram-positive, replication deficient plasmids (carrying a ColE1 replicon and the Ap^r and Cm^r determinants of pBR322 and pC194, respectively) and the ability of the resultant recombinant plasmids to transform *B. subtilis* to Cm^r examined. In the study of Collins *et al.* (1985), two such plasmids were identified. Plasmid pRB7, which carried a 2.0 kb *Sau*3A fragment derived from pCB102, was shown by southern blot analysis of the *B. subtilis* Cm^r transformants to have undergone integration into the host chromosome. Although it was established that more than one copy of pRB7 became integrated, the chromosomal location of the site(s) of integration was not ascertained. It followed that transformation of *B. subtilis* with pRB7 was Rec-dependant, and once integrated the plasmid was stably maintained during growth for several generations without antibiotic selection.

In contrast, the plasmid pRB1 appeared to be established as an autonomously replicating unit in transformed *B. subtilis* cells. Thus transformation was Rec-independent and Cm^s segregants arose at high frequencies in the absence of antibiotic selection. Although the presence of plasmid DNA could not be visualized by standard plasmid isolation techniques, southern hybridization experiments supported the view that pRB1 existed within the cells as an autonomous plasmid, confirmatory evidence being supplied by transfer of the plasmid back to *E. coli*. These observations were interpreted as reflecting either weak replication of the plasmid or inaccurate partitioning of plasmid molecules at cell division. These results were extended by the determination of the complete nucleotide sequence of the *Sau*3A fragment carried by pRB1 within our laboratory. The fragment was shown to be 3.48 kb in length, to be extremely A+T-rich (74 per cent), and to carry four major ORFs, one of which began outside the cloned region. The sequence obtained did not resemble any other characterized replicon, with no regions capable of coding for RNA transcripts with significant secondary structure (i.e., replicons of ColE1 and R1, and gram-positive plasmids such as pC194, pT181 and pC221; Scott, 1984; Brenner and Shaw, 1985) nor were there any regions with extensive repeat structures (i.e., as in the F replicon; Scott, 1984). Subsequent subcloning experiments indicated that the pCB101 replicon could be more accurately defined as being located within a 2.5-kb region of this *Sau*3A fragment. This particular subfragment carries only two ORFs, encoding proteins of mol.wt 27 000 and 43 000. Removal of nine codons from the larger of these two ORFs destroyed the replicative ability of this region (Brehm *et al.*, 1986). The two proteins encoded by these ORFs demonstrated no significant primary sequence homology with proteins encoded by any of the currently sequenced gram-positive plasmids.

The question of whether the cloned region carries the replicative function of pCB101 awaits further experiments in a clostridial host. However, we have now determined the complete nucleotide sequence of pCB101 (Minton, Brehm and Barstow, unpublished data) and have established that there are no regions carried by the plasmid which bear any resemblance to any known replication regions. Furthermore, the complete nucleotide sequence of pCB102 has also been determined (Thompson, Brehm and Minton, unpublished data) and a similar lack of an obvious replication region is apparent.

An alternative procedure for mobilizing small non-conjugative plasmids from *B. subtilis* to *C. acetobutylicum* using pAMβ1 has recently been developed (Oultram, Davies and Young; unpublished data). In this case transfer occurs via the formation of a plasmid cointegrate. A series of plasmids was constructed (pOD1–3) which carry a gram-negative replicon (ColE1), an antibiotic-resistance determinant (Ap^r) and two gram positive antibiotic-resistance determinants, Cm^r (pC194) and Em^r (pAMβ1). Although unable to replicate in *B. subtilis*, transformation into cells carrying pAMβ1, with selection for Cm^r, allows the establishment of the plasmid as a cointegrate with pAMβ1 by virtue of the homologous Em^r regions present on both plasmids. The cointegrates formed were subsequently shown to be capable of conjugal transfer to *C. acetobutylicum*. These vectors are currently being employed to introduce cloned genes into *C. acetobutylicum*.

Guiney (1984) tested plasmids from 14 different incompatibility groups in *E. coli* for transfer to *B. fragilis*. Their lack of success suggested a substantial barrier to transfer between *E. coli* and *Bacteroides*. To determine the nature of this barrier they constructed two shuttle vectors, pDP1 and pDP1.1, composed of pCP1, a ColE1 replicon and *oriT* of RK2 (Guiney *et al.*, 1984c). Both plasmids were mobilized into *B. fragilis* by the RK2 helper plasmid pRK231. Neither shuttle vector could replicate in an *E. coli polA* strain, indicating that pCP1 cannot replicate in *E. coli*. In addition, as transfer of pRK231 to *B. fragilis* was not observed, the RK2 replicon cannot function in *Bacteroides*. Analysis of pDP1 deletion variants established that the region essential for the replication of pCP1 was within a 5 kb *Eco*R1 fragment (Guiney *et al.*, 1984c; Mathews and Guiney, 1986). It was also apparent that none of the three markers carried by pDP1 (Cc^r, Ap^r and *Tc^r) could express in both hosts. Thus the barriers to plasmid exchange between *E. coli* and *Bacteroides* exist at the level of replication and drug resistance expression.

Similar shuttle vectors were utilized by Shoemaker *et al.* (1985) in their analysis of Tn*4351*. In this case one vector, pSS-2, carried a 33-kb *Pst*1 fragment of pBF4, while the other, pE5-2, carried the 4.4-kb cryptic plasmid, pB8-51, isolated from *B. eggerthii*. Both vectors contained RSF1010. Although pE5-2 was mobilized by R751 to strains of *B. thetaiotaomicron* and *B. uniformis*, no transfer to *B. fragilis* V479C was observed. These experiments illustrate that it cannot be assumed that *E. coli* IncP conjugal systems will function in all *Bacteroides* strains. They also suggest that *Bacteriodes* are the first gram-negative bacteria that do not maintain IncP (i.e., R751 and RK2) and IncQ (i.e., RSF1010) plasmids. In later constructions, Shoemaker *et al.* (1986b) demonstrated that the presence of *E. coli* mobilization regions on shuttle vectors was not necessary for their mobilization by R751 from *E. coli* to *Bacteroides* spp. Sequences present on pBFTM10 or pB8-51 were sufficient for both IncP mediated mobilization from *E. coli* to *Bacteroides* spp. In addition, the Tc^r ERL mobilized the same vectors between *Bacteroides*, and from *Bacteroides* to *E. coli*. Mapping studies localized the Mob region of pBFTM10 adjacent to the replication function determined by Guiney *et al.* (1984c). The type of shuttle vectors described in these studies provide a useful addition to the classic shuttle vectors constructed by Smith (1985a). Conjugation may be employed to transfer these vectors to *Bacteroides* spp. that are not yet transformable (*B. ovatus*, *B. thetaiotaomicron*, *B. uniformis*), and may be particularly useful in transferring cloned DNA from *E. coli* to the strictly anaerobic species of *Bacteroides* from the human mouth and the bovine rumen. The shuttle vectors have the additional advantage of being able to transfer larger pieces of cloned DNA (>15 kb), and will circumnavigate any problems with recipient restriction modification systems. The usefulness of conjugation as a vector delivery system has recently been demonstrated by Guthrie and Salyers (1986). They used a mobilizable *Bacteroides* suicide vector (capable of replication in *E. coli* but not *Bacteroides*) to deliver a nonfunctional portion of the cloned chondroitin lyase II gene into *B. thetaiotaomicron*, where insertional inactivation of the homologous chromosomal gene occurred.

Gene cloning

Reports on the cloning and nucleotide sequence analysis of genes from anaerobic bacteria have proliferated in the last few years (Table 4.2). It is apparent, however, that the bulk of these genes are of a clostridial origin rather than *Bacteroides*. This is probably a reflection of the difficulties encountered in the genetic analysis of *Bacteroides* spp. *in situ*, compared to the elegant genetic systems now developed for *Clostridium* spp., necessitating the isolation of genes of interest in *E. coli*.

Table 4.2 Genes cloned from *Clostridium* and *Bacteroides* in *Escherichia coli*

Species	Gene	Reference
B. nodosus	Pilin genes	Elleman et al. (1984; 1986)
		Anderson et al., (1984)
B. thetaiotaomicron	Chondroitin lyase II	Guthrie et al. (1985)
	Chondro-4-sulphatase	Guthrie and Salyers (1987)
B. fragilis	RecA-like gene	Goodman et al. (1986)
	Cc-Em determinant (*Tc^r)	Guiney et al. (1984b)
B. succinogenes	Xylanase	Daeschel and Klaenhammer (1987)
C. pasteurianum	Galactokinase (galK)	Daldal and Appelbaum (1985)
	Nitrogenase (nifH)	Chen et al. (1986)
	Ferredoxin	Graves et al. (1985)
	Molybdenum-pterin binding protein (mop)	Hinton and Freyer (1986)
C. acetobutylicum	Glutamine synthetase (glnA)	Usdin et al. (1986)
	Endoglucanases/cellobiases	Zappe et al. (1986)
	LeuB, ProA, HisG ArgG, LeuC	Efstathiou and Truffaut (1986)
		Zappe et al. (1986)
		This laboratory (unpublished data)
C. thermocellum	Endoglucanases/cellobiases	Cornet et al. (1983)
		Millet et al. (1985)
		Schwarz et al. (1986)
	beta-glucosidases	Schwarz et al. (1986)
C. butyricum	Hydrogenase (hyd)	Karube et al. (1983)
	LeuB	Ishii et al. (1983)
C. tetani	Tetanus toxin	Eisel et al. (1986)
		Fairweather et al. (1986)
C. perfringens	Bacteriocin (bcn)	Garnier and Cole (1986)
	Tc^r determinant (tet)	Heefner et al. (1984)
		Abraham and Rood (1985b)

Bacteroides

To date, only a handful of genes have been cloned from *Bacteroides* spp. These are the pilin gene, from two different serogroups of *B. nodosus*—the causal agent of ovine footrot (Anderson *et al.*, 1984; Elleman *et al.*, 1984, 1986)—the genes encoding chondroitin lyase II and chondro-4-sulphatase from *B. thetaiotaomicron* (Guthrie *et al.*, 1985; Guthrie and Salyers, 1987), a gene from *B. fragilis* which complements the *recA* gene of *E. coli* (Goodman *et al.*, 1986) and a xylanase gene from *B. succinogenes* (Dalschel and Klaenhammer, 1987). In contrast to the situation with the *Bacteroides* R-factor genes encoding Cc–Em resistance, all these genes appear to express in *E. coli*, indicating that, at least in the case of these chromosomal genes, a barrier to the expression of *Bacteroides* genes does not exist in *E. coli*.

Nucleotide sequence determination of the serogroup A and H pilin genes (Elleman and Hoyne, 1984; Elleman *et al.*, 1986) has demonstrated that the coding regions are preceded by sequences complementary to the 3' end of 16S RNA of *E. coli* and that they carry AT-rich regions further upstream, preceded by potential recognition sites for *E. coli* RNA polymerase. At the 3' end of the gene potential rho-independent transcriptional terminators are present. Furthermore, the codon usage is reminiscent of highly expressed *E. coli* genes (Grosjean and Fiers, 1982), with a bias towards the use of U in the wobble position of quartet codons, the preferential use of A in favour of G in both duet and quartet codons and avoidance of *E. coli* modulator codons. These observations help to explain why pilin genes express in *E. coli* at equivalent levels to those observed in *Bacteroides*. The relevance, however, of any of these sequences to expression of the genes in *Bacteroides* is yet to be determined. Indeed, Guthrie and Salyers (1987) have recently obtained evidence that the promoter from which the cloned chondroitin lyase II gene is expressed in *E. coli* may not be the promoter utilized by *B. thetaiotaomicron*.

The pilin genes were cloned to enable the production of a cheaper vaccine by recombinant DNA. However, although it is clear that efficient production of polypeptides may be elicited in *E. coli*, it was shown that the unusual short-signal peptide was not removed (Elleman and Hoyne, 1984). Consequently, attempts are now being made to express pili from *B. nodosus* on the surface of *Pseudomonas aeruginosa*, as the pili of this organism contain similar sequences at their N-termini (Elleman *et al.*, 1986).

A particularly powerful use for cloned DNA is as a DNA probe in the detection and enumeration of bacterial species within clinical samples. A number of such probes have been described for *Bacteroides* spp. These include specific probes for *B. fragilis*, *B. thetaiotaomicron*, *B. uniformis*, *B. distasonis*, *B. ovatus*, *B. vulgatus*, and *Bacteroides* group 3452–A (Salyers *et al.*, 1983; Kuritza and Salyers, 1985; Kuritza *et al.*, 1986a,b). These probes represent a single, more rapid test than the traditional time-consuming methods of detection, which rely on gas–liquid chromatography and biochemical tests. The rapid diagnosis of gram-negative anaerobic infections facilitates the choice of drug regime employed in treatment. However, before such probes may be transferred to a clinical setting, a number of problems have to be resolved, such as probe sensitivity and specificity and the requirement for non-radioactive signals.

Clostridium

A considerable number of clostridial genes have now been cloned in *E. coli* (see Table 4.2, page 52) and it is rapidly becoming apparent that, as with other gram-positive genes, no special barriers to their expression exist in *E. coli*. Thus, many of these genes have been detected by their complementation of *E. coli* mutations. Genomic DNA derived from *C. butyricum* has been shown to complement mutants of the *E. coli leuB* locus (Ishii *et al.*, 1983), while recombinant plasmids carrying *C. acetobutylicum* and *C. thermocellum* chromosomal DNA have been also shown to complement *leuB*, proA, *hisG* and *argG* (Efsathiou and Truffaut, 1986; Zappe *et al.*, 1986), and *leuB*, *leuC* and *trpE*, respectively (Cornet *et al.*, 1983). The same DNA fragment of *C. acetobutylicum* carrying the β-isopropyl malate dehydrogenase gene was shown to complement the equivalent locus of *B. subtilis*, *leuC* (Efsathiou and Truffaut, 1986). Studies in our laboratory have shown that many of the recombinant plasmids carrying *C. acetobutylicum* genomic DNA which are capable of complementing *E. coli leuB* mutations also complement *leuC* mutants, indicating linkage of these two genes on the *C. acetobutylicum* chromosome. Complementation of *E. coli* mutants has also been used to clone the hydrogenase gene, *hyd*, of *C. butyricum* (Karube *et al.*, 1983) the galactokinase gene, *galK*, of *C. pasteurianum* (Daldal and Appelbaum, 1985) and the glutamate synthetase gene, *glnA*, of *C. acetobutylicum* (Usdin *et al.*, 1986). In the case of the latter two genes, convincing evidence was obtained that expression of the clostridial *galK* and *glnA* genes occurred from their own promoters.

The cloning of other clostridial genes has been detected on the basis of the acquisition of novel characteristics by *E. coli*. This is particularly true of genes from *C. thermocellum* and *C. acetobutylicum* coding for various enzymes involved in the degradation of cellulose. Thus, using various cellulose and cellobiose-based substrates, Aubert and co-workers have cloned ten distinct fragments of *C. thermocellum* DNA in *E. coli* and demonstrated expression of enzymic activities related to cellulose degradation (Millet *et al.*, 1985). Two of these cloned fragments appeared to carry the previously cloned *celA* and *celB* genes (Cornet *et al.*, 1983), coding for two different endoglucanases, while another five fragments coded for hitherto unidentified endoglucanases. The remaining three clones appeared to carry three different cellobiohydrolase genes. The suggestion that *C. thermocellum* carries the genetic information for an unexpected multiplicity of endo-β-glucanases was further underlined by the work of Schwarz *et al.* (1986), who cloned three more distinct endoglucanases. In the same study, two genes with distinct β-glucosidase activity were obtained, while a fourth endoglucanase was shown to be the previously identified *celA* (Schwarz *et al.*, 1985). Endoglucanase and cellobiase genes have also been cloned from *C. acetobutylicum* (Zappe *et al.*, 1986). The cellulolytic complex (termed 'cellulosome') secreted by *C. thermocellum* has an estimated molecular weight of 2.1×10^6 and is composed of at least 14 polypeptides which include both endoglucanases and cellobiohydrolases (Lamed *et al.*, 1983). As attempts to resolve this cellulosome into its active subunits by biochemical techniques have been unsuccessful, the cloning and expression of the genes coding for the individual components in *E. coli* offers a unique opportunity to unravel the complexities of this system.

The fact that clostridial genes in general express in *E. coli* suggests that the transcriptional and translational signals of these genes are efficiently recognized by *E. coli* transcription/translation machinery. Confirmatory evidence has been obtained in several instances by nucleotide sequence analysis. To date, all the sequenced genes appear to possess sequences exhibiting a high degree of conformity to the ribosome binding site of gram-positive bacteria. These include the ferredoxin (Fd) gene (Graves *et al.*, 1985), the *mop* gene (Hinton and Freyer, 1986), the multiple *nifH* genes of *C. pasteurianum* (Chen *et al.*, 1986), the *celA*, *celB* and *celD* genes of *C. acetobutylicum* (Beguin *et al.*, 1985; Grepinet and Beguin, 1986; Joliff *et al.*, 1986), the tetanus toxin of *C. tetani* (Eisel *et al.*, 1986), the major ORFs identified on the plasmids pIP404 (Garnier and Cole, 1986) and pCB101 and pCB102 (Minton, Brehm, Thompson and Barstow, unpublished data).

A feature frequently seized upon when analysing structural gene nucleotide sequences is the preference for codons ending in A/U or G/C. Not altogether unexpectedly, the structural genes from *Clostridium* spp. with a low (30 per cent) G+C content exhibit a marked preference for codons ending in A or T, i.e., 89 per cent in the *bcn* gene of *C. perfringens*, 86.5 per cent, 82.1 per cent, 84.2 per cent and 85.7 per cent in the *mop*, Fd, *nifH1* and nifH2 genes of *C. pasteurianum*, respectively, and 86.3 per cent in the tetanus gene of *C. tetani*. In contrast, such a preference is not so marked in the *celA,B,D* genes (60.4, 56, and 60.6 per cent respectively) of *C. thermocellum* which has a higher G+C content (58 per cent). As a bias towards codons ending in A or T is normally associated with weakly expressed *E. coli* genes (Grosjean and Fiers, 1982), this preference has been used to explain the low expression of tetanus polypeptides (Eisel *et al.*, 1986) and the total lack of expression of the *bcn* gene product (Garnier and Cole, 1986) in *E. coli*. In this latter case expression of a polypeptide was not observed even when the structural gene was fused to the *lacZ'* gene on pUC9 to provide both transcriptional and translational signals. A similar fusion construct of the *celD* gene resulted in the production of the encoded protein to a level representing 15 per cent of the cell's soluble protein (Joliff *et al.*, 1986). However, even in *E. coli* the evidence that codon usage can so dramatically affect the production of proteins is conflicting (Andersson *et al.*, 1984; Holm, 1986), i.e., the pseudomonad carboxypeptidase G_2 gene (92.8 per cent A/T in the wobble position; Minton *et al.*, 1984) can be expressed at 16 per cent of the cell's soluble protein (Chambers and Minton, unpublished data). Thus, the fact that the *C. pasteurianum mop* protein is expressed at 5 per cent of the cell's soluble protein, even though there is a 86.5 per cent preference for codons ending in A or T, is not surprising (Hinton and Freyer, 1986).

With regard to transcriptional signals, the majority of the sequenced genes contain GC-rich regions, of dyad symmetry, 3' to coding sequence. These regions are capable of forming hairpin loop structures characteristic of *E. coli* rho-independent terminators. Upstream of the translational initiation codon, sequences reminiscent of *E. coli*-35 and -10 promoter sequences have been noted. However, the assignment of promoter sequences in this manner is fraught with difficulties, and in the two instances where the sequence determinants controlling transcription initiation have been experimentally defined, additional sequences to those predicted have been discovered. In the *celA* gene, S1 mapping experiments and primer extension procedures on mRNA isolated from *E. coli* indicated transcription from a region exhibiting close conformity to the *E. coli* promoter consensus sequence (Beguin *et al.*, 1986). However, although mRNA initiating from this site was also isolated from *C. thermocellum*, its proportions were relatively minor in comparison to a major transcript initiating further downstream from a sequence conforming to the *Bacillus* σ^{28} promoters. The presence of two promoters for *celA* suggests regulation of expression by co-ordinate promoters, one or both sites being utilized under defined conditions.

It has also been shown that the RNA polymerase from *C. pasteurianum* (in *vivo*), *E. coli* (*in vivo* and *in vitro*) and *B. subtilis* (*in vitro*) each produce a transcript of similar size in response to the *C. pasteurianum* Fd gene, suggesting that the same transcriptional signals are used (Graves and

Rabinowitz, 1986). The sequences upstream of the initiation nucleotide (P1) corresponded to the *E. coli* consensus promoter sequence. However, *E. coli* RNA polymerase was also shown to recognize a second promoter (P2), at equal efficiency. At a total length of 255 nucleotides the Fd monocistronic mRNA represents one of the smallest mRNAs yet described.

Features of note are the A+T-rich regions located 5' to the translational initiation codons of the *nifH1* and Fd genes of *C. pasteurianum*, the *celA,B,D* genes of *C. thermocellum* and the *bcn* gene of *C. perfringens*, which suggest that the upstream regulatory regions may be characterized by exceptionally high A + T content, as is the case with other gram-positive promoters (see Pero *et al*., 1982). It is also apparent that the 5' non-coding regions of clostridial genes commonly contain palindromic sequences, which presumably have regulatory significance.

Conclusion

The last decade has seen considerable progress in the genetic analysis of the obligate anaerobes *Clostridium* and *Bacteroides*. With the glaring exception of transduction, the principal means of genetic exchange appear to be operating in both genera. It is now apparent that, as with the streptococci, the rapid emergence of antibiotic resistance in both genera, particularly in *Bacteroides*, is most likely due to chromosomally located conjugal transposable elements, rather than autonomous plasmids. The involvement of plasmids in toxigenicity is beginning to emerge, while nucleotide sequencing of cloned genes is providing valuable information on gene structure. More importantly, rudimentary host vector systems have been developed. One should not, however, be lulled into a false sense of security. It is abundantly clear that the systems devised for one member of a genus will not be immediately applicable to other members of the same genus. This is particularly obvious in *Clostridium* where, although *C. perfringens* appears readily amenable to genetic analysis, the same cannot be said for the saccharolytic clostridia. In the light of their biotechnological potential, it is these latter clostridia that currently command the attention of the industrial microbiologist. Until reliable methods of transformation are developed, a conjugal cointegrate delivery system, based on pAMβ1 and related plasmids, will play a key role in these organisms.

Overall progress to date has relied to a great extent on the considerable advances made in recombinant DNA methodology. Although such techniques are extremely powerful, they are not the sole answer to the genetic analysis of a micro-organism. There is still ample room for the development of classic means of genetic analysis. The necessary 'tools' for undertaking genetic mapping of the genomes are as yet unavailable. The isolation of transducing phages and development of transposon mutagenesis systems would be especially beneficial. In this regard the conjugal transposable elements that appear to be widespread in both *Bacteroides* and *Clostridium* warrant more detailed analysis. However, the genetic systems now available to the microbiologist form a sound basis for future progress, and we can no doubt expect a proliferation of interest in these anaerobic genera over the next decade.

References

Abraham L J, Rood J I. (1985a). Molecular analysis of transferable tetracycline resistance plasmids for *Clostridium perfringens*. *Journal of Bacteriology* **161**: 636–640.
(1985b). Cloning and analysis of the *Clostridium perfringens* tetracycline resistance plasmid pCW3. *Plasmid* **13**: 155–162.

Allcock E R, Reid S J, Jones D T, Woods D R. (1982). *Clostridium acetobutylicum* protoplast formation and regeneration. *Applied and Environmental Microbiology* **43**: 710–721.

Anderson B J, Bills M M, Egerton J A, Mattick J S. (1984). Cloning and expression in *Escherichia coli* of the gene encoding the structural subunit of *Bacteroides nodosus* fimbriae. *Journal of Bacteriology* **160**: 748–754.

Andersson S G E, Buckingham R H, Kurland C G. (1984). Does codon composition influence ribosome function? *EMBO Journal* **3**: 91–94.

Arai T, Kusakabe A, Nakashio S, Nakamura M. (1984). A survey of plasmids in *Clostridium difficile* strains. *Kitasato Archives of Experimental Medicine* **57**: 285–288.

Bawden R E, Crane L W, Palchaudhuri S. (1982). Antibiotic resistance in anaerobic bacteria: molecular biology and clinical aspects. *Reviews of Infectious Diseases* **4**: 1075–1095.

Beguin P, Cornet P, Aubert J. (1985). Sequence of a cellulase gene of the thermophilic bacterium *Clostridium thermocellum*. *Journal of Bacteriology* **162**: 102–105.

Beguin P, Rocancourt M, Chebrou M, Aubert J. (1986). Mapping of mRNA encoding endoglucanase A from *Clostridium thermocellum*. *Molecular and General Genetics* 202: 251–254.

Beul H A *et al.* (1985). Characterisation of cryptic plasmids in clinical isolates of *Bacteroides fragilis*. *Journal of Medical Microbiology* 20: 39–48.

Bibb M J, Ward J M, Hopwood D A. (1978). Transformation of plasmid DNA into *Streptomyces* at high frequency. *Nature* 274: 398–400.

Blaschek H P, Solberg M. (1981). Isolation of a plasmid responsible for caseinase activity in *Clostridium perfringens* ATCC 3626B. *Journal of Bacteriology* 147: 262–266.

Brefort G, Magot M, Ionesco H, Sebald M. (1977). Characterisation and transferability of *Clostridium perfringens* plasmids. *Plasmid* 1: 52–66.

(1978). Characterisation and transferibility of *Clostridium perfringens* plasmids. In *Microbiology-1978*, pp. 242–245. Edited by Schlessinger D. American Society for Microbiology, Washington DC.

Brehm J K, Barstow, D A, Pennock A, Young M, Minton N P. (1986). Characterisation of a replication origin from a cryptic plasmid of *Clostridium butyricum*. P.G3–22:242. Abstracts of the XIV International Congress of Microbiology. Manchester.

Brenner D G, Shaw W V. (1985). The use of synthetic oligonucleotides with universal templates for rapid DNA sequencing: results with staphylococcal replicon pC221. *EMBO Journal* 4: 561–568.

Burt S J, Woods D R. (1976). R factor transfer to obligate anaerobes from *Escherichia coli*. *Journal of General Microbiology* 93: 405–409.

Burt S J, Woods D R. (1977). Transfection of the anaerobe *Bacteroides thetaiotaomicron* with phage DNA. *Journal of General Microbiology* 103: 181–187.

Callihan D R, Young F E, Clark V L. (1983). Identification of three homology classes of small cryptic plasmid in *Bacteroides* species. *Plasmid* 9: 17–30.

Chang S, Cohen S N. (1979). High frequency transformation of *Bacillus subtilis* protoplasts by plasmid DNA. *Molecular and General Genetics* 168: 111–115.

Chen K C, Chen J S, Johnson J L. (1986). Structural features of multiple *nifH*-like sequences and very biased codon usage in nitrogenase genes of *Clostridium pasteurianum*. *Journal of Bacteriology* 166: 162–172.

Clarke D J, Morris J G. (1976). Partial purification of a DCCD-sensitive membrane ATPase complex from the obligately anaerobic bacterium *Clostridium pasteurianum*. *Biochemical Journal* 154: 725–729.

Clarke D J, Robson R M, Morris J G. (1975). Purification of two *Clostridium* bacteriocins by procedures appropriate to hydrophobic proteins. *Antimicrobial Agents and Chemotherapy* 7: 256–264.

Collins M E, Oultram J D, Young M. (1985). Identification of restriction fragments from two cryptic *Clostridium butyricum* plasmids that promote the establishment of a replication-defective plasmid in *Bacillus subtlis*. *Journal of General Microbiology* 131: 2097–2105.

Cornet P, Tronik D, Millet J, Aubert J. (1983). Cloning and expression in *Escherichia coli* of *Clostridium thermocellum* genes coding for amino acid synthesis and cellulose hydrolysis. *FEMS Microbiology Letters* 16: 137–141.

Cue D, Beckler G S, Reeve J N, Konisky J. (1985). Structure and sequence divergence of two archaebacterial genes. *Proceedings of the National Academy of Sciences of the USA* 82: 4207–4211.

Daeschel M A, Klaeenhammer T R. (1985). Association of a 13.6-megadalton plasmid in *Pediococcus pentosaceus* with bacteriocin activity. *Applied and Environmental Microbiology* 50: 1538–1541.

Daldel F, Applebaum J. (1985). Cloning and expression of *Clostridium pasteurianum* galactokinase gene in *Escherichia coli* K-12 and nucleotide sequence analysis of a region affecting the amount of the enzyme. *Journal of Medical Biology* 186: 533–545.

Donkersloot J A, Cisar J O, Wax M E, Harr R J, Chassy B M. (1985). Expression of *Actinomyces viscosus* antigens in *Escherichia coli*: cloning of a structural gene (*fim*A) for type 2 fimbriae. *Journal of Bacteriology* 162: 1075–1078.

Duncan C L, Rokos E A, Christenson C M, Rood J I. (1978). Multiple plasmids in different toxigenic types of *Clostridium perfringens*: possible control of beta-toxin production. In *Microbiology-1978*, pp. 246–248. Edited by Schlessinger D. American Society for Microbiology, Washington DC.

Efstathiou I, Truffaut N. (1986). Cloning of *Clostridium acetobutylicum* genes and their expression in *Escherichia coli* and *Bacillus subtlis*. *Molecular and General Genetics* 204: 317–321.

Eisel U *et al.* (1986). Tetanus toxin: primary structure, expression in *Escherichia coli*, and homology with botulinum toxins. *The EMBO Journal* 5: 2495–2502.

Eklund M W, Poysky F T. (1974). Interconversion of type C and D strains of *Clostridium botulinum* by specific bacteriophages. *Applied Microbiology* 27: 251–258.

Eklund M W, Poysky F T, Meyers J A, Pelroy G A. (1974). Interspecies conversion of *Clostridium botulinum* type C to *Clostridium novyi* type A by bacteriophage. *Science* 186: 456–458.

Elleman T C, Hoyne P A. (1984). Nucleotide sequence of the gene encoding pilin of *Bacteroides nodosus*, the causal organism of ovine footrot. *Journal of Bacteriology* 160: 1184–1187.

Elleman T C, Hoyne P A, Emery D L, Stewart D J, Clark B L. (1984). Isolation of the gene encoding pilin of *Bacteroides nodosus* (strain 198), the causal organism of ovine footrot. *FEBS Letters* 173: 103–107.

Elleman T C, Hoyne P A, McKern N M, Stewart D J. (1986). Nucleotide sequence of the gene encoding the two-subunit pilin of *Bacteroides nodosus* 265. *Journal of Bacteriology* 167: 243–250.

Fairweather N F, Lyness V A. (1986). The complete

nucleotide sequence of tetanus toxin. *Nucleic Acids Research* 14: 7809–7812.

Fairweather N F, Lyness V A, Pickard D J, Allen G, Thomson R O. (1986). Cloning, nucleotide sequencing, and expression of tetanus toxin fragment C in *Escherichia coli*. *Journal of Bacteriology* 165: 21–27.

Finn C W, Silver R P, Habig W H, Hardegree M C. (1984). The structural gene for tetanus neurotoxin is on a plasmid. *Science* 224: 881–884.

Fornari C S, Kaplan S. (1982). Genetic transformation of *Rhodopseudomonas sphaeroides* by plasmid DNA. *Journal of Bacteriology* 154: 1513–1515.

Franke A E, Clewell D B. (1981). Evidence for a chromosome-borne resistance transposon (Tn916) in *Streptococcus feacalis* that is capable of 'conjugal' transfer in the absence of a conjugative plasmid. *Journal of Bacteriology* 145: 494–502.

Garnier T, Cole S T. (1986). Characterisation of a bacteriocinogenic plasmid from *Clostridium perfringens* and molecular genetic analysis of the bacteriocin encoding gene. *Journal of Bacteriology* 168: 1189–1196.

Gawron-Burke C, Clewell D B. (1982). A transposon in *Streptococcus faecalis* with fertility properties. *Nature* 300: 281–284.

Goodman H J, Southern J A, Parker J, Woods D R. (1986). Cloning and expression of a *recA*-like gene from the obligate anaerobe *Bacteroides fragilis* in *Escherichia coli*. P.B32–3:143. Abstracts of the XIV International Congress of Microbiology, Manchester.

Graves M C, Mullenbach G T, Rabinowitz J C. (1985). Cloning and nucleotide sequence determination of the *Clostridium pasteurianum* ferredoxin gene. *Proceedings of the National Academy of Sciences of the USA* 82: 1653–1657.

Graves M C, Rabinowitz J C. (1986). *In vivo* and *in vitro* transcription of the *Clostridium pasteurianum* ferredoxin gene. *Journal of Biological Chemistry* 261: 11409–11415.

Grepinet O, Beguin P. (1986). Sequence of the cellulase gene of *Clostridium thermocellum* coding for endoglucanase B. *Nucleic Acid Research* 14: 1791–1799.

Grosjean H, Fiers W. (1982). Preferential codon usage in procaryotic genes: the optimal codon-anticodon interaction energy and the selective codon usage in efficiently expressed genes. *Gene* 18: 199–209.

Guiney D G. (1984). Promiscuous transfer of drug resistance in gram negative bacteria. *Journal of Infectious Diseases* 149: 320–329.

Guiney D G, Davis C E. (1978). Identification of a conjugative R plasmid in *Bacteroides ochraceus* capable of transfer to *Escherichia coli*. *Nature* 274: 181–182.

Guiney D G, Davis C E. (1982). Incompatibility and host range of pGD10 from *Capnocytophaga ochraceus*, formerly *Bacteroides ochraceus*. *Plasmid* 7: 196–198.

Guiney D G, Hasegawa P, Stalker P, Davis C E. (1983). General genetic analysis of clindamycin resistance in *Bacteroides* species. *Journal of Infectious Diseases* 147: 551–558.

Guiney D G, Hasagawa P, Davis C E. (1984a). Homology between clindamycin resistance plasmids in *Bacteroides*. *Plasmid* 11: 268–271.

(1984b). Expression in *Escherichia coli* of cryptic tetracycline resistance genes from *Bacteroides* R plasmids. *Plasmid* 11: 248–252.

(1984c). Plasmid transfer from *Escherichia coli* to *Bacteroides fragilis*: differential expression of antibiotic resistance phenotype. *Proceedings of the National Academy of Sciences of the USA* 81: 7203–7206.

Guthrie E P, Salyers A A. (1986). Use of targeted insertional mutagenesis to determine whether chondroitin lyase II is essential for chondroitin sulfate utilisation by *Bacteroides thetaiotaomicron*. *Journal of Bacteriology* 166: 966–971.

Guthrie E P, Salyers A A. (1987). Evidence that the *Bacteroides thetaiotaomicron* chondroitin lyase II gene is adjacent to the chondro-4-sulfatase gene may be part of the same operon. *Journal of Bacteriology* 169: 1192–1199.

Guthrie E P, Shoemaker N B, Salyers A A. (1985). Cloning and expression in *Escherichia coli* of a gene coding for a chondroitin lyase from *Bacteroides thetaiotaomicron*. *Journal of Bacteriology* 164: 510–515.

Hara T, Matsuda M, Yoneda M. (1977). Isolation and some properties of non-toxigenic derivatives of a strain of *Clostridium tetani*. *Biken Journal* 20: 105–115.

Hayter P M, Dale J W. (1984). Detection of plasmids in clinical isolates of *Clostridium difficile*. *Microbios Letters* 27: 151–156.

Heefner D L, Squires C H, Evans R J, Kopp B J, Yarus M J. (1984). Transformation of *Clostridium perfringens*. *Journal of Bacteriology* 159: 460–464.

Hinton S M, Freyer G. (1986). Cloning, expression and sequencing the molybdenum-pterin binding protein *(mop)* gene of *Clostridium pasteurianum* in *Escherichia coli*. *Nucleic Acids Research* 14: 9371–9380.

Holloway B W. (1978). Plasmids that mobilise bacterial chromosome. *Plasmid* 2: 1–19.

Holm L. (1986). Codon usage and gene expression. *Nucleic Acids Research* 14: 3075–3087.

Hongo M A, Murato A, Ogato S, Kono K, Kato F. (1968). Characterisation of a temperate phage and four bacteriocins produced by nonpathogenic *Clostridium* species. *Agricultural and Biological Chemistry* 32: 773–780.

Horodiceanu T, Bougueleret L, Bieth G. (1981), Conjugative transfer of multiple-antibiotic resistance markers in beta-hemolytic group A, B, F and G streptococci in the absence of extrachromosomal deoxyribonucleic acid. *Plasmid* 5: 127–137.

Inamine J M, Burdett V. (1985). Structural organisation of a 67-kilobase streptococcal conjugative element mediating antibiotic resistance. *Journal of Bacteriology* 161: 620–626.

Ionesco H. (1980). Transfert de la resistance a la tetracycline chez *Clostridium difficile*. *Annales de Microbiologie (Institut Pasteur)* **131A**: 171–179.

Ionesco H, Bieth G, Dauguet C, Bouanchaud D. (1976). Isolement et identification de deux plasmides d'une souche bacteriocinogene de *Clostridium perfringens*. *Annales de Microbiologie (Institut Pasteur)* **127B**: 283–294.

Ionesco H, Bouanchaud D H. (1973). Production de bacteriocine liee a la presence d'un plasmide chez *Clostridium perfringens* type A. *Comptes Rendus-Academie Sciences Paris* **276** Serie D: 2855–2857.

Ishii K, Kudo T, Honda H, Horikoshi L. (1983). Molecular cloning of β-isopropylmalate dehydrogenase from *Clostridium butyricum* M588. *Agricultural and Biological Chemistry* **47**: 2313–2317.

Joliffe G, Beguin P, Aubert J. (1986). Nucleotide sequence of the cellulase gene *celD* of *Clostridium thermocellum*. *Nucleic Acids Research* **14**: 8605–8613.

Jones D T, Jones W A, Woods D R. (1985). Production of recombinants after protoplast fusion in *Clostridium acetobutylicum* P262. *Journal of General Microbiology* **131**: 1213–1216.

Jong E C, Ko H L, Pulverer G. (1975). Studies on bacteriophages of *Propionibacterium acnes*. *Medical Microbiology and Immunology* **161**: 263–271.

Karube I, Urano N, Yamada T, Hirochika H, Sakaguchi K. (1983). Cloning and expression of the hydrogenase gene from *Clostridium butyricum* in *Escherichia coli*. *FEBS Letters* **158**: 119–122.

Kawata T, Takumi K, Sato S, Yamashita H. (1968). Autolytic formation of spheroplasts and autolysis of cell walls in *Clostridium botulinum* type A. *Japanese Journal of Microbiology* **12**: 445–455.

Klebe R J, Harriss J V, Sharp Z D, Douglas M G. (1983). A general method for polyethylene-glycol-induced genetic transformation of bacteria and yeasts. *Gene* **25**: 333–341.

Kleckner N. (1981). Translocatable elements in procaryotes. *Annual Review of Genetics* **15**: 341–404.

Knowlton S, Ferchak J D, Alexander J K. (1984). Protoplast regeneration in *Clostridium tertium*: isolation of derivatives with high frequency regeneration. *Applied and Environmental Microbiology* **48**: 1246–1247.

Konisky J. (1978). The Bacteriocins. In *The Bacteria*, vol VI. pp. 71–136. Edited by Gunlans I C. Academic Press, New York.

Kuritza A P, Getty C E, Shaughnessy P, Hesse R, Salyers A A. (1986a). DNA probes for identification of clinically important *Bacteroides* species. *Journal of Clinical Microbiology* **23**: 343–349.

Kuritza A P, Salyers A A. (1985). Use of species-specific DNA hybridisation probe for enumerating *Bacteroides vulgatus* in human faeces. *Applied and Environmental Microbiology* **50**: 958–964.

Kuritza A P, Shaughnessy P, Salyers A A. (1986b). Enumeration of polysaccharide-degrading *Bacteroides* species in human feces by using species-specific DNA probes. *Applied and Environmental Microbiology* **51**: 385–390.

Laird W J, Aaronson W, Silver R P, Habig W H, Hardegree M C. (1980). Plasmid-associated toxigenicity in *Clostridium tetani*. *Journal of Infectious Diseases* **142**: 623.

Lamed R, Setter E, Kenig R, Bayer E A. (1983). The celluosome. A discrete cell surface organelle of *Clostridium thermocellum* which exhibits separate antigenic, cellulose-binding and various cellulolytic activities. *Biotechnology and Bioengineering Symposium* **13**: 163–181.

Li A W, Krell P J, Mahoney D E. (1980). Plasmid detection in a bacteriocinogenic strain of *Clostridium perfringens*. *Canadian Journal of Microbiology* **26**: 1018–1022.

Lin Y, Blaschek H P. (1984). Transformation of heat-treated *Clostridium acetobutylicum* with pUB110 plasmid DNA. *Applied and Environmental Microbiology* **48**: 737–742.

Luczak H, Schwarzmoser H, Staudenbauer W L. (1985). Construction of *Clostridium butyricum* plasmids and transfer to *Bacillus subtilis*. *Applied Microbiology and Biotechnology* **23**: 114–122.

Macrina F L, Mays T D, Smith C J, Welch R A. (1981). Non-plasmid associated transfer of antibiotic resistance in *Bacteroides*. *Journal of Antimicrobial Chemotherapy* **8** Suppl D: 77–86.

Magot M. (1983). Transfer of antibiotic resistances from *Clostridium innocuum* to *Clostridium difficile* in the absence of detectable plasmid DNA. *FEMS Microbiology Letters* **18**: 149–151.

(1984). Physical characterisation of the *Clostridium perfringens* tetracycline-chloramphenical resistance plasmid pIP401. *Annales de Microbiologie (Institut Pasteur)* **135B**: 269–282.

Magot M, Fayolle F, Privitera G, Sebald M. (1981). Transposon-like structures in the *Bacteroides fragilis* MLS plasmid pIP410. *Molecular and General Genetics* **181**: 559–561.

Mahony D E. (1974). Bacteriocin susceptibility of *Clostridium perfringens*: a provisional typing scheme. *Applied and Environmental Microbiology* **28**: 172–176.

Mahony D E, Clark G A, Stringer M F, MacDonald M C, Duchesse D R, Mader J A. (1986). Rapid extraction of plasmids from *Clostridium perfringens*. *Applied and Environmental Microbiology* **51**: 521–523.

Mahony D E, Li A. (1978). Comparative study of ten bacteriocins of *Clostridium perfringens*. *Antimicrobial Agents and Chemotherapy* **14**: 886–892.

Mahony D E, Moore T I. (1976). Stable L-forms of *Clostridium perfringens* and their growth on glass surfaces. *Canadian Journal of Microbiology* **22**: 953–959.

Malamy M H, Tally F P. (1981). Mechanisms of drug-resistance transfer in *Bacteroides fragilis*. *Journal of Antimicrobial Chemotherapy* **8**: 59–75.

Mancini C, Behme R J. (1977). Transfer of multiple antibiotic resistance from *Bacteroides fragilis* to *Escherichia coli*. *Journal of Infectious Diseases* **136**: 597–600.

Marsh P K, Malamy M H, Shimell M J, Tally F P. (1983).

Sequence homology of clindamycin resistance determinants in clinical isolates of *Bacteroides* species. *Antimicrobial Agents and Chemotherapy* **23**: 726–730.

Martinez-Suarez J V, Barquez F, Reig M, Perez C. (1985). Transferable plasmid-linked chloramphenical acetyltransferase conferring high-level resistance in *Bacteroides fragilis. Antimicrobial Agents and Chemotherapy* **28**: 113–117.

Matthew B G, Guiney D G. (1986). Characterisation and mapping of regions encoding clindamycin resistance, tetracycline resistance, and a replication function on the *Bacteroides* R plasmid pCP1. *Journal of Bacteriology* **167**: 517–521.

Mays T D, Johnson J L. (1979). Plasmid DNA homology studies of *Bacteroides fragilis* and related species. *Plasmid* **2**: 299–30.

Mays T D, Smith C J, Welch R A, Delfino C, Macrina F L. (1982). Novel antibiotic resistance transfer in *Bacteroides. Antimicrobial Agents and Chemotherapy* **21**: 110–118.

Milhelc V A, Duncan C L, Chambliss G H. (1978). Characterisation of a bacteriocinogenic plasmid in *Clostridium perfringens* CW55. *Anti-microbial Agents and Chemotherapy* **14**: 771–779.

Millet J, Petre D, Beguin P, Raynaud O, Aubert J. (1985). Cloning of ten distinct DNA fragments of *Clostridium thermocellum* coding for cellulases. *FEMS Microbiology Letters* **29**: 145–149.

Minton N P, Atkinson T, Bruton C J, Sherwood R F. (1984). The complete nucleotide sequence of the *Pseudomonas* gene coding for carboxypeptidase G_2. *Gene* **31**: 31–38.

Minton N P, Morris J G. (1981). Isolation and partial characterisation of three cryptic plasmids from strains of *Clostridium butyricum. Journal of General Microbiology* **127**: 325–331.

(1983). Regeneration of protoplasts of *Clostridium pasteurianum* ATCC 6013. *Journal of Bacteriology* **155**: 432–434.

Muldrow L L, Archibald E R, Nunez-Montiel O L, Sheehy R J. (1982). Survey of the extrachromosomal gene pool of *Clostridium difficile. Journal of Clinical Microbiology* **16**: 637–640.

Muller B, Allmansberger R, Klein A. (1985). Termination of a transcription unit comprising highly expressed genes in the archaebacterium *Methanococcus voltae. Nucleic Acids Research* **13**: 6439–6445.

Nieves B M, Gil F, Castillo F J. (1981). Growth inhibition activity and bacteriophage and bacteriocin-like particles associated with different species of *Clostridium. Canadian Journal of Microbiology* **27**: 216–225.

O'Brien R W, Morris J G. (1971). The ferredoxin-dependent reduction of chloramphenicol by *Clostridium acetobutylicum. Journal of General Microbiology* **67**: 265–271.

Ogata S, Choi K, oshino S, Hayashida S. (1981). Studies on sucrose-induced autolysis of clostridial cells. *Journal of the Faculty of Agriculture, Kyushi University* **25**: 201–222.

Ogata S, Hongo M. (1979). Bacteriophages of the genus *Clostridium. Advances in Applied Microbiology* **25**: 241–273.

Oultram J D, Young M. (1985). Conjugal transfer of plasmid pAMB1 from *Streptococcus lactis* and *Bacillus subtilis* to *Clostridium acetobutylicum. FEMS Microbiology Letters* **27**: 129–134.

Pan-Hou H S K, Hosono M, Imura N. (1980). Plasmid-controlled mercury biotransformation by *Clostridium cochlearium* T-2. *Applied and Environmental Microbiology* **40**: 1007–1011.

Pero J, Lee G, Moran C P, Lang N, Losick R. (1982). Promoters controlled by novel sigma factors in *Bacillus subtilis*. In: *Promoters; structure and function*, pp. 283–292. Edited by Rodriguez R L, Chamberlin M J. Praeger, New York.

Popoff M R, Truffaut N. (1985). Survey of plasmids in *Clostridium butyricum* and *Clostridium beijerinkii* strains from different origins and different phenotypes. *Current Microbiology* **12**: 151–156.

Privitera G, Dublanchet A, Sebald M. (1979a). Transfer of multiple antibiotic resistance between subspecies of *Bacteroides fragilis. Journal of Infectious Diseases* **139**: 97–101.

Privitera G, Sebald M, Fayolle F. (1979b). Common regulatory mechanisms of expression and conjugative ability of a tetracycline resistance plasmid in *Bacteroides fragilis. Nature* **278**: 657–659.

Pugsley A P. (1984). The ins and outs of colicins. Part I: production and translocation across membranes. *Microbiological Sciences* **1**: 168–175.

Rashtchian A, Dubes G R, Booth J. (1982). Transferable resistance to cefoxitin in *Bacteroides thetaiotaomicron. Antimicrobial Agents and Chemotherapy* **22**: 701–703.

Rauhut R, Gabius H, Kuhn W, Cramer F. (1984). Phenylalanyl-tRNA synthetase from the archaebacterium *Methanosarcina barkeri. Journal of Biological Chemistry* **259**: 6340–6345.

Reid S J, Allcock E R, Jones D T, Woods D R. (1983). Transformation of *Clostridium acetobutylicum* protoplasts with bacteriophage DNA. *Applied and Environmental Microbiology* **45**: 305–307.

Reysset G, Sebald M. (1985). Conjugal transfer of plasmid-mediated antibiotic resistance from *Streptococcus* to *Clostridium acetobutylicum. Annales de Microbiologie (Institut Pasteur)* **136B**: 275–282.

Roberts I, Holmes M, Hylemon P B. (1986). Modified plasmid isolation method for *Clostridium perfringens* and *Clostridium absonum. Applied and Environmental Microbiology* **52**: 197–199.

Robillard N J, Tally F P, Malamy M H. (1985). Tn4400, a compound transposon from *Bacteroides fragilis* functions in *Escherichia coli. Journal of Bacteriology* **164**: 1248–1255.

Rogers P. (1986). General genetics and biochemistry of

Clostridium relevant to development of fermentation processes. *Advances in Applied Microbiology* **31**: 1–60.

Rokos E A, Rood J I, Duncan C I. (1978). Multiple plasmids in different types of *Clostridium perfringens*. *FEMS Microbiology Letters* **4**: 323–326.

Romaniec M P M, Clarke N G E, Orpin C G, Hazlewood G P. (1986). Cloning and expression in *Escherichia coli* of cellulase genes from rumen anaerobic bacteria, *Ruminococcus albus* and *Butyrivibrio* sp. Proceedings of the XIV International Congress of Microbiology, pp. 101.

Rood J I. (1983). Transferable tetracycline resistance in *Clostridium perfringens* strains of porcine origin. *Canadian Journal of Microbiology* **29**: 1241–1246.

Rood J I, Maher E A, Somers E B, Campase E, Duncan C L. (1978a). Isolation and characterisation of multiply antibiotic-resistance *Clostridium perfringens* strains from porcine feces. *Antimicrobial Agents and Chemotherapy* **13**: 871–880.

Rood J I, Scott V N, Duncan C L. (1978b). Identification of a transferable tetracycline resistance plasmid (pCW3) from *Clostridium perfringens*. *Plasmid* **1**: 563–570.

Rotimi V O, Duerden B I, Hafiz S. (1981). Transferable plasmid-mediated antibiotic resistance in *Bacteroides*. *Journal of Medical Microbiology* **14**: 359–370.

Rowbury R J. (1977). Bacterial plasmids with particular reference to their replication and transfer properties. *Progress in Biophysics and Molecular Biology* **31**: 271–317.

Salaki J S, Black R, Tally F P, Kislak J W. (1976). *Bacteroides fragilis* resistant to clindamycin. *Journal of Medicine* **60**: 426–428.

Salyers A A. (1984). *Bacteroides* of the human lower intestinal tract. *Annual Review of Microbiology* **38**: 293–313.

Salyers A A, Lynn S P, Gardener J F. (1983). Use of randomly cloned DNA fragments for identification of *Bacteroides thetaiotaomicron*. *Journal of Bacteriology* **154**: 287–293.

Satija K C, Narayan K J. (1980). Passive bacteriocin typing of strains of *Clostridium perfringens* type A causing food poisoning for epidemiologic studies. *Journal of Infectious Diseases* **142**: 899–902.

Schallehn G, Eklund M W. (1980). Conversion of *Clostridium novyi* type D to alpha toxin production by phages of *Clostridium novyi* type A. *FEMS Microbiology Letters* **7**: 83–86.

Schwarz W, Bronnonmeier K, Staudenbaum W L. (1985). Molecular cloning of *Clostridium thermocellum* genes involved in β-glucan degradation in bacteriophage lambda. *Biotechology Letters* **7**: 859–864.

Schwarz W, Grabnitz F, Staudenbaum W L. (1986). Properties of a *Clostridium thermocellum* endoglucanase produced in *Escherichia coli*. *Applied and Environmental Microbiology* **51**: 1293–1299.

Scott J R. (1984). Regulation of plasmid replication. *Microbiological Reviews* **48**: 1–23.

Scott V N, Duncan C L. (1978). Cryptic plasmids in *Clostridium botulinum* and *C. botulinum* like organisms. *FEMS Microbiology Letters* **4**: 55–58.

Sebald M, Bouanchaud D, Bieth G. (1975). Nature plasmidique dela resistance a plusieurs antibiotiques chez *Clostridium perfringens* type A. *Comptes Rendus-Academie des Sciences, Paris Serie D* **280**: 2401–2404.

Sebald M, Brefort G. (1975). Transfert du plasmide tetracycline–chloramphenical chez *Clostridium perfringens*. *Comptes Rendus-Academie des Sciences Paris, Serie D* **281**: 317–319.

Sell T L, Schaberg D R, Fekety F R. (1983). Bacteriophage and bacteriocin typing scheme for *Clostridium difficile*. *Journal of Clinical Microbiology* **17**: 1148–1152.

Sgorbati B, Scardovi V, Leblanc D J. (1982). Plasmids in the genus *Bifidobacterium*. *Journal of General Microbiology* **128**: 2121–2131.

Shimell M J, Smith C J, Tally F P, Macrina F L, Malamy M H. (1982). Hybridisation studies reveal homologies between pBF4 and pBFTM10, two clindamycin–erythromycin resistance transfer plasmids of *Bacteroides fragilis*. *Journal of Bacteriology* **152**: 950–953.

Shoemaker N B, Getty C, Gardner J F, Salyers A A. (1986a). Tn4351 transposes in *Bacteroides* species and mediates the integration of plasmid R751 into the *Bacteroides* chromosome. *Journal of Bacteriology* **165**: 929–936.

Shoemaker N B, Getty C, Guthrie E P, Salyers A A. (1986b). Regions in *Bacteroides* plasmids pBFTM10 and pB8-51 that allow *Escherichia coli-Bacteroides* shuttle vectors to be mobilised by IncP plasmids and by a conjugative *Bacteroides* tetracycline resistance element. *Journal of Bacteriology* **166**: 959–965.

Shoemaker N B, Guthrie E P, Salyers A A, Gardner J F. (1985). Evidence that the clindamycin-erythromycin resistance gene of *Bacteroides* plasmid pBF4 is on a transposable element. *Journal of Bacteriology* **162**: 626–634.

Shoemaker N B, Smith M D, Guild W R. (1979). Organisation and transfer of heterologous chloramphenicol and tetracycline resistance genes in pneumococcus. *Plasmid* **3**: 432–441.

Smith C J. (1985a). Development and use of cloning systems for *Bacteroides fragilis*: cloning of a plasmid-encoded clindamycin resistance determinant. *Journal of Bacteriology* **164**: 294–301.

(1985b). Polyethylene glycol-facilitated transformation of *Bacteroides fragilis* with plasmid DNA. *Journal of Bacteriology* **164**: 466–469.

Smith C J, Gonda M A. (1985). Comparison of the transposon-like structures encoding clindamycin resistance in *Bacteroides* R-plasmids. *Plasmid* **13**: 182–192.

Smith C J, Macrina F L. (1984). Large transmissible clindamycin resistance plasmid in *Bacteroides ovatus*. *Journal of Bacteriology* **158**: 739–741.

Smith C J, Markowitz S M, Macrina F L. (1981). Transferable tetracycline resistance in *Clostridium difficile*. *Antimicrobial Agents and Chemotherapy* **19**: 997–1003.

Smith C J, Welch R A, Macrina F L. (1982). Two independent conjugal transfer systems operating in *Bacteroides fragilis* V479–1. *Journal of Bacteriology* **151**: 281–287.

Soong T W, Ho B, Thong T W, Ng H C, Ng S C, Yap E H. (1984). Characterisation of plasmids from clinical isolates of *Bacteroides*. *Singapore Medical Journal* **25**: 146–150.

Soutschek-Bauer E, Hartl L, Staudenbauer W L. (1985). Transformation of *Clostridium thermohydrosulfuricum* DSM 568 with plasmid DNA. *Biotechology Letters* **7**: 705–710.

Squires C H, Heefner D L, Evans R J, Kopp B J, Yarus M J. (1984). Shuttle plasmids for *Escherichia coli* and *Clostridium perfringens*. *Journal of Bacteriology* **159**: 465–471.

Stiffler P W, Keller R, Traub N. (1974). Isolation and characterisation of several cryptic plasmids from clinical isolates of *Bacteroides fragilis*. *Journal of Infectious Diseases* **130**: 544–548.

Strom M S, Eklund M W, Potsky F T. (1984). Plasmids in *Clostridium botulinum* and related *Clostridium* species. *Applied and Environmental Microbiology* **48**: 956–963.

Tally F P, Snydman D R, Gorbach S L, Malamy M H. (1979). Plasmid-mediated transferable resistance in *Bacteroides fragilis*. *Journal of Infectious Diseases* **139**: 83–88.

Tally F P, Snydman D R, Shimell M J, Malamy M H. (1982). Characterisation of pBFTM10 a clindamycin–erythromycin resistance factor from *Bacteroides fragilis*. *Journal of Bacteriology* **151**: 686–691.

Tinnell W H, Macrina F L. (1976). Extrachromosomal elements in a variety of strains representing the *Bacteroides fragilis* group of organisms. *Infection and Immunity* **14**: 955–964.

Totsuka M. (1976). Studies on Veillonellophages isolated from washings of human oral cavity. *Bulletin of the Tokyo Medical and Dental University* **23**: 261–273.

Truffaut N, Sebald M. (1983). Plasmid detection and isolation in strains of *Clostridium acetobutylicum* and related species. *Molecular and General Genetics* **189**: 178–180.

Tylenda C A, Calvert C, Kolenbrander P E, Tylenda A. (1985a). Isolation of *Actinomyces* bacteriophage from human dental plague. *Infection and Immunity* **49**: 1–6.

Tylenda C A, Enriquez E, Kolenbrander P E, Delisle A L. (1985b). Simultaneous loss of bacteriophage receptor and coaggregation mediator activities in *Actinomyces viscosus* MG-1. *Infection and Immunity* **48**: 228–233.

Urano N, Karube I, Suzuki S, Yamada T, Hirochika H, Sakaguicha K. (1983). Isolation and partial characterisation of large plasmids in hydrogen evolving bacterium *Clostridium butyricum*. *European Journal of Applied Microbiology and Biotechnology* **17**: 349–354.

Usdin K P, Zappe H, Jones D T, Woods D R. (1986). Cloning, expression, and purification of glutamine synthetase from *Clostridium acetobutylicum*. *Applied and Environmental Microbiology* **52**: 413–419.

Varley J M, Boulnois G J. (1984). Analysis of a cloned colicin 1b gene: complete nucleotide sequence and implications for regulation of expression. *Nucleic Acids Research* **12**: 6727–6739.

Vandenbergh P A, Syed S A, Gonzalezm C F, Loesche W J, Olsen R H. (1982). Plasmid content of some oral microorganisms isolated from subgingival plaque. *Journal of of Dental Research* **61**: 497–501.

Vijayakumar M N, Priebe S D, Guild W R. (1986). Structure of a conjugative element in *Streptococcus pneumoniae*. *Journal of Bacteriology* **166**: 978–984.

Wallace B L, Bradley J E, Rogolsky M. (1981). Plasmid analyses in clinical isolates of *Bacteroides fragilis* and other *Bacteroides* species. *Journal of Clinical Microbiology* **14**: 383–388.

Weickert M J, Chambliss G H, Sugiyama H. (1986). Production of toxin by *Clostridium botulinum* type A strains cured of plasmids. *Applied and Environmental Microbiology* **51**: 52–56.

Welch R A, Jones K R, Macrina F L. (1979). Transferable lincosamide–macrolide resistance in *Bacteroides*. *Plasmid* **2**: 261–268.

Welch R A, Macrina F L. (1981). Physical characterisation of *Bacteroides fragilis* R plasmid pBF4. *Journal of Bacteriology* **145**: 867–872.

Wich G, Jarsch M, Bock A. (1984). Apparent operon for a 5S ribosomal RNA gene and for tRNA genes in the archaebacterium *Methanococcus vannielii*. *Molecular and General Genetics* **196**: 146–151.

Wust J, Hardegger U. (1983). Transferable resistance to clindamycin, erythromycin, and tetracycline in *Clostridium difficile*. *Antimicrobial Agents and Chemotherapy* **23**: 784–786.

Yongsmith B, Cole J A. (1986). Cloning and expression of propionibacterial genes in *Escherichia coli*. Proceedings of the XIV International Congress of Microbiology, p. 244.

Yoshino S, Ogata S, Hayashida S. (1984). Regeneration of protoplasts of *Clostridium saccheroperbutylacetonicum*. *Agricultural and Biological Chemistry* **48**: 249–250.

Yu P, Pearce L E. (1986). Conjugal transfer of streptococcal antibiotic resistance plasmids into *Clostridium acetobutylicum*. *Biotechnology Letters* **80**: 469–474.

Zappe H, Jones D T, Woods D R. (1986) Cloning and expression of *Clostridium acetobutylicum* endoglucanase, cellobiose and amino acid biosynthesis genes in *Escherichia coli*. *Journal of General Microbiology* **132**: 1367–1372.

Note added in proof

Since completion of this Chapter, further advancements have been made in the rapidly expanding field of anaerobe genetics. Attention is therefore drawn to the following more recent reviews on the subject:– Salyers A A, Shoemaker N B, Guthrie E P. (1987). Recent advances in *Bacteroides* genetics. *Critical Reviews in Microbiology* **14**: 49–71; Young M, Staudenbauer W L, Minton N P. (1989). Genetics of *Clostridium*. In *Clostridia*, pp. 63–103. Edited by Minton N P, Clarke D J. Plenum Publishing Co., New York, and; Young M, Minton N P, Staudenbauer W L. (1989). Recent advances in the genetics of the clostridia. *FEMS Microbiology Reviews* **63**: 301–326.

5

Bacteroides and *Fusobacterium*: classification and relationships to other bacteria

H N Shah and S E Gharbia

Current status of classification
    Fusobacterium
    Bacteroides
Isolation and Identification
    Routine laboratory methods for isolation
    Identification
Pathogenesis of infections and the nature of virulence
    Studies of virulence of *F. nucleatum, B. fragilis, B. asaccharolyticus* and *B. gingivalis*
References

The inspired work of Veillon and Zuber (1898) which resulted in the isolation of gram-negative, non-sporing, obligately anaerobic bacteria from infections at various sites, led to an enthusiastic search for anaerobic bacteria by other workers around the turn of this century. Noteable examples include the work of Tissier (1908), who reported that the predominant micro-organisms of the human colon were anaerobes and described five 'types'. Tunnicliff (1919, 1923) reported 'fusiform' bacteria from various sources while Krumwiede and Pratt (1913) isolated several strains of fusiform bacteria from different sites and classified them into two groups on the basis of carbohydrate fermentation patterns. At first, the recognition of *Bacteroides* and *Fusobacterium* as distinct entities was not immediately apparent, and the early nomenclature includes a number of generic names to describe these bacteria: *Fusiformis, Sphaerophorus, Bacteroides, Necrobacterium, Pseudobacterium, Ristella, Zuberella, Corynebacterium, Actinomyces, Bacterium, Bacillus, Fusobacterium, Leptothrix* and *Leptotrichia*.

The non-fusiform, gram-negative, anaerobic rods were named *Bacteroides* by Castellani and Chalmers (1919). Knorr (1923) suggested the generic name *Fusobacterium* for obligately anaerobic, gram-negative bacilli that were fusiform, he proposed three species, *F. nucleatum F. polymorphum* and *F. plauti-vincenti*. The family name Bacteroidaceae was subsequently used by Pribram (1929) for both genera. Eggerth and Gagnon (1933) reported one of the first detailed systematic studies of these bacteria in which 18 species were delineated on the basis of morphology and the fermentation of 20 carbohydrates. Surprisingly, no black-pigmented bacteroides were included in this classification, although they had been reported earlier in the microbial flora of the mouth, faeces, urine and in infected wounds (Oliver and Wherry, 1921).

The classic work of Prévot and his colleagues (1966) extended over three decades. These were the first attempts to employ more reliable taxonomic criteria, such as acid end product analysis and the quantitation of purines and pyrimidines. Furthermore, accurate measurements of oxidation–reduction potentials for various species, as well as their pathogenic potential in various experimental animal models, were reported.

Prévot's system was adopted, and in the 6th edition of *Bergey's Manual of Determinative Bacteriology* (Breed *et al.*, 1957) all non-sporing gram-negative anaerobes with rounded ends were included in the genus *Bacteroides*; 30 species were recorded, most of which readily fermented glucose, and only rarely fermented lactose and sucrose. The species considered to be proven pathogens were *B. fragilis*, *B. serpens* and *B. melaninogenicus*. In the same edition of the Manual, *Fusobacterium* was included in the family Bacteroidaceae along with *Bacteroides* and *Sphaerophorus*. However, in the subsequent edition emphasis shifted from morphology to metabolic end products as a more realiable criterion for classification of these microorganisms. Data from several sources, particularly Beerens *et al.* (1962), Sebald (1962) Werner (1972a, b) and Holdeman and Moore (1974) were used to redefine the Bacteroidaceae and, following these redefinitions, the genus *Fusobacterium* comprised species whose major metabolic end product was butyric acid. Fusiform bacilli that produced lactic acid were placed in the genus *Leptotrichia*, while physiologically related organisms which conformed to neither pattern were assigned to the genus *Bacteroides*. Currently, the genus *Bacteroides* comprises some 39 species, *Fusobacterium* eight species and *Leptotrichia* remains a monospecific genus *L. buccalis* (*Bergey's Manual of Systematic Bacteriology*, 9th edition; Holdeman *et al.*, 1984). However, the number of *Bacteroides* and *Fusobacterium* species proposed after the publication of the 9th edition of *Bergey's Manual* (1984) has increased rapidly, and most are beyond the scope of this chapter. Many of these new species fit only loosely into the description of the above genera, and will undoubtedly be reclassified in the future. Examples include *B. gracilis*, *B. cellulosolvens*, *B. mycoides*, *B. polypragmatus* and *B. symbiosus*.

Phylogenetically, the genera *Bacteroides* and *Fusobacterium* are most closely related to the *Flavobacterium* and *Cytophaga* genera as defined by their 16S rRNA sequences (Weisburg *et al.*, 1985) and oligonucleotide catalogues (Paster *et al.*, 1985). The relationship, although not close, is nevertheless specific such that the organisms are placed into one of the 10 eubacterial phyla of Woese *et al.* (1985). Within this phylum, the *Bacteroides* and *Fusobacterium* constitute one subdivision, while the flavobacteria and cytophagas form another subphylum (Paster *et al.*, 1985). However, incoherent groups were reported within the *Bacteroides–Fusobacterium* subphylum (Paster *et al.*, 1985), e.g., *B. amylophilus* belonged to the phylogenetic unit defined by the purple photosynthetic bacteria, whereas *B. succinogenes* was unrelated to any of the 10 defined phyla of Woese *et al.* (1985). These inconsistencies highlight some major phylogenetic problems which still exists in the classification of these microorganisms.

Current status of classification

Fusobacterium

The genus *Fusobacterium* was defined mainly by morphological criteria (Knorr, 1922) until it was recognized that butyric acid, the major metabolic end product of its member species, could be used to help redefine the genus. This redefinition resulted in the delineation of the genus *Leptotrichia* to contain the single species, *L. buccalis* which is morphologically similar to the fusobacteria, but which produces lactic acid as a major metabolic end product. Several species assigned to the genus *Fusobacterium* are no longer recognized, or have been transferred to other genera, e.g., *F. bullosum* and *F. symbiosum* were shown to produce spores and are now included in the genus *Clostridium*. *F. polysaccharolyticum* was also transferred to the genus *Clostridium* on the basis of its heat resistance. *F. aquatile*, *F. glutinosum* and *F. stabile* have been deleted from the Approved Lists of Bacteriological Names because strains were no longer extant, while *F. plauti* was shown to have a gram-positive type of cell wall (Hofstad and Aasjord, 1982) and was transferred to the genus *Eubacterium*. Similarly, a DNA base composition of G + C 52–57 mol per cent for *B. prausnitzii* was incompatible with membership of the genus *Fusobacterium*. Characteristics useful in the differentiation of gram-negative, non-sporing rods are given in Table 5.1, and species which are currently recognized as members of the genus *Fusobacterium* are listed in Table 5.2.

Despite the apparent homogeneity of species

Table 5.1 Biochemical and chemical characteristics useful in the differentiation of gram-negative, non-spore forming, rods

Taxon	DNA base composition ($G + C$ mol%)	Dibasic amino acid of the peptidoglycan*	Dehydrogenase enzymes present†	Predominant cell fatty acid‡	Major menaquinone	References
B. fragilis group e.g., *B. fragilis* (type species)	41–48	DAP	G-6-PDH, 6-PGDH MDH, GDH	aiC$_{15:0}$	MK-10 MK-11	Shah and Collins (1980); Shah and Collins (1983)
B. splanchnicus	40	DAP	G-6-PDH, 6-PGDH MDH, GDH	iC$_{15:0}$	MK-9	Shah and Collins (1980); Shah and Collins (1983a)
Pigmented species:						
a) Non-fermentative e.g., *B. gingivalis*	48–54	DAP/Lys	MDH, GDH	iC$_{15:0}$	MK-9	Shah *et al.* (1976); Shah and Collins (1980); Shah and Collins (1983)
b) Fermentative	40–51	DAP	MDH, GDH	aiC$_{15:0}$	MK-10 MK-11	Shah *et al.* (1976); Shah and Collins (1980); Shah and Collins (1983)
Non-pigmented saccharolytic spp. e.g., *B. oralis*	40–52	DAP	MDH, GDH	aiC$_{15:0}$ iC$_{15:0}$ C$_{15:0}$	MK-10 MK-13	Shah *et al.* (1976); Shah and Collins (1980); Shah and Collins (1983)
Non-pigmented non-fermentative spp. e.g., *B. coagulans*	28–37	NT	GDH	C$_{16:0}$ C$_{18:1}$	—	Collins and Shah (1987)

Organism	% G+C	Cell wall†	Enzymes†	Fatty acids‡	Menaquinone	Reference
Anaerorhabdus furcosus	34	NK (Not DAP)	G-6-PDH	$C_{16:1}$; $C_{18:0}$	—	Shah and Collins, (1986a)
Megamonas hypermegas	32–35	DAP	G-6-PDH, 6-PGDH, MDH	$C_{15:0}$	—	Shah *et al.* (1983); Shah and Collins (1982a)
Sebaldella termitidis	30–32	—	—	$C_{16:0}$; $C_{18:1}$	—	Shah and Collins (1986b)
Tissierella praeacuta	28	mDAP	6-PGDH	$iC_{15:0}$	—	Collins and Shah (1986b)
B. capillosus	60	NT	GDH	$C_{14:0}$; $C_{16:0}$	—	Collins and Shah (1987)
Mitsuokella multiacidus	56–58	mDAP	MDH	$C_{16:1}$	—	Shah and Collins (1982b); Shah *et al.* (1983)
Rikenella microfusus	60–61	DAP	MDH	$iC_{15:0}$	MK-8	Collins *et al.* (1985)
Fusobacterium	28–32	Lan or DAP or both	GDH	$C_{16:0}$; $C_{14:0}$; 3-OH-$C_{14:0}$	—	Jantzen and Hofstad (1981); Gharbia and Shah (1986)
Leptotrichia	25	DAP	MDH	$C_{16:0}$; $C_{18:1}$; 3-OH-$C_{14:0}$	NT	Hofstad and Jantzen (1982); Page and Krywolap (1976)

* DAP, diaminopimelic acid; Lys, Lysine; Lan, lanthionine;
† G-6-PDH, glucose-6-phosphate dehydrogenase; 6-PGDH, 6-phosphogluconate dehydrogenase; MDH, malate dehydrogenase; GDH, glutamate dehydrogenase;
‡ $C_{14:0}$, tetradecanoic acid; $C_{16:0}$, hexadecanoic acid; $C_{16:1}$, hexadecenoic acid; $C_{18:1}$, octadecenoic acid; $iC_{15:0}$, 13-methyltetradecanoic acid; $aiC_{15:0}$, 12-methyltetradecanoic acid; 3-OH-$C_{14:0}$, 12-methyltridecanoic acid; NT, not tested; NK, not known.

Table 5.2 Currently recognized species of the genus *Fusobaterium*

Current name	Previous epithet
*F. nucleatum**	*F. fusiforme* (Werner), *F. fusiformis*, *F. polymorphum*, *F. plauti-vincenti* (Boe), *F. polymorphus*
*F. necrophorum**	*Fusiformis necrophorus*, *Bacilus necrophorus*, *Sphaerophorus necrophorus*
*F. gonidiaformans**	*Bacillus gonidiaformans*, *Sphaerophorus gonidiaformans*
*F. varium**	*Bacteroides varius*, *Sphaerophorus varius*
*F. necrogenes**	*Bacillus necrogenes*, *Sphaerophorus necrogenes*
*F. perfoetens**	*Coccobacillus perfoetens*, *Ristella perfoetens*, *Sphaerophorus perfoetens*
*F. naviforme**	*Bacillus naviforme*, *Ristella naviforme*
*F. russii**	*Bacteroides russii*
*F. mortiferum**	*Bacillus mortiferus*, *Sphaerophorus mortiferus*
*F. prausnitzii**	*Bacteroides prausnitzii*
F. simiae	New species (Slots and Potts 1982)
F. periodonticum	New species (Slots *et al.* 1982)
F. alocis	New species (Cato *et al.* 1985)
F. sulci	New species (Cato *et al.* 1985)
F. ulcerans	New species (Adriaans and Drasar, 1987; Adriaans and Shah, 1988)

*Listed in the 9th edition of *Bergey's Manual of Systematic Bacteriology*

within the genus *Fusobacterium*, the application of chemotaxonomic analysis to the species listed in Table 5.2 indicates considerable heterogeneity.

Based on our studies (Gharbia and Shah, 1988) and the characteristics given by Hofstad (1979, 1981) and Holdeman *et al.* (1984), the following criteria for defining the genus *Fusobacterium* are suggested:

obligately anaerobic, gram-negative, non-sporing rods; some species, but not all, possess fusiform-shaped cells;

produce major amounts of butyric acid as a metabolic end product;

contain glutamate dehydrogenase;

DNA base composition within the range G+C 28–32 mol per cent;

dibasic amino acid of the peptidoglycan either diaminopimelic acid or lanthionine or both;

cell fatty acids consist mainly of straight-chain and monounsaturated acid types (methyl branched and cyclopropane acids are absent);

lack respiratory quinones (which are present in most other eubacteria).

Bacteroides

Gram-negative, obligately anaerobic, non-sporing rods, which cannot be assigned to the genera *Fusobacterium* or *Leptotrichia* are frequently deposited in the genus *Bacteroides*. Lack of good criteria for defining the *Bacteroides* is partly responsible for the genus accumulating such a large and heterogeneous collection of species. More than 50 species are currently listed in *Bergey's Manual of Systematic Bacteriology*, 9th edition (Holdeman *et al.*, 1984) and in the Approved Lists of Bacteriological Names. Species from a range of animals that includes cats, dogs, pigs and marine animals and their environments, insects and plants, as well as man, have been reported recently (e.g., Love *et al.*, 1986). Poor definition of the genus and the unacceptably wide range of DNA base composition of G + C (28–61 mol per cent; Holdeman *et al.*, 1984) among *Bacteroides* spp. allows isolates which share only marginal similarities to be assigned to the genus *Bacteroides*. Thus the number of species incorrectly placed in the genus *Bacteroides* continues to increase. It has been suggested (Shah and Collins, 1983; Collins and Shah, 1987) that the genus should be restricted to those species whose DNA base composition is within the range G + C 40–48 mol per cent and which are biochemically and chemically related to the type species *Bacteroides fragilis*. Thus, the genus *Bacteroides sensu stricto* should comprise the following species:

B. fragilis (type species);
B. caccae;
B. distasonis;
B. eggerthii;

B. merdae;
B. ovatus;
B. stercoris;
B. thetaiotaomicron;
B. uniformis;
B. vulgatus.

These species are biochemically and chemically fairly homogeneous and share all the characteristics which are suggested to facilitate clearer circumscription of the genus *Bacteroides* (below). Thus, the genus *Bacteroides* should be restricted to those species which fulfil the following criteria:

obligately anaerobic, gram-negative, non-sporing rods;

saccharolytic, produce major amounts of acetate and succinate as metabolic end products;

contain enzymes of the pentose phosphate pathway such as glucose-6-phosphate dehydrogenase (G-6-PDH) and 6-phosphogluconate dehydrogenase (6-PGDH) in addition to malate dehydrogenase (MDH) and glutamate dehydrogenase (GDH);

have a DNA base composition within the range G + C 40–48 mol per cent;

possess sphingolipids;

contain a mixture of long chain fatty acids with predominantly straight chain saturated, anteiso-methyl branched and isomethyl branched acids;

possess menaquinones with MK-10 and MK-11 as major components;

peptidoglycan contains *meso*-diaminopimelic acid as the diamino acid.

Table 5.3 Proposed taxonomic revision of the genus *Bacteroides* with particular reference to human species*

Species	Taxonomic status/comments
'B. fragilis group'	
B. fragilis, B. caccae, B. distasonis, B. eggerthii, B. merdae, B. ovatus, B. stercoris, B. thetaiotaomicron, B. uniformis, B. vulgatus	*Bacteroides*
'B. oralis–melaninogenicus group'	
B. bivius, B. buccae, B. buccalis, B. corporis, B. denticola, B. intermedius, B. melaninogenicus, B. disiens, B. loescheii, B. oralis, B. oulorum, B. oris, B. ruminicola, B. veroralis, B. heparinolyticus, B. zoogleoformans	New genus†
'Pigmented asaccharolytic group'	
B. asaccharolyticus, B. endodontalis, B. gingivalis	*Porphyromonas*
B. amylophilus	*Ruminobacter*
B. capillosus	New genus
B. coagulans	New genus
B. furcosus	*Anaerorhabdus*
B. gracilis	*Wolinella*
B. hypermegas	*Megamonas*
B. levii	Uncertain (possibly related to pigmented asaccharolytic group)
B. macacae	
B. microfusus	*Rikenella*
B. multiacidus	*Mitsuokella*
B. praeacutus	*Tissierella*
B. putredinis	Uncertain (possibly related to *B. macacae* and pigmented asaccharolytic group)
B. pneumosintes	Uncertain (not *Bacteroides*)
B. splanchnicus	Uncertain (not *Bacteroides*)
B. succinogenes	*Fibrobacter*
B. terminitidis	*Sebaldella*
B. ureolyticus	*Campylobacter* or *Wolinella*

*Adapted from Collins and Shah, 1987
†Recently reclassified as 'Prevotella' Shah and Collins, 1990

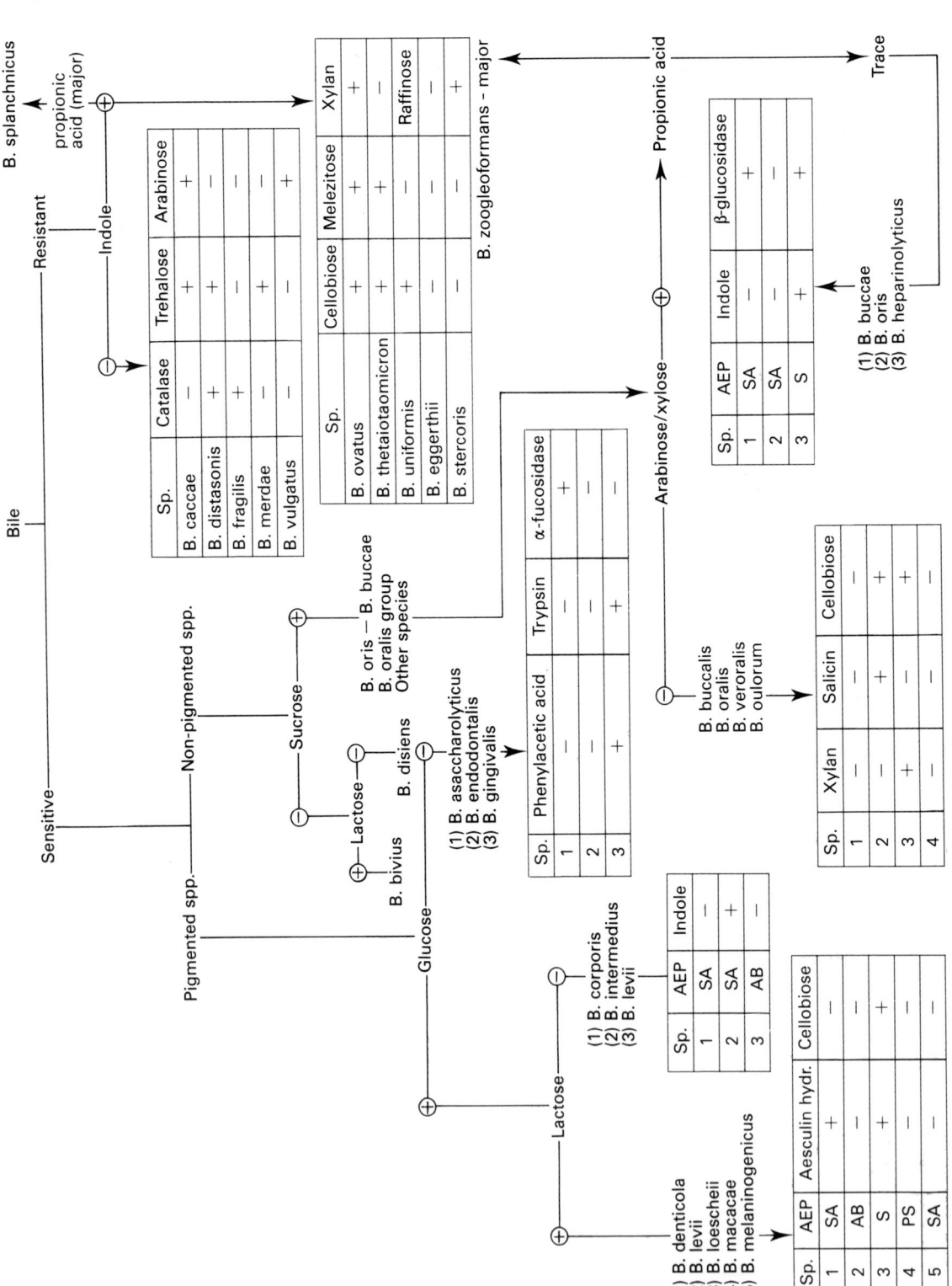

Fig. 5.1 Identification scheme for *Bacteroides* and bacteroides-like species (Adapted from Holdeman *et al.*, 1984 and Sutter *et al.*, 1985)

AEP, acid end product (major); A, acetic acid; B, butyric acid; P, propionic acid; S, succinic acid.

On the basis of the above criteria and earlier data (Shah and Collins, 1983; Collins and Shah, 1987) we proposed the following revision of the genus *Bacteroides* (Table 5.3). The table includes many of the older non-human species to show their present taxonomic status, but largely omits many of the recent non-human species such as *B. helcogenes, B. suis* and *B. pyogenes* (Benno *et al.*, 1983), which extend beyond the scope of this chapter. The newly revised genus *Bacteroides* which has been validly published (Shah and Collins, 1989) contains only saccharolytic, non-pigmented species. The growth of all species is stimulated by bile and, with the exception of *B. splanchnicus*, they are the only gram-negative, anaerobic, non-spore-forming rods which possess high levels of both G-6-PDH and 6-PGDH (see Table 5.1). Although, superficially, *B. splanchnicus* strains appear to form a homogeneous group with the '*B. fragilis*' group, biochemical, chemical and genetic data indicate that they are taxonomically very diverse. *B. splanchnicus* differs markedly from the '*B. fragilis*' group in producing propionic acid as a major metabolic end product, possessing menaquinones with nine isoprene units and containing mainly 13-methyl-tetradecanoic acid (Shah and Collins, 1983). Subsequent rRNA homology studies by Johnson and Harich (1986) revealed less than 20 per cent relatedness between *B. splanchnicus* and the '*B. fragilis*' group, and indicate that *B. splanchnicus* should be excluded from the genus *Bacteroides sensu stricto*.

The newly included species in the '*B. fragilis*' group, *B. caccae, B. merdae* and *B. stercoris* (Johnson *et al.*, 1986) are the DNA homology groups '3452', 'T4-1' and 'ss. a' respectively of Johnson and Ault (1978). Delineation of species within the '*B. fragilis*' group is achieved by conventional bacteriological tests (Holdeman *et al.*, 1984; Sutter *et al*, 1985), and is summarized in Fig. 5.1.

The group of species referred to as the '*B. oralis–melaninogenicus*' group (see Table 5.3) contains many of the species commonly isolated from the oral cavity; these species have a DNA G+C composition between 40 and 50 mol per cent. However, when the criteria given above are used to define the genus *Bacteroides sensu stricto*, it is evident that the group constitutes a new genus, unrelated to *B. fragilis*. Most of these species are moderately saccharolytic and produce mainly acetic and succinic acids in a glucose medium. The pigmented species, such as *B. melaninogenicus, B. loescheii, B. corporis, B. intermedius* and some strains of *B. denticola*, produce both protohaem and protoporphyrin pigments (Fig. 5.2). *B. melaninogenicus, B. loescheii* and the pigmented strains of *B. denticola* fluoresce brilliantly under ultraviolet radiation (365 nm) (Fig.

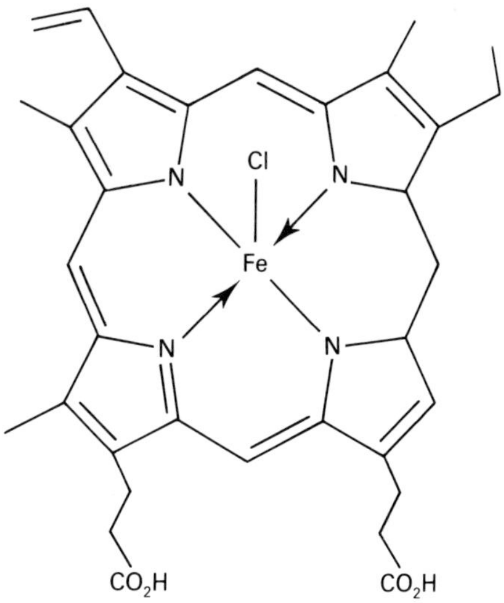

(a) Protohaem

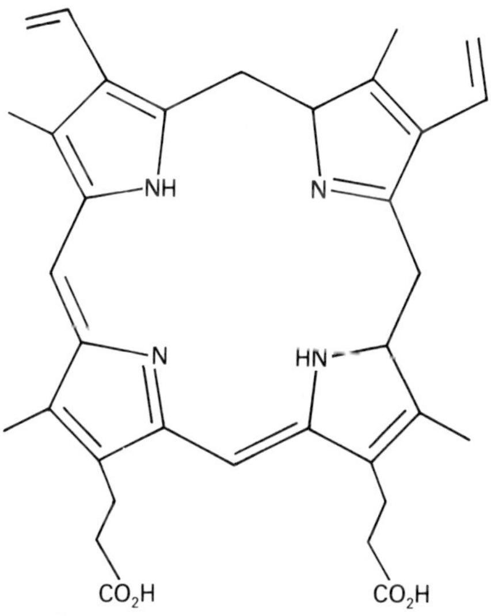

(b) Protoporphyrin

Fig. 5.2 Porphyrins of *Bacteroides* species

5.3) due to the production of mainly protoporphyrin. These species have the ability to demetallate the iron-containing protohaem molecule leaving a highly conjugated structure which fluoresces under UV radiation (Shah et al., 1979). The darker pigmented colonies of B. intermedius and B. corporis are due to an excess of protohaem production over protoporphyrin. Before these colonies become entirely black, however, the non-pigmented younger cells at the centre of the colonies also fluoresce (Fig. 5.3). This property is useful for recognizing at least some species of the 'B. oralis–melaninogenicus' group, and has been of major importance in elucidating the taxonomic inter-relationships of this group of species. Other members of this large cluster include B. bivius, B. buccae, B. buccalis, B. disiens, B. oralis, B. oulorum, B. oris, B. ruminicola and B. veroralis. These species differ from the B. fragilis group in that they are sensitive to bile and lack the enzymes G-6-PDH and 6-PGDH. Although these species have been proposed against a background of DNA-DNA homology studies, heterogeneity is nevertheless evident in some species, such as B. intermedius and B. loescheii. Biochemical and chemical data indicate that the rumen species, B. amylophilus and B. succinogenes, are unrelated to this group and to B. fragilis, and it has been suggested (Shah and Collins, 1983) that both species should be excluded from the genus Bacteroides. Details of the taxonomic position of the rumen species is discussed by Collins and Shah (1987). It is interesting that most of the species which contribute to the B. oralis–melaninogenicus group are of oral origin (Shah and Collins, 1981; Sundqvist, 1976). Of the physiological reactions used to delineate species of this group, the fermentation of xylose, arabinose and cellobiose are of major importance (see Fig. 5.1). The pigmented, asaccharolytic species, B. asaccharolyticus, B. gingivalis and B. endodontalis, comprise another tight cluster of species that differ so significantly from B. fragilis and the B. oralis–melaninogenicus group that they have recently been reclassified in a new genus, Porphyromonas (Shah and Collins, 1988). These species produce major amounts of protohaem on blood agar plates, lack G-6-PDH and 6-PGDH, produce high levels of butyric acid and possess predominantly isomethyl branched long chain fatty acids.

B. levii and B. macacae resemble the asaccharolytic pigmented group in producing mainly protohaemin and possessing a similar dehydrogenase pattern (Shah and Collins, 1983). The similarity in metabolic end products (butyric and acetic acids) and a compatible G+C content of 45–48 mol. per cent further reinforces the similarity between B. levii and the asaccharolytic species. Like B. macacae, however, B. levii is weakly saccharolytic. B. macacae differs further from the rest of this group in producing mainly propionic and succinic acids as major metabolic end products and possessing a lower G+C content of 42–44 mol per cent. Two broad groups of species are discernible by DNA base composition: a high G+C group, whose mean base composition is in the range 56–61 mol per cent, and a low G+C group of 28–37 mol per cent. The high G+C group, which comprises B. multiacidus (G+C 56–58 mol per cent) B. microfusus (G+C 60–61 mol per cent) and B. capillosus (G+C 60 mol per cent) share little interspecies DNA homology and can be readily excluded from the genus Bacteroides. B. multiacidus has been reclassified as Mitsuokella multiacidus (Shah and Collins, 1982b) and B. microfusus is now Rikenella microfusus (Collins et al., 1985), but B. capillosus remains unclassified. Recently, a second species, M. dentalis, has been added to the genus Mitsuokella (Haapasalo et al., 1986).

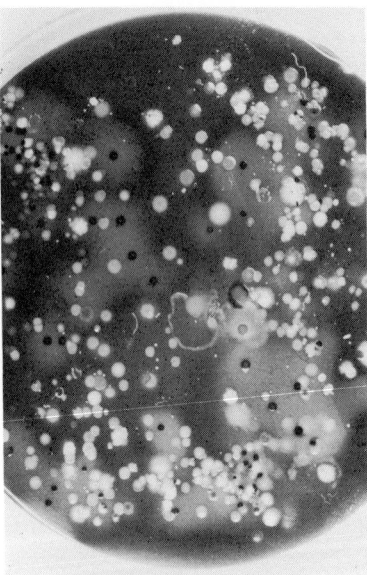

Fig. 5.3 Sample of dental plaque on blood agar plate. UV-fluorescent colonies of B. melaninogenicus and B. denticola (entire colonies fluoresce) and B. intermedius (only the centre of these colonies fluoresce)

Species whose DNA base composition falls within the range G+C 28–37 mol per cent are phenotypically and genetically so different from *B. fragilis* as to justify their exclusion from the genus *Bacteroides*. With the exception of *B. hypermegas* and *B. termitidis*, all species are non-fermentative. *B. hypermegas* was reclassified in a new genus, *Megamonas*, as *Megamonas hypermegas* (Shah and Collins, 1982a). More recently, *B. furcosus, B. praeacutus* and *B. termitidis* have been reclassified in separate genera as *Anaerorhabdus furcosus* (Shah and Collins, 1986) *Tissierella praeacuta* (Collins and Shah, 1986b) and *Sebaldella termitidis* (Collins and Shah, 1986a), respectively. Although *B. coagulans* and *B. ureolyticus* have not yet been formally reclassified, biochemical and chemical characteristics so far carried out indicate that they warrant separate generic status. Paster and Dewhirst (1988) used partial 16S ribosomal ribonucleic acid sequence data to show that *B. ureolyticus* and another 'corroding bacteroides', *B. gracilis*, 'are not true bacteroides', thus confirming our previous proposals (Shah and Collins, 1983; Collins and Shah, 1987). Paster and Dewhirst (1988) showed that these species have a very high 16S rRNA affinity with the genus *Campylobacter* and suggested that they should be transferred to this genus.

The phenotypic properties of *B. pneumosintes* are incompaticable with membership of the genus *Bacteroides* but since strains are unavailable for biochemical and chemical characterization, the taxonomic position of this species remains uncertain.

Isolation and identification

Routine laboratory methods for isolation

Methods available for specimen collection, transport and anaerobic culture techniques have been dealt with in detail by Sutter *et al*. (1985), Holdeman *et al*. (1984), Dowell and Hawkins (1974) and in various chapters in this book. The following media have been selected from various recommended media (Sutter *et al*., 1985) for the primary isolation of anaerobic bacteria, with particular reference to *Bacteroides*, *Fusobacterium* and *Leptotrichia* spp.

For total counts

Brucella 5 per cent sheep blood agar supplemented with vitamin K_1 1 mg per cent and haemin 0.5 mg per cent.

For selective isolation and presumptive identification of *B. fragilis* and related species

Bacteroides Bile Esculin (BBE) agar trypticase soy agar 4 g, oxgall 2 g per cent esculin 0.1 g per cent, ferric ammonium citrate 0.05 g per cent, haemin 1.0 mg per cent and gentamicin 10 mg per cent.

For the selection of pigmented and other *Bacteroides* spp

Kanamycin–vancomycin laked blood (KVLB) Brucella Agar (BBL) 4.3 g per cent haemin 0.5 mg per cent, vitamin K_1 mg per cent, kanamycin 7.5 mg per cent, vancomycin 0.75 mg per cent and laked sheep blood 5 per cent v/v. A separate medium has been described for *B. gingivalis* (Hunt *et al*., 1986). It was shown previously (Shah *et al*., 1976; van Winkelhoff and de Graaff, 1983) that vancomycin was inhibitory for the oral asaccharolytic species at these concentrations. However, a concentration of vancomycin 2.5 μg/ml was successfully employed by Duerden (1980) for isolating *Bacteroides* spp. from the gingival flora.

For the isolation of *Fusobacterium* and the inhibition of *Bacteroides* spp. and most other faecal micro-organisms

Rifampicin Blood Agar (RBA), containing rifampicin 5 mg per cent in Brucella blood agar is generally used.

For the selective isolation of *Fusobacterium*, *Veillonella* and *Leptotrichia*

Fusobacterium Egg-yolk Agar (FEA) consists of Brucella agar base with disodium hydrogen phosphate 0.5 g per cent, potassium dihydrogen phosphate 0.1 g per cent, magnesium sulphate 0.01 g per cent, haemin 0.05 mg per cent (all w/v) and polysorbate 80 (BBL) 0.1 per cent v/v. The medium is adjusted to pH 7.6, autoclaved and the following additions made before pouring: vancomycin 0.5 mg per cent, neomycin, 10 mg per cent and josamycin 0.3 mg per cent, and egg yolk emulsion (Difco) 5 per cent v/v.

Table 5.4 Presumptive identification of gram-negative, anaerobic, non-sporing, rods*

	Kanamycin (1mg) susceptibility	Colistin (10µg) susceptibility	Growth in 20% bile	Indole production	Nitrate reduction	Urease production	Pitting colonies	Pigment/ fluorescence
B. fragilis group	R	R	+	V	–	–	–	–
Pigmented species	R	V	–	V	–	–	–	+/+–
Non-pigmented species	R	V	–	V	–	–	–	–
B. ureolyticus -like group	S	S	–	–	+	V	V	–
Fusobacterium– Leptotrichia	S	S	–+	V	–+	–	–	–

R, resistant; S, susceptible; V, variable; +–, most strains positive, few negative; –+, most strains negative, few positive
*Adapted from Sutter *et al.* (1985)

Identification

When successfully isolated, organisms can be maintained by weekly subculture on blood agar plates. Cultures for long term storage are best kept at −70°C in liquid nitrogen, or in freeze-dried ampoules.

Identification

Gram's stain, cell and colonial morphology, susceptibility to special antibiotic disks placed on the first quadrant of a streaked plate, together with simple tests such as nitrate reduction, growth in bile and indole production, enable a preliminary grouping of isolates (Table 5.4) into the following major categories:

B. fragilis group;
pigmented group;
non-pigmented group;
B. ureolyticus-like group;
Fusobacterium–Leptotrichia group.

Figures 5.1 (page 68) and 5.4 give more detailed flow charts for the routine identification of major groups of species. Members of the *B. ureolyticus*-like group, which includes *B. gracilis* and *Wolinella* spp., are all stimulated to grow by the presence of

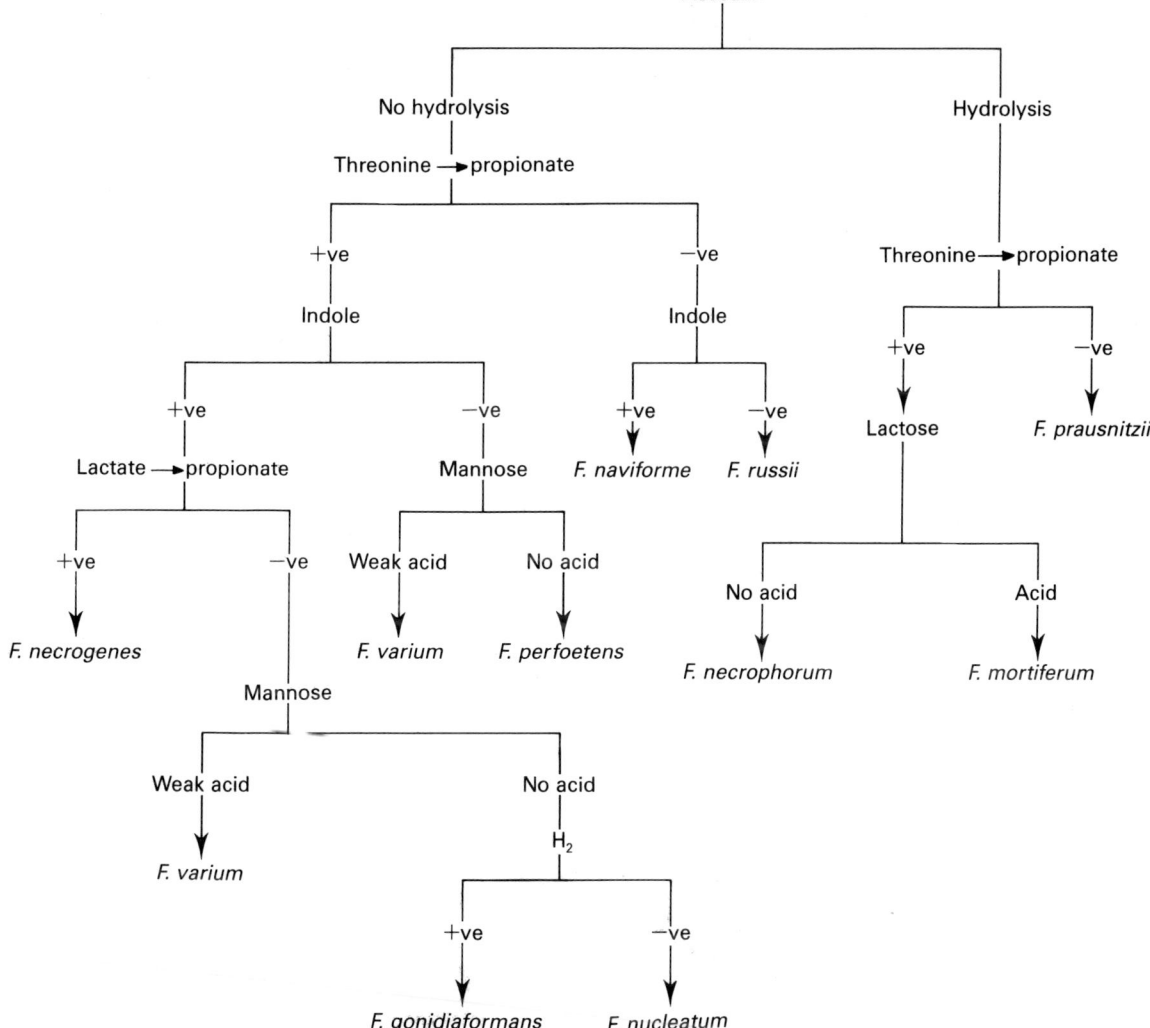

Fig. 5.4 Identification scheme for *Fusobacterium* species (Adapted from Holdeman and Moore, 1984 and Sutter *et al.*, 1985)

formate and fumarate in the media (Sutter et al., 1985). These three taxa can be distinguished from each other by the presence of urease activity in *B. ureolyticus* and the motility of *Wolinella* spp.

Several commercially available multitest systems (e.g., API 20A, RAP-ID ANA etc) can be used for most of the taxa listed above (Hofstad, 1980). However, the data base for many of these systems is still inadequate and cannot be relied upon entirely. Gas–liquid chromatography (GLC) of the metabolic end products of isolates is necessary for accurate indentification to the species level, and is employed in all current identification schemes. Some other methods for determining metabolic end products have been described recently (Shah et al., 1985).

Identification methods available to the specialist laboratory

The application of chemical methods for provision of characters of value in microbial classification have so far led to gradual improvements in the circumscription of taxa within the Bacteroidaceae. As stated earlier, the genus *Bacteroides* is poorly defined in the 9th edition of *Bergey's Manual of Systematic Bacteriology* (Holdeman et al., 1984) and contains species whose DNA base compositions are in the range G+C 28–61 mol per cent. Application of chemotaxonomic methods has led to a more restricted definition of the genus *Bacteroides* (see above), enabled the identification of hitherto unrecognizable species, such as *B. gingivalis* and *B. veroralis*, and led to the establishment of several new genera (see Table 5.3, page 67). Details of these analyses and the chemical composition of the components analysed have already been given (Shah et al., 1983). Technical advances to simplify and improve these analyses are occurring rapidly and many of the techniques which are at present beyond the limits of the diagnostic laboratory will soon become available. A brief summary of the techniques used, together with some examples of their value, is shown in Table 5.5.

Pathogenesis of infections and the nature of virulence

Anaerobic bacteria involved in infections are usually derived from the normal flora of the skin, oral cavity, nasopharynx, large intestine or the female genital tract. Infections are therefore usually polymicrobial and often pyogenic. It is generally accepted that a preceding infection by an anaerobic or facultative organism predisposes a site to invasion by the indigenous anaerobic flora. Other predisposing fac-

Table 5.5 Techniques currently in use to study the systematics of gram-negative, nonsporing, anaerobic rods

Method	Example of components analysed	Reference (example)
Thin layer chromatography (TLC) (including reverse phase TLC)	Lipids, e.g., menaquinones, polyamines	Collins and Shah (1987)
High-performance liquid chromatography (HPLC)	DNA base composition Menaquinones Amino acids	Shah and Collins (1983) Tamaoka and Komagata (1984)
Capillary gas chromatography	Cell fatty acids	Jantzen and Hofstad (1981)
Electrophoresis and blotting techniques	Enzyme–protein patterns DNA, RNA-hybridization DNA-probes Restriction enzymes RNA-cataloging Rna-sequencing	Shah and Williams (1982) Johnson (1978) Kurita, et al. (1986) Bradbury et al. (1985) Poxton and Brown (1986) Paster et al. (1985)
Spectrophotometry	DNA-base composition DNA/DNA-reassociation cytochromes	Shah et al. (1976) van Steenbergen et al. (1982) Reddy and Bryant (1977)

tors include disruption of the cutaneous or mucosal barriers, tissue injury and tissue necrosis, impairment of the blood supply, obstruction of hollow viscus (e.g., tracheobronchial tree) and the presence of a foreign body (Finegold et al., 1985). However, once initiated, it is clear that anaerobes play an active role in the infectious process such that treatment with antibiotics such as metronidazole or clindamycin, which are active only against the anaerobic component, is usually successfully (Thadepalli et al., 1973; Gorbach, 1975). Furthermore, it has been demonstrated that certain gram-negative anaerobes can protect facultative bacteria against phagocytosis and serum killing (Ingham et al., 1977, 1981). Anaerobes are thought to be only poorly invasive (Onderdonk et al., 1976), but once a suitable environment has been attained, they proliferate and elaborate a host of substances that function as potential virulence factors (Table 5.6).

Within the *Bacteroides* and *Fusobacterium* spp., research into pathogenicity and virulence has been focused on three main areas:

- *B. fragilis* of the *B. fragilis* group, particularly in pyogenic mixed anaerobic and facultative/aerobic infections;
- *B. gingivalis*, *B. asaccharolyticus* and *B. intermedius* of the pigmented species, particularly in periodontal diseases;
- *F. necrophorum* and *F. nucleatum* in various oral and soft tissue infections and abscesses.

These species all possess a number of potential virulence factors, one or more of which may be elaborated *in vivo*.

Table 5.6 Possible virulence determinants of *Bacteroides* and *Fusobacterium*

Property	Potential effect	Comments	Example
Adhesins	Cell attachment Colonisation Initiation of disease	Several surface attachment structures reported, e.g., fimbriae, filamentous, appendages, outer slime layers	Gibbons and van Houte (1971); Onderdonk et al. (1978); Slots and Gibbons (1978); Falkler et al. (1979); Falkler and Burger (1981); Haapasalo et al. (1985)
Capsule	Antiphagocytic Inhibits macrophage migration Potentiates abscess formation	Effects demonstrated by comparison of capsulated and non-capsulated strains	Onderdonk et al. (1977); Babb and Cummins (1978); Lindberg et al. (1979); Sundqvist et al. (1982); Sundqvist and Carlsson (1984)
Lipopolysaccharide	Stimulates bone resorption and inflammation in animal experiments. Reduce opsonic activity of serum complement	Most of the effects so far studied relate to periodontal diseases. Role of LPS in suppurative anaerobic infections still unknown	Hausmann et al. (1972); Hofstad et al. (1977); Hofstad and Sveen (1979); Sveen and Skaug (1980); Dahlen et al. (1981); Iino and Hopps (1984); Jones and Gemmell (1986)
Hydrolytic enzymes, hyaluronidase, elastase, chondroitin sulphatase, neuraminidase, protease, heparinase, fibrinolysin, phosphatase, gelatinase, collagenase, deoxyribonuclease, ribonuclease, lecithinase, lipase, penicillinase	Potential damage to host tissue, may act as spreading factors and impair blood supply to infected areas	Many species of *Bacteroides* and *Fusobacterium* possess these enzymes. Unlike some *Clostridium* spp. the role of these enzymes as virulence factors is largely speculative	Gibbons and MacDonald (1961); Muller and Werner (1970); Kaufman et al. (1972); Porschen and Sonntag (1974); Rudek and Haque (1976); Pulverer et al. (1977); Fraser and Brown (1981); Steffen and Hentges (1981); Robertson et al. (1982); Okuda et al. (1985); Smalley and Birss (1987)

Table 5.6 Continued.

Property	Potential effect	Comments	Example
Other enzymes Immunoglobulin proteases (IgA, IgG and IgM)	Induce degradation of immunoglobulins	Several patterns of degradation demonstrated. Good correlation between suspected pathogens and enzyme potency	Kilian (1981); Kilian et al. (1983)
Superoxide dismutases and catalase	Reduce the toxic effect of superoxide radicals and hydrogen peroxide	Present in both putative pathogens and species which are regarded as normal commensals	Carlsson et al. (1977); Tally et al. (1978); Gregory et al. (1978); Rolfe et al. (1978)
Soluble metabolites and products Leukotoxic substances, e.g., volatile fatty acids and volatile sulphur compounds chemotactic inhibitors	Cytopathic and cytotoxic for a variety of cell types Interference of neutrophil chemotaxis	These products appear to be more common to asaccharolytic species but are poorly understood	Roberts (1967); Karmen (1981); Touw et al. (1982); van Steenbergen et al. (1982a; 1985)
Growth factors Menadione, haemin, succinate, amino acids, peptides and steroids	Stimulates bacterial growth of some species thus conferring a selective advantage	Haemin shown to affect the virulence of some bacteroides. Asparagine and glutamine deamidation yield toxic ammonium ions	MacDonald et al. (1963); Lev et al. (1971); Lev (1980); Mayrand and McBride (1980); Kornman and Loesche (1982); McKee et al. (1986); Shah and Williams (1987); Gharbia and Shah (1988)

Studies of the virulence of F. nucleatum, B. fragilis, B. asaccharolyticus and B. gingivalis

The clinical importance of *F. nucleatum* has been regularly reported (Hofstad, 1974; Gorbach, 1975; Duerden et al., 1976, 1980; Sveen et al., 1977; Hofstad and Fredriksen, 1979; Moore et al., 1983; Bennett and Duerden, 1985). However, unlike *F. necrophorum*, which is predominantly an animal pathogen causing occasional severe infection in man, and which has been studied in detail, the virulence mechanisms of *F. nucleatum* remain largely unknown. *F. nucleatum* is present in only very low numbers in supragingival plaque of healthy individuals but its numbers increase dramatically in gingivitis, sometimes in children but most noticeably in adults, and it becomes one of the predominant species (Moore et al., 1982, 1984). Its co-occurrence with spirochaetes in acute necrotizing ulcerative gingivitis, has been recognized for some time (Listgarten, 1965; Loesche et al., 1982) but the nature of this association is unknown.

Filamentous appendages have not been reported in *F. nucleatum*. However, these organisms adhere to oral epithelial cells (Gibbons and van Houte, 1971) and, by means of an arginine 'bridge' (Dehezya and Coles, 1980, 1982), to a galactose-containing receptor on certain oral bacteria or human erythrocytes (Falkler and Hawley, 1977; Falkler et al., 1979; Mongiello and Falkler, 1979; Falkler and Burger, 1981). Adherence is crucial for the initiation of some diseases, thus the ability of *F. nucleatum* and *B. gingivalis* to selectively adhere to crevicular epithelium suggests a prominent role for these micro-organisms in the initiation and maintenance of inflammatory periodontal diseases (Hofstad, 1984).

Apart from these features, *F. nucleatum* possesses no obvious virulence factors which may help elucidate its role in disease. The energy metabolism of this species is, however, very similar to *B. gingivalis* (Gharbia and Shah, 1988b). Its production of hydrolytic enzymes (Table 5.6 page 75) is minimal and its lipopolysaccharide (LPS) possesses no unique chemical features (Hofstad, 1984). However, current work (Gharbia and Shah, unpublished data) indicates considerable heterogeneity of LPS chemotypes, as revealed by silver staining in polyacrylamide gels. Metabolically, *Fusobacterium* spp. utilize various dibasic amino acids to produce an array of toxic amines which are presently being characterized (Gharbia and Shah, unpublished results). When cultured in a mixed microbial community in a chemostat, *F. nucleatum* resisted the acidic conditions of pH 4.1 for 16 generations and returned to its original level after the pH was readjusted to 7.0 (McDermid *et al.*, 1986). These properties reflect the tremendous versatility and resilience of this species and clearly indicate that further studies are needed.

Unlike *F. nucleatum*, considerable research has been focused on *B. fragilis* and *B. gingivalis* because of their clinical significance. Both species produce various extracellular enzymes which may be important virulence determinants (Table 5.6). Both species also possess many similar putative virulence factors but differ in their sites of infection. *B. fragilis* represents only about 0.5 per cent of the total human intestinal flora, but it can reach up to 50 per cent of the anaerobic flora in certain infections related to the intestine (Polk and Kasper, 1977). The ability of *B. fragilis* to adhere to specific sites, such as the peritoneal mesothelium, correlates strongly with the presence of a capsule (Onderdonk, 1978), which has been carefully analysed (Kasper, 1976; Kasper *et al.*, 1980, 1983). *B. fragilis* resists and impairs phagocytosis, although the exact mechanism is still uncertain, but Onderdonk *et al.* (1977), Kasper *et al.* (1979) and more recently Connolly *et al.* (1984) have all demonstrated that the capsular polysaccharide (CPS) inhibits phagocytosis. However, this is refuted by Wade *et al.* (1983) and, more recently, by Jones and Gemmell (1986) who reported that the LPS, and not the CPS, reduced the opsonic activity of serum complement. It has not been possible to demonstrate any major differences in LPS chemotypes amongst clinical isolates of *B. fragilis* (Shah *et al.*, unpublished work). However, there is little doubt that the capsule is an important virulence determinant (Lindberg *et al.*, 1982), especially in promoting abscess formation.

Some strains of *B. fragilis* can increase their molar growth yield in the presence of haemin, and, like *B. gingivalis* (McKee *et al.*, 1986), their virulence is possibly enhanced. *B. fragilis* strains can be cultured without haemin in batch or in continuous culture, but the molar growth yield is increased in the presence of haemin by amounts which are strain-dependent (Al-Jalili and Shah, 1988). This data, together with the considerable enzyme polymorphism recently detected amongst strains of *B. fragilis*, reflect the heterogeneity of the species (Shah *et al.*, 1987) and may partly explain the frequent conflicting results pertaining to *B. fragilis* strains.

B. gingivalis (Coykendall *et al.*, 1980) was first recognized as one of the two G+C mol per cent and enzyme clusters of *B. melaninogenicus* ss. *asaccharolyticus* (Shah *et al.*, 1976). These organisms are now believed to play an important role in the aetiology of periodontal diseases. Despite the apparent genetic homogeneity of the species (Naito *et al.*, 1985), phenotypic heterogeneity is still evident (Shah, unpublished work) and may explain the wide variation in biological activities reported amongst strains of *B. gingivalis*. Strains W50 and W83, which were originally used to show the heterogeneity of the group (Shah *et al.*, 1976) appear to be the most virulent biotypes of the species. *B. gingivalis* strains possess several pathogenic attributes which may partly explain the nature of their virulence (Okuda *et al.*, 1981; Slots, 1982; Yamamoto *et al.*, 1982). High resistance to phagocytosis and killing was reported (Sundqvist *et al.*, 1982) possibly due to the composition of the capsule (Mansheim and Coleman, 1980). The most virulent strains require complement for leucocyte activation. However, these strains elaborate products which are not only non-chemotactic, but which compete with chemotactic peptides and heat-labile opsonins of the complement system and block the chemotactic receptors on the polymorphonuclear leucocytes (PMNL) (Van Dyke *et al.*, 1982) resulting in a reduced activity of the PMNLs. Intact cells of *B. gingivalis*, and particularly its LPS, activate both the classic complement pathway (Okuda *et al.*, 1978; Tofte *et al.*, 1980) and the alternative complement pathway (Sundqvist and Johansson, 1980) but the presence of a capsule decreases this activity (Okuda and Takazoe, 1983). Furthermore, the ability of *B. gingivalis* to degrade IgA, IgG, IgM and a wide array of plasma proteins (Kilian, 1981;

Kilian et al., 1983; Carlsson et al., 1984) must seriously impair the host defences. The effect of these proteolytic enzymes *in vivo* has been elegantly demonstrated by Sundqvist and Carlsson (1984) who correlated the enzyme activity around implanted tissue cages in guinea pigs with purulent tissue breakdown, degradation of immunoglobulins and C3 complement protein around the implants. Virulent strain W83 required serum to activate the leukocytes, unlike avirulent strain 380 (van Steenbergen and de Graaff, 1986), thus degradation of the plasma opsonins may impair the host leucocyte defences (Sundqvist and Carlsson, 1984).

The pathogenic role of *B. asaccharolyticus* is less clear. This species colonizes the gastrointestinal tract and genitalia and has not been isolated from dental plaque (Slots and Genco, 1979). It is often reported in mixed anaerobic infections of the gastrointestinal tract (Holdeman et al., 1974; Finegold et al., 1985).

In a study of men with balanoposthitis and nongonococcal urethritis, *B. asaccharolyticus* was the major single species isolated. *B. asaccharolyticus* is thought to be a significant pathogen in several types of superficial, necrotizing lesions and in diabetic gangrene and perineal lesions (Duerden, 1987). Several years ago it was demonstrated that the pigmented bacteroides were a necessary component for inducing polymicrobial infection in animals (MacDonald et al., 1963; Socransky and Gibbons, 1965). These studies were, however, carried out before the taxonomy of the black pigmented bacteroides was clarified, and thus the role of *B. asaccharolyticus* in these studies remains unclear. More recent studies by Mayrand *et al.* (1980) and van Steenbergen *et al.* (1982b) indicate that *B. asaccharolyticus* was usually less virulent in animal model experiments. Studies relating to the activity of specific virulence factors such as LPS, capsules or lytic enzymes of the pigmented bacteroides have so far been aimed at species other than *B. asaccharolyticus*. *B. asaccharolyticus* can, however, degrade IgA, IgG, IgM and a wide array of plasma proteins (Werner and Muller, 1971; Kilian, 1981) and may thus have the capacity to impair the host defences. Furthermore, *B. asaccharolyticus*, being a non-fermenting species, has the ability to utilize amino acids, peptides and substances such as mucin, as sources of energy. As a result, any hydrolytic activity at the site of infection will stimulate the growth of *B. asaccharolyticus* and must enhance the disease process.

B. gingivalis, *B. fragilis* and, to a lesser extent, *B. asaccharolyticus*, produce an array of hydrolytic enzymes which probably enhance their pathogenic potential (see Table 5.6). For example there was a marked increase in the level of collagenases in pathogenic strains of *B. gingivalis* compared to avirulent isolates (Grenier and Mayrand, 1987). The virulence of *B. gingivalis* strain W50 has also been found to be markedly enhanced by the presence of haemin in the medium (McKee et al., 1986) but although much data concerning aspects of the pathogenic potential of *B. gingivalis* strains W83 and W50 is known, the virulence determinants of other strains remain largely uncertain. Haem may also affect the virulence of *B. fragilis*, but within this species the requirement for haem varies considerably (Al-Jalili and Shah, 1988).

Recently, vesicles and also layers external to the outer membrane (S layer) have been reported in several *Bacteroides* spp. (Shah et al., 1976; Kornman and Holt, 1981; Haapasalo et al., 1985). These vesicles are high in proteolytic activity (Smalley and Birss, 1987) and may be useful in adsorbing opsonins *in vivo* (Haapasalo, 1986). The possibility that these structures may be important virulence factors for *Bacteroides* and *Fusobacterium*, is currently under investigation, in our laboratory and elsewhere.

References

Adriaans B, Drasar B S. (1987). The isolation of fusobacteria from tropical ulcers. *Journal of Epidemiology and Infection* 99: 361–372.

Adriaans B, Shah H N. (1988). *Fusobacterium ulcerans* sp. nov. from tropical ulcers. *International Journal of Systematic Bacteriology* 38: 447–448.

Al-Jalili T A R, Shah H N. (1988). Protoheme, a dispensable growth factor for *Bacteroides fragilis* grown by batch and continuous culture in a basal medium. *Current Microbiology* 17: 13–18.

Babb J L, Cummins C S. (1978). Encapsulation of *Bacteroides* species. *Infection and Immunity* 19: 1088–1091.

Beerens H, Castel M M, Fievez L. (1962). Classification des Bacteroidaceae. Abstracts of the VIIIth International Congress for Microbiology, Montreal, p. 120.

Bennett K W, Duerden B I. (1985). Identification of fusobacteria in a routine diagnostic laboratory. *Journal of Applied Bacteriology* 59: 171–181.

Benno Y, Watabe J, Mitsuoka T. (1983). *Bacteroides pyogenes* sp. nov., *Bacteroides suis* sp. nov., and *Bacteroides helcogenes* sp. nov., new species from abscesses and feces of pigs. *Systematic and Applied Microbiology* **4**: 396–407.

Bradbury W C, Murray R G E, Mancini C, Morris U L. (1985). Bacterial chromosomal restriction endonuclease analysis of the homology of *Bacteroides* species. *Journal of Clinical Microbiology* **21**: 24–28.

Breed R S, Murray E G D, Smith N R (eds.) (1957). *Bergey's Manual of Determinative Bacteriology*, 7th edn. Williams and Wilkins, Baltimore.

Carlsson J, Hofling J F, Sundqvist G. (1984). Degradation of albumin haemopexin, haptoglobin and transferrin by black-pigmented *Bacteroides* species. *Journal of Medical Microbiology* **18**: 39–46.

Carlsson J, Wrethen J, Beckman G. (1977). Superoxide dismutase in *Bacteroides fragilis* and related *Bacteroides* species. *Journal of Clinical Microbiology* **6**: 280–284.

Castellani A, Chalmers A J. (1919). *Manual of Tropical Medicine*, 3rd edn. Ballière, Tindall and Cox, London.

Cato E P, Moore L V, Moore W E. (1985). *Fusobacterium alocis* sp. nov. and *Fusobacterium sulci* sp. nov. from the human gingival sulcus. *International Journal of Systematic Bacteriology* **35**: 475–477.

Collins M D, Shah H N. (1986a). Reclassification of *Bacteroides termitidis* Sebald (Holdeman and Moore) in a new genus, *Sebaldella*, as *Sebaldella termitidis* comb. nov. *International Journal of Systematic Bacteriology* **36**: 349–350.

(1986b). Reclassification of *Bacteroides praeacutus* Tissier (Holdeman and Moore) in a new genus *Tissierella*, as *Tissierella praeacuta* comb. nov. *International Journal of Systematic Bacteriology* **36**: 461–463.

(1987). Recent advances in the chemotaxonomy of the genus *Bacteroides*. In *Recent Advances in Anaerobic Bacteriology; Proceedings of the Fourth Anaerobe Discussion Group Symposium*, pp. 249–258. Martinus Nijhoff Publishers, Dordrecht.

Collins M D, Shah H N, Mitsuoka T. (1985). Reclassification of *Bacteroides microfusus* (Kaneuchi and Mitsuoka) in a new genus *Rikenella*, as *Rikenella microfusus* comb. nov. *Systematic and Applied Microbiology* **6**: 79–81.

Connolly J C, McLean C, Tabaqchali S. (1984). The effect of capsular polysaccharide and lipopolysaccharide of *Bacteroides fragilis* on polymorph function and serum killing. *Journal of Medical Microbiology* **17**: 1259–1271.

Coykendall A L, Kaczmarek F S, Slots J. (1980). Genetic heterogeneity in *Bacteroides asaccharolyticus* (Holdeman and Moore 1970) Finegold and Barnes 1977 (Approved Lists, 1980) and proposal of *Bacteroides gingivalis* sp. nov. and *Bacteroides macacae* (Slots and Genco) comb. nov. *International Journal of Systemic Bacteriology* **30**: 559–564.

Dehazya P, Coles R S. (1980). Agglutination of human erythrocytes by *Fusobacterium nucleatum*: factors influencing hemagglutination and some characteristics of the agglutinin. *Journal of Bacteriology* **143**: 205–211.

(1982). Extraction and properties of hemagglutinin from cell wall fragments of *Fusobacterium nucleatum*. *Journal of Bacteriology* **152**: 298–305.

Dahlen G, Magnusson BC, Moller A. (1981). Histological and histochemical study of the influence of lipopolysaccharide-endotoxin extracted from *Fusobacterium nucleatum* on the periapical tissues in the monkey *macaca fasicularis*. *Archives of Oral Biology* **26**: 591–598.

Dowell V R, Hawkins T M. (1974). *Laboratory methods in anaerobic bacteriology*. CDC Publication No. 77-8272. Center for Disease Control, Atlanta.

Duerden B I. (1980). The isolation and identification of *Bacteroides spp.* from the normal human gingival flora. *Journal of Medical Microbiology* **13**: 89–101.

(1987). Anaerobes in non-gonococcal urethritis, balanoposthitis and genital ulcers. In *Recent Advances in Anaerobic Bacteriology; Proceedings of the Fourth Anaerobe Discussion Group Symposium*, pp. 231–248. Martinus Nijhoff Publishers, Dordrecht.

Duerden B I, Holbrook W P, Collee J G, Watt B. (1976). The characterisation of clinically important gram-negative anaerobic bacilli by conventional bacteriological tests. *Journal of Applied Bacteriology* **40**: 163–188.

Eggerth A H, Gagnon B H. (1933). The bacteroides of human feces. *Journal of Bacteriology* **25**: 389–413.

Falkler W A, Burger B W. (1981). Microbial surface interactions: reduction of the haemagglutination activity of the oral bacterium *Fusobacterium nucleatum* by absorption with *Streptococcus* and *Bacteroides*. *Archives of Oral Biology* **26**: 1015–1025.

Falkler W A, Hawley C E. (1977). Hemagglutinating activity of *Fusobacterium nucleatum*. *Infection and Immunity* **13**: 230–238.

Falkler W A, Mongiello J R, Burger B W. (1979). Haemagglutination inhibition and aggregation of *Fusobacterium nucleatum* by human salivary mucinous glycoproteins. *Archives of Oral Biology* **24**: 483–489.

Finegold S M, George W L, Mulligan M E. (1985). Anaerobic infections, part 1. *Disease-a-month* **xxx1 (10)**: 17–69.

Fraser A G, Brown R. (1981). Neuraminidase production by Bacteroidaceae. *Journal of Medical Microbiology* **14**: 63–76.

Gharbia S E, Shah H N. (1986). Biochemical properties of *Fusobacterium* species. XIV International Congress of Microbiology. Abstract P.GA-10.

(1988a). Characteristics of glutamate dehydrogenase, a new diagnostic marker for the genus *Fusobacterium*. *Journal of General Microbiology* **134**: 327–332.

(1988b). Growth responses to glucose and protein hydrolysates by *Fusobacterium* species. *Current Microbiology* **17**: 229–234.

Gibbons R J, MacDonald J B. (1961). Degradation of collagenous substrates by *Bacteroides melaninogenicus*. *Journal of Bacteriology* **81**: 614–621.

Gibbons R J, van Houte J. (1971). Selective bacterial adherence to oral epithelial surface and its role as an ecological determinant. *Infection and Immunity* **3**: 567–573.

Gorbach S L. (1975). Management of anaerobic infections: intraabdominal sepsis. *Annals of Internal Medicine* **83**: 377–384.

Gregory E M, Moore W E C, Holdeman L V. (1978). Superoxide dismutase in anaerobes. *Applied and Environmental Microbiology* **35**: 988–991.

Grenier D, Mayrand D. (1987). Selected characteristics of pathogenic and nonpathogenic strains of *Bacteroides gingivalis*. *Journal of Clinical Microbiology* **25**: 738–740.

Haapasalo M. (1986). The genus *Bacteroides* in human dental root canal infections. Ph.D thesis, University of Helsinki, Finland.

Haapasalo M, Lounatmaa K, Ranta H, Shah H N, Ranta K. (1985). Ultrastructures of *Bacteroides capillus*, *Bacteroides buccae*, *Bacteroides pentosaceus*, *Bacteroides oris*, *Bacteroides oralis*, *Bacteroides veroralis* and pentose sugar fermenting *Bacteroides* species from humans with periapical osteitis: occurrence of external proteinaceous cell wall layer. *International Journal of Systematic Bacteriology* **35**: 65–75.

Haapasalo M, Ranta H, Shah H N, Ranta K, Lounatmaa K, Kroppenstedt R M. (1986). *Mitsuokella dentalis* sp. nov. from dental root canals. *International Journal of Systematic Bacteriology* **36**: 566–568.

Hausmann E, Weinfeld N, Miller W A. (1972). Effects of lipopolysacchides on bone resorption in tissue culture. *Calcified Tissue Research* **9**: 272–282.

Hofstad T. (1974). Antibodies reacting with lipopolysaccharides from *Bacteroides melaninogenicus*, *Bacteroides fragilis* and *Fusobacterium nucleatum* in serum from normal human subjects. *Journal of Infectious Diseases* **129**: 349–352.

(1979). Serological responses to antigens of *Bacteroidaceae*. *Microbiological Reviews* **43**: 103–115.

(1980). Evaluation of the API ZYM system for identification of *Bacteroides* and *Fusobacterium* species. *Medical Microbiology and Immunology* **168**: 173–177.

(1981). The Genus *Fusobacterium*. In *The Prokaryotes. A handbook on habitats, isolation, and identification of bacteria*, pp. 1464–1474. Edited by Starr M P *et al*. Springer-Verlag, Berlin-Heidelberg, New York.

(1984). Pathogenicity of anaerobic gram-negative rods: possible mechanisms. *Reviews of Infectious Diseases* **6**: 189–199.

Hofstad T, Aasjord P. (1982). *Eubacterium plautii* (Seguin 1928) comb nov. *International Journal Systematic Bacteriology* **32**: 346–349.

Hofstad T, Fredriksen G. (1979). Immunochemical studies of partially hydrolyzed lipopolysaccharide from *Fusobacterium nucleatum* Fev1. *European Journal of Biochemistry* **94**: 59–64.

Hofstad T, Jantzen E. (1982). Fatty acids of *Leptotrichia buccalis*: taxonomic implication. *Journal of General Microbiology* **128**: 151–153.

Hofstad T, Sveen K. (1979). The chemotactic effect of *Bacteroides fragilis* lipopolysaccharide. *Reviews of Infectious Diseases* **1**: 42–46.

Hofstad T, Sveen K, Dahlen G. (1977). Chemical composition, serological reactivity and endotoxicity of lipopolysaccharides extracted in different ways from *Bacteroides fragilis*, *Bacteroides melaninogenicus* and *Bacteroides oralis*. *Acta Pathologica et Microbiologica Scandinavica* **69**: 543–548.

Holdeman L V, Cato E P, Moore W E C. (1974). Current classification of clinically important anaerobes. In *Anaerobic bacteria: role in disease*. Edited by Balows A *et al*. Charles C. Thomas, Springfield.

Holdeman L V, Kelley R W, Moore W E C. (1984). Genus *Bacteroides* (Castellani and Chalmers 1919). In *Bergey's Manual of Systematic Bacteriology*, vol 1, 9th edn. pp. 604–631. Edited by Krieg M R, Holt J G. William and Wilkins, Baltimore.

Holdeman L V, Moore W E C. (1974). Genus I. *Bacteroides* (Castellani and Chalmers 1919). In *Bergey's Manual of Determinative Bacteriology*, 8th edn. p. 385. Edited by Buchanan R E, Gibbons N E. William and Wilkins, Baltimore.

Hunt D E, Jones J V, Dowell V R. (1986). Selective medium for the isolation of *Bacteroides gingivalis*. *Journal of Clinical Microbiology* **23**: 441–445.

Iino Y, Hopps R M. (1984). The bone resorbing activities in tissue culture of lipopolysaccharides from the bacteria *Actinobacillus actinomycetemcomitans*, *Bacteroides gingivalis* and *Capnocytophaga ochracea* isolated from the human mouth. *Archives of Oral Biology* **29**: 59–63.

Ingham H R, Sisson P R, Middleton R L, Narang H K, Codd A A, Selkon J B. (1981). Phagocytosis and killing of bacteria in aerobic and anaerobic conditions. *Journal of Medical Microbiology* **14**: 391–399.

Ingham H R, Sisson P R, Tharagonnet D, Selkin J B, Codd A A. (1977). Inhibition of phagocytosis *in vitro* by obligate anaerobes. *Lancet* **2**: 1252–1254.

Jantzen E, Hofstad T. (1981). Fatty acids of *Fusobacterium* species: taxonomic implications. *Journal of General Microbiology* **123**: 163–171.

Johnson J L. (1978). Taxonomy of *Bacteroides*, I. Deoxyribonucleic acid homologies among *Bacteriodes fragilis* and other saccharolytic *Bacteroides*. *International Journal of Systematic Bacteriology* **28**: 245–256.

Johnson J L, Ault D A. (1978). Taxonomy of the *Bacteroides*, II. Correlation of phenotypic characteristics with deoxyribonucleic acid homology groupings for *Bacteroides fragilis* and other saccharolytic *Bacteroides* species. *International Journal of Systematic Bacteriology* **28**: 257–268.

Johnson J L, Harich B. (1986). Ribosomal ribonucleic acid

homology among species of the genus *Bacteroides*. *International Journal of Systematic Bacteriology* **36**: 71-79.

Johnson J L, Moore W E C, Moore L V H. (1986). *Bacteroides caccae* sp. nov., *Bacteroides merdae* sp. nov. and *Bacteroides stercoris* sp. nov. isolated from human feces. *International Journal of Sysematic Bacteriology* **36**: 499-501.

Jones G R, Gemmell C G. (1986). Effects of *Bacteroides asaccharolyticus* cells and *B. fragilis* surface components on serum opsonisation and phagocytosis. *Journal of Medical Microbiology* **22**: 225-229.

Karmen P R. (1981). The effects of bacterial sonicates on human keratinizing stratified squamous epithelium *in vitro*. *Journal of Periodontal Research* **16**: 323-330.

Kasper D L. (1976). The polysaccharide capsule of *Bacteroides fragilis* subspecies *fragilis*: immunochemical and morphologic definition. *Journal of Infectious Diseases* **133**: 79-87.

Kasper D L, Onderdonk A B, Polk B F, Bartlett J G. (1979). Surface antigens as virulence factors in infection with *Bacteroides fragilis*. *Reviews of Infectious Diseases* **1**: 278-288.

Kasper D L, Onderdonk A B, Reinap B G, Lindberg A A. (1980). Variations of *Bacteroides fragilis* with in vitro passage: presence of an outer membrane-associated glycan and loss of capsular antigen. *Journal of Infectious Diseases* **142**: 750-756.

Kasper D L, Weintraub A, Lindberg A A, Lonngren J. (1983). Capsular polysaccharide and lipopolysaccharide from two *Bacteroides fragilis* reference strains: chemical and immunochemical characterization. *Journal of Bacteriology* **199**: 991-997.

Kaufman E J, Mashimo P A, Hausmann E, Hanks C T, Ellison S A. (1972). Fusobacterial infection: enhancement by cell free extracts of *Bacteroides melaninogenicus* possessing collagenolytic activity. *Archives of Oral Biology* **17**: 577-580.

Kilian M. (1981). Degradation of immunoglobulins A1, A2 and G by suspected principal periodontal pathogens. *Infection and Immunity* **34**: 757-765.

Kilian M, Thomsen B, Petersen T E, Bleeg H S. (1983). Occurrence and nature of bacterial IgA proteases. *Annals of the New York Academy of Science* **409**: 612-624.

Knorr M. (1922). Uber die fusospirillare Symbiose, die Gattung *Fusobacterium* (K. B. Lehmann) und *Spirillum sputigenum*. Die Gattung *Fusobacterium*, I. Mitteilung, Die Epidemiologie der fusospirillaren Symbiose, besonders des Plaut-Vincentschen Angina. *Zentralblatt fur Bakteriologie, Parasitenkunde und Hygiene, Abteilung 1, Originale* **87**: 536-545.

(1923). Uber die fusospirillare Symbiose, die Gattung *Fusobacterium* (K. B. Lehmann) and *Spirillum sputigenum*. II. Mitteihung. Die Gattung *Fusobacterium*. *Zentralblatt fur Bakteriologie, Parasitenkunde and Infektionskrankheiten, Abteilung 1, Originale* **89**: 4-22.

Kornman K S, Holt S C. (1981). Physiological and ultrastructural characterisation of a new *Bacteroides* sp (*Bacteroides capillus*) isolated from severe localised periodontitis. *Journal of Periodontal Research* **16**: 542-555.

Kornman K S, Loesche W J. (1982). Effects of estradiol and progesterone on *Bacteroides melaninogenicus* and *Bacteroides gingivalis*. *Infection and Immunity* **35**: 256-263.

Krumweide C, Pratt J. (1913). *Fusiform bacilli*, isolation and cultivation. *Journal of Infectious Diseases* **12**: 199-201.

Kurita A P, Getty C E, Shaughnessy P, Hesse R, Salyers A A. (1986). DNA probes for the identification of clinically important *Bacteroides* species. *Journal of Clinical Microbiology* **23**: 343-349.

Lev M, Keudell K C, Milford A F. (1971). Succinate as a growth factor for *Bacteroides melaninogenicus*. *Journal of Bacteriology* **108**: 175-178.

Lev M. (1980). Glutamine-stimulated amino acid and peptide incorporation in *Bacteroides melaninogenicus*. *Journal of Bacteriology* **26**: 301-304.

Lindberg A A, Berthold P, Nord C E, Weintraub A. (1979). Encapsulated strains of *Bacteroides fragilis* in clinical specimens. *Medical Microbiology and Immunology* **167**: 29-36.

Lindberg A A, Weintraub A, Kasper D L, Lonngren J. (1982). Virulence factors in infections with *Bacteroides fragilis*: isolation and characterization of capsular polysaccharide and lipopolysaccharide. *Scandinavian Journal of Infectious Diseases Supplement*. **35**: 45-52.

Listgarten M A. (1965). Electron-microscopic observations on the bacterial flora of acute necrotizing ulcerative gingivitis. *Journal of Periodontology* **36**: 328-339.

Loesche W J, Syed S A, Laughon B E, Stoll J. (1982). The bacteriology of acute necrotizing ulcerative gingivitis. *Journal of Periodontology* **53**: 223-230.

Love D N, Johnson J L, Jones R F, Bailey M, Calverley A. (1986). *Bacteroides tectum* sp. nov. and characteristics of other nonpigmented *Bacteroides* isolates from soft tissue infections from cats and dogs. *International Journal of Systematic Bacteriology* **36**: 123-128.

MacDonald J B, Socransky S S, Gibbons R J. (1963). Aspects of the pathogenesis of mixed anaerobic infections of mucous membranes. *Journal of Dental Research* **42**: 529-544.

Mansheim B J, Coleman S E. (1980). Immunochemical differences between oral and nonoral strains of *Bacteroides asaccharolyticus*. *Infection and Immunity* **27**: 589-596.

Mayrand D, McBride B C. (1980). Ecological relationships of bacteria involved in a simple, mixed anaerobic infection. *Infection and Immunity* **27**: 44-50.

Mayrand D, McBride B C, Edwards T, Jensen S. (1980). Characterization of *Bacteroides asaccharolyticus* and *B. melaninogenicus* oral isolates. *Canadian Journal of Microbiology* **26**: 1178-1183.

McDermid A S, McKee A S, Ellwood D C, Marsh P D. (1986). The effect of lowering the pH on the composition and metabolism of a community of nine oral bacteria grown in a chemostat. *Journal of General Microbiology* **132**: 1205–1214.

McKee A S, McDermid A S, Baskerville A, Dowsett A B, Ellwood D C, Marsh P D. (1986). Effect of hemin on the phsiology and virulence of *Bacteroides gingivalis* W50. *Infection and Immunity* **52**: 349–355.

Mongiello J R, Falkler W A. (1979). Sugar inhibition of oral *Fusobacterium nucleatum* haemagglutination and cell binding. *Archives of Oral Biology* **24**: 539–545.

Moore W E C et al. (1982). Bacteriology of experimental gingivitis in young adult humans. *Infection and Immunity* **38**: 1137–1148.

(1983). Bacteriology of moderate (chronic) periodontitis in mature adult humans. *Infection and Immunity* **42**: 510–515.

(1984). Bacteriology of experimental gingivitis in children. *Infection and Immunity* **46**: 1–6.

(1985). Comparative bacteriology of juvenile periodontitis. *Infection and Immunity* **48**: 507–519.

Muller H E, Werner H. (1970). Die Neuraminidase als pathogenetischer Faktor bein einem durch *Bacteroides fragilis* bedingten Abscess. *Zeitschrift fur Medizinische Microbiologie und Immunologie* **156**: 98–106.

Naito Y, Okuda K, Kato T, Takazoe I. (1985). Monoclonal antibodies against surface antigens of *Bacteroides gingivalis*. *Infection and Immunity* **50**: 231–235.

Oliver W W, Wherry W B. (1921). Notes on some bacterial parasites of the human mucous membranes. *Journal of Infectious Diseases* **28**: 341–345.

Okuda K, Kato T, Shiozu J, Takazoe I, Nakamura T. (1985). *Bacteroides heparinolyticus* sp. nov. isolated from humans with periodontitis. *International Journal of Systematic Bacteriology* **35**: 438–442.

Okuda K, Slots J, Genco R J. (1981). *Bacteroides gingivalis, Bacteroides asaccharolyticus*, and *Bacteroides melaninogenicus* subspecies: cell surface morphology and adherence to erythrocytes and human buccal epithelial cells. *Current Microbiology* **6**: 7–12.

Okuda K, Takazoe I. (1983). Antiphagocytic effects of the capsular structure of a pathogenic strain of *Bacteroides melaninogenicus*. *Bulletin of Tokyo Dental College* **14**: 99–104.

Okuda K, Yanagi K, Takazoe I. (1978). Complement activation by *Propionibacterium acnes* and *Bacteroides melaninogenicus*. *Archives of Oral Biology* **23**: 911–915.

Onderdonk A B, Bartlett J G, Louie T, Sullivan-Seigler N, Gorbach S L. (1976). Microbial synergy in experimental intra-abdominal abscess. *Infection and Immunity* **13**: 22–26.

Onderdonk A B, Kasper D L, Cisneros R L, Bartlett J G. (1977). The capsular polysaccharide of *Bacteroides fragilis* as a virulence factor: comparison of the pathogenic potential of encapsulated and unencapsulated strains. *Journal of Infectious Diseases* **136**: 82–89.

Onderdonk A B, Moon N E, Kasper D L, Bartlett J G. (1978). Adherence of *Bacteroides fragilis in vivo*. *Infection and Immunity* **19**: 1083–1087.

Page L R, Krywolap G N. (1976). Determination of deoxyribonucleic acid composition and deoxyribonucleic acid-deoxyribonucleic acid hybridisation of *Fusobacterium fusiforme, Fusobacterium polymorphum* and *Leptotrichia buccalis*: taxonomic considerations. *International Journal of Systematic Bacteriology* **26**: 301–304.

Paster B J, Dewhirst F E. (1988). Phylogeny of campylobacters, wolinellas, *Bacteroides gracilis*, and *Bacteroides ureolyticus* by 16S ribosomal ribonucleic acid sequencing. *International Journal of Systematic Bacteriology* **38**: 56–62.

Paster B J et al. (1985). A phylogenetic grouping of the bacteroides, cytophagas and certain flavobacteria. *Systematic and Applied Microbiology* **6**: 34–42.

Polk B F, Kasper D I. (1977). *Bacteroides fragilis* subspecies in clinical isolates. *Annals of Internal Medicine* **86**: 569–571.

Porschen R K, Sonntag S. (1974). Extracellular deoxyribonuclease production by anaerobic bacteria. *Applied Microbiology* **27**: 1031–1033.

Poxton I, Brown R. (1986). Immunochemistry of the surface carbohydrate antigens of *Bacteroides fragilis* and definition of a common antigen. *Journal of General Microbiology* **132**: 2475–2481.

Prévot A R. (1966). *Manual for the classification and determination of the anaerobic bacteria*, 1st American edn, translated by V. Fredette. Lea and Febiger, Philadelphia.

Pribram E. (1929). A contribution to the classification of microorganisms. *Journal of Bacteriology* **18**: 361–394.

Pulverer G, Ko H L, Wegrzynowicz Z, Jeljaszewicz J. (1977). Clotting and fibrinolytic activities of *Bacteroides melaninogenicus*. *Zentralblatt fur Bakteriologie Mikrobiologie und Hygiene* (A) **239**: 510–513.

Reddy C A, Bryant M P. (1977). Deoxyribonucleic acid base composition of certain species of the genus *Bacteroides*. *Canadian Journal of Microbiology* **23**: 1252–1256.

Roberts D S. (1967). The pathogenic synergy of *Fusiformis necrophorus* and *Corynebacterium pyogenes*. I. Influence of the leucocidal exotoxin of *F. necrophorus*. *British Journal of Experimental Pathology* **48**: 665–673.

Robertson P B, Lantz M, Marucha P T, Kornman K S, Trummel C L, Holt S C. (1982). Collagenolytic activity associated with *Bacteroides* species and *Actinobacillus actinomycetemcomitans*. *Journal of Periodontal Research* **17**: 275–283.

Rolfe R D, Hentges D J, Cambell B J, Barrett J T. (1978). Factors related to the oxygen tolerence of anaerobic bacteria. *Applied and Environmental Microbiology* **36**: 306–313.

Rudek W, Haque R-U. (1976). Extracellular enzymes of the

genus *Bacteroides. Journal of Clinical Microbiology* 4: 458–460.

Sebald M. (1962). Etude sur les bacteries anaerobies gram-negatives asporulees. These de l'Universite Paris.

Shah H N, Bonnet R, Mateen B, Williams R A D. (1979). The porphyrin pigmentation of subspecies of *Bacteroides melaninogenicus. Biochemistry Journal* 180: 45–50.

Shah H N, Collins M D. (1980). Fatty acid and isoprenoid quinone composition in the classification of *Bacteroides melaninogenicus* and related taxa. *Journal of Applied Bacteriology* 48: 75–87.

(1981). *Bacteroides buccalis* sp. nov. *Bacteroides denticola*, sp. nov. and *Bacteroides pentosaceus*, sp. nov., new species of the genus *Bacteroides* from the oral cavity. *Zentralblatt fur Bakteriologie, Mikrobiologie und Hygiene I. Abteilung, Originale C* 2: 235–241.

(1982a). Reclassification of *Bacteroides hypermegas* (Harrison and Hansen) in a new genus *Megamonas* as *Megamonas hypermegas* comb. nov. *Zentralblatt fur Bacteriologie und Hygiene I. Abteilung, Originale C* 3: 394–398.

(1982b). Reclassification of *Bacteroides multiacidus* (Mitsuoka, Terada, Watanabe, Uchida) in a new genus *Mitsuokella*, as *Mitsuokella multiacidus* comb. nov. *Zentrablatt fur Bakteriologie. I. Abteilung Originale C* 3: 491–494.

(1983). Genus *Bacteroides*: A chematoxonomical perspective. *Journal of Applied Bacteriology* 55: 403–416.

(1986a). Reclassification of *Bacteroides furcosus* (Holdeman and Moore) in a new genus *Anaerorhabdus* as *Anaerorhabdus furcosus. Systematic and Applied Microbiology* 8: 86–88.

(1986b). Reclassification of *Bacteroides termitidis* (Sebald) in a new genus *Sebaldella*, as *Sebaldella termitidis* comb. nov. *International Journal of Systematic Bacteriology* 36: 349–350.

(1988). Proposal for reclassification of *Bacteroides asaccharolyticus, Bacteroides gingivalis*, and *Bacteroides endodontalis* in a new genus, *Porphyromonas. International Journal of Systematic Bacteriology* 38: 128–131.

(1989). Proposal to restrict the genus *Bacteroides* (Castellani and Chalmers) to *Bacteroides fragilis* and closely related species. *International Journal of Systemic Bacteriology* 39: 85–87.

(1990). Prevotella, a new genus to include *Bacteroides melaninogenicus* and related species formerly classified in the genus *Bacteroides. International Journal of Systematic Bacteriology* 40: 205–208.

Shah H N, Collins M D, Watabe J, Mitsuoka T. (1985). *Bacteroides oulorom* sp. nov., a nonpigmented saccharolytic species from the oral cavity. *International Journal of Systematic Bacteriology* 35: 193–197.

Shah H N, Elhag K M, Al-Jalili T A R, Mundegar Z R. (1987). Glucose-6-phosphate dehydrogenase and malate dehydrogenase enzyme electrophoretic patterns amongst strains of *Bacteroides fragilis. Journal of General Microbiology* 133: 1975–1981.

Shah H N, Nash R A, Whiley R A, Hardie J M, Collins M D. (1983). *Practical workshop on chemical methods for bacterial classification and identification.* The London Hospital Medical College, Department of Oral Microbiology, University of London.

Shah H N, Williams R A D. (1982). Dehydrogenase patterns in the taxonomy of *Bacteroides. Journal of General Microbiology* 128: 2955–2965.

(1987). Utilization of glucose and amino acids by *Bacteroides intermedius* and *Bacteroides gingivalis. Current Microbiology* 15: 241–246.

Shah H N, Williams R A D, Bowden G H, Hardie J M. (1976). Comparison of the biochemical properties of *Bacteroides melaninogenicus* from human dental plaque and other sites. *Journal of Applied Bacteriology* 41: 473–492.

Slots J. (1982). Importance of black-pigmented bacteroides in human periodontal disease. In *Host–parasite interactions in periodontal diseases*, pp. 27–45. Edited by Genco R J, Mergenhagen S F. American Society for Microbiology, Washington DC.

Slots J, Genco R J. (1979). Direct hemagglutination technique for differentiating *Bacteroides asaccharolyticus* oral strains from non-oral strains. *Journal of Clinical Microbiology* 10: 371–373.

Slots J, Gibbons R J. (1978). Attachment of *Bacteroides melaninogenicus* subsp. *asaccharolyticus* to oral surfaces and its possible role in colonization of the mouth and of periodontal pockets. *Infection and Immunity* 19: 254–264.

Slots J, Potts T V. (1982). *Fusobacterium simiae* a new species from monkey dental plaque. *International Journal of Systematic Bacteriology* 32: 191–194.

Slots J, Potts T V, Mashimo P A. (1983). *Fusobacterium periodonticum* new species from the human oral cavity. *Journal of Dental Research* 62: 960–963.

Smalley J W, Birss A J. (1987). Trypsin-like activity of the extracellular membrane vesicles of *Bacteroides gingivalis* W50. *Journal of General Microbiology* 133: 2883–2894.

Socransky S S, Gibbons R J. (1965). Required role of *Bacteroides melaninogenicus* in mixed anaerobic infections. *Journal of Infectious Diseases* 113: 247–253.

Steffen E K, Hentges D J. (1981). Hydrolytic enzymes of anaerobic bacteria isolated from human infections. *Journal of Clinical Microbiology* 14: 153–156.

Sundqvist G K. (1976). Bacteriological studies of necrotic dental pulps. Odontological dissertation No. 7. University of Umea, Sweden.

Sundqvist G, Bloom G D, Enberg K, Johansson E. (1982). Phagocytosis of *Bacteroides melaninogenicus* and *Bacteroides gingivalis in vitro* by human neutrophils. *Journal of Periodontal Research* 17: 113–121.

Sundqvist G, Carlsson J. (1984). *In-vitro* and *in-vivo* studies on the pertubation of host defense by black-pigmented *Bacteroides* species. In *Models of anaerobic infection*, pp. 129–138. Edited by Hill M J. Martinus Nijhoff Publishers, Dordrecht.

Sundqvist G, Johansson E. (1980). Bactericidal effect of pooled human serum on *Bacteroides asaccharolyticus* and *Actinobacillus actinomycetemcomitans*. *Scandinavian Journal of Dental Research* 90: 29–36.

Sutter V L, Citron D M, Edelstein M A C, Finegold S M (1985). *Wadsworth anaerobic bacteriology manual*, 4th edn. Star Publishing Company, California.

Sveen K, Hofstad T, Milner K C. (1977). Lethality for mice and chick embryos, pyrogenicity in rabbits and ability to gelate lysate from amoebocytes of *Limulus polyphemus* by lipopolysaccharides from *Bacteroides*, *Fusobacterium* and *Veillonella*. *Acta Pathologica et Microbiologica Scandinavica* 85: 388–396.

Sveen K, Skaug N. (1980). Bone resorption stimulated by lipopolysaccharides from *Bacteroides*, *Fusobacterium* and *Veillonella*, and by the lipid A and the polysaccharide part of *Fusobacterium* lipopolysaccharide. *Scandinavian Journal of Dental Research* 88: 535–542.

Tally F P, Goldin B R, Jacobus N V, Gorbach S L. (1977). Superoxide dismutase in anaerobic bacteria of clinical significance. *Infection and Immunity* 16: 20–25.

Tamaka J, Komagata K. (1984). Determination of base composition by reversed-phase high-performance liquid chromatography. *FEMS Microbiology Letters* 25: 125–128.

Thadepalli H, Gorbach S L, Kiethl. (1973). Anaerobic infections of the female genital tract bacteriologic and therapeutic aspects. *American Journal of Obstetrics and Gynecology* 117: 1034–1040.

Tissier H. (1908). Recherches sur la flore intestinale normale des enfants eges d'un an a cinq ans. *Annales de L'Institut Pasteur (Paris)* 22: 189–208.

Tofte R W, Peterson P K, Schmeling D, Bracke J, Kim Y, Quie P G. (1980). Opsonisation of four *Bacteroides* species: role of the classical complement pathway and immunoglobulin. *Infection and Immunity* 27: 784–792.

Touw J J A, van Kampen G P S, van Steenbergen T J M, Veldhuizen J P, de Graaff J. (1982). The effect of culture filtrates of oral strains of black-pigmented bacteroides on the matrix production of chick embryo cartilage cells *in vitro*. *Journal of Periodontal Research* 17: 351–357.

Tunnicliff R. (1919). The microscopic appearances in ulceromembranous tonsillitis (Vincents Angina). *Journal of Infectious Diseases* 25: 132–134.

(1923). The life cycle of *Bacillus fusiformis*. *Journal of Infectious Diseases* 33: 147–154.

Van Dyke T E, Bartholomew E, Genco R J, Slots J, Levine M J. (1982). Inhibition of neutrophil chemotaxis by soluble bacterial products. *Journal of Periodontology* 53: 502–508.

van Steenbergen T J M, de Graaff J. (1986). Proteolytic activity of black-pigmented *Bacteroides* strains. *FEMS Microbiology Letters* 33: 219–227.

van Steenbergen T J M, den Ouden M D, Touw J J A, de Graaff J. (1982). Cytotoxic activity of *Bacteroides gingivalis* and *Bacteroides asaccharolyticus*. *Journal of Medical Microbiology* 15: 253–258.

van Steenbergen T J M, Kastelein P, Touw J J A, de Graaff J. (1982). Virulence of black-pigmented bacteroides strains from periodontal pockets and other sites in experimentally induced skin lesions in mice. *Journal of Periodontal Research* 17: 41–49.

van Steenbergen T J M, Namavar F, de Graaff J. (1985). Chemiluminescence of human leukocytes by black-pigmented bacteroides strains from dental plaque and other sites. *Journal of Periodontal Research* 20: 58–71.

van Winkelhoff A J, de Graaff J. (1983). Vancomycin as a selective agent for isolation of *Bacteroides* species. *Journal of Clinical Microbiology* 18: 1282–1284.

Veillon M H, Zuber H. (1898). Recherches sur quelques microbes strictement anaerobies et leur role en pathologie. *Archives de Medicine Experimentale et d'Anatomie Pathologique* 10: 517–545.

Wade B H, Kasper D L, Mandell G L. (1983). Interactions of *Bacteroides fragilis* and phagocytes: studies with whole organisms, purified capsular polysaccharide and clindamycin-treated bacteria. *Journal of Antimcirobial Chemotherapy* 12 Suppl C: 51–52.

Weisburg W G, Oyaizu Y, Oyaizu H, Woese C R. (1985). Natural relationship between *Bacteroides* and flavobacteria. *Journal of Bacteriology* 164: 230–236.

Werner H. (1972a). A serological study of strains belonging to *Sphaerophorus necrophorus*, *Sphaerophorus varium* and *Sphaerophorus freundii*. *Medical Microbiology and Immunology* 157: 315–324.

(1972b). Anaerobierdifferenzierung durch gaschromatographische Staffnechselanalysen. *Zentralblatt fur Bakteriologie, Parasitenkunde, Infectionskrankheiten und Hygiene, I. Abteilung Originale, Reiche A* 220: 446–451.

Werner H, Muller H E. (1971). Immunoelektrophoretische Untersuchungen uber die Einwirkung von *Bacteroides*, *Fusobacterium*-, *Leptotrichia*- und *Spaerophorus*-Arten auf menschliche Plasmaproteine. *Zentralblatt fur Bakteriologie I Abteilung Originale* 216: 96–113.

Woese C R, Stackebrandt E, Macke T J, Fox G E. (1985). A phylogenetic definition of the major eubacterial taxa. *Systematic and Applied Microbiology* 6: 133–142.

Yamamoto A, Takahashi M, Takamori K, Sasaki T. (1982). Ultrastructure of the outer membrane surface of black-pigmented bacteroides isolated from the human oral cavity. *Bulletin of Tokyo Dental College* 23: 47–60.

6

The clostridia

G Hobbs

Current status of classification
    Morphology and biochemical tests
    Toxin production and pathogenicity
    Other serological tests
    Cell composition
    Nucleic acid studies
    Numerical analyses
Isolation of clostridia
    Anaerobic conditions
    Treatment of samples
    Enrichment of cultures
    Plating of enrichment media
    Purification
Identification procedures
    Characterization of toxins
    Other serological tests
    Morphological and biochemical properties
Conclusions
References

With a few exceptions, the genus *Clostridium* embraces all the anaerobic spore-forming bacteria. Historically, the occurrence of endospores has been regarded as a feature of prime taxonomic importance; bacteria producing them are separated into different genera from those which do not. Further separation of spore-forming bacteria into genera and species is based on relatively few selected criteria. The groups established in this way embrace bacteria with a wide range of properties, and it is not surprising that the existing classification schemes are not entirely compatible with modern views on taxonomy.

The most generally accepted classification scheme is that in *Bergey's Manual of Systematic Bacteriology* (Cato *et al.*, 1986) where most of the spore-forming bacteria are in two genera, *Bacillus* and *Clostridium*, which are distinguished according to whether or not they are strictly anaerobic. Other spore-forming genera are separated from these two by a few simple criteria: *Sporolactobacillus* because it is microaerophilic, does not produce catalase and has a homofermentative metabolism of sugars; *Desulfotomaculum* by the exclusion from the genus *Clostridium* of those anaerobic spore-forming rods able to reduce sulphate; *Sporosarcina* by the microscopic appearance of its coccoid-shaped cells and strict anaerobic metabolism. The aerobic spore-forming actinomycetes producing aerial hyphae are classified in the genus *Thermoactinomyces*.

There have been several attempts to subdivide the anaerobic spore-forming bacteria further, the most notable of which is the classification scheme of Prévot *et al.* (1967). In this scheme they are separated into two orders, four families and nine genera. In addition, a number of varieties are introduced for some species. Because of its complexity and lack of convenience this scheme has not found widespread acceptance. It places too much emphasis on trivial differences and needlessly complicates the nomenclature.

Because of their anaerobic metabolism the clostridia have presented particular problems in methodology, equipment and media which are not encountered in other branches of bacteriology. In much of the early work these problems were not always recognized and, as a result, studies were carried out with mixed culture. Another major problem resulted from the publication of incomplete and inadequate species descriptions; many descriptions originated from industrial research being created simply to secure commercial advantage or patent cover.

Nevertheless, most of the pathogenic clostridia had been recognized by the end of the nineteenth century and, by the early part of the twentieth century, it was apparent that they were, for the most part, toxin-producing bacteria. World War I gave a great impetus to the serious study of anaerobic infections in man and the development of the anaerobic jar provided a ready means of obtaining pure cultures.

The normal habitat for clostridia is soil and human and animal intestines, and they are widespread in soil and aquatic sediments throughout the world. Because of the somewhat arbitrary definition of the genus it embraces organisms with a wide range of properties. They differ in tolerance of oxygen; three species (*C. carnis*, *C. histolyticum*, *C. tertium*) will in fact grow aerobically, although they do not sporulate and die rapidly under these conditions. Others, such as *C. novyi* type B and *C. haemolyticum* are strict anaerobes and will not tolerate much exposure to air during manipulation. Some species have very simple growth requirements while others require most of the amino acids and several vitamins in the growth medium.

The majority of clostridia, including all the pathogenic species, are mesophilic, with optimum growth temperatures in the region of 30–37°C, although there are psychrophilic and thermophilic species.

Pathogenic clostridia cause a variety of diseases in animals and man. During war, wound infections become a major factor, but in peacetime clostridial diseases in man are generally restricted to occasional incidents of tetanus, botulism, gas gangrene and necrotic enteritis, with comparatively frequent incidents of food poisoning due to *C. perfringens* type A. In addition, in recent years *C. difficile* has been recognized as a major cause of antibiotic-induced pseudomembranous colitis, and possibly some non-antibiotic related bowel diseases. For the most part, symptoms result from the action of toxins produced during growth of the organism concerned. The pathogenic clostridia produce a wide range of toxins which affect the production of particular diseases in various ways. They are proteins synthesized during growth of the vegetative organisms, or associated with the sporulation process, and by and large they are secreted into the surrounding medium. In many cases the organism may no longer be present, for instance in the case of botulism. In others, such as gangrene, the toxins are produced by the organism growing *in situ* in a particular lesion. Diagnosis of clostridial diseases is usually, therefore, a matter of correlating symptoms with the presence of specific toxins or organisms capable of producing them. Morphological and biochemical characterization of the organism is, when used, usually for confirmation of a diagnosis.

There are a number of standard texts and laboratory manuals dealing with the isolation and identification of pathogenic anaerobes and these should be consulted for detailed descriptions of culture procedures and media. (Willis, 1969; Sterne and Batty, 1975; Holdeman *et al.*, 1977; Dowell and Hawkins, 1977; Sutter *et al.*, 1980; Mitsuoka, 1980; Gall and Riely, 1981; Smith and Williams, 1984). Moreover, the methods for isolation and identification of pathogenic clostridia have been the subject of two recent reviews (Crowther and Baird-Parker, 1983; Hardie, 1986).

Current status of classification

It has already been stated that classification schemes are based on considerations of convenience rather than sound taxonomic principles, and no serious taxonomic study has been carried out on the whole group of anaerobic spore-forming bacteria, although attempts have been made with isolated groups of related species. The main difficulties in identification and classification of clostridia have recently been discussed in detail (Hobbs, 1986). These difficulties arise for a variety of reasons. Some species have been studied much more closely than others, so that detailed characterization of serotypes is available in some cases, but barely adequate species descriptions are all that exist for others. Many species are represented by one strain, others by hundreds. There are genuine difficulties in obtaining reproducible results for some of the biochemical activities as there are real capabilities for switching metabolic pathways during growth. Taxonomic problems start, however, with the use of very few criteria to define the genus. Subsequent descriptions of species have not all used a uniform approach, and in the case of pathogenic clostridia the over-riding criterion is the ability to produce particular toxins.

Of the 83 species described in the most recent edition of *Bergey's Manual of Systematic Bacteriology* (Cato et al., 1986), some 18 are generally regarded as pathogenic, most producing extracellular toxins lethal to experimental animals (usually mice). A further 21 species are reported to be present in or have been cultured from, pathological material and may or may not have been involved in the pathological process. These species are listed in Table 6.1. Other species have been isolated from healthy human beings. Classification of the pathogenic clostridia is based primarily on the production of particular toxins, and species are frequently divided into serotypes on the basis of the serological- or substrate-specificity of the toxins. Morphological and biochemical tests are used primarily for confirmation. The classification of the non-pathogenic clostridia is based entirely on morphological and biochemical criteria. While more recent taxonomic criteria, such as cell composition, DNA and RNA hybridization and protein electrophoresis patterns have all been applied to some clostridia, no studies have considered the genus as a whole and this kind of data is at best only used as an aid to identification. In a few instances it has helped to clarify classification of related organisms.

Morphology and biochemical tests

By definition members of the genus *Clostridium* are rod-shaped, produce endospores and are strictly anaerobic. Historically, much has been made of the shape of the vegetative cell and the shape and position of the spore in the cell. In practice most clostridia are pleomorphic, and the morphology is often affected quite considerably by the cultural conditions. In a classification scheme, the only reliable morphological criteria that are of general use are whether the organism is rod-shaped and whether the spores are terminal. Many species with subterminal spores will show apparently terminal spores in the same field; these should be considered as subterminal species. With experience it is, of course, often possible to recognize the morphology, albeit pleomorphic, which is characteristic of particular species; this is extremely useful in practice. However, it is difficult to incorporate experienced judgement into a determinative scheme.

A number of interesting observations on the spores of clostridia have been made by electron-microscopy. Some species have characteristic

Table 6.1 Pathogenic and pathology-associated clostridia

Pathogenic	*Occur in pathological conditions*
C. absonum	C. barati
C. botulinum	C. bifermentans
C. carnis	C. cadaveris
C. chauvoei	C. clostridiforme
C. colinum*	C. cochlearium
C. difficile	C. ghoni
C. fallax	C. glycolicum
C. haemolyticum	C. hastiforme
C. histolyticum	C. indolis
C. limosum	C. innocuum
C. malenominatum	C. irregulare
C. novyi	C. oroticum
C. perfringens	C. paraputrificum
C. ramosum	C. putrificum
C. septicum	C. sphenoides
C. sordellii	C. sporogenes
C. spiriforme	C. subterminale
C. tetani	C. sporosphaeroides
	C. symbrosum
	C. tertium
	C. villosum*

*not in human disease

appendages and an exosporium, or extra membrane, surrounding the mature spore. (Hodgkiss et al., 1966, 1967; Rode et al., 1971; Rode and Smith, 1971). Further work on other species is required before the full potential of these observations can be assessed, but they can be useful for some species.

Gram's stain is of limited value in the classification of clostridia. The genus is defined as gram-positive and most species generally are, especially in young cultures. However, a lot of species rapidly become gram-negative with age and after incubation for a few days most cultures will be largely gram-negative. There are a few species in which gram-positive cells have not been seen; however, when these species are examined they all have a cell wall structure considered typical of gram-positive organisms.

The fact that three species can grow aerobically has already been mentioned. These are *C. carnis, C. histolyticum* and *C. tertium*. They are retained in the genus *Clostridium* because they grow better anaerobically, they lose viability rapidly after aerobic growth and they do not sporulate under aerobic conditions.

For most clostridia, and certainly for all the pathogenic species, a fairly standard list of biochemical tests is used for classification. Not all of these tests are always necessary for identification of particular species. The list includes the morphology and reaction to Gram's stain, which have already described, motility, growth and reaction on blood agar and egg-yolk agar, liquefaction of gelatin, digestion of casein, production of indole from tryptophan, reduction of nitrate and production of acid from a range of carbohydrate substrates. In addition, characterization of the fermentation products resulting from growth in a suitable glucose-containing medium is useful. Because of their anaerobic metabolism, clostridia do not break down the substrates in fermentation reactions completely. This leads to the situation, useful in practice, in which a variety of products accumulate in the growth medium, the particular products depending on the metabolism of the species. These patterns of fermentation products can then be used in identification and classification (Holdeman et al., 1977).

Many clostridia can obtain their energy from the fermentation of various amino acids. Attempts have been made to use this information for identification and taxonomic purposes (Mead, 1971; Elsden and Hilton, 1979). While the results of this work look promising, these tests are not yet widely used.

The use of any criteria to establish classification schemes requires that, for the most part, the criteria are consistent and stable, but although the morphological criteria of clostridia are reasonably stable, the same cannot be said for some of the biochemical criteria. Reference to the discussion by Hobbs (1986) shows that while some criteria, such as egg-yolk reactions, proteolysis, indole production, nitrate reduction and urease activity present few problems, carbohydrate fermentation reactions are frequently unreliable. Since species identity often depends on these criteria it is perhaps not surprising that there are frequent difficulties in the identification of new isolates and, at times, in persuading named strains to give the 'correct' reactions. It is inevitable, therefore, that taxonomic or classification schemes set up using these criteria remain a matter chiefly of convenience and require a liberal interpretation. Historically, there has been no generally agreed basis for separating two strains of clostridia into different species other than the diseases they cause.

Toxin production and pathogenicity

The pathogenic clostridia have historically been identified by virtue of the diseases they cause and classification schemes have been set up to ensure that clostridia causing the same disease are grouped together. From the clinical point of view this has obvious advantages, although it does lead to a number of irreconcilable differences with the requirements of taxonomic principles.

The pathogenic species are defined primarily on the basis of the lethal toxins they produce, hence suitable antitoxins are required for identification. These are generally commercially available and are frequently used to identify toxins without isolating the organisms. The earliest studies on pathogenic clostridia recognized that not all toxins causing particular symptoms were serologically identical. This led to the setting up of serotypes, the species still being defined by the symptoms produced by the toxin. There are seven serotypes of *C. botulinum*, four of *C. novyi* and five of *C. perfringens*, and evidence that these numbers may well need to be expanded (Gimenez and Ciccarelli, 1972; Gimenez, 1984). These species illustrate the difficulties of

Table 6.2 Biochemical groups of C. botulinum

Character	Metabolic group			
	I	II	III	IV
Proteolysis	+	−	−	+
Lipolysis	+	+	+	−
Acid from glucose	+	+	+	−
Acid from mannose	−	+	+	−
Ferment amino acids	+	−	−	+
Minimum growth temperature (°C)	10	3.3	10	12
Minimum inhibitory NaCl concentration (%)	10	5	3	6
Inhibitory pH	4–4.5	5–5.5	?	5.5
Toxins	A,B,F	B,E,F	C_1,C_2,D	G

reconciling identification requirements for clinical purposes with taxonomic requirements. Table 6.2, modified from Smith (1977), shows the four metabolic groups of C. botulinum and their relationship to toxin types. In the absence of toxins these metabolic groups would have merited separate species status and, in some cases, would have been identified as other species. This situation has been complicated further by the recent isolation of two organisms producing C. botulinum toxins but which would otherwise have been identified as C. butyricum and C. barati (Hall et al., 1985; McCroskey et al., 1986). It has always been recognized that the ability to produce toxins is readily lost by most pathogenic clostridia. One reason for this easy loss is the fact that at least some toxin production is mediated by bacteriophages or other transferable genetic material (Crowther and Baird-Parker, 1983; Strom et al., 1984). In one case transfer of toxin production from C. botulinum to C. novyi has been demonstrated (Eklund et al., 1974). In a taxonomic scheme designed to embrace all clostridia, the ability to produce toxin is therefore as unreliable as many biochemical tests. In an identification scheme for clinical use, toxin production remains of primary importance. What is not clear is the significance of non-toxic strains isolated from clinical situations.

Other serological tests

Many attempts have been made to use the somatic, flagellar and spore antigens of clostridia to study taxonomic relationships and set up classification schemes (Baba, 1969; Smith and Williams, 1984). In practice little of general application to the whole genus has emerged from these studies. There are important applications however; somatic antigens can be used for identification of C. septicum and C. chauvoei (Sterne and Batty, 1975), and capsular polysaccharide antigens can be used to serotype strains of C. perfringens type A (Stringer et al., 1980). A serotyping scheme using spore antigens along with entertoxins has also been proposed, this scheme claims to type over 90 per cent of food poisoning strains of C. perfringens (Chakrabarty and Narayan, 1979).

From the point of view of classification, there is little correlation between groups based on spore and vegetative cell antigens and the species defined on toxin or biochemical criteria.

Cell composition

Various aspects of the chemical composition of the organisms themselves have been studied with a view to providing data of value in classification. These include cell wall amino acids and sugars (Novotny, 1969; Cummins, 1970; Cummins and Johnson, 1971; Paine and Cherniak, 1975) cell lipids and fatty acids (Moss and Lewis, 1967; Farshy and Moss, 1970; Fugate et al., 1971; Hobbs et al., 1971; Vaczi et al., 1971; Elsden et al., 1980) products of pyrolysis (Oyama and Carle, 1967; Cone and Lechowich, 1970; Yemetsev et al., 1971; Reiner and Bayer, 1978; Gutteridge et al., 1980) and electrophoresis of cell proteins (Cato et al., 1982; Bom et al., 1986).

Most of these studies have been performed on relatively few species and there are insufficient data available to assess their value in the genus as a whole. Where data are available there is no separation of toxic organisms from related non-toxic strains. Electrophoresis of protein patterns has been applied most widely and there is evidence that preliminary identification of many species may be possible using this method.

Nucleic acid studies

DNA base ratios have become available for most clostridia and it is clear now that the genus embraces species with base ratios ranging from 21 to 55 moles per cent guanine + cytosine (G+C). The majority have values of between 24 and 32 per cent and about 14 species have higher values (Hobbs, 1986).

Hybridization studies have been performed on small groups of related clostridia (Lee and Riemann, 1970a, b; Cummins and Johnson, 1971; Wu *et al.*, 1972; Nakamura *et al.*, 1973, 1977, 1979, 1983; Johnson, 1973). As far as the pathogenic clostridia are concerned, for instance *C. botulinum*, these studies have shown primarily that toxic and non-toxic strains are closely related, whereas the different metabolic groups are not, and hence there is no correlation with toxin types. One study has examined RNA homologies among 56 species of clostridia (Johnson and Francis, 1975). They found three main groups, with Group I being divided into 10 subgroups. These homology groups do not correspond to any grouping set up on the basis of biochemical properties or toxin types (Hobbs, 1986).

The evidence from nucleic acid studies suggests that the genus *Clostridium*, as presently defined, should be separated into a number of genera. In doing so, however, consideration should be given to whether the genus is properly constituted in the first place. It may well be that better genera can be constituted if a wider range of properties is considered. Nucleotide sequencing of 16S ribosomal RNA is regarded by some as a useful reflection of evolutionary relationships. In studies including clostridia (Fox *et al.*, 1980; Tanner *et al.*, 1981), the species examined became intermixed with species of a number of other genera including *Eubacterium*, *Peptococcus*, *Ruminococcus*, *Sarcina*, *Acetobacterium* and *Mycoplasma*. These studies however still grouped together species with high and low GC ratios.

Numerical analyses

Numerical analytical techniques have been applied to a few small groups of species using conventional morphological and biochemical data. *C. perfringens* and some related organisms were studied by Nakamura *et al.*, (1973) and *C. botulinum* and related organisms were studied by Kiritani *et al.* (1973). While these studies clarified some close relationships, they had no impact on the genus as a whole.

Isolation of clostridia

Most authors on this subject have pointed out that no single method or medium can be relied upon to isolate all the species of clostridia that may be present in samples from widely different environmental conditions. It is useful at the outset of any investigation to ascertain what is required; different approaches are necessary to isolate any clostridia which *might* be present in a particular sample and to find any clostridia which are *active* in a particular environment. In the latter case it is possible to match isolation conditions (e.g., of pH, temperature, salt level, etc.) to those existing in the original sample.

In many circumstances, especially with the pathogenic clostridia, some knowledge of the sample and likely organisms exists. A sound knowledge of the varying habits and properties of the different clostridia will then be useful for successful isolation. Success often in fact depends on the exploitation of specific properties of the organism sought.

Anaerobic conditions

Anaerobic bacteria vary considerably in their reaction to the presence of oxygen. Amongst the clostridia there are species which will grow in the presence of air (see p. 88) and there are others which require the exclusion of oxygen and the provision of reduced media. In general, however, clostridia can be readily isolated without resorting to the more extreme measures to exclude all traces of oxygen which are required by some anaerobes.

The ability of many clostridia to survive exposure to oxygen is undoubtedly due to the fact that they produce endospores, the vegetative cells of which are often oxygen-sensitive. In view of the fact that clostridia are not always present as the spore form it is prudent to minimize exposure of samples and cultures to air during isolation procedures.

Over the years many methods have been devised to attain anaerobic culture conditions; these have been reviewed by Willis (1969). Today clostridia are mostly isolated and cultivated in anaerobic jars or chambers on conventional media and using methods adapted for work in anoxic conditions. These are described in detail in the standard texts already referred to.

Anaerobic jars and cabinets are generally operated with a gax mixture of nitrogen, carbon dioxide and hydrogen (commonly 85:5:10 per cent). A catalyst is incorporated into the jar or chamber to combine any remaining oxygen with hydrogen to form water. The earlier McIntosh and Fildes jars used a palladium catalyst which required heat, either by direct heating before closure of the lid or by passing an electric current afterwards. These have been largely superseded by palladium-coated alumina catalysts which are active at ambient temperature.

An alternative to this is a system which uses a disposable hydrogen and carbon dioxide generating system, commercial versions of which are readily available. The inclusion of carbon dioxide in the gas mix, or bicarbonate in the medium, has been shown to improve germination of spores and, often, the growth of clostridia (Hobbs et al., 1971).

To achieve anaerobic conditions it is also necessary to reduce the oxidation–reduction potential of the growth medium. With liquid media this is readily achieved by boiling and cooling immediately before inoculation. Agar media are normally reduced immediately after pouring and the use of freshly poured agar media is frequently adequate. However, it is usual to add reducing agents to the medium; in some cases normal medium components are active reducing agents, e.g., reducing sugars. The most useful and generally added reducing agent is cysteine, which will allow the growth of the most exacting clostridia, especially if used in conjunction with dithiothreitol (Collee et al., 1971).

Most media quickly lose their reducing properties if stored in contact with air. Liquid media can, as stated above, be largely restored by boiling, solid media, however, must be stored under anaerobic conditions to retain their efficiency.

Given the use of reduced media and anaerobic jars or cabinets, clostridia can be successfully handled and isolated by otherwise conventional laboratory methods.

Treatment of samples

The vegetative cells of many clostridia are sensitive to oxygen and, although many samples may themselves be highly reduced, it is sensible to expose them as little as possible to oxygen. If storage is necessary before processing this should be done under anaerobic conditions.

Before specimens are treated in any way samples for direct examination should be taken. In most cases a direct microscopic examination will be required, with or without specific staining techniques. Second, especially in the case of suspect food samples, an extract will be made to attempt a direct demonstration of toxins. Third, a portion of the sample will be used for inoculating enrichment cultures. The sample may or may not be treated before inoculating into enrichment media.

It is common to treat samples for isolation of clostridia by heat or chemicals. These treatments assume that the organism is present as spores, and should only be carried out in addition to the examination of untreated samples. The spores of many clostridia will survive heating at 80°C for 10 min, although survival is also a function of the numbers present. This treatment will kill all vegetative cells present.

Various chemicals have been used to kill vegetative organisms. Ethyl alcohol is one of the most successful and has been used successfully where thermolabile spores such as those of *C. botulinum* type E are sought (Johnson et al., 1964).

Enrichment cultures

In many samples clostridia are likely to be outnumbered by facultative organisms. With a few exceptions, for example in investigations of infant botulism (Glasby and Hatheway, 1985), direct plating of the sample is of limited value. However, the tendency of many clostridia to spread on the surface of agar media can be exploited. Isolation of *C. tetani* by inoculating one spot on a blood agar plate is a long-established method. After inoculation a pure culture can usually be obtained from a part of the plate remote from the inoculation site.

For many food samples, which are themselves good growth media, enrichment may simply consist of incubating the sample anaerobically, for instance, by vacuum packaging in oxygen-impermeable material.

By far the most commonly used enrichment medium used is Robertson's cooked meat broth, which can be improved for the isolation of saccharolytic species by the addition of 0.5–1.0 per cent w/v glucose.

The addition of antibiotics or other selective agents to enrichment media has been suggested on a number of occasions (see Hobbs et al., 1971; Smith and Williams, 1984). In general these are not necessary but may be advantageous in particular circumstances (Levett, 1985).

Because of the widely differing growth rates of clostridia, no single recommendation can be made concerning the incubation period of enrichment cultures before plating out. For fast growing species such as *C. perfringens* subcultures should be made after incubation for 18 h, or even sooner, at 30–37°C. Slower growing species such as *C. botulinum* or *C. tetani* are best obtained by plating out after 2–4 days, or longer.

The pathogenic clostridia have optimum growth temperatures in the region of 35–37°C. However, this may not be the optimum temperature for toxin production, and in practise the enrichment culture will be tested for the presence of toxin before proceeding further. Many *C. botulinum* strains produce toxin best at 25°C, and toxin is gradually destroyed at 37°C. A temperature of 30°C is usually preferred for this organism. In some cases a higher incubation temperature may be useful. *C. perfringens*, for instance, still grows rapidly at 45°C whereas most accompanying organisms will not grow at this temperature. Incubation at 45°C has been successfully used to isolate *C. perfringens* over many years (see Hobbs *et al.*, 1971).

Plating of enrichment media

Isolation of clostridia is often facilitated by the recognition of specific reactions on solid media. The two most common and most useful agar media are fresh blood agar and some form of egg-yolk agar, and in practise both media are usually used together. The reactions produced on these media are a result of the action of the toxins, or other soluble antigens (enzymes), which are produced and excreted by the clostridia, particularly the pathogenic species.

The haemolysins produced by different clostridia show varying activity against the red blood cells of different animal species. For example, only the θ-toxin of *C. perfringens* is active against horse erythrocytes, whereas both the α- and θ-toxins of this organism are active against sheep erythrocytes. *C. chauvoei* is strongly haemolytic on sheep-blood agar but weakly so on horse-blood agar.

There are several egg-yolk media in common use some, such as that described by Willis and Hobbs (1958, 1959) incorporate protein and lactose with a pH indicator, thus enabling several biochemical reactions to be determined at the same time. Egg-yolk media are primarily designed to recognize two reactions, that caused by a lecithinase C, such as the α-toxin of *C. perfringens*, and that caused by a lipase, such as produced by *C. botulinum*. The lecithinase reaction results in a wide zone of opacity in the medium, surrounding and extending well beyond the colonies. The precipitate is chiefly composed of protein rendered insoluble by the breakdown of lecithin in the egg-yolk. A lipase reaction results in a precipitation of free fatty acids in the medium together with a coextensive irridescent film (or pearly layer) covering the colony. The fatty acids are derived from neutral lipids present in the egg-yolk and are mainly insoluble in the aqueous phase of the medium.

Since the reactions on these two media are produced by toxins or other soluble antigens, they can often be inhibited by appropriate antitoxins, thus providing useful identification criteria. Antitoxin can be spread over half of the plate before inoculation (see Hobbs *et al.*, 1971).

The capacity of many clostridia to spread on agar media has already been mentioned as being useful for isolation. However, it is often more useful to inhibit the spreading and to allow single colonies to be isolated. The easiest way to prevent spreading is by incorporating 3–4 per cent agar into media instead of the more usual 1–1.5 per cent. Another method is to incorporate specific agglutinating antisera into the medium. This assumes some knowledge of the clostridia being sought, but has been used successfully for the isolation of *C. tetani* and *C. septicum* from mixtures with other anaerobes (Willis and Williams, 1970).

For routine use, most pathogenic clostridia will grow and produce typical reactions on blood agar and egg-yolk agar in 48 h. *C. perfringens* and *C. bifermentans* will normally be easily recognized after 24 h, although some may require three days. Daily examination of the plates is generally recommended, allowing colonies to be picked off for purification at the appropriate time.

Purification

The use of anaerobic jars and cabinets permits conventional, single colony isolation for purification of cultures. As has already been stated, exposure of the cultures to air should be kept to a minimum and the use of anaerobic cabinets provides an obvious advantage over opening anaerobic jars in a laboratory atmosphere.

Purification of clostridia by single colony selection, while being the method of choice, does have some peculiar difficulties. Many clostridia, especially the pathogenic species, have very similar colonies and, with most, colony variation is commonplace. One of the most common reasons for colony variation is the proportion of spores in a colony. When a high proportion of spores is present colonies are usually less irregular in shape and are opaque. When a few spores are present, colonies tend to be more irregular or rhizoid and translucent. Also, colony variation tends to be more obvious in older cultures. Despite these variations the presence of different colonies in a culture should always be regarded as potentially indicating a mixture, and the colonies should be examined separately. It will usually be apparent on the next subculture whether the colonies are different organisms or not. On selecting single colonies for subculture it is usually more profitable to transfer them directly to an agar medium again, rather than a liquid medium. Even so, mixed cultures have been known to persist through several subcultures.

Once pure cultures have been obtained a stock culture should be set aside in Robertson's cooked meat broth and a second culture in this medium should be used to inoculate the various media required for identification.

Identification procedures

Characterization of the pathogenic clostridia may be required for two main reasons. First, and most obviously, when the causative organism of a particular disease is sought. Second, when an investigation into the incidence and role of a particular clostridium in nature is required. In the first case there will often be some pathology which will permit some sort of preliminary diagnosis. In the second case some kind of enrichment culture will generally be set up to start the investigation. In both cases the initial investigation will be to look for toxin in the pathological material, food or enrichment culture. The toxin will then be characterized with specific antitoxins; after isolation of the organism biochemical tests are used to confirm a diagnosis. Non-toxic clostridia have to be identified solely from morphological and biochemical data.

Characterization of toxins

The exact procedure to be followed will depend on the type of sample under investigation. Whether the sample itself or a subsequent enrichment culture is examined, a test for toxin will usually precede any isolation or purification procedure. In a few cases the blood of the patient may well contain circulating toxin and this can be examined directly. For precise details of the protocols for toxin testing, the standard texts referred to earlier, particularly Sterne and Batty (1975), should be consulted.

In the first instance a sample of the suspect sample, or of the enrichment culture, is injected into an experimental animal to determine whether toxin is present. Some clostridial toxins are produced in an inactive or prototoxin form, which requires activation by proteolytic enzymes. These may or may not be produced by the organism and hence a duplicate test should be set up with the suspect material incubated with trypsin for 30 min before inoculation. For example, the lethal toxins of the non-proteolytic *C. botulinum* strains and the ε-toxin of *C. perfringens* are activated by proteolytic enzymes. On the other hand the β-toxin of *C. perfringens* is inactivated by trypsin. Once toxin has been demonstrated, identification is achieved by further animal tests, this time by protection tests using appropriate diagnostic antitoxins.

In most cases the experimental animal of choice is the mouse, and intraperitoneal injection the preferred route of administration. However, with *C. perfringens* and *C. novyi* strains there are some advantages in using toxin and neutralization tests in the skin of guinea-pigs. More information can be obtained from this test because the appearances of the dermonecrotic responses are characteristic of the different toxins.

In most toxin tests it is helpful to clarify and filter-sterilize the sample to be injected. This is to avoid invasive organisms confusing the interpretation of the results of the tests. With the skin-tests just described this could be a problem, and with mouse tests for *C. botulinum* and *C. tetani* centrifugation is usually considered sufficient. However, difficulties can also arise here, and there have been reports of non-specific deaths in experimental animals resulting from gram-negative organisms being present in, for instance, incubated food samples. This problem can be overcome by the addition of antibiotics to the sample inoculated, as well as by using filter-sterilization. There have been other unidentified causes of non-specific deaths in mice during toxin tests with food or other environmental samples. These are usually overcome by 10-fold dilution of the sample and, in one case, addition of bovine serum was found to be effective (Solberg et al., 1985).

Antisera for recognizing the main lethal toxins produced by clostridia are commercially available. Characterization of other soluble antigens requires specific antitoxins which are not available and which would have to be prepared. This is not necessary for most diagnostic purposes but to differentiate, for instance, all the varieties of *C. perfringens*, antisera to the proteinases K and λ, the δ-haemolysin and others are required.

The enterotoxin of *C. perfringens* is not detected in the tests described above. It is quite distinct from the lethal toxins and is not neutralized by the diagnostic antitoxins described. A specific antitoxin is now available and a variety of serological procedures can be used to detect this toxin (Crowther and Baird-Parker, 1983). *C. difficile* produces two toxins, one is lethal for a variety of laboratory animals and the other is primarily cytotoxic. Toxin is normally detectable in the patients' faeces and the most reliable and sensitive assay at present is for cytotoxicity (Bartlett, 1981). A number of *in vitro* serological methods have been described for detection of these toxins (Crowther and Baird-Parker, 1983; Smith and Williams, 1984; Kamiya et al., 1986; Rautenberg et al., 1986; Lyerly and Wilkins, 1986) and more work is clearly necessary before satisfactory reagents are commercially available.

Antisera are not available for a few species, these are *C. absonum*, *C. carnis*, *C. limosum*, *C. spiriforme*. Identification of these species has to depend on morphological and biochemical data from purified cultures.

Other serological tests

There have been many attempts to devise alternative tests for pathogenic clostridia. For the most part, these have been designed to measure some antigen other than the toxin which should be present in toxic strains, but not in non-toxic clostridia. In many cases individual enzymes have been sought and, in others, more generalized tests such as agar diffusion precipitin tests have been tried. In most cases these tests have failed because the correlation with toxicity was not good, although in some they are helpful for isolation (Lilly et al., 1984). Also, serological tests which could be used as alternatives to experimental animals have been sought for the toxins themselves. While these are feasible, in most cases antisera prepared against crude toxin preparations have been used and cross-reactions with non-toxic species usually occur. It became obvious many years ago that this approach would be successful only if antitoxins which were monospecific, i.e., prepared against purified toxin, were available. A number of such tests have recently been described, e.g., for *C. botulinum* toxins (Kozaki et al., 1979; Notermans et al., 1979; Shone et al., 1985).

Cell and spore antigens are, with few exceptions, of little value in the identification of clostridia. Their uses for *C. septicum*, *C. chauvoci* and serotypes of *C. perfringens* type A have already been described.

Morphological and biochemical properties

The morphological and biochemical properties of bacteria are only of value for identification if pure cultures are used. Once a pure culture has been achieved the various test media are inoculated and incubated at the appropriate temperature, for the pathogenic clostridia this is usually 37°C. Details of media and methods can be found in the various texts referred to at the beginning of the chapter.

If a positive toxin test has been found then the purpose of morphological and biochemical tests is to confirm, or not, a preliminary diagnosis. In most situations where food and pathological material is examined more than one clostridium will be isolated, and it is often desirable to identify them all. Indeed, some may be involved in the pathology as secondary infections.

Tables of diagnostic tests are available in a number of publications—*Bergey's Manual of Systematic Bacteriology* (Cato et al., 1986), the *VPI Anaerobe Laboratory Manual* (Holdeman et al., 1977) and

Table 6.3 Pathogenic clostridia producing a reaction on egg-yolk media

Character	C. absonum	C. botulinum (proteolytic)	C. botulinum (non-proteolytic)	C. haemolyticum	C. limosum	C. novyi type A	C. novyi type B	C. perfringens	C. sordellii
Motility	−	+	+	+	+	+	+	−	+
Lecithinase	+	−	−	+	+	+	+	+	+
Lipase	−	+	+	−	−	+	−	−	−
Gelatin liquefaction	+	+	+	+	+	+	+	+	+
Casein digestion	−	+	−	−	+	−	v	−	+
Haemolysis	+	+	+	+	+	+	+	v	v
Indole	−	−	−	+	−	−	−	−	+
Nitrate reduction	+	−	−	−	+	v	v	v	v
Acid from cellobiose	+	−	−	−	−	−	−	v	−
Acid from glucose	+	+	+	+	−	+	+	+	+
Acid from lactose	+	−	−	−	−	−	−	+	−
Acid from maltose	+	v	v	−	−	v	+	+	+
Acid from mannitol	−	−	−	−	−	−	−	−	−
Acid from mannose	+	−	+	−	−	−	v	+	v
Acid from ribose	−	−	v	−	−	v	v	v	v
Acid from salicin	+	−	−	−	−	−	−	v	−
Acid from starch	+	−	v	−	−	−	−	+	−
Acid from sucrose	+	v	v	−	−	v	−	+	−
Acid from trehalose	+	−	v	−	−	−	−	v	−

v = Variable

C. absoncum is differentiated from *C. perfringens* by not producing H$_2$S, its relative lack of toxicity for mice and by the fact that its lecithinase activity is not completely inhibited by *C. perfringens* type A antitoxin.

Hobbs and Cross (1984) being three of the most recent. These publications should be consulted for full descriptions of all clostridia. The reactions of the main pathogenic clostridia are reproduced in Tables 6.3 and 6.4. By the time a pure culture has been obtained and a toxicity test has been performed, there will already be a considerable amount of diagnostic information available. The morphology, colony characteristics, motility and reactions on blood agar and egg-yolk agar, as well as toxicity, will all be known. Sensible use of the diagnostic tables will usually mean that few further tests are necessary in many cases.

In addition to the tests listed in the diagnostic tables, it has now become general practice to examine the fermentation products after growth of anaerobic bacteria in a suitable glucose medium (for details see Holdeman *et al.*, 1977). A relatively simple gas–chromatographic procedure, which is generally available to most anaerobe laboratories, is required. The results of this test are usually of more value in the identification of non-sporing anaerobes, but the information is still useful for clostridia.

Conclusions

It can be concluded that classification of the anaerobic spore-forming bacteria is not soundly based on modern taxonomic thinking, and that more new data are required before this could be done. The present classification schemes evolved from a need to characterize those clostridia which caused disease and, to a lesser extent, those which accumulated

Table 6.4 Pathogenic clostridia not producing a reaction on egg-yolk media

Character	C. carnis	C. chauvoei	C. difficile	C. fallax	C. histolyticum	C. malenominatum	C. ramosum	C. septicum	C. spiroforme	C. tetani
Motility	+	+	+	+	+	+	−	+	−	+
Gelatin liquefaction	−	+	+	−	+	−	−	+	−	+
Casein digestion	−	−	−	−	+	−	−	−	−	−
Haemolysis	+	+	−	−	+	−	−	+	−	+
Indole	−	−	−	−	−	+	−	−	−	v
Nitrate reduction	−	+	−	−	−	+	−	v	−	−
Acid from cellobiose	+	v	v	−	−	−	+	+	v	−
Acid from glucose	+	+	+	+	−	−	+	+	+	−
Acid from lactose	v	+	−	−	−	−	+	+	+	−
Acid from maltose	+	+	−	+	−	−	+	+	−	−
Acid from mannitol	−	−	+	−	−	−	v	−	−	−
Acid from mannose	+	+	+	+	−	−	+	+	+	−
Acid from ribose	v	+	v	v	−	−	v	v	−	−
Acid from salicin	+	−	v	−	−	−	+	v	v	−
Acid from starch	−	−	−	+	−	−	−	−	−	−
Acid from sucrose	+	+	−	−	−	−	+	−	+	−
Acid from trehalose	−	−	v	−	−	−	+	v	−	−
Spore position	T	S	S	S	S	T	T	S	T	T

V, variable; T, terminal location of spores; S, sub-terminal location of spores

useful metabolic products. In as much as pathogenic clostridia are generally identified by virtue of the toxins they produce, and that these relate to the symptoms of the disease and the treatment required, this remains a satisfactory solution. Many clostridia, not recognized as pathogenic, are present in normal and diseased intestines and tissues, and these may or may not have a role in the pathology. The present classification schemes are inadequate for studies into this kind of situation and this is borne out by the large number of 'unidentifiable' isolates found.

References

Baba T. (1969). Analytical serology of bacillaceae. In: *Analytical serology of microorganisms*, vol 2. pp. 609–642. Edited by Kwapinski J B G. Interscience Publishers, New York.

Bartlett J G. (1981). Laboratory diagnosis of antibiotic-associated colitis. *Laboratory Medicine* 12: 347–351.

Bom I J, Smelt J P P M, Kersters K, Verrips C T. (1986). Identification and grouping of *Clostridium botulinum* strains by numerical analysis of their electrophoretic protein patterns. *Journal of Applied Bacteriology* 60: 483–490.

Cato E P, George W L, Finegold S M. (1986). Genus *Clostridium*. In *Bergey's Manual of systematic bacteriology*, vol. 2. pp. 1141–1200. Edited by Sneath P H A et al. Williams and Wilkins, Baltimore.

Cato E P, Hash D E, Holdeman L V, Moore W E C. (1982). Electrophoretic study of *Clostridium* species. *Journal of Clinical Microbiology* 15: 688–702.

Chakrabarty A K, Narayan K G. (1979). A proposed serogrouping scheme for epidemiological investigation of food poisoning due to *Clostridium perfringens* type A. *Zentralblatt fur Bakteriologie, Parasitenkunde, Infektionskrankheiten und Hygiene. Abteilung I. Originale Reihe A.* 245: 114–122.

Collee J G, Rutter J M, Watt B. (1971). The significantly viable particle: a study of the subculture of an exacting sporing anaerobe. *Journal of Medical Microbiology* 4: 271–288.

Cone R D, Leckowich R V. 1970. Differentiation of *Clostridium botulinum* types, A, B and E by pyrolysis gas–

liquid chromatography. *Applied Microbiology* **19**: 138–145.

Crowther J S, Baird-Parker A C. (1983). The pathogenic and toxigenic spore-forming bacteria. In *The bacterial spore*, vol. 2. pp. 275–311. Edited by Hurst A, Gould G W. Academic Press, London.

Cummins C S. (1970). Cell-wall composition in the classification of gram-positive anaerobes. *International Journal of Systematic Bacteriology* **20**: 413–419.

Cummins C S, Johnson J L. (1971). Taxonomy of the clostridia wall composition and DNA homologies in *Clostridium butyricum* and other butyric acid-producing clostridia. *Journal of General Microbiology* **67**: 33–46.

Dowell V R, Hawkins T M. (1977). *Laboratory methods in anaerobic bacteriology*. Publication No. 78–8272, Center for Disease Control, Public Health Service, US Department of Health, Education and Welfare, Atlanta.

Eklund M W, Poysky F T, Meyers J A, Pelroy G A. (1974). Interspecies conversion of *Clostridium botulinum* type C to *Clostridium novyi* type A by bacteriophage. *Science* **186**: 456–458.

Elsden S R, Hilton M G. (1979). Amino acid patterns in clostridial taxonomy. *Archives of Microbiology* **123**: 137–141.

Elsden S R, Hilton M G, Parsley K R, Self R. (1980). The lipid fatty acids of proteolytic clostridia. *Journal of General Microbiology* **118**: 115–123.

Farshy D C, Moss C W. (1970). Characterization of clostridia by gas chromatography: differentiation of species by tri-methyl silyl derivatives of whole cell hydrolysates. *Applied Microbiology* **20**: 78–84.

Fox G E *et al.* (1980). The phylogeny of prokaryotes. *Science* **209**: 457–463.

Fugate K J, Hansen L B, White O. (1971). Analysis of *Clostridium botulinum* types A, B and E for fatty and carbohydrate content. *Applied Microbiology* **21**: 470–475.

Gall L S, Riely P E. (1981). *Manual for the determination of the clinical role of anaerobic microbiology*. CRC Press Inc, Boca Raton.

Gimenez D F. (1984). *Clostridium botulinum* sub type Ba. *Zentralblatt für Bakterologie, Parasitenkunde, Infektionskrankheiten und Hygiene Abteilung I. Originale Reihe A.* **257**: 68–73.

Gimenez D F, Cicearelli A. S. (1972). Antigenic variations in botulinal toxins of type F. Proposed definitions for the serological typing and classification of *Clostridium botulinum*. *Medicina* **32**: 596–606.

Glasby C, Hatheway C L. (1985). Isolation and enumeration of *Clostridium botulinum* by direct inoculation of infant faecal specimens on egg-yolk agar and *Clostridium botulinum* isolation media. *Journal of Clinical Microbiology* **31**: 264–266.

Gutteridge C S, Mackey B M, Norris J R. (1980). A pyrolysis gas–liquid chromatography study of *Clostridium botulinum* and related organisms. *Journal of Applied Bacteriology* **49**: 165–174.

Hall J D, McCroskey L M, Pincomb B J, Hatheway C L. (1985). Isolation of an organism resembling *Clostridium barati* which produces type F botulinal toxin from an infant with botulism. *Journal of Clinical Microbiology* **21**: 654–655.

Hardie J M. (1986). Methods for the isolation and identification of anaerobes. In *Anaerobic bacteria in habitats other than man*, SAB Symposium Series No. 13, pp. 397–410. Edited by Barnes E M, Mead G C. Blackwell Scientific Publications, Oxford.

Hobbs G. (1986). Some problems in the identification and taxonomy of clostridia. In *Anaerobic bacteria in habitats other than man*, SAB Symposium No. 13, pp. 23–36. Edited by Barnes E M, Mead G C. Blackwell Scientific Publications, Oxford.

Hobbs G, Cross T. (1984). Identification of endospore-forming bacteria. In *The bacterial spore*, vol. 2. pp. 49–78. Edited by Hurst A, Gould G W E. Academic Press, London.

Hobbs G, Hardy R, Mackie P R. (1971). Characterization of *Clostridium* species by means of their lipids. *Journal of General Microbiology* **68**: ii–iii.

Hobbs G, Williams K, Willis A T. (1971). Basic methods for the isolation of clostridia. In *Isolation of anaerobes*, SAB Technical Series No. 5, pp. 2–23. Edited by Shapton D A, Board R G. Academic Press, London.

Hodgkiss W, Ordal Z J, Cann D C. (1966). The comparative morphology of spores of *Clostridium botulinum* type E and the spores of the 'OS mutant'. *Canadian Journal of Microbiology* **12**: 1283–1284.

Hodgkiss W, Ordal Z J, Cann D C. (1967). The morphology and ultrastructure of the spore and exosporium of some *Clostridium* species. *Journal of General Microbiology* **47**: 213–225.

Holdeman L V, Cato E P, Moore W E C. (1977). *Anaerobe laboratory manual*, 4th edn., Anaerobe Laboratory, Virginia Polytechnic Institute and State University, Blacksburg.

Johnson J L. (1973). Use of nucleic acid homologies in the taxonomy of anaerobic bacteria. *International Journal of Systematic Bacteriology* **23**: 308–315.

Johnson J L, Francis B S. (1975). Taxonomy of the clostridia: ribosomal ribonucleic acid homologies among the species. *Journal of General Microbiology* **88**: 229–244.

Johnson R, Harmon S, Kautter D A. (1964). Method to facilitate the isolation of *Clostridium botulinum* type E. *Journal of Bacteriology* **88**: 1521–1522.

Kamiya S, Nakamura S, Yamakawa K, Nishida S. (1986). Evaluation of a commercially available latex immunoagglutination test kit for detection of *Clostridium difficile* D-1 toxin. *Microbiology and Immunology* **30**: 177–181.

Kiritani K, Mitsui N, Nakamura S, Nishida S. (1973). Numerical taxonomy of *Clostridium botulinum* and *Clostridium sporogenes* strains and their susceptibilities to induced lysins and to mitomycin C. *Japanese Journal of Microbiology* **17**: 361–372.

Kozaki S, Dufrenne J, Hagenaars A M, Notermans S. (1979). Enzyme linked immunoabsorbent assay (ELISA) for the detection of *Clostridium botulinum* type B toxin. *Japanese Journal of Medical Science and Biology* **32**: 199–205.

Lee W H, Riemann H. (1970a). Correlation of toxic and non-toxic strains of *Clostridium botulinum* by DNA composition and homology. *Journal of General Microbiology* **60**: 117–123.

Lee W H, Riemann H. (1970b). The genetic relatedness of proteolytic *Clostridium botulinum* strains. *Journal of General Microbiology* **93**: 27–35.

Levett P N. (1985). Effect of antibiotic concentration in a selective medium on the isolation of *Clostridium difficile* from faecal specimens. *Journal of Clinical Pathology* **38**: 223–234.

Lilly T, Kautter D A, Lynt R K, Solomon H M. (1984). Immunodiffusion detection of *Clostridium botulinum* colonies. *Journal of Food Protection* **47**: 868–870.

Lyerly D M, Wilkins T D. (1986). Commercial latex test for *Clostridium difficile* toxin A does not detect toxin A. *Journal of Clinical Microbiology* **23**: 622–623.

McCroskey L M, Hatheway C L, Fenicia L, Pasolini B, Aureli P. (1986). Characterization of an organism that produces type E botulinal toxin but which resembles *Clostridium butyricum* from the feces of an infant with type E botulism. *Journal of Clinical Medicine* **23**: 201–202.

Mead G C. (1971). The amino-acid fermenting clostridia. *Journal of General Microbiology* **67**: 47–56.

Mitsuoka T. (1980). *A colour atlas of anaerobic bacteria.* Sobunsha, Tokyo.

Moss C W, Lewis V J. (1967). Characterization of clostridia by gas chromatography. Differentiation of species by cellular fatty acids. *Applied Microbiology* **15**: 390–397.

Nakamura S, Kimura I, Yamakawa K, Nishida S. (1983). Taxonomic relationships among *Clostridium novyi* types A and B, *Clostridium haemolyticum* and *Clostridium botulinum* type C. *Journal of General Microbiology* **125**: 1473–1479.

Nakamura S, Okado I, Abe T, Hishida S. (1979), Taxonomy of *Clostridium tetani* and related species. *Journal of General Microbiology* **113**: 29–35.

Nakamura S, Okado I, Nakashio S, Nishida S. (1977). *Clostridium sporogenes* isolates and their relationship to *Clostridium botulinum* based on deoxyribonucleic acid reassociation. *Journal of General Microbiology* **100**: 395–401.

Nakamura S, Shimamura T, Hayase M, Nishida S. (1973). Numerical taxonomy of saccharolytic clostridia, particularly *Clostridium perfringens*-like strains: description of *Clostridium absonum* sp. nov. and *Clostridium paraperfringens*. *International Journal of Systematic Bacteriology* **23**: 419–429.

Notermans S, Dufrenne, J, Kozaki S. (1979). Enzyme linked immunoabsorbent assay for detection of *Clostridium botulinum* type E toxin. *Applied and Environmental Microbiology* **37**: 1173–1175.

Novotny P. (1969). Composition of cell walls of *Clostridium sordellii* and *Clostridium bifermentans* and its relation to taxonomy. *Journal of Medical Microbiology* **2**: 81–100.

Oyama V I, Carle G C. (1967). Pyrolysis gas chromatography applications to life detection and chemotaxonomy. *Journal of Gas Chromatography* **5**: 151–154.

Paine C M, Cherniak R. (1975). Composition of the capsular polysaccharides of *Clostridium perfringens* as a basis for their classification by chemotypes. *Canadian Journal of Microbiology* **21**: 181–185.

Prévot A R, Turpin A, Kaiser P. (1967). *Les bacteries anaerobes*. Dunod, Paris.

Rautenberg P, Stender F, Ullmann U. (1986). Detection of *Clostridium difficile* toxin A by immunoblottiny. *Zentralblatt fur Bakteriologie, Parasitenkunde, Infektionskrankheiten und Hygiene Abteilung I. Originale Reihe A.* **261**: 29–42.

Reiner E, Bayer F L. (1978). Botulism: a pyrolysis gas–liquid chromatographic study. *Journal of Chromatographic Science* **16**: 623–629.

Rode L J, Pope L, Filip C, Smith L D S. (1971). Spore appendages and taxonomy of *Clostridium sordellii*. *Journal of Bacteriology* **108**: 1384–1389.

Rode L J, Smith L D S. (1971). Taxonomic implications of spore fine structure in *Clostridium bifermentans*. *Journal of Bacteriology* **105**: 349–354.

Shone C *et al*. (1985) Monoclonal antibody-based immunoassay for Type A *Clostridium botulinum* is comparable to the mouse bioassay. *Applied and Environmental Microbiology* **50**: 63–67.

Smith L D S. (1977). *Botulism. The organism, its toxins, the disease* p. 19. Thomas, Springfield.

Smith L D S, Williams B L. (1984). *The pathogenic anaerobic bacteria*, 3rd edn. Thomas, Springfield.

Solberg M, Post L S, Furgang D, Graham C. (1985). Bovine serum eliminates rapid non-specific toxic reactions during bioassay of stored fish for *Clostridium botulinum* toxin. *Applied and Environmental Microbiology* **49**: 646–649.

Sterne M, Batty I. (1975). *Pathogenic clostridia*. Butterworths, London.

Stringer M F, Turnbull P C B, Gilbert R J. (1980). Application of serological typing to the investigation of outbreaks of *Clostridium perfringens* food poisoning 1970–1978. *Journal of Hygiene* **84**: 443–456.

Strom M S, Eklund M W, Poysky F T. (1984). Plasmids in *Clostridium botulinum* and related *Clostridium* species. *Applied and Environmental Microbiology* **8**: 956–963.

Sutter V L, Citron D M, Finegold S M. (1980). *Wadsworth anaerobic bacteriology manual*, 3rd edn. C V Mosby, St Louis.

Tanner R S, Stackebrandt E, Fox G E, Woese C R. (1981). A

phylogenetic analysis of *Acetobacterium woodii*, *Clostridium barkeri*, *Clostridium butyricum*, *Clostridium lituseburense*, *Eubacterium limosum* and *Eubacterium tenne*. Current Microbiology **5**: 35–38.

Vaczi L, Surjan M, Kiss J M, Fust G, Redai I. (1971). Bacterial cardiolipins and their serological activity. Acta Microbiologica Academiae Scientiarum Hungaricae, Budapest **18**: 159–165.

Willis A T. (1969). *Clostridia of wound infection*. Butterworths, London.

Willis A T, Hobbs G. (1958). A medium for the identification of clostridia producing opalescence in egg-yolk emulsions. Journal of Pathology and Bacteriology **75**: 299–305.

Willis A T, Hobbs G. (1959). Some new media for the isolation and identification of clostridia. Journal of Pathology and Bacteriology **77**: 511–521.

Willis A T, Williams K. (1970). Some cultural reactions of *Clostridium tetani*. Journal of Medical Microbiology **3**: 291–301.

Wu J I J, Riemann H, Lee W H. (1972). Thermal stability of the DNA hybrids between the proteolytic strains of *Clostridium botulinum* and *Clostridium sporogenes*. Canadian Journal of Microbiology **18**: 97–99.

Yemetsev V T, Dzadzaia T D, Murzakov B G, Gostenkow V F. (1971). Taxonomic differentiation of *Clostridium pasteurianum* and *Clostridium acetylbutylicum* by the method of pyrolysis–gas chromatography. Izvestiia Akademii Nank SSSR. Seriia Biologicheskaia **3**: 465–468.

7

Anaerobic cocci

B Watt

The nature of anaerobic cocci
    Morphological metabolic and cultural characteristics
Isolation
Classification
    The development of classification schemes
    Characterization of anaerobic cocci in the diagnostic laboratory
    Laboratory diagnosis of infections due to anaerobic cocci
Ecology
    Role of anaerobic cocci in disease
    Antibiotic susceptibility of anaerobic cocci
    Treatment of infections due to anaerobic cocci
References

The increasing interest in anaerobic microbiology in recent years has largely been confined to the clostridia and the non-sporing gram-negative anaerobic bacilli, to the neglect of the anaerobic cocci. It was not until the 1970s that the relevance of the anaerobic cocci to human infections was appreciated (Pien et al., 1972; Lambe et al., 1974). Yet these organisms are part of the commensal flora in man and animals, and are associated with a variety of human infections, often in combination with other organisms.

Thus, our knowledge of the anaerobic cocci is limited; we can culture the organisms in the laboratory but cannot agree on a straightforward classification scheme for the diagnostic laboratory. We can isolate the organisms from cases of human infection but have little idea of their pathogenic role.

The nature of anaerobic cocci

There is no taxonomically acceptable definition of anaerobic cocci, and several workers have included microaerophilic or capnophilic strains under the heading 'anaerobic' cocci, usually because such strains appear to have a requirement for anaerobic conditions on primary isolation (because most anaerobic atmospheres include about 10 per cent CO_2). In an effort to remove this confusion, Watt and Jack (1977) proposed a working definition of anaerobic coci as 'cocci that grow well under satisfactory conditions of anaerobiosis and do not grow on suitable solid media in CO_2 10 per cent in air even after incubation for seven days at 37°C'. We found that such strains were sensitive to metronidazole (MICs of 5 mg/l or less), whereas strains that grew in CO_2 10 per cent were not sensitive (MICs >500 mg/l). It must be stressed, however, that although the separation of true anaerobic cocci from 'non-anaerobic' cocci is helpful for the purposes of characterization, it does not necessarily reflect differences in pathogenicity.

The anaerobic cocci isolated from man are generally considered to belong to one of three genera—*Peptococcus* or *Peptostreptococcus* (both gram-positive) and *Veillonella* (gram-negative). The classification of these anaerobic cocci is discussed later.

Morphological, metabolic and cultural characteristics

Peptococcus and *Peptostreptococcus*

Members of these two genera are non-motile and non-sporing cocci, with diameters in the range 0.4–2.5 μm. They may appear in clusters or chains of varying length in gram-stained smears, but cell arrangement and morphology are not constant characteristics, being influenced by the composition of the medium. They appear to derive energy from peptone and amino acids; carbohydrates are not an essential energy source and many strains are asaccharolytic. Some strains will only ferment (i.e., produce acid from) carbohydrates if a fatty acid or Tween 80 is added to the basal medium (Parker, 1983). Other strains produce gas by fermentation of carbohydrates only when provided with a sulphur source, such as thiosulphate (Wildy and Hare, 1953). Some strains may retain Gram's methyl violet stain poorly or irregularly. Peptostreptococci and peptococci are obligate anaerobes and are only moderately oxygen-sensitive. They will grow readily on a variety of 'rich' conventional media (e.g., Columbia agar, Mueller–Hinton agar, brain–heart infusion agar); most workers find that addition of 7–10 per cent human or horse blood greatly enhances growth, but no information is available on the relative effects of different species of blood. The growth of many strains is enhanced by addition of CO_2 10 per cent to the incubation environment. Surface colonies on blood-containing media are non-haemolytic, without distinctive characters, low convex, dull grey-white and 0.5–1.5 mm in diameter after incubation for 48 h. One species (*Peptococcus niger*) forms a black 'pigment' after 5–7 days incubation in liquid media.

Veillonella

These are small non-motile, non-sporing obligately anaerobic gram-negative cocci, 0.3–0.5 μm in diameter, which are arranged in pairs, clusters or short chains (Rogosa, 1964). Studies of the cell wall of *Veillonella* spp. suggest that its composition is similar to that of *Neisseria* spp., but that it contains more mucopeptide. The organisms grow poorly on ordinary media but grow well on media with added trypticase (Wilson, 1983). Metabolically, they are rather inert and require incubation for at least 48 h before visible colonies develop. Growth appears to be enhanced by the presence of CO_2 10 per cent in the incubation atmosphere. The physiology of these organisms is well reviewed by Delwiche *et al.* (1985).

Isolation

Anaerobic cocci grow readily on conventional media with added blood (see above) and in liquid media such as thioglycollate or cooked meat broths. Many strains do not grow reliably in small volumes of liquid media and this may give rise to problems with identification systems such as AP1 20A or Minitek (B Watt and F Brown, unpublished data).

Selective media

Anaerobic cocci are often isolated from mixed culture of specimens from human or animal sources, and several workers have developed selective media to facilitate the isolation of these organisms from mixed infections. Anaerobic cocci can be isolated from mixed cultures by use of neomycin blood agar (final concentration 70 mg/l) or gentamicin blood agar (20–25 mg/l) but there may be some quantitative reduction in isolation even at these concentrations (B Watt, unpublished data). Wren (1978) found that a nalidixic acid–Tween medium gave the best recovery of anaerobic cocci from clinical samples. These selective agents suffer from the disadvantage that they do not inhibit the *Bacteroides* spp. which are often present in association with anaerobic cocci in clinical samples; a medium containing bicozamycin (500 mg/l) and neomycin (30–40 mg/l) has given encouraging preliminary results in this regard (Watt and Brown, 1983).

Classification

The development of classification schemes

The first isolation of an anaerobic coccus was reported by Veillon in 1893, from a case of Bartholinitis. Taylor (1929), reviewing this and other reports, noted that there had been many subsequent reports of isolations of anaerobic cocci from human infection and that these isolates had been given various names by different investigators. The first attempts at a detailed classification scheme were by Prévot who, in a series of papers from 1924 onwards, developed a scheme based on morphological and biochemical tests (Prévot, 1933; Prévot and Fredette, 1966). He and his co-workers divided anaerobic cocci into eight genera, on the basis of morphological criteria, and subdivided these genera into species on the basis of tests such as carbohydrate fermentation reactions, growth in litmus milk, formation of indole and liquefaction of gelatin. Although other workers made important contributions to this area of anaerobic microbiology (e.g., Harris and Brown, 1929; Stone, 1940; Foubert and Douglas, 1948), it was not until the 1950s that an alternative scheme was devised. Hare *et al.* (1952) criticized Prévot's scheme on the basis of the small numbers of strains examined. From a study of nearly 100 clinical isolates (many from the respiratory and genital tracts) they proposed a classification scheme based on the formation of gas from carbohydrate and inorganic acid substrates and the fermentation of a range of carbohydrates. The scheme originally contained six groups—these were later expanded to ten (Thomas and Hare, 1954). (It should be noted that both of the above schemes included microaerophilic cocci.) In 1971 Rogosa placed the gram-positive anaerobic cocci in a new family, which he named Peptococcaceae. The family was considered to contain four genera—*Peptococcus*, *Peptostreptococcus*, *Ruminococcus* and *Sarcina*, although other genera, such as *Coprococcus*, have been included by Holdeman and Moore (1974). These genera have been used as the basis of the more recent classification schemes, but different workers have applied different criteria to the development of species names. For a detailed review of the confused status of classification in the 1970s the reader is referred to the reviews of Wells and Field (1976) and Watt *et al.* (1986). Wells and Field noted that the same organism may appear under different names in different schemes—for example, they considered that the organism named *Peptococcus asaccharolyticus* in the scheme of Holdeman and Moore (1974) was equivalent to the organism given the name *Peptostreptococcus* CDC Group I by Dowell and Hawkins (1974). Most workers in this field have used the scheme published in the *Manual of the Virginia Polytechnic Institute* (V.P.I.) (Holdeman and Moore, 1972; Holdeman *et al.* 1977). The advantage of this scheme is the considerable amount of detail which is given about the biochemical reactions and GLC profiles of each of the described species, but its disadvantage is that, as there is no acceptable test to differentiate *Peptococcus* from *Peptostreptococcus*, and as many species are biochemically rather inert, classification is suspect.

In recent years, several workers have applied various modern analytical methods to the classification of the anaerobic cocci. These include analysis of cell-wall peptidoglycan structure and whole-cell fatty acids; determination of G+C content; DNA-DNA homology and polyacrylamide gel electrophoresis (PAGE) of whole-cell proteins. This subject is excellently reviewed by Taylor (1984). The use of such techniques has tended to cast doubt on established classification schemes, so much so that Ezaki *et al.* (1983) showed that, whereas the G+C content of the type strain of the genus *Peptococcus* (*Pc. niger*) was 51 mol per cent, that of other species within the genus *Peptococcus* ranged from 29 to 34 mol per cent. The content for *Pstr. anaerobius* (the type strain of the genus *Peptostreptococcus*) was 33 mol per cent. Also, these authors showed that the DNA-DNA homology and cell fatty acid profiles of four strains of *Peptococcus* were closer to those of *Pstr. anaerobius* than *Pc. niger*. For these reasons, they proposed that *Pc. asaccharolyticus*, *Pc. indolicus*, *Pc. prevotii* and *Pc. magnus* be reassigned to the genus *Peptostreptococcus* and that the 'group' of organisms previously named *Gaffkya anaerobia* be renamed *Pstr. tetradius*. These proposals have now been accepted (Moore *et al.*, 1985). As *Pstr. constellatus*, *Pstr. intermedius* and *Pstr. morbillorum* were transferred to the genus *Streptococcus* in 1974 (Holdeman and Moore) and *Pstr. parvulus* in 1983 (Cato), and as *Pc. saccharolyticus* has now been transferred to the genus *Staphylococcus* (Killper-Balz and Schleifer, 1981), the genus *Peptococcus* now contains only one species—*Pc. niger*. Another species, *Pc. heliotrinreducens*, was isolated by Lanigan (1976) from the sheep rumen but is of no clinical significance in man. The present situation is outlined in Table 7.1. It should be noted, however, that the present taxonomic status of *Coprococcus*, *Ruminococcus* and *Sarcina* is unclear, and that some workers (e.g., Huss *et al.*, 1984) have cast

Table 7.1 Present classification status of the gram-positive anaerobic cocci*

Accepted name	Previous name(s)
Peptococcus niger	Peptococcus niger
Peptococcus heliotrinreducens	new species
Peptococcus asaccharolyticus	Peptostreptococcus asaccharolyticus
Peptostreptococcus micros	Peptococcus glycinophilus
Peptostreptococcus indolicus	Peptococcus indolicus
Peptostreptococcus magnus	Peptococcus magnus
Peptostreptococcus prevotii	Peptococcus prevotii
Peptostreptococcus tetradius	Gaffkya anaerobia

*From Moore et al. (1985)

doubt on the present classification of *Peptococcus* and *Peptostreptococcus*.

Taxonomy of gram-negative anaerobic cocci

Rogosa (1964) proposed that the genus *Veillonella*, defined by Prévot in 1933 to include two species, *V. alcalescens* and *V. parvula*, be enlarged to include three subspecies of *V. parvula* (*parvula, rodentium* and *atypica*) and four subspecies of *V. alcalescens* (*alcalescens, ratti, criceti* and *dispar*). These proposals were made on the basis of serological tests and biochemical characteristics which included decomposition of hydrogen peroxide and absolute requirements for putrescine or cadaverine. However, Mays *et al.* (1982), after studying the DNA-DNA homologies of 116 strains of *Veillonella*, suggested that the type strains of *V. parvula* ss. *parvula* and *V. alcalescens* ss. *alcalescens*, which showed a high degree of homology, be renamed *V. parvula*, and that the other subspecies be given species status. The present situation is summarized in Table 7.2. Differentiation of these strains is extremely difficult by biochemical tests, and serological tests give incomplete results without DNA-DNA homology studies.

Characterization of anaerobic cocci in the diagnostic laboratory

If the classification of anaerobic cocci has presented problems for specialized laboratories, it has become a nightmare for the diagnostic laboratory with limited resources and experience. Several workers have tried to produce simplified characterization schemes, using conventional tests, to allow diagnostic laboratories to attempt some kind of labelling of clinical isolates, rather than simply 'anaerobic cocci'. Following the working definition of anaerobic cocci proposed by Watt and Jack (1977), Wren *et al.* (1977) suggested that the genus *Peptococcus* could be separated from *Peptostreptococcus* by the use of a 5 µg novobiocin disc—peptococci are resistant whereas peptostreptococci show a zone of inhibition. However, other workers (Watt *et al.*, 1986) have found this test to be fallible, as is the use of inhibition of growth by discs containing liquoid (sodium polyanethol sulphonate) to specifically identify *Pstr. anerobius*. The use of antibiotic resistogram patterns, so useful in the classification of the gram-negative anaerobic bacilli (Finegold *et al.*, 1971), has also proved to be unreliable. Similarly, the use of commercially produced microsystems (e.g., API 20A and API ZYM) may be of some use in characterizing biochemically active species, but is of little or no use for less reactive strains. Also, many strains may fail to grow in the small volumes of liquid media used in such microsystems (B. Watt and F. Brown, unpublished data). It is therefore not surprising that Taylor (1984) was able to reliably identify only 39 of 87 isolates examined by conventional methods.

We have proposed (Watt *et al.*, 1984) a simple identification scheme (Figure 7.1), suitable for the diagnostic laboratory, in which anaerobic cocci can be divided into four groups on the basis of simple biochemical tests and liquoid susceptibility. This scheme, while not a formal identification scheme,

Table 7.2 Present classification status of the gram-negative anaerobic cocci*

Accepted name	Previous name(s)
Veillonella criceti	Veillonella alcalescens ss. criceti
Veillonella dispar	Veillonella alcalescens ss. dispar
Veillonella ratti	Veillonella alcalescens ss. ratti
Veillonella atypica	Veillonella parvula ss. atypica
Veillonella caviae	new species
Veillonella rodentium	Veillonella parvula ss. rodentium
Veillonella parvula	Veillonella alcalescens ss. alcalescens / Veillonella parvula ss. parvula

*From Moore et al. (1985)

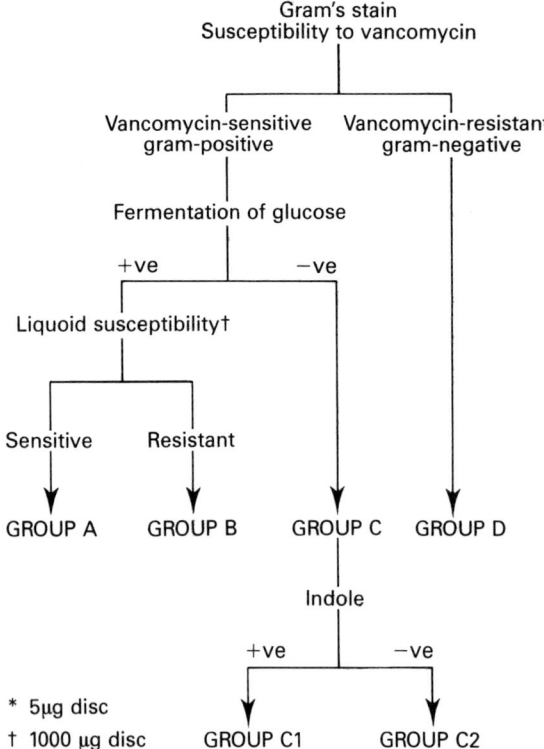

Fig. 7.1 Simple identification scheme for anaerobic cocci (Adapted from Watt *et al.*, 1984)

nevertheless allows the diagnostic laboratory to attempt some labelling of clinical isolates and should be of use in epidemiological studies.

Laboratory diagnosis of infections due to anaerobic cocci

Anaerobic cocci are not demanding anaerobes, they survive well in swabs (Smith *et al.* 1984) and will grow well on blood agar media, provided that CO_2 10 per cent is present in the incubation atmosphere and that anaerobic incubation is continued uninterrupted for at least 48 h. (In the case of *Pc. niger*, incubation for up to one week is necessary for the development of the characteristic black 'pigment'.) Anaerobic cocci will also grow well in liquid media such as cooked meat broth or thioglycollate-containing broth. Subsequent subculture and sensitivity testing can be performed on similar blood agar media.

Ecology

Anaerobic cocci have been isolated from various sites in healthy people; they occur as part of the skin flora and are present in large numbers in the mouth, bowel and genital tracts. Their occurrence is dealt with in more detail in Chapter 10. There have been few studies covering large numbers of healthy people, so that the normal distribution of species is not known for certain. A study by Neut *et al.* (1985) confirmed that gram-positive anaerobic cocci were found as part of the normal flora of the mouth, vagina and intestinal tract but indicated that the distribution of species was different at different sites. Thus they found *Pstr. anaerobius* and *Pstr. micros* to be the most frequently isolated species from the mouth; *Pstr. magnus* and *Pc. asaccharolyticus* (sic) were most frequently found in the vagina and *Pstr. prevotii* in the faecal flora. Much more information is needed in this area—but the lack of a suitable characterization scheme for the diagnostic laboratory is an obstacle to progress.

Role of anaerobic cocci in disease

Anaerobic cocci have been isolated from many different infections. Although they are occasionally isolated in pure culture—from cases of bacteraemia (Topiel and Simon, 1986), pericarditis and mediastinitis (Phelps and Jacobs, 1985) and osteomyelitis (Ziment *et al.*, 1968)—they usually occur in mixed culture, often with gram-negative anaerobes. They are relatively frequent isolates from cases of brain abscess (together with aerobes and other anaerobes); in a survey of the literature, Finegold (1977) showed that from 142 cases of brain abscess, there was a total of 248 isolates of anaerobic bacteria, of which 54 (22 per cent) were anaerobic cocci. They are found as components of mixed infections in abscesses of dental origin (Sklavounos *et al.*, 1986). Other infections in which anaerobic cocci occur in mixed culture include a wide variety of wound infections, lung abscesses and soft-tissue infections, as well as pilonidal sinuses and sebaceous cysts. This latter group of soft-tissue infections (which includes anaerobic cellulitis, infected varicose ulcers and anaerobic streptococcal myonecrosis) are well reviewed by Finegold (1977). The clinical observations of Meleney (1931) led to the discovery of a condition which was called progressive synergistic gangrene, in which microaerophilic streptococci or anaerobic cocci, usually in combination with *Staphylococcus aureus*, cause a progressive gangrene of the skin. The fact that anaerobic cocci can act synergically with other organisms is borne out by the work of Brook and his colleagues (Brook and Walker, 1984). Using a subcutaneous abscess model in mice, they showed that the presence of anaerobic cocci in the abscess significantly increased the size and severity of the abscess, which was initially produced by other aerobic and anaerobic bacteria.

Anaerobic cocci are not associated with specific or characteristic infections and they appear to be of limited pathogenicity. Indeed, their isolation from clinical samples is often difficult to interpret—pathogen or commensal? Several reviewers have noted that the most common species isolated from clinical material is *Pstr. anaerobius*. This may simply reflect the fact that this organism is a common constituent of the commensal flora, and not that it possesses specific pathogenic potential. Anaerobic cocci do not appear to possess specific virulence factors, although the demonstration of hyaluronidase production by oral peptostreptococci (Tam and Chan, 1985) is of interest. More work is required to assess the 'epidemiology' of different species of anaerobic cocci, together with detailed studies to look for virulence factors.

Antibiotic susceptibility of anaerobic cocci

The detailed susceptibility patterns of the obligately anaerobic cocci are discussed in Chapter 24. In general, they are much more susceptible to antibiotics than are the gram-negative anaerobes (Watt et al., 1979), the vast majority of strains being sensitive to the penicillins and the cephalosporins. (We have only encountered two strains that produce detectable amounts of β-lactamase.) All strains are sensitive to metronidazole (Watt and Jack, 1977), while many strains are resistant to tetracycline and to bicozamycin. However, there does not appear to be any correlation between susceptibility patterns and identification, and the use of 'antibiogram' patterns has proved to be of little or no value as an aid to the identification of these organisms.

Veillonella spp. have received even less attention than the gram-positive anaerobic cocci. They are less sensitive to penicillin but are otherwise rather antibiotic-sensitive (see Chapter 24).

Treatment of infections due to anaerobic cocci

Infections involving anaerobic cocci are usually polymicrobial in nature. Such infections are usually treated with a combination of metronidazole and a broad-spectrum antibiotic such as a cephalosporin, but there have been no adequately documented trials of antibiotic therapy in infections due to anaerobic cocci. It should be noted that a proportion of strains are resistant to the newer agents such as ciprofloxacin (Watt and Brown, 1986) and to newer cephalosporin derivatives (Fuss and Helsel, 1986).

Anaerobic cocci remain an enigma. Their identification, their role in human disease and their molecular biology and genetics remain rather obscure. Much more needs to be done to unravel the secrets of these interesting organisms.

References

Brook I, Walker R I. (1984), Pathogenicity of anaerobic gram-positive cocci. *Infection and Immunity* 45: 320–324.

Cato E P. (1983). Transfer of *Peptostreptococcus parvulus* (Weinberg, Nativella and Prévot 1937) Smith 1957 to the genus *Streptococcus: Streptococcus parvulus* (Weinberg, Nativella and Prévot 1937). Comb. nov. nom. rev., *International Journal of Systematic Bacteriology* 33: 82–84.

Delwiche E A, Pestka J J, Tortorello M L. (1985). The veillonellae: gram-negative cocci with a unique physiology. *Annual Review of Microbiology* 39: 175–193.

Dowell V R, Hawkins T M. (1974). *Laboratory methods in anaerobic bacteriology*. Communicable Disease Centre Laboratory Manual, Atlanta: H E W Publication No. (CDC) 77–8272.

Ezaki T, Yamamoto N, Ninomiya K, Suzuki S, Yabbuchi E. (1983). Transfer of *Peptococcus indolicus*, *Peptococcus asaccharolyticus*, *Peptococcus prevotii* and *Peptococcus magnus* to the genus *Peptostreptococcus* and proposal of *Peptostreptococcus tetradius* sp. nov. *International Journal of Systematic Bacteriology* 33: 683–698.

Finegold S M, Sugihara P T, Sutter V L. (1971). Use of selective media for isolation of anaerobes from humans. In *Isolation of anaerobes*, pp. 99–108. Edited by Shapton D A, Board R G. Society for Applied Bacteriology Technical Series No. 5. Academic Press, London.

Foubert E L, Douglas H C. (1948). Studies on the anaerobic micrococci. 1. Taxonomic considerations. *Journal of Bacteriology* 56: 25–34.

Fuss R J, Helsel V. (1986). In vitro activities of RO19–5247 and RO15–8074, new oral cephalosporins. *Antimicrobial Agents and Chemotherapy*, 30: 429–434.

Hare R, Wildy P, Billet F S, Twort D N. (1952). The anaerobic cocci: gas formation, fermentation reactions, sensitivity to antibiotics and sulphonamides. Classification. *Journal of Hygiene* 50: 295–319.

Harris J W, Brown J H. (1929). A clinical and bacteriological study of 113 cases of streptocccal puerperal infection. *Bulletin of Johns Hopkins Hospital* 44: 1–31.

Holdeman L V, Moore W E C. (1972). *Anaerobe laboratory manual*, 2nd ed. Anaerobe Laboratory, Virginia Polytechnic Institute and State University, Blacksburg.

Holdeman L V, Moore W E C. (1974). New genus *Coprococcus*. Twelve new species, and emended descriptions for four previously-described species of bacteria from human faeces. *International Journal of Systematic Bacteriology* 24: 260–277.

Holdeman L V, Cato E P, Moore W E C. (1977). *Anaerobe laboratory manual*, 4th edn. Anaerobe Laboratory. Virginia Polytechnic Institute and State University, Blacksburg.

Huss V A R, Festl H, Schleifer K H. (1984). Nucleic acid hybridization studies and deoxyribonucleic acid base compositions of anaerobic gram-positive cocci. *International Journal of Systematic Bacteriology* 34: 95–101.

Killper-Balz R, Schleifer K H. (1981). Transfer of *Peptococcus saccharolyticus* Foubert and Douglas to the genus *Staphylococcus: Staphylococcus saccharolyticus* (Foubert and Douglas) comb. nov. *Zentralblatt fur Bakteriologie, Parasitenkunde, InfectionsKrankheiten und Hygiene Abteilung* 2: 324–331.

Lambe D W, Vroon D H, Rietz C W. (1974). Infections due to anaerobic cocci. In *Anaerobic bacteria, role in disease*. pp. 585–599. Edited by Barlows A *et al.* Thomas, Springfield.

Lanigan G W. (1976). *Peptococcus heliotrinreducans* sp. nov. a cytochrome-producing anaerobe which metabolizes pyrrolizadime alkaloids. *Journal of General Microbiology* 94: 1–10.

Mays T D, Holdeman L V, Moore W E C, Rogosa M, Johnson J L. (1982). Taxonomy of the genus *Veillonella* Prévot. *International Journal of Systematic Bacteriology* 32: 28–36.

Meleney F L. (1931). Bacterial synergism in disease process with a confirmation of the synergistic bacterial aetiology of a certain type of progressive gangrene of the stomach wall. *Annals of Surgery* 19: 961–968.

Moore W E C, Cato E P, Moore L V H. (1985). Index of three bacterial and yeast nomenclatural changes published in the *International Journal of Systematic Bacteriology* since the 1980 Approved Lists of bacterial names (1 January 1980 to 1 January 1985). *International Journal of Systematic Bacteriology* 35: 382–407.

Neut C, Lesieur V, Ramond C, Beerens H. (1985). Analysis of gram-positive anaerobic cocci in oral, faecal and vaginal flora. *European Journal of Clinical Microbiology* 4: 435–437.

Parker M T. (1983). *Staphylococcus* and *Micrococcus*; the anaerobic cocci. In *Topley and Wilson's Principles of bacteriology, virology and immunity*, p. 239. Edited by Wilson G S *et al.* Edward Arnold, London.

Phelps R, Jacobs R A. (1985). Purulent pericarditis and mediastinitis. *Journal of the American Medical Association* 254: 947–948.

Pien F D, Thompson R L, Martin W J. (1972). Clinical and bacteriological studies of anaerobic cocci. *Mayo Clinic Proceedings* 47: 251–257.

Prévot A R. (1933). Etudes de systematiques bacterienne. I. Lois generales. II. Cocci anaerobies. *Annales des Science Naturelles Botanique* 15: 23–260.

Prévot A R, Fredette V. (1966). *Manual for the classification and identification of the anaerobic bacteria*. Lea and Febiger, Philadelphia.

Rogosa M. (1964). The genus *Veillonella* IV. General cultural, ecological and biochemical considerations. *Journal of Bacteriology* 87: 162–170.

Rogosa M. (1971). Peptococcaceae, a new family to include the gram-positive, anaerobic cocci of the genera *Peptococcus*, *Peptostreptococcus* and *Ruminococcus*. *International Journal of Systematic Bacteriology* 21: 234–237.

Sklavounos A, Legakis N J, Ioannidou H, Patrinkou A. (1986). Anaerobic bacteria in dentoalveolar abscesses. *International Journal of Oral and Maxillofacial Surgery* 15: 288–291.

Smith G L F, Cumming C G, Ross P W. (1984). Survival of gram-positive anaerobic cocci on swabs and their isolation from the mouth and vagina. *Journal of Clinical Pathology* 36: 93–98.

Stone M L. (1940). Studies on the anaerobic streptococcus 1. Certain biochemical and immunological properties of anaerobic streptococci. *Journal of Bacteriology*, 39: 559–582.

Tam Y C, Chan E C. (1985). Purification and characterisation of hyaluronidase from oral *Peptstreptococcus* species. *Infection and Immunity* 47: 508–513.

Taylor A L. (1929). The anaerobic streptococci. In *A system of bacteriology in relation to medicine*, vol 2. pp. 136–142. HMSO, London.

Taylor E. (1984). Taxonomy of asaccharolytic anaerobic gram-positive cocci. pp. 54–85. *PhD. Thesis*, University of London.

Thomas C G A, Hare R. (1954). The classification of anaerobic cocci and their isolation in normal human beings and pathological processes. *Journal of Clinical Pathology* 7: 300–304.

Topiel M S, Simon G L. (1986). Peptococcaceae bacteraemia. *Diagnostic Microbiology and Infectious Diseases* 4: 109–117.

Watt B, Jack E P. (1977). What are anaerobic cocci? *Journal of Medical Microbiology* 10: 461–468.

Watt B, Young O, McCurdy G. (1979). The susceptibility of anaerobic cocci from clinical samples to six antimicrobial agents. *Journal of Infection* 1: 143–149.

Watt B, Brown F V. (1983). A selective agent for anaerobic cocci. *Journal of Clinical Pathology* **36**: 605–606.

Watt B, Bushell A C, Wallace E T. (1984). Characterisation of anaerobic cocci in the diagnostic laboratory. *Journal of Clinical Pathology* **37**: 1197.

Watt B, Brown F V. (1986). Is ciprofloxacin active against clinically important anaerobes? *Journal of Antimicrobial Chemotherapy* **17**: 605–613.

Watt B, Wallace E T, Bushell A C. (1986). Characterisation of anaerobic cocci. In *Anaerobic bacteria in habitats other than man*, Society of Applied Bacteriology Symposium No. 13. Edited by Barnes E M, Mead G C. Blackwells, Oxford.

Wells C L, Field C R. (1976). Long-chain fatty acids of peptococci and peptostreptococci. *Journal of Clinical Microbiology* **4**: 515–521.

Wildy P, Hare R. (1953). The effect of fatty acids on the growth, metabolism and morphology of the anaerobic cocci. *Journal of General Microbiology* **9**: 216–225.

Wilson G S. (1983). *Veillonella*. In *Topley and Wilson's Principles of bacteriology, virology and immunity*, p. 479. Edited by Wilson G S *et al*. Edward Arnold, London.

Wren M W D, Eldon C P, Dakin G H. (1977). Novobiocin and the differentiation of peptococci and peptostreptococci. *Journal of Clinical Pathology* **30**: 620–622.

Wren M W D. (1978). A new selective medium for the isolation of non-sporing anaerobic bacteria from clinical specimens. *Medical Laboratory Sciences* **35**: 371–378.

Ziment I, Miller L G, Finegold S M. (1968). Nonsporulating anaerobic bacteria in osteomyelitis. *Antimicrobial Agents and Chemotherapy* 77.

8
The spirochaetes

M J Hudson

Morphology and ultrastructure
Ecology
Classification
   The order Spirochaetales
Isolation techniques
   Anaerobic culture
   Enrichment and isolation techniques
Characterization and typing methods
   Morphological and ultrastructural methods
   Serological methods
   Cultivation and metabolism
   Structural components
The treponematoses
   Diagnostic aspects
   Treatment
   Syphilis and Human Immunodeficiency Virus (HIV) coinfection
Spirochaetes in oral and periodontal diseases
   Spirochaetes and acute ulcerative gingivitis (AUG)
   Spirochaetes and the periodontal diseases
Intestinal spirochaetosis
   Historical perspective
   Prevalence studies
   Pathological aspects of intestinal spirochaetosis
Miscellaneous disease
   Multiple sclerosis
   Tropical ulcer
   Non-syphilitic genital lesions
Future developments
References

There is an intrinsic elegance in the symmetry of the spiral or helix, and micro-organisms with this morphology immediately attract the attention of the microscopist, even when present as minor components of a complex microbiota. Indeed, spiral microbes were amongst the first protists seen and described by the pioneer seventeenth century microscopist, Antonie van Leeuwenhoek, who

Table 8.1 Examples of non-spirochaetal spiral and helical bacteria

Genus	Brief description
*Spiroplasma**	Spiral mycoplasma
Methanospirillum	Spiral methanogen
Spirillum	Marine and freshwater spirilla
Aquaspirillum	Marine and freshwater spirilla
Rhodospirillum	Photosynthetic spirilla
*Anaerobiospirillum**	Digestive tract anaerobes
*Clostridium spireforme**	Toxigenic intestinal clostridium
*Helicobacter pylori**	Gastric spiral of man
Selenomonas	Digestive tract anaerobes

*Includes species pathogenic for man, animals or plants

observed motile, spiral 'little animalcules' in pond water, dental plaque and in diarrhoeic faeces (Dobell, 1932).

Species with spiral or helical morphology occur in various genera, both gram-positive and gram-negative (Table 8.1). However, the presence of internal (periplasmic) flagella enclosed within an outer sheath serves to distinguish the spirochaetes from other spiral-shaped bacteria (Holt, 1978). The general term spirochaetes is used hereafter to include the thin, flexuous, spiral- or helical-shaped, motile, gram-negative microbes with internal flagella and chemoheterotrophic metabolism.

The classic spirochaetal diseases of man are:
1. Syphilis and the endemic treponematoses (yaws, pinta, bejel or non-venereal syphilis), caused by *Treponema pallidum* and closely related species;
2. Tick-borne and louse-borne relapsing fevers, caused by some 13 species of *Borrelia*;
3. Lyme disease and erythema chronicum migrans (ECM), caused by *Borrelia burgdorferi* and transmitted by tick-bite;
4. Leptospirosis and Weil's disease, caused by *Leptospira interrogans*.

There are few other human infections directly attributable to spirochaetes although it has long been recognized that the spirochaetes, as part of the normal microflora, may play a significant role in opportunist and mixed infections, such as in Vincent's disease.

The rekindled interest in the spirochaetes is largely a result of the application of the new techniques of molecular biology to the study of the pathogenic spirochaetes, and also of the discovery of 'new' spirochaetal diseases, such as Lyme disease in man and swine dysentery in pigs. These have occurred in parallel with recent advances in taxonomic techniques, and interest in the ecology, physiology and metabolism of the free-living and host-associated spirochaetes.

Morphology and ultrastructure

The electron microscope has greatly assisted the study of the spirochaetes, both in differentiating them from other bacteria and as a tool of taxonomy. There is considerable variation in the number and arrangement of the internal flagella, and this has taxonomic importance. It is appropriate that these structures be referred to as internal flagella, since they are composed of flagellin-like proteins and arise from insertion structures similar to those of typical external flagella.

The major features of a generalized spirochaete are shown in Fig. 8.1. The protoplasmic cylinder encloses the cytoplasm, containing ribosomes, the bacterial chromosome, and (in some species) longitudinal arrays of cytoplasmic tubules. The internal flagella or periplasmic fibrils (synonyms: endoflagella, axial filaments, axial fibrils or axiostyle) are inserted subterminally at each end into basal anchorage structures similar to those of 'normal' flagella. The internal flagella wind around the spiral protoplasmic cylinder and, except in the leptospires, overlap and interdigitate in the middle of the cell. The protoplasmic cylinder and flagella are enclosed within an outer sheath (or cell envelope). The basic morphology of spirochaetes is helical, with spiral ends and supercoils; the shape is a function of several factors, including peptidoglycan structure and the action of the internal flagella.

The number of flagella arising from each end, a feature of taxonomic significance, is commonly recorded in shorthand—one flagellum arising from each end and overlapping in the middle of the cell is referred to as a 1-2-1 arrangement and five arising from each end and overlapping as 5-10-5. Alternatively, the total number of flagella present in the spirochaete is recorded.

SPIROCHAETES: SCHEMATIC ANATOMY

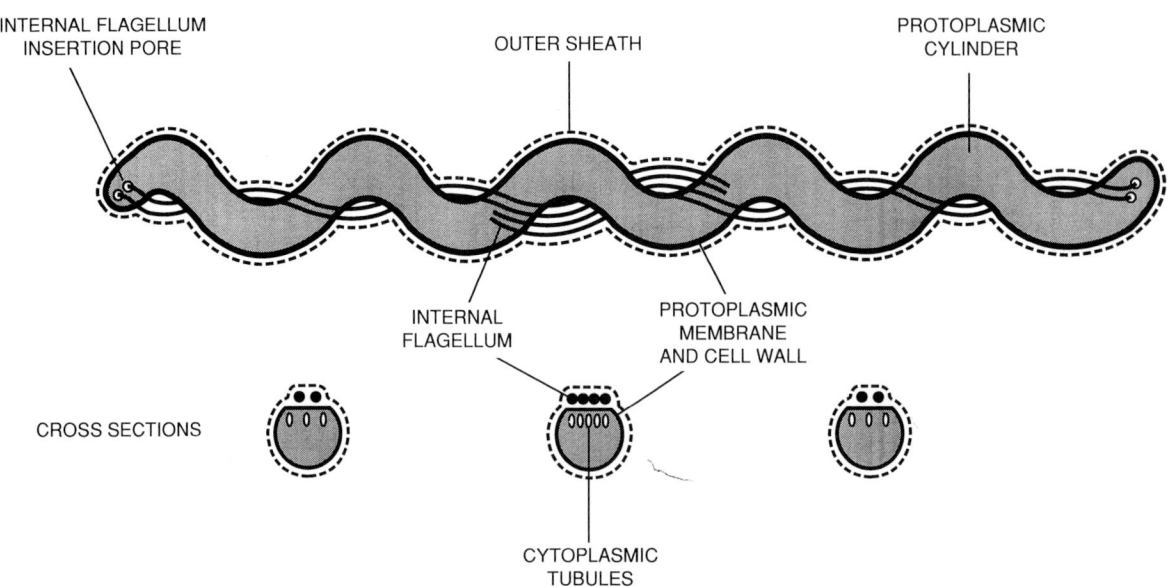

Fig. 8.1 Simplified diagrammatic representation of a spirochaete

The precise mechanism(s) of spirochaetal motility is unclear (Canale-Parola, 1978). They lack external flagella but are capable of motility in free solution and this motility is enhanced in viscous media. Motility can be flexuous, corkscrew or serpentine, darting or languid; some spirochaetes may also show a gliding motility when in contact with surfaces. Intact internal flagella, but not cytoplasmic tubules, are an absolute requirement for motility.

Cytoplasmic tubules or fibrils of approximately 7 nm diameter are present in most treponemes, in *Spirochaeta zuelzerae* and in *Leptonema (Leptospira) illini* (Hovind-Hougen, 1976; Holt, 1978). Their function is unknown, but it is unlikely that they play a significant role in either motility or morphology since many spirochaetes lack them. These structures are quite distinct from the much larger (21–24 nm diameter) cytoplasmic microtubules observed in large, non-cultivable spirochaetes and gliding bacteria found in the hindguts of termites and other xylophagous insects (Margulis *et al.*, 1981). These resemble eukaryote microtubules in size and composition, containing immunoreactive tubulin but not actin (Margulis *et al.*, 1978). In contrast, the 7 nm cytoplasmic tubules of treponemes are composed of neither tubulin nor actin (Margulis *et al.*, 1978; Paster and Kaine, 1984).

Ecology

The spirochaetes are found in diverse habitats and ecology is used as an important criterion in spirochaete taxonomy (Harwood and Canale-Parola, 1984). Some are strictly free-living species that inhabit fresh and salt waters, muds and even hot springs. Most treponemes are host-associated commensals of man and various animals, living in the digestive tract and on mucosal surfaces; some may be opportunist pathogens. Other spirochaetes are commensals of invertebrates such as molluscs, insects and protozoa. However, some spirochaetes are obligate pathogenic parasites of man and animals, e.g., the syphilis treponeme. Other human and animal pathogens follow a more complex ecology involving intermediate hosts and vectors; man may be an obligatory intermediate in the life cycle, as in the relapsing fevers, or may be accidentally infected, as in Lyme Disease.

Classification

Morphology is often used as a basic criterion of microbial systematics; therefore, the spirochaetes have always been classified separately from other groups. This morphological approach to taxonomy was later refined so that the spirochaetes include only those microbes with internal flagella. However, given the diversity of ecologies, phenotypes and wide range of DNA base compositions (G+C 25–65 mol per cent) this is clearly an heterogenous group of microbes with a shared characteristic. Does this diversity arise from some protospirochaete ancestor or from parallel evolution of various microorganisms?

The comparison of ribosomal RNA (rRNA) oligonucleotide catalogues and sequence data is arguably the most accurate phylogenetic technique currently available (Woese, 1987), and this technique is as appropriate for the study of high-level taxa (kingdoms) as for low-level taxa (species and genera). Such studies have been greatly facilitated by recent advances in the rapid sequencing of nucleic acids and statistical comparison of primary and secondary sequences. In such studies the spirochaetes cluster together as one of ten distinct eubacterial phyla defined from rRNA data (Paster et al., 1984; Woese et al., 1985). Within the spirochaete group there are clear sub-groups (branches) approximating to genera previously described on the basis of phenotypic data and DNA-DNA homologies. It is rare for classic phenotypic and modern phylogenetic groupings to correspond so well.

Thus, the spirochaetes constitute a distinct and relatively deep (ancient) evolutionary branch of eubacterial prokaryotes, presumably arising from some common ancestor. They comprise the order Spirochaetales, although a higher taxon (e.g., phylum or division) may be more appropriate (Woese et al., 1985).

The order Spirochaetales

The families Leptospiraceae and Spirochaetaceae are the major subdivisions of the order (Table 8.2). DNA-DNA hybridization and rRNA sequence data have shown the leptospires to be only distantly related to other spirochaetes, supporting their separation at this level (Johnson and Faine, 1984; Paster et al., 1984). Ultrastructural and chemical studies show major differences between the families, particularly in cell wall composition and

Table 8.2 The taxonomy of the spirochaetes

Order	Family	Genera
Spirochaetales	Leptospiraceae	Leptospira, Leptonema*
	Spirochaetaceae	Spirochaeta, Cristispira, Borrelia, Treponema, Brachyspira*

The pillotinas (large spirochaetes from termite gut) are not included in this simplified taxonomic scheme
*Genera not included in Bergey's Manual of Systematic Bacteriology

Table 8.3 Major chemical and ultrastructural differences between the families Leptospiraceae and Spirochaetaceae*

Character	Leptospiraceae	Spirochaetaceae
Peptidoglycan dibasic amino acid	Diaminopimelic acid	Ornithine
Cell wall 4-O-methylmannose	Yes	No
Flagella at each end	Single	Multiple (single)†
Sheathed flagella	Yes	Yes (no)‡
Flagella overlapping	No	Yes
Flagella insertion body	Gram-negative type	Gram-positive type

*Data from Holt (1978); Yanagihara et al. (1984)
† Most Spirochaeta have a single flagellum at each end
‡ The flagella of borrelia are not sheathed

arrangement of internal flagella (Table 8.3). Classification within the Spirochaetaceae is complex and unsatisfactory. There are five genera and all are defined on the basis of type-strains which have yet to be cultured *in vitro*; two (*Spirochaeta* and *Treponema*) include species with a very wide range of DNA base composition. Fortunately, cultivability is not a prerequisite for phylogenetic analyses of rRNA sequences generated by cloning strategies or direct primer-initiated reverse transcription, and such techniques should help to clarify many of the uncertainties in spirochaete taxonomy.

Those obligately anaerobic spirochaetes implicated in human disease fall within the genera *Treponema* and *Brachyspira*. However, it is useful to review the general properties and ecology of all of the spirochaetes.

Little is known of the pillotinas—large, non-cultivable and ultrastructurally complex spirochaetes found as commensals in the hindgut of termites and xylophagous insects (Margulis *et al.*, 1981) and symbiotes of protozoa found therein. These may constitute a third family, Pillotaceae, with several genera, including *Hollandia*, *Pillotina*, *Diplocalyx* and *Clevelandina*.

The family Leptospiraceae

The leptospires are obligate aerobes which use long chain fatty acids or alcohols as carbon and energy sources. With the exception of a single species, they comprise a relatively well defined taxon with a DNA G+C content of 35–41 mol per cent (Johnson and Faine, 1984). Cells are 0.1 μm in diameter by 6–12 μm in length, tightly coiled with a hooked end. The two internal flagella, inserted one at each end, rarely overlap. Two species are recognized—*L. interrogans* (pathogenic strains) and *L. biflexa* (free-living, non-pathogenic strains). The two species are further subdivided into serogroups and serovars. Differentiation of the species is of great importance in human and veterinary medicine and can be made using the relatively few simple tests which are summarized in Table 8.4.

The outlier in the group and '*species incertae sedis*' is *L. illini*. It broadly resembles *L. biflexa* but differs in being able to grow in serum-free media, has 7 nm cytoplasmic tubules and a DNA G+C of 51–53 per cent. Furthermore, its internal flagella are inserted into basal bodies of gram-positive type, in contrast to the gram-negative type found in the true leptospires. 16S rRNA data (Paster *et al.*, 1984) support the proposed transfer of this species to a second genus within the family Leptospiraceae, as *Leptonema illini*, comb. nov. (Hovind-Hougen, 1979).

Table 8.4 Differential characteristics of pathogenic and non-pathogenic species of *Leptospira*

Character	*L. interrogans*	*L. biflexa*
Pathogenicity	+	−
Growth at 13°C	−	+
Growth on C14.0	−	+
8-azaguanine (225mg/l)	S	R
Lipase	+/−	+
DNA G+C mol%	35–40	38–41

The family Spirochaetaceae

This family includes four major genera of obligate or facultatively anaerobic and microaerophilic spirochaetes (Canale-Parola, 1984a). Many species, indeed all type species and even an entire genus are non-cultivable. Those members of the family available for study use either carbohydrates or amino acids as carbon and energy sources. L-Ornithine is the cross-linking dibasic amino acid present in the peptidoglycan of this family.

The genus *Spirochaeta*

Spirochaeta spp. are free-living, obligate or facultatively anaerobic spirochaetes found widely distributed in marine and fresh waters, muds and sediments (Canale-Parola, 1984b). Most of the species have a single internal flagellum inserted at each end, overlapping centrally (1-2-1 arrangement). However, the type strain, *S. plicatilis*, has not been grown in pure culture and is atypically large (0.75 μm by 80–250 μm) and has a bundle of 18–20 flagella arising from each end; most *Spirochaeta* spp. are 0.2–0.5 μm by 5–40 μm in length. DNA G+C contents vary between 51 and 65 mol per cent (that of *S. plicatilis* is not known).

A simplified classification of the genus *Spirochaeta* based on oxygen requirement (Canale-Parola, 1984b) is not supported by 16S rRNA oligonucleotide cataloguing (Paster *et al.*, 1984). Two obligately anaerobic species (*S. stenostrepta* and *S. zuelzerae*) group with the treponemes whereas two halotolerant obligately anaerobic species (*S. litoralis* and *S. isovalerica*) are more closely affiliated to the facultative spirochaeta. The oxygen requirement of *S. plicatilis* is not known.

The genus *Cristispira*

The cristispires are very large (0.5–3 μm by 30–180 μm) commensal spirochaetes found in the crystalline style and digestive tract of freshwater and marine molluscs. They have not been cultivated *in vitro*. Their characteristic feature, other than size, is a bundle of more than 100 internal flagella, inserted at each end, which can form a distinct ridge (crista) within the loose outer membrane (Kuhn, 1981; Breznak, 1984). Microaerobic or anaerobic conditions are required for sustained motility *in vitro*. The genus is represented by the single species *Cristispira pectinis*; the DNA base content is not known. It is possible that the pillotinas and cristispires are related.

The genus *Borrelia*

In the past, spirochaetes were classified on the basis of their visualization with common aniline dyes (including Gram's stain); the larger, easily-stained spirochaetes seen in relapsing fever and Vincent's disease (*B. vincentii*) were classed as borrelia; this practice is still in common use. As presently defined (Kelly, 1984) the genus is restricted to arthropod-borne pathogenic spirochaetes. Size varies from 0.2 to 0.5 μm wide by 3 to 20 μm in length; 15–20 internal flagella are inserted at each end, overlapping in the middle. Some borrelia, but not the type species, *B. anserina*, have been cultivated *in vitro* and are microaerophiles.

Interest in the borrelias has been rekindled with the application of the new techniques of molecular biology to the study of antigenic shift in the relapsing fever borrelias (Barbour and Hayes, 1986) and the discovery of the agent of erythema chronicum migrans (ECM) and Lyme disease, *B. burgdorferi*.

Some 16 species of borrelia are described and are all agents of tick-borne and louse-borne relapsing fevers and other infections of man, animals and birds. Species are commonly named according to the arthropod vector and probably reflect geographic distribution of the vector rather than true genotypic species (Barbour and Hayes, 1986); a more reliable method for speciation is required. Ribosomal RNA studies and a relatively narrow range of DNA G+C content (27–32 mol per cent) confirm the borrelia as a distinct branch within the main spirochaete group (Paster *et al.*, 1984).

The genus *Treponema*

Treponemes are commensal, pathogenic spirochaetes found in the digestive tract and genital mucosae of man and animals (Smibert, 1984). Cell size varies from 0.1 to 0.4 μm wide by 5 to 20 μm in length. Cultivable species are obligate anaerobes. The DNA G+C content ranges from 25 to 54 mol per cent. Species vary in their ability to use carbohydrate or amino acids as carbon and energy sources, and also in their growth requirements, such as long or short chain fatty acids and cofactors.

The type species for the genus, *T. pallidum*, is not cultivable *in vitro*, but can be grown in coculture with mammalian cells.

Phylogenetic relationships determined from rRNA analysis (Paster *et al.*, 1984) show that the majority of treponemes form a group well separated from the spirochaetes, leptospires and borrelias. An important exception is *T. hyodysenteriae*, which is only distantly related to other spirochaete taxa, including the main group of treponemes. Two free-living spirochaetes (*S. stenostrepta* and *S. zuelzerae*) group with the treponemes.

The *Treponema pallidum* 'complex'

The *T. pallidum* complex describes several very closely related spirochaetes which cause specific cutaneous and systemic disease in man, non-human primates and mammals. They are variously allocated to subspecies or species level, but they are genetically indistinguishable and, for practical purposes, may be regarded as a single genospecies.

T. pallidum is the type species for the genus but is unrepresentative in many respects, notably its high DNA G+C content (52–54 mol per cent) and its microaerobicity. Numerous studies of *T. pallidum ss. pallidum* and *T. pallidum ss. pertenue* have shown that motility, infectivity, protein synthesis and metabolism all require oxygen at low concentration. Cytochromes b and c are present and molecular

Table 8.5 General characteristics of the *Treponema pallidum* complex and their associated diseases

Species (disease)	Disease type		Sexual transmission	Geographic distribution
	systemic	cutaneous		
T. pallidum ss. *pallidum* (venereal syphilis)	+	−	+	Worldwide
T. pallidum ss. *pertenue* (yaws, frambesia)	−	+	−	Tropics
T. pallidum ss. *endemicum* (endemic/non-venereal syphilis, bejel)	−	+	−	Old-world
T. carateum (pinta, carate)	−	+	−	New-world
T. paraluiscuniculi (rabbit syphilis)	−	+	+	NR

NR, Not relevant

oxygen is required for the uptake and metabolism of glucose and pyruvate. The internal flagellar arrangement is 3-6-3 and a bundle of cytoplasmic tubules is present within the cell. *T. pallidum* is typically tightly coiled and thinner than other treponemes. It is 0.1–0.2 μm in diameter by 5–15 μm in length; the longer forms are probably formed prior to binary transverse fission.

In the absence of species- and subspecies-specific markers the differentiation of the non-cultivable pathogenic treponemes is extremely difficult and relies on the ability to infect particular laboratory animals experimentally and produce pathological changes and subtle differences in the distribution of lesions.

Members of the *T. pallidum* complex have evolved characteristic modes of infectivity and virulence. Phylogenic classifications have been proposed based on historical and geographical occurences of the diseases and their clinical manifestations (Table 8.5) (Rothwell, 1981).

In most experimental studies the Nichols strain of *T. pallidum*, which was first isolated in 1912 and maintained by intratesticular passage in *T. paraluiscuniculi*-free rabbits, has been used.

The cultivable treponemes

The remaining treponemes are obligately anaerobic host-associated spirochaetes inhabiting the digestive tract (mouth, rumen, intestine) and genital mucosae

Table 8.6 The cultivable treponemes

CHO-fermenting	Site	DNA G+C mol%	Non-CHO-fermenting	Site	DNA G+C mol%
T. phagedenis	G	38–39	*T. denticola*	O	37–38
T. succinifaciens	I	36	*T. vincentii*	O	44
T. bryantii	R	36	*T. scoliodontum*	O	NA
T. hyodysenteriae	I	26	*T. refringens*	G	39–43
T. innocens	I	26	*T. minutum*	G	37
T. macrodentium	O	39			
T. orale	O	37			
*T. socranskii**	O	50			
*T. pectinovorum**	O	39			
*T. saccharophilum**	R	54			

G, genital; I, intestinal; O, oral; R, rumen; NA, data not available
*Species not included in *Bergey's Manual of Systematic Bacteriology*

of man and various animals. Many such spirochaetes have been observed but not yet cultured; it is assumed that these are oxygen-sensitive or nutritionally fastidious species. The cultivable treponemes fall into two categories, defined according to the preferred carbon source (Table 8.6). This list is by no means exhaustive, as many treponemes have been isolated from various sites and animal species but not formally classified. Groupings are most often made on the basis of size, morphology and the number and arrangement of internal flagella.

The treponemes are speciated on the basis of size, biochemical reactions, tolerance tests and determination of metabolic end products. The reactions of the species are found in standard textbooks (Holdeman et al., 1977; Smibert, 1984) and relevant publications of new species (Smibert and Burmeister, 1983; Smibert et al., 1984).

Table 8.6 shows three distinct groups based on DNA G+C content: the 'high' G+C group (50–54 mol per cent) which includes *T. socranskii*, *T. saccharophilum* and *T. pallidum*; a 'very low' G+C group (of about 26 mol per cent), which includes *T. hyodysenteriae* and *T. innocens*; and the remaining named species spanning the range 36–44 mol per cent G+C. It is clear that there are at least three new or redefined genera here.

Apart from genetic differences, the pig intestinal pathogen *T. hyodysenteriae* is also unusual in its requirement for and metabolism of cholesterol (Stanton, 1987) and low levels of oxygen (Stanton, 1989). The ability to tolerate and indeed to utilize low levels of oxygen could be an important pathogenic mechanism for tissue invasion, and has intriguing parallels with other microaerobic spirochaete pathogens such as *T. pallidum* and the borrelias.

Genetic studies comparing the large intestinal treponemes of man with the morphologically similar pig organisms, *T. hyodysenteriae* and *T. innocens*, show that they also fall into the same low G+C content group (Coene et al., 1989). On the basis of DNA-DNA hybridization data, many of the human strains and both pig species belong to a single species (>70 per cent homology) with, perhaps, a separate subspecies group of human strains (<50 per cent homology). *T. innocens* (Kinyon and Harris, 1979) was defined on the basis of studies with relatively few strains of non-enteropathogenic, weakly β-haemolytic intestinal treponemes from pig and dog. The human strains are also weakly β-haemolytic. With rare exceptions, *T. hyodysenteriae* is strongly β-haemolytic and this trait is stable even after attenuation by extensive laboratory passage (Taylor et al., 1980).

Given such a high degree of genetic relatedness within this group of intestinal spirochaetes, the species *T. hyodysenteriae* and *T. innocens* will have to be carefully redefined with a view to combining genetic, phenotypic and ecological data.

The genus *Brachyspira*

This genus was proposed by Hovind-Hougen et al. (1982) to describe a single species, *Brachyspira aalborgi*, a slow-growing anaerobic spirochaete found in the large intestine of man. Cells are 0.2 μm wide by 1.7–6 μm long and have four flagella inserted at each end, overlapping at the middle (4-8-4 arrangement). In contrast to the treponemes, B. aalborgi lacks cytoplasmic tubules and, by analogy with *Lept. illini* (see page 112), this appears to be an important taxonomic feature. Neither the DNA G+C content nor any nucleic acid hybridization data have been reported.

Isolation techniques

Anaerobic culture

The cultivable treponemes are all obligate anaerobes and many are extremely oxygen-sensitive. The isolation and maintenance of treponemes demands care, perseverance and good anaerobic technique. They all grow slowly and many are also nutritionally fastidious. Sensitivity to oxygen is usually most evident upon initial isolation; strains often develop a degree of oxygen-tolerance on subculture. Smibert (1981) has summarized many of the media and isolation techniques necessary for their growth.

The Hungate technique

The best methods for the culture of obligate anaerobes are based on the techniques described by Hungate (1950). With the exception of high-integrity anaerobic chambers, these are the only reliable methods for the cultivation of the exquisitely oxygen-sensitive methanogens.

The Hungate technique is conceptually simple. Media are prepared, inoculated and incubated under a stream or head-space of a dense, deoxygenated gas (nitrogen, carbon dioxide or argon) thus ensuring that air is always excluded and media remain highly reduced. Growth continues unchecked through each passage and can be monitored without disturbing anaerobiosis. However, the method is manipulatively difficult, although it can be quickly mastered; the major drawback is that media preparation is labour-intensive. The many ingenious adaptations of the basic methods permit the use of such techniques as 'streaking out' and the use of opaque media (Bryant, 1972; Macy et al., 1972; Holdeman et al., 1977; Mitsuoka, 1980).

Anaerobic chambers

The major advantage of anaerobic chambers (or glove-boxes) is that most conventional 'bench' techniques can be used inside them with little modification. There is a wide choice of excellent chambers available from commercial sources, including high-integrity (and expensive) chambers which guarantee oxygen levels of <1ppm. Most commercial chambers double as incubators and some offer a 'bare-hands' facility to improve comfort; these have not been properly assessed for suitability when working with extremely oxygen-sensitive anaerobes, for which oxygen-impervious gloves are used.

For the culture of oxygen-sensitive treponemes, media must be prepared, inoculated and incubated entirely within the anaerobic chamber. It is important to store plastic ware in the anaerobic chamber for at least 24 h before use to remove adsorbed oxygen.

Bench anaerobic techniques

The less oxygen-sensitive treponemes, such as the large intestinal treponemes, can be isolated and subcultured by good technique on the open bench and by reliable anaerobic transfer. The sensitivity and reproducibility of such methods are maximized by attention to detail and by minimizing exposure of cultures to air and oxidized medium. Media should be freshly prepared or stored anaerobically before use; if possible, plates should not be surface dried before inoculation and the anaerobic atmosphere should be kept humid. Colonial growth is favoured by low gelling concentrations of agar and growth in broth is often improved by the addition of a small amount of agar.

The anaerobic atmosphere in the jar should contain sufficient hydrogen to maintain anaerobiosis (in conjunction with an active deoxygenating catalyst) and CO_2 5–10 per cent to improve growth; this can be achieved by either evacuation–replacement systems or gas-generating kits; the latter have the advantages of simplicity and of maintaining a humid atmosphere.

Enrichment and isolation techniques

Media

The isolation of many treponemes is facilitated by the use of habitat-simulating media, devised to mimic the natural nutritional environment of the organism. The habitat of free-living intestinal treponemes differs in many respects from that of tissue-associated species; whereas blood or serum may be required for the growth of the latter, species resident in the gut lumen usually form part of a complex food chain and are adapted to growth with only low levels of protein nitrogen, various complex carbohydrates and end products of microbial metabolism such as short chain fatty acids. Habitat-simulating media tend to be relatively simple basal media, buffered with bicarbonate, and containing vitamins and minerals, a mixture of complex carbohydrates and a little yeast extract and peptone. This is supplemented with short chain fatty acids, serum or sterilized liquor prepared from the habitat under investigation, e.g., faeces, rumen fluid or digestor waste.

Human intestinal treponemes, like *T. hyodysenteriae* and the weakly haemolytic non-pathogenic large treponemes of the pig intestine, grow well on most types of nutrient agar supplemented with 5–10 per cent blood or serum. While Columbia, tryptone soya and brain–heart infusion agars all support good growth of these spirochaetes, earlier and markedly more luxuriant growth may be obtained with media specifically formulated for anaerobes, such as Fastidious Anaerobe Agar (FAA, Lab M). However, since the enhanced surface growth obtained with the latter medium tends to obscure smaller colonial variants and slower-growing species, for routine isolation, a combination of media based on FAA and tryptone soya agars, should be used.

Improved media for the culture of spirochaetes will result from the identification of specific nutritional requirements. Suzuki and Loesche (1989) have shown that several purified serum proteins, particularly ceruloplasmin, may be substituted for whole serum in media for *T. denticola* and *T. vincentii*. Some oral and rumen spirochaetes have very specific requirements for a carbon source, e.g., pectin and glucuronic acids in the human oral species *T. pectinovorum*. In continuous culture the growth of the treponeme component of dental plaque microflora is dependent upon the presence of mucin glycoproteins (Glenister *et al.*, 1988); it is not clear whether this growth stimulation is attributable to native or degraded mucin, or to cross-feeding by other bacteria.

Physical methods

Two attributes of treponemes can be used to aid their isolation from complex microflora—their flexuous and narrow cell size and their motility. Slender spirochaetes can be partially purified (with other small micro-organisms) by filtration through a single or graded membrane filter, with either positive or negative pressure. The filtrate is then cultured as above. In the well-in-plate method, an inoculum is placed in a shallow well cut into the agar medium; spirochaetes and other small motile micro-organisms (e.g., anaerobic curved rods) migrate through the gel, away from the inoculum, presumably by chemotaxis along a nutrient gradient. In a modification of this phenomenon, the inoculum is placed on the surface of a sterile membrane filter (pore sizes between 0.1 and 0.65 µm) lying on the agar. The inoculum is contained within a well constructed from a nylon ring on a seal of paraffin oil; for small inocula, the ring of paraffin oil alone is sufficient. Alternatively, a membrane filter with a hydrophobic margin may be used. The spirochaetes migrate through the membrane pores, into and then through the agar, usually beyond the periphery of the filter.

In either technique a plug of agar is taken from the margin of growth, examined for purity by microscopy and subjected to a repeat procedure for purification or cloned by streaking or serial dilution. For most rapid migration of the spirochaete front through the gel, low agar concentrations should be used. The type of spirochaete isolated will also depend on the pore size of the chosen membrane filter; larger pore sizes (>0.45 µm) are less selective for spirochaetes but may be necessary depending on the species sought.

A drawback of these physical methods for isolation is that they are not particularly effective for the selective enrichment and recovery of the larger spirochaetes.

Selective culture

There are several antibiotics with little activity towards treponemes which can be used to reduce the growth of competing flora. These include polymyxin B, nalidixic acid, cycloserine, and spectinomycin; the MIC of each is >500 mg/l. Spirochaetes also appear to have an innate resistance to rifampicin and the inclusion of up to 50 mg/l (but usually 1–2 mg/l) in media greatly facilitates the isolation of spirochaetes from mixed cultures (Stanton and Canale-Parola, 1979). Porcine and human intestinal spirochaetes are commonly isolated on media with spectinomycin 400 mg/l as the selective agent (Songer *et al.*, 1976), with or without polymyxin B 5–80 mg/l. This medium can be incubated at 37°C for the selective isolation of *T. innocens* and *B. aalborgi*, but the sensitivity and selectivity for *T. hyodysenteriae* are improved by incubation at 42–43°C. Polymyxin B has been used as the sole selective agent for the isolation of murine caecal spirochaetes (Lee and Phillips, 1978).

Interaction and growth of T. pallidum with cell cultures

T. pallidum can be grown in coculture with mammalian cells, although the treponemes rapidly lose motility and virulence in cell culture medium alone. *T. pallidum* attaches to various cell lines cultured in chemically reduced tissue culture media at 33–36°C and under microaerobic atmospheres (Fitzgerald, 1981). Although multiplication and retention of motility by the treponemes has been demonstrated independently by several groups of workers, neither multiple serial passage nor prolonged culture has been achieved (Fieldsteel et al., 1982; Norris and Edmondson, 1986).

The culture of *T. pallidum in vitro* requires high levels of serum in the growth medium and undefined growth factor(s) present in testis extract. Typical lag, exponential and stationary phases of growth and multiplication can be demonstrated in such coculture systems (Table 8.7) but the factors required for active growth have yet to be elucidated.

The concentrations of glucose, dissolved oxygen and reduced dithiothreitol all fall in the culture medium; there is a small fall in pH and a small rise in Eh. The cell line achieves confluency after about one week. Whether the physicochemical variables influence the metabolism of the treponemes or of the cell line, or both, is not known. Continuous culture of the feeder cells may permit prolonged and higher yields of viable, virulent treponemes.

Characterization and typing methods

Morphological and ultrastructural methods

Microscopy has always been of fundamental importance in the study of the spirochaetes and it is the routine diagnostic technique for the differentiation of *T. pallidum* ss. *pallidum* from commensal treponemes in exudates from genital ulcers. Cell shape, motility (and therefore viability) and approximate dimensions can all be gleaned from examination of wet films by dark-field, phase-contrast or differential interference-contrast microscopy. Silver stains can be used, but only the larger spirochetes stain well with aniline dyes. Jennings et al. (1983) found that spirochaetes bound avidin strongly and so could be visualized by an avidin-biotinylated peroxidase staining technique. Electron microscopy yields more detailed morphological data including accurate cell dimensions, the number and arrangement of internal flagella and the presence of cytoplasmic fibrils and surface arrays (Hovind-Hougen, 1976).

Serological methods

The spirochaetes share several common antigens, but can often be readily differentiated with absorbed polyclonal sera (Meyer and Hunter, 1967). This is the basis of the FTA-ABS (Fluorescent Treponemal Antibody-Absorbed) test; the specificity of anti-*T. pallidum* antiserum is increased by prior absorbtion with *T. phagedenis* (Reiter).

Western blotting with polyclonal sera has been used to demonstrate the extent of shared antigens among the *T. pallidum* complex (Baker-Zander and Lukehart, 1984). Monoclonal antibodies have been selected for their high specificity (Simonson et al., 1988) although they could equally be selected for genus-specific epitopes. Chatfield et al. (1988) have used western blotting to identify the major specific antigens of *T. hyodysenteriae*. Tall and Nauman (1986) used a microagglutination test (as used for leptospires) to differentiate *T. denticola*, *T. pectinovorum* and *T. vincentii*. By this method, no antigens shared between the species were identified; however, both *T. denticola* and *T. vincentii* showed marked heterogeneity.

Cultivation and metabolism

Most studies of the cultivable treponemes have compared species on the basis of growth characteristics *in vitro*, the fermentation or utilization of various

Table 8.7 Coculture of *Treponema pallidum* and Sf1Ep cells*

Days incubation	Number of T. pallidum per culture	Doubling time (h)	Motility (%)	Sf1Ep cells per culture
0	5.2×10^6	—	100	1.5×10^5
3	11.0×10^6	67	90	3.0×10^5
6	33.1×10^6	45	96	7.3×10^5
9	111.0×10^6	42	97	8.8×10^5†
12	132.0×10^6	288	73	8.3×10^5†

*Data from Norris and Edmondson, 1986
†Conflent growth

carbohydrates and amino acids, and the end products of metabolism. This is not only of taxonomic interest; the type of metabolism and the end products may aid our understanding of the micro-ecology of the spirochaetes. Tables of the fermentation reactions and end products of the cultivable treponemes are given by Smibert (1984).

Since many treponemes grow poorly or slowly in broth media, identifications based on fermentation reactions can be both time consuming and inconclusive. It is now common to assay cultures of fastidious micro-organisms, such as spirochaetes, for preformed enzyme activity against a battery of selected chromogenic substrates. The obvious advantage is speed and ease of assay (4 h, aerobic incubation) but the technique requires a good growth of cells for the standardized inoculum and the range of tests is a compromise and may be poorly discriminatory.

Hunter and Wood (1979) and Taylor *et al.* (1980) have used API-ZYM test strips as an aid in the differentiation of *T. hyodysenteriae* and *T. innocens*. Human intestinal spirochaetes resembling these pig species have also been characterized in the same way (Hovind-Hougen, *et al.*, 1982; Tompkins, *et al.*, 1986; Coene *et al.*, 1989). Oral treponemes can also be identified to species level by this approach. Loesche and his colleagues have investigated the use of both API-ZYM and the RapID-ANA system (Laughon *et al.*, 1982; Syed *et al.*, 1988). This latter system includes several tests, not present in the API-ZYM gallery, which subdivide strains of *T. denticola*.

Structural components

Proteins

The comparison of electrophoretic profiles of soluble microbial proteins is a rapid and highly discriminatory technique for microbial identification and differentiation (Moore *et al.*, 1980). Sodium dodecyl 1-1 sulphate-polyacrylamide gel electrophoresis (SDS-PAGE) of soluble proteins from *T. denticola* and *T. pectinovorum* (Tall and Nauman, 1986) and *T. hyodysenteriae* (Chatfield *et al.*, 1988) yielded protein profiles which are highly conserved but characteristic for each species, with relatively minor differences between strains in the intensity or molecular weight of some polypeptides. In contrast, strains of *T. vincentii* (Tall and Nauman, 1986) and *T. innocens* (Chatfield *et al.*, 1988) were far more heterogenous. Whether such variation is indicative of strain, sub-species or species differences can be answered only by nucleic acid hybridization data.

Lipids and cellular fatty acids

Treponemes, like all spirochaetes, are lipid-rich organisms (15–20 per cent by weight) and the distribution of polar and neutral lipids can be used to differentiate families and genera (Livermore and Johnson, 1974; Matthews *et al.*, 1979). The major phospholipid is phosphatidylcholine, with phosphatidylethanolamine present in some species; cardiolipin and phosphatidylglycerol are present in smaller amounts. *T. pallidum* is unique among the treponemes (and spirochaetas) in lacking a monoglycosyldiglyceride; there are also differences between treponeme species in the substituent sugar in this lipid. The lipids of *T. hyodysenteriae* and *T. innocens* have been studied in detail (Matthews and Kinyon, 1984); they are unusual in the predominance of plasmalogens (alk-1-enyl ether phopholipids) and phosphatidylglycerol and few differences were found between the two species.

The analysis of total cellular long chain fatty acids or the differential analyses of polar lipid fractions has been applied to the treponemes (Livermore and Johnson, 1974; Matthews *et al.*, 1979; Matthews and Kinyon, 1984; Tompkins *et al.*, 1986). Treponemes readily incorporate serum fatty acids into lipid and the analysis of structural fatty acids is thus sensitive to differences in media. In the case of the plasmalogens of *T. hyodysenteriae* and *T. innocens*, a further pitfall is the cochromatography of fatty acids and aldehydes derived from the hydrolysis and esterification of normal and ether phospholipids, respectively.

Nucleic acid techniques

There are few reports of the systematic use of plasmid profiling of restriction endonuclease digests of chromosomal DNA for classification and typing of spirochaetes. Coene *et al.* (1989) compared the activities of various endonucleases on DNA from human and porcine intestinal treponemes as a classification system and also to determine the distribution of modified bases. Such an approach may be a useful general technique for typing spirochaetes. It is clear from rRNA studies that the spirochaetes are only distantly related to other bacteria. This should simplify the design of specific oligonucleotide probes (against rRNA) for rapid and sensitive detection of slow growing or non-cultivable treponemes, particularly known and putative pathogens.

The treponematoses

The *T. pallidum* complex of species includes the aetiological agents of the classic treponemal diseases of man: venereal syphilis, yaws, pinta and endemic syphilis (bejel). There are no species- or subspecies-specific serodiagnostic reagents with which to differentiate between these diseases. Epidemiological studies have shown some cross-protection between the treponematoses; indeed, the local incidence of venereal syphilis can soar in the aftermath of eradication programmes for the endemic treponematoses. Although the incidence of yaws and pinta has been greatly diminished in the Americas, the prevalence of bejel and yaws is increasing in areas of the Middle East, the Far East and Africa (Anon, 1984; Perine *et al.*, 1984).

Diagnostic aspects

The isolation and identification of *T. pallidum* is technically difficult and costly. Spirochaetes are cultured by intratesticular passage in *T. paraluiscuniculi*-free rabbits and identified by the ability to produce characteristic lesions in various laboratory animals.

For serological screening it is still usual to perform an agglutination or flocculation assay against reagin (cardiolipin), followed by confirmation of positive results by a specific antitreponemal assay. Since *T. pallidum* contains cardiolipin (Matthews *et al.* 1979), the reagin-based assays may not be as non-specific as previously thought. Haemagglutination and microhaemagglutination assays (with erythrocytes sensitized with treponemal extracts) are now available as alternatives to reagin-based assays; they are more costly but they provide an accuracy approaching that of the FTA-ABS test. The most specific assay is still the *T. pallidum* immobilization (TPI) test (Bradford and Larsen, 1985).

The development of cheap, automatable, sensitive and species-specific serodiagnostic assays (such as ELISA-based techniques) for syphilis and the endemic treponematoses is greatly hampered by our inability to culture *T. pallidum* in bulk *in vitro*. Suspensions of *T. pallidum* prepared from infected rabbit testicular tissues are difficult to obtain in high yields without contamination from adsorbed host components. Tissue-culture techniques have not yet achieved sufficiently high yields of viable, virulent *T. pallidum* to meet these needs.

An alternative strategy is to use molecular immunology and recombinant DNA technology to identify and produce species- and subspecies-specific antigens. Several surface-expressed antigens, identified by immunoprecipitation and partially characterized with monoclonal antibodies, have been cloned and successfully expressed in *Escherichia coli*, including a species-specific, immunodominant, hydrophobic 47-kda surface protein (Chamberlain *et al.*, 1988). Minor differences in polypeptides have been observed between *T. pallidum* ss. *pallidum*, *T. pallidum* ss. *pertenue* and *T. paraluiscuniculi* (Baker-Zander and Lukehart, 1983, 1984) which suggest that it should be possible to produce subspecies-specific monoclonal antibodies.

Treatment

The treatment of choice for the treponematoses is a long-acting penicillin such as procaine benzylpenicillin or benzathine benzylpenicillin. Penicillin-resistant strains of *T. pallidum* have not been recorded. Treatment of late disease is necessarily more aggressive and probenecid may be used to sustain high tissue levels of antibiotic. Alternative drugs with good CNS penetration such as amoxycillin and doxycycline may be indicated in neurosyphilis. In the penicillin-hypersensitive patient, oral tetracycline or erythromycin stearate are effective, although treatment failure with erythromycin has been reported. Some third-generation cephalosporins, such as ceftriaxone, also show **promise** and penetrate the CNS readily.

Response to treatment is monitored with repeated serological tests; reagin tests are sensitive indicators of disease activity. Recrudescence is not uncommon in inadequately treated patients and there is always a risk of reinfection.

Antimicrobial resistance

The possibility that *T. pallidum* and related organisms may acquire or develop resistance to penicillins is a continuing threat, and is the driving force in the search for alternative treatments. Future acquisition of antimicrobial resistance is a real possibility; cryptic plasmid DNA has been detected in the Nichols strain (Norgard and Miller, 1981) and β-lactamases are produced by some strains of intestinal spirochaetes (Tompkins *et al.*, 1987).

Several techniques are used for the assay of antimicrobial activity against *T. pallidum*, including inhibition of treponemal motility and treatment of experimentally infected laboratory animals, but these may not accurately reflect clinical efficacy. Stamm *et al.* (1988) have recently developed an elegant *in vitro* assay for antitreponemal activity of antimicrobials based on the inhibition of new protein synthesis (measured as uptake of radiolabelled methionine) by suspensions of *T. pallidum*. Although most appropriate for the assay of antimicrobials which inhibit ribosome function, it also appears to be a sensitive indicator of the bacteriostatic or bactericidal activity of penicillins. In this assay, protein synthesis in *T. pallidum ss. pallidum* (Nichols and 'street strain 14') and *T. pallidum ss. pertenue* (Gauthier) was unaffected by cycloheximide, rifampicin or streptomycin. The Nichols strain of *T. pallidum ss. pallidum* and *T. pallidum ss. pertenue* were sensitive to chloramphenicol, tetracycline and erythromycin and partially so to penicillin. In contrast, the recent syphilis isolate 'street strain 14' was totally resistant to erythromycin (and roxithromycin). This strain was subsequently found to have been isolated from a penicillin-hypersensitive patient who developed active secondary ulcers during intensive, but unsuccessful, treatment with erythromycin. Thus, the laboratory findings and clinical observations were entirely consistent.

In addition to a specific macrolide resistance, 'street strain 14' showed reduced sensitivity to a wide range of antimicrobials, perhaps due to a non-specific alteration in permeability. Studies with further isolates are required to determine how typical 'street strain 14' is of contemporary *T. pallidum* strains.

Syphilis and Human Immunodeficiency Virus (HIV) coinfection

Syphilis may follow an altered or accelerated clinical course in patients with concomitant HIV infection. Lukehart *et al.* (1988) described several such cases in which there was poor response to standard treatment and rapid progression to neurosyphilis, confirmed by demonstration of spirochaetes in the central nervous system (CNS). This combination of sexually transmitted diseases probably warrants early and aggressive antimicrobial therapy, ensuring adequate treponemacidal levels in the CNS. This dual infection is likely to increase as the prevalence of HIV increases. It will also be important to monitor for possible alterations in the pathogenesis of the endemic treponematoses that might result from concomitant HIV infection. As an indication of what may happen, it may be possible to produce a model of these interactions in primates experimentally coinfected with Simian Immunodeficiency Virus (SIV) and syphilis.

Spirochaetes in oral and periodontal diseases

Oral treponemes are members of the indigenous microflora of the gingival crevice and supra- and subgingival plaque; in the absence of gingival and peridontal diseases they are present in relatively low numbers. The periodontopathic potential of the oral treponemes remains largely unknown, although most studies agree that their presence in large numbers is, at the very least, indicative of active disease. Critical evaluation of a general or species-specific role for treponemes in gingival and periodontal disease is limited, partly by inadequate laboratory methods for quantitation and identification. Microscopy remains the most commonly used method although this limits classification to morphological variants. The design of selective or differential media for the oral spirochaetes is hampered by their slow growth rates and fastidious requirements. The use of rifampicin as a selective agent in complex media goes some way towards quantitative selective isolation. Treponemes are also physically fragile and tend to fragment during commonly used plaque dispersion techniques (Salvador *et al.*, 1987).

A promising alternative to culture is the direct

measurement of spirochaete biomass in subgingival plaque by an enzyme-linked immunosorbent assay (ELISA); species- or biotype-specific antibodies would be the preferred reagents in this kind of assay although a serovar-specific monoclonal antibody has been used to quantitify populations of *T. denticola* (Simonson et al., 1988).

Spirochaetes and acute ulcerative gingivitis (AUG)

Acute ulcerative gingivitis (synonyms: acute necrotizing ulcerative gingivitis, Vincent's disease, Vincent's gingivostomatitis, Trench mouth) is a non-communicable infectious disease affecting the gingiva and buccal mucosae, cited in standard text books as the classic fusospirochaetal infection. Diagnosis is confirmed by microscopic examination of a simple gram-stained smear prepared from the ulcerated gingiva. The condition responds well to prompt treatment with an antibiotic (penicillin or a nitroimidazole such as metronidazole) and rigorous oral hygiene.

There are many factors which appear to predispose to AUG, both physiological (poor oral hygiene, viral infections, smoking, puberty, pregnancy) and psychological (stress). In many developing countries malnourished children debilitated by recent infectious disease are prone to a particularly aggressive form of AUG, cancrum oris (Noma), with deep and spreading necrotic ulceration of the adjoining facial tissues, bone exposure leading to disfigurement, and risk of superinfection.

Gram-stained smears of ulcerated tissue in early AUG and cancrum oris show a characteristic fusospirochaetal flora and treponemes can be found invading the gingival epithelium and underlying connective tissues. Loesche *et al.* (1982) showed that the predominant flora of plaque associated with affected gingival tissue in AUG was also abnormal and complex. Treponemes comprised 30 per cent and fusobacteria only 5 per cent of the microscopical count with other spiral bacteria (mostly selenomonads) contributing 3 per cent. Culture, however, showed the black-pigmented *Bacteroides intermedius* to predominate (Table 8.8), present as 24 per cent of the cultivable flora compared with less than 3 per cent for fusiform organisms.

Treponemes, fusobacteria and *B. intermedius* were all markedly reduced in plaque following treatment, a finding consistent with involvement of this abnormal microflora in active ulcerative disease but not necessarily implying an aetiological role for any particular species or combination. The risk factors outlined above appear to compromise local host defence mechanisms to some critical degree which then permits a shift in the microflora, particularly of treponemes (*T. vincentii* and *T. denticola*) and gram-negative anaerobes. An animal model of AUG can be produced in experimental rats and dogs, stressed or otherwise compromised with systemic corticosteroids. In man the response of the abnormal plaque flora to specific antianaerobe chemotherapy is rapid and specific, with concomitant gingival healing; counselling in oral hygiene is vital if recrudescence of AUG is to be avoided.

These studies have altered our perceptions of the changes in the microflora associated with AUG and need to be corroborated and extended with particular emphasis on the culture and speciation of the treponemes, fusiforms and bacteroides involved.

Table 8.8 Comparisons of plaque flora in eight patients with acute ulcerative gingivitis*

Organisms present	Percentage of total flora by	
	Microscopy	Culture
Treponemes	30	—
Fusiforms (including capnocytophagas)	5	3
Other spirals and curved rods	7	—
Bacteroides intermedius	—	24

*Adapted from Loesche *et al.* (1982)

Spirochaetes and the periodontal diseases

Simple dark-ground or phase-contrast microscopy of a wet film of crevicular fluid or subgingival plaque from a diseased periodontal site commonly shows a flora teeming with actively motile rods, spirochaetes and curved rods; the ratio of non-motile to motile bacteria is about one, in contrast to a 50-fold excess of non-motile cells in the absence of disease (Listgarten and Hellden, 1978). The increased frequency, proportion and types of spirochaetes present in various periodontal diseases is well documented (Table 8.9) (Loesche *et al.*, 1985; Moore WEC *et al.*, 1985, 1987; Westergaard and Fiehn, 1987).

Table 8.9 Frequency of detection (percentage) of various treponeme species in healthy controls and in patients with different periodontal diseases*

Treponemes detected	Periodontitis			Gingivitis		Healthy	
	Juvenile	Moderate	Severe	Adult	Child	Adult	Child
By microscopy							
Total	84	65	88	40	11	0	7
Large treponeme	48	11	63	6	0	0	0
By culture							
T. denticola	19	25	21	3	0	0	0
T. vincentii	6	4	2	0	0	0	0
T. socranskii	58	59	60	31	43	26	17
ss. buccalis				11	3		
ss. paredis	0	1	2	3	29	0	3
ss. socranskii				0	6		
T. pectinovorum	19	6	12	9	3	0	0
Treponema sp. D	16	9	14	11	3	0	0
Treponema sp. F	0	4	2	3	3	0	0

*Data adapted from Moore WEC et al. (1987)

However, the role of spirochaetes in the aetiology and pathogenesis of gingivitis and periodontal disease remains controversial (Loesche and Laughon, 1981); many believe that they are passive colonizers of diseased sites and play no part in the initiation of the lesion or its progression. Although electronmicroscopy often shows pallisades of spirochaetes aligned at the leading edge of the plaque–tissue interface, this might well be expected of a highly motile organism following a nutrient gradient by chemotaxis.

In addition to the technical difficulties of culture, quantitation and identification of the various spirochaetes that are present in a sample from a single site, there is also evidence of considerable variation in the microbial flora between patients with the same clinical disease (Moore et al., 1985). This may contribute to differences in findings between investigating centres.

Of course, an association of treponemes with periodontal disease might only be an epiphenomenon related to altered local environment and nutrient availability. Loesche et al. (1983) showed that strict anaerobes such as the spirochaetes and bacteroides are more commonly found in deep periodontal pockets, with a low pO_2 (<15 mmHg, 2 per cent O_2), whereas microaerobes such as capnocytophagas and actinomyces are more common in medium pockets with higher pO_2 (>15 mmHg). In deep pockets which bled, the pO_2 was raised and the proportion of intermediate-sized treponemes decreased; small and large treponemes appeared to be tolerant of the raised pO_2, perhaps due to some protective factor or nutrient in the blood.

In addition to isolation and quantitative studies, it is imperative that treponemes from diseased sites be examined for possible virulence characteristics, including mechanisms of evading or modifying host-defence mechanisms. This is the only strategy which might clarify the role of treponemes in the aetiology and pathogenesis of the periodontal diseases and gingivitis. Screening for enzyme activities may also aid differentiation of species and biovars.

Putative virulence factors in oral spirochaetes

Few candidate virulence factors have been identified in oral treponemes. Given the difficulty of isolating and growing treponemes in bulk it is not surprising that there are relatively few studies published in this field compared with, for example, the virulence factors of the proteolytic, black-pigmented bacteroides.

T. denticola produces an array of proteases active against substrates specific for trypsin-like and collagenase-like activities (Makinen et al., 1986); it also produces an extracellular fibrinolysin (Nitzan et al.,

1978). There are, of course, many other oral spirochaetes (and other bacteria) to be surveyed for similar activities. It would be particularly interesting to try to match the specificities and properties (pH optima and pI) of peptidases in crevicular fluids taken from patients with active gingival and periodontal diseases with those produced by *T. denticola*.

Both *T. denticola* and *T. vincentii* readily adhere to cultured epithelial cells with subsequent reduced cell proliferation and cytotoxicity, evident by vacuolization and rounding of the cells. This activity is only loosely cell-bound and both culture supernate and spirochaete washings induce detachment of the cultured epithelial cells (Reijntjens *et al.*, 1986).

Intestinal spirochaetosis

Historical perspective

Motile spiral-shaped 'little animalcules' were observed by Antonie van Leeuwenhoek in diarrhoeic faeces although, given the limitations in resolution of his relatively simple microscopes, it is difficult to be confident of the exact nature of the microbes he observed. Theodor Escherich (1884, 1886) first made detailed observations of faecal spirochaetes in relation to cholera and diarrhoea, and his observations were confirmed by others. Many of these early publications have also been cited in support of early observations of *Campylobacter jejuni* enteritis (Kist, 1985) and it is probable that both spirochaetes and campylobacters were observed in some early studies; campylobacters often exhibit spiral morphology, although they are easily distinguished by their darting motility. Coincidentally, the first case of human intestinal spirochaetosis brought to my attention (in 1982) was recognized when stools were being screened for campylobacters by direct microscopy of wet films.

Ruane *et al.* (1989) have elegantly reviewed much of the literature concerning intestinal spirochaetal infections in man. In the early part of the century, systematic microscopical screening of stool samples from patients and healthy subjects in temperate and tropical countries gave a confusing picture of widely different carriage rates of spirochaetes; they were more commonly found in patients from local populations in the tropics and most commonly associated with diarrhoea and dysentery (McFie, 1917; Hogue, 1922; Parr, 1923). The large spiral organism observed was named *Spirochaeta eurygyrata* (Werner, 1909) and was the subject of many learned tracts, but the organism was not successfully cultivated and the name is no longer considered legitimate.

After the 1920s, reports of intestinal spirochaetosis were only sporadic. Shera (1962) described 52 patients in whom he demonstrated a fusobacterial–spirochaetal infection of the rectal mucosa, this appeared, by sigmoidoscopy, as a 'strawberry lesion'; the condition was attributed to ascorbate deficiency and was treated successfully with both a vitamin C supplement and an organic arsenical drug with antitreponemal activity.

Human intestinal spirochaetosis was revived as a pathological entity by Harland and Lee (1967) and Lee *et al.* (1971). They were unable to match any particular symptoms to the heavy overgrowth of spirochaetes attached to the epithelium of rectal biopsies and resected appendix specimens. Their description of the typical basophilic epithelial margin in colorectal biopsy specimens was timely, and served to alert pathologists to the condition.

Recent culture studies of the intestinal spirochaetes suggest that more than one species is present. In most European studies (Sanna *et al.*, 1984; Cooper *et al.*, 1986; Tompkins *et al.*, 1986) the predominant isolate has been phenotypically similar to *T. innocens*—large treponemes found as commensals in the pig and dog (Kinyon and Harris, 1979). The organism is relatively fast-growing (3–7 days) and the spreading veil of surface growth is only partially or weakly β-haemolytic, like *T. innocens*, but in contrast to the strong β-haemolysis of the pig enteropathogen, *T. hyodysenteriae*. A smaller, slow-growing spirochaete *Brachyspira aalborgii* has also been described. This is weakly or non-haemolytic (Hovind-Hougen *et al.*, 1982). Both the human spirochaetes have four or five internal flagella inserted at each end (4-8-4), whereas *T. innocens* usually has more than eight.

It is puzzling that there should be such a dearth of studies which consider both organisms; no studies report the isolation of both species in parallel and there are no studies on comparative physiology and metabolism. Genetic studies on the human *T. innocens*-like strains show considerable homology with *T. hyodysenteriae* (Coene et al., 1989), but no comparable data are available for *B. aalborgii*.

Prevalence studies

There have been few recent surveys of the faecal carriage rate of spirochaetes in the general populations of developed countries. Most data comes from specimens of faeces or biopsy tissue taken as part of diagnostic procedures. However, Tompkins et al. (1986) included culture for spirochaetes as part of the routine screen for stool pathogens in 995 people in Bradford, one-third of whom were from the local Asian community. Intestinal spirochaetes were isolated from only 16 subjects (1.6 per cent) but all of these were Asian and 12 had visited the Indian subcontinent during the preceding year. Phenotypic and biochemical profiles showed similarities with *T. innocens* (weakly β-haemolytic). Spirochaetes isolated from several members of the same household had different biochemical profiles.

Benno et al. (1984) cultured 'borrelias' from seven of ten healthy highlanders in Papua New Guinea, at counts of $\log_{10}$ 9.5 cfu/g of faeces, representing about 10 per cent of the total cultivable flora. None was observed in faeces from a matched Japanese control population. This was not easily explained as simply due to environmental contamination; in terms of the number of species present, the faecal microflora of the highlanders was far more simple than that of the Japanese subjects.

Goossens et al. (1983) compared, by microscopy and culture, the prevalence of large treponemes in faeces submitted for examination in Brussels (Belgium) and from healthy control subjects and patients with diarrhoea in Butare (Rwanda). The results are summarized in Table 8.10.

In a recent survey of the normal population and hospital patients in the Sultanate of Oman, intestinal spirochaetes were cultured from 114 of 1000 specimens of faeces (11.4 per cent) submitted from hospital patients (S P Barrett, personal communication). When compared with randomly matched case controls the isolation of spirochaetes was unrelated to symptoms but was significantly more common in Omani nationals. Spirochaetes were cultured from 78 (26.7 per cent) of 292 faecal samples from volunteer healthy Omani nationals.

Biopsy material from Scandinavian patients investigated for various gastrointestinal diseases was screened by microscopy for colorectal spirochaetosis (Henrik-Nielsen et al., 1983). Fifteen (5 per cent) of 300 patients had spirochaetosis, which was cleared by a course of neomycin and bacitracin but without any change in symptoms. Infection was similarly common in patients with alternating diarrhoea and constipation (6 of 65) as in those without gastrointestinal disease (4 of 53). Spirochaetosis was not found in 84 patients with idiopathic inflammatory bowel disease (IBD). We have also failed to observe or culture intestinal spirochaetes in a series of more than 120 IBD patients, with IBD of varied disease activities (M J Hudson and M G Hartley, unpublished data).

Spirochaetosis has been sought microscopically in appendix specimens in several surveys, two of which are summarized in Table 8.11. A notable feature is the lack of spirochaetosis in acute appendicitis and the contrast between this lack and the higher prevalence in patients who presented with 'pseudoappendicitis' but who had a histologically normal appendix, other than spirochaetosis.

Thus, in summary, in otherwise unselected patient populations, colorectal spirochaetosis is not associated with any particular disease or symptoms, with the exception of 'pseudoappendicitis' in which

Table 8.10 Prevalence of intestinal spirochaetes in two populations

Country	n	Number (%) of subjects with spirochaetosis		
		Total	With diarrhoea	Without diarrhoea
Belgium	1679	21 (1.2%)	17 (1%)	4 (0.24%)
Rwanda	222	39 (17.6%)	18 (14%)*	21 (22.6%)†

*Of 129 patients with diarrhoea
†Of 93 healthy control subjects

Table 8.11 Spirochaetosis in resected appendix specimens

Source of specimens	Number of specimens with spirochaetosis	
	Lee et al (1981)	Henrik-Nielsen et al. (1985)
All appendicectomies	62/790 (7.8%)	18/627 (2.6%)
Acute appendicitis	7/160 (4.4%)	3/414 (0.7%)
Pseudoappendicitis	51/523 (9.8%)	13/106 (12.3%)
Incidental resection	4/107 (3.7%)	2/107 (1.9%)

the incidence of spirochaetosis is significantly increased. Recent surveys support historical observations of considerable geographical variation in the incidence of intestinal spirochaetosis and confirm that, in several developing countries, intestinal spirochaetosis is not uncommon and is largely asymptomatic.

The pathological role of spirochaetes in colorectal disease, especially in 'pseudoappendicitis', cannot be determined without reliable isolation methods and comparative studies between intestinal treponemes isolated from patients with and without gastrointestinal disease, and from different countries. No serological data are available on patients with colorectal spirochaetosis.

Prevalence studies in homosexual/bisexual men

McMillan and Lee (1981) surveyed rectal biopsies from 100 homosexual men and found spirochaetosis in 36 per cent whereas the prevalence in 67 presumed heterosexual men of comparable age was only 3 per cent, this difference was highly significant. Spirochaetosis was present in four out of 10 men with early syphilis and eight out of 18 with rectal gonorrhoea. Among the 70 patients in whom no other recognized pathogen was detected, 23 (32.9 per cent) had spirochaetosis. Sexually active homosexual and bisexual men appear to be at particular risk of polymicrobic infection of the rectum and anal margin, a condition referred to as the Gay Bowel Syndrome (GBS). In addition to recognized pathogens such as the gonococcus, Herpes simplex viris and *Chlamydia trachomatis*, other opportunist pathogens may colonize the rectal mucosa, perhaps secondarily to trauma. These include protozoa, unusual campylobacters, and yeasts. Intestinal spirochaetosis can be added to this list.

Treponemes isolated from homosexual patients with rectal spirochaetosis are phenotypically similar to *T. innocens* (Cooper *et al.*, 1986; Tompkins *et al.*, 1986). No differences were noted between strains from homosexual patients and those from heterosexual subjects. In a series of eight homosexual patients, proctitis was present in three—all had rectal spirochaetosis and no other pathogen; overgrowth with spirochaetes also appeared to result in a depletion of colonocyte microvilli (Cooper *et al.*, 1986). The spirochaetes were eliminated with a course of metronidazole which also resulted in symptomatic improvement.

While it seems unlikely that intestinal spirochaetosis is associated with any symptomatic disease complex in heterosexual patients, except perhaps 'pseudoappendicitis', in the practising male homosexual patient with GBS the high incidence of infection in the absence of other pathogens suggests that these spirochaetes may be able to exert some limited pathology in the colorectum. It would seem sensible to test this theory by a controlled trial of the effect of metronidazole on gastrointestinal symptoms and proctitis in homosexual patients with GBS. If metronidazole does have a therapeutic effect but spirochaetes are not significantly involved, some other (anaerobic) candidate pathogen may be involved. Nothing is known about the possibility of transfer of spirochaetes between partners or whether the organisms survive on, and perhaps colonize, the oral or genital mucosae.

The infectious diseases associated with the Acquired Immunodeficiency Syndrome (AIDS) and the AIDS-related complex (ARC) in homosexual males have rather displaced microbiological studies on GBS itself. The types of opportunist infection are similar but not identical. The isolations of intestinal spirochaetes made in a large study reported by

Laughon *et al.* (1988) of the prevalence of recognized and potential intestinal pathogens in the faeces of 388 homosexual and bisexual males in relation to diarrhoea (including GBS) and AIDS are summarized in Table 8.12. Intestinal spirochaetes were cultured from five (42 per cent) of 12 patients with diarrhoea in whom no other candidate intestinal pathogens were identified; this was a significant association at the 5 per cent level. Although the high incidence of spirochaetosis in this population of homosexual men was similar to that reported in other series of homosexual men without AIDS or ARC, the question of how much spirochaetosis is attributable to homosexual activity, and how much to the depressed immune response seen in AIDS, ARC and, to a lesser extent, in GBS still remains. The question can be answered only by comparative studies with heterosexual patients with HIV infection acquired through blood products, etc. It would also be interesting to survey heterosexual AIDS and ARC patients in developing countries where there is a naturally high incidence of intestinal spirochaetosis.

Pathological aspects of intestinal spirochaetosis

Intestinal spirochaetosis does not seem to produce marked pathological changes in the underlying mucosa. Tissue invasion is rarely seen and, apart from depleted microvilli (Cooper *et al.*, 1986), there are few specific changes in the colonocytes and only limited inflammatory infiltration of the submucosa.

Gebbers *et al.* (1987) described two patients with mild intestinal symptoms in whom rectal spirochaetosis was the only abnormal finding. Unusually, spirochaetes were observed within epithelial cells and macrophages and there were degranulating mast cells and an excess of IgE plasma cells in the lamina propria. In one patient the symptoms settled after treatment with metronidazole.

Mathan and Mathan (1985) proposed the term 'tropical colonopathy' to describe morphological abnormalities seen in rectal mucosal biopsies from otherwise well patients in India. Electron-dense bodies and vesicles were present in colonocytes and crypt cells; microvilli were stunted and patchy. Spirochaetosis was present in nine of 14 subjects with 'tropical colonopathy' and, in four subjects with heavy infestation, abnormalities were more pronounced. Since spirochaetosis was not observed in five of the subjects and infection with spirochaetes can be patchy, multiple biopsies from the length of the colorectum are required to exclude spirochaetosis.

Intestinal spirochaetes and disease in other animals

Spirochaetes form part of the indigenous microflora of the large intestine of many non-human primates and other animals. They have rarely been associated with intestinal abnormality or disease other than in the pig. Indeed, in rodents, spiral bacteria including treponemes form part of the stable autochthonous microflora of the intestine, and may play a role in excluding pathogens (Lee and Phillips, 1978).

Swine dysentery is an infectious mucohaemorrhagic colitis and caecitis of pigs caused by *T. hyodysenteriae*, a large spirochaete found only in the colon and faeces of affected or carrier pigs. *T. hyodysenteriae* is strongly β-haemolytic on blood agar and is enteropathogenic for pigs, mice and

Table 8.12 Faecal spirochaetes in homosexual and bisexual males in relation to intestinal symptoms and AIDS*

Patient group	Number of patients with spirochaetes		
	Without diarrhoea	With diarrhoea	With proctitis
Asymptomatic	34/243 (16%)	NA	NA
Gay bowel syndrome	NA	8/30 (27%)	9/31 (29%)
AIDS or ARC	3/26 (12%)	4/37 (11%)	NA

NA, not available
*Data from Laughton *et al.* (1988)

rabbits. Studies with germ-free animals have shown that the spirochaete cannot initiate the disease alone but, in the presence of other intestinal microorganisms, it is able to invade the mucosa. Non-pathogenic commensal spirochaetes of similar size and biochemical properties found in healthy pigs (and dogs) are not enteropathogenic and only weakly β-haemolytic; a separate species, *T. innocens*, describes such strains. Spirochaetes resembling *T. innocens* may proliferate and attach to the colorectal mucosa in some pigs, and this may be associated with stress and nutritional changes at the time of weaning. Whereas *T. hyodysenteriae* and *T. innocens* appear to share many antigens, strains of the latter species are heterogenous with respect to electrophoretic polypeptide profile (Chatfield *et al.*, 1988).

Natural Tyzzer's disease in the guinea pig (McLeod *et al.*, 1977; Zwicker *et al.*, 1978) is unusual in being associated with intestinal spirochaetosis. Necrotic and inflammatory lesions were found in the ileum, caecum and colon; spirochaetosis was most commonly found associated with crypt cells containing intracellular '*Bacillus piliformis*'.

Miscellaneous diseases

Multiple sclerosis

No studies have been published to test the hypothesized role for oropharyngeal treponemes in the aetiology and pathogenesis of multiple sclerosis (MS) advanced by Gay and Dick (1986). Briefly, this hypothesis suggests that if the sphenoidal sinus were innately weak, or damaged by sinusitis, or both, it would present little obstacle to penetration by oral treponemes, leading to local infection as plaques at the base of the brain. In this protected location a lipid-rich organism, such as an oral treponeme, might elicit an abnormal immune response to the organism, nerve tissues and myelin sheath material. This is an intriguing, attractive and testable hypothesis; treponemes should be sought in early plaque material and the oligoclonal immunoglobulins in CSF examined in western-blots against oral treponemes and other members of the oropharyngeal flora, with appropriate controls.

Tropical ulcer

Treponeme-like microbes have been observed as part of the synergic flora associated with tropical ulcer (Adriaans *et al.*, 1987). Anaerobic culture yielded facultative gram-negative bacilli and *Fusobacterium ulcerans* in addition to spirochaetes. The latter were not fully characterized and their contribution to the lesion remains undetermined.

Non-syphilitic genital lesions

Non-*T. pallidum* spirochaetes have long been recognized as the normal flora of the genital mucosae. Early attempts to culture the treponeme of syphilis were dogged by growth of 'saprophytic' species such as *T. phagedenis* and *T. refringens*. In a recent study (Piot *et al.*, 1986), non-*T. pallidum* spirochaetes of various sizes were observed in large numbers associated with ulcerative balanoposthitis in 41 (12 per cent) of 344 uncircumcised men with genital ulcer disease. In 14 (34 per cent) of these patients, no other microbial pathogen was identified, although the exudate was clearly polymicrobic; the foul smell of the discharge indicated that other obligate anaerobes were present.

Future developments

Several areas of research need to be pursued to clarify the role of anaerobic spirochaetes in human disease. Studies of the physiology and metabolism of treponemes will lead not only to improved isolation media but also to improved methods for the rapid and precise quantitation of slow-growing, fastidious treponemes. This may be aided by monoclonal antibody-based assays or by DNA hybridization with specific oligonucleotide probes.

The taxonomy of the treponemes is confused and most isolates differ from the recognized type strains, indicating that they should be allocated to new genera. The role of rRNA sequence data as an accurate measure of phylogeny within the spirochaete cluster, or merely as a reflection of a fast or uneven evolutionary clock, requires further evaluation. The significance of genera that are defined on the basis of ecology, such that free-living forms are spirochaetes, arthropod-borne species borrelias, host-associated anaerobes treponemes, and on the taxonomic value of features such as internal flagella arrangement and cytoplasmic tubules also requires further consideration.

In diagnostic microbiology, the production of sensitive, specific tests for syphilis and the treponematoses (e.g., ELISA) are largely dependent upon improved *in vitro* culture of *T. pallidum* or production of recombinant species-specific treponemal antigens. Amongst other putative spirochaetal diseases, the role and significance of detecting either the treponemes associated with periodontal and gingival disease or those associated with intestinal disease remains far from clear. Even if the overgrowth of oral and intestinal spirochaetes are epiphenomena, there remains confusion about the conditions that allow proliferation and no concensus about whether they are reasonable markers of dysfunction or disease activity.

The list of unanswered questions is extensive. The spirochaetes are a diverse and important group of micro-organisms which share a particular morphology. They are able to compete with other microbes in diverse ecosystems. The classic spirochaetal diseases, such as syphilis, leptospirosis and the borrelial diseases, are among the most common and most serious infectious diseases in the world, and much is still to be learnt about the pathogenesis and control of these infections. The investigation of the pathogenic potential of spirochaetes normally resident in the flora is less dramatic but will provide valuable insights into the delicate balance of tolerance between host and microflora.

References

Adriaans B, Hay R, Lucas S, Robinson D C. (1987). Light and electron microscopic features of tropical ulcer. *Journal of Clinical Pathology* 40: 1231–1234.

Anon. (1984). Surveillance of treponematoses: yaws and other endemic treponematoses. *WHO Weekly Epidemiological Record* 59: 377–380.

Baker-Zander S A, Lukehart S A. (1983). Molecular basis of immunological cross-reactivity between *Treponema pallidum* and *Treponema pertenue*. *Infection and Immunity* 42: 634–638.

Baker-Zander S A, Lukehart S A. (1984). Antigenic cross-reactivity between *Treponema pallidum* and other pathogenic members of the family Spirochaetaceae. *Infection and Immunity* 46: 116–121.

Barbour A G, Hayes S F. (1986). Biology of *Borrelia* species. *Microbiological Reviews* 50: 381–400.

Benno Y, Takahashi M, Kametaka M, Rikimaru T, Koishi H, Mitsuoka T. (1984). The fecal flora of Papua New Guinea highlanders and its protein synthesis. In *Intestinal flora and dietary factors*. Edited by Mitsuoka T, Japan Scientific Societies Press, Tokyo.

Bradford L L, Larsen S A. (1985). Serologic tests for syphilis. In *Manual of clinical microbiology*, 4th edn. Edited by Lennete E H *et al.* American Society for Microbiology, Washington DC.

Bresnak J A (1984) *Cristispira*, Gross 1910, 44[AL]. In *Bergey's Manual of systematic bacteriology*, vol 1. Edited by Krieg, N R, Holt J R. Williams and Wilkins, Baltimore.

Bryant M P. (1972). Commentary on the use of the Hungate technique for culture of anaerobic bacteria. *American Journal of Clinical Nutrition* 25: 1324–1328.

Canale-Parola E. (1978). Motility and chemotaxis of spirochetes. *Annual Reviews in Microbiology* 32: 69–99.

Canale-Parola E. (1984a). Spirochaetales, Buchanan 1917, 163[AL]. In *Bergey's Manual of systematic bacteriology*, vol 1. Edited by Krieg N R, Holt J R. Williams and Wilkins, Baltimore.

Canale-Parola E. (1984b). Spirochaetaceae, Swellengrebel 1907, 581[AL]. In *Bergey's Manual of systematic bacteriology*, vol 1. Edited by Krieg N R, Holt J R. Williams and Wilkins, Baltimore.

Canale-Parola E. (1984c). Spirochaeta, Ehrenberg 1835, 313[AL]. In *Bergey's Manual of systematic bacteriology*, vol 1. Edited by Krieg N R, Holt J R. Williams and Wilkins, Baltimore.

Chamberlain N R, Radolf J D, Hsu P-L, Sell S, Norgard M V. (1988). Genetic and physicochemical characterization of the recombinant DNA-derived 47-kilodalton surface immunogen of *Treponema pallidum* subsp. *pallidum*. *Infection and Immunity* 56: 71–78.

Chatfield S N, Fernie D S, Penn C, Dougan G. (1988). Identification of the major antigens of *Treponema hyodysenteriae* and comparison with those of *Treponema innocens*. *Infection and Immunity* 56: 1070–1075.

Coene M *et al.* (1989). Comparative analysis of the genomes of intestinal spirochetes of human and animal origin. *Infection and Immunity* 57: 138–145.

Cooper C, Cotton D W K, Hudson M J, Kirkham N, Wilmott F E W. (1986). Rectal spirochaetosis in homosexual men: characterisation of the organism and pathophysiology. *Genitourinary Medicine* 62: 47–52.

Dobell C. (1932). *Antonie van Leeuwenhoek and his 'Little Animals'*. Harcourt Brace Co, New York.

Escherich T. (1884). Klinisch-therapeutische beobachtungen aus der cholera-epidemie in Neapel. *Arztliche Intelligentia* 33: 561–564.

Escherich T. (1886). Beitrage zur kenntnis der darmbakterien. *Munchnener Medezinische Wochenschrift* 38: 815–833.

Fieldsteel A H, Cox D L, Moeckli R A. (1982). Further studies on the replication of virulent *Treponema pallidum* in tissue cultures of Sf1Ep cells. *Infection and Immunity* **35**: 449–455.

Fitzgerald T J. (1981). In-vitro culture of *Treponema pallidum*: a review. *Bulletin of the World Health Organization* **59**: 787–812.

Gay D, Dick G. (1986). Is multiple sclerosis caused by an oral spirochaete? *Lancet* **2**: 75–77.

Gebbers J-O, Ferguson D J P, Mason C, Kelly P, Jewell D P. (1987). Spirochaetosis of the human rectum associated with an intraepithelial mast cell and IgE plasma cell response. *Gut* **28**: 588–593.

Glenister D A, Salamon K E, Smith K, Beighton D, Keevil C W. (1988). Enhanced growth of complex communities of dental plaque in mucin-limited continuous culture. *Microbial Ecology in Health and Disease* **1**: 31–38.

Goossens H, Lejour M, De Mol P, De Boeck M, Verhaeghen G, Dekegel D, Butzler J P (1983) Isolation of *Treponema hyodysenteriae* from human faecal specimens: a curiosity or a new enteropathogen? Abstracts of the *First European Congress of Clinical Microbiology*, Bologna, October 17–21, 1983. Poster 267.

Harland W A, Lee F D. (1967). Intestinal spirochaetosis. *British Medical Journal* **2**: 718–719.

Harwood C S, Canale-Parola E. (1984). Ecology of spirochetes. *Annual Reviews in Microbiology* **38**: 161–192.

Henrik Nielsen R, Orholm M, Pederson J O, Hovind-Hougen K, Teglbjaerg P S, Thaysen E H. (1983). Colorectal spirochaetosis: clinical significance of the infestation. *Gastroenterology* **85**: 62–67.

Henrik-Nielsen R, Lundbeck F A, Teglbjaerg P S, Ginnerup P, Hovind-Hougen K. (1985). Intestinal spirochaetosis of the vermiform appendix. *Gastroenterology* **88**: 971–977.

Hogue M J. (1922). *Spirochaeta eurygyrata*: a note on its life history and cultivation. *Journal of Experimental Medicine* **36**: 617–626.

Holdeman L V, Cato E P, Moore W E C. (eds) (1977). *Anaerobe laboratory manual*, 4th edn. Virginia Polytechnic Institute and State University, Blacksburg.

Holt S C. (1978). Anatomy and chemistry of spirochetes. *Microbiological Reviews* **42**: 114–160.

Hovind-Hougen K. (1976). Determination by means of electron microscopy of morphological criteria of value for classification of some spirochetes, in particular treponemes. *Acta Pathologica et Microbiologica Scandinavica, Section B* Suppl **255**: 1–41.

Hovind-Hougen K. (1979). Leptospiraceae, a new family to include *Leptospira* Noguchi 1917 and *Leptonema* gen. nov. *International Journal of Systematic Bacteriology* **29**: 245–251.

Hovind-Hougen K *et al.* (1982). Intestinal spirochetosis: morphological characterization of the spirochete *Brachyspira aalborgi* gen.nov, sp.nov. *Journal of Clinical Microbiology* **16**: 1127–1136.

Hunter D, Wood T. (1979). An evaluation of the API-ZYM system as a means of classifying spirochaetes associated with swine dysentery. *Veterinary Record* **104**: 383–384.

Hungate R E. (1950). The anaerobic mesophilic culloltic bacteria. *Bacteriological Reviews* **14**: 1–49.

Jennings B R, Mincer H, Turner J, Baselski V, Kelly R T. (1983). Use of avidin-biotinylated horseradish peroxidase complex for the visualization of spirochaetes. *Journal of Clinical Microbiology* **18**: 1250–1251.

Johnson R C, Faine S. (1984). Leptospiraceae, Hovind-Hougen 1979, 245[AL]. In *Bergey's Manual of systematic bacteriology*, vol 1. Edited by Krieg N R, Holt J R. Williams and Wilkins, Baltimore.

Kelly R T. (1984). Borrelia, Swellengrebel 1907, 582[AL] In *Bergey's Manual of systematic bacteriology*, vol 1. Edited by Krieg N R, Holt J R. Williams and Wilkins, Baltimore.

Kinyon J M, Harris D L. (1979). *Treponema innocens*, a new species of intestinal bacteria, and emended description of the type strain of *Treponema hyodysenteriae* Harris *et al. International Journal of Systematic Bacteriology* **29**: 102–109.

Kist M. (1985). The historical background to campylobacter infection: new aspects. In *Campylobacter III*. Edited by Pearson A D *et al.* Public Health Laboratory Service, London.

Kuhn D A. (1981). The genus *Cristispira*. In *The Prokaryotes; a handbook on habitats, isolation, and identification of bacteria*, vol 1. Edited by Starr M P *et al.* Springer-Verlag, Berlin.

Laughon B E, Syed S A, Loesche W J. (1982). API-ZYM system for identification of *Bacteroides* spp., *Capnocytophaga* spp., and spirochetes of oral origin. *Journal of Clinical Microbiology* **15**: 97–102.

Laughon B E *et al.* (1988). Prevalence of enteric pathogens in homosexual men with and without acquired immunodeficiency syndrome. *Gastroenterology* **94**: 984–993.

Lee A, Phillips M. (1978). Isolation and cultivation of spirochetes and other spiral-shaped bacteria associated with the cecal mucosa of rats and mice. *Applied and Environmental Microbiology* **35**: 610–613.

Lee F D, Kraszewski A, Gordon J, Howie J G R, McSeveney D, Harland W A. (1971). Intestinal spirochaetosis. *Gut* **12**: 126–133.

Listgarten M A, Hellden L. (1978). Relative distribution of bacteria at clinically healthy and periodontally diseased sites in humans. *Journal of Clinical Periodontology* **5**: 115–132.

Livermore B P, Johnson R C. (1974). Lipids of the Spirochaetales: comparison of the lipids of several members of the genera *Spirochaeta*, *Treponema*, and *Leptospira*. *Journal of Bacteriology* **120**: 1268–1273.

Loesche W J, Laughon B E. (1981). Role of spirochetes in periodiontal disease. In *host-parasite interactions in periodontal diseases*, pp. 62–75. Edited by Genco R J, Mergenhage S E. American Society for Microbiology, Washington DC.

Loesche W J, Syed S A, Laughon B E, Stoll J. (1982). The bacteriology of acute necrotizing ulcerative gingivitis. *Journal of Periodontology* **53**: 223–230.

Loesche W J, Gusberti F, Mettraux G, Higgins T, Syed S A. (1983). Relationship between oxygen tension and subgingival bacterial flora in untreated human periodontal pockets. *Infection and Immunity* **42**: 659–667.

Loesche W J, Syed S A, Schmidt E, Morrison E C. (1985). Bacterial profiles of subgingival plaques in periodontitis. *Journal of Periodontology* **56**: 447–456.

Lukehart S A, Hook E W, Baker-Zander S A, Collier A C, Critchlow C W, Handsfield H H. (1988). Invasion of the central nervous system by *Treponema pallidum*: implications for diagnosis and treatment. *Annals of Internal Medicine* **109**: 855–862.

Macy J M, Snellen J E, Hungate R E. (1972). Use of syringe method for anaerobiosis. *American Journal of Clinical Nutrition* **25**: 1318–1323.

McFie J W S. (1917). The prevalence of *Spirochaeta eurygyrata* in europeans and natives in the Gold Coast. *Lancet* **1**: 336–340.

McLeod C G, Stookey J L, Harrington D G, White J D. (1977). Intestinal Tyzzer's disease and spirochetosis in a guinea pig. *Veterinary Pathology* **14**: 229–235.

McMillan A, Lee F D. (1981). Sigmoidoscopic and microscopic appearance of the rectal mucosa in homosexual men. *Gut* **22**: 1035–1041.

Makinen K K, Syed S A, Makinen P-L, Loesche W J. (1986). Benzoylarginine peptidase and iminopeptidase profiles of *Treponema denticola* strains isolated from the human periodontal pocket. *Current Microbiology* **14**: 85–89.

Margulis L, To L, Chase D. (1978). Microtubules in prokaryotes. *Science* **200**: 118–24.

Margulis L, To L P, Chase D G. (1981). The genera *Pillotina*, *Hollandia* and *Diplocalyx*. In *The Prokaryotes; a handbook on habitats, isolation, and identification of bacteria*, vol 1. Edited by Starr M P *et al*. Springer-Verlag, Berlin.

Mathan M M, Mathan V I. (1985). Rectal mucosal morphologic abnormalities in normal subjects in southern India: a tropical colonopathy? *Gut* **26**: 710–717.

Matthews H A, Kinyon J M. (1984). Cellular lipid comparisons between strains of *Treponema hyodysenteriae* and *Treponema innocens*. *International Journal of Systematic Bacteriology* **34**: 160–165.

Matthews H M, Yang T-K, Jenkin H M. (1979). Unique lipid composition of *Treponema pallidum* (Nicols virulent strain). *Infection and Immunity* **24**: 713–719.

Meyer P E, Hunter E F. (1967). Antigenic relationships of 14 treponemes demonstrated by immunofluorescence. *Journal of Bacteriology* **93**: 784–789.

Mitsuoka T. (1980). *A color atlas of anaerobic bacteria*. Sobunsha Publishers, Tokyo.

Moore L V H *et al*. (1987). Bacteriology of human gingivitis. *Journal of Dental Research* **66**: 989–995.

Moore W E C, Hash D E, Holdeman L V, Cato E P. (1980). Polyacrylamide slab gel elecrophoresis of soluble proteins for studies of bacterial floras. *Applied and Environmental Microbiology* **39**: 900–907.

Moore W E C *et al*. (1984). Variation in periodontal floras. *Infection and Immunity* **46**: 720–726.

Moore W E C *et al*. (1987). Comparative bacteriology of juvenile periodontitis. *Infection and Immunity* **48**: 507–519.

Nitzan D, Sperry J F, Wilkins T D. (1978). Fibrinolytic activity of oral anaerobic bacteria. *Archives of Oral Biology* **23**: 465–470.

Norgard M V, Miller J N. (1981). Plasmid DNA in *Treponema pallidum* (Nichols): potential for antibiotic resistance by syphilis bacteria. *Science* **213**: 553–555.

Norris S J, Edmondson D G. (1986). Factors affecting the multiplication and subculture of *Treponema pallidum* subsp. *pallidum* in a tissue culture system. *Infection and Immunity* **47**: 534–539.

Parr L W. (1923). Intestinal spirochetes. *Journal of Infectious Disease* **33**: 369–383.

Paster B J, Kaine B P. (1984). Absence of actin genes in spirochetes. *Current Microbiology* **11**: 285–288.

Paster B J, Stackebrandt E, Hespell R B, Hahn C M. (1984). The phylogeny of spirochetes. *Systematic and Applied Microbiology* **5**: 337–351.

Perine P L, Hopkins D R, Niemel P L A, St John R K, Causse G, Antal G M. (1984). *Handbook of endemic treponematoses*. World Health Organization, Geneva.

Piot P, Duncan M, Van Dyck E, Ballard R C. (1986). Ulcerative balanoposthitis associated with non-syphilitic spirochaetal infection. *Genitourinary Medicine* **62**: 44–46.

Reijntjens F M J, Mikx F H M, Wolters-Lutgerhorst J M L, Maltha J C. (1986) Adherence of oral treponemes and their effects on the morphological damage and detachment of epithelial cells *in vitro*. *Infection and Immunity* **51**: 642–647.

Rothwell D. (1981). Microevolutionary change in the human pathogenic treponemes: an alternative hypothesis. *International Journal of Systematic Bacteriology* **31**: 82–87.

Ruane P J, Nakata M M, Reinhardt J F, George W L. (1989). Spirochete-like organisms in the human gastrointestinal tract. *Reviews of Infectious Diseases* **11**: 184–196.

Salvador S L, Syed S A, Loesche W J. (1987). Comparison of three dispersion procedures for quantitative recovery of cultivable species of subgingival spirochaetes. *Journal of Clinical Microbiology* **25**: 2230–2232.

Sanna A *et al*. (1984). Studies of treponemes isolated from human gastrointestinal tract. *L'Igiene Moderna* **81**: 959–973.

Shera A G. (1962). Specific granular lesions associated with intestinal spirochaetosis. *British Journal of Surgery* **50**: 68–77.

Simonson L G, Goodman C H, Bial J J, Morton H E. (1988). Quantitative relationship of *Treponema denticola* to severity of periodontal disease. *Infection and Immunity* **56**: 726–728.

Smibert R M. (1981). The genus *Treponema*. In *The Prokaryotes; a handbook on habitats, isolation, and identification of bacteria*, vol 1. Edited by Starr M P *et al*. Springer-Verlag, Berlin.

Smibert R M. (1984). Treponema, Schaudinn 1905, 1728[AL]. In *Bergey's Manual of systematic bacteriology*, vol 1. Edited by Krieg N R, Holt J R. Williams and Wilkins, Baltimore.

Smibert R M, Burmeister J A. (1983). *Treponema pectinovorum* sp.nov. isolated from humans with periodontitis. *International Journal of Systematic Bacteriology* **33**: 852–856.

Smibert R M, Johnson J L, Ranney R R. (1984). *Treponema socranskii* sp.nov., *Treponema socranskii* subsp. *socranskii* subsp.nov., *Treponema socranskii* subsp. *buccale* subsp.nov., *Treponema socranskii* subsp. *paredis* subsp.nov. isolated from the human periodontia. *International Journal of Systematic Bacteriology* **34**: 457–462.

Songer J G, Kinyon J M, Harris D L. (1976). Selective medium for the isolation of *Treponema hyodysenteriae*. *Journal of Clinical Microbiology* **4**: 57–60.

Stamm L V, Stapleton J T, Bassfore P J. (1988). In vitro assay to demonstrate high level resistance of a clinical isolate of *Treponema pallidum*. *Antimicrobial Agents and Chemotherapy* **32**: 164–169.

Stanton T. (1987). Cholesterol metabolism by *Treponema hyodysenteriae*. *Infection and Immunity* **55**: 309–313.

Stanton T. (1989). Glucose metabolism and NADH recycling by *Treponema hyodysenteriae*, the agent of swine dysentery. *Applied and Environmental Microbiology* **55**: 2365–2371.

Stanton T B, Canale-Parola E. (1979). Enumeration and selective isolation of rumen spirochaetes. *Applied and Environmental Microbiology* **38**: 965–973.

Suzuki M, Loesche W. (1989). Ceruloplasmin can substitute for rabbit serum in stimulating the growth of *Treponema denticola*. *Infection and Immunity* **57**: 643–644.

Syed S A, Salvador S L, Loesche W J. (1988). Enzyme profiles of oral spirochetes in RapID-ANA system. *Journal of Clinical Microbiology* **26**: 2226–2228.

Tall, B D, Nauman R K. (1986). Microscopic agglutination and polyacrylamide gel electrophoresis analyses of oral anaerobic spirochetes. *Journal of Clinical Microbiology* **24**: 282–287.

Taylor D J, Simmons J R, Laird H M. (1980). Production of diarrhoea and dysentery in pigs by feeding pure cultures of a spirochaete differing from *Treponema hyodysenteriae*. *Veterinary Record* **106**: 323–370.

Tompkins D S, Foulkes S, Godwin P G R, West A P. (1986). Isolation and characterisation of intestinal spirochaetes. *Journal of Clinical Pathology* **39**: 535–541.

Tompkins D S, Millar M R, Heritage J, West A P. (1987). β-lactamase production by intestinal spirochaetes. *Journal of General Microbiology* **133**: 761–765.

Werner H. (1909). Ueber befunde von darmspirochaten beim menschen. *Zentralblatt für Bakteriologie und Parasitologie Abteilung 1 Originale* **52**: 241–243.

Westergaard J, Fiehn N–E. (1987). Morphological distribution of spirochetes in subgingival plaque from advanced marginal periodontitis in humans. *Acta Pathologica, Microbiologica et Immunologica Scandinavica: Section B* **95**: 49–55.

Woese C R. (1987). Bacterial evolution. *Microbiological Reviews* **51**: 221–271.

Woese C R, Stackebrandt E, Macke T J, Fox G E. (1985). A phylogenetic definition of the major eubacterial taxa. *Systematic and Applied Microbiology* **6**: 143–151.

Yanagihara Y *et al*. (1984). Chemical compositions of cell walls and polysaccharide fractions of spirochaetes. *Microbiology and Immunology* **28**: 535–544.

Zwicker G M, Dagle G E, Adee R R. (1978). Naturally occurring Tyzzer's disease and intestinal spirochetosis in guinea pig. *Laboratory Animal Science* **28**: 193–198.

9
Actinomyces and *Arachnia*

G H Bowden

The nature of *Actinomyces* and *Arachnia*
    Colonial and cellular morphology
    Habitat
    Physiology and metabolism
Classification
Isolation
    Routine laboratory methods
Identification
    Routine laboratory methods
    Reference or research laboratory methods
Pathogenicity
References

Although these genera share only a similarity in habitat and morphology, being quite distinct in many taxonomic characters, it is convenient to consider them together. One reason for this is that pathogenic species or strains of both genera cause human infections which may be indistinguishable on the basis of clinical symptoms and pathogenesis. The excellent sections on these genera in the current *Bergey's Manual of Systematic Bacteriology* (Schaal, 1986a,b) give extensive details on some aspects of these organisms and should be consulted by anyone interested in the biology of *Actinomyces* and *Arachnia*.

The nature of *Actinomyces* and *Arachnia*

Colonial and cellular morphology

Actinomyces spp. produce a variety of colony forms (Schaal, 1986b; Slack and Gerencser, 1975) ranging from relatively simple smooth colonies with entire edges to those which are heaped and rough. In a limited way the colony form can be used to suggest the species of *Actinomyces*. Thus, as a general rule, strains of *A. bovis, A. viscosus, A. naeslundii, A. odontolyticus, A. denticola, A. howellii, A. meyerii* and *A. israelii* serovar 2 produce smooth colonies 1–2 mm diameter after growing on blood agar under anaerobic conditions. The colonies of certain species have other distinguishing characteristics: *A. odontolyticus* produces a red–brown pigment on blood agar; some strains of *A. naeslundii* can produce a rough, crumbly, irregular colony; *A. pyogenes* is β-haemolytic and strains of *A. israelii* serovar 1 usually produce a heaped, rough colony resembling a 'molar tooth'. The morphology of microcolonies of *Actinomyces* may also provide useful information as filamentous or smooth microcolonies may be seen with different species. Details of microcolony appearance can be found in Schaal (1986b) and Slack and Gerencser (1975). Broth cultures of

Actinomyces reflect the type of growth on solid media; those species which produce smooth regular colonies grow as smooth suspensions. Strains of *A. viscosus*, and sometimes *A. naeslundii*, also produce ropy sediments, presumably related to the production of extracellular polymers. *Actinomyces* which produce rough colonies will often produce granular deposits in broth and a completely clear supernate. Typical of this type of granular growth is *A. israelii* serovar 1, which grows as filamentous masses which are difficult to suspend.

Cellular pleomorphism, ranging from short coccoid cells to branching filaments, is typical of members of *Actinomyces* and *Arachnia*. Strains of *A. israelii* serovar 1, the rough forms of *Arachnia propionica* and strains of *A. viscosus* and *A. naeslundii*, often produce branching filaments. The other species produce filaments less readily and it is more common for the majority of the cells to be short, pleomorphic rods. However, careful examination of a gram-stained smear of most species will reveal branching cells with occasional filaments. Branching of the cells of *Actinomyces* and *Arachnia* can most easily be seen with the cell wall staining method described by Hale (1953).

Habitat

As far as we know, most *Actinomyces* and *Arachnia* are limited in their habitat to man and animals. More specifically, the oral cavity and tonsils appear to be the major habitat for these genera (Bowden and Hardie, 1973; Slack and Gerencser, 1975), where they are members of the autochthonous flora Table 9.1. Strains of both genera have been isolated occasionally from the gut or the female genital tract, but it seems unlikely that they represent part of the indigenous flora. The distribution of species of *Actinomyces* among their hosts may vary (Slack and Gerencser, 1975). Evidence for this has been obtained by Dent and Williams (1984a,b; 1986), who have isolated three new species from the mouths of cattle, and Buchanan *et al.* (1984) who identified *A. hordeovulneris* as a pathogen in dogs. More extended studies of other oral habitats will probably also reveal new species. Little is known of the presence of *Arachnia* in the oral cavity of animals.

Both *Actinomyces* and *Arachnia* grow in the mouths of man and animals (Dent, 1979; Dent and Marsh, 1981) as components of the bacterial communities which develop in association with the tooth surface (Bowden *et al.*, 1979; Bowden and Hardie, 1978). These communities grow either above the gingiva or subgingivally, and representatives of *Actinomyces* may be isolated from either site. *A. meyerii* (Cato *et al.*, 1984) has been isolated from subgingival pockets (Moore *et al.*, 1982).

Physiology and metabolism

There have been relatively few detailed studies on the biochemistry of *Actinomyces* (Howell and Pine,

Table 9.1 Species currently placed in the genera *Actinomyces* and *Arachnia*

Species	Serovars	Habitat	Reference
Actinomyces bovis	2	Oral cavity (animals)	Slack and Gerencser (1975)
Actinomyces israelii	2 (others*)	Oral cavity (human) Tonsils Female genital tract	Schaal, (1986b)
Actinomyces naeslundii	4	Oral cavity (human)	Schaal (1986b)
Actinomyces odontolyticus	2	Oral cavity (human)	Schaal (1986b)
Actinomyces viscosus	2	Oral cavity (human)	Schaal (1986b)
Actinomyces denticolens	ND	Oral cavity (cattle)	Dent and Williams (1984a)
Actinomyces howellii	ND	Oral cavity (cattle)	Dent and Williams (1984b)
Actinomyces slackii	ND	Oral cavity (cattle)	Dent and Williams (1986)
Actinomyces meyerii	1	Oral cavity (human)	Cato *et al.* (1984)
Actinomyces hordeovulneris	ND	Not known	Buchanan *et al.* (1984)
Actinomyces pyogenes	1	Mucosa (human/animal)	Reddy *et al.* (1982)
Arachnia propionica	2	Oral cavity	Schaal (1986a)

*Schaal (1986b); ND, not defined

Table 9.2 Semi-defined medium for *Actinomyces*

Basic solution	Volume 950 ml (g)	Solution 1	Volume 250 ml (mg)
Possatium phosphate monobasic	6	Para-amino benzoic acid	50
Potassium phosphate dibasic	9	Thiamine	50
Calcium chloride (anhydrous)	20	Riboflavin	50
Sodium acetate $3H_2O$	0.2	Nicotinic acid	50
Glucose	0.3	Pyriodoxal HCl	50
L-cysteine HCl	0.2	Inositol	50
Glutathione	0.05	Calcium pantothenate	50
L-asparagine	0.1		
L-tryptophane	0.04		
L-glutamic acid (Na salt)	0.5		
Casein hydrolysate*	2		

Solution 2	Volume 25 ml (mg)	Solution 3	Volume 100 ml (mg)
DL-thioctic acid	2.5	Ferrous sulphate $7H_2O$	400
Biotin	2.5	Mangenese sulphate $2H_2O$	15
Haemin†	2.5	Sodium molybdate $2H_2O$	15
Folic acid	5.0		

Add 10 ml of solution 1, 1 ml of solution 2 and 1 ml of solution 3 to 950 ml of the basic solution. Adjust to pH 7.0, warm to dissolve, filter, sterilize or autoclave at 110°C for 30 min

* The casein hydrolysate may be either acid- or enzyme-hydrolysed. If the former, no peptide will be included in the medium
† Dissolve the haemin by adding a drop of 0.880 ammonia

1956; Christie and Porteus, 1962a,b; Kiel and Porteus, 1962; Buchanan and Pine, 1965, 1967; Kiel and Tanzer, 1977; Reddy and Cornell, 1982; Hamilton and Ellwood, 1983). Studies on the growth requirements of *A. israelii* strains were made by Christie and Porteus (1962a,b,c) and Kiel and Porteus (1962), who developed a fully defined medium for this species. The medium was complex, with a wide range of growth factors and amino acids. Strains of *A. bovis*, *A. viscosus*, *A. naeslundii*, *A. odontolyticus* and *Arachnia* grow well in this medium or in its modification (Bowden et al., 1976) (Table 9.2). Consequently, *Actinomyces* spp. could be regarded as having complex nutritional requirements. Studies on the fermentation balances of *A. israelii* (called *A. naeslundii* in the publication) have been made (Buchanan and Pine, 1965, 1967) and the ratios of end products of glucose metabolism vary with the growth conditions. Generally, *Actinomyces* spp. growing anaerobically with CO_2 produce formic, acetic, lactic and succinic acids. Varying the atmospheric requirements alters the end products (Buchanan and Pine, 1965, 1967) and CO_2 is important in the production of succinate. Almost nothing is known of the activities of *Actinomyces* in continuous culture although some aspects of the carbohydrate metabolism of a strain of *A. viscosus* have been reported by Hamilton and Ellwood (1983). The strain was a human isolate which fitted the characteristics of Cluster 1 (Fillery et al., 1978). A glucose phosphoenol pyruvate phosphotransferase system was demonstrated, and it was also suggested that a second glucose transport system existed in the organism. This second system may be based on proton motive force, as has been shown for oral strains of *Streptococcus* (Hamilton, 1987). The products of glucose metabolism included formic, acetic, lactic and succinic acids and ethanol. Amino acid utilization was measured and growth was shown to be limited by asparagine. Detailed studies on the growth requirements of *A. meyeri* have also been made by Reddy and Cornell (1982) and Reddy *et al.* (1982).

In general, *Actinomyces* and *Arachnia* spp. show little capacity to degrade proteins, although some species hydrolyse starch, Tween 40 and 60 and aesculin, and *A. pyogenes* degrades casein and gelatine. Carbohydrates, including mono-, di- and trisaccharides and sugar alcohols are all fermented and, generally, *Actinomyces* and *Arachnia* spp. show a wide fermentative capacity (Schaal, 1986a,b). Variations in the metabolism of glucose and glycogen by different strains of *A. viscosus* and *A. naeslundii* have been described by Komiyama *et al.* (1986).

In tests for enzyme activities of whole cells, with the API enzyme tests (API, Bio Merieux) *Actinomyces* spp. have been shown to produce leucine arylamidase, β-galactosidase, α-glucosidase and β-xylosidase (Killian, 1978; Schofield and Schaal, 1981; Dent and Williams, 1984a).

Actinomyces spp. cannot be regarded as strictly obligate anaerobes; indeed, the majority of the species are facultative, growing well under aerobic conditions with CO_2. Strains of *A. bovis*, *A. israelii* and *A. meyerii* are among those that require anaerobic conditions. However, given the addition of CO_2, many strains of *A. israelii* will grow in the presence of air. *Actinomyces* could therefore be considered a genus of facultative anaerobes with a requirement for CO_2 (Slack and Gerencser, 1975). *Arachnia* spp. are generally considered to be facultative anaerobes and they do not show a need for, or growth enhancement in, CO_2.

Certain *Actinomyces* spp. produce significant amounts of extracellular carbohydrate polymers and surface proteins. The studies on the presence and nature of such materials have generally been limited to strains identified as *A. viscosus* or *A. naeslundii* (Van der Hoeven, 1974; Ellen *et al.*, 1978; Masuda *et al.*, 1983; Birdsell *et al.*, 1984; Cisar *et al.*, 1984). The reason for the interest in these species is their role in dental plaque and an association with periodontal disease (Bowden *et al.*, 1979; Ellen, 1982). It is well known that the adherence and retention of oral bacteria to their habitat (the tooth surface) can involve relatively specific interactions between enamel coated with salivary pellicle and the bacterial surface (Qureshi and Gibbons, 1981; Ellen, 1982; Ellen and Sivendra, 1985). Strains of *A. viscosus* produce 'levan sucrases' (Warner and Miller, 1978) which may be free or cell-bound. In addition to this, a strain of *A. viscosus* (Ny 1) isolated from rats was shown to produce a complex extracellular polysaccharide (Van der Hoeven, 1974) and a similar polysaccharide has also been demonstrated in human strains (Rosan and Hammond, 1974). Strains of *A. naeslundii* and *A. viscosus* also produce fibrillar proteins which facilitate their attachment to surfaces and to other oral bacteria (Cisar *et al.*, 1978, 1984; Ellen *et al.*, 1978, 1980; Cisar and Vatter, 1979; Birdsell *et al.*, 1984; Ellen and Sivendra, 1985; Gabriel *et al.*, 1985). Musada *et al.* (1983) have shown that variation in fimbriae can be related to serological and numerical taxonomic clusters within *A. naeslundii* and *A. viscosus*.

When strains of *A. viscosus* are used in studies of their adherence and metabolic properties it is important to designate the source, as the animal strains (serovar 1) differ distinctly from the human strains (serovar 2). Relatively few well defined animal strains of *A. viscosus* are available and most studies include *A. viscosus* serovar 1 ATCC 15987, Ny 1 or WVU 440 (Fillery *et al.*, 1978); human isolates of *A. viscosus* serovar 2 are much more easily obtained. Little is known of any surface structures associated with the other species of *Actinomyces*. In the case of several of the species, no attempt has been made to describe or isolate extracellular polymers or surface components. However, electronmicroscopy, which reveals extensive fimbriae on strains of *A. viscosus* and *A. naeslundii*, does not show these structures on cells of *A. israelii* (Slack and Gerencser, 1975; Musada *et al.*, 1983; Birdsell *et al.*, 1984). Recently, attempts to remove surface proteins from strains of *A. israelii* with an alkaline buffer (Jenkinson *et al.*, 1981) have given very low yields of material (Bowden, unpublished data). The same technique applied to *A. viscosus* ATCC 15987 gave a high yield of fimbriae. This would suggest that *A. israelii* may not produce very extensive surface structures or extracellular polymers. Little is known on the surfaces of strains of *Arachnia*, although they appear to be relatively smooth (Lai and Listgarten, 1980).

Classification

The classification of *Actinomyces* remained somewhat confused until the development of chemotaxonomic methods (Goodfellow and Minnikin, 1985).

Studies of the cell wall composition of *Actinomyces* by Cummins and Harris (1958) and Cummins (1962) provided concrete evidence for the classification of *Actinomyces* as bacteria. Since this time, chemotaxonomic methods have made a considerable impact upon decisions on the classification of *Actinomyces* and *Arachnia* (Bowden and Hardie, 1978; Schaal, 1986a,b). Coupled to the definition of the two genera, chemotaxonomy, particularly as DNA homology, has contributed to the definition of new species of *Actinomyces* (Coykendal and Munzenmaier, 1979; Dent and Williams, 1984b,c, 1986). Given the relatively firm definition of species as a result of chemotaxonomy, serology, at both the species and serovar level, has been particularly valuable in aiding the classification of *Actinomyces* and *Arachnia* (Gerencser, 1979). Finally, several extensive numerical taxonomic studies of *Actinomyces* and *Arachnia*, involving traditional tests, chemotaxonomy and serology, have confirmed some established species and suggested the potential for subdivision within other taxa (Holmberg and Nord, 1975; Holmberg *et al.*, 1975; Fillery *et al.*, 1978; Gerencser, 1979; Schaal and Schofield, 1981; Schofield and Schaal, 1981; Schaal and Gatzer, 1985). Table 9.1 (page 134) shows those species and the accepted serovars which are currently included in *Actinomyces* and *Arachnia*. Following the early studies of Cummins and Harris (1958) and Cummins (1962), chemotaxonomic examination of the *Actinomyces* and *Arachnia* has kept pace with progress in the definition of species and genera of other bacteria. In general, DNA homology has been used as the final arbiter of species rank. However, other chemotaxonomic markers have proved extremely valuable for defining the genera and species.

A definitive study of the murein types of bacteria by Schleifer and Kandler (1972) included data on *A. israelii*, *A. naeslundii*, *A. viscosus* and *A. odontolyticus* and showed that the murein contained the amino acid ornithine. In contrast, *A. bovis* and *A. pyogenes* contain lysine as the dibasic amino acid. Quantitative data and details of the structure of *Actinomyces* murein can be found in several recent references (Dent and Williams, 1984b, 1985; Schleifer and Seidle, 1985). Few data are available on the composition of the murein of *Arachnia* which contains LLDAP (Schleifer and Kandler, 1972; Bowden and Hardie, 1978).

The carbohydrate composition of the walls of *Actinomyces* and *Arachnia* spp. has been studied by several workers (Slack and Gerencser, 1975; Bowden *et al.*, 1976; Bowden and Fillery, 1978; Buchanan *et al.*, 1984; Dent and Williams, 1984a,b, 1986) and can be useful in supporting the separation of species (Table 9.3). Combinations of various carbohydrates have been identified in *Actinomyces* and include the relatively unusual methyl pentose 6-deoxytalose. However, it seems likely that strains currently included in a single species may vary their cell wall carbohydrate composition (Dent and Williams 1984a, 1986). Differences in cell wall carbohydrate composition supported the separation of *A. denticolens* (Dent and Williams, 1984a), *A. howellii* (Dent and Williams, 1984b), *A. slackii* (Dent and Williams, 1986) and *A. hordeovulneris* (Buchanan *et al.*, 1984) from *A. viscosus* and *A. naeslundii*. These last two species are reported to contain 6-deoxytalose (Bowden *et al.* 1976; Dent and Williams, 1984a) whereas none of the previous species contain this carbohydrate (see Table 9.3). However, strains identified as *A. viscosus* (Buchanan *et al.*, 1984) or *A. naeslundii* (strain TF11) (Dent and Williams, 1984a)

Table 9.3 Carbohydrates present in the cell walls of *Actinomyces* and *Arachnia*

Species	Glucose	Galactose	Mannose	Rhamnose	6-deoxytalose	Glucosamine	Galactosamine
Actinomyces							
A. israelii							
serovar 1	T	+	−	−	−	+	−
serovar 2	T	+	−	+	−	+	+
A. bovis	T	−	−	+	+	+	−
A. naeslundii*	T	−	−	+	+	+	+
A. viscosus*	+	T	T	+	+	+	+
A. denticolens	−	−	+	−	−	−	−
A. howellii	+	−	−	+	−	ND	ND
A. slackii	+	+	−	+	−	ND	ND
A. odontolyticus	+	−	+	+	+	−	−
A. hordeovulneris	+	+	−	−	−	ND	ND
A. meyerii			NT				
A. pyogenes	+	−	T	+	−	ND	ND
Arachnia							
A. propionica	+	+	+	−	−	ND	ND

T, trace; ND, not determined; NT, not tested
*Strains may vary in composition

have been shown to lack 6-deoxytalose. It is possible, therefore, that 6-deoxytalose may not be a good marker to identify *A. viscosus* and *A. naeslundii* and that strains lacking 6-deoxytalose may be different species. It is apparent that further examination of the carbohydrate content of the cell wall of the facultative *Actinomyces* spp. is necessary before definitive statements on the relationship of the wall carbohydrate composition and species can be made. Whether these variations in carbohydrate can be translated into serovar differences is not known, as specific serovar antigens have not been precisely identified (Bowden *et al.* 1976; Bowden and Fillery, 1978). Table 9.3 lists those carbohydrates that have been detected in *Actinomyces* cell walls.

Studies on the composition of carbohydrates extracted from the cell walls of *Actinomyces* are few (Bowden *et al.*, 1976; Bowden and Fillery, 1978; Fillery *et al.* 1978). Trichloracetic acid-extracted carbohydrate generally represents between 5 and 24 per cent by weight of *Actinomyces* walls (Bowden, 1976; Bowden *et al.*, 1976; Bowden and Fillery, 1978). Strains of *Arachnia* have been reported to contain galactose, or glucose plus galactose (Boone and Pine, 1968; Johnson and Cummins, 1972; Bowden 1976) (Table 9.3) although Johnson and Cummins (1972) reported that some strains contained mannose.

A variety of other chemotaxonomic tests have been used to support the classification of *Actinomyces* and *Arachnia* at the generic level. These include various lipid analyses (Amdur *et al.*, 1978; Kropenstadt and Kutzner, 1978; Pandhi and Hannovand, 1978; Kropenstadt, 1985). In general, lipid analyses of *Actinomyces* have identified n-saturated and n-unsaturated fatty acids with no odd-numbered, branched or hydroxy acids. Similar profiles are found in *Arachnia*. Isoprenoid quinones have been identified in *Actinomyces* spp. (Collins and Jones, 1981; Kropenstadt, 1985).

Another chemotaxonomic method which has proved valuable in demonstrating variations in *Actinomyces* species and also identifying new species is the analysis of whole-cell proteins by polyacrylamide gel electrophoresis (PAGE) (Jackman, 1985). Studies by Dent and Williams (1984a,b, 1985, 1986), Cato *et al.* (1984) and McCormick *et al.* (1985) have shown the potential for differentiation within the *Actinomyces* by this technique. Distinct patterns are produced by different species and separation can be based on visual examination or some calculation of similarities (Dent and Williams 1984a,b, 1985, 1986; McCormick *et al.*, 1985). It is possible to use PAGE in a relatively sophisticated way with densitometry and computer analysis (Jackman, 1985) in taxonomic studies.

Acid end product analysis is valuable in differentiating between *Actinomyces* and *Arachnia*; the latter produce propionic acid (Bowden and Hardie, 1978).

The current status of *Actinomyces* and *Arachnia*, together with their taxonomy, are discussed by Schaal (1986a,b) and Schaal and Gatzer (1985). It is most likely that the genus *Actinomyces* is more complex than is currently accepted. Despite the apparent definition of such well documented species as *A. israelii* there are indications that these species represent groups of organisms (Schaal and Gatzer, 1985) which could be proposed as 'sub-species'. Two subdivisions of *A. israelii* are represented by the two accepted serovars, apparently two further serovars can be detected (Schaal and Gatzer, 1985). This confirms previous observations of some heterogenicity within *A. israelii* serotype 1 (Gerencser and Slack, 1975; Bowden *et al.*, 1976; Bowden and Fillery, 1978). Moreover, separation of these four serovars can also be made on the basis of biochemical characters (Schaal and Gatzer, 1985).

Strains which fall into the species *A. naeslundii* and *A. viscosus* can also be subdivided (Fillery *et al.*, 1978; Schaal and Schofield, 1981; Schofield and Schaal, 1981). Fillery *et al.* (1978) identified seven clusters within *A. viscosus* and *A. naeslundii*. Some clusters could be regarded as being 'typical' while others appeared to be intermediate between the typical clusters of *A. viscosus* and *A. naeslundii*. Cluster 1 contained 'typical' human *A. viscosus* (serovar 2) isolates and cluster 5 typical *A. naeslundii*. Cluster 7 included all the strains of *A. viscosus* of animal origin (serovar 1). The other clusters included 'intermediate' strains. Masuda *et al.* (1983) were able to identify clusters 1, 3 and 5 defined by Fillery *et al.* (1978) by using absorbed antisera to surface fimbriae. In addition to this, these authors suggested that cluster 3 (Fillery *et al.*, 1978) represented *A. naeslundii* serovar 3 (Slack and Gerencser, 1975; Gerencser, 1979). The relationship of a serovar designation to a true taxonomic position may be difficult to assess as relatively little is known of the nature of the antigens that define the serovars. Recent studies on the distribution of epitopes on strains of *A. viscosus* and *A. naeslundii* revealed by monoclonal antibodies have shown that strains possess a 'mosaic' of determinants, several of which are shared between strains of both species (Firtel and Fillery, 1986).

Studies on the DNA homologies within *A. naeslundii* and *A. viscosus* (Coykendall and Munzenmaier, 1979) generally support the observations taken from numerical taxonomy. The three recently-described species, *A. denticolens*, *A. howellii* and *A. slackii*, resemble *A. naeslundii* and *A. viscosus* but can be differentiated from these species (Dent and Williams 1984 a,b,c, 1985, 1986) by chemotaxonomic methods. Serological relationships between these species and *A. viscosus* and *A. naeslundii* have not been examined in any detail. However, Schaal (1986b) reports that *A. denticolens* showed little cross-reaction with other species of *Actinomyces* and appeared to be serologically distinct. The relatively recent isolation from animals of new species which are closely similar to *A. viscosus* and *A. naeslundii* (see above) suggests that there may be several species associated with the oral cavities of different animals.

It seems certain that further division will also be possible within *A. odontolyticus*. Relatively few strains of this species have been examined, except by Batty (1958). To date, two serovars have been described (Slack and Gerencser, 1975) and serovar 2 cross reacts with *A. pyogenes* (Schaal and Gatzer, 1985), but little more than this is known. It will be necessary to isolate and examine a large number of strains of *A. odontolyticus*, and also of *A. bovis*, using chemotaxonomic tests, before more definite statements on diversity within these species can be made. Cato *et al.* (1984) reported that 16 strains of *A. meyerii* formed a homogenous group when examined by routine biochemical tests, serology and PAGE. Study of *Actinomyces*, with the objective of better definition of the taxonomy of this genus, is necessary and would provide a fertile field for research.

It seems likely that the two serovars of *Arachnia* may represent separate species (Slack and Gerencser, 1975; Schaal and Schofield, 1981; Schaal and Gatzer, 1985). As with some of the *Actinomyces* spp., detailed studies of larger numbers of *Arachnia* strains are required.

Isolation

A review by Schaal (1984) gives detailed information on the isolation and identification of members of the Actinomycetaceae from clinical samples. This review includes methods which have been effective at the Institute of Hygiene in Cologne, one of the main centres for the processing of clinical specimens from actinomycete infections in Europe. This review should be read in conjunction with the following sections.

Routine laboratory methods

Media for isolation and cultivation

In general most species of *Actinomyces* and *Arachnia* will grow well on a good quality nutrient agar such as Blood Agar Base No. 2 CM271 (Oxoid) or Brain–Heart Infusion Agar (Difco) supplemented with 5 per cent v/v defibrinated horse or sheep blood (Bowden and Hardie 1973; Slack and Gerencser, 1975). *A. hordeovulneris* requires the addition of fetal calf serum to media (Buchanan *et al.*, 1984) and media for *A. meyerii* should include vitamin K_1 and Tween 80 (Cato *et al.*, 1984). Various broth media can be used, and again media such as Trypticase Soy Broth (BBL) or Brain–Heart Infusion Broth (Difco) are adequate. These broths may be enriched with laked horse blood (Oxoid) 0.5 per cent v/v to enhance the growth of some species. Semi-solid broth formulations with reducing components can be useful in primary isolation, as can cooked meat media. Schaal (1984, 1986b) describes several other non-selective media for isolation of *Actinomyces*. It is certain that these media are also effective and probably the decision on which media to use should rest with the individual and the availability of media. It is not correct to believe that *Actinomyces* and *Arachnia* are difficult to cultivate; most good quality standard media supplemented with blood or serum will support their growth.

Some selective media have been described for isolation of *Actinomyces* spp. or actinomycetes generally. Such media have been designed to select the genera from samples from the oral cavity and are not for use with clinical samples, although there seems no reason why they should not be useful in this respect. However, until detailed studies on the effectiveness of selective media relative to non-selective media for the isolation of *Actinomyces* spp. from clinical samples have been made, selective media should be used with care. Cultivating clinical samples on both selective and non-selective media could be employed as a routine. Schaal (1984) recommends using three or four different media.

The selective media include those for oral actinomycetes (Beighton and Colman, 1976) or for *A. viscosus* and *A. naeslundii* (Ellen and Balcerzak-Raczkowski, 1975; Kornman and Loesche, 1978; Zylber and Jordan, 1982). The selective agents employed include colistin, sodium fluoride and cadmium sulphate, and it is possible that individual strains of *Actinomyces* and *Arachnia* are inhibited by different concentrations of these agents.

Strains of *A. israelii, A. viscosus, A. naeslundii, A. odontolyticus* and *Arachnia* spp. grow on the defined media described by Christie and Porteus (1962a,b). A 'semi-defined' modification of this medium, in which several of the amino acids are replaced by an acid hydrolysate of casein, has been described by Bowden et al. (1976). The modified medium is very useful for studies on the physiology and antigenic structure of *Actinomyces* as it can be standardized and all components are dialysable. The formula for this medium is given in Table 9.2 (page 135). Generally, filter sterilization of the entire medium is preferred, but in the absence of a suitable filter system the medium can be prepared in bulk and sterilized by autoclaving at a pressure of $10 + 6.9$ KDA (10 lb/sq in) for a sufficient period, depending on the volume; Amounts of 1 litre can be autoclaved with a holding time of 20 min, larger volumes, of up to 45 litres, should be autoclaved for 40–60 min without glucose and the phosphate buffer, which should be filter-sterilized and added aseptically to the autoclaved base. The cells for inoculum should be grown in a non-defined complex medium and 1 litre of the semi-defined medium should be inoculated with cells from 20 ml of complex medium. As far as is known, the more recently described species of *Actinomyces* have not been tested for growth in this medium.

Isolation of Actinomyces and Arachnia

Isolation of both of these genera is relatively simple and requires no apparatus or media which would be unique to a specialist laboratory (Slack and Gerencser, 1975; Schaal, 1984). As mentioned above, the organisms will grow on most good quality standard laboratory media with blood or serum as an enrichment. Incubation in an anaerobic environment with added CO_2 is important, but this can be achieved in a simple candle jar or 'Fortner' plate (Schaal, 1984), in the absence of more sophisticated equipment. There is a suggestion (Schaal, 1984) that strict anaerobic conditions may suppress the growth of some catalase-positive *Actinomyces*. If a candle jar is not available the organisms can be grown from samples inoculated directly to reduced media such as thioglycollate broth with added sodium carbonate (1–2 per cent of sodium carbonate solution 1 per cent w/v).

The methods of sample taking and transport generally used for anaerobes are suitable for *Actinomyces* and care must be taken to avoid drying the specimens. Samples of pus from lesions can be taken into sterile syringes which are then sealed in tubes for transport, or commercial anaerobic sample tubes can be used. Granules of *A. israelii* or *Arachnia* may be isolated from pus by diluting a small sample with sterile saline, when they will fall to the bottom of the container.

As a routine, samples suspected of containing *Actinomyces* or *Arachnia* spp. should be cultured in a suitable broth and also plated on to an enriched, non-selective medium. One of the selective or enrichment media could also be employed. Fundamental to the isolation of *Actinomyces* and *Arachnia* is incubation for a sufficient period in an atmosphere containing additional CO_2 (see above). Relatively long incubation periods (7–14 days) should be given and the plates or broths examined every 48–72 h during this time. In the case of the more rapidly growing *Actinomyces* spp., e.g., *A. viscosus, A. naeslundii* and *A. odontolyticus*, cultures should be positive within three days.

The most effective method for isolating *Actinomyces* and *Arachnia* from oral samples is to plate dilutions of the samples on to non-selective media (Bowden et al., 1975, 1976; Fillery et al., 1978; Dent, 1979). The selective media described above can also be used, although the possibility of inhibition of some strains must be borne in mind. A major problem in isolating *Actinomyces* and *Arachnia* spp. from oral samples plated on to non-selective media is the recognition of their colonies among those of the large numbers of other bacterial species in the sample (Bowden and Hardie, 1973; Bowden et al.,

1979). The selection of 'suspect' colonies becomes easier with experience, and if studies of these genera in oral samples are contemplated, standard type–culture strains should be obtained, in order to become familiar with their colonial characters. Suspect colonies from plates should be subcultured on blood agar (see above) and subjected to a series of tests for identification.

Identification

Routine laboratory methods

Identification of *Actinomyces* and *Arachnia* spp. can be made relatively easily in routine laboratories. However, it must be recognized that there is a limit to the precision of identification of species and, obviously, antisera have to be available to define the serovars. Certain tests which can be run routinely, e.g., PAGE and the demonstration of enzyme activities by API-ZYM tests, can be useful in identification of *Actinomyces* species (Kilian, 1978; Dent and Williams, 1985). Although such tests may be employed, it should be noted that the number of times that a routine clinical laboratory needs to identify *Actinomyces* spp. may be too few to warrant their use.

The most effective of the routine tests for identification of the genus is the analysis of acid end products (Bowden and Hardie, 1978; Sweetman *et al.*, 1985). Suspect colonies can be subcultured into a suitable broth and incubated anaerobically with CO_2 for four to seven days, when the supernate can be analysed by gas–liquid chromatography (Holdeman *et al.*, 1977; Sweetman *et al.*, 1985). Differentation of *Actinomyces* and *Arachnia* from similar genera is shown in Table 9.4.

It is fairly common to use biochemical tests to support the identification of species of *Actinomyces* and *Arachnia*. However, despite the amount of data available (Slack and Gerencser, 1975; Schaal, 1984; Schaal, 1986a,b) the tests in isolation may be of relatively little value. Certainly, in a routine laboratory the large variety of substrates necessary will not always be available. Reference to Schaal's data (1984, 1986a,b) shows that many of the tests have little discriminating value. However, individual tests may be employed to help in defining a particular species. Schaal (1984) gives tables of data that include tests which are of most value in identifying members of the Actinomycetaceae. Many of the test results in these tables are listed as variable and, if those tests which are definitely positive or negative are used, differentiation of a species may depend on only one or two tests. Table 9.5 shows some tests which may be useful, but it should be remembered that test results can vary with the media used (Slack and Gerencser, 1975).

In practice, colony and cellular morphology, atmospheric growth requirements and the catalase test, coupled with acid end product analysis, prove to be particularly valuable (Tables 9.4 and 9.5). When these tests have been made the addition of certain selected biochemical tests help to support the allocation of an isolate to a species. For example, *A. israelii* could be expected to ferment amygdalin, while the fermentation of glycogen and the hydrolysis of starch would support the identification of an

Table 9.4 Differentiation of *Actinomyces* and *Arachnia* from similar genera*

Character	*Actinomyces*	*Arachnia*	*Bacterionema* (*Corynebacterium*)	*Rothia*	*Propionibacterium*	*Bifidobacterium*
Growth						
$O_2 + CO_2$	V	+	+	+	−	−
Catalase	V	−	+	+	+	−
Acid products						
Acetic	+	+	+	+	+	+
Propionic	−	+	+	−	+	−
Lactic	+	+	+	+		+
Succinic	+			−		
Cell wall						
Ornithine	+†	−	−	−	−	+
DL-DAP	−	−	+	−	−	−
LL-DAP	−	+	−	−	+	−
Galactose‡	+	+	+	+	+	+
Arabinose	−	−	+	−	−	−
Rhamnose	V	−	−	−	−	V
6-deoxytalose	V	−	−	−	−	−

V, variable; † *A. bovis* has lysine as the dibasic amino acid; ‡ not all carbohydrates are shown
*Data from Bowden and Hardie (1978); Schaal (1986a,b) Dent and Williams (1986)

Table 9.5 Selected biochemical and physiological characteristics of *Actinomyces* and *Arachnia*

Character	*Actinomyces bovis*	*Actinomyces israelii*	*Actinomyces naeslundii*	*Actinomyces viscosus*	*Actinomyces denticola*	*Actinomyces bouellii*	*Actinomyces slackii*	*Actinomyces bordeovuli*	*Actinomyces odontolyticus*	*Actinomyces meyerii*	*Actinomyces pyogenes*	*Arachnia propionica*
Growth												
O_2	−	−	+	+	V	NK	+	NK	+	−	+	+
$O_2 + CO_2$	V	V	+	+	+	+	+	+	+	V	+	+
$AnO_2 + CO_2$	+	+	+	+	+	+	+	+	+	+	+	+
Catalase	−	−	−	+	−	+	+	+	−	−	−	−
Fermentation of												
Glucose	+	+	+	+	+	+	+	+	+	+	+	+
Amygdalin	−	+	V	−	−	−	−	NK	−	V	−	V
Cellobiose	−	+	V	−	−	−	−	+	−	−	V	−
Glycogen	+	−	V	V	NK	NK	NK	NK	V	V	+	−
Mannitol	−	V	−	−	V	−	−	−	−	−	−	+
Melibiose	V	+	+	V	NK	+	NK	+	−	−	−	V
Raffinose	−	+	+	+	+	+	+	+	−	−	−	+
Rhamnose	−	V	−	−	−	−	−	V	−	−	−	−
Sorbitol	−	V	−	−	−	NK	−	NK	−	−	V	V
Trehalose	−	V	+	V	−	+	V	+	V	−	V	NK
Arabinose	−	V	−	−	−	+	−	−	V	V	V	V
Salicin	−	+	V	V	+	−	−	NK	V	−	−	NK
Hydrolysis of												
Starch	+	−	−	−	NK	NK	NK	NK	V	−	V	−
Aesculin	V	+	+	+	−	+	NK	−	+	V	−	−
Urea	−	−	+	+	V	NK	NK	NK	−	−	V	−
Reduction of												
NO_3	−	V	+	V	+	NK	NK	−	−	−	−	NK
NO_2	−	−	V	−	−	NK	NK	NK	−	−	−	NK

NK, result is not known or is unclear; V, variable

anaerobic catalase-negative strain as *A. bovis*. Dent and Williams (1984a) have suggested that the inability of *A. denticolens* to ferment trehalose separates it from strains of *A. naeslundii*.

A technique which could be valuable in establishing the species of *Actinomyces* and *Arachnia* on a routine basis is the determination of the protein profile of a strain (Dent and Williams, 1984a,b, 1985, 1986). All that is necessary is equipment to run polyacrylamide gels.

In the past this equipment was relatively bulky; gels have been 16 × 16 cm, or more. Recently, smaller units have been introduced and, with these, a gel can be run in 1 h. PAGE has been applied to the identification and analysis of *Actinomyces* (Dent and Williams, 1984a,b, 1985; McCormick *et al.*, 1985). The preparation of samples may involve cell disruption and centrifugation (Dent and Williams, 1985) and McCormick *et al.* (1985) have used a novel system in which urea acts to liberate protein from whole cells. This method would be particularly applicable to cells for electrophoresis in a routine laboratory. Although sophisticated methods have been proposed for the analysis of protein profiles of bacteria in taxonomic studies (Jackman, 1985), visual comparisons (Dent and Williams, 1984a,b, 1985) are perfectly adequate to identify differences in profiles on a routine basis. PAGE will probably provide a useful adjunct to the basic tests described above in identifying *Actinomyces* and *Arachnia* spp.

Obviously, it is only possible to identify *Actinomyces* or *Arachnia* spp. serologically if standard sera are available. Such sera are usually only used in reference laboratories, or when *Actinomyces* and *Arachnia* form the basis of major areas of study (Slack and Gerencser, 1975; Gerencser, 1979; Schaal and Gatzer, 1985; Schaal, 1986a,b). Sera can be used in various standard techniques to identify the species and serotype of *Actinomyces* and *Arachnia* (see below). In most routine laboratories, the isolation of a given serovar may have little clinical significance. However, in studies of oral disease it may be necessary to identify to the level of the serovar. Moore *et al.* (1987) have produced data which show that while *A. naeslundii* serovar 1 is associated with healthy gingivae, *A. naeslundii* serovar 3 increases in gingivitis. These authors also reported isolating strains with the biochemical characteristics of *A. israelii, A. odontolyticus* and *A. naeslundii*, but which could not be identified as any of the currently accepted serovars.

Reference or research laboratory methods

Actinomyces and *Arachnia* spp. can be identified in more sophisticated laboratories by the range of techniques which are applicable to any other bacteria (Slack and Gerencser, 1975; Bowden and Hardie, 1978; Minnikin and Goodfellow, 1981; Schaal and Schofield, 1981; Dent and Williams, 1985; Schleifer and Seidle, 1985).

Chromatographic techniques for analysis of cell wall composition are easily applied and both the amino acid and carbohydrate composition (Table 9.3, page 137) of *Actinomyces* and *Arachnia* cell walls are valuable in identification. The distinct nature of the murein of the *Actinomyces* and *Arachnia* (Schleifer and Kandler, 1972; Bowden and Hardie, 1978; Dent and Williams, 1984a,b, 1985) has been used to support the placing of an isolate into *Actinomyces*. The techniques used for these analyses vary but in the main the standard cell wall material is produced by mechanical disruption followed by digestion with proteolytic enzymes (Cummins, 1962; Dent and Williams, 1984a; Schleifer and Seidle, 1985). The simplest machine for cell disruption is the 'Mickle Disintegrator' (Mickle Engineering Co., Surrey, UK) although the Braun homogenizer (Braun, Melsungen, Federal Republic of Germany) will take larger volumes of cells, as will the 'Bead Beater' (Biospec. Products, Bartlesville, Oklahoma, USA). Cell walls prepared in the above manner contain murein and associated cell wall carbohydrate and, therefore, can be used both for amino acid and carbohydrate analysis. To determine the amino acid composition accurately, or to quantitate the ratios of amino acids, it is necessary to remove the carbohydrate. This can be done most easily by extraction of the cell walls with trichloracetic acid (Perkins, 1963; Schleifer, 1975) although formamide and sodium hydroxide have also been proposed (Schleifer and Seidle, 1985). A particularly valuable method of producing murein for amino acid analysis is that of Schleifer and Kandler (1972) which was used for *Actinomyces* by Dent and Williams, (1984a,b, 1986). The analysis of the amino acid composition of whole cell walls or purified murein can be made by routine methods (Schleifer and Seidle, 1985). Automatic amino acid analysis is used for quantitative studies.

It is apparent that acid end product analysis, protein profiles and cell wall analysis can form a firm basis for identifying *Actinomyces* and *Arachnia*. This is evidenced by the work of Dent and Williams (1984a,b,c, 1986), in which these tests were the foundation of the identification of new species.

However, as with all bacteria, the analysis of DNA is necessary to confirm differences between strains which are thought to be different species. Isolation of DNA from *Actinomyces* generally requires some pretreatment of cells before extraction. Dent and Williams (1984a) treated growing cells with glycine or benzyl penicillin (Dent and Williams, 1984b, 1986); these methods aid lysis of the cells. Cell lysis can also be achieved by pretreatment with lysozyme or a proteinase followed by sodium dodecyl sulphate (SDS) (Dent and Williams, 1984b, 1986). Buchanan *et al.* (1984) used a sarcosyl buffer (sodium lauryl sarcosinate) and glass beads to isolate DNA from *A. hordeovulneris*. Standard techniques are used to measure the DNA homology of isolates of *Actinomyces* (Coykendal and Munzenmaier, 1979; Dent and Williams, 1986). Very few studies have been made on the DNA from strains of *Arachnia* (Johnson and Cummins, 1982).

Several other chemataxonomic methods can be used to support the identification of a strain as a member of the genera *Actinomyces* and *Arachnia* (see section on classification) but the methods given above are probably the most valuable and easily applicable.

One method for identification of *Actinomyces* has been in use for many years. Before the advent of many of the chemotaxonomic methods, *Actinomyces* were being identified serologically. Serology is perhaps one of the earliest of all the chemotaxonomic tests which involve the identification of specific components of the bacterial cell. The most common of the methods in use is fluorescent antibody recognition of *Actinomyces* and *Arachnia* in clinical specimens or pure culture (Gerencser, 1979). Gerencser (1979) states 'Since biochemical identification presents problems, serological identification has been attempted for many years'. As a result of this, most reference laboratories use serology as an integral part of their identification schemes (Schaal, 1984). The use of serology has allowed the identification of *Actinomyces* species, but also has indicated serovar variation within species (see Table 9.1, page 134).

Discussion of the definition of *Actinomyces* spp. and serovars and their cross-reactivities is presented by Gerencser (1979) and Schaal and Gatzer (1985). Cross-reactions between species can be relatively extensive at low dilutions of antisera. However, as might be expected, at high dilutions the sera become more specific and absorbtion can produce relatively specific antisera. Those cross-reactions which have been detected within *Actinomyces* spp. are described by Gerencser (1979) and Schaal and Gatzer (1985) and cross-reactions may be reciprocal or only one way (Gerencser, 1979). Double diffusion in agar gels (Bowden *et al.*, 1976; Bowden and Fillery, 1978; Schaal and Gatzer, 1985) can be used in identification and Holmberg *et al.* (1975) described standardized crossed immunoelectrophoretic techniques for *A. israelii*. Generally, diffusion techniques support the separation of serovars identified by fluorescent antibodies. The preparation of antisera and reagents for fluorescent antibody identification and the techniques are described in detail by Gerencser (1979) and Schaal (1984).

Despite the extensive use of serology in the identification of *Actinomyces* and *Arachnia* spp., relatively little is known of the antigens responsible for species- and serovar-specific reactions. There seems little doubt that, in common with most gram-positive bacteria, the cell wall carbohydrate polymers are significant antigens (Bowden *et al.*, 1976; Bowden and Fillery, 1978). In addition to the carbohydrates, a group of polypeptide antigens (Bowden *et al.*, 1976) and amphipathic antigens (Wicken *et al.*, 1978) exist and will be involved in the reactions of strains with fluorescent antibody. The carbohydrate antigens of the cell wall have been shown to be responsible for the species and serovar definition of *A. israelii* (Bowden and Fillery, 1978) and strains which may represent new serovars were identified on the basis of the wall carbohydrates (Bowden and Fillery, 1978). This observation is supported by Gerencser (1979) and Schaal and Gatzer (1985). These last authors found four serological groupings within *A. israelii*; two of these were the accepted serovars 1 and 2, the others were represented by subclusters 1c and 1d, which had been defined in a previous numerical taxonomic study (Schofield and Schaal, 1981). Recent isolates were used to produce sera to subclusters 1c and 1d, and culture-collection strains ATCC 10048 and ATCC 12102 were used to produce sera against *A. israelii* serotype 1. Bowden and Fillery (1978) had noted previously that culture-collection strains may lack the wall carbohydrate antigens present in fresh isolates. One of the strains of *A. israelii* 1 examined by Bowden and Fillery seemed to be distinct from other recent isolates. It is not known whether the variant strains in *A. israelii* 1 demonstrated by Bowden and Fillery (1978) are the same as subclusters 1c and 1d of Schaal and Gatzer (1985). However, if they are, it suggests that wall carbohydrate could be a useful antigen to define serovars.

As has been stated earlier, almost nothing is known of the wall carbohydrate or the other antigens which define the serovars of most *Actinomyces* or *Arachnia*. Definition and standardization of the serovar antigens could add to the value of serology in the taxonomy of *Actinomyces* and *Arachnia* by giving a more precise picture of relationships between species and serovars.

Pathogenicity

The pathogenesis and clinical significance of infections due to *Actinomyces* and *Arachnia* spp. has been reviewed relatively recently (Beaman, 1984; Schaal and Beaman, 1984; Schaal and Pulverer, 1984). In general, strains of *Actinomyces* and *Arachnia* cannot be considered as virulent pathogens. Most infections involving these genera are mixed and the roles played by other organisms in the lesions are not known in detail. However, some studies have suggested nutritional relationships (Mayrand and McBride, 1980) or a protective role for the *Actinomyces* (Jordan and Kelly, 1982). It is possible that *Actinomyces* may have the capacity to modify the immune response of the host (Bowden, 1984) by inducing suppression (Fitzgerald and Birdsell, 1982; Burckhardt, 1984).

Despite their relatively low virulence, *Actinomyces* and *Arachnia* will produce lesions in experimental animals (Slack and Gerencser, 1975; Beaman *et al.*, 1979; Beaman, 1984; Bowden, 1984; Burckhardt, 1984). The most common laboratory animal used for experimental infections is the mouse, although successful infections have been produced in guinea-pigs, hamsters and rabbits (Slack and Gerencser, 1975; Beaman, 1984; Bowden, 1984). It is therefore possible to study infections in the laboratory. Recently, studies of mixed infections including *Actinomyces* in guinea-pigs have been undertaken by Grenier and Mayrand (1983). Although *A. israelii* was a component of a mixture of organisms which produced a transmissible infection, it was not essential to infections.

There is no doubt that the infected host mounts an immune response to *Actinomyces*; from the data of Colebrook (1920) and Holmberg (1981) it can be seen that antibodies to *A. israelii* are present in the sera of patients. However, relatively few attempts have been made to use serology in the diagnosis of infection with *Actinomyces* and *Arachnia* spp. (Schaal, 1984). Serological diagnosis might be complicated by the presence of *Actinomyces* as components of the normal oral flora, and almost nothing is known of the nature of the antibody response to oral *Actinomyces* in man. Together with antibody responses, some attempts have been made to measure cell-mediated responses in patients and uninfected persons. Skin tests in patients are apparently negative (Matheson *et al.*, 1935; Emmons, 1938) although skin responses to extracts of *Actinomyces* were positive in uninfected persons (Matheson *et al.*, 1935). This latter observation has been confirmed by Nissengard (1977) in studies of the relationships of *Actinomyces* to periodontal disease in normal persons.

The source of *Actinomyces* and *Arachnia* in infections is most likely to be the oral cavity and the route of spread is haematogenous (Bowden and Hardie, 1973; Bowden, 1984). However, infections involving intrauterine contraceptive devices (Schaal and Pulverer, 1984) suggest that *Actinomyces* and *Arachnia* may colonize the uterine cavity. Nevertheless, Pulverer and Schaal (1984), in studies of over 3000 cases of actinomycosis in man, note that bacteria commensal in the oral cavity were most commonly associated with the infection. This observation supports the suggestion that the oral cavity is the source of infection in most cases of the disease. It is of interest to note that *Actinobacillus* (*Haemophilus*) *actinomycetemcomitans* is often associated with *A. israelii* and that these infections are among the most severe (Pulverer and Schaal, 1984). *Actinobac. actinomycetemcomitans* has been implicated as a significant organism in a specific type of periodontal disease, and may enhance the virulence of other bacteria, including *Actinomyces* (Bowden, 1984).

One area which deserves attention in this section is the suggestion that *Actinomyces* are significantly involved in the production of periodontal diseases (Jordan *et al.*, 1972; Guggenheim and Schroeder, 1974; Loesche and Syed, 1978; Syed and Loesche, 1978; Ellen, 1982) and that they also contribute to the development of root surface caries lesions (Jordan and Sumney, 1973; Syed *et al.*, 1975; Behbehani *et al.*, 1983). There is no doubt that the oral flora is extremely complex, examination of only one of the publications by Moore *et al.* (1987) on the subgingival flora will make this abundantly clear. Studies on the supragingival flora also demonstrate the diversity of species in the community (Bowden *et al.*, 1979). *Actinomyces* contributes a major portion of the supragingival flora, and strains can be isolated from most tooth surfaces. The initial studies (Jordan and Sumney, 1973) of an association of *Actinomyces* with root surface lesions have not been confirmed by more recent and extensive surveys (Ellen *et al.*, 1985). However, certain strains of *A. viscosus* do increase in numbers relative to *A. naeslundii* in the developing lesions of nursing caries (Milnes and Bowden, unpublished data). This observation, coupled to that of Komiyama *et al.* (1986), suggests that certain strains of *Actinomyces* may be able to establish in an acid environment. Further studies are required in order to define the ecology of *Actinomyces* in the oral cavity.

A. viscosus can produce periodontal destruction in experimental infections of hamsters and rats (Jordan *et al.*, 1972; Guggenheim and Schroeder,

1974). Consequently, strains of *A. viscosus* have been used to test the responses of lymphocytes from patients with periodontal disease (Burkhardt, 1984). *A. viscosus* has also been shown to have the capacity to act as a mitogen for B cells (Clagett and Engel, 1984).

During the 1970s considerable effort was expended in an attempt to define the host immune response to *A. viscosus*, and also to relate this response to periodontal disease (Baker *et al.*, 1976; Lang and Smith, 1977a,b; Lehner *et al.*, 1974). Detailed discussion of this aspect of *Actinomyces* pathogenicity is beyond the scope of this chapter. However, those interested will find the articles by Burckhardt (1984) and Ellen (1982) useful reading. It should be noted, however, that microbiological studies by Socransky *et al.* (1982) indicate that *Actinomyces* spp. are associated with healthy gingiva. Given the diversity of the oral flora, the potential for different strains of *A. viscosus* and *A. naeslundii*, and the effect that normal carriage of these bacteria may have on the immune response, it seems probable that identifying a unique and major role for *A. viscosus* in periodontal destruction will be difficult.

References

Amdur B H, Szabo E I, Socransky S S. (1978). Fatty acids of gram-positive bacterial rods from human dental plaque. *Archives of Oral Biology* 23: 23–29.

Baker J J, Chan S P, Socransky S S, Oppenheimer J J, Mergenhagen S E. (1976). Importance of *Actinomyces* and certain gram-negative organisms in the transformation of lymphocytes from patients with periodontal disease. *Infection and Immunity* 13: 1363–1368.

Batty I. (1958). *Actinomyces odontolyticus*, a new species of actinomycete regularly isolated from deep carious dentine. *Journal of Pathology and Bacteriology* 75: 455–459.

Beaman B L. (1984). Actinomycete pathogenesis. In *The biology of the actinomycetes*, pp. 457–479. Edited by Goodfellow M *et al.* Academic Press, London.

Beaman B L, Gershwin M E, Maslan S. (1979). Infectious agents in immunodeficient murine models:pathogenicity of *Actinomyces israelii* serotype 1 in congenitally athymic (nude) mice. *Infection and Immunity* 24: 583–585.

Behbehani M J, Jordan H V, Heeley J D. (1983). Oral colonisation and pathogenicity of *Actinomyces israelii* in gnotobiotic rats. *Journal of Dental Research* 62: 69–74.

Beighton D, Colman G. (1976). A medium for the isolation and enumeration of oral Actinomycetaceae from dental plaque. *Journal of Dental Research.* 55: 875–878.

Birdsell D C, Callihan D R, Powell J T, Fischschweiger W. (1984). Adherence of *Actinomyces viscosus* to teeth and its role in pathogenesis. In *Biological, biochemical and biomedical aspects of actinomycetes.* pp. 33–46. Edited by Ortiz-Ortiz L *et al.* Academic Press, New York.

Boone C J, Pine L. (1968). Rapid method for characterisation of actinomycetes by cell wall composition. *Applied Microbiology* 16: 279–284.

Bowden G H. (1976). MPhil Thesis, University of London.

Bowden G H. (1984). Pathogenesis of *Actinomyces israelii* infections. In *Biological, biochemical and biomedical aspects of actinomycetes*, pp. 1–10. Edited by Ortiz-Ortiz L *et al.* Academic Press, New York.

Bowden, G H W, Ellwood D E, Hamilton I R. (1979). Microbial ecology of the oral cavity. In *Advances in microbial ecology*, vol. 3. pp. 135–217. Edited by Alexander M Plenum Press, New York.

Bowden G H W, Fillery E D. (1978). Wall carbohydrate antigens of *A. israelii*. In *Advances in experimental medicine and biology*, vol. 107. pp. 685–693. Edited by McGhee J H *et al.* Plenum Press, New York.

Bowden G H W, Hardie J M. (1973). Commensal and pathogenic *Actinomyces* species in man. In *Actinomycetales. Characteristics and practical importance.* pp. 277–295. Edited by Sykes, Skinner Academic Press, London.

Bowden G H, Hardie J M. (1978). Oral pleomorphic (coryneform) gram-positive rods. In *Coryneform bacteria.* pp. 235–263. Edited by Bousefield I J, Callely A G. Academic Press, London.

Bowden G H W, Hardie J M, Fillery E D. (1976). Antigens from *Actinomyces* species and their value in identification. *Journal of Dental Research* 55 Special Issue A: A192–A204.

Bowden G H, Hardie J M, Slack G L. (1975). Microbial variations in approximal dental plaque. *Caries Research* 9: 253–277.

Buchanan B B, Pine L. (1965). Relationship of carbon dioxide to aspartic acid and glutamic acid in *Actinomyces naeslundii*. *Journal of Bacteriology* 89: 729–733.

Buchanan B B, Pine L. (1967). Path of glucose breakdown and cell yields of a facultative anaerobe *Actinomyces naeslundii*. *Journal of General Microbiology* 46: 225–236.

Buchanan A M, Scott J L, Gerencser M A, Beaman B L, Jang S, Biberstein E L. (1984). *Actinomyces hordeovulneris* sp. nov. An agent of canine actinomycosis. *International Journal of Systematic Bacteriology* 34: 439–443.

Burckhardt J J. (1984). Lymphocyte interactions in host responses to oral infection caused by *Actinomyces viscosus*. In *Biological, biochemical and biomedical aspects of actinomycetes*, pp. 47–60. Edited by Ortiz-Ortiz L *et al.* Academic Press, New York.

Cato E P, Moore W E C, Nygaard G, Holdeman L V. (1984).

Actinomyces meyeri sp. nov. Specific epithet rev. *International Journal of Systematic Bacteriology* **34**: 487–489.

Christie A O, Porteus J W. (1962a). The cultivation of a single strain of *Actinomyces israelii* in a simplified and chemically defined medium. *Journal of General Microbiology* **28**: 487–489.

Christie A O, Porteus J W. (1962b). The growth factor requirements of the Wills strain of *Actinomyces israeii* growing in a chemically defined medium. *Journal of General Microbiology* **28**: 455–460.

Christie A O, Porteus J W. (1962c). Growth of several strains of *A. israelii* in chemically defined medium. *Nature* **195**: 408–409.

Cisar J O, Sandberg A L, Mergenhagen S E. (1984). The function and distribution of different fimbriae on strains of *Actinomyces viscosus* and *Actinomyces naeslundii*. *Journal of Dental Research* **63**: 393–396.

Cisar J O, Vatter A E. (1979). Surface fibrils (fimbriae) of *Actinomyces viscosus* T. 14V. *Infection and Immunity* **24**: 523–531.

Cisar J O, Vatter A E, McIntire F C. (1978). Identification of the virulence associated antigen on the surface fibrils of *Actinomyces viscosus* T.14. *Infection and Immunity* **19**: 312–319.

Clagett J A, Engel L D. (1984). Polyclonal B-cell activation in response to *Actinomyces viscosus*—its nature and genetics. In *Biological, biochemical and biomedical aspects of actinomycetes*, pp. 61–71. Edited by Ortiz-Ortiz L et al. Academic Press, New York.

Colebrook L. (1920). The mycelial and other microorganisms associated with human actinomycosis. *British Journal of Experimental Pathology* **1**: 197–212.

Collins M D, Jones D. (1981). Distribution of isoprenoid quinone structural types in bacteria and their taxonomic implications. *Microbiological Reviews* **45**: 316–354.

Coykendall A L, Munzenmaier A J. (1979). Deoxyribonucleic acid hybridization among strains of *Actinomyces viscosus* and *Actinomyces naeslundii*. *International Journal of Systematic Bacteriology* **29**: 234–240.

Cummins C S. (1962). Chemical composition and antigenic studies of cell walls of *Corynebacterium*, *Mycobacterium*, *Nocardia*, *Actinomyces* and *Arthrobacter*. *Journal of General Microbiology* **39**: 35–50.

Cummins C S, Harris H. (1958). Studies on the cell wall composition and taxonomy of *Actinomycetales* and related groups. *Journal of General Microbiology* **18**: 173–189.

Dent V E. (1979). The bacteriology of dental plaque from a variety of zoo-maintained mammalian species. *Archives of Oral Biology* **24**: 277–282.

Dent V E, Marsh P O. (1981). Evidence for a basic plaque microbial community on the tooth surface of animals. *Archives of Oral Biology* **26**: 171–179.

Dent V E, Williams R A D. (1984a). *Actinomyces denticolens* sp. nov.: a new species from the dental plaque of dairy cattle. *International Journal of Systematic Bacteriology* **34**: 316–320.

Dent V E, Williams R A D. (1984b). *Actinomyces howellii*, a new species from the dental plaque of dairy cattle. *International Journal of Systematic Bacteriology* **34**: 316–320.

Dent V E, Williams R A D. (1984c). Deoxyribonucleic acid reassociation between *Actinomyces denticolens* and other *Actinomyces* species from dental plaque. *International Journal of Systematic Bacteriology* **34**: 501–502.

Dent V E, Williams R A D. (1985). A combined biochemical approach to the taxonomy of gram positive rods. In *Chemical methods in bacteriological systemics*, Society for Applied Bacteriology, Technical Series No. 20, pp. 341–357. Edited by Goodfellow M, Minnikin D E. Academic Press, London.

Dent V E, Williams R A D. (1986). *Actinomyces slackii* sp. nov. from dental plaque of dairy cattle. *International Journal of Systematic Bacteriology* **36**: 392–395.

Ellen R P. (1982). Oral colonization by gram positive bacteria significant to periodontal diseases. In *Host-parasite interactions in periodontal diseases*, pp. 98–111. Edited by Genco R J, Mergenhagen S E. American Society for Microbiology, Washington DC.

Ellen R P, Balcerzak-Raczkowski R P. (1975). Differential medium for detecting dental plaque bacteria resembling *Actinomyces viscosus* and *Actinomyces naeslundii*. *Journal of Clinical Microbiology* **2**: 305–310.

Ellen R P, Balcerzak-Raczkowski R P, Fillery E D, Chan K H, Grove D A. (1980). Sialidase-enhanced lectin like mechanism for *Actinomyces viscosus* and *Actinomyces naeslundii* haemagglutination. *Infection and Immunity* **27**: 335–343.

Ellen R P, Banting D W, Fillery E D. (1985). Longitudinal microbiological investigation of a hospitalized population of older adults with a high root surface caries risk. *Journal of Dental Research* **64**: 1377–1381.

Ellen R P, Fillery E D, Chan K H, Grove D A. (1980). Sialidase-enhanced lectin like mechanism for *Actinomyces viscosus* and *Actinomyces naeslundii* haemagglutination. *Infection and Immunity* **27**: 335–343.

Ellen R P, Sivendra R. (1985). In vitro attachment, salivary agglutination and surface fibril density of fresh *Actinomyces* isolates from two distinct oral surfaces. *Journal of Dental Research* **64**: 799–803.

Ellen R P, Walker D L, Chan K H. (1978). Association of long surface appendages with adherence related functions of the gram positive species *Actinomyces naeslundii*. *Journal of Bacteriology* **134**: 1171–1175.

Emmons C W. (1938). The isolation of *Actinomyces bovis* from tonsillar granules. *Public Health Report* **53**: 1967–1975.

Fillery E D, Bowden G H, Hardie J M. (1978). A comparison of strains of bacteria designated *Actinomyces viscosus* and *Actinomyces naeslundii*. *Caries Research* **12**: 299–312.

Firtel M, Fillery E D. (1986). The distribution of antigenic determinants between *Actinomyces viscosus* and *Actinomyces naeslundii*. *Journal of Dental Research* **65**: Abstract No. 40, 535.

Fitzgerald J E, Birdsell H E. (1982). Systemic immune response to oral colonisation. *Journal of Periodontal Research* **17**: 237–246.

Gabriel O, Heeb M, Hinrichs M. (1985) Interaction of the surface adhesions of the oral *Actinomyces* spp. with mammalian cells. In *Molecular basis of oral microbial adhesion*, pp. 45–52. Edited by Mergenhagen S E, Rosan B. American Society for Microbiology, Washington DC.

Gerencser M A. (1979). The application of fluorescent antibody techniques to the identification of *Actinomyces* and *Arachnia*. In *Methods in microbiology*, No. 13, pp. 187–321. Edited by Bergan T, Norris J R. Academic Press, London.

Goodfellow M, Minnikin D E. (1985). Introduction to chemosystematics. In *Chemical methods in bacterial systematics*, Society for Applied Bacteriology Technical Series Vol 20, pp. 1–15. Edited by Goodfellow M, Minnikin D E. Academic Press, London.

Grenier D, Mayrand D. (1983). Etudes d'infections mixtes anaerobies comportant *Bacteroides gingivalis*. *Canadian Journal of Microbiology* **29**: 612–618.

Guggenheim B, Schroeder H E. (1974). Reactions in the periodontium to continuous antigenic stimulation in sensitised gnotobiotic rats. *Infection and immunity* **10**: 565–572.

Hale C M F. (1953). The use of phosphomolybdic acid in the demonstration of bacterial cell walls. *Laboratory Practice* **2**: 115–118.

Hamilton I R. (1987). Effects of changing environment on sugar transport and metabolism by oral bacteria. In *Sugar transport and metabolism in gram positive bacteria*, pp. 94–113. Edited by Reizer A, Peterkofsky A. Ellis Horwood, Chichester.

Hamilton I R, Ellwood D C. (1983). Carbohydrate metabolism by *Actinomyces viscosus* growing in continuous culture. *Infection and Immunity* **42**: 19–26.

Holdeman L V, Cato E P, Moore W E C. (1977). *Anaerobe laboratory manual*, 4th edn. Virginia Polytechnic Institute and State University Blacksburg.

Holmberg K. (1981). Immunodiagnosis of human actinomycosis. In Actinomycetes. *Zentrallblatt fur Bakteriologie, Mikrobiologie und Hygiene. 1 Abteilung*, Suppl 11, pp. 259–261. Edited by Schaal K P, Pulverer G. Gustav Fischer, Verlag, Stuttgart.

Holmberg K, Nord C E. (1975). Numerical taxonomy and laboratory identification of *Actinomyces* and *Arachnia* and some related bacteria. *Journal of General Microbiology* **91**: 17–44.

Holmberg K, Nord C E, Wadstrom I. (1975). Serological studies of *A. israelii* by crossed immunoelectrophoresis: standard antigen-antibody system for *A. israelii*. *Infection and Immunity* **12**: 387–397.

Howell A, Pine L. (1956). Studies on the growth of species of *Actinomyces* I. Cultivation in a synthetic medium with starch. *Journal of Bacteriology* **71**: 47–53.

Jackman P J H. (1985). Bacterial taxonomy based on electrophoretic whole cell protein profiles. In *Chemical methods in bacterial systematics*, pp. 115–129. Edited by Goodfellow M, Minnikin D E. Academic Press, London.

Jenkinson H F, Sawyer W D, Mandelstam J. (1981). Synthesis and order to assembly of spore coat proteins in *Bacillus subtilis*. *Journal of General Microbiology* **123**: 1–16.

Johnson J L, Cummins C S. (1972). Cell wall composition and deoxyribonucleic acid similarities among the anaerobic coryneforms, chemical propionibacteria and strains of *Arachnia propionica*. *Journal of Bacteriology* **109**: 1047–1066.

Jordan H V, Kelly D M. (1982). Persistence of *Eikenella corrodens* and *Actinobacillus actinomycetemcomitans* in mixed actinomycotic lesions in mice. *Journal of Dental Research* **61**: 231 Abstract No. 478.

Jordan H V, Keyes P H, Bellack S. (1972). Periodontal lesions in hamsters and gnotobiotic rats infected with *Actinomyces* of human origin. *Journal of Periodontal Research* **7**: 21–28.

Jordan H V, Sumney D L. (1973). Root surface caries: a review of the literature and significance of the problem. *Journal of Periodontology* **44**: 158–163.

Kiel R A, Porteus J W. (1962). The amino acid requirements of a single strain of *Actinomyces israelii* growing in a clinically defined medium. *Journal of General Microbiology* **28**: 193–201.

Kiel M A, Tanzer J M. (1977). Regulation of invertase of *Actinomyces viscosus*. *Infection and Immunity* **17**: 510–12.

Kilian M. (1978). Rapid identification of Actinomycetaceae and related bacteria. *Journal of Clinical Microbiology* **8**: 127–133.

Komiyama K, Khandelwal R L, Duncan D E. (1986). Glycogen synthetic abilities of *Actinomyces viscosus* and *Actinomyces naeslundii* freshly isolated from dental plaque and root surface caries lesions and non-carious sites. *Journal of Dental Research* **65**: 899–902.

Kornman K S, Loesche W J. (1978). New medium for isolation of *Actinomyces viscosus* and *Actinomyces naeslundii* from dental plaque. *Journal of Clinical Microbiology* **7**: 514–518.

Kropenstadt R M. (1985). Fatty acid and menaquinone analysis of actinomycetes and related organisms. In *Chemical methods in bacterial systematics*, pp. 173–199. Edited by Goodfellow M, Minnikin D E. Academic Press, London.

Kropenstadt R M, Kutzner H J. (1978). Biochemical taxonomy of some problem actinomycetes. *Zentrallblatt fur Bakteriologie, Parasitenkunde, Infektionskrankheiten und Hygiene. I. Abteilung*, Suppl 6, pp. 125–133.

Lai C H, Listergarten M A. (1980). Comparative ultrastructure of certain *Actinomyces* species, *Arachnia*, *Bacterionema* and *Rothia*. *Journal of Periodontology* **51**: 136–154.

Lang N P, Smith F N. (1977a). Lymphocyte blastogenesis to plaque antigens in human periodontal disease. *Journal of Periodontal Research* **12**: 298–309.

Lang N P, Smith F N. (1977b). Lymphocyte blastogenesis to plaque antigens in human periodontal disease II. The relationship to clinical parameters. *Journal of Periodontal Research* **12**: 310–317.

Lehner T, Wilton J M A, Challacombe S J, Ivanyi L. (1974). Sequential cell mediated immune responses in experimental gingivitis in man. *Clinical and Experimental Immunology* **16**: 481.

Loesche W J, Syed S A. (1978). Bacteriology of human experimental gingivitis: effect of plaque and gingivitis score. *Infection and Immunity* **21**: 830–839.

Mathieson D R, Harrison R, Hammond C, Henrici A T. (1935). Allergic reactions of actinomycetes. *American Journal of Hygiene* **21**: 405–421.

Mayrand D, McBride B C. (1980). Ecological relationships of bacteria involved in a simple, mixed anaerobic infection. *Infection and Immunity* **27**: 44–50.

McCormick S S, Mengoli H F, Gerencser M A. (1985). Polyacrylamide gel electrophoresis of whole cell preparations of *Actinomyces* sp. *International Journal of Systematic Bacteriology* **35**: 429–433.

Minnikin D E, Goodfellow M. (1981). Lipids in the classification of actinomycetes. *Zentralblatt fur Bakteriologie, Parasitenkunde, Infektionskrankheiten und Hygiene 1. Abteilung*, Suppl 11, pp. 99–109.

Moore W E C, Ranney R R, Holdeman L V. (1982). Subgingival microflora in periodontal disease: cultural studies. In *Host-parasite interactions in periodontal diseases*, pp. 13–26. Edited by Genco R J, Mergenhagen S E. American Society for Microbiology, Washington DC.

Moore L V H *et al.* (1987). Bacteriology of human gingivitis. *Journal of Dental Research* **66**: 989–995.

Musada N, Ellen R P, Fillery E D, Grove D A. (1983). Chemical and immunological comparison of surface fibrils of strains representing six taxonomic groups of *Actinomyces viscosus* and *Actinomyces naeslundii*. *Infection and Immunity* **39**: 1325–1333.

Nissengard R J. (1977). The role of immunology in periodontal diseases. *Journal of Periodontology* **48**: 505–516.

Pandhi P N, Hammond B F. (1978). The polar lipids of *Actinomyces viscosus*. *Archives of Oral Biology* **23**: 17–21.

Perkins H R. (1963). A polymer containing glucose and amino hexuronic acid isolated from the cell walls of *Micrococcus lysodeikticus*. *Biochemical Journal* **86**: 475.

Pulverer G, Schaal K P. (1984). Medical and microbiological problems in human actinomycetes. In *Biological, biochemical and biomedical aspects of actinomycetes*, pp. 161–170. Edited by Ortiz-Ortiz L *et al.* Academic Press, New York.

Qureshi V, Gibbons R J. (1981). Differences in the absorption rate of human strains of *Actinomyces viscosus* and *Actinomyces naeslundii*. *Infection and Immunity* **131**: 261–266.

Reddy C A, Cornell C P. (1982). Physiological and nutritional features of *Corynebacterium pyogenes*. *Journal of General Microbiology* **128**: 2851–2855.

Reddy C A, Cornell C P, Fraga A M. (1982). Transfer of *Corynebacterum pyogens* (Glage) Eberson to the genus *Actinomyces* as *Actinomyces pyogenes* (Glage) comb. nov. *International Journal of Systematic Bacteriology* **32**: 419–429.

Rosan B, Hammond B F. (1974). Extracellular polysaccharides of *Actinomyces viscosus*. *Infection and Immunity* **10**: 304–308.

Schaal K P. (1984). Laboratory diagnosis of actinomycete disease. In *The biology of the actinomycetes*, pp. 425–456. Edited by Goodfellow M *et al.* Academic Press, London.

Schaal K P. (1986a). Genus *Arachnia*. In *Bergey's Manual of systematic bacteriology*, vol 2, pp. 1332–1342. Edited by Sneath P H A *et al.* Williams and Wilkins, Baltimore.

Schaal K P. (1986b). Genus *Actinomyces*. In *Bergeys' Manual of systematic bacteriology*, vol 2, pp. 1383–1418. Edited by Sneath P H A *et al.* Williams and Wilkins, Baltimore.

Schaal K P, Beaman B L. (1984). Clinical significance of actinomycetes. In *The biology of the actinomycetes*, pp. 389–424. Edited by Goodfellow M *et al.* Academic Press, London.

Schaal K P, Gatzer R. (1985). Serological and numerical phenetic classification of clinically significant fermentative actinomycetes. In *Filamentous microorganisms, biomedical aspects*, pp. 85–109. Edited by Arai A. Japan Scientific Societies Press, Tokyo.

Schaal K P, Pulverer G. (1984). Epidemiologic, etiologic, diagnostic and therapeutic aspects of endogenous actinomycetes infections. In *Biological, biochemical and biomedical aspects of actinomycetes*, pp. 13–32. Edited by Ortiz-Ortiz L *et al.* Academic Press, New York.

Schaal K P, Schofield G M. (1981). Current ideas on the taxonomic status of the Actinomycetaceae. In *Actinomycetes, Proceedings of the 4th International symposium on actinomycete biology. Zentralblatt fur Bakteriologie, Mikrobiologie und Hygiene* Suppl 11, pp. 67–78. Edited by Schaal K P, Pulverer G. Gustave Fischer, Verlag, Stuttgart.

Schleifer K H. (1975). Chemical structure of the peptidoglycan; its modifiability and relation to the biological activity. *Zeitschrift fur Immunitatsforschung* **149**: 157–164.

Schleifer K H, Kandler O. (1972). Peptidoglycan types of bacterial cell walls and their taxonomic implications. *Bacteriological Reviews* **36**: 407–477.

Schleifer K H, Seidl P H. (1985). Chemical composition

and structure of murein. In *Chemical methods in bacterial systematics*, pp. 201–219. Edited by Goodfellow M, Minnikin D E. Academic Press, London.

Schofield G M, Schaal K P. (1981). A numerical taxonomic study of members of the Actinomycetaceae and related taxa. *Journal of General Microbiology* **127**: 237–259.

Slack J M, Gerencser M A. (1975). *Actinomyces filamentous bacteria. Biology and Pathogenicity*. Burgess, Minneapolis.

Socransky S S, Tanner A C R, Haffajee A D, Hillman J D, Goodson J M. (1982). Present status of studies on the microbial etiology of periodontal diseases. In *Host-parasite interactions in periodontal diseases*, pp. 1–12. Edited by Genco R J, Mergenhagen S E. American Society for Microbiology, Washington DC.

Sweetman D A, Geddes D A, MacFarlane T W, Shah H N, Nash K A, Hardie J M. (1985). Detection of acidic end products of metabolism of anaerobic gram negative bacteria. In *Chemical methods in bacterial systematics*, pp. 317–340. Edited by Goodfellow M, Minnikin D E. Academic Press, London.

Syed S A, Loesche W J. (1978). Bacteriology of human experimental gingivitis: effect of plaque age. *Infection and Immunity* **21**: 821.

Syed S A, Loesche W J, Pape H L. (1975). Predominant cultivable flora isolated from human root surface caries plaque. *Infection and Immunity* **11**: 727–731.

Van der Hoeven J S. (1974). A slime producing microorganism in dental plaque of rats selected by glucose feeding: chemical composition of extracellular slime elaborated by *Actinomyces viscosus* strain Nyl. *Caries Research* **8**: 183–210.

Warner T N, Miller C H. (1978). Cell-associated levan of *Actinomyces viscosus*. *Infection and Immunity* **19**: 711–719.

Wicken A J, Broody K W, Evans J D, Knox K W. (1978). New cellular and extracellular amphypathic antigens from *Actinomyces viscosus* Nyl. *Infection and Immunity* **22**: 615–616.

Zylber L J, Jordan H V. (1982). Development of a selective medium for detection and enunciation of *Actinomyces viscosus* and *Actinomyces naeslundii* in dental plaque. *Journal of Clinical Microbiology* **15**: 253–259.

10

Bifidobacteria

A K Roberts

The nature of bifidobacteria
Classification
Isolation
Identification
   Routine laboratory methods
   Specialist laboratory methods
Ecology of bifidobacteria in the intestine
   Occurrence
   Interaction with other organisms
   Effects of bifidobacteria on the host
Conclusions
References

Bifidobacteria were first discovered in the faeces of breast-fed infants by Tissier in 1899 (cited by Poupard *et al.*, 1973) who called the organisms *Bacillus bifidus*. In 1924 Orla-Jensen suggested that the organism should be classified in a separate genus, *Bifidobacterium*, but this classification was not generally accepted and researchers continued to classify the organisms as lactobacilli, actinomycetes or corynebacteria, etc. Even into the 1970s, bifidobacteria were referred to as *Lactobacillus bifidus* (Bullen and Willis, 1971; Willis *et al.*, 1973). However, since the 1960s a lot of time has been devoted to the classification of the species and of the biotypes of *Bifidobacterium* isolated from the intestinal flora of man and animals, and from extraenteral sources (Reuter, 1971; Scardovi *et al.*, 1971; Poupard *et al.*, 1973; Scardovi, 1981, 1986; Biavati *et al.*, 1982, 1986; Mitsuoka, 1984). Further attempts have been made to study the ecology of bifidobacteria in various habitats (Mitsuoka and Kaneuchi, 1977; Resnick and Levin, 1981a,b; Mitsuoka, 1982; Yoshioka *et al.*, 1983; Benno *et al.*, 1984; Biavati *et al.*, 1984; Mevissen-Verhage *et al.*, 1987).

In this chapter, the isolation, identification and ecology of bifidobacteria from the human intestine will be discussed.

The nature of bifidobacteria

Bifidobacteria are gram-positive, non-sporing obligately anaerobic rods. They may develop a degree of tolerance to air, but this usually occurs after the strain has been subcultured in the laboratory several times. Cultural conditions greatly affect cell morphology, but after anaerobic growth on blood agar they are thin rods 0.25 μm wide and up to 3 μm long, straight or slightly curved, and usually regular or with slight bulges at one pole. Gram-stained smears of infant faeces also reveal bifidobacteria with this morphology. In less ideal conditions the rods tend to branch and to have very large bulbous extremities. The name *Bifidobacterium* suggests that this bifurcated form is their normal morphology; this can be very misleading in the initial identification of isolates. One species, *B. bifidum*, branches to a greater or lesser extent under most cultural conditions.

Bifidobacteria obtain energy from the fermentation of various carbohydrates via a unique glycolysis pathway, the fructose-6-phosphate shunt. The key enzyme of this pathway, fructose-6-phosphate phosphoketolase, splits hexose phosphate to erythrose-4-phosphate and acetyl phosphate; the latter compound is converted to acetic acid. The erythrose-4-phosphate and an additional fructose-6-phosphate undergo conjugation and rearrangement, until they are finally converted to acetic and lactic acids, in the theoretical ratio of 3:2. The formation of formic acid and ethanol often alters this fermentation balance (Scardovi, 1981).

Ammonium can be used as a sole nitrogen source and most species require riboflavin, pantothenate and cysteine for growth. Other compounds, e.g., N-acetylglucosamine-containing saccharides and certain peptides, have been shown to enhance the growth of *B. bifidum in vitro*. The significance of these 'bifidus factors' is not yet understood (Scardovi, 1986; Petschow and Talbott, 1990).

Nine named species of *Bifidobacterium* have been isolated from human faeces (infant and adult) or the vagina (Table 10.1). Ten other species have been found in the intestines and faeces of various animals, e.g., pigs, cattle, chickens and rabbits. Three species are unique to the honey bee intestine and two species have been found only in sewage. *B. dentium* is the only species which has been tentatively implicated in pathogenicity; it has been isolated from lung abscesses, pleural fluid and dental caries as well as from adult faeces and the vagina, but a pathogenic role has not been proven (Scardovi, 1986; Lauer, 1990).

Members of the genus *Lactobacillus* are also gram-positive, non-sporing rods which, when isolated from faeces, can be confused morphologically with bifidobacteria. They are often microaerophilic, but some strains are anaerobic on isolation and will not grow in aerobic conditions. Lactobacilli and bifidobacteria are usually distinguished from each other by the fatty acid end products of fermentation. Lactobacilli produce primarily lactic acid via the Embden–Meyerhof pathway, although some species also produce detectable amounts of acetic acid.

Table 10.1 Human species of *Bifidobacterium* and their habitats

Species	Habitat
B. bifidum	Faeces of infant and adult
	Vagina
B. adolescentis	Faeces of infant and adult
	Sewage
B. infantis	Faeces of infant
	Vagina
B. breve	Faeces of infant
	Vagina
	Sewage
B. longum	Faeces of infant and adult
	Vagina
	Sewage
B. catenulatum	Faeces of infant and adult
	Vagina
	Sewage
B. dentium	Faeces of adult
	Vagina
	Dental caries; pleural fluid
B. angulatum	Faeces of adult
	Sewage
B. pseudocatenulatum	Faeces of infant
	Sewage
**B. gallicum*	Faeces of adult

*Only one strain isolated (Dsm 20093) (Lauer, 1990).

Table 10.2 The major classification schemes of human species of the genus *Bifidobacterium*

Eggerth, 1935	Dehnert, 1957	Reuter, 1963–64	Scardovi et al., 1971	Scardovi et al., 1974, 1979
I	I, II	*B. bifidum* a *B. bifidum* b		*B. bifidum*
	III	*B. breve* a *B. breve* b *B. parvulorum* a *B. parvulorum* b	*B. breve*	*B. breve*
	IV	*B. infantis* *B. liberorum* *B. lactentis*	*B. infantis*	*B. infantis*
II	V	*B. longum* a *B. longum* b		*B. longum*
		B. adolescentis a *B. adolescentis* b *B. adolescentis* c *B. adolescentis* d		*B. adolescentis*
				B. catenulatum *B. dentium* *B. angulatum* *B. pseudocatenulatum*

Although the amount of acetic acid produced is less than the concentrations produced by the bifidobacteria, it may, in practice, lead to confusion over the identification of the two genera.

Classification

In the current edition of *Bergey's Manual of Systematic Bacteriology* (Scardovi, 1986), 24 species of *Bifidobacterium* are listed; they have been isolated from a variety of animal habitats and from sewage. Before 1935, *B. bifidum* was the only recognized species of the genus. Eggerth (1935 cited by Poupard *et al.* 1973) realized that these organisms were not a homogeneous group and divided his strains into two groups based on the fermentation patterns of 12 carbohydrates; group I contained strains from infants and group II consisted of strains from adults. Dehnert (1957, cited by Poupard *et al.*, 1973) continued this work; on the basis of results from 24 carbohydrates, he split Eggerth's group I into four groups, giving a total of five groups of bifidobacteria.

However, Reuter (1964, cited by Poupard *et al.*, 1973) recognized eight species and several biotypes for strains isolated from human infants and adults.

In the early 1970s, with the help of DNA homology studies, Scardovi and colleagues established eleven new species of bifidobacteria, these came from diverse habitats, including the human intestine. These workers confirmed existing species, and also proposed that some of Reuter's species be merged into a single species (Poupard *et al.*, 1973) (Table 10.2).

The discovery of unique species in waste waters suggests that there may be an extraenteral source of bifidobacteria. With methods available to study genotypic characteristics, e.g., DNA homology plus more sophisticated chemotaxonomy, it should not be too long before more new species of bifidobacteria are described from enteral and extraenteral sources.

Isolation

A large variety of media have been devised for isolating or enumerating the bifidobacteria in faeces and sewage. All the media contain a carbohydrate source such as glucose, lactose, fructose or maltose. Various peptones are also included, e.g., trypticase and phytone (Scardovi, 1981), peptone and casamino acids (Resnick and Levin, 1981a). In some recipes manufactured complex substrates were used instead, e.g., reinforced clostridial medium (Wiel-Korstanje and Winkler, 1970; Willis *et al.*, 1973) or Columbia blood agar base (Borriello *et al.*, 1978). Other ingredients have also been added to the basic media to encourage the growth of all species of bifidobacteria, e.g., Yeast extract (Borriello *et al.*, 1978; Resnick and Levin, 1981a), liver digest (Simhon *et al.*, 1982), blood (Wiel-Korstanje and Winkler, 1970), haemin, tomato juice and vitamin K_1 (Borriello *et al.*, 1978).

To improve selectivity or enable differentiation, or both, antibiotics and other compounds have been added, e.g., kanamycin and nalidixic acid (Borriello *et al.*, 1978), nalidixic acid and neomycin sulphate (Resnick and Levin, 1981a). Dyes such as china blue (Wiel-Korstanje and Winkler, 1970) and bromocresol green (Resnick and Levin, 1981a) have been included to help in the differentiation of bifidobacteria from other bacteria present in the samples that also grow on the media. Lowering the pH of a medium has also been used for selectivity (Willis *et al.*, 1973) (Table 10.3).

However, selective media have not been very successful in the cultivation and identification of bifidobacteria. Other gram-positive, relatively acid-tolerant organisms also occur in faeces, e.g., enterococci and lactobacilli. These bacteria often grow profusely on *Bifidobacterium* selective media. As a result of this, all colony types must be examined by microscopy, and selective media do not significantly reduce the work load in the identification of bifidobacteria.

The formulation and use of selective media for isolating bifidobacteria from a particular ecosystem requires a knowledge of the various species and biotypes of bifidobacteria that may be present in that habitat. As this information is not complete it is probably better to devise a medium which permits satisfactory growth of the largest number of species and biotypes known. This medium should contain peptone, a universal carbohydrate (glucose or fructose), yeast extract (for vitamin B) and cysteine, plus various minerals and reducing agents. All manipulations should be performed in an anaerobic environment, e.g., in an anaerobic cabinet. The plates should then be incubated anaerobically (H_2 80 per cent, CO_2 20 per cent) at 37°C for four days to allow the growth of bifidobacteria. For subculture, the isolates need only be incubated for 48 h and the strains can be manipulated on the open bench.

Identification

Routine laboratory methods

The phenotype an organism displays is intrinsically connected with the environment in which it is

Table 10.3 Components of media used in the isolation and enumeration of *Bifidobacterium* species

Components	Wiel-Korstanje and Winkler (1970)	Willis et al. (1973)	Borriello et al. (1978)	Resnick and Levin (1981a)	Scardovi (1981)
Carbohydrate	Glucose		Maltose	Lactose	Glucose
Peptone	Reinforced clostridial agar	Reinforced clostridial medium	Columbia agar base	Peptone + casamino acids	Trypticase + phytone
Selective agents		pH 5.0	Kanamycin + nalidixic acid	Neomycin sulphate + nalidixic acid	
Differentiation agents	China blue	Cotton blue		Bromocresol green	
Enrichment agents	Horse blood		Haemin Vitamin K_1 Yeast extract Tomato juice	Yeast extract	Yeast extract Tween 80

Table 10.4 Characteristics used to identify subspecies and biotypes in the genus *Bifidobacterium*

Species and biotype	Fermentation of							
	Sucrose	Melibiose	Xylose	Aesculin	Mannitol	Sorbitol	Glycogen	Mannose
B. bifidum a	(+)	(+)						
B. bifidum b	–	–						
B. breve ss. breve					+	+		
B. breve ss. parvulorum					–	–		
B. infantis ss. infantis			–	–	–			
B. infantis ss. liberorum			+	+	–			
B. infantis ss. lactentis			+	–	+			
B. longum ss. longum a							(+)	
B. longum ss. longum b							+	
B. adolescentis a					+	+	+	
B. adolescentis b					+	–	+	
B. adolescentis c					–	+	d	
B. adolescentis d					–	–	–	

+, positive; –, negative; (+), slow reaction; d, different reactions for different strains

metabolizing. Factors such as the composition of medium, incubation conditions and the process of subculturing can grossly effect the phenotype. Colonial and cell morphology cannot be used reliably to differentiate between the species of bifidobacteria, or to identify a particular strain as definitely belonging to the genus.

In the routine anaerobic bacteriology laboratory, bifidobacteria are identified to genus level by the analysis of the short chain volatile and non-volatile fatty acids produced during the fermentation of glucose. During fermentation, bifidobacteria produce large amounts of acetic acid (usually 20–40 mmol/l) and smaller amounts of lactic acid. In practice, both acids need to be measured to avoid confusion with *Lactobacillus*.

The most widely used method of identifying the various species of bifidobacteria in a sample is a study of carbohydrate fermentation patterns. An array of carbohydrate substrates (usually 15–20) are inoculated with each strain of *Bifidobacterium* to be speciated. After anaerobic incubation, the pH of each substrate culture is measured to determine whether the carbohydrate has been metabolized. A pH drop of greater than one unit in comparison with the control substrate is classed as strong acid production; a drop of 0.5–1.0 pH unit is classed as weak acid production (Holdeman *et al.*, 1977). The fermentation pattern obtained from this array of carbohydrates is used for species identification. This method is subjective and there would undoubtedly be reader-to-reader variation if the same strains were given to different workers to identify. These patterns can also be used to define biotypes within a species. Therefore, a fermentation pattern must be interpreted bearing in mind the ecosystem from which the organism was isolated (Tables 10.4 and 10.5).

To complicate matters further, some *Bifidobacterium* spp. isolated from man have the same fermentation patterns as other species isolated from human faeces (Table 10.4). There is no way of distinguishing *B. dentium*, *B. catenulatum* and *B. angulatum* from the various biotypes of *B. adolescentis* in the routine laboratory. These species were only discovered by examining the genotype of these organisms in DNA-DNA homology studies.

Specialist laboratory methods

The most reliable assignment of a bacterial strain to the genus *Bifidobacterium* is based on the demonstration in cellular extracts of fructose-6-phosphate phosphoketolase, the key enzyme in hexose metabolism by bifidobacteria (Scardovi, 1981). The methodology is not difficult, but it is time consuming

156 Bifidobacteria

Table 10.5 Fermentation patterns of bifidobacteria isolated from human material.

Species	Acid from								
	D-ribose	L-arabinose	lactose	cellobiose	melezitose	raffinose	sorbitol	starch	gluconate
B. bifidum	−	−	+	−	−	−	−	−	−
B. breve	+	−	+	d	d	+	d	−	−
B. infantis	+	−	+	−	−	+	−	−	−
B. longum	+	+	+	−	+	+	−	−	−
B. adolescentis	+	+	+	+	+	+	d	+	+
B. catenulatum	+	+	+	+	−	+	+	−	d
B. dentium	+	+	+	+	+	+	−	+	+
B. angulatum	+	+	+	−	−	+	d	+	d
B. pseudocatenulatum	+	+	+	d	−	+	d	+	d

+, 90% of strains positive; −, 90% of strains negative; d, different reactions for different strains

and, therefore, not particularly suited to the routine laboratory. The method is definitive and is more reliable than the analysis of fatty acid end products of metabolism.

Techniques used to speciate bifidobacteria in the specialist laboratory can be divided into chemotaxonomic and genetic methods; the latter are more specific. All the methods require a data base which should include the results of the various species of *Bifidobacterium* isolated from the habitats to be examined, as well as type strains from culture collections. The isolates should be subjected to the same subculturing and storage techniques as the reference strains used in the data base. Furthermore, experience in each of the methods is needed to interpret the results. Prior to these specialist methods the strains will probably have been examined in routine fermentation tests to reduce the number of strains needing special methods. The work load involved in establishing and routinely performing these techniques is high.

Cell proteins

Electrophoretic patterns of about 30 bacterial proteins have been used to confirm the species identity of strains of bacteria by running them simultaneously with type strains and other isolates of that species. This technique has also been used to identify unlabelled strains (Moore *et al.*, 1980; Biavati *et al.*, 1982, 1986). At least three specific protein bands for each species are needed, plus a trained eye and a good data base. This method is reported to distinguish between closely related species which can be confused when using fermentation reactions. Homologous strains within a species have identical, or nearly identical protein patterns.

Isozymes of certain enzymes, e.g., transaldolase, occur in bifidobacteria. The electrophoretic mobility varies between these isozymes. This mobility can be used to differentiate closely related species. Isozymes of transaldolase belong to four immunologically distinct groups, representing strains from man, animals, bees and the environment (Scardovi, 1986).

Cell wall murein

In gram-positive bacteria, the cell wall murein accounts for 40–90 per cent of the dry weight of the cell wall. Murein is composed of chains of two amino sugars, N-acetylglucosamine and N-acetyl muramic acid, alternating within a strand. A peptide containing four amino acids is linked to each N-acetyl muramic acid. Two parallel strands of amino sugars are then linked together by a peptide bond between the amino acid chains in each strand. There is usually a direct peptide link between the chains, but in some organisms a cross-linking peptide containing up to five amino acids may be employed. The composition of these cross-links can be useful in species identification in the bifidobacteria. However, closely related species may have the same cross-links, e.g., *B. longum* and *B. infantis* (Kandler and Lauer, 1974 cited by Scardovi, 1986). Other chemotaxonomic studies, such as the analysis of cellular fatty acids and phosphoglycerides have been unfruitful for species identification (Exterkate *et al.*, 1971; Veerkamp, 1971).

Guanine + cystosine content of DNA

The genetic character of an organism appears to be more definitive than its phenotype, which can vary widely depending on growth conditions. Simple analysis of DNA from various species of bifidobacteria shows little inter-species variation. For example the percentage of guanine plus cytosine (G+C mol per cent) in species of bifidobacteria isolated from man is between 55 and 61 mol per cent. *B. bifidum*, *B. longum*, *B. breve* and *B. adolescentis* have indistinguishable G+C contents of 58 mol per cent. The other human species have values in the range 55–61 per cent (Gasser and Mandel, 1968, cited by Scardovi, 1986).

Although bifidobacteria are homogeneous in their base composition, the arrangement of these bases within the genome varies from species to species.

DNA-DNA homology

The similarity or homology of the DNA polynucleotide sequences has been determined in DNA-DNA hybridization competition experiments. The closer the homology of the DNA, the more closely related the species or strains of bifidobacteria (Scardovi *et al.*, 1971, Biavati *et al.*, 1984). This method has been used to confirm the differences between species isolated from various habitats, and to show the similarity of biotypes within the same species. *B. bifidum* and *B. adolescentis* form distinct genetic groups. *B. bifidum* biotypes a and b are closely related to each other, as are the four biotypes of *B. adolescentis*. Four species, *B. dentium*, *B. angulatum*, *B. catenulatum* and *B. pseudocatenulatum*, with fermentation patterns which are indistinguishable from *B. adolescentis*, are seen to be distinct from each other, and from *B. adolescentis*, using DNA-DNA homology. *B. longum* and *B. infantis* are the most closely related species, showing about 50 per cent homology; *B. breve* shows 40–50 per cent homology with the *B. infantis* biotypes (Scardovi, 1981).

Thus, it appears that, in the evolution of human species of bifidobacteria, *B. infantis* may have been the source of *B. breve* and *B. longum* and that *B.*

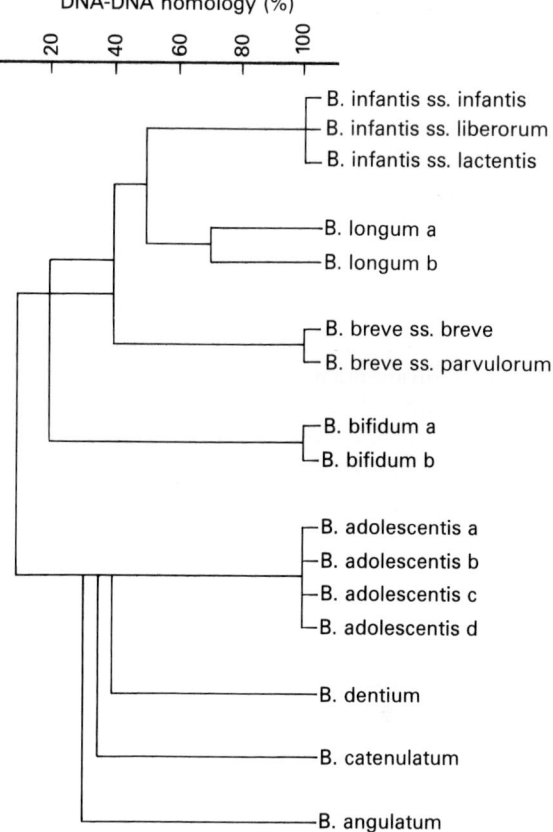

Fig. 10.1 DNA-DNA homology of human species of the genus *Bifidobacterium* (Adapted from Scardovi *et al.*, 1971)

bifidum evolved from a separate line, as did *B. adolescentis*. The remaining human species probably descended from the independent *B. adolescentis* line (Figure 10.1).

Ecology of bifidobacteria in the intestine

Occurrence

The gut is sterile before birth, but various organisms can be isolated from the faeces of infants on the first day of life and bifidobacteria can be detected by day two. There is some controversy about the source of these bifidobacteria. Maternal faeces have been implicated as a potential source during parturition.

This may be true away from modern medical care, but in large hospitals, with high levels of hygiene, this source becomes less likely. Bifidobacteria from the vagina and from breast milk have also been suggested as the source of this early inoculum, but bifidobacteria have been isolated from the vagina in only 2–3 per cent of pregnant women (Goplerud et al., 1976) and in low numbers from the colostrum of some women. Furthermore, there is doubt about the identification of the bifidobacteria in the colostrum studies (Poupard et al., 1973).

Epidemiological studies suggest that there is a 'local' environmental source of bifidobacteria for each maternity clinic in modern hospitals, because breast-fed, formula-fed, vaginally delivered and caesarian section term babies can all become colonized with the same species of bifidobacteria (Long and Swenson, 1977; Neut et al., 1985). The acquisition of this organism may be delayed in some babies and, in others, it may not appear in the faecal flora even at two months of age (Lundequist et al., 1985). Therefore, an individual 'baby factor' must also operate. This may be related to gut maturity, as a colonization delay has been observed in preterm infants (Stark and Lee, 1982a; Sakata et al., 1985) or the babies may simply not have been exposed to bifidobacteria.

B. infantis, B. longum, B. bifidum, B. breve and *B. adolescentis* are the most common species isolated from infant faeces. Colonization by these species appears to be independent of age and diet (Mitsuoka and Kaneuchi, 1977; Beerens et al., 1980; Benno et al., 1984; Biavati et al., 1984; Mitsuoka, 1984; Mevissen-Verhage et al., 1987). Babies often have two or three different species of bifidobacteria in their faeces, but all the babies in each particular clinic harbour the same predominant species. This suggests that one *Bifidobacterium* strain may be transmitted to all the infants in a clinic via the hands of the nurses (Mitsuoka and Kaneuchi, 1977) (Table 10.6) and that individual hospital practices and policies in maternity units are factors influencing the ecology of bifidobacteria. Species do not persist in the intestine for long periods and species can disappear and reappear in the next sample; thus the home environment and host factors are also important (Mevissen-Verhage et al., 1987).

Once established, bifidobacteria became the dominant members of the faecal flora of both breast- and formula-fed infants, reaching levels of 10^{10}–10^{11} colony forming units (cfu)/g of faeces. In purely breast-fed babies, bifidobacteria predominate, and they significantly outnumber other obligate and facultative anaerobes in the climax faecal flora at approximately one month of age (Stark and Lee, 1982b; Yoshioka et al., 1983; Benno et al., 1984; Roberts et al., 1985, Roberts, 1988). In formula-fed infants, facultative anaerobes often occur in larger numbers in the climax faecal flora than in their breast-fed counterparts. Thus bifidobacteria, enterobacteria and enterococci may be equally dominant in the faeces of formula-fed babies.

At weaning, there are rapid changes in faecal bacterial populations. Facultative anaerobes can increase in number by 1000-fold in breast-fed infants and the *Bacteroides* spp. are detected in all babies at

Table 10.6 *Bifidobacterium* species from the faeces of infants

Study	Country	Predominant species
Mitsuoka and Kaneuchi (1977)	Japan: Clinic A	*B. longum*
	Clinic B	*B. infantis* ss. *infantis* b
	Clinic C	*B. infantis* ss. *infantis* a
Beerens et al. (1980)	France, Lille	*B. bifidum*
		B. longum
Benno et al. (1984)	Japan	*B. breve*
Biavati et al. (1984)	Italy	*B. breve*
		B. infantis
Mitsuoka (1984)	Japan	*B. infantis*
Mevissen-Verhage et al. (1987)	The Netherlands	*B. longum*
		B. breve
		B. adolescentis
Roberts (1987) (unpublished data)	Italy, Milan	*B. breve*
		B. infantis

10^7–10^{10} cfu/g. An adult-type faecal flora is not fully established until at least one year of age. In this adult faecal flora, bifidobacteria remain at similar levels to those found in infant faeces, but they are now co-dominant with *Bacteroides* spp. and *Eubacterium* spp. (Mitsuoka and Kaneuchi, 1977; Stark and Lee, 1982b). However, with the transition from a milk diet to a predominantly solid diet there is a shift in the species of *Bifidobacterium* detected in the faeces. *B. infantis*, commonly isolated from infants, is not found in the faeces of adults and children, and *B. breve* is rarely isolated. In Japan, the most ubiquitous species in the faeces of children and adults is *B. adolescentis* (mainly biotypes a, b and c). *B. longum* is common in adults, but biotype a predominates over biotype b, the latter being more common in infants (Mitsuoka and Kaneuchi, 1977; Mitsuoka, 1982). Biavati *et al.* (1986) have found that *B. pseudocatenulatum* is the most frequently isolated species from adult faeces in Italy. These workers identified their isolates by the electrophoretic analysis of cell proteins. Mitsuoka (1982) used sugar fermentation patterns in his study. However, with this technique *B. pseudocatenulatum* cannot be separated from *B. adolescentis* and Mitsuoka's strains may therefore have contained species which were phenotypically identical to *B. adolescentis*, but which were genetically distinct. Both studies suggest that '*B. adolescentis*-type' strains are dominant in children and adults.

The change in diet from a high carbohydrate liquid food to a solid diet rich in protein does not affect the numbers of bifidobacteria in the faeces. The transition favours the dominance of *B. adolescentis*, and appears to lead to the demise of the *B. infantis* and *B. breve* present in the infant faeces. The solid diet may select for species that can utilize various carbohydrate sources in their metabolism, e.g., *B. adolescentis* and phenotypically similar species.

Data on microbial populations of various regions of the gastrointestinal tract of adults reveal that bifidobacteria are unable to colonize the motile upper reaches of the gut. In the terminal ileum, where motility is reduced, bifidobacteria have been detected at 10^7 cfu/g of contents. On reaching the caecum, bifidobacteria multiply to the levels observed in the faeces (Mitsuoka, 1982). There are no data on the species of bifidobacteria isolated from the individual regions of the intestine.

Interaction with other organisms

The organic acids produced by bifidobacteria may exert an inhibitory effect on other intestinal bacteria *in vitro* and *in vivo*. In the early 1970s, studies of infant faecal flora showed that the levels of *Escherichia coli* were higher in formula-fed babies than in breast-fed infants. The converse was true for bifidobacteria (Bullen and Willis, 1971). It was thought that the unbuffered acid from the bifidobacteria in the breast-fed infant inhibited the growth of *E. coli*. *In vitro* experiments confirmed these results. However, it was then revealed that *E. coli* numbers were also low in breast-fed babies that did not harbour bifidobacteria (Willis *et al.*, 1973). Thus, unbuffered acid from the whole bacterial population appeared to control the growth of *E. coli*. Formula feeds, with their much higher buffering capacity, did not allow the accumulation of free acid. Modern studies of infant faecal flora, which were conducted with low buffering capacity, 'humanized' formula feeds, substantiate this theory. *E. coli* is now found at similar concentrations in breast-fed and formula-fed infants, whether bifidobacteria are present or not.

Breast-fed babies have significantly lower numbers of enterococci in their faeces than formula-fed infants (Yoshioka *et al.*, 1983; Benno *et al.*, 1984; Roberts *et al.*, 1985; Roberts, 1988). This may be due to competition between the two organisms for a growth factor(s), e.g., B vitamins, that are present in lower concentrations in breast-milk than in modern formulae.

Research groups have reported bacterial antagonism to *Clostridium difficile*, *in vitro*, by species of bifidobacteria. However this does not appear to be the case *in vivo*, particularly in infants, who are often substantially colonized by both *C. difficile* and bifidobacteria (Rolfe *et al.*, 1981; Rolfe, 1984). Bifidobacteria may prevent the outgrowth of these organisms in the faecal population.

Apart from their possible role in bacterial antagonism, bifidobacteria may also form part of the food web in the intestine. Organic acids produced by bifidobacteria can be metabolized by other organisms and some strains secrete amino acids that can then be fermented by the flora.

Effects of bifidobacteria on the host

Certain species appear to be specifically adapted to live in the human intestine, e.g., *B. catenulatum*, *B.*

angulatum. The role of these, and other species of *Bifidobacterium* in the intestine is not known, but they are thought to be involved in colonization resistance and pathogen exclusion. They are one of the dominant members of the faecal flora and they probably contribute to colonization resistance by occupying the potential bacterial binding sites along the large intestine and possibly in the terminal ileum.

The volatile fatty acids produced by intestinal bacteria represent about 70 per cent of the energy available from the initial carbohydrate consumed by the host; 80–90 per cent of these short chain acids are absorbed by the host, mainly in the large intestine. Absorbed acetate probably passes into the circulation and is then taken up by the tissues. Other acids have been shown to be an important energy source for the colonic mucosa. These acids also stimulate absorbtion of salt and water in the colon, thus retaining the water and mineral homeostasis of the body (Cummings, 1983; Hoverstad, 1986). Bifidobacteria contribute significantly to the short chain acids secreted into the gut.

Some strains of bifidobacteria are able to degrade human colonic mucin (Hoskins *et al.*, 1985). The monosaccharides released from the mucin could help to support the growth of other faecal bacteria which are unable to degrade mucin themselves. Degradation could also impair the lubricating and protective functions of mucin in the gut, and thus be detrimental to the host. Approximately 1 per cent of normal human faecal bacteria are mucin degraders. If the proportion increased it could lead to 'holes' in the mucus layer, leaving the intestinal mucosa susceptible to attack by bacterial, physical and chemical agents.

Conclusions

Bifidobacteria are ubiquitous in the intestine of man from infancy to adult life. They probably play a passive role in colonization resistance and pathogen exclusion. Their production of lactic and acetic acids may inhibit the growth of certain organisms.

Research on the growth requirements of known species of bifidobacteria has allowed appropriate media to be devised for their isolation. However, selective media are not particularly successful. Identification in the routine laboratory is not an easy task, but DNA-DNA homology studies in specialist laboratories may help to ease the problem in the future.

References

Beerens H, Romond C, Neut C. (1980). Influence of breast-feeding on the bifid flora of the new born intestine. *American Journal of Clinical Nutrition* **33**: 2434–2439.

Benno Y, Sawada K, Mitsuoka T. (1984). The intestinal microflora of infants: composition of faecal flora in breast-fed and bottle-fed infants. *Microbiology and Immunology* **28**: 975–986.

Biavati B, Castagnoli P, Crociani F, Trovatelli L D. (1984). Species of the genus *Bifidobacterium* in the faeces of infants. *Microbiologica* **7**: 341–345.

Biavati B, Castagnoli P, Trovatelli L D. (1986). Species of the genus *Bifidobacterium* in the faeces of human adults. *Microbiologica* **9**: 39–45.

Biavati B, Scardovi V, Moore W E C. (1982). Electrophoretic patterns of proteins in the genus *Bifidobacterium* and proposal of four new species. *International Journal of Systematic Bacteriology* **32**: 358–373.

Borriello P, Hudson M, Hill M. (1978). Investigation of the gastrointestinal bacterial flora. *Clinics in Gastroenterology* **7**: 329–349.

Bullen C L, Willis A T. (1971). Resistance of breast-fed infants to gastroenteritis. *British Medical Journal* **3**: 338–343.

Cummings J H. (1983). Fermentation in the human large intestine: evidence and implications for health. *Lancet* **1**: 1206–1209.

Exterkate F A, Otten B J, Wassenberg H W, Veerkamp J H. (1971). Comparison of the phospholipid composition of *Bifidobacterium* and *Lactobacillus* strains. *Journal of Bacteriology* **106**: 824–829.

Goplerud C P, Ohm M J, Galask R P. (1976). Aerobic and anaerobic flora of the cervix during pregnancy and the puerperium. *American Journal of Obstetrics and Gynecology* **126**: 858–868.

Holdeman L V, Cato E P, Moore W E C. (1977). *Anaerobe laboratory manual*, 4th edn. Virginia Polytechnic Institute and State University, Blacksburg.

Hoskins L C, Agustines M, McKee W B, Boulding E T, Kriaris M, Niedermeyer G. (1985). Mucin degradation in human colon ecosystems. *Journal of Clinical Investigation* **75**: 944–953.

Hoverstad T. (1986). Studies of short-chain fatty acid absorption in man. *Scandinavian Journal of Gastroenterology* **21**: 257–260.

Lauer E. (1990). *Bifidobacterium gallicum* sp. nov. Isolated from human faeces. International Journal of Systematic Bacteriology. **40**: 100–102.

Long S S, Swenson R M. (1977). Development of anaerobic faecal flora in healthy new born infants. *Journal of Pediatrics* **91**: 298–301.

Lundequist B, Nord C E, Winberg J. (1985). The composition of the faecal microflora in breast fed and bottle fed infants from birth to eight weeks. *Acta Paediatrica Scandinavica* 74: 45–51.

Mevissen-Verhage E A E, Marcelis J H, de Vos, M N, Harmsen-van Amerongen W C M, Verhoef J. (1987). *Bifidobacterium*, *Bacteroides*, and *Clostridium* spp. in faecal samples from breast-fed and bottle-fed infants with and without iron supplement. *Journal of Clinical Microbiology* 25: 285–289.

Mitsuoka T. (1982). Recent trends in research on the intestinal flora. *Bifidobacteria and Microflora* 1: 3–24.

Mitsuoka T. (1984). Taxonomy and ecology of bifidobacteria. *Bifidobacteria and Microflora* 3: 11–28.

Mitsuoka T, Kaneuchi C. (1977). Ecology of the bifidobacteria. *American Journal of Clinical Nutrition* 30: 1799–1809.

Moore W E C, Hash D E, Holdeman L V, Cato E P. (1980). Polyacrylamide slab gel electrophoresis of soluble proteins for studies of bacterial floras. *Applied and Environmental Microbiology* 39: 900–907.

Neut C, Romond C, Beerens H. (1985). Bacterial interactions. *Scandinavian Journal of Infectious Diseases*, Suppl 46: 37–46.

Orla-Jensen M. (1924). La classification des bacteries lactiques. *Lait* 4: 468–474.

Petschow B M, Talbott R D. (1990). Growth promotion of *Bifidobacterium* species by whey and casein fractions from human and bovine milk. *Journal of Clinical Microbiology* 28: 287–292.

Poupard J A, Husain I, Norris R F. (1973). Biology of the bifidobacteria. *Bacteriological Reviews* 37: 136–165.

Resnick I G, Levin M A. (1981a). Quantitative procedure for enumeration of bifidobacteria. *Applied and Environmental Microbiology* 42: 427–432.

Resnick I G, Levin M A. (1981b). Assessment of bifidobacteria as indicators of human faecal pollution. *Applied and Environmental Microbiology* 42: 433–438.

Reuter G. (1971). Designation of type strains for *Bifidobacterium* species. *International Journal of Systematic Bacteriology* 21: 273–275.

Roberts A K. (1988). *The development of the infant faecal flora*. PhD Thesis. Council for National Academic Awards.

Rolf R D. (1984). Role of volatile fatty acids in colonisation resistance to *Clostridium difficile*. *Infection and Immunity* 45: 185–191.

Rolfe R D, Helebian S, Finegold S M. (1981). Bacterial interference between *Clostridium difficile* and normal faecal flora. *Journal of Infectious Diseases* 143: 470–475.

Sakata H, Yoshioka H, Fujita K. (1985). Development of the intestinal flora in very low birth weight infants compared to normal full-term new borns. *European Journal of Pediatrics* 144: 186–190.

Scardovi V. (1981). The genus *Bifidobacterium*. In *The Prokaryotes*, vol 2. pp. 1951–1961. Edited by Starr M P *et al*. Springer-Verlag, New York.

Scardovi V. (1986). *Bifidobacterium*. In *Bergey's Manual of systematic bacteriology*, vol 2, pp. 1418–1434. Edited by Sneath P H A *et al*. Williams and Wilkins, Baltimore.

Scardovi V, Crociani F. (1974). *Bifidobacterium catenulatum*, *B. dentium* and *B. angulatum*: three new species and their deoxyribonucleic acid homology relationships. *International Journal of Systematic Bacteriology*. 24: 6–20.

Scardovi V, Trovatelli L D, Biavati B, Zani G. (1979). *Bifidobacterium cumculi*, *B. choerinum*, *B. boum* and *B. pseudocatenulatum*: four new species and their deoxyribonucleic acid homology relationships. *International Journal of Systematic Bacteriology*. 29: 291–311.

Scardovi V, Trovatelli L D, Zani G, Crociani F, Matteuzzi D. (1971). Deoxyribonucleic acid homology relationships among species of the genus *Bifidobacterium*. *International Journal of Systematic Bacteriology* 21: 276–294.

Simhon A, Douglas J R, Drasar B S, Soothill J F. (1982). Effect of feeding on infants' faecal flora. *Archives of Disease in Childhood* 57: 54–58.

Stark P L, Lee A. (1982a). The bacterial colonisation of the large bowel of pre-term low birth weight neonates. *Journal of Hygiene* 89: 59–67.

Stark P L, Lee A. (1982b). The microbial ecology of the large bowel of breast-fed and formula-fed infants during the first year of life. *Journal of Medical Microbiology* 15: 189–203.

Veerkamp J H. (1971). Fatty acid composition of *Bifidobacterium* and *Lactobacillus* strains. *Journal of Bacteriology* 108: 861–867.

Wiel-Korstanje J A A Van der, Winkler K C. (1970). Medium for differential count of the anaerobic flora in human faeces. *Applied Microbiology* 20: 168–169.

Willis A T, Bullen C L, Williams K, Fagg C G, Bourne A, Vignon M. (1973). Breast milk substitute: a bacteriological study. *British Medical Journal* 4: 67–72.

Yoshioka H, Iseki K-I, Fujita K. (1983). Development and differences of intestinal flora in the neonatal period in breast-fed and bottle-fed infants. *Pediatrics* 72: 317–321.

11

Anaerobes in the normal flora of man

B S Drasar and B I Duerden

Intestine
   The intestinal bacteria
Skin
Mouth
   Tongue
   Saliva
   Gingival crevice and dental plaque
Genitourinary tract
   Vaginal microflora
   Anaerobes in the urethra
References

Man has an enormous resident microbial flora– approximately 10^{14} bacteria per human being. Despite the fact that man is an aerobic organism, a very large proportion of this normal flora is anaerobic. Even in apparently aerobic situations, such as in the mouth and on the skin, there are anaerobic microenvironments which provide conditions suitable for anaerobes to flourish, and even in these seemingly inimical areas the anaerobes predominate in the normal bacterial flora. However, anaerobes rarely comprise the whole bacterial microflora at any particular site, and the facultative or aerobic species may play a role in scavenging any free oxygen that may be present, thus protecting the strict anaerobes and maintaining conditions of low Eh. The importance of the normal flora has been recognized since the beginning of bacteriology. The earlier literature was comprehensively reviewed by Rosebury (1962).

Anaerobes of many various species are found on the skin and the mucosal surfaces. The population often includes species which, given suitable initiating conditions such as trauma or a reduction in the blood supply and oxygenation, are able to cause infection either in the local vicinity or at a distant site if carried there by the bloodstream. The major sites of colonization by anaerobic bacteria are: intestine, skin, mouth and genitourinary tract.

Intestine

During the past 15 years, our knowledge of the intestinal flora has been greatly expanded. This development can be traced in a series of books (Drasar and Hill, 1974; Clarke and Bauchop, 1977; Hentges, 1983; Drasar and Barrow, 1985). However, our knowledge of the composition and functions of the flora remain far from complete. A list of the major bacterial species isolated from the intestine can be presented but, undoubtedly, many species await description.

In a medical context, the intestinal flora is important (a) as a source of bacteria able to cause infection;

(b) as a means of excluding pathogens from the intestine (colonization resistance) and (c) in terms of its metabolic potential. However, a description of the distribution of bacteria in the intestine and explanations of their effects on the body, or of the mechanisms by which they are controlled, remains possible only in outline.

Samples of intestinal contents have been obtained from the proximal small intestine of people resident in Western Europe and North America. These contain few cultivable bacteria; indeed, most investigators have reported samples from which they were unable to isolate bacteria despite careful and reliable techniques. The viable count of bacteria seldom exceeds 10^4/ml of intestinal contents. For a period after a meal a greater number of bacteria can be isolated from jejunal samples. These bacteria are probably transients from the mouth which have been swallowed with the food and which are protected in the food from gastric acidity. Although few bacteria can be cultured, the microscopic examination of duodenal or jejunal juice always reveals the presence of large numbers of bacteria. The microscopic count usually exceeds 10^6/ml of intestinal contents but lower counts are observed after a meal—the opposite results to those obtained by culture. Gram-positive bacteria predominate both among those bacteria isolated and those observed microscopically. Thus, in the healthy residents of Western Europe and North America, the concentration of cultivable bacteria is lowest in the duodenum and jejunum and greatest in the terminal ileum. The importance of those bacteria that can be seen but not grown is still to be determined. Very few anaerobes have been isolated from samples from the small intestine. Bacteroides and gram-positive anaerobes are occasionally found in small numbers, most often in ileal samples.

These results differ markedly from those obtained in studies of the small intestinal flora in developing countries. These studies suggest that a richer and more permanent normal flora is found in the small intestine of people living in these areas. Studies in south India (Bhat *et al.*, 1972) have shown that non-sporing anaerobes are a major part of the proximal small intestinal flora in normal people. Geographical differences have also been demonstrated with respect to numbers of bacteria of different genera present in faeces (see below).

The relationship of this flora to nutritional status, nutrient utilization and subclinical disease may have important practical consequences. The flow rate of the gut contents contributes to controlling the colonization of some sites. Flow rate is at its greatest at the top of the small intestine, where microbial multiplication in the contents cannot usually overcome the rate at which organisms are removed by peristalsis. In the terminal ileum, the flow rate is slower and the colonizing flora resembles that of the caecum.

Despite intensive investigation, our knowledge of the bacterial flora of the large intestine of man is limited. The major problem of these studies is obtaining adequate samples. Many studies on the intestinal flora have been concerned with the bacterial content of faeces. Such studies are relevant to discussions of the significance of the intestinal flora in so far as the findings reflect the bacterial flora of the large intestine. Those few studies that have been made on bowel contents suggest that the contents of the large bowel support a flora qualitatively similar to that of faeces. It seems likely, however, that there is also a specific, mucosally-associated flora and that, although the bacterial species concerned can be isolated from faeces, the functional association with the intestinal wall is of greater ecological importance. Whatever the problems in interpreting the results of studies of faeces, our knowledge of the bacterial species inhabiting the intestine has been built up, to a large extent on these studies.

The numbers of bacteria in faeces are very large —10^{10}–10^{11} bacteria/gram can readily be isolated —and anaerobic bacteria predominate. Some 30–40 species make up 99 per cent of the bacterial mass.

The intestinal bacteria

When intestinal bacteria are mentioned generally this most often refers to bacteria isolated from faeces. For many years it was generally accepted that a very limited range of bacteria could be isolated from faeces. Prominent amongst these were *Escherchia coli*, *Enterococcus faecalis* and *Clostridium perfringens*. Although Eggerth and Gagnon (1933) had demonstrated the importance of non-sporing anaerobes, their findings were largely ignored for many years, but subsequent studies on the faecal flora of man have confirmed that these organisms are predominant. It is estimated that the human intestine contains some 400–500 species of bacteria, mostly non-sporing anaerobes, and it is likely that the intestines of other animals harbour similar numbers of species. At present it is not known if the same bacterial species are present in all animal species but it is likely that each animal species has a corresponding group of bacterial commensals. Even if the same species are present, it is likely that the types associated with a particular animal species will be unique to that species.

Very few investigators have attempted a systematic investigation of the intestinal bacteria and any list of the species present in the gut must be provisional. The tables presented in this chapter are based on systematic studies of the faecal flora of man. These investigations were performed in an attempt to elucidate the interaction of diet, gut bacteria and cancer. When faced with complexity of the gut bacteria it is useful to consider the categories that must be represented. These include (a) contaminants or transients from the environment and (b) indigenous bacteria, i.e., bacteria that have colonized the individuals under investigation (these may include bacterial species 'normal' or 'common' in the study group but not present in all members of the species). Special mention must be made of the so called autochthenous flora. This group of organisms may be considered to have undergone parallel evolution with the host and to represent the true intestinal organisms.

It is not possible, with any certainty, to assign individual species of intestinal anaerobes to any of these categories. Thus, more practically, it is important to distinguish between the species isolated from the intestine in terms of their source. The system used for the tables in this chapter includes:

organisms occurring in large numbers in faeces —the most common faecal organisms;

organisms that have been isolated from faeces and for which faeces is probably the primary source, but which are also commonly isolated from non-faecal sources;

organisms commonly isolated from non-faecal sources and for which faeces is a minor source. (The bacteria of the rumen are among the best studied groups of non-faecal, non-sporing anaerobes and this site is usually identified as a major source of anaerobic bacteria.)

Most of the bacteria growing in the intestine are non-sporing anaerobes; they include members of the genera *Bacteroides*, *Bifidobacterium*, *Eubacter-*

Table 11.1 *Bacteroides* spp. isolated from human faeces*

Species	Presence in group†			
	(i)	(ii)	(iiia)	(iiib)
B. amyophilus		+		
B. asaccharolyticus				+
B. caccae‡		+		
B. capillosus				+
B. coagulans				+
B. distasonis‡	+			
B. eggerthii‡	+			
B. fragilis‡				+
B. hypermegas		+		
B. melaninogenicus				+
B. merdae‡		+		
B. multiacidus		+		
B. oralis				+
B. ovatus‡	+			
B. pneumosintes				+
B. praeacutus				+
B. putredinis		+		
B. ruminicola ss. brevis			+	
B. ruminicola ss. ruminicola			+	
B. splanchnicus				+
B. stercoris‡		+		
B. succinogenes			+	
B. thetaiotaomicron‡	+			
B. uniformis‡	+			
B. ureolyticus				+
B. vulgatus‡	+			

†(i), common faecal organisms; (ii), commonly isolated from faeces and non-faecal sources; (iiia), faeces minor source—rumen major source; (iiib), faeces minor source—other (non-rumen) sources major, ‡Fragilis group of bile- and penicillin-resistant *Bacteroides* spp

*Data from Drasar and Hill (1974), Holdeman and Moore (1974), Moore and Holdeman (1974), Holdeman *et al.* (1976), Finegold *et al.* (1975, 1977, 1983), Johnson *et al.* (1986)

ium and many others. Clostridia are also represented, although they are far outnumbered by non-sporing anaerobes. It has been suggested that the flora becomes more diverse with age, but definitive data to support this hypothesis are lacking. Numerically, the most important genus of intestinal bacteria in animals and man is the *Bacteroides* group of gram-negative anaerobic bacilli (Table 11.1). Saccharolytic bacteroides of the *Bacteroides fragilis* group are predominant. At one time it was considered that all saccharolytic, bile salt-resistant bacteroides that produce succinic acid as a major metabolic product could be identified as *B. fragilis*, but improvements in classification have resulted in the delineation of a number of separate species in this group. These include the former subspecies of '*B. fragilis*'—*B. distasonis*, *B. fragilis* (*sensu stricto*), *B. vulgatus*, *B. uniformis*, *B. ovatus* and *B. thetaiotaomicron* (Cato and Johnson, 1976)—together with new species such as *B. caccae*, *B. merdae* and *B. stercoris* (Johnson *et al.*, 1986). Among the fusobacteria, only *Fusobacterium prauznitzii*, which should not really be considered to be a member of the genus *Fusobacterium* (see Chapter 5), occurs regularly in large numbers (Table 11.2) and can thus be regarded as a primary member of the normal flora. Various other fusobacteria and, indeed, other gram-negative non-sporing anaerobes, have also been isolated.

Table 11.2 Fusobacteria and other gram-negative non-sporing anaerobes isolated from faeces*

Species	Presence in group†		
	(i)	(ii)	(iii)
F. gonidiaformans		+	
F. mortiferum		+	
F. naviforme			+
F. necrogenes		+	
F. necrophorum		+	
F. nucleatum			+
F. plauti			+
F. prausnitzii	+		
F. russii		+	
F. symbiosum			+
F. varium		+	
Butyrivibrio fibrisolvens			+
Desulfomonas pigra	+		
Leptotrichia buccalis			+
Oscillospira guillermondii			+
Selenomonas ruminatium			+
Succinimonas amylolytica		+	
Succinivibrio dextrinosolvens			+

†(i), common faecal organisms; (ii), commonly isolated from faeces and non-faecal sources; (iii), faeces minor source
*Data from Drasar and Hill (1974), Holdeman and Moore (1974), Moore and Holdeman (1974), Holdeman *et al.* (1976), Finegold *et al.* (1975, 1977, 1983)

Table 11.3 *Bifidobacterium* spp. isolated from faeces*

Species	Presence in group†
	(i)
B. adolescentis	+
B. angulatum	
B. bifidum	+
B. breve	+
B. catenulatum	
B. dentium	
B. infantis	+
B. longum	+
B. pseudolongum	

†(i), common faecal organisms
*Data from Drasar and Hill (1974), Holdeman and Moore (1974), Moore and Holdeman (1974), Holdeman *et al.* (1976), Finegold *et al.* (1975, 1977, 1983)

Several genera of gram-positive, non-sporing rods are numerically important in the gut. *Bifidobacterium* spp., including *Bifidobacterium bifidum* and *B. infantis*, are prominent in the faeces of breast-fed infants. Other species notably, *B. adolescentis* and *B. longum*, are commonly isolated from adult faeces (Table 11.3). The bacteria included in this genus can be regarded as characteristic intestinal bacteria and they are the most obvious candidates for autochthenous status. All have been isolated from the intestine of infants or adults. In breast-fed infants, and in some adults, they are numerically the dominant component of the flora.

Very many other species of gram-positive non-sporing anaerobes have also been isolated from faeces. *Eubacterium* spp., particularly *Eubacterium aerofaciens* and *E. biforme*, are prominent (Table 11.4). Assessment of the importance of this genus is complicated by its uncertain taxonomic status. The genus includes all those organisms that cannot be assigned to other genera of gram-positive non-sporing anaerobes. The group is metabolically diverse and includes bacteria important in the metabolism of bile acids, cholesterol and digoxin.

Actinomyces and propionibacteria have also been isolated from faeces (Table 11.5) but these genera are usually considered to be characteristic of other microecological locations.

Facultative and obligately anaerobic gram-positive cocci are numerically important in the gut. The strict anaerobes include *Coprococcus*, *Peptostreptococcus*, *Ruminococcus* and *Sarcina ventriculi* (Table 11.6). Gram-negative anaerobic cocci include

Table 11.4 *Eubacterium* spp. isolated from human faeces*

Species	Presence in group†		
	(i)	(ii)	(iii)
E. aerofaciens	+		
E. alactolyticum			+
E. biforme	+		
E. budayi			+
E. cellulosolvens			+
E. combesii		+	
E. contortum		+	
E. cylindroides		+	
E. dolichum		+	
E. eligens		+	
E. formicigenerans		+	
E. hallii		+	
E. lentum		+	
E. limosum			+
E. moniliforme			+
E. multiforme			+
E. nitritogenes			+
E. ramulus	+		
E. rectale	+		
E. ruminantium			+
E. saburream			+
E. siraeum		+	
E. tenue			+
E. tortuosum			+
E. ventriosum		+	

†(i), common faecal organisms; (ii), commonly isolated from faeces and non-faecal sources; (iii), faeces minor source
*Data from Drasar and Hill (1974), Holdeman and Moore (1974), Moore and Holdeman (1974), Holdeman *et al.* (1976), Finegold *et al.* (1975, 1977, 1983)

Table 11.5 Other gram-positive anaerobic rods occasionally isolated from faeces*

Species
Actinomyces naeslundii
Actinomyces odontolyticus
Lachnospira multiporus
Propionibacterium acnes
Propionibacterium avidum
Propionibacterium granulosum
Propionibacterium jensenii

*Data from Drasar and Hill (1974), Holdeman and Moore (1974), Moore and Holdeman (1974), Holdeman *et al.* (1976), Finegold *et al.* (1975, 1977, 1983)

Table 11.6 Gram-positive cocci isolated from faeces*

Species	Presence in group†		
	(i)	(ii)	(iii)
Copro. eutactus	+		
Copro. catus	+		
Copro. comes	+		
Pstr. asaccharolyticus			+
Pstr. magnus			+
Pstr. prevotii			+
Pstr. anaerobius			+
Pstr. micros			+
Pstr. parvulus			+
Pstr. productus			+
Rumino. albus			+
Rumino. bromii	+		
Rumino. flavefaciens			+
Rumino. lactaris	+		
Rumino. obeum	+		
Rumino. torques	+		
Sarcina ventriculi			+

†(i), common faecal organisms; (ii), commonly isolated from faeces and non-faecal sources; (iii), faeces minor source
*Data from Drasar and Hill (1974), Holdeman and Moore (1974), Moore and Holdeman (1974), Holdeman *et al.* (1976), Finegold *et al.* (1975, 1977, 1983)

Veillonella, *Acidaminococcus* and the budding bacterium *Gemmiger formiculis* (Table 11.7).

Several types of spore-forming rods are normal inhabitants of the gut; the genus *Clostridium* is probably ubiquitous (Table 11.8). The variety of species isolated probably reflects the efficiency of the procedures used rather than the relative abundance of the different types. Many, such as *C. perfringens* and *C. paraputrificum*, occur regularly in moderate numbers. Others, such as *C. ramosum*, may be dominant members of the flora. The sul-

Table 11.7 Gram-negative cocci isolated from faeces*

Species
Acidaminococcus fermentans
Gemmiger formiculus
Megasphera elsdenii
Veillonella parvula

*Data from Drasar and Hill (1974), Holdeman and Moore (1974), Moore and Holdeman (1974), Holdeman *et al.* (1976), Finegold *et al.* (1975, 1977, 1983)

phate-reducing *Desulfotomaculum* group has seldom been demonstrated but is undoubtedly present.

Several types of spirochaete can be seen attached to the mucosa and in the gut contents of healthy animals, but their status in the human gut flora is uncertain.

Balance within the population and exclusion of pathogens

Except when successful invasion by a pathogen occurs, the gut flora is probably a stable, self-regulating system. There is a natural resistance to the alteration of the flora by the introduction of new bacteria, even if they are of species commonly encountered in the intestine. This resistance to change is understandable, the bacteria that comprise the flora are adapted to life in the intestine, whereas freshly introduced strains require time to adapt. The role of the flora in excluding pathogenic bacteria has important consequences in the context of acute intestinal disease. This phenomenon has been referred to variously as bacterial antagonism, bacterial interference and colonization resistance, and is examined in detail in Chapter 20. It can most usefully be regarded as a special case of those ecological interactions that occur in the gut.

The human colon harbours a diverse collection of bacteria. That these organisms do not grow and engulf the body is a tribute to the host's control systems, and yet the way in which the flora is controlled, its constitution determined and its activity expressed is not completely understood. Intestinal physiology and host defence mechanisms must play an important role in preventing the flora over-running the host, and in determining the distribution of the flora within the intestine. It seems likely that environmental factors such as diet also influence the flora.

In any confined environment, such as the large intestine, space and nutrients are limited, and the bacteria that are able to transform nutrients into bacterial cells fastest, under the prevailing conditions, will be present in the greatest numbers. Bacteria which are able to utilize nutrients that are not used by other bacteria will also possess an ecological advantage. Interactions between the bacteria must play an important role in determining both the total number and the relative frequency of

Table 11.8 *Clostridium* spp. and other spore-forming anaerobic rods isolated from faeces*

Species

C. acetobutylicum
C. aminovalericum
C. aurantibutyricum
C. barati
C. barkeri
C. beijerinkii
C. bifermentans
C. butyricum
C. cadaveris
C. carnis
C. celatum
C. cellobioparum
C. chauvoei
C. clostridiiforme
C. cochlearium
C. difficile
C. fallax
C. felsineum
C. ghoni
C. glycolicum
C. haemolyticum
C. indolis
C. innocuum
C. irregulare
C. lentoputrescens
C. limosum
C. leptum
C. malenominatum
C. mangenoti
C. nexile
C. oxeanicum
C. oroticum
C. paraputrificum
C. pasteurianum
C. perfringens
C. plagarum
C. pseudotetanicum
C. putrefaciens
C. ramosum
C. sartagoformum
C. septicum
C. sordellii
C. sporogenes
C. sporosphaeroides
C. subterminale
C. tertium

Desulfotomaculum nigrificans
Desulfotomaculum ruminis
Desulfotomaculum orientis

*Data from Drasar and Hill (1974), Holdeman and Moore (1974), Moore and Holdeman (1974), Holdeman *et al.* (1976), Finegold *et al.* (1975, 1977, 1983)

the various species. Metabolic inhibitors such volatile fatty acids and H_2S, produced by members of the flora, mediate these interactions and act as significant ecological determinents.

Metabolic significance of the gut bacteria

The intestinal bacteria have been studied because of the importance of their contribution to the host's general metabolism, their role in drug metabolism and their relationship to large bowel cancer.

Man, and mammalian species in general, are not thought to secrete enzymes that can breakdown polysaccharides other than starch, and so the complete degradation of a heterogeneous polysaccharide to its monomeric constituents by a bacterium in the gastrointestinal tract requires that the bacterium produces a range of hydrolytic enzymes. Human intestinal bacteria primarily synthesize cell associated polysaccharidases which initiate polymer breakdown by hydrolysing the polysaccharide backbone. Glycosidases hydrolyse the polysaccharide side chains and also further break down the backbone oligosaccharides. The substrates for these activities include the non-absorbed carbohydrate residues from the diet, the intestinal mucins and dietary fibre.

The principal end products of the fermentation of polysaccharides by the gut bacteria are short chain fatty acids. These are absorbed from the gut and, once absorbed, are potential substrates for energy metabolism in the colonic mucosa. Thus, fermentation of carbohydrate in the large gut may be said to contribute to the host's nutrition. There is also some evidence of a contribution to the available pool of vitamins.

The role of the gut flora in the fermentation of protein and the conservation of intestinal nitrogen is probably less benign. Indeed, some of the products of amino acid fermentation have been implicated in cancer. The metabolic importance of colonic fermentation in man was recently reviewed (Cummings *et al* 1989) and is considered further in Chapter 24.

The varied studies highlight the importance of improving our knowledge of the composition of the gut flora and our understanding of the mechanisms controlling the gut bacteria and their metabolism. The solution of these problems becomes increasingly important as more evidence is accumulated for the role of the large bowel flora in carcinogenesis and in drug metabolism. These themes were discussed at a major symposium (Banbury Report, 1981) and in recent books (Hill 1986a,b). The theme of drug metabolism by intestinal bacteria is a recurrent one and has been reviewed by Scheline (1968), Goldman (1973, 1981), Goldin (1986) and Rowland (1988).

Geographical variation in the intestinal flora

The total number of samples of intestinal contents or faeces that have been investigated cannot be regarded as an adequate sampling of the human population of the world. However, because of interest in large bowel disease, and in particular, cancer, some attempts have been made to compare the bacteria isolated from faecal samples collected from people living in different parts of the world. The results of these investigations are discussed in Chapter 24. In this section the results of these investigations are considered in a bacteriological context.

The intestine may contain as many as 500 species of bacteria; if this is the case, then the flora of each individual could be regarded as a unique ecosystem. Thus, this very diversity would preclude general conclusions. However, very few studies have been undertaken that might reveal the degree of detail.

The examination of the bacterial flora of faeces from people living in widely different circumstances in different parts of the world have revealed differences in both the numbers and frequency of isolation of groups of bacteria, particularly non-sporing anaerobes. The reasons for these differences are not known. Consideration of the data from individuals indicates that the differences may be a reflection of the number of individuals in each group whose flora is dominated by a particular bacterial group (Peach *et al.*, 1974). The data presented from the study of Japanese–Hawaiians (Moore and Holdeman, 1974) can also be analysed in this way. Both of the studies showed that bacteroides, bifidobacteria or eubacteria may be dominant in a particular individual or, indeed, that no genus need be dominant.

Bacteroides are less prominent among African populations than among other groups (Hill *et al.*, 1971; Koornhoff *et al.*, 1979; Drasar *et al.*, 1986). This must reflect the importance of other bacterial groups in these populations.

Table 11.9 Comparison of the faecal flora of people living in various countries

Country	Carriage rates* expressed as a percentage of the bacterial groups named		
	Bacteroides	Bifido-bacterium	Eubacterium
USA	85	60	42
Scotland	93	46	37
Denmark	100	94	40
England	90	70	50
Hong Kong	89	53	51
Finland	100	74	57
India	30	56	52
Japan	63	75	60
Nigeria	15	100	—
Uganda	37	80	28

*A bacterial group must constitute at least 10 per cent of the flora to be detected

Differences between groups are often minimized by the way the results are presented; comparison of $\log_{10}$ of the viable counts are often employed. The original data from various published studies (Hill et al., 1971; Crowther et al., 1976; IARC, 1977; Tomkins et al., 1981; Drasar et al., 1986) has been presented in the form of carriage rates (Table 11.9), this illustrates the wide differences that can be demonstrated. The reasons for these findings is not known, but the role of diet has been widely canvassed.

Skin

Although the skin is one of the most accessible sites with a normal microbial flora, the patterns of colonization and the interactions between members of the mixed population are not well established. Most of the work that has been done has concentrated upon the (easier to handle) aerobic species, despite the fact that even at this site, where the surface is constantly exposed to air, the anaerobic skin flora outnumbers the aerobic flora by between 10:1 and 100:1 (Evans et al., 1950; Haenel. 1958).

The anaerobes are mostly the 'anaerobic diphtheroids', i.e., propionibacteria, and include the lipolytic species *Propionibacterium acnes*, which is implicated in the pathogenesis of acne. Three species of *Propionibacterium* inhabit the skin—*P. acnes*, *P. granulosum* and *P. avidum*. *P. acnes* is the most numerous and is found in all postpubertal subjects. *P. granulosum* is found in numbers 100–1000-fold less than *P. acnes* and can be isolated from only 10–12 per cent of normal subjects. Both species are found on skin with a high sebum content. *P. avidum* is found in the axillae; it requires a skin environment with a high water content rather than sebum (Noble, 1984). Sampling of the anaerobic skin flora is particularly difficult, but Holland and Roberts (1974) devised a technique of harvesting bacteria from pilosebaceous ducts by treating an area of skin within a Teflon ring with cyanoacrylate.

The resident skin flora interferes with the implantation of fresh organisms. This may be due to the production of bacteriocins or antibiotics. Perhaps more likely, metabolites produced by the anaerobes may be inhibitory to other bacteria—particularly the propionic acid and other free fatty acids liberated from sebum as a result of metabolism by propionibacteria. This may also be an important component of the 'self-disinfecting' mechanism, by which bacteria implanted on the normal skin are rapidly removed (Marples, 1969; Woodroffe and Shaw, 1974). See Chapter 17 for further discussion of the skin microflora.

Mouth

The mouth is a collection of microenvironments with quite different resident groups of bacteria. In the neonate, the oral flora is rapidly established and is mostly aerobic. Anaerobes appear only after the eruption of the early teeth, which provide the appropriate conditions for anaerobes in the gingival crevice.

Tongue

The surface flora of the tongue is made up mostly of *Streptococcus salivarus* and anaerobic gram-negative cocci of the genus *Veillonella*; there are very few bacteroides or fusobacteria.

Saliva

The salivary glands and saliva, as it is initially secreted, are sterile, but the saliva in the mouth rapidly gains a bacterial flora from the surrounding oral surfaces over which it flows. These organisms are primarily streptococci, bacteroides, fusobacteria and veillonellae. The gram-negative anaerobic bacilli are derived from the flora of the gingival crevice, the streptococci from plaque and the veillonellae from the tongue. Hadi and Russell (1968, 1969) found viable counts of salivary fusobacteria of 2.72×10/ml in normal subjects (counts were higher in patients with gingivitis).

Gingival crevice and dental plaque

Anaerobes are concentrated as a major part of the bacterial population in the gingival crevice and the subgingival plaque. This is a highly reduced, very anaerobic environment between the gingival epithelium and the tooth surface, and is capped over by the supragingival plaque, made up mostly of streptococci. The anaerobes that contribute to this subgingival flora are a mixture of gram-positive and gram-negative cocci and bacilli. These gingival anaerobes often have demanding growth requirements and are difficult to isolate. As a result, quantitative bacteriological studies generally underestimate the numbers present. Gibbons et al. (1963, 1964) and Loesche et al. (1972) found that gram-negative anaerobic bacilli constituted about 16 per cent of the cultivable flora of the gingival crevice, but only about 4 per cent of that of dental plaque. The proportion contributed by anaerobes varies with the standards of dental hygiene of test populations; the same workers found that gram-negative anaerobic bacilli represented 17 per cent of the cultivable flora in plaque from institutionalized subjects. This may represent a shift towards a greater degree of periodontal disease in these subjects. The predominant cultivable microbial flora from healthy periodontal sites comprises gram-positive cocci and bacilli (a mixture of aerobic, facultative and anaerobic species) with the gram-negative anaerobes occupying a minority position, as indicated above. The presence of periodontal disease (see Chapter 15) is associated with an increase in the numbers of gram-negative anaerobic species and a change in the gram-positive:gram-negative ratio (Slots, 1977, 1982; Listgarten and Hellden, 1978; Newman et al., 1978).

Quantitative studies of the gingival microbial flora must be interpreted with caution because of the difficulties encountered in isolating some (perhaps many) of the species present. Not only are some very oxygen-sensitive and require a high standard of anaerobic technique, but they are often nutritionally demanding as well. Some grow only in mixed cultures and cannot be maintained in pure culture on solid media. This problem has been recognized with fusobacteria and with some pigmented and non-pigmented strains of the melaninogenicus–oralis group of bacteroides, which will grow only in mixed cultures and in which satellitism may be demonstrated. Growth of some strains may be improved by the addition of 5–10 per cent CO_2 to the anaerobic atmosphere (Watt, 1973; Stalons et al., 1974) and by growth factors such as metal ions (Caldwell and Arcand, 1974), haemin (Gibbons and MacDonald, 1960; Gilmour and Poole, 1970), menadione (vitamin K) (Lev, 1959) or its precursors (Robins et al., 1973) and succinate, but some strains still require other, as yet unidentified, substances that are provided in mixed cultures.

Anaerobic species that contribute to the normal gingival flora

Gram-positive cocci

A range of anaerobic gram-positive cocci is found in the gingival crevice. Their identity is obscure because they have not been speciated in many studies, but Neut et al. (1985) found that *Peptostreptococcus anaerobius* and *Pstr. micros* were the most frequently identified species in the mouth.

Gram-positive bacilli

Members of the genera *Actinomyces*, *Lactobacillus* and *Eubacterium* are normally present in the gingival flora and constitute a large proportion of the normal flora. The *Eubacterium* species have not been studied in as great detail as some other species, but are undoubtedly present. Lactobacilli, both anaerobic and aerobic species, have long been recognized in dental plaque. They are found in healthy mouths but are present in large numbers in the presence of caries, although they are not now considered to have a major role in the pathogenesis of this common condition.

The oral gram-positive anaerobic bacilli that have received most attention from microbiologists are the actinomyces. They are normal commensals of the gingival crevice and form an important part of dental plaque; some species are also capable of causing severe, invasive infection. *Actinomyces israelii* is the most common pathogen, causing typical actinomycosis in man (see Chapters 9 and 15). It is also found almost universally as a commensal. *A. eriksonii* is found in similar commensal situations, but less commonly, and is only occasionally a cause of typical actinomycosis. *A. odontolyticus* is found mostly in saliva and in carious teeth; there is considerable doubt about its pathogenicity. *A. naeslundii* is also found as a commensal at similar sites in the mouth and is generally thought to be non-pathogenic, although it may contribute to lesions of the gums and teeth. The role of *A. viscosus* is less clear. It is found in the oral cavity of man and hamsters, and has been shown to contribute to dental plaque and periodontal disease in the latter. Generally, it is reasonable to regard the actinomyces as important components of the normal flora that contribute to the development of dental plaque and, while recognizing that *A. israelii*, the most common commensal, can cause invasive and widespread disease, the other species are generally less invasive and their pathological role appears to be restricted to a contribution to the mixed flora of local lesions of the gums and teeth.

Gram-negative cocci

Veillonellae are reported to be commensals of the surface of the tongue and can be found in moderate numbers in saliva. They do not appear to contribute significantly to the mixed anaerobic flora of the gingival crevice and they have no known pathological role.

Gram-negative bacilli

The Bacteroidaceae—gram-negative non-sporing anaerobic bacilli—make a major contribution to the normal gingival flora and are implicated in a wide range of infections related to the teeth and gums. They can be isolated in large numbers from the gingival crevice and from mature plaque, although in smaller numbers from early dental plaque in which streptococci and other gram-positive species predominate.

The species found in the normal gingival flora belong to the melaninogenicus–oralis group of bacteroides and the fusobacteria. From the earlier reports of gingivitis by Plaut (1894) and Vincent (1896), there has been great interest in the role of fusobacteria in the gingival flora. Fusiform organisms are readily seen in smears of subgingival plaque, even from healthy subjects, although their numbers are increased in patients with gingivitis. They are more difficult to isolate in culture, but of those that can be isolated, most that can be reliably identified are *F. nucleatum*. This is the only species that is consistently fusiform in shape when examined by microscopy and, because of this, it is generally assumed that most of the gingival fusobacteria are *F. nucleatum*, although only a minority of those seen are cultivable and some of those that are isolated cannot be identified by current schemes (Bennett and Duerden, 1985).

Although more readily noticed in stained smears, the fusiform organisms are greatly outnumbered by the small gram-negative anaerobic bacilli, the various species of *Bacteroides*. These belong almost entirely to the melaninogenicus–oralis group. *B. fragilis* is rarely isolated from the normal mouth and asaccharolytic bacteroides are present in only small numbers, if at all.

In qualitative cultures of the normal gingival flora, pigmented strains of *B. melaninogenicus* and *B. intermedius* are commonly found and readily isolated. They are easy to see and are frequently reported. However, in quantitative cultures the pigmented strains are generally greatly outnumbered by non-pigmented species of the same group. These all used to be known by the general specific name *B. oralis*, with some described as *B. ruminicola* (human strains), but more detailed taxonomic studies have shown that there are distinct differences between *B. oralis* isolates. This species has now been divided into at least four recognized species (*B. oralis, B. buccalis, B. veroralis* and *B. zoogleoformans*) and *B. ruminicola* (human strains) into two species (*B. buccae* and *B. oris*) (Holdeman et al., 1982) (see Chapter 5). The names given to these species reflect, with one exception, their particular association with the mouth.

Genitourinary tract

The mucosal surfaces of the genitourinary tract are colonized by a complex mixture of bacterial species, many of which are unique to these sites and are not found elsewhere in the body. This is particularly true of the vaginal lactobacilli, but also applies to the most common vaginal bacteroides, *B. bivius* and *B. disiens*, which belong to the melaninogenicus–oralis group and are similar in many respects to oral strains but which have sufficient stable distinguishing characteristics to be clearly distinct species. There are two distinct mucosal surfaces to be considered in women (vagina and urethra) and only one in men (urethra). The contribution of anaerobes to the normal flora at these sites, and their roles there, is the subject of considerable debate and reports of studies by different groups of workers using different techniques have come to widely different and sometimes frankly opposing conclusions. Most work has been done on the normal vaginal flora. Little attention has been paid to the female urethra, which will share some of the vaginal flora but may have a distinctive pattern of colonization. There has been more interest in recent years in the normal urethral flora in men as part of the investigation of non-gonococcal non-chlamydial urethritis.

Vaginal microflora

Although the vaginal flora is generally considered to be a single entity, the vagina does not provide a uniform environment throughout its length from the introitus to the cervix, and the balance of species differs with these different environments. The flora of the lower vagina is really a mixture of organisms from the upper vaginal mucosa, and from the cervical and upper vaginal secretions, and organisms from the vulva and perineum. Sampling from the lower vagina gives a much more variable pattern of colonization than is seen in the upper vagina because of the variable nature of the admixture of organisms from the perineal skin and faecal contamination. In this lower vaginal flora there are staphylococci, propionibacteria and various coryneform (diphtheroid) organisms from the perineal skin flora, together with variable numbers of *E. coli* and other enterobacteria from the faeces. Intestinal anaerobes, particularly the fragilis group of bacteroides, may be present at the introitus and in the lower vagina as part of this faecal contamination, but they are not found higher up the vagina in the 'true' vaginal flora. It is usual, therefore, to study the flora of the cervix and fornices, as this will more closely represent this true vaginal flora. Above the cervix, the upper genital tract is normally sterile and the presence of bacteria indicates some abnormality, i.e., a pathological process.

Having defined the site, however, the vaginal environment is far from constant. It changes markedly with the various hormonal changes of the human reproductive cycle. For obvious ethical reasons, definitive studies on the female prepubertal vaginal flora are not possible, but the examination of vaginal swabs taken for various reasons from prepubertal girls shows that the flora found in adult women does not develop until puberty. From puberty until the menopause, the state of the mucosa and secretion changes with each menstrual cycle and with pregnancy, and it would be expected that the microbial flora might change with this changing environment. The menopause again marks a major change in the physiological environment of the vagina. Finally, superimposed upon these changes are the effects of sexual intercourse; not only may bacteria be introduced at this time, but seminal fluid is alkaline. This buffers the normal acidity and causes a temporary rise in pH, essential for the sperm but possibly also affecting the vaginal flora.

However, despite these interacting variables, a general pattern of bacterial colonization of the normal adult vagina is emerging. The flora is predominantly anaerobic and the environmental pH is low. Some aerobic/facultative species are present, mostly staphylococci, streptococci and coryneforms, but they do not appear to play a significant role in the normal ecology or in disease. The anaerobic bacteria present in the normal vagina form a varied and variable group, and different groups of workers do not agree on the roles of several reported genera. There is general agreement that lactobacilli, many of which are obligate anaerobes, others facultative organisms, are a major part of the normal vaginal flora, but the significance of *Bacteroides* spp. and anaerobic gram-positive cocci is much more controversial. Although there have been some reports of *Clostridium difficile* in the vagina, clostridia are probably not a major part of the normal flora.

Lactobacilli

Vaginal lactobacilli were first described by Doderlein in 1894, and his view, that the vaginal flora was generally homogeneous, consisting of lactobacilli and a few other irrelevant organisms, was generally accepted until the 1970s, despite reports of the isolation of a more varied flora, including many other anaerobes. However, this almost exclusive concentration on the lactobacilli reflected the fact that they are a major component of the normal vaginal flora in healthy non-pregnant women. They are the most common anaerobic gram-positive bacilli found in vaginal cultures and are clearly evident in gram-stained films of vaginal secretions.

There are facultative and obligately anaerobic species in the genus *Lactobacillus*, but all grow better in microaerophilic or anaerobic conditions. Both groups are found in the vagina. The classification of the lactobacilli is uncertain and debatable, but the species most commonly isolated from the vagina is *L. acidophilus*, a name that reflects its presumed acid-producing role in the normal vaginal flora. The metabolic activity of lactobacilli has long been considered to make a major contribution towards maintaining the normal vaginal environment at a pH below 7.0. They are acidophilic organisms, growing best at pH values below 6, and they are homofermentative, breaking down glucose with the production of at least 95 per cent lactic acid and only minimal amounts of CO_2 and acetic acid. The vaginal epithelial cells contain glycogen, and it has been generally accepted since the mid-1930s that this is metabolized, first to glucose and then to lactic acid, by the vaginal lactobacilli, maintaining the low pH and selectively favouring colonization by acid-tolerant organisms such as the lactobacilli themselves (Cruickshank and Sharman, 1934). Gas–liquid chromatography analysis of short chain fatty acids in vaginal secretions has confirmed that lactic acid is the main acid in normal secretions, and the lactobacilli are the only members of the normal flora that produce lactate as their major metabolic product. However, Wilks and Tabaqchali (1987) have cast some doubt on the simplicity of this explanation. They found that the vagina does not provide a constant acidic environment, the pH falling from a mean value of 6.5 around the onset of menstruation to 4–6 during weeks two to four of the cycle. They suggested that this could have been due to metabolism by lactobacilli of glycogen in shed epithelial cells, but they could detect no increase in numbers of lactobacilli associated with this change. They also pointed out that Brown (1982) had produced no unequivocal proof that vaginal lactobacilli could produce acid from glycogen.

Despite these indications that the interactions between lactobacilli, other bacteria and the host environment may be more complex than first thought, they do seem to have a role in maintaining the normal vaginal ecology, and this seems to be an important protective mechanism against infection with potential pathogens. Other than during menstruation, when the shed blood and cells may buffer the acidity, the normal vaginal pH of 4.5 or less is inhibitory to many potential pathogens.

Lactobacilli or their products have been shown to inhibit a range of aerobic, facultative and anaerobic organisms. This effect has been ascribed variously to the pH or to the lactic acid itself, or to production of inhibitory protein or 'lactocidins'. Diffusible and dialysable products of *Lactobacillus* cultures that lower the pH, i.e., probably lactic acid itself, inhibit the bacteria associated with bacterial vaginosis, *Bacteroides* spp., *Mobiluncus* spp., *Gardnerella vaginalis* and anaerobic cocci (Mårdh, 1983; Bosu *et al.*, 1987).

L. acidophilus has no known pathogenic activities.

Anaerobic cocci

Early reports of the contribution of anaerobic cocci to the normal vaginal flora were sporadic and inconclusive, but recent studies have established that they are a regular part of the flora. Neut *et al.* (1985) reported that anaerobic gram-positive cocci were generally present as commensals of the normal vagina, and that the common species were *Pstr. magnus* and *Pstr. asaccharolyticus*. Bartlett *et al.* (1977) found that 'peptococci' were the most frequent isolates in a quantitative study of the vaginal flora; they were present in 65 per cent of subjects.

Gorbach *et al.* (1973) found a similar proportion of women to be colonized by gram-positive anaerobic cocci and Ohm and Galask (1975) found them in 75 per cent of cervical samples taken from women about to undergo hysterectomy, who may not be regarded as an entirely normal group. However, Masfari *et al.* (1986) isolated gram-positive anaerobic cocci from only 27 per cent of control (asymptomatic) subjects. Between these figures, Wilks *et al.* (1984) found various species of anaerobic cocci, including *Peptococcus magnus*, *Pc. asaccharolyticus*, *Pstr. micros* and *Pstr. anaerobius*, in half of 20 control (asymptomatic) subjects, and in a subsequent quantitative study Wilks and Tabaqchali

(1987) isolated species that included *Pstr. anaerobicus*, *Pc. prevotii*, *Pstr. productus*, *Pc. asaccharolyticus* and *Pstr. magnus* from six out of 10 normal subjects at different stages of the menstrual cycle.

It is clear from these varied results that anaerobic gram-positive cocci can be isolated from about 50–70 per cent of normal women, and that they are probably a regular part of the normal flora. It is likely that any group of organisms detected in 60 per cent of subjects when there is such a complex mixed flora is generally present in most, if not all, subjects.

Bacteroides spp.

The reported isolation rates of gram-negative anaerobic bacilli from the vagina of normal healthy women have varied considerably. Some workers have found *Bacteroides* spp. in only about 5 per cent of their subjects, whereas others report isolation rates of 60–70 per cent. The presence of bacteroides in the vagina was first reported by Burdon (1928) who isolated '*B. melaninogenicus*' from 28 out of 35 normal women. Mead and Louria (1969) and Suzuki and Ueno (1971) also found high colonization rates. *Bacteroides* spp. have been isolated from cervical cultures in 57 per cent of 30 normal women (Gorbach *et al.*, 1973) in 65 per cent of 26 women (Sanders *et al.*, 1975) and from the fornices in 65 per cent of 20 women (Duerden, 1980). However, Neary *et al.* (1973) found bacteroides in only 8.6 per cent of 246 preoperative gynaecological patients, and in 4.6 per cent of 500 women attending a family planning clinic. Similarly, Leigh (1976) found only 5 per cent of 200 gynaecology out-patients to have positive cultures and, in a quantitative study, Lindner *et al.* (1978) isolated *Bacteroides* spp. from only 4 per cent of 20 normal subjects. However, in other quantitative studies *Bacteroides* spp. were reported in 41 per cent of normal women by Bartlett *et al.* (1977), 12 out of 20 and six out of 10 women by Wilks *et al.* (1984) and Wilks and Tabaqchali (1987); Levison *et al.* (1979) found them to be part of the predominant flora.

There are probably several reasons for these wide variations in the reported frequency of *Bacteroides* spp. in the normal vaginal flora. Some of the differences may reflect differences in the subject populations, but these are probably insignificant in comparison with differences in methodology. Methods of sampling and transport, systems of anaerobic culture and the nature of the study (qualitative or quantitative) could all affect the isolation rate. When swabs are used, particularly if they are not placed in a suitable anaerobic transport medium, many gram-negative anaerobes can be lost from a combination of oxygen toxicity, desiccation and adherence to the swab; much more reliable results would be expected with direct plating or the collection of secretion directly into liquid transport medium. Rapid transport to the laboratory and an efficient anaerobic system are essential for optimal isolation of vaginal bacteroides, which are mostly of the melaninogenicus–oralis group and are more fastidious in terms of media, growth factors and anaerobic conditions than the intestinal bacteroides (fragilis group); moreover, some of them appear to grow only in mixed cultures, and are lost when subcultured on to pure plates. They also grow more slowly, and prolonged incubation is required for optimum isolation rates. Differences in carriage rates between qualitative and quantitative studies, even from the same groups of workers, may be a reflection of the quantitative sensitivity of the sampling and culture methods. In quantitative methods, secretions are diluted before plating out known volumes to obtain single colonies for counting and identification; the lower limit of sensitivity of such methods is about 10^5 cfu/ml. In qualitative studies, however, a more concentrated inoculum is used and, particularly if the colonies are easy to see in a mixture (e.g., pigmented bacteroides), the limit of sensitivity may be much lower.

It would appear from the consensus of these studies that *Bacteroides* spp. are a regular part of the normal vaginal flora. There is also general agreement on the species that make up this part of the microflora. In most studies in which isolates have been identified to species level, most vaginal isolates have been found to belong to the melaninogenicus–oralis group, and the most common species is *B. bivius*. The similar species, *B. disiens*, is also a common isolate, but is less frequently reported than *B. bivius*. These two species are particularly associated with the vagina, being common at this site and rarely found elsewhere in the body. Two species of pigmented saccharolytic bacteroides, *B. melaninogenicus* and *B. intermedius*, are also commonly found, *B. melaninogenicus* being more frequently isolated than *B. intermedius*. These pigmented strains are usually present in smaller numbers than the non-pigmented species, but they are easily seen after prolonged incubation of mixed cultures. Asaccharolytic bacteroides, principally the pigmented *B. asaccharolyticus*, and less commonly *B. ureolyticus*, are also found in vaginal samples, but they are much more common in pathological states and it is more doubtful whether they should be regarded as normal commensals. The fragilis group is not part of the normal vaginal flora.

Qualitative and quantitative approaches to the normal vaginal flora

The inconsistencies in the reports of the isolation of different genera from the normal vaginal flora probably result from a combination of methodological differences and actual differences in a constantly changing environment. The lower limit of detection in qualitative tests, particularly of fastidious anaerobes, is of the order of 10^4–10^5 cfu/ml of secretions, and it is reasonable to assume that bacteria isolated from 60 per cent of subjects under these circumstances are 'always' present, i.e., are part of the normal flora. On this basis, aerobic and anaerobic lactobacilli, coryneforms, coagulase-negative staphylococci, *Bacteroides* spp. (of the melaninogenicus–oralis group) and gram-positive anaerobic cocci are the major constituents of the normal flora.

However, this qualitative recording of the species does not distinguish between 'major' constituents present at approximately 10^{8-9} cfu/ml of secretions, and relatively 'minor' components present at 10^4–10^5 cfu/ml, i.e., outnumbered by 10 000:1, nor does it give any indication of changes in the balance with the physiological changes of the reproductive cycle. Furthermore, some of these components of the normal flora, e.g., bacteroides and anaerobic cocci, are also associated with infective pathological conditions of the genitourinary tract (abcesses, ulcers, vaginosis, deep pelvic infection, etc.; see Chapter 14).

In attempts to define more clearly the roles of the different species, particularly the anaerobes, in cyclical changes and in the balance between health and disease, several workers have attempted detailed quantitative studies of the vaginal flora. Various methods have been used, and their variety reflects the difficulties inherent in accurately measuring the amount of vaginal secretions in a test sample. There have been three main approaches:

1. The double swab method in which duplicate swabs, one preweighed, are used to collect samples of secretions; one is placed in transport media and the preweighed one is reweighed as an indirect measure of the weight of sample tested (Onderdonk et al., 1977);
2. Sampling of secretions by volume, by collecting defined volumes of secretions either by calibrated pipette (Lindner et al., 1978) or standard loops (Levison et al., 1979);
3. Sampling by weight, by collecting samples with a loop into preweighed transport medium and reweighing (Wilks et al., 1982; Masfari et al., 1986).

The double swab method of Onderdonk et al. (1977) has given unreliable results in other hands and the volumetric methods run into problems because the viscosity of secretions varies greatly; sampling by weight appears to give more reliable results.

From these various quantitative studies, a reasonable consensus of the quantitative relationships in the normal vaginal flora has emerged. The total count reported by most is 10^{8-10} cfu/ml or g of secretions and anaerobes (10^{9-10} cfu/ml) outnumber aerobic/facultative bacteria (10^8 cfu/ml) by a ratio of 10–100:1. The three main groups of anaerobes have been found in more or less equal numbers if the findings are averaged, but the order differs in different studies. Bartlett et al. (1977) found that anaerobic cocci ($10^{8.7}$ cfu/ml) outnumbered bacteroides ($10^{7.6}$ cfu/ml), whereas Levison et al. (1979) found the opposite. Wilks et al. (1984) also found greater numbers of bacteroides ($10^{7.3}$ cfu/g) than anaerobic cocci ($10^{6.7}$ cfu/g), but their figures were generally about 1 $\log_{10}$ lower than the earlier reports, as were those of Masfari et al. (1986), who used the same method and found equal numbers of bacteroides and anaerobic cocci (10^7 cfu/g) with lactobacilli 10-fold higher (10^8 cfu/g).

Variation of the flora with the reproductive cycle

The physiological environment of the vagina changes during each menstrual cycle and, even more markedly, with pregnancy. Although the hormonal factors have long been held to influence the vaginal flora (Cruickshank and Sharman, 1934), most studies have been based on single specimens taken at arbitrary points in the cycle. Bartlett et al. (1977) found constant mean levels of anaerobes throughout the cycle, but fewer aerobes in the premenstrual specimens from five women, but the opposite effect was found by Sautter and Brown (1980), who found that lactobacilli decreased in numbers from $10^{9.4}$ cfu/g before menstruation to $10^{6.7}$ cfu/g during menstruation. Wilks and Tabaqchali (1987) did quantitative studies of the vaginal flora in ten asymptomatic women at weekly intervals through a normal cycle and found a marked fall in pH from 6.5 in week one to about pH 4.5 in weeks two to four. This fall in pH was associated with a fall in the number of anaerobic lactobacilli and a small rise in the number of aerobic lactobacilli; there were no significant changes in the frequency or numbers of other anaerobes. These results are confusing and underline the complexity of the bacteria–host interactions and the need for further study.

The physiological changes in the vagina in pregnancy are dramatic, as are the changes in the bacterial flora. The amount of glycogen in the mucosa is much lower, the pH of secretions much less acidic and the normally anaerobic flora is replaced by a preponderance of aerobic/facultative species (Hurley et al., 1974). The lactobacilli, bacteroides and anaerobic cocci generally disappear during pregnancy but return and re-establish the normal, non-pregnant, patterns soon after delivery (Hite et al., 1947; Willis, 1977).

Anaerobes in the urethra

There have been few detailed studies of the anaerobic flora of the urethra. In women, most emphasis has been on the vaginal flora and little is known of the urethra. In men, most interest has centred upon the problems of urethritis, and investigation of the normal flora has only come about with the need to establish the role of organisms found in samples from patients with non-gonococcal non-chlamydial urethritis.

Most workers who have examined controls as part of their study of urethritis have isolated anaerobes from urethral swabs from asymptomatic men; these are considered to be part of the normal flora (Sullivan et al., 1972; Smith, 1975). The predominant anaerobic species are gram-positive cocci (Weinberg, 1974); bacteroides are found less commonly and in smaller numbers. Bowie et al. (1977) isolated anaerobes from 91 per cent of normal males. Fontaine et al. (1982) isolated anaerobes from 80 per cent of men without urethritis (normal controls) and gram-positive cocci were predominant, being present in 67 per cent of specimens. However, Masfari et al. (1983, 1985) isolated anaerobes from only 6 (21 per cent) of 28 asymptomatic controls; they concluded that the normal flora of the male urethra was predominantly aerobic and that the presence of anaerobes in large numbers indicated some abnormal or pathological condition. As with the vaginal flora, the differences may reflect differences in sampling methods or in quantitative sensitivity. As others had found, Masfari et al. (1983, 1985) isolated anaerobes in large numbers from most patients with any type of urethritis, suggesting that the presence of a largely anaerobic flora was associated with some pathological disorder.

References

Banbury Report. (1981). *Banbury Report 7. Gastrointestinal cancer: endogenous factors.* Cold Spring Harbor Laboratory, Cold Spring Harbor.

Bartlett J G, Onderdonk A B, Drude E, Goldstein C, Anderka M, Alpert S, McCormack W M. (1977). Quantitative bacteriology of the vaginal flora. *Journal of Infectious Diseases* 136: 271–277.

Bennett K W, Duerden B I. (1985). Identification of fusobacteria in a routine laboratory. *Journal of Applied Bacteriology* 59: 171–187.

Bhat P, Shantakumari S, Rajan D, Mathan V I, Kapadia C R, Swamabi C, Baker S J. (1972). Bacterial flora of the gastrointestinal tract in south India control subjects and patients with tropical sprue. *Gastroenterology* 62: 11.

Bosu W, Duerden B I, Bennett K W. (1987). Metabolic interactions of vaginosis-associated bacteria. *Journal of Medical Microbiology* 23: xii.

Bowie W R, Wang S-P, Alexander E R et al. (1977). Etiology of non-gonococcal urethritis. Evidence for *Chlamydia trachomatis* and *Ureaplasma urealyticum*. *Journal of Clinical Investigation* 59: 735–742.

Brown W J. (1982). Variation in the vaginal bacterial flora: a preliminary report. *Annals of Internal Medicine* 96: 931–934.

Burdon K L. (1928). *Bacterium melaninogenicum* from normal and pathological tissue. *Journal of Infectious Diseases* 42: 161–171.

Caldwell D R, Arcand C. (1974). Inorganic and metalorganic growth requirements of the genus *Bacteroides*. *Journal of Bacteriology* 120: 322–333.

Cato E P, Johnson J L. (1976). Reinstatement of species rank for *Bacteroides fragilis*, *B. ovatus*, *B. distasonis*, *B. thetaiotaomicron*, and *B. vulgatus*. Designation of neotype strains for *Bacteriodes fragilis* (Veillon and

Zuber) Castellani and Chalmers and *Bacteroides thetaiotaomicron* (Distaso) Castellani and Chalmers. *International Journal of Systematic Bacteriology* 26: 230.

Clarke R T J, Bauchop T (eds). (1977). *The microbial ecology of the gut*. Academic Press, London.

Crowther J S, Drasar B S, Hill M J, MacLennan R, Magnin D, Peach S, Teoh-Chan. (1976). Faecal steroids and bacteria and large bowel cancer in Hong Kong by socioeconomic groups. *British Journal of Cancer* 34: 191–198.

Cruickshank R, Sharman A. (1934). The biology of the vagina in the human subject; glycogen in the vaginal epithelium and its relation to ovarian activity. *Journal of Obstetrics and Gynaecology* 41: 190–207.

Cummings J H, MacFarlane G T, Drasar B S. (1989). The gut microflora and its significance. In *Gastrointestinal and oesophogeal pathology*, pp. 201–219. Edited by Whitehead R. Churchill Livingstone, Edinburgh.

Doderlein A. (1894). Die scheidensekretuntersuchungen. *Zentralblatt fur Gynakologie* 18: 10–14.

Drasar B S, Barrow P A. (1985). *Intestinal microecology*. Van Nostrand Reinhold, Wokingham.

Drasar B S, Hill M J. (1974). *Human intestinal flora*. Academic Press, London.

Drasar B S, Montgomery F, Tomkins A M. (1986). Diet and faecal flora in three dietary groups in rural northern Nigeria. *Journal of Hygiene* 96: 59–65.

Duerden B I. (1980). The isolation and identification of *Bacteroides* spp. from the normal human vaginal flora. *Journal of Medical Microbiology* 13: 79–87.

Eggerth A H. Gagnon B H. (1933). The *Bacteroides* of human faeces. *Journal of Bacteriology* 25: 389–413.

Evans C A, Smith W M, Johnston E A, Giblett E R. (1950). Bacterial flora of the normal human skin. *Journal of Investigative Dermatology* 15: 305.

Finegold S M, Flora J J, Attebury H R, Sutter V L. (1975). Faecal bacteriology of colonic polyp patients and control patients. *Cancer Research* 35: 3407–3417.

Finegold S M, Sutter V L, Sugihara P T, Edler H A, Lehmann S M, Phillips R L. (1977). Faecal flora of Seventh Day Adventist populations and control subjects. *American Journal of Clinical Nutrition* 30: 1718–1792.

Finegold S M, Sutter V L, Mathisen G E. (1983). Normal and indigenous intestinal flora. In *Human intestinal microflora in health and disease*, pp. 3–31. Edited by Hentges D J. Academic Press, New York.

Fontaine E A R, Taylor-Robinson D, Hanna N F, Coufalik E D. (1982). Anaerobes in men with urethritis. *British Journal of Venereal Diseases* 58: 321–326.

Gibbons R J, MacDonald J B. (1960). Hemin and vitamin K compounds as required factors for the cultivation of certain strains of *Bacteroides melaninogenicus*. *Journal of Bacteriology* 80: 164–170.

Gibbons R J, Socransky S S, de Araujo W C, van Houte J. (1964). Studies of the predominant cultivable microbiota of dental plaque. *Archives of Oral Biology* 9: 365–370.

Gibbons R J, Socransky S S, Sawyer S, Kapsimalis B, MacDonald J B. (1963). The microbiota of the gingival crevice area of man. II: The predominant cultivable organisms. *Archives of Oral Biology* 8: 281–289.

Gilmour M N, Poole A E. (1970). Growth stimulation of the mixed microbial flora of human dental plaques by haemin. *Archives of Oral Biology* 15: 1343–1353.

Goldin B R. (1986). *In situ* bacterial metabolism and colon mutagens. *Annual Review of Microbiology* 40: 367–393.

Goldman P. (1973). Therapeutic implications of the intestinal microflora. *New England Journal of Medicine* 289: 623–628.

Goldman P. (1981). The metabolism of xenobiotics by the intestinal flora. In *Banbury Report 7*, pp. 25–39. Cold Spring Harbor Laboratory, Cold Spring Harbor.

Gorbach S L, Menda K B, Thadepalli H, Keith L. (1973). Anaerobic microflora of the cervix in healthy women. *American Journal of Obstetrics and Gynaecology* 117: 1053–1055.

Hadi A W, Russell C. (1968). Quantitative estimations of fusiforms in saliva from normal individuals and cases of acute ulcerative gingivitis. *Archives of Oral Biology* 73: 1371–1376.

Hadi A W, Russell C. (1969). Fusiforms in gingival material. Quantitative estimations from normal individuals and cases of periodontal disease. *British Dental Journal* 126: 82–84.

Haenel H. (1958). Experimentelle Untersuchungen uber die Zusammensetzung der Darmflora. *Zentralblatt fur Bakteriologie, Parasitenkunde, Infecktionskrankheiten und Hygiene I. Abteilung Origale* 170: 323–326.

Hentges D J. (1983). *Human intestinal microflora in health and disease*. Academic Press, New York.

Hill M J. (1986a). *Microbial metabolism in the digestive tract*. CRC Press, Boca Raton.

Hill M J. (1986b). *Microbes and human carcinogenesis*. Edward Arnold, London.

Hill M J, Crowther J S, Drasar B S, Hawksworth G, Aries V, Williams R E O. (1971). Bacteria and the aetiology of cancer of the large bowel. *Lancet* i: 95–100.

Hite K E, Hesseltine H C, Goldstein L. (1947). Study of bacterial flora of normal and pathologic vagina and uterus. *American Journal of Obstetrics and Gynecology* 53: 233–240.

Holdeman L V, Good I J, Moore W E C. (1976). Human faecal flora: variation in composition within individuals and a possible effect of emotional stress. *Applied and Environmental Microbiology* 31: 359–375.

Holdeman L V, Moore W E C. (1974). New genus, *Coprococcus* twelve new species, and amended descriptions of four previously described species of bacteria from human faeces. *International Journal of Systematic Bacteriology* 24: 260–277.

Holdeman L V, Moore W E C, Churn P J, Johnson J L. (1982). *Bacteroides oris* and *Bacteroides buccae*, new species from human periodontitis and other human infections. *International Journal of Systematic Bacteriology* **32**: 125–131.

Holland K T, Roberts C D. (1974). A technique for sampling micro-organisms from the pilo-sebaceous ducts. *Journal of Applied Bacteriology* **32**: 289–296.

Hurley R, Stanley V C, Leask B G C, de Louvois J. (1974). Microflora of the vagina during pregnancy. In *The normal microbial flora of man*. Society for Applied Bacteriology Symposium Series 3, pp. 155–185. Edited by Skinner F A, Carr J G. Academic Press, London.

IARC Microecology Group. (1977). Dietary fibre transit time, faecal bacteria, steroids and colon cancer in two Scandinavian populations. *Lancet* **ii**: 207–212.

Johnson J L, Moore W E C, Moore L V H. (1986). *Bacteroides caccae* sp.nov., *Bacteroides merdae* sp.nov., and *Bacteroides stercoris* sp.nov. isolated from human faeces. *International Journal of Systematic Bacteriology* **36**: 499–501.

Koornhoff H J, Richardson N J, Wall D M, Moore W E C. (1979). Faecal bacteria in South African rural blacks and other population groups. *Israel Journal of Medical Science* **15**: 335–340.

Leigh D A. (1976). Bacteroides infections: diagnosis and treatment. In *Selected topics in clinical bacteriology*, p. 129. Edited by de Louvois J. Balliere Tindall, London.

Lev M. (1959). The growth promoting activity of compounds of the vitamin K group and analogues for a rumen strain of *Fusiformis nigrescens*. *Journal of General Microbiology* **20**: 697–703.

Levison M E, Corman L C, Carrington E R, Kay C D. (1979). Quantitative microflora of the vagina. *American Journal of Obstetrics and Gynecology* **127**: 80–85.

Lindner J G E M, Plantema F A F, Hoogkamp-Korstranje J A A. (1978). Quantitative studies of the vaginal flora of healthy women and of obstetric and gynaecological patients. *Journal of Medical Microbiology* **11**: 233–241.

Listergarten M A, Hellden L. (1978). Relative distribution of bacteria at clinically healthy and periodontally diseases sites in humans. *Journal of Clinical Periodontology* **5**: 115–132.

Loesche W J, Hocket R N, Syed S A. (1972). The predominant cultivable flora of tooth surface plaque removed from institutionalised subjects. *Archives of Oral Biology* **17**: 1311–1325.

Mårdh P A, Soltesz L V. (1983). Interaction between lactobacilli and other organisms occurring in the vaginal flora. *Scandinavian Journal of Infectious Diseases* Suppl 40: 47–51.

Marples R R. (1969). Diphtheroids of normal human skin. *British Journal of Dermatology* Suppl 81: 1–47.

Masfari A N, Kinghorn G R, Duerden B I. (1983). Anaerobes in genitourinary infections in men. *British Journal of Venereal Diseases* **59**: 255–259.

Masfari A N, Kinghorn G R, Hafiz S, Barton I G, Duerden B I. (1985). Anaerobic bacteria and herpes simplex virus in genital ulceration. *Genitourinary Medicine* **61**: 109–113.

Masfari A N, Duerden B I, Kinghorn G R. (1986). Quantitative studies of vaginal bacteria. *Genitourinary Medicine* **62**: 256–263.

Mead P B, Louria D B. (1969). Antibiotics in pelvic infections. *Clinical Obstetrics and Gynecology* **12**: 219–239.

Moore W E C, Holdeman L V. (1974). Human fecal flora: the normal flora of 20 Japanese-Hawaiians. *Applied Microbiology* **27**: 961–979.

Neary M P, Allen J, Okubadejo O A, Payne D J H. (1973). Preoperative vaginal bacteria and postoperative infections in gynaecological patients. *Lancet* **ii**: 1291.

Neut C, Lesieur V, Romond C, Beerens H. (1985). Analysis of gram-positive anaerobic cocci in oral, fecal and vaginal flora. *Journal of Clinical Microbiology* **4**: 435–437.

Newman M G, Grinenco V, Weiner M, Angel I, Karge H, Nisengard R. (1978). Predominant microbiota associated with periodontal health in the aged. *Journal of Periodontology* **49**: 553–559.

Noble W C. (1984). Skin microbiology: coming of age. *Journal of Medical Microbiology* **17**: 1–12.

Ohm M J, Galask R P. (1975). Bacterial flora of the cervix from 100 pre-hysterectomy patients. *American Journal of Obstetrics and Gynecology* **122**: 6823–6826.

Onderdonk A B, Polk B F, Moon N E, Goren B, Bartlett J G. (1977). Methods for quantitative vaginal flora studies. *American Journal of Obstetrics and Gynecology* **128**: 777–781.

Peach S, Fernandez F, Johnson K, Drasar B S. (1974). The non-sporing anaerobic bacteria in human faeces. *Journal of Medical Microbiology* **7**: 213–221.

Plaut H C. (1894). Studien zur Bakteriellen Diagnostik der Diphtherie und der Anginen. *Deutsche Medizinische Wochenscrhirft* **20**: 920–923.

Robins D J, Yee R B, Bentley R. (1973). Biosynthetic precursors of vitamin K as growth promoters for *Bacteroides melaninogenicus*. *Journal of Bacteriology* **14**: 965–971.

Rosebury T. (1962). *Microorganisms indigenous to man*. McGraw-Hill, New York.

Rowland I R, (ed). (1988). *The role of the gut flora in toxicity and cancer*. Academic Press, London.

Sanders C V, Mickal A, Lewis A C, Torres J. (1975). Anaerobic flora of the endocervix in women with normal and abnormal Papanicolaou (Pap) smears. *Clinical Research* **23**: 30A.

Sautter R L, Brown W J. (1980). Sequential vaginal cultures from normal young women. *Journal of Clinical Microbiology* **11**: 479–484.

Scheline R R. (1968). Drug metabolism by intestinal micro-organisms. *Journal of Pharmaceutical Science* **57**: 2021–2037.

Slots J. (1977). Microflora in the healthy gingival sulcus in man. *Scandinavian Journal of Dental Research* **85**: 247–254.

Slots J. (1982). Importance of black-pigmented *Bacteroides* in human periodontal disease. In *Host-parasite interactions in periodontal diseases*, pp. 27–45. Edited by Genco R J, Mergenhagen S E. American Society for Microbiology, Washington DC.

Smith L DS. (1975). *The pathogenic anaerobic bacteria*, 2nd edn. Thomas, Springfield.

Stalons D R, Thornsberry C, Dowell V R. (1974). Effect of culture medium and carbon dioxide concentration on growth of anaerobic bacteria commonly encountered in clinical specimens. *Applied Microbiology* 27: 1098–1104.

Suzuki S, Ueno K. (1971). Methods of isolation and identification of anaerobes from clinical material. In *First Symposium on anaerobic bacteria and their infectious diseases*, p. 81. Edited by Ishiya et al. Esai Co, Tokyo.

Tomkins A M, Bradley A F, Oswald S, Drasar B S. (1981). Diet and faecal microflora of infants children and adults in rural Nigeria and urban UK. *Journal of Hygiene*, 86: 285–293.

Vincent M H. (1896). Sur l'etiologie et sur les lesions anatomopathologiques de la pourriture d'hopital. *Annales de l'Institut Pasteur* 10: 488–510.

Watt B. (1973). The influence of carbon dioxide on the growth of obligate and facultative anaerobes on solid media. *Journal of Medical Microbiology* 6: 307–314.

Wilks M, Thin R N, Tabaqchali S. (1982). Quantitative methods for studies on vaginal flora. *Journal of Medical Microbiology* 15: 141–147.

Wilks M, Tabaqchali A. (1987). Quantitative bacteriology of the vaginal flora during the menstrual cycle. *Journal of Medical Microbiology* 24: 241–245.

Wilks M, Thin R N, Tabaqchali S. (1984). Quantitative studies of the vaginal flora in genital disease. *Journal of Medical Microbiology* 18: 217–231.

Willis A T. (1977). *Anaerobic bacteriology: clinical and laboratory practice*, 3rd edn. Butterworth, London.

Woodroffe R C, Shaw D A. (1974). *Natural control and ecology of microbial populations on skin and hair*. Society for Applied Bacteriology, Symposium Series 3: 13–34.

12

Laboratory diagnosis of anaerobic infection

M W D Wren

Sites from which anaerobes may be isolated
    Head, neck and CNS
    Thoracic sites
    Abdominal sites
    Female genital tract
    Superficial sites
Types of specimen
    Pus
    Tissue
    Dressings
Transport of clinical material
Direct examination of clinical material
    Gram's stain
    Fluorescent antibody staining
    Ultraviolet light
    Gas–liquid chromatography
Anaerobic techniques
    Anaerobic jars
    Anaerobic chambers
Hungate (roll tube) techniques
    Holding jars
    The role of carbon dioxide
    Length of incubation
Isolation media
    Basal media
    Supplements
    Selective media
    Broth enrichment
Isolation procedure
Preliminary identification
    Comments on the preliminary groupings
    Commercial identification kits
    Identification of anaerobes with preformed enzymes
    Gas–liquid chromatography
Appendix
References

Sites from which anaerobes may be isolated

It is now generally recognized that anaerobic bacteria may be involved in most human bacterial infections that follow any form of surgery or are related to those body sites that have a large anaerobic population. Infections that occur as a result of trauma, especially where the integrity of a mucous surface is interrupted, where penetration of a foreign body has occurred or where 'spillage' of anaerobes from the normal flora occurs, frequently harbour a mixed anaerobic flora, with or without aerobes. An example of this involvement would be when anaerobes from the oral cavity are responsible for brain abscess, chronic sinusitis, chronic otitis media, lung abscess, aspiration pneumonia and empyema. Anaerobes must therefore be sought in a wide variety of clinical specimens.

Head, neck, and central nervous system

Specimens from patients with chronic sinusitis and chronic otitis media may yield anaerobes in up to 52 and 56 per cent of specimens, respectively. Orofacial surgery for infections of dental origin is a rich source of anaerobes, with recovery rates of up to 96 per cent recorded (Chow et al., 1978). Pus from brain abscesses of otitic origin frequently contain anaerobes, (Chapter 16; Ingham et al., 1978) whereas those with sinusitis as the predisposing factor are generally caused by *Streptococcus milleri*, although we have found such abscesses where this organism was accompanied by a mixture of anaerobes. Attention has recently been drawn to necrobacillosis, a disease caused by *Fusobacterium necrophorum*, the portal of entry for which is often a tonsillar infection, (Moore-Gillon et al., 1984).

Thoracic sites

Specimens of pus from empyema yield anaerobic bacteria in 76–80 per cent of samples (personal observation). Similar figures for aspiration pneumonia and lung abscess have been reported by other workers (Bartlett et al., 1974a,b).

Abdominal sites

Most of the many different types of intra-abdominal infections involve a wide range of anaerobes (Gorbach, 1975). In particular, appendicitis with perforation results in anaerobic infection and may lead to the formation of intra-abdominal abscesses, which always contain anaerobes as part of a mixed flora.

Female genital tract

The vagina harbours large numbers of anaerobes, so it is not surprising that infections of the female genital tract and the surrounding tissues involve anaerobic bacteria. Infections of particular importance include postabortal and puerpeural sepsis, tubo-ovarian abscess, Bartholin's gland infections, postoperative gynaecological infections and infections of the uterine wall secondary to intrauterine contraceptive devices. In female genital tract infections anaerobes are frequently accompanied by microaerophilic streptococci.

Superficial sites

Superficial abscesses and ulcers frequently yield anaerobic bacteria on culture. Such sites include breast and axillary abscesses, ischiorectal and pilonidal abscesses, and infected sebaceous cysts. Pressure sores, chronic leg ulcers and ischaemic ulcers of the extremities of diabetics are frequently infected with anaerobes. Bite wounds, especially those from humans, are particularly destructive and we have isolated anaerobes from at least half of those studied.

More rarely, anaerobes may be cultured from paronychia, bone and joint infections and ophthalmic injuries occurring as a result of a penetrating foreign body.

Types of specimen

Proper and adequate collection and transport are the main links in the chain of events leading to a meaningful result in anaerobic bacteriology. Generally, the best specimens, from any site, are samples of fluid or pus, because these have the advantage of being their own best transport medium and are more representative of the lesion from which they come. A small survey in this laboratory has shown that pus is superior to swabs for the recovery of anaerobes from clinical material (Table 12.1).

Table 12.1 Isolation of anaerobes from pus and swabs

Organism	Number of isolates	
	Pus	Swabs
B. fragilis group	8	6
B. melaninogenicus group	6	2
Bacteroides spp.	14	8
Fusobacteria	6	1
Anaerobic cocci	8	4
Clostridia	3	3
NSGPR*	6	3
Total	51	27

*NSGPR, non-sporing, gram-positive rods

Pus

A universal container filled to exclude most, if not all the air, will generally suffice for direct examination and culture of material from large abscesses. Small amounts of pus are probably better taken into a small syringe or, if available, a gassed-out tube. Swabs are always a poor alternative to liquid pus. However, should a swab be the only specimen available at the time, the type of swab used, the material it is made from and the transport method employed will affect the success of recovery of anaerobic bacteria from the sample.

Tissue

Pieces of tissue are good specimens if they are available. Immediate delivery to the laboratory is essential because there may be some difficulty in transporting tissue specimens under good anaerobic conditions, although Sutter *et al.* (1980) have suggested a method of doing so.

Dressings

The inner parts of surgical dressings are also good sources of clinical material but are all too frequently disposed of by the ward or the out-patient department. They can be a rich source of certain types of anaerobe (e.g., actinomyces) and ward and other staff should be encouraged to remove the inner portion of the dressing into a sterile container and send it to the laboratory without delay.

Transport of clinical material

It always used to be stated that any method of transportation of specimens for anaerobic bacteriology must aim at maintaining a moist atmosphere, minimal exposure to oxygen, preventing desiccation and maintaining the original balance between the numbers and types of organisms in the original inoculum. However, recent studies have shown that not only should these traditional assumptions be reviewed but that there is also a general lack of agreement on what system to use and, where swabs are taken, which transport medium gives the best results. (Helstad *et al.*, 1977; Gargan and Phillips, 1979; Smith *et al.*, 1986).

A comparison between the transportation of small volumes of pus in gassed-out tubes and the collection of material on swabs has been made (Gargan and Phillips, 1979). Results showed that gassed-out tubes were only marginally better than a swab in a reduced transport medium, provided that the interval between loading the swab and culture onto plates was no longer than 6 h. It is likely that, with small samples of pus at least, the longer the time interval before culture the greater the advantage of a gassed-out tube (Helstad *et al.*, 1977). Furthermore, apparent loss of bacteria from swabs may be due to retention of organisms on the swab rather than to death of the organisms themselves (Collee *et al.*, 1974).

If commercially produced gassed-out tubes are needed, we have found the Port-a-Cul tube (Becton Dickinson) successful for the transportation of brain abscess material. Generally, most workers advocate the use of a reduced transport medium when swabs are taken. The types of medium recommended include Stuarts, Cary and Blair, Amies and a modified Stuarts medium. Our preference is for Cary and Blair medium in 13 × 100 mm screw-capped glass tubes filled almost to the top of the tube. Even so, the use of such media may still not result in a recovery superior to that obtained with even a dry swab, at least with some anaerobic species (Human and Jones, 1986; Smith *et al.*, 1986).

The coating of cotton wool or alginate swabs with haemoglobin and glycerol has also been used successfully in laboratory tests and clinical studies (Smith and Ferguson, 1977). However, experience in this laboratory with commercially prepared transport media shows them to be inferior to those prepared in the laboratory. Some success has been achieved with the vacutainer transport device (Becton Dickinson) provided that ward and clinical staff are trained to use it properly. One way to avoid these problems of specimen transport would be to plate the specimen on to appropriate media at the bedside, and to use an anaerobic bag system (Bio-Bag System, Marion Scientific) for transport. This method of transport generates an anaerobic atmosphere immediately and is based on the work of Rosenblatt and Stewart (1975). The bags also make good transport devices for tissues and dressings but they are expensive to use.

The use of nutrient (broth) media in broth form for the transport of fluids and pus should be strongly discouraged. Overgrowth by facultative bacteria may occur if any delay ensues and, importantly, oxidized thioglycollate media may become toxic to some anaerobes and thereby impede their growth.

Direct examination of clinical material

Direct examination of clinical material can lead to a rapid presumptive diagnosis of anaerobic infection. Preliminary examination may merely require a good look at the specimen and an acute sense of smell when the lid of the container is removed. Pus samples taken from patients with an anaerobic infection possess a foul odour. Occasional infections with *Proteus* spp. also have a bad smell, but as a general rule, if the specimen smells offensive then anaerobic bacteria may be presumed to be present. Conversely, some anaerobic infections that produce very little foul odour also occur, and in these cases other preliminary examinations may be of help. The presence of 'sulphur granules' (actinomyces microcolonies) may suggest actinomycosis. Further examinations should include examination of a gram-stained smear, fluorescent antibody staining, examination for fluorescence under ultraviolet light and the use of gas–liquid chromatography for the presence of short chain fatty acids.

Gram's stain

Although many anaerobic bacteria have similar morphologies to aerobes, there are some that have a typical morphology (e.g., some fusobacteria). One particular problem with anaerobes is their capacity to become decolourized very easily. This can be avoided if Kopeloff's modification of Gram's stain is used (see Appendix, page 195). We have found that this is the best modification to use when anaerobes are present because it does not over-decolourize the anaerobes.

Fluorescent antibody staining

These methods can be very useful for obtaining a rapid diagnosis. Fluorescent antibodies can be either bought commercially, which is very expensive, or made in the laboratory. They provide an accurate and reliable technique for the identification of some anaerobes and can be a very useful laboratory tool, e.g., in the examination of blood cultures. Although good correlation with culture has been reported (Willis *et al.*, 1982) it would be very costly to use fluorescent antibodies routinely. The commercial products (for *Bacteroides fragilis* and *B. melaninogenicus*) also give good correlation with the results of cultures for these two organisms (Labbe *et al.*, 1980).

Ultraviolet light

Brick red fluorescence of a specimen under long wave ultraviolet light indicates the presence of pigmented bacteroides. In one study this appearance was noted in nearly one-third of specimens. However, this property is not absolute; in the same study several specimens that failed to fluoresce yielded pigmented bacteroides on culture (Phillips, 1982).

Gas–liquid chromatography (GLC)

Direct gas–liquid chromatography of a pus sample detects short chain fatty acids present as a result of the metabolism of anaerobic bacteria. The presence of one or several of these acids (particularly *iso*-butyric, butyric, *iso*-valeric, valeric, and *iso*-caproic) is a good indication that anaerobes are present. False-negative results (i.e., a negative result on GLC but growth of anaerobes on culture) are relatively rare and may occasionally be due to infections in which only one anaerobe is present, or where the bacterial population is very low. Figure 12.1 shows typical positive and negative GLC traces.

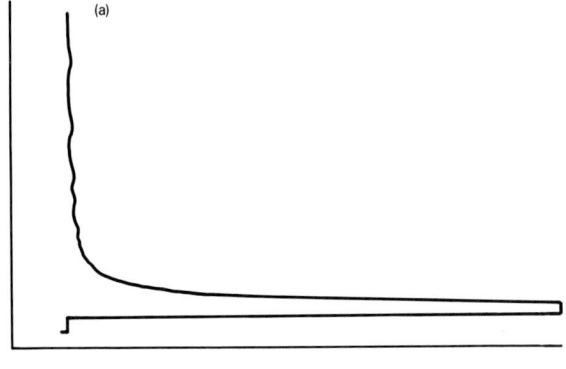

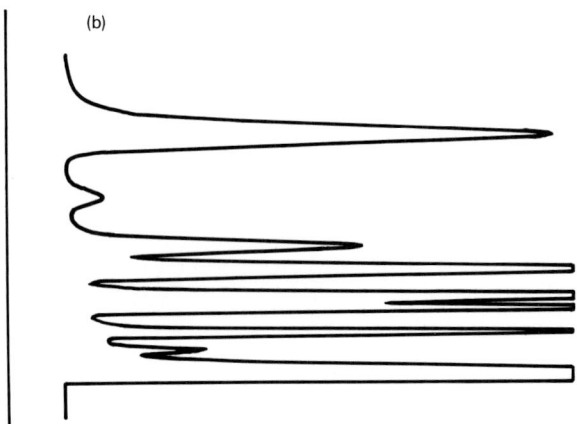

Fig. 12.1 Gas–liquid chromatographic analysis of pus from (a) an abscess caused by *Staphylococcus aureus* and (b) an empyema caused by multiple anaerobes

Anaerobic techniques

Considerable debate has surrounded the use of different systems for anaerobic culture. However, with whichever system is used, cultures must be placed into an anaerobic atmosphere immediately after inoculation with the specimen. In the UK two methods of achieving anaerobiosis are in common use. The most common method is the use of an anaerobic jar; an alternative method, growing in acceptance, is the use of an anaerobic chamber. Either of these methods will give satisfactory results provided that careful procedures are followed. A third technique, which is not commonly used, the Hungate or roll-tube technique, (Hungate, 1966; Moore, 1969) will also be discussed briefly.

Anaerobic jars

There are two types of anaerobic jar—those which require evacuation of the air by vacuum pump and refilling with a gas mixture suitable for the growth of anaerobic bacteria, and those which rely on the generation of gas from a commercially prepared sachet with removal of all oxygen by catalysis. Suitable jars of the evacuation–replacement type are commercially available (e.g., through Don Whitley Scientific or Oxoid Ltd). Those which rely on commercial sachets are produced by Becton Dickinson (Gas Pak® system) and Oxoid Ltd, who produce the Oxoid jar, which can be used with the Oxoid Gas Kit® envelopes.

Evacuation–replacement jars

The use of these jars must also involve the use of a standard anaerobic procedure such as that suggested by Collee *et al.* (1972). The attainment of a secondary vacuum in these jars within about 10 min after setting up is the quickest way to assess whether the jar is functioning properly or not. An alternative to this is the methylene blue indicator system; this is easy to prepare and simple to use (Fildes and McIntosh, 1921) and may be bought as commercially prepared indication strips that are placed in the jar with the cultures. Alternatively, biological indicators may be used. Failure of a fairly exacting anaerobe, such as *Clostridium oedematiens* type A or *C. tetani*, to grow on a blood agar plate or, conversely, the growth of *Pseudomonas aeruginosa* on nutrient agar or Simmon's citrate agar would be an indication that the jar had failed. However, total reliance on biological indications is not recommended since failure of the jar will be indicated only after at least 18 h, by which time precious cultures will be irretrievably lost.

Commercial gas producing envelopes

Jars in which commercial sachets are used rely on the generation of H_2 after addition of water to the sachet. The sachet is then placed in the jar with the cultures, catalyst and indicators and the lid is closed. If a catalyst is present the volume of hydrogen released is enough to remove all the oxygen in the jar.

Comparison of the above two methods of jar

operation has shown them to be equally effective (Collee et al., 1972; Ferguson et al., 1976), provided that each is carefully controlled.

Maintenance of anaerobic jars

Each type of jar will enable the recovery of all clinically significant anaerobes provided that they are well maintained.

Continuous checks are required to minimize the risk of jar failure. Particular attention should be paid to setting the jar up and to checking the development of a secondary vacuum, which shows satisfactory catalyst activity.

Investigation of jar failure

If no secondary vacuum develops the catalyst should be replaced and the jar put up again. If the jar still fails there will probably be a leak, either in the jar, or in the valves.

Checking for leaks

The catalyst packet should be removed from the lid and the plate rack removed from the jar. A small paper towel is then soaked in acetone and placed in the jar and the lid is replaced tightly. If the jar is then immersed in a bucket of warm water, the acetone will evaporate and force air out through any leaks in the jar. When the leak has been detected the appropriate action may be taken. A common cause of failure is the 'O' ring, which should be reseated properly. The ring itself, and the channel in which it sits, should be cleared and faulty valves should be replaced. When the corrections have been done the catalyst packet should be replaced and the jar set up to check satisfactory performance. This procedure must be done in a well ventilated room and away from naked lights.

Sources of gas

The safest mixture to use for anaerobic jars (and anaerobic chambers) consists of N_2 80 per cent, H_2 10 per cent and CO_2 10 per cent. Hydrogen concentrations higher than 10 per cent involve a greater hazard. The safest way of delivering this gas to the anaerobic jar is to use a reducing valve at the cylinder head and to pass the gas from this valve into a rubber football bladder. The bladder is then connected to the jar and the vacuum in the jar is allowed to suck the gas in from the bladder. Suitable cylinders of gas are available commercially.

Catalyst

The most commonly used catalyst is small pellets of cold catalyst D (Engelhard Industries Ltd) which should be used in the ratio of 1.0 g of catalyst for every litre of jar volume (Willis and Phillips, 1983). For large jars more than one catalyst sachet may be used, or a larger catalyst sachet employed. The catalyst is inactivated by moisture but this may be reversed by heating the pellets to a temperature of 160°C for 60–90 min. Irreversible inactivation occurs when the active surface becomes saturated with gases such as hydrogen sulphide. Batches of catalyst should be regularly checked for such irreversible inactivation and should be replaced when they have reached the end of their efficiency. Catalyst activity may be prolonged by incorporating sachets of Anotox® (Don Whitley Scientific) in the jar; this is a form of activated charcoal which adsorbs poisonous gases on to its surface.

Anaerobic chambers

For those laboratories with a large throughput of specimens for anaerobic culture it may be much more convenient to use an anaerobic chamber rather than 30–40 anaerobic jars. Although it has been shown that anaerobic chambers provide no advantage over anaerobic jars for clinical bacteriology (Wren, 1980a) they do offer certain advantages, one of the most important being the early examination of cultures in a continuous anaerobic atmosphere.

Chambers may be made of flexible plastic or may be of a rigid, gas-tight construction. Operation within the chamber is accomplished through the use of gloves sealed to the wall of the chamber, although a number of chambers now have a 'bare hands' facility which makes manipulation within the chamber easier. Materials are passed into and out of the chamber via an air lock. This is purged of air, which is replaced with oxygen-free gas, and the material is then passed into the main body of the chamber. Anaerobiosis within the chamber is maintained by circulating the atmosphere continually over catalyst with an electric fan. The atmosphere is kept free of toxic products by using a cannister of Anotox® as in an anaerobic jar, and the whole chamber is run at a slightly positive pressure so that, should a leak occur, the chamber atmosphere is pushed out, rather than air sucked in. Leaks may be detected either by noting that the chamber is making an unusually heavy demand for gas or by using a leak detector (Don Whitley Scientific). The amount of gas used is also determined by the amount of use of the air lock system.

Hungate (roll-tube) techniques

Techniques based on roll tubes with PRAS (pre-reduced anaerobically sterilized) media are more fully described by Holdeman and Moore (1977). Basically, the technique involves isolating anaerobes by streaking material on to the inner surface of agar-coated glass tubes. The streaking is performed by rotating the glass tube on a rotor while the tube is flushed with oxygen-free gas from an inserted cannula.

The roll-tube culture media and, all other tubed media, liquid and solid, are prepared in O_2-free gas and contain cysteine hydrochloride as a further reducing agent which has been added before autoclaving. The tubes are then autoclaved within a metal press to prevent the butyl rubber stoppers from blowing during the sterilizing cycle.

Resazurin is used as an indicator of anaerobiosis. Because the tubes are inoculated and maintained with an O_2-free internal atmosphere they can be incubated in racks in the ordinary incubator and do not require much space. After incubation growth is examined by placing the tubes under a stereoscopic microscope and colonies are picked and subcultured under O_2-free gas. The techniques are not difficult to master, but do need some degree of technical expertise. Although roll-tubes are not used very widely for clinical anaerobic bacteriology, there can be no question of their value for the recovery of extremely oxygen-sensitive anaerobic species.

Holding jars

When large numbers of specimens have to be cultured anaerobically care has to be taken that plates of media already inoculated are not exposed to the air for too long. This may be avoided by processing the specimens in small batches or by the use of a holding jar.

A holding jar is constructed in a similar way to an anerobic jar but is without a catalyst and instead of valves the lid contains two vents. Inoculated plates may be placed in the jar, the lid replaced and one of the vents connected by rubber tubing to a cylinder of O_2-free gas (preferably CO_2) which is continuously run into the jar as a gentle stream to exclude air. Once enough cultures have been processed, the plates can be removed from the holding jar and put up anaerobically in the usual way. The technique has been described fully by Martin (1971) and by Ellner et al. (1973) and has been shown to be successful.

The role of carbon dioxide

The inclusion of CO_2 in the anaerobic atmosphere is important. CO_2 enhances the speed of growth and increases the colony size of anaerobes growing on solid media (Watt, 1973). A concentration of 7–10 per cent seems to be optimal, although many species have been shown to require a lower concentration (0.25–1.0 per cent) than this for initiation of growth (O'Reilly, 1980).

Length of incubation

Work in our laboratory (Wren, 1980a) has shown that uninterrupted incubation of cultures for 48 h followed by a further incubation period of three days is necessary for the maximum recovery of anaerobes from clinical material. It was shown that up to 20 per cent of non-sporing anaerobes may not be isolated until five days of anaerobic incubation have elapsed. Organisms suffering most from early oxygen exposure are the asaccharolytic bacteroides, some fusobacteria and the anaerobic cocci. Indeed, we have shown that premature opening of anaerobic jars and exposure of cultures to the atmosphere may alter the final bacteriological picture. It is extremely important, therefore, that if anaerobic jars are to be used, the minimum incubation period before they are opened and the cultures examined is 48 h. Difficulties arise when cultures need to be examined the following day; a duplicate set of plates should be streaked and set up in a separate jar and opened after 24 h. This problem will not arise in those laboratories using an anaerobic chamber where plates may be continuously examined without removal from anaerobic conditions.

Isolation media

It must be stated at the beginning of this section that a poorly transported specimen subsequently plated on to inadequate media will fail to give growth of an anaerobe even if the strictest anaerobic atmosphere is provided. There is no substitute for freshly made, adequately supplemented media. Media stored in the cold for more than two days, with or without reducing agents, do not give optimal results. Quite often our laboratory practices (i.e., bulk pouring and storage of media) preclude us from achieving a good growth of many clinically important anaerobes.

Basal media

Various basal media have been formulated or adapted for clinical anaerobic work. Those which have been shown to give good results over many years of study include Brucella Agar (Difco), Columbia Agar (Oxoid), Brain–Heart Infusion Agar (Difco), Wilkins Chalgren Agar (Oxoid) and Fastidious Anaerobe Agar (Lab M). When these media are made according to the manufacturer's instructions, and 5–10 per cent of horse blood is added, a good growth of most, if not all, clinically important anaerobes can be expected when the medium is used freshly poured.

Supplements

Many supplements have been tried over the years to stimulate better anaerobic growth. The more common supplements that have been found to be useful for this purpose include haemin, vitamin K_3 or K_1, sodium bicarbonate, Tween 80, cysteine hydrochloride, potassium nitrate, dithiothreitol, glucose, sodium succinate and catalase. Of this list those that have proved to be the most useful in our hands, in the routine clinical situation, are haemin, the vitamin K group, sodium succinate and Tween 80. The cysteine–dithiothreitol plates devised by Moore (1968) are required for the recovery of exacting strains of *C. oedematiens*, but we have no experience of this.

The use of a mixture of formate and fumarate improves the growth of *B. ureolyticus* and similar organisms both on solid and in liquid media (Holdeman and Moore, 1977; Fontaine et al., 1984). Tween 80 stimulates growth of anaerobic gram-positive cocci and sodium succinate improves the growth of some anaerobic gram-negative rods.

In the author's experience, the use of cysteine hydrochloride has no beneficial effect in media used in clinical bacteriology, provided that the media are used when freshly poured. During aerobic storage of media this reducing agent can become toxic and the medium may even fail to grow some anaerobes. Moreover, great care must be taken when mixtures of supplements are added. We have found that some mixtures of Tween 80 and cysteine hydrochloride are highly toxic for strains of anaerobic cocci and for some clostridia. Care is needed, therefore, before using supplement mixtures, and any intended mixture should be tested in comparison with media known to give good results before such a mixture is added to the media in routine use.

Selective media

As has been previously stated, most anaerobic infections involve mixtures of different species and, therefore, optimal separation of such a mixture can only be expected if selective media are employed. There is still no single all-purpose selective medium for anaerobes and it is unlikely that one will be devised. Therefore, separation depends on the use of several selective agents which may be used singly or together.

Aminoglycosides

Media incorporating aminoglycosides have formed the traditional basis of selective anaerobic bacteriology for many years. Their effectiveness in facilitating the recovery of anaerobes from specimens by inhibiting aerobic and facultative gram-negative bacilli and staphylococci is well recognized. The formulation of these and other selective media discussed below is given in the appendix.

Neomycin agar (NA)

The major use of neomycin agar is for the recovery of anaerobic gram-positive cocci and clostridia. Additional advantages include the incorporation of neomycin into egg-yolk agar plates for the rapid detection of *C. perfringens* by a direct Nagler reaction in certain circumstances.

Kanamycin agar (KA)

Bifidobacteria are said to grow well on this medium (Finegold *et al.*, 1971) but we have little experience of isolating these organisms on this medium. However, findings in this laboratory suggest that kanamycin is an excellent selective agent for the recovery of clostridia from mixed cultures, but fusobacteria fail to grow on this medium.

Kanamycin–vancomycin agar (KVA)

This medium gives good recovery of the bacteroides group of organisms. However, in our experience, not all of the bacteroides grow equally well and diminishing the vancomycin concentration to one-third of the original concentration improves the isolation rate without compromising its selectivity. It cannot be recommended for use in general clinical work, when all anaerobes should be looked for, (Wren, 1980b). Again, fusobacteria fail to grow on this medium.

Neomycin–vancomycin agar (NVA)

This medium is useful for selective isolation of most of the gram-negative anaerobes. Made in exactly the same way as the kanamycin–vancomycin medium, it is slightly less inhibitory than KVA and will allow fusobacteria to grow.

Media containing nalidixic acid

The use of nalidixic acid as an agent for the selective isolation of anaerobic bacteria is not new. Investigations with this agent have been performed since 1970 (Finegold *et al.*, 1971). However, its use in media at concentrations of 50–100 µg/ml has, in our experience, been far too inhibitory. A suitable concentration of this agent is 10 µg/ml, which is less inhibitory to non-sporing anaerobes than the aminoglycoside media and yet is just as selective. However, its failure to enable selective isolation of clostridia makes it a poor choice when searching for members of this genus. Some problems may also be encountered when searching for anaerobic gram-positive cocci because aerobic cocci will also grow. Substitution of oxolinic acid in place of nalidixic acid may help prevent the growth of aerobic cocci. (D Petts, personal communication). Nalidixic acid media are useful alternatives to aminiglycoside media for the gram-negative anaerobes (Wren 1978; 1980b).

Nalidixic acid–Tween agar (NAT)

NAT medium contains Tween 80, which stimulates the growth of the anaerobic cocci in particular. Tween causes the gradual lysis of the blood in the medium during the incubation period, thus giving rise to early pigmentation of the *B. melaninogenicus* group of organisms. In studies with the medium in this laboratory, superior isolation rates to those for non-selective media were obtained for anaerobic gram-positive cocci, non-sporing gram-positive rods (NSGPR) and the fusobacteria (Table 12.2).

Nalidixic acid—vancomycin agar (NAV)

NAV medium also contains sodium succinate as a source of energy for some bacteroides organisms (Lev *et al.*, 1971). The medium is inhibitory for all gram-positive bacteria and can be regarded as being

Table 12.2 Isolation using nalidixic acid-Tween agar (NAT)

Species	Number of strains	Number isolated on		
		NAT	NA†	KA‡
Peptococcus spp.	462	437	377	232
Peptostreptococcus spp.	316	301	186	154
Clostridia	32	19	32	32
NSGPR*	91	91	53	27

*NSGPR, non-sporing, gram-positive rods
†Neomycin agar
‡Kanamycin agar

Table 12.3 Isolation using nalidixic acid–vancomycin agar (NAV)

Species	Number of strains	Number isolated on		
		NAV	NEO-V*	KVA†
B. fragilis group	455	455	455	455
Bacteroides spp.	362	312	252	226
Fusobacteria	189	189	122	0
Veillonellae	32	28	19	0

*Neomycin–vancomycin agar
†Kanamycin–vancomycin agar

selective for anaerobic gram-negative bacteria. This medium is extremely successful at recovering fusobacteria and asaccharolytic bacteroides from many clinical infections (Table 12.3).

Rifampicin agar

Rifampicin, at a concentration of 50 µg/ml, will permit selective isolation of *F. mortiferum*, *F. varium* and some clostridia. It can be useful when looking for these particular anaerobes in normal flora studies. However, its strong selective activity prevents its use in clinical work.

Bicozamycin agar

Bicozamycin is an antimicrobial agent with selective properties for anaerobic gram-positive cocci (Watt and Brown, 1983a). Preliminary studies indicate that this medium is a good choice for the isolation of these organisms (Watt and Brown, 1983b; Smith et al., 1986).

Selective media are also used for actinomyces and *C. difficile*. Readers are referred to the relevant chapters in this volume for the media for these organisms.

Broth enrichment

The remaining question regarding media is the value of broth enrichment. If the specimen is shown to contain a large number of bacterial types by Gram's stain, little benefit will be gained by enriching the specimen in broth before plating. Generally, the more rapid growers in the population will grow up first at the expense of those that grow more slowly. Under such conditions the more sensitive species fail to do well and are rarely recovered. Broth enrichment may be used beneficially if there are scanty numbers of organisms in the specimen or when looking for spore-forming organisms. The broth may be heated (or 'heat-shocked') to kill all the vegetative forms and to allow the resulting suspension of heat-resistant spores to germinate subsequently. For heat-shocking, about 1 ml of fluid or 1 g of tissue is placed in a tube of Robertson's cooked meat medium and heated to 80°C for 10 min. It is then cooled and incubated at 37°C before the medium is then plated out, over a period of time, onto solid media for aerobic and anaerobic incubation. This allows for the isolation of both rapid and slowly growing clostridia. A fuller discussion of this is given by Willis (1977). Heating in a boiling water bath for one hour is generally unnecessary and rather harsh for some of the spore bearers involved in clinical infections, heat-resistant food poisoning strains of *C. perfringens* being the exception (see Chapter 22).

Isolation procedure

On receipt of the specimen it should be examined macroscopically and smears made of the material for Gram's stain (and fluorescent antibody staining if required). Table 12.4 gives a list of clues to anaerobic infection that may be useful pointers.

A sample may be prepared for direct gas–liquid chromatography (see page 193). The material should then be inoculated on to media suitable for aerobic bacteria and on to the following for anaerobic culture:
1. Blood agar (non-selective);
2. Nalidixic acid–Tween agar (NAT);

Table 12.4 Bacteriological clues that indicate possible anaerobic infection

1. Morphology of some organisms in gram-stained film
2. Failure to obtain aerobic growth of some morphological types seen in the original direct smear
3. Foul odour to the specimen
4. 'Typical' GLC pattern on examination of clinical material
5. Growth on nalidixic acid–vancomycin media
6. Growth on aminoglycoside-containing media incubated anaerobically

3. Nalidixic acid–vancomycin agar (NAV);
4. Kanamycin or neomycin agar (KA or NA);
5. A tube of cooked meat medium (presteamed to drive off any air); this should only be used if it is thought that small numbers of organisms are present. This cooked meat medium may be heated to 80°C for 10 min if the *Clostridium* spp. in the specimen are being sought particularly.

The anaerobic cultures are incubated for a minimum of 48 h before examination if anaerobic jars are being used, but may be examined daily in an anaerobic chamber. Any colonies are then picked off, using a straight wire, on to a blood agar plate for anaerobic incubation and, preferably, onto chocolate agar in air plus CO_2 (10 per cent). A third picking is made if possible and is smeared over the whole surface of another blood agar plate. This can be used with disc testing for preliminary grouping. A Gram's stain may also be performed at this stage, or may be done from the anaerobic purity plate later on. All colonial types should be subcultured, since the practice of subculturing only those colonies which fail to appear on the primary aerobic plate can result in several anaerobes which are present not being followed up. The original cultures should then be incubated for a further three days and re-examined on the fifth day for any additional growth.

Preliminary identification

It is beyond the scope of this chapter to discuss fully the identification of anaerobic isolates and readers are referred to the relevant chapters for this. However, it would be useful if most routine clinical laboratories could at least make preliminary identification of the common important clinical anaerobes. The tests reported below should enable the identification of the more common anaerobes.

The following paper discs may be placed on a plate which has been uniformly seeded with a colony. For gram-positive organisms:
vancomycin (5 µg);
sodium polyanethol sulphonate (liquoid) (SPS) (1000 µg);
novobiocin (5 µg);
metronidazole (5 µg).
For gram-negative organisms:
vancomycin (5 µg);
gentian violet (1 in 100 000);
metronidazole (5 µg);
phosphomycin (300 µg);
penicillin (2 units).
Readers are referred to papers by Sutter and Finegold, 1971; Duerden *et al.*, 1976; Bennett and Duerden, 1985, for details. Plates are incubated for 24 to 48 h and the tests are read for zones of inhibition of growth in the usual way. It should be noted that these disc resistance tests are solely for identification and the results are not used to recommend therapy.

Once growth is obtained on the second anaerobic plate that was plated out for purity, it is used for further tests. Colonial morphology, production of any pigment, haemolysis, fluorescence under UV light and pitting of the agar are observed. Growth is removed for a catalase test, a spot indole test and for a Gram's stain, if this has not already been performed from the original colony. Growth is also placed into fastidious anaerobic broth (FAB; Lab M) with 0.1 per cent Tween 80 for phase contrast microscopy (if it is a gram-positive coccus). Further colonies are then placed into tubes of FAB for gas–chromatography, and also on to an egg-yolk agar plate for lecithinase and lipase production. For gram-negative rods, a colony may also be picked and subcultured to a bacteroides bile aesculin plate, which will identify members of the *B. fragilis* 'group', because only these organisms will grow on, and blacken, the medium. Tables 12.5, 12.6 and 12.7 list a number of preliminary identifications that can be made with these tests.

Colonial morphology and pigmentation

The colonial size and shape are important, as are such features as internal 'flecking' of the colony,

Table 12.5 Provisional identification of anaerobic gram positive cocci

Species	Vancomycin	SPS	Metronidazole	Novobiocin	Pigment	Fluorescence	Catalase	Indole	Lecithinase	Lipase	Morphology (FAB + 0.1% Tu80)
Peptococcus niger	S	R	S	R	+	−	+	−	−	−	M
Peptococcus spp.	S	R	S	R*	−	−	V	V	−	−	M
Peptostrept. anaerobius	S	S	S	S	−	−	−	−	−	−	C
Peptostrept. spp.	S	R	S	S*	−	−	−	−	−	−	C

*Some strains of Pc. magnus and Pstr. micros may give anomalous results
S, sensitive; R, resistant; V, variable results; M, growth in masses; C, growth in chains

whether the colonies are translucent or opaque and whether they are mucoid. Black or brown pigment is produced by members of the B. melaninogenicus 'group' B. asaccharolyticus and Peptococcus niger. Pink pigment is produced by Actinomyces odontolyticus and a brownish pigment by some strains of A. naeslundi.

Haemolysis

Typical haemolysis patterns are produced by C. perfringens and F. nucleatum. C. perfringens produces a 'target' type of haemolysis consisting of an inner zone of complete haemolysis and an outer zone of partial haemolysis. Exposure of cultures of F. nucleatum to the atmosphere results in a greening around the growth.

Pitting of the agar

Many strains of B. ureolyticus will pit the agar surface as they grow. The longer the time of incubation the further the edge of growth extends. It appears that the pitting gets shallower nearer to the colony edge.

Table 12.6 Provisional identification of gram-negative anaerobes

Species	Vancomycin	Gentian violet	Metronidazole	Phosphomycin	Penicillin	Pigment	Haemolysis	Fluorescence	Pitting of agar	Catalase	Indole	Lecithinase	Lipase	Growth	Blackening	BBAA*
B. fragilis group	R	S	S	R	R	−	−	−	−	V	V	−	−	+	+	
Pigmented group	V	S	S	R	V	+	(+)	+	−	−	V	−	V	−	−	
B. ureolyticus	R	S	S	S	S	−	−	−	+	−	−	−	−	−	−	
Bacteroides spp.	R	S	S	R	S	−	−	−	−	V	V	−	−	−	−	
F. mortiferum	R	R	S	S	R	−	−	−	−	−	−	−	−	−	−	
F. varium	R	R	S	S	R	−	−	−	−	−	+	−	−	−	−	
F. necrophorum	R	R	S	S	S	−	W	−	−	−	+	−	+	−	−	
Fusobacterium spp.	R	R	S	S	S	−	−	V	−	V	−	−	−	−	−	
Veillonella	R	NA	S	NA	S	−	−	−	−	−	−	−	−	−	−	

*Bacteroides bile aesculin agar
R, resistant; S, sensitive; V, variable results; NA, not available; W, weak reaction

Table 12.7 Provisional identification of gram-positive rods

Species	Spores	Vancomycin	Metronidazole	Pigment	Haemolysis	Fluorescence	Pitting of agar	Catalase	Indole	Lecithinase	Lipase	'Nagler reaction' positive
C. perfringens	−	S	S	−	+†	−	−	−	−	+	−	+
Other Nagler +ve clostridia	+	S	S	−	−	−	−	−	+	+	−	+
Clostridium spp.	+	S	S	−	V	V	−	−	V	V	V	−
P. acnes	−	S	R	−	−	−	−	V	V	−	−	−
Propionibacterium spp.	−	S	S	V	−	−	−	V	V	−	V	−
Actinomyces spp.	−	S	V	V	−	−	−	V	−	−	V	−
NSGPR*	−	S	V	−	−	−	−	V	V	−	−	−

*NSGPR, non-sporing, gram-positive rods
†double zone
S, sensitive; R, resistant; V, variable

Fluorescence under UV light

Traditionally, members of the pigmented group of bacteroides were said to fluoresce brick red under UV light. However, it is now known that some colonies of other bacteroides, bifidobacteria and clostridia will also sometimes fluoresce, particularly in the presence of fermentable carbohydrate. Readers are referred to the papers by Brazier (1986a,b) for a fuller discussion.

Phase contrast microscopy of FAB–Tween 80 cultures

Gram-positive cocci produce their typical morphological characteristics in this medium. A wet preparation is made when growth has occurred (the earlier the better) and examined under phase contrast microscopy with the ×40 objective; peptococci grow in masses and peptostreptococci in long chains. This is a useful test for differentiating this difficult group of organisms, although their classification is now in serious doubt (see Chapter 7).

Comments on the preliminary groupings

B. fragilis group

All members of this group grow in the presence of 20 per cent bile and hydrolyse aesculin, giving the typical appearance on bacteroides bile aesculin agar. They are also typically resistant to a 2-unit penicillin disc although this property taken alone does not imply a member of the fragilis group because other bacteroides, on occasions, are also penicillin-resistant. However, a gram-negative rod that grows on 20 per cent bile, hydrolyses aesculin and is resistant to penicillin and phosphomycin may be identified as a member of the B. fragilis group with some confidence. All are sensitive to 1 in 100 000 gentian violet.

B. melaninogenicus group

This group generally produces colonies which produce brick red fluorescence under UV light (but see above) and produce brown or black pigmented colonies on prolonged incubation. They are of variable sensitivity to both penicillin and vancomycin but are resistant to phosphomycin. Indole is produced by B. asaccharolyticus, and B. intermedius; the latter may also produce a lipase. All are sensitive to 1 in 100 000 gentian violet.

B. ureolyticus

Found commonly in soft tissue infections, this slender bacillus produces transparent colonies which pit the surface of the agar (see p. 191). It is sensitive to both phosphomycin and gentian violet.

Fusobacteria

This group of organisms can be differentiated from most bacteroides by its resistance to gentian violet and its sensitivity to phosphomycin. *F. necrophorum* is the only member of the genus which produces a lipase. Penicillin-resistant species include *F. mortiferum* and *F. varium*, which may be further differentiated by indole production. The most common species, *F. nucleatum*, is typically indole-positive and produces needle-like (fusiform) cells when examined in gram-stained smears.

The anaerobic cocci

This is a very difficult group of organisms to differentiate (see Chapter 7) but differentiation of the two main gram-positive genera, *Peptococcus* and *Peptostreptococcus*, by a combination of novobiocin sensitivity and morphology in FAB–Tween broth, has proved useful in routine microbiological studies. Novobiocin testing is not foolproof, however, because *Pc. magnus* and *Pstr. micros* may give equivocal results. Sensitivity to a disc containing 1000 μg of liquoid (SPS) will indicate that the organism is *Pstr. anaerobius*. Black pigment would indicate the presence of *Pc. niger*, although this property is often lost on subculture of the organism.

The clostridia

The production, on egg-yolk agar, of lecithinase (phospholipase C) which is specifically neutralized by *C. perfringens* type A antitoxin, in an organism giving a double zone of haemolysis on blood agar and a negative spot indole test, indicates *C. perfringens*. A similar, 'Nagler reaction', but on non-haemolytic, indole-positive organisms would suggest *C. bifermentans* or *C. sordellii*.

Other clostridia are difficult to differentiate with the preliminary tests given in this chapter.

Non-sporing gram-positive bacilli

Differentiation of this group of difficult organisms cannot be achieved using these tests, although a gram-positive rod not showing spores that produce both catalase and indole is likely to be a *Propionibacterium acnes*.

Commercial identification kits

There are a several commercial identification kits available for the routine identification of anaerobic isolates. However, it is our experience that many anaerobes fail to give clear-cut colour reactions in many of these systems. Furthermore, some strains fail to grow well in the media supplied with the kits. With these kits, correct identification can only be expected with 50–70 per cent of isolates.

Identification of anaerobes with preformed enzymes

The detection of preformed enzymes for the identification of anaerobic bacteria has theoretical advantages, particularly because many clinically important anaerobes are asaccharolytic. A further advantage is that only 4 h are required for incubation. Commercial systems utilizing enzyme substrates are available (the API ZYM 32A® System is one example), or substrates may be made in the laboratory. Work so far with these methods shows some promise for the future. Interested readers are referred to the papers published by Tharagonnet *et al.* (1977), Marler *et al.* (1984), Levett (1985) and Wren and Silles (1986).

Gas–liquid chromatography (GLC)

Full identification of many anaerobes involves the use of gas–liquid chromatography for the detection of the acid end products of carbohydrate metabolism. Measurements of both volatile (acetic, propionic, *iso*-butyric, butyric, *iso*-valeric, and *iso*-caproic) and non-volatile (lactic and succinic) acids may be performed.

GLC is necessary for the identification of anaerobic cocci, and some clostridia, and to separate the genera of non-spore forming anaerobic gram-positive bacilli. Separation of bacteroides and fusobacteria is also achieved efficiently by the use of GLC.

The use of a machine with flame ionization detectors is common in the UK. Such a system measures the metabolic acids produced either by analysing acid–ether extracts of the culture supernate (or pus) or from the material injected straight into the column.

A detailed description of the equipment used is prohibited by space, but has been published adequately elsewhere (Sutter et al., 1985; Willis and Phillips, 1988).

Column packing materials found to be useful for this technique include Carbowax 20M TPA, AT1200, PFG 20M and Chromosorb 101. A typical system would be a Pye Unicam series 304 (or 204) chromatograph fitted with a 1.5 m glass column packed with Carbowax 20MTPA on gas chrom Q and flame ionization detection. The injection port is operated at a temperature of 165°C, the oven at 150°C, and the detector at 200°C. Nitrogen, the carrier gas, flows through the column at 60 ml/min. Using such a system, a volatile fatty acid analysis takes approximately 12 min and the non-volatile analysis about 6–7 min.

Sample preparation

Volatile fatty acids

A pure culture of the organism is grown in a suitable broth culture (e.g., Robertson's cooked meat medium, or FAB, LAB M Ltd) for 48 h. A 1.0 ml sample of this culture is transferred to a glass tube and acidified with 0.2 ml of 50 per cent aqueous sulphuric acid. To this is added 1.0 ml of diethyl ether and the whole is mixed for 20–30 s. Two liquid phases are separated by centrifugation. The ether layer (upper layer) is removed and 1 μl is injected into the column.

Pus specimens may be treated in exactly the same way, but it may be necessary to increase the amount of ether to 1.5 or even 2.0 ml.

Non-volatile fatty acids.

A further 1.0 ml of culture is transferred into another glass tube (preferably stoppered) and 2.0 ml of methanol and 0.5 ml of 50 per cent sulphuric acid are added. The tubes are placed in a water bath of 56°C for 30 min before 1.0 ml water and 0.5–1.0 ml of chloroform are added. After mixing for 30 s, the tubes are centrifuged to separate the two phases. The chloroform layer (lower layer) is removed and 1.0 μl is injected into the column.

Finding such fatty acids in clinical material, or directly by injection of the broth phase of an anaerobic blood culture showing growth, provides the clinician with a rapid presumptive diagnosis of anaerobic infection.

Detection of alcohols may be helpful in some circumstances (for details see Willis and Phillips, 1983).

Appendix

Bacteroides bile aesculin agar

agar 40 g;
oxgall 20 g;
ferric ammonium citrate 0.5 g;
aesculin 1 g;
haemin (500 μg/ml) 10 ml;
distilled water 1 litre;
pH 7.0.
After dissolving the solids by heating, dispense and autoclave at 121°C for 15 min. Cool to 50°C and add gentamicin to a final concentration of 100 μg/ml and pour plates.

Identification discs

Whatman AA discs are sterilized by autoclaving at 121°C for 15 min, impregnated with the substances below and dried in the incubator. They are stored ready for use over desiccant at 4°C.
gentian violet 1 in 100 000;
phosphomycin 300 μg;
vancomycin 5 μg;
SPS (sodium polyanethol sulphonate) 1000 μg;
novobiocin 5 μg;
metronidazole 5μg;
penicillin 2 units.

Selective media

These media should be prepared with a nutritious basal medium to which haemin 5 μg/ml, vitamin K_1 0.5 μg/ml and horse blood 5–10 per cent have been added. The necessary media concentrations of antibiotics are given below.
neomycin agar—neomycin 100 μg/ml;
kanamycin agar—kanamycin 75 μg/ml;
neomycin-vancomycin agar—neomycin 100 μg/ml vancomycin 2.5 μg/ml;
kanamycin–vancomycin agar—kanamycin 75 μg/ml vancomycin 2.5 μg/ml;
nalidixic acid–Tween agar—nalidixic acid 10 μg/ml;

nalidixic acid–vancomycin agar—nalidixic acid 10 μg/ml vancomycin 2.5 μg/ml;
rifampicin agar—rifampicin 50 μg/ml;
bicozamycin agar—bicozamycin 300 μg/ml.

Gram stain (Kopeloff method)

1. Flood a heat-fixed smear with a solution of crystal violet (10 g/l) and *immediately* add seven drops of a 5 per cent solution of sodium hydrogen carbonate;
2. Leave for 3 min;
3. Wash in tap water and flood with iodine solution (Gram's iodine);
4. Leave the iodine on for 3 min;
5. Decolourize with a solution containing alcohol and acetone (700 ml of 95 per cent ethanol; 300 ml acetone);
6. Counterstain with 2 per cent safranin;
7. Leave for 3 min. Wash well in tap water and dry.

References

Bartlett J G, Gorbach S L, Finegold S M. (1974a). The bacteriology of aspiration pneumonia. *American Journal of Medicine* **56**: 202–207.

Bartlett J G, Gorbach S L, Tally F P, Finegold S M. (1974b). Bacteriology and treatment of primary pulmonary lung abscess. *American Review of Respiratory Diseases* **109**: 510–518.

Bennett K W, Duerden B I. (1985). Identification of fusobacteria in a routine diagnostic laboratory. *Journal of Applied Bacteriology* **59**: 171–181.

Brazier J S. (1986a). A note on ultra-violet red fluorescence of anaerobic bacteria *in vitro*. *Journal of Applied Bacteriology* **60**: 121–126.

(1986b). Yellow fluorescence of fusobacteria. *Letters in Applied Microbiology* **2**: 125–126.

Collee J G, Watt B, Fowler E B, Brown R. (1972). An evaluation of the Gas-Pak system in the culture of anaerobic bacteria. *Journal of Applied Bacteriology* **35**: 71–82.

Collee J G, Watt B, Brown R, Johnstone S. (1974). The recovery of anaerobic bacteria from swabs. *Journal of Hygiene* **72**: 1.

Chow A W, Rosner S M, Brady F A. (1978). Orofacial odontogenic infections. *Annals of Internal Medicine* **88**: 392–402.

Duerden B I, Holbrook W P, Collee J G, Watt B. (1976). Characterisation of clinically important gram-negative anaerobic bacilli by conventional bacteriological tests. *Journal of Applied Bacteriology* **40**: 163–188.

Ellner P D, Granato P A, Many C B. (1973). Recovery and identification of anaerobes. A system suitable for the routine clinical laboratory. *Applied Microbiology* **26**: 904–913.

Ferguson I R, Phillips K D, Willis A T. (1976). An evaluation of the 'Gaskit' disposable hydrogen and carbon dioxide generator for the culture of anaerobic bacteria. *Journal of Applied Bacteriology* **41**: 433–437.

Fildes P, McIntosh J. (1921). An improved form of McIntosh and Fildes anaerobic jar. *British Journal of Experimental Pathology* **2**: 153–154.

Finegold S M, Sugihara P T, Sutter V. (1971). Use of selective media for isolation of anaerobes from humans. In *Isolation of anaerobes*, pp. 99–108. Edited by Shapton D A, Board R G. Academic Press, London.

Fontaine EAR, Borriello S P, Taylor-Robinson D, Davies H A. (1984). Characteristics of a gram-negative anaerobe isolated from men with non-gonococcal urethritis. *Journal of Medical Microbiology* **17**: 129–140.

Gargan R A, Phillips I. (1979). A comparison of three methods for the transport of clinical specimens containing anaerobes. *Medical Laboratory Sciences* **36**: 159–169.

Gorbach S L. (1975). Treatment of intraabdominal sepsis. *Annals of Internal Medicine* **83**: 377–379.

Helstad A G, Kimball J L, Maki G D. (1977). Recovery of anaerobic, facultative and aerobic bacteria from clinical specimens in three anaerobic transport systems. *Journal of Clinical Microbiology* **5**: 564–569.

Holdeman L V, Moore W E C. (1977). *Anaerobe laboratory manual*. Virginia Polytechnic Institute and State University, Blacksburg.

Human R P, Jones G A. (1986). Survival of bacteria in swab transport packs. *Medical Laboratory Sciences* **34**: 247–258.

Hungate R. (1966). A roll tube method for the cultivation of strict anaerobes. In *Methods in microbiology*, vol. 3B. p. 117. Edited by Norris J R, Gibbons D L. Academic Press, London.

Ingham H R, Selkon J B, Roxby C M. (1978). The role of *Bacteroides fragilis* in abscesses of the central nervous system: implications for therapy. *Journal of Antimicrobial Chemotherapy* **4**: 283–294.

Labbe M, Delaware M, Pepesack F, Crokaert F, Yourassowsky F (1980). Detection of *B. fragilis* and *B. melaninogenicus* by direct immunofluorescence. *Journal of Clinical Pathology* **33**: 1189–1192.

Lev M, Kendall K L, Milford D F. (1971). Succinate as a growth factor for *Bacteroides melaninogenicus*. *Journal of Bacteriology* **108**: 175–178.

Levett P N. (1985). Identification of *Clostridium difficile* using the API-ZYM system. *European Journal of Clinical Microbiology* **4**: 505–507.

Marler L, Allen S, Siders J. (1984). Rapid enzymatic characterisation of clinically encountered anaerobic bacteria with the API-ZYM system. *European Journal of Clinical Microbiology* **3**: 294–300.

Martin W J. (1971). Practical method for isolating anaerobic bacteria in the clinical laboratory. *Applied Microbiology* **22**: 1168–1171.

Moore W E C. (1968). Solidified media suitable for the cultivation of *Clostridium novyii* type D. *Journal of General Microbiology* **53**: 415.

— (1969). Techniques for the culture of fastidious anaerobes. *International Journal of Systematic Bacteriology* **16**: 173.

Moore-Gillon J, Lee T H, Eykyn S, Phillips I. (1984). Necrobacillosis, a forgotten disease. *British Medical Journal* **288**: 1526–1527.

O'Reilly S. (1980). The carbon dioxide requirements of anaerobic bacteria. *Journal of Medical Microbiology* **13**: 573–579.

Phillips I. (1982). Laboratory management of anaerobic infection. *Journal of Antimicrobial Chemotherapy*, **10** Suppl. A: 7–10.

Rosenblatt J E, Stewart P R. (1975). Anaerobic bag culture method. *Journal of Clinical Microbiology* **1**: 527–530.

Smith G L F, Cumming C G, Ross P W. (1986). Survival of gram-positive anaerobic cocci on swabs and their isolation from the mouth and the vagina. *Journal of Clinical Pathology* **39**: 93–98.

Smith L L, Ferguson I R. (1977). Modified bacteriological swabs for the transport of anaerobes in clinical specimens. *Medical Laboratory Sciences* **34**: 247–258.

Sutter V L, Citron D M, Edelstein M A C, Finegold S M. (1985). *Wadsworth anaerobic bacteriology manual*, 4th edn. pp. 71–78. Star, Belmont.

Sutter V L, Citron D M, Finegold S M. (1980). *Wadsworth anaerobic bacteriology manual*, 3rd edn. Mosby, St Louis.

Sutter V L, Finegold S M. (1971). Antibiotic disc susceptibility tests for rapid presumptive identification of gram-negative anaerobic bacilli. *Applied Microbiology* **21**: 13–20.

Tharagonnet D, Sisson P R, Roxby C M, Ingham H R, Selkon J B. (1977). The API-ZYM System in the identification of gram-negative anaerobes. *Journal of Clinical Pathology* **30**: 505–509.

Watt B. (1973). The influence of carbon dioxide on the growth of obligate and facultative anaerobes in solid media. *Journal of Medical Microbiology* **6**: 307–314.

Watt B, Brown F V. (1983a). A selective agent for anaerobic cocci. *Journal of Clinical Pathology*, **36**: 605–606.

— (1983b). The in-vitro activity of biozamycin against anaerobic bacteria of clinical interest. *Journal of Antimicrobial Chemotherapy* **12**: 549–553.

Willis A T. (1977). *Anaerobic bacteriology; clinical and laboratory practice*. Butterworths, London.

Willis A T, Taylor E, Pantosta A, Phillips I, Tabaqchali S. (1982). Comparison of antisera in the fluorescent antibody test for detection of *Bacteroides* species in clinical specimens. *Journal of Clinical Pathology* **35**: 304–308.

Willis A T, Phillips K D. (1983). *Anaerobic infections*. PHLS Monograph No. 3, 2nd edn. HMSO.

Willis A T, Phillips K D. (1988) *Anaerobic infections; clinical and laboratory practice*, pp. 128–139. Public Health Laboratory Service, London.

Wren M W D. (1978). A new selective medium for the isolation of non-sporing anaerobes from clinical specimens. *Medical Laboratory Sciences* **35**: 371–378.

— (1980a). Prolonged primary incubation in the isolation of anaerobic bacteria from clinical specimens. *Journal of Medical Microbiology* **13**: 257–263.

— (1980b). Multiple selective media for the isolation of anaerobic bacteria from clinical specimens. *Journal of Clinical Pathology* **33**: 61–65.

Wren M W D, Silles J M. (1986). Enzyme profiles as an aid in the identification of anaerobic bacteria. In *Recent advances in the identification of anaerobes*. Martinus Nijhoff, Amsterdam.

13

Abdominal sepsis

A Trevor Willis

Anaerobes as normal flora
    Infant gut flora
    Stomach
    Small Bowel
    Colon
    Biliary tract
Pathogenesis of abdominal infections
    Predisposing factors to anaerobic infections
    Development and progression of abdominal anaerobic infections
Clinical considerations
    Surgical wound infections
    Intra-abdominal anaerobic infections
Treatment
    Surgical management
    Antimicrobial therapy
Prophylaxis
    Postsurgical endogenous infections
    Prevention of surgical sepsis
References

It was in the setting of the digestive tract that Antonie van Leeuwenhoek in 1683 pre-empted modern concepts of nonclostridial anaerobic infections in man. He noted that 'there are more animals in the uncleaned matter on the teeth in one's mouth than there are men in a whole kingdom, ... owing to which such a stench comes from the mouth of many that one can hardly bear talking to them. Many call this a stenching breath, but actually it is in most cases a stinking mouth'. The 'animalcules' described by Leeuwenhoek were not only the first microorganisms to be described, they were also the first recognized anaerobes, and their clinical association with the malodour of oral sepsis pointed out a common and prominent feature of non-clostridial anaerobic infections some three centuries before its general recognition in the 1960s.

Nor was the association of non-sporing anaerobes and malodour with intra-abdominal sepsis a discovery of the twentieth century. It was discovered as early as 1848, when Henry Hancock performed the first successful operation to treat peritonitis due to appendicitis in a 30-year-old lady; on opening into the abdomen a quantity of excessively offensive turbid serum with fibrinous flocculi poured out, mixed with bubbles of gas and patches of false membrane (Goldman, 1966). Towards the end of the nineteenth century, Veillon (1893) noted the association of foetid suppuration with anaerobic infections, and subsequently Veillon and Zuber (1898) made the first recorded isolations of anaerobic cocci, fusobacteria and other gram-negative anaerobic bacilli from a variety of septic foci. Despite these observations on the participation of non-sporing anaerobes in common infective processes their clinical significance was not immediately appreciated and they slipped into obscurity, where they largely remained until the 'anaerobe revolution' of the mid-twentieth century.

Unlike the clostridia, very little attention was paid to the non-sporing anaerobes during the first two decades of the twentieth century. All the bacilli were uncritically lumped together in the single genus *Bacteroides*, until Eggerth and Gagnon (1933) and Weiss and Rettger (1937) recognized the need for their rational generic subdivision. The anaerobic cocci fared no better; for many years they were virtually ignored by microbiologists and later proposals, such as those of Stone (1940), were at best only interim expedients in a taxonomic jungle. Happily, the non-clostridial anaerobic infections of man did not meet with the same degree of neglect. It became increasingly clear that these infections, of which there was a considerable variety, where characterized by foul, localized, rather indolent suppuration, and were prone to progress to bacteraemia. They were frequently recognized in local infections of the appendix (Altemeier, 1938a,b, 1941; Gunn, 1956) and of the female genital tract (Schottmuller, 1911; Harris and Brown, 1927; Colebrook, 1930; Altemeier, 1940).

It is remarkable how successful the early studies of intra-abdominal sepsis were in defining the cardinal features of non-clostridial anaerobic sepsis in general. Regarding their endogenous origin, Schottmuller (1911) positively asserted that the source of '*Streptococcus putridus*' in puerperal infections of the uterus was the vagina. In his report on the involvement of anaerobic streptococci in tuboovarian abscess, Altemeier (1940) not only supported Schottmuller's contention but also postulated their mode of entry. He suggested that the bacteria present in the vagina may either accompany the gonococcus in its ascent to the tubes, or may alone be capable of setting up a state of prolonged pelvic inflammation and suppuration. In the same classic paper Altemeier further referred to the characteristically putrid odour of the pus from these infections, and gave the lie to laboratory explanations for its 'bacteriological sterility'. Two years earlier Altemeier (1938a,b) had demonstrated the importance of mixed anaerobes in acute perforated appendicitis with peritonitis, in contrast to the popular concept that *Escherichia coli* and/or the faecal streptococci alone were the essential aetiological agents. He also corrected the erroneous belief that *E. coli* was responsible for the offensive odour of the pus. These classic papers were followed by two others (Altemeier, 1941, 1942) in which he restated the *polymicrobial* nature of appendicitis peritonitis, and introduced the idea of synergic pathogenicity in this setting.

As medicine moved into the antibiotic era, early studies on the antimicrobial sensitivities of fusobacteria and bacteroides set the pattern of prescribing practice that was to successfully span the period of the 'anaerobe renaissance' of the 1960s. While *Fusobacterium necrophorum* was susceptible to penicillin (and to chloramphenicol and tetracyclines), *Bacteroides* spp. derived from intra-abdominal sepsis were almost always resistant to penicillin (but sensitive to tetracyclines, chloramphenicol and erythromycin). All groups were resistant to aminoglycosides (Alston, 1955; Gillespie and Guy, 1956).

During the 1960s came the clear recognition that the indigenous non-clostridial anaerobes of the gastrointestinal and female genital tracts were common participants in intra-abdominal sepsis, a conclusion which was consistent both with the observed ecological dominance of anaerobes in the normal indigenous floras of these sites and with the typically opportunist character of intra-abdominal infections, which almost always present as sepsis secondary to some preceding local compromising event. Laboratory studies identified the frequent participation of non-sporing anaerobes in a wide variety of infections associated with the gastrointestinal tract including 'primary' intra-abdominal sepsis, infections complicating abdominal trauma, postsurgical intra-abdominal and wound infections, and bacteraemic states related to abdominal lesions (Finegold *et al.*, 1985). The final cornerstone of evidence for the major aetiological association of endogenous anaerobes with intra-abdominal sepsis in man was put in place by the observation that serious postoperative sepsis, so commonly encountered as a complication of appendicectomy and of operations on the colon and female genital tract, could be largely prevented (or specifically treated) by the prophylactic (or therapeutic) use of specifically antianaerobic antimicrobials such as metronidazole (Study Group, 1974, 1975, 1976, 1977; Goldring *et al.*, 1975; Eykyn *et al.*, 1979). The impact of these observations on gastrointestinal and gynaecological surgery was considerable, for they finally and firmly established the validity of rational antimicrobial prophylaxis in surgical practice.

Although clinical patterns of anaerobic sepsis in experimental animals show marked differences in detail from those seen in the corresponding human infections, subsequent animal model studies utilizing restricted spectrum antimicrobials such as metronidazole, clindamycin and gentamicin as 'therapeutic probes' fully substantiated the clinical experience, in relation both to the aetiology and the prevention of intra-abdominal sepsis (Weinstein *et al.*, 1975; Wilkins *et al.*, 1977; Onderdonk *et al.*, 1979; Nichols *et al.*, 1979).

Anaerobes as normal flora

Infections due to the non-sporing anaerobes are almost always endogenous, because most of these organisms are obligate parasites that form part of the normal bacterial flora of the mucosal surfaces of the oronasopharynx, and of the alimentary and lower female genital tracts. While strains of *Bacteroides* are normally present in all of these situations, *Fusobacterium* spp. are most commonly encountered in the oropharynx, and to a lesser extent in the female genital tract. The anaerobic cocci are also prevalent in the upper respiratory and genital tracts, but are relatively sparse in the intestinal tract. Abdominal anaerobic infections are almost always due to organisms derived from the gastrointestinal or female genital tract. In this chapter the concern is chiefly with infections of gastrointestinal origin; genital tract infections and the associated bacterial pathogens are considered in Chapter 14.

The gastrointestinal tract, and especially the colon, affords a natural environment for various species of *Clostridium*, of which *C. perfringens* is dominant among the potential pathogens, while *C. sporogenes* and *C. bifermentans* are common non-pathogenic inhabitants (see Chapter 11).

Infant gut flora

Bacterial colonization of the gastrointestinal tract commences at the time of delivery with aspiration of the maternal genital tract flora (Brook *et al.*, 1979); its subsequent composition is determined by the nature of the infant feed. In purely breast-fed infants the colonic flora is dominated by bifidobacteria (Willis *et al.*, 1973; Bullen *et al.*, 1976; Tomkins *et al.*, 1981; Stark and Lee, 1982), but other anaerobes, such as *B. fragilis*, anaerobic cocci and clostridia, are sometimes also present in smaller numbers. In infants receiving a formula-feed, either as a 'total' feed or as a supplement to breast-feeding, the anaerobic bacterial faecal flora tends to resemble that of the adult (Rotimi and Duerden, 1981). Thus, bifidobacteria become much less numerous and are usually outnumbered by both *Bacteroides* spp. and anaerobic streptococci by about 3 to 1. Moreover, the distribution of the different anaerobic genera is profoundly different for the two groups of infants. Thus, while bifidobacteria are present and dominant in virtually 100 per cent of breast-fed infants and weanlings, they are absent from about 25 per cent of adults. Bacteroides dominate in 100 per cent of adults, while other anaerobes such as peptostreptococci and clostridia are present in a much higher proportion of adults than breast-fed infants. Weanlings show a transitional bacterial flora (Mata *et al.*, 1969).

Stomach

Because the stomach is constantly seeded with the bacteria of the oropharyngeal flora, it is by no means always 'sterile'. There appears to be no resident indigenous flora of the normal stomach, but only a small transient one in which anaerobes are scantily represented. During periods of fasting the normal stomach may be regarded as virtually sterile (Drasar *et al.*, 1969; Nelson and Mata, 1970).

Small bowel

The bacterial flora of the small bowel is sparse and probably largely transitory, and contains few anaerobic organisms (Hayward, 1963; Borriello *et al.*, 1978). In a study of North American subjects, Thadepalli *et al.* (1979) found that the duodenum, jejunum and ileum were sterile in 82 per cent, 69 per cent and 55 per cent of cases, respectively; both clostridia and bacteroides were rare.

Colon

Distal to the ileocaecal valve there are astronomical numbers of bacteria, comprising the normal colonic or faecal flora. In this setting, as has been noted, anaerobes are dominant and outnumber facultative components by about 1000 to 1. Organisms of the *B. fragilis* group not only predominate numerically but are found in 100 per cent of faecal samples (Finegold and Miller, 1968) (Table 13.1). It is of some interest that while *B. vulgatus*, *B. thetaiotaomicron* and *B. distasonis* predominate in the normal faecal flora both in terms of numbers and ubiquity, the great majority of isolates recoverable from *pathological* material are strains of *B. fragilis* (Werner, 1974; Moore and Holdeman, 1975; Holland *et al.*, 1977; Duerden, 1980) (Table 13.2). This fortuity of pathogenicity is put into perspective by the detailed systematic observations by Moore and Holdeman (1974) on the microflora of faecal specimens from 20 individuals. They isolated 113 different kinds of bacteria, and estimated that as many as 500 species may be present in this habitat; nearly 70 per cent of their isolates were represented by only ten anaerobic species, among which only '*B. fragilis*' was a recognized pathogen.

In the total anaerobic bacterial population of the colon, the clostridia are very much a minority group, showing mean viable counts of only 10^3–10^4 (Drasar, 1967; Collee, 1974). The most frequently

Table 13.1 Adult human faecal flora (from Finegold and Miller, 1968)

Species	Count/g of faeces	Percentage of 31 subjects colonized
B. fragilis group	10^{10}	100
Bifidobacterium spp.	10^9	68
Fusobacterium spp.	10^9	58
Clostridium spp.	10^8	42
Anaerobic cocci	10^8	39
E. coli	10^7	97
Enterococci	10^7	80
Lactobacilli	10^7	50
B. melaninogenicus group	10^4	45

Table 13.2 Incidence of *Bacteroides* spp. in faecal flora and in pathological material*

Species	Distribution in normal faeces (%)	Distribution in pathological material (%)
B. vulgatus	43	2
B. thetaiotaomicron	29	17
B. fragilis	13	81

*After Werner (1974)

encountered species in human faeces are *C. perfringens* (a constant inhabitant of the gut), *C. bifermentans*, *C. tertium*, *C. sporogenes*, *C. ramosum* and *C. paraputrificum* (Drasar *et al.*, 1976). Among infants, there is a characteristically high faecal carriage rate of *C. difficile* (Larson *et al.*, 1978; Stark *et al.*, 1982). A study by Holst *et al.* (1981) of children aged from two weeks to 15 years revealed a peak incidence of 64 per cent carriage in children aged 1–18 months, and 4 per cent in children aged above and below these limits. In contrast to this high prevalence in infants, *C. difficile* is both sparse and uncommon in healthy adults (George *et al.*, 1979; Marrie *et al.*, 1982), although it makes a significant reappearance in the adult colon, not as part of the normal flora, but as the putative pathogen in antibiotic-associated diarrhoea and pseudomembranous colitis.

Unlike the breast-fed infant gut, once the adult colonic flora has been established it appears to be remarkably stable. Neither drastic changes in diet nor long periods of fasting have any appreciable effect on the bacterial load or its composition (Aries *et al.*, 1971; Moore *et al.*, 1969; Attebury *et al.*, 1974; Peach *et al.*, 1974; Axelsson and Justesen, 1977). While attempts to 'sterilize' the colon by mechanical and/or antimicrobial methods have met with singularly little success, therapy with some antimicrobial agents may cause serious disturbances of the gut ecology and predispose the patient to either suprainfection or antibiotic-associated diarrhoea or pseudomembranous colitis. Abnormal physiological or pathological events, on the other hand, greatly favour massive colonization of regions of the gut which are normally free from anaerobes. Thus, obstruction of the small bowel by mechanical means, by ileus or by the presence of surgical blind loops or diverticulae, is associated with rapid retrograde luminal invasion by a faecal flora, while malignant disease of the stomach, pernicious anaemia, medication with H$_2$-receptor antagonists such as cimetidine, and other events associated with reduced gastric acidity favour colonization of the stomach by the oropharyngeal flora so that overgrowth by *Bacteroides* spp., peptostreptococci and clostridia is common (Drasar *et al.*, 1969; Goldstein *et al.*, 1969; Browning *et al.*, 1974; Sykes *et al.*, 1976a,b; Muscroft *et al.*, 1981a,b; Lancet, 1981).

Biliary tract

Although bile is normally sterile (Lou *et al.*, 1977; Dye *et al.*, 1978; Peel, 1980), a variety of bacteria is recoverable from it, notably in cases of cholecystitis, cholelithiasis, local malignant disease and obstruction. Most workers agree that aerobes are the most common isolates, among which *E. coli* is most frequently encountered. Anaerobes account for only about 10 per cent of the total flora, and are represented chiefly by *C. perfringens* and anaerobic streptococci. *Bacteroides* species are infrequently present (Fukunaga, 1973; Sapala *et al.*, 1975; Keighley, 1977; Lou *et al.*, 1977; Willis *et al.*, 1984). However, in contrast to these findings, Nielsen and Justesen (1976), Schimada *et al.* (1977), England and Rosenblatt (1977) and Marne *et al.* (1986) found that *B. fragilis* was commonly present in biliary tract disease; indeed, it was the single most commonly isolated anaerobe in England and Rosenblatt's study, accounting for 21 per cent of all positive cultures, while *C. perfringens* represented only 6 per cent of the total isolates. These apparently contradictory findings may merely reflect differences in underlying pathology in the patients studied. Thus, Finegold (1979) noted that *B. fragilis* more often occurred in patients who had previous biliary surgery (especially biliary–intestinal anastomosis), postoperative stricture or some other obstructive lesion.

The route by which bacteria gain access to the gall bladder is uncertain, although it is believed by many that they are transported to the liver from the gastrointestinal tract via the portal circulation and then excreted with the bile into the biliary tract. Ascending colonization from the duodenum is a logical route of entry and seems likely to occur, for example, in cases of biliary tract obstruction.

Some of the pathogenic anaerobes encountered in infections associated with the gastrointestinal tract are listed in Table 13.3.

Table 13.3 Some anaerobic pathogens in gastrointestinal-associated infections

Common isolates	Uncommon isolates
B. fragilis	B. melaninogenicus
B. thetaiotaomicron	B. intermedius
F. nucleatum	B. asaccharolyticus
P. anaerobius	B. ruminicola
C. perfringens	B. ureolyticus
	F. necrophorum
	C. septicum
	C. difficile

Pathogenesis of abdominal infections

Although there are fundamental differences between clostridial and non-clostridial infections, in terms of the common source of the organisms and the mode and site of 'transmission' and host-challenge, it is generally true to say that anaerobic bacterial infections of man are typically associated with locally compromised tissues and also with a systemically compromised host.

The anaerobic pathogens are themselves endowed with a diversity of factors which enable them to progressively colonize and infect contaminated and debilitated tissues, and synergic effects of one bacterial species upon another may be a significant factor in many polymicrobial infections. Moreover, once an anaerobic infection is established, production of toxic substances by the infecting organisms may lead to a progressive weakening or neutralization of various host defence mechanisms. There is thus always an interplay between the components of bacterial pathogenicity on the one hand and the elements of host resistance on the other.

Predisposing factors to anaerobic infections

Bacterial determinants

Despite intensive studies during recent years, the precise mechanisms of pathogenicity of the non-clostridial anaerobes remain largely unknown.

Microbial adherence to various types of cells is

exhibited by a number of species. Thus, *F. nucleatum* and species of black pigmented *Bacteroides* agglutinate human erythrocytes, and show varying degrees of adhesion to the crevicular epithelium and the vagina. Such selective adherence may be an enabling prerequisite for oral anaerobes in the pathogenesis of inflammatory periodontal disease (Moore *et al.*, 1982). In the same way, the virulence of *B. fragilis* in gastrointestinal tract-associated infections may be due to a high capacity to penetrate the local host defences and to adhere to the mucosa (Hofstad, 1984). Onderdonk *et al.* (1978) noted that capsulate *B. fragilis* strains were more apt to adhere to rat peritoneal mesothelium than were noncapsulate variants of the *B. fragilis* group. In similar clinical settings, potential determinants of bacterial pathogenicity include the production of extracellular and cell-bound enzymes such as collagenase, fibrinolysin, hyaluronidase, deoxyribonuclease and neuraminidase, some or a number of which are produced by a variety of pathogenic species of *Bacteroides*; *Fusobacterium* spp. produce DNAase, but are otherwise comparatively inactive (Porschen and Sonntag, 1974; Hofstad, 1984). The role played by these different enzymes is somewhat speculative as virulent and avirulent *Bacteroides* do not appear to differ in their *in vitro* production of these enzymes.

It has been suggested that the superoxide dismutase present in some *Bacteroides* and *Fusobacterium* spp. may protect colonizing organisms from the toxic effects of oxygen until conditions are propitious for their multiplication and tissue invasion (Tally *et al.*, 1977; Gregory *et al.*, 1978). Production of β-lactamases by all strains of the *B. fragilis* group, and by ever increasing numbers of isolates of *B. melaninogenicus*, *B. oralis*, *B. ruminicola* and *B. disiens*, provides a direct mechanism of pathogenesis in patients receiving therapy with β-lactam agents. Moreover, a similar mechanism of pathogenesis may operate indirectly by virtue of the presence of β-lactamase-producing commensals protecting fully sensitive pathogens from exposure to β-lactam agents (Hackman and Wilkins, 1976; O'Keefe *et al.*, 1978; Maddocks, 1980; Brook *et al.*, 1983b).

Although the role played by lipopolysaccharide (LPS; 'endotoxin') in suppurative sepsis due to nonclostridial anaerobes is speculative, there is some evidence to suggest that the LPS of *F. nucleatum* contributes to the pathogenesis of dentoalveolar (periapical) inflammation. On the other hand, capsular polysaccharide of *B. melaninogenicus* and *B. fragilis* has been clearly identified as a virulence factor which inhibits both the migration and the phagocytic activity of macrophages (Okuda and Takazoe, 1973; Onderdonk *et al.*, 1977; Kasper *et al.*, 1977, 1979a,b; Brook *et al.*, 1983a, 1984a). Moreover, crude capsular polysaccharide from *B. fragilis* induces abscess formation in experimental animals in the absence of viable organisms (Onderdonk *et al.*, 1977), while immunization of animals with capsular polysaccharide protects them against challenge with *B. fragilis*. Protection against abscess formation appears to be T cell-mediated and does not require the presence of serum antibody (Onderdonk *et al.*, 1982).

Bacterial synergy

As long ago as 1928, Weinberg *et al.* drew attention to the role of bacterial synergy in the aetiology of acute appendicitis, and showed that certain combinations of intestinal organisms are lethal in smaller doses than the pure cultures alone. At about the same time, Brewer and Meleney (1962a,b) and Meleny (1931) recorded cases of progressive gangrene of the abdominal wall (Meleney's progressive synergistic (sic) gangrene), following drainage of appendical abscesses. In these, *Staphylococcus aureus* and an anaerobic coccus were isolated from the gangrenous margin of the lesion, and it was shown that a similar gangrenous infection could be produced in dogs and guinea pigs only when the two organisms were injected together; pure cultures of the individual strains had little or no effect. Similarly, by introducing infected emboli into the lungs of rabbits, Meleney showed that severe and widespread infection resulted only when the staphylococcus and the anaerobic coccus were present together (see also Mergenhagen *et al.*, 1958; Brook *et al.*, 1984b).

Hite *et al.* (1949) demonstrated a synergic action between *F. necrophorum* and *S. liquefaciens* in producing necrotic abdominal wall lesions in mice. Roberts (1967) showed that infective bulbar necrosis of sheep is produced by the synergic association of *F. necrophorum* and *Corynebacterium pyogenes*. The latter organism produces a nutrient necessary for the growth of the fusobacterium, while the fusobacterium produces a leucocidal toxin which protects both organisms from phagocytosis. Similarly, in actinomycotic lesions in man, *Actinomyces israelii* is commonly accompanied by anaerobic cocci, *B. melaninogenicus* and staphylococci. It has been suggested that secretion of collagenases and hyaluronidases by these organisms may facilitate invasion of tissue either by disrupting its mechanical structure or by providing split products which may be essential for the nutrition of *A. israelii* (Smith and Holdeman, 1968).

Other clear examples of synergic interrelations between micro-organisms in the setting of mixed anaerobic–aerobic sepsis relate to oxidation–reduction (OR) potential, and to enzymatic protection against antimicrobial agents. Thus, it is accepted that facultative bacteria assist the growth of anaerobes by utilizing oxygen, by diminishing the OR potential, or by supplying catalase. Protection of anaerobes against the activity of β-lactam agents results from β-lactamase production by the concomitant facultative flora, a phenomenon that might more correctly be considered as 'indirect synergy' (Tally, 1979; Hamilton-Miller, 1981).

With the increasing recognition of the importance of anaerobic bacteria in many clinical settings, and their frequent association with facultative organisms, workers have been prompted to re-examine the pathogenicity of pure and mixed cultures of these organisms. For this work, various sophisticated and elegant animal model studies have been developed, most of which have been concerned with intra-abdominal sepsis.

Evolution of mixed anaerobic–aerobic infections

Since septic complications of colonic perforation may involve multiple bacterial species derived from the intestinal flora, Weinstein and his colleagues (Onderdonk *et al.*, 1974; Weinstein *et al.*, 1974, 1975) developed an experimental model in rats to study the quantitative bacteriology and antimicrobial therapy of intra-abdominal sepsis. Following implantation of colonic contents into the pelvic region of the experimental animals, they observed quantitative differences in bacterial populations according to the stage of the disease. During the initial, often fatal stage of acute peritonitis, *E. coli*, enterococci and *B. fragilis* were always present; blood cultures at this time were uniformly positive, *E. coli* being the principal isolate. Animals that survived this early bacteraemic stage of acute peritonitis developed indolent intra-abdominal abscesses. The major isolates from abscess contents were *B. fragilis* and *F. necrophorum*; *E. coli* and enterococci were also present but in lower concentrations. Thus, in the early acute stage of the infection the two aerobic species were dominant, while in the secondary chronic phase the anaerobes dominated. Of equal importance was the observation that during the infective process a major simplification of the original faecal inoculum occurred, although the subsequent infection apparently remained bacteriologically complex.

Further studies, which made use of selective antimicrobial therapy as 'therapeutic probes'— gentamicin as a specific anti-*E. coli* agent, and clindamycin against anaerobes—confirmed that coliforms caused the early mortality, while anaerobes were primarily responsible for the late complication of intra-abdominal abscess formation. Thus, both aerobes and anaerobes are pathogens in this model, a fundamental observation which not only offered rational guidelines to antimicrobial selection for patients with intra-abdominal sepsis, but also indicated that similar experimental techniques may be relevant for assessing mixed infections at other anatomical sites (Bartlett, 1978; Nulsen *et al.*, 1983).

At about the same time that Weinstein and co-workers were carrying out their definitive studies, two other highly significant observations were made by Ingham *et al.* (1977) and Kelly (1978, 1980):

First, Ingham and his colleagues noted that, in *in-vitro* studies, most obligate anaerobes inhibited the phagocytosis of aerobic bacteria. They referred to the magnitude of this effect as the 'inhibitory index', which, in descending order, was greatest for *B. melaninogenicus*, *B. fragilis* and *F. necrophorum*. This antiphagocytosis effect, which was specifically abolished by the addition of metronidazole to the *in vitro* system, was not exhibited by any of 36 aerobes tested. These important observations, which were subsequently further refined and developed by Ingham *et al.* (1980; 1981), were confirmed and extended by Jones and Gemmell (1982) and Wade *et al.* (1983). They provide strong evidence that recognized obligately anaerobic pathogens may act synergically in mixed infections with aerobic species to inhibit phagocytosis and destruction of other bacteria, which when present alone may not be overtly pathogenic. This synergy, taken in concert with the biphasic nature of mixed aerobic/anaerobic sepsis as demonstrated by Weinstein and colleagues, provides a compelling scenario for the general pathogenesis of these infections. The findings of subsequent studies from various centres are in general agreement that *Bacteroides* species are able to impair opsonization, chemotaxis, phagocytosis and intraleucocytic killing of other bacteria, although the precise mechanisms for these effects have not been fully elucidated (Simon et al., 1982; Namavar et al., 1984, 1986; MacLaren et al., 1984; Vel et al., 1985).

Second, Kelly (1978, 1980) produced infections with standardized mixtures of *E. coli* and *B. fragilis* in wounds in the shoulders of guinea-pigs. These wounds were designed to mimic surgical wounds of the human abdominal wall. From histological and quantitative bacteriological studies of these infected lesions, Kelly showed that, whereas quantitatively similar subinfective doses of pure *E. coli* or pure *B. fragilis* failed to produce inflammation, an inoculum of the same volume, containing the same total number of *mixed* organisms (half *E. coli* and half *B. fragilis*) produced marked inflammation with formation of frank pus; here is a clear demonstration of pathogenic synergy *in vivo*.

These questions of bacterial synergy in mixed aerobic–anaerobic infections have been briefly reviewed in the Lancet (1980) and by McGowan and Gorbach (1981).

The interplay of the various mechanisms of synergy which operate in the aerobic–anaerobic setting has been referred to by Gorbach (1982) as a "cascade effect" whereby the infection occurs in a series of stages, each being fostered by specific organisms in the flora. In the case of abdominal sepsis, coliforms promote the peritonitis and septic shock phase, which is followed by intra-abdominal abscess related to anaerobes. The anaerobic abscess itself is a protective environment that fosters a mixed infection.'

In vivo studies by Brook *et al.* (1984b), using lethality and production of subcutaneous abscesses in mice as the animal model markers, demonstrated the potential for synergy between a wide range of aerobic and anaerobic bacteria which are commonly encountered in mixed infections in man. Their findings not only confirm those of other workers, outlined above, but also extend the observed range of organisms among which synergy occurs. Thus, synergy was demonstrated between *Bacteroides* on the one hand and *S. aureus*, *Pseudomonas aeruginosa*, *Klebsiella pneumoniae* and *Proteus mirabilis* on the other. Similar energy was demonstrated between anaerobic streptococci and *P. aeruginosa* or *S. aureus*. There was also demonstrable synergy between anaerobes themselves—between anaerobic cocci and *B. fragilis* or *B. asaccharolyticus*.

Comment

One of the important practical consequences of the developing knowledge of synergy is the impact it has upon the rational use of antimicrobials for both the prevention and treatment of mixed infections. Synergy between anaerobes and aerobes implies that inhibition of either component of the infecting flora should have a beneficial effect, since the other component would then experience a less favourable environment. There is some persuasive clinical evidence for this view, deriving particularly from the use of metronidazole (which has virtually exclusively anti-anaerobic activity) and of macrolides, which are active against aerobic cocci and some anaerobes, but are inactive against enterobacteria. In a study of experimental peritonitis in rabbits, Pressey *et al.* (1983) found that large doses of erythromycin suppressed not only the sensitive aerobic bacterial flora (gram-positive cocci and bacilli), but also significantly reduced the numbers of insensitive anaerobic gram-negative bacilli. Moreover, the addition of gentamicin, itself inactive against anaerobes, resulted in the total suppression of anaerobic gram-negative bacilli. These relationships, especially the further inhibition of the anaerobic flora by an agent with no *in vitro* action against anaerobes, add further support to the proposal that bacterial synergy is important in the genesis of mixed anaerobic sepsis. Not withstanding these considerations, the anaerobic components in mixed infections frequently play a dominant role (Dunn and Simmons, 1984) and specific anti-anaerobic therapy should never be omitted from the treatment regimen.

Although synergy is by no means essential to infection *per se* (single organism experimental and clinical infections are not uncommon) when synergy between organisms is suspected antimicrobial treatment should be directed at each of the synergic elements.

Valuable commentaries on pathogenetic mechanisms of non-sporing anaerobes were published by Kasper and Finegold (1979), Tally (1979), Tally and Gorbach (1979), Collee (1980), Ingham *et al.* (1980), Bjornson (1982), Hofstad (1984) and Rotstein *et al.* (1985).

Host determinants

Some major predisposing factors to anaerobic infections are listed in Tables 13.4 and 13.5.

Any condition which interferes with the systemic defence mechanisms of the host, and thus predisposes to infection in general, increases the incidence of anaerobic infection also. There is no clear evidence, however, that diabetes, corticosteroid therapy, cytotoxic therapy, hypogammaglobulinaemia, leucopenia or splenectomy are associated with a higher incidence of anaerobic, as opposed to aerobic, infection; any apparent increase in incidence in recent years is probably attributable to their correct clinical recognition.

It is generally true to say that anaerobic abdominal infections occur secondary to mechanical disruption of anatomical barriers; spontaneous infections are relatively uncommon. In non-clostridial anaerobic infections the disrupted barriers are commonly mucosal surfaces, where the offending organisms exist as the dominant part of normal indigenous bacterial floras. Examples of this type of disruption include accidental and surgical trauma to the gastrointestinal and female genital tracts and a variety of pre-existing pathological states, ranging from infection by aerobes or viruses to ulcerating malignancies. Disturbance of mucosal barriers also follows upon obstruction and stasis such as may occur in diverticular disease, appendicitis and blind loop syndrome.

Antimicrobial therapy may alter the balance of organisms in the normal floras and enhance the growth of anaerobic elements. Thus, the use of oral aminoglycosides for the preoperative 'sterilization' of the large bowel may predispose to postoperative anaerobic sepsis following colorectal surgery, since the anaerobes are resistant to this group of drugs. The causal relationship of a variety of antimicrobials, in particular clindamycin and cephalosporins, to *C. difficile* antibiotic–associated pseudomembranous colitis is well established (see Chapter 21).

The most important single factor which enables anaerobes to flourish in the tissues is a low oxygen tension, which is usually the result in the first instance of a compromised blood supply and poor tissue perfusion. A number of the determinants listed in Table 13.4 may operate in concert,

Table 13.4 Anaerobic infections—host-related determinants*

1. Compromised blood supply
2. Disruption of epithelial barriers
3. Compromised defence mechanisms (cellular and humoral)
4. Previous antimicrobial therapy
5. Bacterial contamination (exogenous and endogenous)

*From Willis (1985)

Table 13.5 Anaerobic infections—factors which reduce tissue Eh

Compromised blood supply	Other factors
Accidental trauma	Necrotic tissue
Surgery	Blood clot
Malignant disease	Foreign bodies
Obliterative arterial disease	Facultative bacteria
Shock	
Anaemia	
Pressure	
Vasoconstrictors	

*From Willis (1985)

Table 13.6 Abdominal anaerobic infections—some predisposing factors

Disruption of epithelial barriers	Compromised defence mechanisms
Gastrointestinal perforation	Diabetes
Accidental trauma	Malignant disease (including leukaemia)
Surgery	Immunosuppressive therapy
Malignant disease	Cytotoxic therapy
	Steroid therapy
Antecedent infections	Alcoholism
	Drug addiction
	Leucopenia
	Hypogammaglobulinaemia
	Splenectomy
	Liver disease
	Hypovolaemia

while others may be brought into play sequentially after the initial challenge.

Some clinical determinants of abdominal anaerobic (chiefly non-clostridial) infections are summarized in Table 13.6.

Postoperative infections

Non-sporing anaerobes are the most common cause of postoperative sepsis following surgery of the gastrointestinal tract and female genital tract (Willis et al., 1981). In addition to the many general and local host factors that may increase the patient's susceptibility to these infections, is a number of important surgical and operative determinants (Table 13.7). Almost all of the predisposing factors that are brought into play as a result of surgery may be prevented or minimized by a meticulous surgical technique (Altemeier, 1979).

Anaesthesia and the trauma of surgery are associated with a variety of metabolic, hormonal and immunological disturbances that contribute to temporary systemic defects in host resistance.

The length of the operative procedure has a direct effect on postoperative infection rates, which are also influenced by the type and extent of the surgery. Thus, emergency operations on the large bowel carry a higher sepsis rate than similar elective procedures. Anterior resection and left-sided colon operations are higher risk procedures than right-sided surgery. Again, while gastric surgery for non-malignant states is occasionally complicated by postoperative anaerobic sepsis, surgery for gastro-oesophageal malignancy or obstruction carries a high risk, and usually involves organisms derived from the oropharynx (Nichols et al., 1975). Some of this increased susceptibility to anaerobic sepsis is attributable to mechanical opening of bacterial barriers and the subsequent 'spilling' of endogenous luminal flora into the surrounding tissues—the peritoneal cavity, parietal wounds and anastomoses. Marked obesity, which predisposes to wound sepsis, may prolong operation time but is also a risk factor in its own right because adipose tissue is poorly vascularized and relatively inert, and so is unable to deal with contaminating bacteria. Impaired local circulation invariably accompanies surgical incision; it may be aggravated by various factors such as damage to a large vessel, excessive electrocoagulation, prolonged and traumatic retraction of wounds and closure of wounds under excessive tension.

Foreign bodies in wounds are a well known predisposing factor to sepsis. Sutures embedded in tissues must be regarded as foreign bodies and use of excessive sutures increases the risk factor; multifilament braided sutures and permanent sutures are more hazardous than others. The placement of drains in the wound is controversial among colorectal surgeons, but it seems clear that their use in 'clean–contaminated' wounds increases the risk of postoperative sepsis.

Poor haemostasis and the presence of haematomas or tissue 'dead spaces', are important operative determinants that favour the development of anaerobic infections. Operative blood loss is also a well known general predisposing factor.

Some antimicrobial agents, such as latamoxef and several cephalosporins, cause a high incidence of postoperative bleeding if used during the peroperative period, particularly in elderly or debilitated patients. Far from preventing postoperative sepsis, these agents pave the way for the development of infected haematomas (Smith and Lipsky, 1983; Morris et al. 1984; Bowcock et al., 1986).

Today, the incidence of significant postoperative anaerobic sepsis can be reduced virtually to nil by attention to host systemic determinants and by the application of a meticulous surgical technique supported, where indicated, by relevant antimicrobial prophylaxis (British Medical Journal, 1980a,b). However, in the operative setting the observation made by Lord Moynihan in 1920 is still apposite: 'Every operation in surgery is an experiment in bacteriology'.

Table 13.7 Some surgical determinants of anaerobic infection*

Trauma of surgery; anaesthesia
Length of surgery
Type and extent of surgery
Emergency surgery
Destruction of mechanical barriers
Endogenous contamination
Obesity
Foreign bodies—excessive sutures; drains
Poor haemostasis
Excessive electrocoagulation
Operative blood loss
Inappropriate antimicrobials
Previous irradiation

*From Willis (1985)

Development and progression of abdominal anaerobic infections

The march of events in abdominal infections due to non-clostridial anaerobes may be illustrated by reference to the female genital tract. Almost any condition that causes bleeding or tissue damage in the female genital tract predisposes to invasion of the pelvic viscera by non-sporing anaerobes. Important among these are malignant disease, gynaecological surgery including curettage, cervical cauterization and laparoscopic sterilization, childbirth, self-induced and incomplete abortion and gonococcal and chlamydial infections. By far the most dangerous obstetrical and gynaecological infections are those that commence in the uterus. Septic thrombophlebitis originating in the uterine sinusoids may extend via venous and lymphatic routes into the broad ligaments and retroperitoneal spaces in the pelvic floor. Extending cellulitis may directly involve the adjacent peritoneum, leading to pelvic abscess formation, or adnexal abscesses may rupture into the peritoneal cavity producing generalized peritonitis or multiple intra-abdominal abscesses. Septic thrombophlebitis of the pelvic veins results in a continuous bacteraemia, endocarditis may occasionally develop, and there may be metastatic abscess formation in the lungs, brain, liver, bones and joints, and other organs. In a study of 540 intra-abdominal abscesses in 501 patients, Altemeier et al. (1973) found that primary disease of the genitourinary tract was second only to appendicitis as a source of intra-abdominal abscess; 19 per cent were associated with appendicitis, 18 per cent with the genitourinary tract.

Anaerobic abscess

Abscess formation provides an environment that is highly favourable to the maintenance of anaerobic infection. The structure of the abscess, with its thick fibrous wall and poor blood supply not only ensures that host immunological defence mechanisms are largely negated, but also that delivery of antimicrobial agents to the abscess cavity may be severely impaired. The dense and diverse bacterial population of the abscess commonly elaborates high concentrations of β-lactamases and other enzymes that ensure inactivation of penicillins, cephalosporins and chloramphenicol (O'Keefe et al., 1978; Tally, 1979; Brook, 1986). The low oxidation–reduction potential of the abscess interferes with the activity of aminoglycosides and reduces the ability of polymorphonuclear leucocytes to phagocytose and kill micro-organisms. Moreover, the very high bacterial counts reflect the fact that bacteria within an abscess are in a relatively quiescent growth phase, thus ensuring the inability of many agents to exert an antimicrobial effect, since this is often dependent on logarithmic growth (Gorbach et al., 1981; Gorbach, 1982). Some reasons for the inefficacy of antimicrobials in abscesses are listed in Table 13.8 (McDonald et al., 1984). Interestingly, the activity of metronidazole is not adversely affected by abscess formation since its anti-anaerobic effect requires bacterial metabolic activity, but not replication. Moreover, its pharmacokinetic properties ensure delivery of therapeutic concentrations to the abscess cavity (Joiner et al., 1982) where its activity is not adversely affected either by the low oxygen tension or by enzyme inactivators (Wells and Wilkins, 1980; Bartlett, 1982).

Anaerobic bacteraemia

Any account of the progression of non-clostridial anaerobic sepsis would be incomplete without brief reference to the serious events which may complicate anaerobic bacteraemia. The immediate clinical features of bacteraemia due to gram-negative anaerobes vary little from those due to sepsis caused by any gram-negative organism. There is a sudden onset of hectic fever, with rigors and profuse sweating, and endotoxic shock may develop. Important complications, some of which help serve as distinguishing features of anaerobic (as opposed to

Table 13.8 Reasons for inefficacy of antimicrobials in abscesses*

1. Poor drug penetration
2. Destruction of drug, e.g., bacterial β lactamases
3. Low pH reduces activity, e.g., clindamycin; erythromycin
4. Protein binding, e.g., clindamycin
5. Low oxygen tension, e.g., aminoglycosides are oxygen-dependent; metronidazole requires anaerobiosis
6. Phagocytosed organisms remain viable
7. Quiescent bacteria fail to take up drug
8. High bacterial inoculum

*After McDonald et al. (1984)

aerobic) bacteraemia include a high incidence of jaundice and septic thrombophlebitis, from which septic embolism may derive. Rather less frequently, various haematological disturbances such as bleeding diathesis and disseminated intravascular coagulation are seen. Anaerobic bacterial endocarditis is by no means unknown. Gunn (1956) summarized the common clinical sequence of events in bacteroides bacteraemia very succinctly. '(1) Symptoms and signs of the primary lesion, e.g. appendicitis, (2) a period of apparent recovery, (3) symptoms and signs of a spreading infection at the site of the original lesion, (4) abrupt onset of septicaemia with rigors, profuse sweating, anaemia and icterus, and (5) symptoms and signs of metastatic infective lesions'. Bacteroides bacteraemias most often follow from a failure to recognize the likely anaerobic aetiology of declared pre-existing sepsis. Not uncommonly the predisposing determinants of the initial localized anaerobic infection pass unrecognized and the infection is misleadingly attributed to facultative organisms. There can be little doubt that most anaerobic bacteraemias may be prevented by fostering a knowledge of the natural history of bacteroides and related infections.

Clinical considerations

Surgical wound infections

In the absence of appropriate antimicrobial prophylaxis, abdominal and perineal wound infection by endogenous anaerobes is a common complication of gastrointestinal and female genital tract surgery; it is seen most frequently after appendicectomy, colorectal operations and hysterectomy. Wound infections, which may be associated with deep seated sepsis, usually become evident on the third or fourth postoperative day. Infection may be largely confined to the wound itself, with local inflammation and production of malodorous pus, or there may be a large element of associated cellulitis. Cellulitis commonly commences at the lower end of an abdominal wound and may then spread laterally or distally into the flanks, perineum and thighs. Perineal wound sepsis is often responsible for considerable delay in wound healing; less frequently infection of abdominal wounds may also show frank wound dehiscence. Rarely, but devastatingly, postsurgical wound sepsis may progress to one of the synergic infections—Meleney's gangrene or anaerobic necrotizing fasciitis.

Meleney's gangrene and anaerobic necrotizing fasciitis

Meleney's gangrene most commonly involves the abdominal wall postoperatively. There is a good deal of experimental evidence which supports the concept of a synergic bacterial aetiology for a number of soft tissue anaerobic infections, including Meleney's gangrene and anaerobic necrotizing fasciitis. Meleny's gangrene is an uncommon instance of chronic, superficial, progressive gangrene which is differentiated clinically from other types of superficial gangrene by its slow and relentless progression, its severe local symptoms and the absence of severe systemic symptoms. Bacteriologically, it is typically a synergic infection, and is due to anaerobic cocci mixed with facultative anaerobes, such as *S. aureus*, *Str. pyogenes*, *Proteus* spp. or *P. aeruginosa* (Grainger *et al.*, 1967).

An essentially similar condition to Meleney's gangrene is anaerobic necrotizing fasciitis (Stone and Martin, 1972; British Medical Journal, 1973; Giuliano *et al.*, 1977). This infection is much more acute than Meleney's gangrene, is associated with a high degree of systemic toxicity and a significant mortality rate. Anaerobic necrotizing fasciitis, which is usually caused by anaerobic cocci and/or bacteroides mixed with enterobacteria and aerobic cocci, occurs most commonly in the perineum (Fournier's gangrene) or the thighs, usually in debilitated patients who have diabetes or other arterial insufficiency. In addition to occurring as a complication of major abdominal or pelvic surgery, and of surgical drainage of vulvovaginal and perirectal abscesses, it may develop secondary to occult intra-abdominal pathology (Galbut *et al.*, 1977; Daly *et al.*, 1978; Pruyn, 1978; Bocking *et al.*, 1981; Percival and Hargreaves, 1982). In addition to severe systemic intoxication, the disease is characterized by a fulminating, rapidly progressive spreading necrosis of the skin, subcutaneous and connective tissues, with massive loss of tissue and an associated highly offensive serosanguinous or frankly purulent discharge.

Surgery, with radical excision and full debridement of tissue to the limits of the involved soft tissues, is the most important element of therapy.

Immediate full combination antimicrobial therapy is mandatory, and must be directed against those groups of organisms most likely to be involved —anaerobes, enterobacteria and pyogenic cocci. This is fulfilled by combination therapy employing metronidazole, gentamicin and benzylpenicillin (or amoxycillin).

Intra-abdominal anaerobic infections

Most intra-abdominal infections are due to endogenous bacteria derived from the gastrointestinal tract, although some infections originate in the female genital tract. The dominance of non-sporing anaerobes in these normal floras partly explains their frequent and major participation in intra-abdominal sepsis.

Anaerobic intra-abdominal infection is usually due to a breach in the integrity of a hollow viscus; the main causes are listed in Table 13.9. Injury to the viscus, whether due to trauma, 'spontaneous' perforation or preceding sepsis, leads to contamination of the peritoneal cavity, which in turn results in local or generalized peritonitis and intra-abdominal abscess formation. Nichols (1980) refers to five factors that determine dissemination of infection within the peritoneal cavity:

the location and size of the primary leak;
the nature of the underlying pathology;
the presence of adhesions;
the duration of the precipitating illness;
the efficacy of the patients' defence mechanisms.

Generalized peritonitis commonly follows such acute catastrophies as perforation in obstructive appendicitis and penetrating or blunt abdominal trauma. Localized peritonitis accompanies more slowly-advancing lesions such as non-obstructive appendicitis, when spread of infection may be limited by the greater omentum. Such localization of infection results in the development of intra-peritoneal abscess associated with the infected or damaged viscus, e.g., appendix abscess. Loculated abscesses between loops of bowel (interloop abscesses) result from localization of generalized peritonitis.

The sites of intraperitoneal abscesses are partly determined by the layout of the peritoneal cavity, the mechanical limiting effect of the greater omentum and gravitational effects. Common sites for abscess formation are the subphrenic and subhepatic spaces, the right and left paracolic gutters, the right iliac fossa and the pelvis (Altemeier et al., 1973).

Visceral abscesses develop within the limits of an abdominal organ, and result either from primary disease developing within the affected organ, e.g., tubo-ovarian abscess, or from haematogenous or lymphatic dissemination of sepsis located elsewhere, e.g., intrahepatic abscess derived from appendicitis. Liver abscess is a common consequence of intra-abdominal disease, from which gastrointestinal anaerobes reach the liver by the portal venous circulation. Like the primary infection, these secondary abscesses are usually polymicrobial, containing not only mixed anaerobic species, but also important facultative pathogens such as Str. milleri.

The diagnosis of intra-abdominal infection is a clinical exercise, an account of which is beyond the scope of the present work. Suffice it to note that, in addition to the patient's history and findings of repeated physical examination, a variety of special investigations is of value. Neutrophil leucocytosis and raised erythrocyte sedimentation rate are the rule. Intra-abdominal sepsis is frequently complicated by anaerobic bacteraemia, so that blood culture is relevant. Pathological material for culture may also be obtained by aspiration of peritoneal fluid or abscess content, and from wound exudate and high vaginal swabs. Various imaging techniques are available for recognition and localization of abscesses; these include plain and contrast radiography, ultrasonic scan, computer tomography and radionuclide scanning.

Useful discussions of intra-abdominal sepsis have

Table 13.9 Some common causes of intra-abdominal anaerobic infection

1.	Gastro-oesophageal malignancy
2.	Carcinoma of colon
3.	Diverticular disease
4.	Appendicitis
5.	Gut ischaemia
6.	Crohn's disease
7.	Ulcerative colitis
8.	Blunt and penetrating abdominal trauma
9.	Invasive instrumentation such as laparoscopy
10.	Gastrointestinal and female genital tract surgery
11.	Anastomotic dehiscence
12.	Pelvic inflammatory disease

been published by Thadepalli et al. (1973), Altemeier (1979), Wise and Donovan (1981), Simmons (1982), Nichols et al. (1984), Springall and Gilmore (1985) and Thomas et al. (1986).

Treatment

Surgical management

The treatment of established intra-abdominal anaerobic infections usually involves the application of all three basic principles of therapy:
drainage of pus;
correction of underlying pathology;
antimicrobial therapy.
In gangrenous perforated appendicitis, for example, evacuation of pus and removal of diseased tissue is achieved at laparotomy by formal appendicectomy and peritoneal toilet. Antimicrobial therapy is combined with these surgical measures and not only serves an essential therapeutic role, but also provides systemic prophylaxis aimed at preventing the development of secondary intra-abdominal and haematogenous sepsis, and postoperative infection of the surgical wound.

Thorough surgical drainage was for many years regarded as the single most important procedure for the proper treatment of intra-abdominal sepsis. Recognition of a common polymicrobial aetiology of these infections, and the dominant role frequently played by non-sporing anaerobes, has enabled a more rational and effective use of antimicrobial agents, and this in turn has permitted some modification of the surgical approach. Many cases of intra-abdominal infection which were formerly treated by peritoneal drainage following laparotomy are now successfully managed with peritoneal toilet and primary closure, combined with relevant antimicrobial therapy (Pashby and Mee, 1978). Indeed, one surgical view considers that because drainage is likely to promote both intra-abdominal infection and extension of sepsis to healthy (parietal) tissues, it should be avoided if at all possible (Magarey et al., 1971; Stone et al., 1978; Lancet, 1979). Surgical drainage is well reviewed by Smith and Gilmore (1985).

In the same way, the traditional approach to the treatment of intraperitoneal and hepatic abscesses involves surgical drainage combined with appropriate antimicrobial therapy, the latter being employed primarily to prevent local or haematogenous extension of the infection. Adequate drainage of pus has been considered essential to the successful management of hepatic abscesses, although there are now several accounts of their response to aspiration (Perera et al., 1980; Herbert et al., 1982; Gerzof et al., 1985) and even to medical treatment alone (Back et al., 1978; Lancet, 1978; Maher et al., 1979; Thomas et al., 1982; George et al., 1982). Berger and Osborne (1982) described nine patients with anaerobic liver abscess who were successfully treated with metronidazole combined with percutaneous needle aspiration; all abscesses healed without further intervention. It is significant that some of these patients had multiple abscesses, many of which it was impracticable to aspirate. More recently, Ralph (1984) reported four patients, two with multiple liver abscesses and two with large intra-abdominal abscesses, in whom antimicrobial therapy (metronidazole alone or in combination with other antimicrobials) was effective without drainage or aspiration.

In an elegant *in vitro* study that used an experimental animal model of subcutaneous *B. fragilis* abscesses in mice, Joiner et al. (1982) and Bartlett (1982, 1983) showed that metronidazole was the most effective of a number of anti-anaerobic agents in reducing the bacterial counts of the abscess contents. This activity correlated with the unique activity of the drug *in vitro* (Ralph and Kirby, 1975), its pharmacokinetic profile (good penetration into the infected site and low protein binding) and its insusceptibility to enzymes such as β-lactamases produced by the infecting organisms. These experimental findings are in accord with the reported advantages of nitroimidazole therapy in extra-abdominal anaerobic abscesses. Thus, following the striking success of metronidazole in the treatment of brain abscess by Ingham et al. (1977), Kottas and Smith (1978) confirmed that even encapsulated cerebral abscesses resolve under metronidazole therapy without surgical intervention.

There is more than a hint from these experimental and clinical observations that metronidazole may rewrite the long established surgical principle that collections of pus always require drainage (Lancet, 1978).

Peritoneal lavage

Peritoneal lavage is regarded by many as a beneficial part of peritoneal toilet as it mechanically removes organisms, toxins and particulate matter from patients with generalized peritonitis. However, in cases of local peritonitis, lavage is likely to disseminate the infection throughout the peritoneal cavity. The value of adding antibiotics to the lavage fluid is controversial. Protagonists of antibiotic lavage argue that the antimicrobial agents are likely to reach foci of infection that are inaccessible to agents administered by the systemic route. Others consider that this is unlikely, but believe that significant absorption of drugs from the peritoneal cavity may result in unexpectedly high and even toxic systemic levels. There can be no argument that the best therapeutic (and prophylactic) results for most infections are achieved by the systemic use of relevant antimicrobials, whose pharmacokinetic properties ensure that they reach the target tissues at effective and sustained concentrations; topical (intraperitoneal) antimicrobial therapy is likely to be less effective, and is certainly much more difficult to control (Normann et al., 1975; Stewart, 1978; British Medical Journal, 1979; Browne, 1979).

Antimicrobial therapy

Although antimicrobial agents and their use in the prevention and treatment of anaerobic infections are considered in depth in Chapter 25, some comment on the management of intra-abdominal infections is relevant here.

Because the response of the non-clostridial anaerobes to the various first-line drugs is predictable, there should rarely be any difficulty in choosing the most appropriate agent in particular circumstances. Since chloramphenicol, clindamycin, cefoxitin and metronidazole are all widely active against the non-clostridial anaerobes they may be used empirically, either when an abdominal infection is suspected clinically, or in cases of known anaerobic sepsis before the results of sensitivity tests become available. The choice of drug may be influenced by such factors as its toxicity, the incidence of developing resistance to it, the required route of administration, location and severity of the infection and concomitant aerobic infection.

Twelve years' experience with metronidazole leads me to recommend it as the drug of choice. In the UK, metronidazole is almost invariably added to broad-spectrum regimens if an anaerobic infection is suspected (Drug and Therapeutics Bulletin, 1986). A standard adult course of intravenous therapy is 0.5 g 8-hourly for 24 h, followed by 0.5 g 12-hourly. This is either continued for five to seven days, or oral therapy (400 mg 6-hourly) is substituted as convenient; rectal suppositories are also available.

A prompt and sustained clinical response is the essential requirement for whichever first-line drug is chosen. If significant clinical and bacteriological resolution of the infection does not occur within 24–48 h, a careful review should be made of the chosen antimicrobial therapy (including its dosage and route of administration) and of the possible presence of an undeclared deep-seated abscess. Although there are some notable exceptions, for example, the treatment of liver abscess, appropriate antimicrobial therapy rarely need extend for longer than seven days.

Broad-spectrum vs narrow-spectrum antimicrobials

During the last few years there has been a scramble among drug manufacturers to produce very broad-spectrum antimicrobial agents with the much sought after characteristics of a 'panaceamycin'. Each new product fulfils some, but not all, of the theoretical *in vitro* requirements of such an agent, but it seems clear that a 'panaceamycin' may never be achieved, because the bacteria mutate to resistant strains faster than chemists can modify chemical structure. This has been adequately demonstrated by the old and new β-lactam agents (penicillins and cephalosporins) and, if only for this reason, there is now a need to return to the more logical, more effective and more flexible approach offered by relatively narrow-spectrum antibiotics.

The disadvantages of unthinking broad-spectrum therapy are legion. Prominent among them are the induction of resistant bacteria, an undesirable and even dangerous suppression of normal bacterial floras, leading to iatrogenic complications such as antibiotic-associated diarrhoea and pseudomembranous colitis, and predisposition of the patient to suprainfection, i.e., secondary colonization with opportunist organisms such as *Candida*, *Pseudomonas* and *Serratia* (Selwyn, 1980). Perhaps even more important is the fact that routine broad-spectrum prescribing distorts and discourages clinical judgement. There is almost no infective clinical syndrome or group of organisms for which relatively narrow-spectrum drugs are not a better choice (Reese and Betts, 1983) and for this reason alone single agent broad-spectrum prescription is rightly regarded as the therapy of the diagnostically destitute.

The preferential use of narrow-spectrum antimicrobials is not a new clinical concept; indeed, it has always been the guiding principle of the most successful therapeuticians (British Medical Journal, 1975; Smith, 1979; Tyrrell et al., 1979; Sanderson, 1983). The successful management of infected patients requires informed clinical judgement, which first defines the likely nature of the suspected organism(s) and the need for antimicrobial chemotherapy, and then selects an antibiotic(s) known to be effective, each with the narrowest spectrum of antimicrobial activity (Finland, 1973). In keeping with these guide-lines is the insistence by Lant (1978) that 'routine prescription of broad-spectrum antibiotics must be strongly resisted; all this does is to predispose to infection with resistant bacteria and thereby encourage the spread of problem organisms—often gram negative—within the hospital'.

It is instructive to refer to the recommendations of Reese and Betts (1983) and Sanderson (1983) concerning the antibiotics of choice for common pathogens. These recommendations highlight the major uses of relatively narrow-spectrum agents, and the infrequency of broad-spectrum prescription. Quite properly, Reese and Betts do not regard any cephalosporin as a first choice antimicrobial for any group of organisms, and they comment that 'cephalosporins are seldom the drug of choice for known specific infections . . . They are often over-utilized'. In fact, the most useful antimicrobials remain 'specialist' drugs, with good activity against one or two groups of micro-organisms, but usually with inferior activity against others.

Nowhere has the importance and value of narrow-spectrum agents been more compellingly demonstrated than in the setting of anaerobic bacterial infections. Until the 'anaerobe renaissance' of the late 1960s and early 1970s the frequent occurrence of postsurgical sepsis complicating operations on the gastrointestinal and female genital tracts was thought to be due to facultative bacteria (*E. coli*, staphylococci, enterococci etc.). Prophylactic and therapeutic measures adopted in these surgical patients aimed at the elimination of aerobic pathogens but, despite the use of both broad-spectrum and combination antimicrobial therapy, this met with little success. Recognition of the dominant role of anaerobes in these settings, and the consequent use of relatively specific anti-anaerobic agents, such as metronidazole and clindamycin, finally resolved this vexing clinical problem, and revolutionized the surgical approach to colorectal and gynaecological operations. Single agent, broad-spectrum theory, so ineffective in practice, soon gave way to the pragmatic practicalities of successful patient management.

Combination therapy

The occurrence of intra-abdominal infection is one of those common occasions when urgent empirical treatment is required for severely ill patients, and because the nature of the pathogen(s) is obscure, effective broad-spectrum therapy (as opposed to broad-spectrum antibiotics) is mandatory.

Broad-spectrum therapy is initiated by the use of drug combinations, virtually total effective cover being provided by a combination: metronidazole plus gentamicin plus amoxycillin. A macrolide is an alternative to amoxycillin for the penicillin-hypersensitive patient. When microbiological studies have revealed the identity of the offending pathogen(s), the therapeutic regimen is refined and adjusted to narrow its spectrum and increase its specificity. For example, if laboratory studies on such a patient implicate *Str. milleri* together with *B. fragilis*, gentamicin is withdrawn from the treatment schedule; if, on the other hand, a coliform bacillus only is isolated, gentamicin is the only drug retained in the schedule.

There are many clinical situations in which clear evidence of an underlying focus of infection makes possible a rational choice of antimicrobial agents to cover the organisms most likely to be involved. For example, abdominal sepsis derived from the urinary tract rarely involves anaerobic bacteria, so that metronidazole is not generally indicated; and since gram-negative organisms such as *E. coli* are the most common pathogens in this situation, empirical therapy may be restricted initially to gentamicin. In the same way, sepsis associated with the biliary tract, and following gastro-oesophageal surgery for non-malignant disease, more often involves *E. coli* and entercocci (including *Str. milleri*) than anaerobes such as *B. fragilis*. In contrast, abdominal sepsis derived from the colon and the female genital tract commonly involves anaerobes, enterobacteria and enterococci, thus dictating a requirement for initial therapy against all three groups of organisms (metronidazole, gentamicin and amoxycillin).

The great advantage of combination antimicrobial therapy is that it utilizes the best choice agent for each group of organisms in a fully flexible regimen. Thus, the spectrum of therapy may be expanded, narrowed or reinforced as indicated by the clinical response of the patient and the microbiological findings.

Prophylaxis

Postsurgical endogenous infections

That patients having gastrointestinal and major gynaecological surgery might become infected with components of their own bacterial flora is not a new concept. Indeed, much of the preoperative ritual enacted with these patients, and their peroperative management, is concerned with the prevention of just such an event.

Patients with advanced appendicitis in which the appendix may be gangrenous or perforated are already severely compromised, and administration of antimicrobial agents is properly regarded as *therapeutic* rather than *prophylactic*. The same is also true for various other intra-abdominal infections, for inflammatory bowel disease and for infected gastrointestinal and gynaecological neoplasms. Surgical intervention in these patients, however, creates surgical wounds (abdominal, perineal, anastomotic) for which the administration of antibiotics represents true prophylaxis. In all such settings, it is therefore probably best to regard prophylaxis as the prevention of postsurgical morbidity due to infection, rather than to allow the semantics of '*prophylaxis*' and '*treatment*' to dictate rigid antimicrobial schedules.

Prevention of surgical sepsis

The prevention of postoperative anaerobic infections in gastrointestinal surgery may be considered under four headings:

operative technique;
preoperative mechanical preparation;
intraluminal preoperative antimicrobial preparation;
systemic antimicrobial prophylaxis.

Operative technique

A good operative technique is clearly important in reducing the incidence of endogenous surgical infection. Manipulative skill not only ensures that contamination due to seeding of bacteria from the gut lumen is minimal, but it also takes account of operative factors known to favour infection, such as prolonged operation time, excessive trauma, poor haemostasis and tissue perfusion, and overtight or excessive sutures (Table 13.7, page 206). The division of opinion about the value and dangers of peritoneal drains was referred to earlier. There is more general agreement about the prophylactic value of drainage of the parietes by closed suction drainage, or management by delayed primary suture, especially in the obese and in children (Clark, 1976; Stuart, 1976, 1978; Brown *et al.*, 1977; Everson *et al.*, 1977).

Preoperative mechanical preparation

Despite the most expert surgical technique it is often difficult to prevent spillage of bowel contents, and it is partly for this reason that attempts are usually made to reduce the load of faeces preoperatively by various means in patients having elective colorectal surgery. Mechanical preparation of the colon may follow a standard procedure and a low residue diet, enemas and purgatives, or even the technique of whole bowel irrigation may be employed (Hewitt *et al.*, 1973). Although excellent mechanical cleansing of the gut may be achieved, these procedures have little effect on the bacterial counts (Nichols *et al.*, 1972; Arabi *et al.*, 1978) or on the incidence of postsurgical sepsis. It is probable that any beneficial effect is attributable more to facilitation of operative technique and to the mechanical protection from leakage or dehiscence afforded by an empty gut to intestinal anastomoses (Alexander-Williams, 1980), than to any reduction of the bacterial load.

Intraluminal preoperative antimicrobial preparation

Attempts to reduce the colonic flora by the use of intraluminal antimicrobial agents were begun some 40 years ago by Garlock and Seley (1939), since when a great variety of oral prophylactic antibiotic schedules has been used. These have usually employed non-absorbable drugs, such as aminoglycosides and phthalylsulphathiazole, but have had only marginal success in reducing the incidence of postsurgical sepsis. The use of non-absorbable antimicrobial agents in the gastrointestinal tract may be looked upon as *topical* therapy, in which the drug action is largely local against the luminal microflora. Such conventional prophylactic regimens attempted, if not to 'sterilize' the gut, at least to numerically reduce the total bacterial load, and their likely success was often judged in these terms (Arabi *et al.*, 1978; Brass *et al.*, 1978). Such is the case with the regimen first proposed by Nichols *et al.* (1972; 1973), which utilizes oral neomycin and erythromycin, and is widely employed in the United States as the 'gold standard' for preoperative prophylaxis in colorectal surgery (Condon and Nichols, 1975; Clarke *et al.*, 1977). Although this combination of oral aminoglycoside and macrolide has a profound effect on the gut flora (Bartlett *et al.*, 1978), it seems to provide little protection against the risks of postsurgical sepsis (Weaver *et al.*, 1986).

Systemic antimicrobial prophylaxis

Although a mechanical barrier to the transfer of bacteria from the bowel lumen or the vagina to the surrounding tissues during surgery is not easily erected, it is possible to effect a 'chemical' barrier by the use of systemic antimicrobial prophylaxis. Here, the aim is to provide active levels of an appropriate antibacterial drug in the tissues at the time they are exposed to the challenge of endogenous contamination. Fortunately, well directed prophylaxis of this sort does not require the elimination, or even the reduction, of endogenous bacterial floras, which has often proved to be ineffective and, moreover, may place the patient at risk from antibiotic-induced suprainfection.

The concept of systemic prophylaxis in general surgical patients is not new, but its use has frequently been unsuccessful, and in retrospect it is clear that failures were commonly due to the use of inappropriate antibiotics. The first successful clinical trial in colorectal surgery was conducted by Bernard and Cole (1964), who used prophylactic systemic penicillin G, combined with methicillin and chloramphenicol. With the benefit of hindsight we now know that chloramphenicol was the main contributory factor to the success of their trial because of its (then unappreciated) activity against non-sporing anaerobes.

Excellent results attended the use of systemic lincomycin and clindamycin in the prevention of postoperative anaerobic infection among patients having appendicectomy and colorectal surgery; it is regrettable that use of clindamycin had to be abandoned in view of its association with antibiotic-induced pseudomembranous colitis. Fortunately, as was shown by Smith *et al.* (1980), metronidazole is at least as effective as clindamycin in clinical practice.

In the United Kingdom, most surgeons now employ metronidazole for prophylaxis in appendicectomy and colorectal operations, combined with other agents relevant to the prevention of polymicrobial sepsis (Campbell, 1980; McDonald and Karran, 1982; Wilson *et al.*, 1982; Weaver *et al.*, 1986). As in the case of treatment, drugs of first choice for combination with metronidazole are gentamicin for its activity against enterobacteria, and amoxycillin for the prevention of streptococcal infection.

Timing of systemic prophylaxis

It is fashionable to advise the use of relatively short term perioperative antimicrobial prophylaxis, often restricted in duration to 24 h or even less, irrespective of the nature of the operative procedure, events and findings (Stokes *et al.*, 1974; British Medical Journal, 1980a,b, Higgins *et al.*, 1980; Sandusky, 1982). This approach recognizes the critical importance of prophylaxis covering the operative challenge period, but often regards any justifiable extension of antimicrobial administration as 'therapy'. It possibly derives from earlier erroneous assertions that 'routine systemic antibiotic prophylaxis in general surgery is now widely regarded as malpractice' because of the 'development in the patients' commensal flora of bacteria resistant to whatever drugs are used'. There is, of course, nothing inherently good about short term prophylaxis, especially if it fails; conversely, there is nothing inherently bad about three to five day prophylaxis, especially if it is successful.

Indeed, Hares et al. (1982) and Keighley (1985) recognized the importance of extended prophylaxis (five days) in patients having surgery for inflammatory bowel disease (Crohn's disease and ulcerative colitis) and Seligman (1978) recommended 48-h prophylaxis for hysterectomy patients. Some surgeons take the view that intestinal anastomoses may become infected with organisms derived from the luminal flora with which it is in intimate and continuous contact, and consider that extended antimicrobial protection of these gut wounds significantly reduces the incidence of anastomotic leaks and dehiscence. Clinical evidence for a septic aetiology of anastomotic dehiscence is sparse and conflicting. Herter and Slanetz (1967), Goligher et al. (1970) and Rosenberg et al. (1971) found that preoperative antimicrobial bowel preparation reduced the incidence of anastomotic leaks and dehiscence following colonic resection; Gallagher et al. (1982), on the other hand, found no evidence that dehiscence was preceded by perianastomotic infection. In view of the adverse influence that dehiscence has on the postoperative course, it seems reasonable to employ extended antimicrobial protection of these intestinal wounds until an infective aetiology of dehiscence is clearly excluded (Eykyn et al., 1979). Available evidence suggests that infection is as much a cause of the anastomotic leak as the leak is the cause of infection (Hunt, 1982). In complex reconstructive procedures, too, such as ileal pouch formation following proctocolectomy, prophylaxis extending to the fifth postoperative day was recommended by Parks et al. (1980).

'Pretreatment' with antimicrobial agents

A number of important surgical benefits of 'pretreatment' with anaerobicidal antimicrobials such as metronidazole have been identified in the management of colorectal cancer (Fiddian, 1978; Willis and Fiddian, 1983).

1. Non-sporing anaerobes are responsible for the infective complications of colorectal cancer, which should thus be treated preoperatively with an effective anaerobicide. The response may be so rapid, even in the presence of abscess formation and peritonitis, that surgery can be delayed until sufficient improvement has been achieved for a safer, easier surgical procedure;
2. Obstruction, when associated with infectious complications, may be relieved by treatment with metronidazole;
3. This new-found control of anaerobic sepsis permits radical cancer-curing resection and anastomosis, even in the presence of perforation and abscess formation, with expectation of smooth recovery and clean healing;
4. Following pretreatment, obstructing left colonic cancers with and without infective aspects can be treated by primary resection and anastomosis with only a caecostomy or without proximal decompression altogether, and even without drainage of any sort.

The philosophy of antimicrobial prophylaxis

The decision as to which patients require prophylaxis and which do not is a matter of clinical experience and judgment. Some surgeons prefer to withhold prophylaxis until the abdominal cavity has been explored, arguing, for example, that those appendicectomy patients least in need of prophylaxis are those in whom laparotomy discloses either a diagnostic error (about 10 per cent of cases) or only a mildly infected appendix. Others regard this delay as unacceptable but choose to modify existing protective therapy in the light of operative findings. Yet others prefer to provide antimicrobial prophylaxis to all patients in whom the gastrointestinal tract or vagina are surgically involved, on the basis that postoperative infective morbidity is by no means restricted to those who are infected at the time of operation.

There are three groups of these surgical patients for whom antimicrobial prophylaxis is indicated:
  those having procedures known to be associated with a high risk of postoperative infection;
  those having surgery known to be associated with serious postoperative infections;
  those debilitated or otherwise compromised patients whose resistance to infection is low.

Expected incidence and severity of postsurgical anaerobic sepsis

The incidence of significant postoperative anaerobic infection following all elective colorectal surgery is around 50 per cent, but that following formation of a colostomy only is virtually nil, while that after abdominoperineal or left colon resection is much greater than after right hemicolectomy. Similarly, significant anaerobic pelvic sepsis develops in about 20 per cent of women having hysterectomy, but is uncommon after vaginal repair. The incidence of anaerobic infection following elective small bowel and gastric surgery (except in cases of gastro-oesophageal malignancy) is low, and it rarely develops following biliary tract surgery. Vulnerable patients, in whom postsurgical anaerobic infection is likely to be of greater severity than in a young robust patient, include the elderly, those with general debility, those with malignant or arterial disease, diabetic or anaemic patients and those receiving immuno-suppressive therapy.

Comment

The indiscriminate use of antimicrobial prophylaxis is not condoned in the setting of surgical sepsis; its prudent and thoughtful use in individual patients, and in particular groups of patients who are estimated to be at significant risk from endogenous anaerobic infection, is to be commended. Within the space of a mere decade the remarkable efficacy of metronidazole in the prevention of postoperative sepsis in abdominal surgery has revised the orthodox view of antimicrobial prophylaxis from 'Malpractice if you use it' to 'Malpractice if you don't'.

References

Alexander-Williams J. (1980). Cleaning the gut for colonic surgery: a gentle compromise. *World Medicine* 15: 18–19.
Alston J M. (1955). Necrobacillosis in Great Britain. *British Medical Journal* 2: 1524–1528.
Altemeier W A. (1938a). The bacterial flora of acute perforated appendicitis with peritonitis. A bacteriologic study based upon one hundred cases. *Annals of Surgery* 107: 517–528.
(1938b). The cause of the putrid odour of perforated appendicitis with peritonitis. *Annals of Surgery* 107: 634–636.
(1940). The anaerobic streptococci in tubo-ovarian abscess. *American Journal of Obstetrics and Gynecology* 39: 1038–1042.
(1941). The pathogenicity of the bacteria of appendicitis peritonitis. *Annals of Surgery* 114: 158–159.
(1942). The pathogenicity of the bacteria of appendicitis peritonitis. An experimental study. *Surgery*, 11: 374–384.
(1979). Principles in the management of traumatic wounds and in infection control. *Bulletin of the New York Academy of Medicine* 55: 123–138.
Altemeier W A, Culbertson W R, Fullen W D, Shook C D. (1973). Intra-abdominal abscesses. *American Journal of Surgery* 125: 70–79.
Arabi Y, Dimock F, Burdon D W, Alexander-Williams J, Keighley M R B. (1978). Influence of bowel preparation and antimicrobials on colonic microflora. *British Journal of Surgery* 65: 555–559.
Aries V C, Crowther J S, Drasar B S, Hill M J, Ellis F R. (1971). The effect of a strict vegetarian diet on the faecal flora and faecal steroid concentrations. *Journal of Pathology* 103: 54–56.
Attebury H R, Sutter V L, Finegold S M. (1974). Normal human intestinal flora. In *Anaerobic bacteria: role in disease*, pp. 81–97. Edited by Balows A. *et al*. Thomas, Springfield.
Axelsson C K, Justesen T. (1977). Studies on the duodenal and fecal flora in gastrointestinal disorders during treatment with an elemental diet. *Gastroenterology* 72: 379–401.
Back E, Hermanson J, Wickman M. (1978). Metronidazole treatment of liver abscess due to *Bacteroides fragilis*. *Scandinavian Journal of Infectious Diseases* 10: 152–154.
Bartlett J G. (1978). Antimicrobial therapy in experimental intra-abdominal sepsis. *Journal of Antimicrobial Chemotherapy* 4: 392–394.
(1982). Intraabdominal abscess: pathogenesis and antibiotic selection. In *Topics in intraabdominal surgical infection*, pp. 49–63. Edited by Simmons R L. Appleton-Century-Crofts, Norwalk.
(1983). Recent developments in the management of anaerobic infections. *Reviews of Infectious Diseases* 5: 235–245.
Bartlett J G, Condon R E, Gorbach S L, Clarke J S, Nichols R E, Ochi S. (1978). Impact of oral antibiotic regimen on colonic flora, wound irrigation cultures and bacteriology of septic complications. *Annals of Surgery* 188: 249–254.
Berger L A, Osborne D R. (1982). Treatment of pyogenic liver abscess by percutaneous needle aspiration. *Lancet* i: 132–134.
Bernard H P, Cole W R. (1964). The prophylaxis of surgical infection: the effect of prophylactic antimicrobial drugs on the incidence of infection following potentially contaminated operations. *Surgery* 56: 151–157.
Bjornson H S. (1982). Bacterial synergy, virulence factors, and host defense mechanisms in the pathogenesis of intraabdominal infections. In *Topics in intraabdominal*

surgical infection, pp. 65–78. Edited by Simmons R L. Appleton-Century-Crofts, Norwalk.

Bocking J K, Holliday R L, Duff J H. (1981). Necrotizing anaerobic infections. *Canadian Journal of Surgery* **24**: 453–455.

Borriello P, Hudson M, Hill M. (1978). Investigation of the gastrointestinal bacterial flora. *Clinics in Gastroenterology* **7**: 329–349.

Bowcock S, Mackie I J, Ho D, Moulsdale M, Billings P, Machin S J. (1986). Effects of various doses of latamoxef (moxalactam) on haemostasis. *Journal of Hospital Infection* **8**: 193–199.

Brass C, Richards, G K, Ruedy J, Prentis J, Hinchley E J. (1978). The effect of metronidazole on the incidence of postoperative wound infection in elective colon surgery. *American Journal of Surgery* **135**: 91–96.

Brewer G E, Meleney F L. (1926a). Progressive gangrenous infection of the skin and subcutaneous tissues, following operation for acute perforated appendicitis: a study in symbiosis. *Annals of Surgery* **84**: 438–450.

(1926b). Progressive gangrenous infection of the skin and subcutaneous tissues, following operation for acute perforated appendicitis: a study in symbiosis. *Transactions of the American Surgical Association* **44**: 389–410.

British Medical Journal. (1973). Synergistic bacterial gangrene. **2**: 3.

(1975). Antibiotics at risk. **2**: 582.

(1979). Antibiotic lavage for peritonitis. **2**: 691–692.

(1980a). Sepsis after bowel surgery. **1**: 882–883.

(1980b). Prophylaxis of surgical wound sepsis. **1**: 1063–1064.

Brook I. (1986). Presence of beta-lactamase-producing bacteria and beta-lactamase activity in abscesses. *American Journal of Clinical Pathology* **86**: 97–101.

Brook I, Barrett C T, Brinkman C R, Martin W J, Finegold S M. (1979). Aerobic and anaerobic bacterial flora of the maternal cervix and newborn gastric fluid and conjunctiva: a prospective study. *Pediatrics* **63**: 451–455.

Brook I, Coolbaugh J C, Walker R I. (1984a). Pathogenicity of piliated and encapsulated *Bacteroides fragilis*. *European Journal of Clinical Microbiology* **3**: 207–209.

Brook I, Gillmore J D, Collbaugh J C, Walker R I. (1983a). Pathogenicity of encapsulated *Bacteroides melaninogenicus* group, *B. oralis* and *B. ruminicola* subsp. *brevis* in abscesses in mice. *Journal of Infection* **7**: 218–226.

Brook I, Hunter V, Walker R I. (1984b). Synergistic effect of *Bacteroides*, *Clostridium*, *Fusobacterium*, anaerobic cocci, and aerobic bacteria on mortality and induction of subcutaneous abscesses in mice. *Journal of Infectious Diseases* **149**: 924–928.

Brook I, Pazzaglia G, Collbaugh J C, Walker R I. (1983b). *In vitro* protection of group A beta-haemolytic streptococci from penicillin by beta-lactamase-producing *Bacteroides* species. *Journal of Antimicrobial Chemotherapy* **12**: 599–606.

Brown S E, Allen H H, Robins R N. (1977). The use of delayed primary wound closure in preventing wound infections. *American Journal of Obstetrics and Gynecology* **127**: 713–717.

Browne M K. (1979). Antibiotic lavage for peritonitis. *British Medical Journal* **2**: 1004–1005.

Browning G G, Buchan K A, Mackay C. (1974). The effect of vagotomy and drainage on the small bowel flora. *Gut* **15**: 139–142.

Bullen C L, Tearle P V, Willis A T. (1976). Bifidobacteria in the intestinal tract of infants: an in-vivo study. *Journal of Medical Microbiology* **9**: 325–333.

Campbell W B. (1980). Prophylaxis of infection after appendicectomy: a survey of current surgical practice. *British Medical Journal* **2**: 1597–1600.

Clark A W. (1976). Management of appendicitis. *British Medical Journal* **2**: 881–882.

Clarke J S, Condon R E, Bartlett J G, Gorbach S L, Nichols R L, Ochi S. (1977). Preoperative oral antibiotics reduce septic complications of colon operations: results of prospective, randomized, double-blind clinical study. *Annals of Surgery* **186**: 251–259.

Colebrook L. (1930). Infection by anaerobic streptococci in puerperal fever. *British Medical Journal* **2**: 134–137, 308.

Collee J G. (1974). *Clostridium perfringens* (*Cl. welchii*) in the human gastro-intestinal tract. In *The normal microbial flora of man*, pp. 205–219. Edited by Skinner F A, Carr J G. Academic Press, London.

(1980). Mechanisms of pathogenicity of anaerobic bacteria of clinical interest. *Infection* **8** Suppl 2: S113–S117.

Condon R E, Nichols R L. (1975). The present position of the neomycin-erythromycin bowel prep. *Surgical Clinics of North America* **55**: 1331–1334.

Daly J W, King C R, Monif G R G. (1978). Progressive necrotizing wound infections in postirradiated patients. *Obstetrics and Gynecology* **52**: Suppl 5s–8s.

Drasar B S. (1967). Cultivation of anaerobic intestinal bacteria. *Journal of Pathology and Bacteriology* **94**: 417–427.

Drasar B S, Goddard P, Heaton S, Peach S, West B. (1976). Clostridia isolated from faeces. *Journal of Medical Microbiology* **9**: 63–71.

Drasar B S, Shiner M, McLeod G M. (1969). Studies on the intestinal flora. I. The bacterial flora of the gastrointestinal tract in healthy and achlorhydric persons. *Gastroenterology* **56**: 71–79.

Drug and Therapeutics Bulletin. (1986). Broad-spectrum antibacterial regimens for the seriously ill. **24**: 33–36.

Duerden B I. (1980). The isolation and identification of *Bacteroides* spp. from the normal human faecal flora. *Journal of Medical Microbiology* **13**: 69–78.

Dunn D L, Simmons R L. (1984). The role of anaerobic bacteria in intraabdominal infections. *Reviews of Infectious Diseases* **6**: S139–S146.

Dye M, Macdonald A, Smith G. (1978). The bacterial flora of the biliary tract and liver in man. *British Journal of Surgery* **65**: 285–287.

Eggerth A H. Gagnon B D. (1933). The bacteroides of human faeces. *Journal of Bacteriology* **25**: 389–413.

England D M, Rosenblatt J E. (1977). Anaerobes in human biliary tracts. *Journal of Clinical Microbiology* **6**: 494–498.

Everson N W, Fossard D P, Nash J R, MacDonald R C. (1977). Wound infection following appendicectomy: the effect of extraperitoneal wound drainage and systemic antibiotic prophylaxis. *British Journal of Surgery* **64**: 236–238.

Eykyn S J, Jackson B T, Lockhart-Mummery H E, Phillips I. (1979). Prophylactic peroperative intravenous metronidazole in elective colorectal surgery. *Lancet* **2**: 761–764.

Fiddian R V. (1978). Prophylaxis in colonic surgery. *Journal of Antimicrobial Chemotherapy*, **4** Suppl C: 39–47.

Finegold S M. (1979). Anaerobes in biliary tract infection. *Archives of Internal Medicine* **139**: 1338–1339.

Finegold S M, George W L, Mulligan M E (1985). Anaerobic infections part II. *Disease-a-Month* **31** No 11: 4–18.

Finegold S M, Miller L G. (1968). Normal fecal flora of adult humans. *Bacteriological Proceedings* **93**.

Finland M. (1973). Superinfections in the antibiotic era. *Postgraduate Medicine* **54**: 175–183.

Fukunaga F H. (1973). Gallbladder bacteriology, histology, and gallstones. *Archives of Surgery* **106**: 169–171.

Galbut D L, Gerber D L, Belgraier A H. (1977). Spontaneous necrotizing fasciitis secondary to occult diverticulitis. *Journal of the American Medical Association* **238**: 2302.

Gallagher P, Cade D, Whale K, Schofield P F. (1982). Does infection cause anastomotic dehiscence? *Journal of the Royal College of Surgeons of Edinburgh* **27**: 90–92.

Garlock J H, Seley G P. (1939). The use of sulfonilamide in surgery of the colon and rectum. *Surgery* **5**: 787–790.

George W L, Kirby B D, Sutter V L, Wheeler L A, Mulligan M E, Finegold S M. (1982). Intravenous metronidazole for treatment of infections involving anaerobic bacteria. *Antimicrobial Agents and Chemotherapy* **21**: 441–449.

George W L, Rolfe R D, Mulligan M E, Finegold S M. (1979). Infectious diseases 1979—antimicrobial agent induced colitis: an update. *Journal of Infectious Diseases* **140**: 266–268.

Gerzof S G, Johnson W C, Robbins A H, Nabseth D C. (1985). Expanded criteria for percutaneous abscess drainage. *Archives of Surgery* **120**: 227–232.

Gillespie W A, Guy J. (1956). Bacteroides in intraabdominal sepsis. Their sensitivity to antibiotics. *Lancet*, **1**: 1039–42.

Giuliano A, Lewis F, Hadley K, Blaisdell F W. (1977). Bacteriology of nectorizing fasciitis. *American Journal of Surgery* **134**: 53–57.

Goldman M. (1966). Appendicitis: a historical survey. *Hospital Medicine* **October**: 42–46.

Goldring J, Scott A, McNaught W, Gillespie G. (1975). Prophylactic oral antimicrobial agents in elective colonic surgery. *Lancet* **2**: 997–1000.

Goldstein F, Wirts C W, Salen G. (1969). Diverticulosis of the small intestine. Clinical bacteriological and metabolic observations on a group of 7 patients. *American Journal of Digestive Diseases* **14**: 170–181.

Goligher J C, Graham N G, De Dombal F T. (1970). Anastomotic dehiscence after anterior resection of rectum and sigmoid. *British Journal of Surgery* **57**: 109–118.

Gorbach S L. (1982). The pre-eminent role of anaerobes in mixed infections. *Journal of Antimicrobial Chemotherapy* **10** Suppl A: 1–6.

Gorbach S L, Bartlett J G, Tally F P. (1981). *Biology of Anaerobes.* Upjohn, Kalamazoo.

Grainger R W, MacKenzie D A, McLachlin A D. (1967). Progressive bacterial synergistic gangrene: chronic undermining ulcer of Meleney. *Canadian Journal of Surgery* **10**: 439–444.

Gregory E M, Moore W E C, Holdeman L V. (1978). Superoxide dismutase in anaerobes: survey. *Applied and Environmental Microbiology* **35**: 988–991.

Gunn A A. (1956). Bacteroides bacteraemia. *Journal of the Royal College of Surgeons of Edinburgh* **2**: 41–50.

Hackman A S, Wilkins T D. (1976). Influence of penicillinase production of *Bacteroides melaninogenicus* and *Bacteroides oralis* on penicillin therapy of an experimental mixed anaerobic infection in mice. *Archives of Oral Biology* **21**: 385–389.

Hamilton-Miller J M T. (1981). Indirect pathogenicity. *Journal of Antimicrobial Chemotherapy* **7**: 307–308.

Hares M M, Bentley S, Allan R N, Burdon D W, Keighley M R B. (1982). Clinical trials of the efficacy and duration of antibacterial cover for elective resection in inflammatory bowel disease. *British Journal of Surgery* **69**: 215–217.

Harris J W, Brown J H. (1927). Description of a new organism that may be a factor in the causation of puerperal infection. *Bulletin of Johns Hopkins Hospital* **40**: 203–215.

Hayward N J. (1963). The bacterial flora of healthy human intestine. *Bulletin de l'Office International des Epizooties* **59**: 1401–1410.

Herbert D A, Fogel D A, Rothman J, Wilson S, Simmons F, Ruskin J. (1982). Pyogenic liver abscesses: successful non-surgical therapy. *Lancet* **1**: 134–136.

Herter F P, Slanetz C A. (1967). Influence of antibiotic preparation of the bowel on complications after colonic surgery. *American Journal of Surgery* **113**: 165–172.

Hewitt J, Rigby J, Reeve J, Cox A G. (1973). Whole-gut irrigation in preparation for large-bowel surgery. *Lancet* **2**: 337–340.

Higgins A F, Lewis A, Noone P, Hale M. (1980). Single and multiple dose cotrimoxazole and metronidazole in colorectal surgery. *British Journal of Surgery* **67**: 90–92.

Hite K E, Locke M, Hesseltine H C. (1949). Synergism in experimental infections with nonsporulating anaerobic bacteria. *Journal of Infectious Diseases* **84**: 1–9.

Hofstad T. (1984). Pathogenicity of anaerobic gram-negative rods: possible mechanisms. *Reviews of Infectious Diseases*, **6**: 189–199.

Holland J W, Hill E O, Altemeier W A. (1977). Numbers and types of anaerobic bacteria from clinical specimens since 1960. *Journal of Clinical Microbiology* **5**: 20–25.

Holst E, Helin I, Mardh P A. (1981). Recovery of *Clostridium difficile* from children. *Scandinavian Journal of Infectious Diseases* **13**: 41–45.

Hunt T K. (1982). Anastomotic failure. In *Topics in intraabdominal surgical infection*, pp. 89–106. Edited by Simmons R L. Appleton-Century-Crofts, Norwall.

Ingham H R, Selkon J B, Roxby C M. (1977). A bacteriological study of otogenic cerebral abscesses: the chemotherapeutic role of metronidazole. *British Medical Journal* **2**: 991–993.

Ingham H R, Sisson P R, Middleton R L, Narang H K, Codd A A, Selkon J B. (1981). Phagocytosis and killing of bacteria in aerobic and anaerobic conditions. *Journal of Medical Microbiology* **14**: 391–399.

Ingham H R, Sisson P R, Selkon J B. (1980). Current concepts of the pathogenetic mechanisms of non-sporing anaerobes: Chemotherapeutic implications. *Journal of Antimicrobial Chemotherapy* **6**: 173–179.

Ingham H R, Sisson P R, Tharagonnet D, Selkon J B, Codd A A. (1977). Inhibition of phagocytosis *in vitro* by obligate anaerobes. *Lancet* ii: 1252–1254.

Joiner K, Lowe B, Dzink J, Bartlett J G. (1982). Comparative efficacy of 10 antimicrobial agents in experimental infections with *Bacteroides fragilis*. *Journal of Infectious Diseases* **145**: 561–568.

Jones G R, Gemmell C G. (1982). Impairment of *Bacteroides* species of opsonisation and phagocytosis of enterobacteria. *Journal of Medical Microbiology* **15**: 351–361.

Kasper D L, Finegold S M. (eds). (1979). Virulence factors of anaerobic bacteria. *Reviews of Infectious Diseases* **1**: 246–400.

Kasper D L, Onderdonk A B, Bartlett J G. (1977). Quantitative determination of the antibody response to the capsular polysaccharide of *Bacteroides fragilis* in an animal model of intraabdominal abscess formation. *Journal of Infectious Diseases* **136**: 789–795.

Kasper D L, Onderdonk A B, Crabb J, Bartlett J G. (1979a). Protective efficacy of immunization with capsular antigen against experimental infection with *Bacteroides fragilis*. *Journal of Infectious Diseases* **140**: 724–731.

Kasper D L, Onderdonk A B, Polk B F, Bartlett J G. (1979b). Surface antigens as virulence factors in infection with *Bacteroides fragilis*. *Reviews of Infectious Diseases* **1**: 278–288.

Keighley M R B. (1977). Microorganisms in the bile. A preventable cause of sepsis after biliary tract surgery. *Annals of the Royal College of Surgeons of England* **59**: 328–334.

(1985). Antimicrobial prophylaxis in colorectal surgery. *Southeast Asian Journal of Surgery* **8** Suppl: 184–192.

Kelly M J. (1978). The quantitative and histological demonstration of pathogenic synergy between *Escherichia coli* and *Bacteroides fragilis* in guinea-pig wounds. *Journal of Medical Microbiology* **11**: 513–523.

(1980). Wound infection: a controlled clinical and experimental demonstration of synergy between aerobic (*Escherichia coli*) and anaerobic (*Bacteroides fragilis*) bacteria. *Annals of the Royal College of Surgeons of England* **62**: 52–59.

Kottas M, Smith L G. (1978). A possible new approach to the management of brain abscess. *Infection* **6**: 81–83.

Lancet. (1978). Appendix abscess: time for a trial? ii: 618.

(1979). Peritoneal drainage. ii: 941–942.

(1980). Bacterial synergy in mixed aerobic/anaerobic infections. i: 405–406.

(1981). Bacteria in the stomach. ii: 906–907.

(1982). Management of brain abscess. i: 742–743.

Lant A. (1978). Antimicrobial therapy: guidelines and pitfalls. *Medicine* **7**: 347–350.

Larson H E, Price A B, Honour P, Borriello S P. (1978). *Clostridium difficile* and the aetiology of pseudomembranous colitis. *Lancet* i: 1063–1066.

Lou M A, Mandal A K, Alexander J L, Thadepalli H. (1977). Bacteriology of the human biliary tract and the duodenum. *Archives of Surgery* **112**: 965–967.

MacLaren D M, Namavar F, Vught A M J J V-v, Vel W A C, Kaan J A. (1984). Pathogenic synergy: mixed intraabdominal infections. *Antonie van Leeuwenhoek* **50**: 775–787.

Maddocks J L. (1980). Indirect pathogenicity. *Journal of Antimicrobial Chemotherapy* **6**: 307–309.

Magarey C J, Chant A D B, Rickford C R K, Magarey J R. (1971). Peritoneal drainage and systemic antibiotics after appendicectomy. *Lancet* ii: 179–182.

Maher J A, Reynolds T B, Yellin A E. (1979). Successful medical treatment of pyogenic liver abscess. *Gastroenterology* **77**: 618–622.

Marne C, Pallares R, Martin R, Sitges-Serra A. (1986). Gangrenous cholecystitis and acute cholangitis associated with anaerobic bacteria in bile. *European Journal of Clinical Microbiology* **5**: 35–39.

Marrie T J, Furlong M, Faulkner R S, Sidorov J, Haldane E V, Kerr E A. (1982). *Clostridium difficile*: epidemiology and clinical features. *Canadian Journal of Surgery* **25**: 438–442.

Mata L J, Carillo C, Villatoro F. (1969). Fecal microflora in healthy persons of a preindustrial region. *Applied Microbiology* **17**: 596–602.

McDonald P J, Karran S J. (1982). Preoperative antimicrobial prescribing practise for elective colorectal surgery in Wessex, 1981 *Lancet*, ii: 753–754.

McDonald P J, Watts J McK, Finlay-Jones J J. (1984). The antimicrobial management of gut-derived sepsis complicating surgery and cancer therapy. In *Microbes and infections of the gut*. Edited by Goodwin C S. Blackwell, Melbourne.

McGowan K, Gorbach S L. (1981). Anaerobes in mixed infections. *Journal of Infectious Diseases* **144**: 181–186.

Meleney F L. (1931). Bacterial synergism in disease processes with a confirmation of the synergistic bacterial aetiology of a certain type of progressive gangrene of the abdominal wall. *Annals of Surgery* **94**: 961–981.

Mergenhagen S E, Thonard J C, Scherp H W. (1958). Studies in synergistic infections. I. Experimental infections with anaerobic streptococci. *Journal of Infectious Diseases* **103**: 33–44.

Moore W E C, Cato E P, Holdeman L V. (1969). Anaerobic bacteria of the gastrointestinal flora and their occurrence in clinical infections. *Journal of Infectious Diseases* **119**: 641–649.

Moore W E C, Holdeman L V. (1974). The human fecal flora: the normal flora of 20 Japanese-Hawaiians. *Applied Microbiology* **27**: 961–979.

(1975). Some newer concepts of the human intestinal flora. *American Journal of Medical Technology* **41**: 427–430.

Moore W E C, Ranney R R, Holdeman L V. (1982). Subgingival microflora in periodontal disease: cultural studies. In *Host-parasite interaction in periodontal disease*, pp. 13–26. Edited by Genco R J, Mergenhagen S E. American Society for Microbiology, Washington, DC.

Morris D L, Fabricius P J, Ambrose N S, Scammell B, Burdon D W, Keighley M R B. (1984). A high incidence of bleeding is observed in a trial to determine whether additon of metronidazole is needed with latamoxef for prophylaxis in colorectal surgery. *Journal of Hospital Infection* **5**: 398–408.

Moynihan B J A. (1920). The ritual of a surgical operation. *British Journal of Surgery* **8**: 27–35.

Muscroft T J, Deane S A, Youngs D, Burdon D W, Keighley M R B. (1981a). The microflora of the postoperative stomach. *British Journal of Surgery* **68**: 560–564.

Muscroft T J, Youngs D, Burdon D W, Keighley M R B. (1981b). Cimetidine and the potential risk of postoperative sepsis. *British Journal of Surgery* **68**: 557–559.

Namavar F et al. (1986). Effect of *Bacteroides fragilis* grown in the presence of clindamycin, metronidazole and fusidic acid on opsonization and killing of *Escherichia coli*. *European Journal of Clinical Microbiology* **5**: 324–329.

Namavar F, van Vught A M J J, Vel W A C, Bal M, MacLaren D M. (1984). Polymorphonuclear leucocyte chemotaxis by mixed anaerobic and aerobic bacteria. *Journal of Medical Microbiology* **18**: 167–172.

Nelson D P, Mata L J. (1970). Bacterial flora associated with the human gastrointestinal mucosa. *Gastroenterology* **58**: 56–61.

Nichols R L. (1980). Infections following gastrointestinal surgery: intra-abdominal abscess. *Surgical Clinics of North America* **60**: 197–212.

Nichols R L, Broido P, Condon R E, Gorbach S L, Nyhus L M. (1973). Effect of preoperative neomycin–erythromycin intestinal preparation on the incidence of infectious complications following colon surgery. *Annals of Surgery* **178**: 453–462.

Nichols R L, Condon R E, Gorbach S L, Nyhus L M. (1972). Efficacy of preoperative antimicrobial preparation of the bowel. *Annals of Surgery* **176**: 227–232.

Nichols R L, Miller B, Smith J W. (1975). Septic complications following gastric surgery: relationship to the endogenous gastric microflora. *Surgical Clinics of North America* **55**: 1367–1372.

Nichols R L, Smith J W, Fossendal E N, Condon R E. (1979). Efficacy of parenteral antibiotics in the treatment of experimentally induced intraabdominal sepsis. *Reviews of Infectious Diseases* **1**: 302–309.

Nichols R L et al. (1984). Risk of infection after penetrating abdominal trauma. *New England Journal of Medicine* **311**: 1065–1070.

Nielsen M L, Justesen T. (1976). Anaerobic and aerobic bacteriological studies in biliary tract disease. *Scandinavian Journal of Gastroenterology* **11**: 437–446.

Normann E, Korvald E, Lotveit T. (1975). Perforated appendicitis—lavage or drainage? *Annales Chirurgiae et Gynaecologiae Fenniae* **64**: 195–197.

Nulsen M F, Finlay-Jones J J, Skinnet J M, McDonald P J. (1983). Intra-abdominal abscess formation in mice: quantitative studies on bacteria and abscess-potentiating agents. *British Journal of Experimental Pathology* **64**: 345–353.

O'Keefe J P, Tally F P, Barza M, Gorbach S L. (1978). Inactivation of penicillin G during experimental infection with *Bacteroides fragilis*. *Journal of Infectious Diseases* **137**: 437–442.

Okuda K, Takazoe I. (1973). Antiphagocytic effects of the capsular structure of a pathogenic strain of *Bacteroides melaninogenicus*. *Bulletin of the Tokyo Dental College* **14**: 99–104.

Onderdonk A B, Kasper D L, Mansheim B J, Louie T J, Gorbach S L, Bartlett J G. (1979). Experimental animal models for anaerobic infections. *Reviews of Infectious Diseases* **1**: 291–301.

Onderdonk A B, Kasper D L, Cisneros R L, Bartlett J G. (1977). The capsular polysaccharide of *Bacteroides fragilis* as a virulence factor: Comparison of the pathogenic potential of encapsulated and unencapsulated strains. *Journal of Infectious Diseases* **136**: 82–89.

Onderdonk A B, Markham R B, Zaleznik D F, Cisneros R L, Kasper D L. (1982). Evidence for T cell-dependent immunity to *Bacteroides fragilis* in an intraabdominal abscess model. *Journal of Clinical Investigation* **69**: 9–16.

Onderdonk A B, Moon N E, Kasper D L, Bartlett J G. (1978). Adherence of *Bacteroides fragilis in vivo*. *Infection and Immunity* **19**: 1083–1087.

Onderdonk A B, Weinstein W M, Sullivan N M, Bartlett J G, Gorbach S L. (1974). Experimental intra-abdominal abscess in rats: quantitative bacteriology of infected animals. *Infection and Immunity* **10**: 1256–1259.

Parks A G, Nicholls R J, Belliveau P. (1980). Proctocolectomy with ileal reservoir and anal anastomosis. *British Journal of Surgery* **67**: 533–538.

Pashby N, Mee W M. (1978). Metronidazole prophylaxis and primary wound closure in appendicectomy. *Journal of Antimicrobial Chemotherapy* **4** Suppl C: 25–28.

Peach S, Fernandex F, Johnson K, Drasar B S. (1974). The non-sporing anaerobic bacteria in human faeces. *Journal of Medical Microbiology* **7**: 213–221.

Peel N. (1980).Bacteria and bile. *World Medicine* **15**: 38.

Percival R, Hargreaves A W. (1982). Necrotizing fasciitis: an alternative approach. *Postgraduate Medical Journal* **58**: 756–759.

Perera M R, Kirk A, Noone P. (1980). Presentation, diagnosis and management of liver abscess. *Lancet* **ii**: 629–632.

Porschen R K, Sonntag S. (1974). Extracellular deoxyribonuclease production by anaerobic bacteria. *Applied Microbiology* **27**: 1031–1033.

Pressey A, Watt B, Sharp D, Brown F V, Lambert H P. (1983). Effects of erythromycin and gentamicin in experimental peritonitis. *Journal of Antimicrobial Chemotherapy* **11**: 339–347.

Pruyn S C. (1978). Acute necrotizing fasciitis of the endopelvic fascia. *Obstetrics and Gynecology* **52** Suppl: 2s–4s.

Ralph E D. (1984). Successful antimicrobial therapy of hepatic, intra-abdominal and intrapelvic abscesses. *Canadian Medical Association Journal* **131**: 605–607.

Ralph E D, Kirby W M. (1975). Unique bactericidal action of metronidazole against *Bacteroides fragilis* and *Clostridium perfringens*. *Antimicrobial Agents and Chemotherapy* **8**: 409–414.

Reese R E, Betts R F. (1983). Antibiotic use. In *A practical approach to infectious diseases*, pp. 51–162. Edited by Reese R E, Douglas R G. Little, Brown and Co., Boston.

Roberts D S. (1967). The pathogenic synergy of *Fusiformis necrophorus* and *Corynebacterium pyogenes*. I. Influence of the leucocidal exotoxin of *F. necrophorus*. *British Journal of Experimental Pathology* **48**: 665–673.

Rosenberg I L, Graham N G, De Dombal F T, Goligher J C. (1971). Preparation of the intestine in patients undergoing major bowel surgery mainly for neoplasms of the colon and rectum. *British Journal of Surgery* **58**: 266–269.

Rotimi V O, Duerden B I. (1981). *Bacteroides* species in the normal neonatal faecal flora. *Journal of Hygiene* **87**: 299–304.

Rotstein O D, Pruett T L, Simmons R L. (1985). Mechanisms of microbial synergy in polymicrobial surgical infections. *Reviews of Infectious Diseases* **7**: 151–170.

Sanderson P J. (1983). *Antibiotics for surgical infections*. Research Studies Press, Chichester.

Sandusky W R. (1982). Prophylactic antibiotics in surgery. *The Guthrie Bulletin* **51**: 143–150.

Sapala J A, Ponka J L, Neblett T R. (1975). The bacteriology of the biliary tract. A preliminary report. *Henry Ford Hospital Medical Journal* **23**: 81–86.

Schimada K, Inamatsu T, Yamashiro M. (1977). Anaerobic bacteria in biliary disease in elderly patients. *Journal of Infectious Diseases* **135**: 850–854.

Schottmuller H. (1911). Ueber bakteriologische Untersuchungen und ihre Methoden bei Febris puerperalis. *Munchener Medizinische Wochenschrift, Munich* **1**: 787–789.

Seligman S A. (1978). Metronidazole in obstetrics and gynaecology. *Journal of Antimicrobial Chemotherapy* **4** Suppl C: 51–54.

Selwyn S. (1980). *The beta-lactamase antibiotics: penicillins and cephalosporins in perspective*. Hodder and Stoughton, London.

Simmons R L. (ed). (1982). *Topics in intraabdominal surgical infection*. Appleton-Century-Crofts, Norwalk.

Simon G L, Klempner M S, Kasper D L, Gorbach S L. (1982). Alterations in opsonophagocytic killing by neutrophils of *Bacteroides fragilis* associated with animal and laboratory passage: effect of capsular polysaccharide. *Journal of Infectious Diseases* **145**: 72–77.

Smith C R, Lipsky J J. (1983). Hypothrombinaemia and platelet dysfunction caused by cephalosporin and oxalactam antibiotics. *Journal of Antimicrobial Chemotherapy* **11**: 496–497.

Smith J A, Skidmore A G, Forward A D, Clarke A M, Sutherland E. (1980). Prospective, randomized, double-blind comparison of metronidazole and tobramycin with clindamycin and tobramycin in the treatment of intra-abdominal sepsis. *Annals of Surgery* **192**: 213–220.

Smith L D S, Holdeman L V. (1968). *The pathogenic anaerobic bacteria*. Thomas, Springfield.

Smith S R G, Gilmore O J A. (1985). Surgical drainage. *British Journal of Hospital Medicine* **33**: 308–315.

Smith T. (1979). Narrow range or broad spectrum? *Hospital Doctor February 21*: 10.

Springall R G, Gilmore O J A. (1985). Intra-abdominal sepsis. *Surgery*, **Issue 26** Suppl: 1–6.

Stark P L, Lee A. (1982). The microbial ecology of the large bowel of breast-fed and formula-fed infants during the first year of life. *Journal of Medical Microbiology* **15**: 189–203.

Stark P L, Lee A, Parsonage B D. (1982). Colonization of the large bowel by *Clostridium difficile* in healthy infants: quantitative study. *Infection and Immunity* **35**: 895–899.

Stewart D J. (1978). Antibiotic lavage in the prevention of intraperitoneal sepsis. *Annals of the Royal College of Surgeons of Edinburgh* **60**: 240–243.

Stokes E J, Waterworth P M, Franks V, Watson B, Clark C G. (1974). Short term routine antibiotic prophylaxis in surgery. *British Journal of Surgery* **61**: 739–742.

Stone H H, Hooper C A, Millikan W J. (1978). Abdominal drainage following appendicectomy and cholecystectomy. *Annals of Surgery* **187**: 606–612.

Stone H H, Martin J D. (1972). Synergistic necrotizing cellulitis. *Annals of Surgery* **175**: 702–710.

Stone M L. (1940). Studies on the anaerobic streptococci. I. Certain biochemical and immunological properties of anaerobic streptococci. *Journal of Bacteriology* **39**: 559–582.

Stuart M. (1976). The role of delayed primary closure in the prevention of wound sepsis after appendicectomy. *Medical Journal of Australia* **2**: 421–422.

—— (1978). The value of delayed primary wound closure in colonic and rectal surgery. *Medical Journal of Australia* **1**: 666–667.

Study Group. (1974). Metronidazole in the prevention and treatment of bacteroides infections in gynaecological patients. *Lancet* **ii**: 1540–1543.

—— (1975). An evaluation of metronidazole in the prophylaxis and treatment of anaerobic infections in surgical patients. *Journal of Antimicrobial Chemotherapy* **1**: 393–401.

—— (1976). Metronidazole in prevention and treatment of bacteroides infections after appendicectomy. *British Medical Journal* **1**: 318–321.

—— (1977). Metronidazole in prevention and treatment of bacteroides infections in elective colonic surgery. *British Medical Journal* **1**: 607–610.

Sykes P A, Boulter K, Schofield P F. (1976a). The alterations in small bowel microflora in intestinal obstruction. *Journal of Medical Microbiology* **9**: 13–22.

—— (1976b). The microflora of the obstructed bowel. *British Journal of Surgery* **63**: 721–725.

Tally F P. (1979). Determinants of virulence in anaerobic bacteria. In *Microbiology—1979*, pp. 219–223. Edited by Schlessinger D. American Society for Microbiology, Washington DC.

Tally F P, Goldin B R, Jacobus N V, Gorbach S L. (1977). Superoxide dismutase in anaerobic bacteria of clinical significance. *Infection and Immunity* **16**: 20–25.

Tally F P, Gorbach S L. (1979). Pathogenesis of bacteroides infections. *Journal of Infection* **1**: 5–12.

Thadepalli H, Gorbach S L, Broido P W, Nyhus N J. (1973). Abdominal trauma, anaerobes, and antibiotics. *Surgery, Gynecology and Obstetrics* **137**: 270–276.

Thadepalli H, Lou M A, Bach V T, Matsui T K, Mandl A K. (1979). Microflora of the human small intestine. *American Journal of Surgery* **138**: 845–850.

Thomas C T, Berk S L, Thomas E. (1982). Management of liver abscesses. *Lancet* **i**: 742–743.

Thomas W E G, Virjee J, Leaper D J. (1986). Intra-abdominal abscesses. *Surgery* **1**: 803–807.

Tompkins A M, Bradley A K, Oswald S, Drasar B S. (1981). Diet and the faecal microflora of infants, children and adults in rural Nigeria and urban UK. *Journal of Hygiene* **86**: 285–293.

Tyrrell D A, Phillips I, Goodwin C S, Blowers R. (1979). *Microbial disease: the use of the laboratory in diagnosis, therapy and control.* Edward Arnold, London.

Veillon M A. (1893). Sur un microcoque anaerobie trouve dans des suppurations fetides. *Comptes Rendus des Seances de la Societe de Biologie, Paris, 9 Ser* **5**: 807–809.

Veillon M A, Zuber A. (1898). Researches sur quelques microbes strictement anaerobies et leur role en pathologie. *Archives de Medicine Experimentale d'Anatomie Pathologique, Paris* **10**: 517–545.

Vel W A C, Namavar F, van Vught A M J J, Pubben A N B, MacLaren D M. (1985). Killing of *Escherichia coli* by human polymorphonuclear leucocytes in the presence of *Bacteroides fragilis*. *Journal of Clinical Pathology* **38**: 86–91.

Wade B H, Kasper D L, Mandell G L. (1983). Interactions of *Bacteroides fragilis* and phagocytes: studies with whole organisms, purified capsular polysaccharide and clindamycin-treated bacteria. *Journal of Antimicrobial Chemotherapy* **17**, Suppl C: 51–62.

Weaver M, Burdon D W, Youngs D J, Keighley M R B. (1986). Oral neomycin and erythromycin compared with single-dose systemic metronidazole and ceftriaxone prophylaxis in elective colorectal surgery. *American Journal of Surgery* **151**: 437–442.

Weinberg M, Prévot A R, Davesne J, Renard C. (1928). Flore microbienne des appendicites aigues. *Comptes Rendus des Seances de la Societe de Biologie, Paris* **98**: 749–752.

Weinstein W M, Onderdonk A B, Bartlett J G, Gorbach S L. (1974). Experimental intra-abdominal abscess in rats: development of an experimental model. *Infection and Immunity* **10**: 1250–1255.

Weinstein W M, Onderdonk A B, Bartlett J G, Louie T J, Gorbach S L. (1975). Antimicrobial therapy of experimental intraabdominal sepsis. *Journal of Infectious Diseases* **132**: 282–286.

Weiss J E, Rettger L F. (1937). The gram-negative *Bacteroides* of the intestine. *Journal of Bacteriology* **33**: 423–434.

Wells C L, Wilkins T D. (1980). Bactericidal effect of metronidazole in mixed-flora abscesses containing anaerobic and facultative bacteria. In *Current chemotherapy and infectious disease*, pp. 892–893. American Society of Microbiology, Washington DC.

Werner H. (1974). Differentiation and medical importance of saccharolytic intestinal anaerobes. *Arzneimittel-Forschung* **24**: 340–343.

Wilkins T D, Walker C B, Nitzan D, Salyers A A. (1977). Experimental infections with anaerobic bacteria in mice. *Journal of Infectious Diseases* **135** Suppl: S13–S17.

Willis A T, Bullen C L, Williams K, Fagg C G., Bourne A, Vignon M. (1973). Breast milk substitute: a bacteriological study. *British Medical Journal* **2**: 67–72.

Willis A T, Fiddian R V. (1983). Metronidazole in the prevention of anaerobic infection. *Surgery* **93**: 174–179.

Willis A T, Jones P H, Reilly S. (1981). *Management of anaerobic infections, prevention and treatment.* Research Studies Press, Chichester.

Willis R G, Lawson W C, Hoare E M, Kingston R D, Sykes

P A. (1984). Are bile bacteria relevant to septic complications following biliary surgery? *British Journal of Surgery* **71**: 845–849.

Wilson N I L, Wright P A, McArdle C S. (1982). Survey of antibiotic prophylaxis in gastrointestinal surgery in Scotland. *British Medical Journal* **285**: 871–873.

Wise R, Donovan I A. (1981). Abdominal sepsis. *International Medicine* **1** 114–117.

14

Anaerobes in genitourinary infections

B I Duerden

Anaerobic species in genitourinary infections
    Gram-positive bacilli
    Gram-negative anaerobic bacilli
    Gram-positive cocci
Anaerobic infections in women
    Infections of the external genitalia and perineum
    Vaginal infections
    Infections related to pregnancy
    Pelvic infections in non-pregnant women
    Postoperative infections in gynaecological surgery
Anaerobic infections in men
Conclusion
References

The genitourinary tract is a complex system of mucosa-lined tubes and hollow organs. The male tract comprises the urethra, bladder, ureters, kidneys, the associated glands (of which the prostate is the most significant) and the vas deferens leading from the testes; all have a common pathway to the body surface via the urethra. In women, the anatomy and physiology are more complex and this affects both the pattern of normal bacterial colonization and the susceptibility to different types of infection. The urinary system (kidneys, ureters and bladder) reaches the surface of perineum via a short urethra opening, with the vagina at the introitus. The genital tract comprises the vagina, uterus, fallopian tubes, with their direct connection to the peritoneum and the pelvic sac, and the associated secretory glands.

Infections of the genitourinary tract comprise a wide range of diseases, from the classic venereal or sexually transmitted diseases caused by specific virulent and well recognized pathogens such as *Treponema pallidum* (syphilis) and *Neisseria gonorrhoeae* (gonorrhoea), to a variety of less well defined but none the less common conditions that tend to be labelled 'non-specific', as a general cloak for the lack of clear understanding about their aetiology and pathogenesis. Anaerobes make a significant contribution to this spectrum of disease.

In strict terms, one of the major sexually transmitted diseases, syphilis, belongs within the category of anaerobic infections; *T. pallidum* is an anaerobic spirochaete, although it cannot at present be cultivated in the laboratory. However, the clinical and laboratory aspects of syphilis are covered in depth in many monographs or textbooks of sexually transmitted diseases, and the disease is so specific and not at all related in any clinical, pathological or epidemiological way to other anaerobic genitourinary infections that it will not be considered further in this chapter.

The anaerobic infections which will be considered in this Chapter generally fall within the 'non-specific' group of genital infections, but that is not to say that they do not include a varied group of distinct conditions, although many share some common aetiological or epidemiological features. Like many of the anaerobic infections at other sites in the body, but unlike the specific sexually transmitted diseases, the anaerobic genitourinary infections are usually related to some physiological or anatomical factors affecting the functioning of the system and to the presence of a complex, but predominantly anaerobic, normal bacterial flora in the lower genitourinary tract itself (vagina and urethra) and in close proximity to the gastrointestinal flora contaminating the perineum.

The contribution of anaerobes to the normal flora of the genitourinary tract is described in Chapter 11. In women, the vaginal flora is predominantly anaerobic. Lactobacilli and other anaerobic gram-positive rods dominate the flora but anaerobic cocci are also a major component and gram-negative anaerobic bacilli are generally present, although in smaller numbers. In relation to the pathogenesis of genitourinary infections, the presence of the gram-negative anaerobic rods and the gram-positive cocci is probably the most significant feature. Reports of the incidence of *Bacteroides* spp. in the normal vaginal flora have varied from 4 per cent to 70 per cent; this probably represents variation in sampling technique, anaerobic methodology and the quantitative sensitivity of the culture system (Duerden, 1980a; Wilks *et al*., 1982; Bartlett and Polk, 1984). The latter may be the most important reason for the variation because the presence of 10^6 cfu of *Bacteroides* spp. per gram of secretions, the normal quantitative level of colonization, may not be detected by some of the dilution–culture techniques employed (Bartlett and Polk, 1984; Masfari *et al*., 1986; Wilks and Tabaqchali, 1987a) (Table 14.1). The *Bacteroides* spp. found in the normal vagina mostly belong to the melaninogenicus–oralis group; the non-pigmented species *Bacteroides bivius* and *B. disiens* are the most common species, with *B. melaninogenicus* also isolated frequently (Duerden, 1980a; Hammann, 1982; Wilks and Tabaqchali, 1987b). The fragilis group of faecal bacteroides are not part of the indigenous vaginal flora and there is doubt about the presence of asaccharolytic bacteroides. The pigmented *B. asaccharolyticus* is commonly found in vaginal swabs from patients with some abnormality and sometimes from ostensibly normal subjects (Duerden, 1980a,b; Masfari *et al*., 1986), but their association is more with pathological than with normal states.

In men, anaerobic gram-positive cocci are a major part of the normal flora of the anterior urethra (Fontaine *et al*., 1982). The role of *Bacteroides* spp. is more debatable. Generally, *Bacteroides* of the fragilis and melaninogenicus–oralis groups are not isolated from the urethra of normal males (Masfari *et al*., 1983). However, some confusion surrounds the isolation of the asaccharolytic non-pigmented species *B. ureolyticus*. This species has been associated with non-gonococcal urethritis and with genital ulcers but some investigators have also found it in urethral samples from normal control subjects (see below).

Table 14.1 Quantitative vaginal bacteriology in various clinical conditions*

Diagnosis (number of patients)	Log_{10} mean viable count (cfu/g) of		
	Total anaerobes	Bacteroides spp.	Anaerobic cocci
Normal controls (22)	6.9	6.5	6.7
Candidiasis (16)	7.4	7.2	6.9
Trichomoniasis (2)	8.5	8.4	7.3
Bacterial vaginosis (22)	9.1	9.1	7.6
Chlamydial infection (12)	9.4	9.3	8.4
Gonorrhoea (10)	9.7	9.6	8.3
Gonorrhoea + chlamydial infection (6)	9.8	9.8	8.6

*Adapted from Masfari *et al*. (1986)

Anaerobic species in genitourinary infections

Apart from the anaerobic spirochaete *T. pallidum* in syphilis, representatives of the three main groups of pathogenic anaerobic bacteria are associated with a variety of genitourinary infections—clostridia and other gram-positive bacilli, gram-negative bacilli and gram-positive cocci.

Gram-positive bacilli

Clostridia

Primary clostridial infections of the genitourinary tract are fortunately rare, but *Clostridium perfringens* may cause infection of the uterus after childbirth or abortion, usually when these have been difficult procedures and there has been considerable tissue damage and manipulation, and therefore increased opportunity for implantation of this essentially faecal organism into the uterus (Willis, 1977). *C. perfringens* is not a major part of the normal vaginal flora although it has been isolated from the vagina and cervix of 4 to 10 per cent of healthy women (Gorbach *et al.*, 1973; Ohm and Galask, 1975) and from 19 to 29 per cent of postabortal women (Holtz and Mauch, 1962). It is commonly present on the perineal skin and may readily gain access to the traumatized area during manipulation and instrumentation. The result of infection, gas gangrene of the uterus (see p. 233 and Chapter 18), is a very serious condition; without prompt treatment there is a high mortality. Successful treatment of established infection may entail emergency hysterectomy to remove the focus of infection in the necrotic tissue. Postpartum uterine gas gangrene is associated with prolonged rupture of the membranes and a difficult delivery; postabortal infection is commonly the result of an illegal, unsterile instrumental abortion.

The other clostridial species that has received some attention in relation to the genitourinary tract is *C. difficile*. Before it became linked with pseudomembranous colitis and antibiotic-associated diarrhoea, Hafiz and colleagues had isolated it from 18 per cent of normal women attending a family planning clinic and 72 per cent of women attending a clinic for sexually transmitted diseases with various conditions (Hafiz *et al.*, 1975; Bramley *et al.*, 1981). This has not been confirmed in other studies, but O'Farrell *et al.* (1984) obtained an isolation rate of 11 per cent using an enrichment medium. There is insufficient evidence to implicate *C. difficile* in any specific pathological changes in the genitourinary tract.

Gardnerella vaginalis

This CO_2-dependent microaerophilic species, often described as a gram-variable bacillus, is included here because, although not a strict anaerobe, it appears to have an important place in bacterial (anaerobic) vaginosis, a condition in which, by their metabolic activity, a mixture of anaerobic bacteria produce an altered vaginal environment manifest as a profuse, offensive discharge. The organism was first described by Gardner and Dukes (1954, 1955) as *Haemophilus vaginalis*, but its precise contribution to bacterial vaginosis remains obscure. Clearly not a *Haemophilus* sp., it was first moved to the genus *Corynebacterium* (Dunkelberg *et al.*, 1970) and then, because it appeared to be unrelated to any established genera, it was accorded its own monospecific genus as *Gardnerella vaginalis* by Greenwood and Picket (1980). Structurally it is a gram-positive organism but it has an unusual cell wall structure which does not retain Gram's stain and generally appears gram-variable, or even frankly gram-negative (Sadhu *et al.*, 1989).

Gram-negative anaerobic bacilli

Several species belonging to all four main groups of gram-negative anaerobic bacilli are associated with infections of the genitourinary tract.

Fragilis group

These are intestinal (faecal) commensals and *B. fragilis*, in particular, is the cause of a range of gut-associated wound infections, abscesses and soft tissue infections (see Chapter 13). They are not part of the normal vaginal flora but they readily contaminate the perineum and introitus and are responsible for the anaerobic component of up to half the infections of the uterus and upper genital tract and pelvis in women.

Melaninogenicus–oralis group

Members of this group are the commensal bacteroides of the vagina. They are greatly increased in numbers in bacterial vaginosis, in which they probably contribute to the disturbance of vaginal metabolic activity represented by the clinical condition. *B. bivius*, *B. disiens* and *B. melaninogenicus* can also be found in infections of the upper tract and may cause anaerobic bacteraemia from a pelvic focus.

Asaccharolytic group

The pigmented and strongly proteolytic species *B. asaccharolyticus* is commonly isolated from vaginal swabs. It may be present in small numbers in specimens from normal subjects, but isolation is more common and the density of growth greater in patients with a vaginal discharge (Duerden, 1980a; Masfari *et al.*, 1986). It is also a common isolate from male and female patients with genital ulcers (Masfari *et al.*, 1985). It appears to have a particular association with abscesses and necrotic ulcers of the external genitalia and perineum, in which it appears to play a significant role in the extending tissue damage.

B. ureolyticus, another assacharolytic, proteolytic but non-pigmented gram-negative bacillus, which is uncommon in the normal vaginal flora, is also commonly isolated from men and women with genital ulcers, often with *B. asaccharolyticus* and anaerobic gram-positive cocci as part of a mixed infection (Masfari *et al.*, 1985). This combination appears to have particular pathogenic potential, causing spreading ulceration and gangrene of the external genitalia of the type described by Meleney (1931, 1933) and Fournier (see Randall, 1920). *B. ureolyticus* has also been linked to non-gonococcal nonchlamydial urethritis (Fontaine *et al.*, 1984), but this has not been confirmed and there is some indication that it can be isolated from urethral specimens from normal controls as often as from men with urethritis.

Fusobacteria

These are not common isolates from genital tract specimens, but *Fusobacterium necrophorum* is a recognized cause of postpartum and postabortal uterine sepsis with severe tissue damage, spread to local tissues and spread via a clinically severe bacteraemia or septicaemia to cause metastatic abscesses elsewhere in the body (Finegold, 1977). In the more common type of *F. necrophorum* infection, the normal mode of spread is by the airborne route, initially causing a throat infection, but with subsequent spread to deeper tissues and the bloodstream. Spread to women with a uterus that is susceptible to infection may represent airborne or contact spread similar to the spread of the group A streptococcal infection causing puerperal fever.

Mobiluncus spp.

Curved, motile gram-negative rods of this genus are implicated in bacterial vaginosis. They were first described in vaginal discharge by Curtis (1913) but received little attention until the 1980s, when various workers saw them by microscopy and isolated them in large numbers from many patients with bacterial vaginosis (Phillips and Taylor, 1982; Sprott *et al.*, 1982, 1983; Spiegel *et al.*, 1983). Two species have been described and named—*M. curtisii* (short) and *M. mulieris* (long); *M. curtisii* has been split into two subspecies—*curtisii* and *holmesii* (Spiegel and Roberts, 1984). There is debate about their taxonomic status because, although staining gramnegatively, they do not have the typical cell wall of gram-negative bacteria and share many characteristics with some gram-positive species. Their contribution to the pathogenesis of bacterial vaginosis is not clear, but they appear to be a significant component of the mixed anaerobic flora in some of these patients.

Gram-positive cocci

Anaerobic gram-positive cocci are part of the urethral and vaginal flora and some of them are associated with soft-tissue infections at sites closely related to their normal habitat. They are often isolated along with *Bacteroides* spp. from abscesses and from genital ulcers (Masfari *et al.*, 1985).

Anaerobic infections in women

Anaerobes are either the major component of, or at least contribute to, a wide range of infections of the female genitourinary system (Table 14.2). To give a clearer pattern to the clinical conditions, they may

Table 14.2 Anaerobic infections of the female genitourinary tract

Superficial infections
  Local abscesses, e.g., Bartholins, Skenes
  Genital ulcers
  Synergic gangrene
Vaginal infections
  Trichomoniasis
  Bacterial vaginosis
Pregnancy-related infections
  Amnionitis
  Postpartum (puerperal) uterine infection,
    i.e., endometritis, myometritis
  Gas gangrene of the uterus
  F. necrophorum infection
  Septic abortion
	Pelvic infections (non-pregnant)
	  Endometritis
	  IUCD-associated infections
	  Pyometra
	  Parametritis
	  Adnexal abscess Pelvic
	  Salpingitis inflammatory
	  Tubo-ovarian abscess disease (PID)
	  Pelvic abscess
	  Peritonitis
	Postgynaecological surgery
	  Wound infections
	  Abscesses

be divided into several groups on the basis of site of infection and predisposing factors, although there is some inevitable overlap between the groups. In this review, five categories will be considered:

superficial infections of the external genitalia and perineum;
vaginal infections;
infections of the uterus and associated structures as complications of pregnancy;
infections of the deep pelvic structures in non-pregnant women;
postoperative infections in gynaecological surgery.

There is no mention of urinary tract infections (UTI) in this categorization because there is no convincing evidence for the involvement of anaerobes in UTI except as part of a mixed flora in complex UTI, when there are chronic problems of large calculi, diverticula and tumours, or when obstruction leads to pyonephrosis and renal or perirenal abscesses. Despite repeated searches, anaerobes have not been established as a cause of UTI in any significant proportion of the 50 per cent of women with the symptoms of UTI who do not have the typical aerobic (mostly enterobacterial) infections. Although anaerobes abound at the sites which form the source of infecting bacteria in UTI (perineal and faecal flora) they do not survive well in urine and they do not cause ascending infection. One reason for the failure of anaerobes to grow in urine and cause infection may be that the Eh and the dissolved oxygen content in urine are too high for anaerobes to become established.

Infections of the external genitalia and perineum

These infections comprise a group of local abscesses of the labia and surrounding tissues and the range of superficial ulcerating lesions that come under the general heading of genital ulcers. The abscesses are mainly associated with the local secretory glands and are best characterized by abscesses of Bartholin's and Skene's glands. Although *N. gonorrhoeae* is one of the most common causes of such abscesses, many cases are unrelated to gonorrhoea or other classic sexually transmitted diseases. They are more common in sexually active women in whom minor local trauma, increased secretory activity of the glands and, possibly, changed vaginal flora, may contribute to the development of the abscesses, but generally the aetiology of these abscesses is most closely linked with that of sebaceous cysts and pilonidal abscesses. Blockage of the ducts of the glands, as a result of trauma, inspissated secretions or inadequate hygiene, leads to a build up of secretions and infection with a mixture of bacteria that is generally predominantly anaerobic (Pearson and Anderson, 1970b). The species most commonly isolated from such abscesses include gram-positive anaerobic cocci and *Bacteroides* spp., particularly the asaccharolytic group of bacteroides (Duerden, 1980b; Duerden *et al.*, 1982). The resultant abscesses are acutely painful and the most effective treatment is surgical drainage, with the abscess cavity opened widely to avoid recurrence. Antibacterial agents such as metronidazole are often unnecessary but they are useful if there is inflammation and cellulitis of the surrounding tissue.

Ulcers of the external genitalia and perineum (genital ulcers) are a common problem in women attending genitourinary medicine clinics. The specific ulcerogenic genitourinary pathogens, *T. pallidum* and *Haemophilus ducreyi*, are rare causes in developed countries, although they still represent common causes of genital ulceration in many parts of the world. For example, chancroid, caused by *H. ducreyi*, is responsible for 42–62 per cent of cases of genital ulcers in some parts of Africa (Duncan *et al.*, 1981; Nsanze *et al.*, 1981) but only for about 1 per cent (or less) in the USA and UK (Hammond *et al.*, 1980), where genital herpes (Herpes simplex virus type 2) is the primary cause of 40–80 per cent of ulcers (Kinghorn *et al.*, 1982a; Masfari *et al.*, 1985). Various aerobic pathogens such as *Staphylococcus aureus* and β-haemolytic streptococci have been isolated from a small proportion of patients, but the major part of the bacterial flora of genital ulcers is anaerobic (Masfari *et al.*, 1985). The anaerobes may not be the primary cause of the epithelial damage which leads to the ulcer—that is most commonly herpes virus infection, trauma or poor hygiene—but once there is damaged tissue it is readily colonized by anaerobic bacteria, which then contribute to the progressive, spreading tissue damage. A good example of the multiple aetiology of genital ulcers is provided by the development of many genital herpes lesions. These occur in both men and women and the same considerations apply in both. The initial herpetic lesion, whether it is a primary infection or is a recurrent, reactivated lesion, is a superficial, thin-walled blister which soon breaks down to expose a shallow, red and extremely painful ulcer; bacteria have made no contribution up to this stage. However, as the condition progresses, the ulcer enlarges, becomes deeper, develops an undermined edge and the base becomes covered in a necrotic slough. These latter effects are not due to the herpes virus itself, but to the secondary bacterial colonization, and the predominant group of bacteria in such necrotic ulcers are anaerobes. In a study of 27 women with genital ulcers, Masfari *et al.* (1985) isolated herpes virus in cell cultures from 18; however, the predominant bacterial flora was anaerobic and the pattern of bacterial colonization or infection was the same in herpetic and non-herpetic ulcers. The most common anaerobic species isolated were asaccharolytic bacteroides (*B. asaccharolyticus* and *B. ureolyticus*), members of the melaninogenicus–oralis group (mainly *B. bivius* and *B. melaninogenicus*, which were presumed to come from the vaginal flora) and anaerobic cocci. Apart from the obviously vaginal bacteria, the anaerobic flora of the genital ulcers was clearly similar to that of other superficial, necrotizing, ulcerative conditions in which anaerobes, although not initiators, are often considered to be important colonizers, e.g., decubitus and varicose ulcers (Chow *et al.*, 1977), diabetic gangrene (Fierer *et al.*, 1979), perianal abscesses and sebaceous cysts (Meislin *et al.*, 1977; Duerden, 1980b; Duerden *et al.*, 1982). In most of the conditions the anaerobes do not necessarily cause the initial lesions, but they do appear to be significant in the continued development of the lesion with spreading tissue damage.

Vaginal infections

Vaginal discharge resulting from various types of infection is the most common condition complained of by women attending genitourinary medicine clinics. Some specific infections are clearly recognized from among the several causes—gonorrhoea, chlamydial infection, candidiasis and trichomoniasis —but the most common diagnosis, accounting for up to one-third of all diagnoses made in women attending such clinics, is bacterial vaginosis, the agreed name for the condition previously known as non-specific vaginitis, or sometimes as anaerobic vaginosis. Of these five conditions, two may quite rightly be addressed in this review of anaerobic infections—*Trichomonas vaginalis* is an anaerobic protozoan, and the pathogenesis of bacterial vaginosis is closely linked to changes in the complex pattern of anaerobic bacteria in the vagina.

However, it should also be noted that anaerobic bacteria are also found in increased numbers in vaginal infections other than bacterial vaginosis. Masfari *et al.* (1986) found significant increases of 2–3 $\log_{10}$ units, from 10^6/g in normal women to 10^9/g in those with infections, of anaerobic bacteria. These were predominantly bacteroides of the vaginal types, and were seen in patients with gonorrhoea, chlamydial infection and trichomoniasis, as well as in those with bacterial vaginosis (Table 14.1 page 225). Similar findings were reported by Wilks *et al.* (1984). In these patients it seemed that the increase in anaerobic bacteria was an indication of some pathological change in the vagina, a disturbance of the normal pattern that caused a change in the flora rather than the increase in anaerobes causing the pathological change. One specific infection in which there was no increase in the anaerobic flora was candidiasis.

Trichomoniasis

The anaerobic flagellate protozoon *Tr. vaginalis* has been recognized for a long time as an important and specific cause of vaginitis. The characteristic features of trichomoniasis are a profuse, frothy, greenish discharge, often with an offensive odour, although the odour is not as characteristic as that of bacterial vaginosis. There is often some degree of vaginal and vulval irritation and inflammation, which can cause an acutely painful vaginitis in some cases, and there are increased numbers of polymorphonuclear leucocytes in the discharge.

Infection with *Tr. vaginalis* also leads to an alteration of the bacterial flora of the vagina. The lactobacilli are suppressed, the pH is raised and the flora becomes predominantly anaerobic, with large numbers of bacteroides of the vaginal species (mainly *B. bivius*, *B. disiens* and *B. melaninogenicus*, as in bacterial vaginosis) and anaerobic cocci; unlike bacterial vaginosis, *G. vaginalis* and *Mobiluncus* spp. have not been described in association with trichomoniasis. Whether or not the anaerobic bacteria contribute to the clinical symptoms is not known.

Trichomoniasis is a sexually transmitted disease and infection is often present in the male partners of women with acute disease. In some men *Tr. vaginalis* causes a urethritis, with symptoms of dysuria and a discharge, but in many it is asymptomatic and the man may be a silent carrier. It is important, therefore, to treat both partners to avoid reinfection. By virtue of its anaerobic metabolism, *Tr. vaginalis* is sensitive to metronidazole and the drug was, in fact, first introduced for the treatment of trichomoniasis a decade before its usefulness in anaerobic bacterial infections was appreciated; it remains the drug of choice. Infections in some women are apparently resistant to treatment with metronidazole, although their *Tr. vaginalis* strains remain sensitive to the drug *in vitro*. Treatment of such cases is difficult and many require hospital in-patient treatment with intravenous metronidazole plus tetracycline or clindamycin.

Bacterial vaginosis

A vaginal discharge with an offensive smell is the presenting clinical feature in one-third of women attending genitourinary medicine clinics and is a common problem in general practice. *Tr. vaginalis* causes a small proportion of cases (see above) but the common forms, previously called 'non-specific vaginitis', are now known as bacterial (or anaerobic) vaginosis. This condition is characterized by a proliferation of anaerobic bacteria in the vagina; the bacteria are localized and there is no systemic involvement, but it is an unpleasant and often distressing condition (Spiegel *et al.*, 1980; Blackwell *et al.*, 1983). The primary symptom of bacterial vaginosis is a foul (fishy) smelling, thin, grey, non-purulent discharge (often described as being 'like flour paste') without inflammation or irritation of the vaginal mucosa. Half of the women with bacterial vaginosis are 'asymptomatic', in that they do not specifically complain of symptoms and only admit that something had been wrong after treatment (Gardner and Dukes, 1955; Jones, 1983).

Bacterial vaginosis is not a specific, monobacterial infection, but a synergic mixture of anaerobic, microaerophilic and CO_2-dependent species that are present in small numbers in many normal, asymptomatic women (McCormack *et al.*, 1977) but in large numbers in vaginosis. The normal lactobacillary flora is replaced by a mixture of small bacilli normally inhibited by the lactobacilli: CO_2-dependent *G. vaginalis* and two anaerobic gram-negative groups—*Bacteroides* spp. of the melaninogenicus–oralis group (principally *B. bivius* and *B. disiens*) and curved, motile rods of *Mobiluncus* spp. *G. vaginalis* and *Bacteroides* spp. are present in most cases, *Mobiluncus* spp. in 25–50 per cent. An altered physiological environment allows these organisms to multiply and induce the discharge, probably by production of biologically active metabolites (Spiegel, 1987). *Mycoplasma* spp. are also associated with bacterial vaginosis but their role is uncertain. The vaginal pH, normally ≤ 4.0, rises to ≥ 5.5. The lactate concentration is reduced and the amounts of succinate, acetate, propionate and butyrate (all principally produced by *Bacteroides* spp.) increase. The

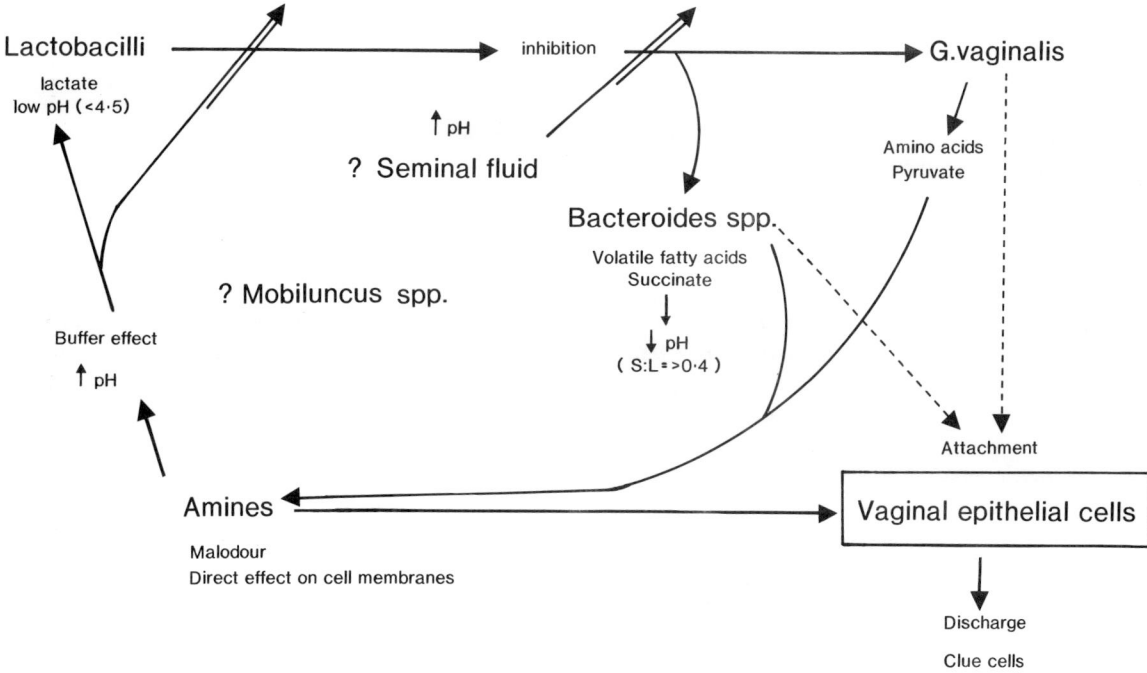

Fig. 14.1 The possible pathogenesis of bacterial vaginosis—a hypothesis

secretions also contain volatile amines—e.g., putrecine, methylamine, cadaverine, etc.—which are products of anaerobic metabolism and cause the fishy smell (Chen et al., 1979, 1982; Sanderson et al., 1983).

Bacterial relationships in the pathogenesis of bacterial vaginosis are not clear, but metabolic interactions may generate active products that cause excessive secretion, e.g. pyruvate and amino acids secreted by G. vaginalis may be decarboxylated to amines by Bacteroides spp. A tentative hypothesis that might, in part, represent the changes recognized in bacterial vaginosis is shown in Figure 14.1. Lactobacilli predominate in the normal vagina, the pH is low and the principal fatty acid product of metabolism is lactate. The vaginosis-associated organisms that are present in relatively small numbers, particularly the Bacteroides spp. and G. vaginalis, are inhibited in vitro by lactic acid and low pH (Bosu et al., 1987). In susceptible women, some factor(s) interferes with the natural protective mechanism by some combination of inhibition of the lactobacilli, increase in pH and buffering of the lactate, and allows the proliferation of G. vaginalis and Bacter-oides spp. The metabolic interactions of these synergic mixtures may then produce active metabolites which induce secretion from the vaginal mucosa while endowing the discharge with its offensive (anaerobic) character.

Most interest in this area has developed around the presence of fatty acids and amines in the vaginal secretion—products of anaerobic metabolism which may be biologically active—but confirmation of the suggested bacterial interaction from in vitro studies is still lacking. There is also no biochemical evidence for a specific pathogenic role for the Mobiluncus spp., although the anaerobic curved rods are clearly a major part of the altered vaginal flora in a large group of patients with bacterial vaginosis (Curtis, 1913; Sprott et al., 1982; Spiegel et al., 1983). Biochemical activities are not the only bacterial characters which might contribute to the pathogenesis of vaginosis. One of the characteristic findings in vaginosis is the presence of vaginal epithelial cells ('clue cells') coated with adherent bacilli, both gram-positive (presumed G. vaginalis) and gram-negative (presumed Bacteroides spp.) and the adhesive properties of the bacteria, which have been well demonstrated in haemagglutination tests, may be important in maintaining the abnormal flora at the mucosal surface where its metabolic activities may be most effective.

Although bacterial vaginosis is not a sexually transmitted disease, it occurs only in sexually active women. The initial change in the vaginal environment is an increase in pH, as occurs with the introduction of seminal fluid and this may be the main initiating factor. In women who suffer repeated episodes of vaginosis, there is a direct relationship between the deposition of semen in the vagina during sexual intercourse and the development of vaginosis. When contraceptive methods that exclude semen from the vagina are used (i.e. when the partner uses a condom) vaginosis does not develop, even in highly susceptible women. The semen may have a dual role in the pathogenesis of vaginosis. It is alkaline and thus raises the pH, inhibits lactobacilli, reduces lactate production and so allows the proliferation and metabolic activity of the vaginosis-associated organisms. It may also provide some substrate(s) for the production of active metabolites. In an *in vitro* model system, mixtures of *Bacteroides* spp., *G. vaginalis* and *Mobiluncus* spp. were damaging in a tissue-culture model, producing potentially active metabolites and causing cytopathic effect, only in the presence of seminal fluid (B M Jones, personal communication). However, this does not explain why only some women are susceptible; hormonal and physiological factors may also be important.

The diagnosis of bacterial vaginosis is essentially clinical, confirmed by simple tests on the secretions (Jones, 1983).
1. The pH (measured with indicator papers) is ≥ 5.5 and microscopy shows characteristic features.
2. In wet films, masses of small bacilli are seen, and *Mobiluncus* spp. are recognized by their active motility. 'Clue cells', vaginal epithelial cells with many attached bacteria, and which are often regarded as pathognomonic of bacterial vaginosis, are seen in wet films, but are probably best visualized in gram-stained films as large epithelial cells coated with small gram-positive and -negative ('gram-variable') rods.
3. The amine test: a drop of 10 per cent KOH added to a saline emulsion of secretions produces the fishy odour of volatile amines. Laboratory confirmation of this diagnosis is usually unnecessary but if required, a vaginal swab in transport medium (e.g., Amies) can be sent for culture.

Although *G. vaginalis* and *Bacteroides* spp. may be sought specifically in specialized laboratories, a heavy growth of mixed anaerobes generally provides sufficient evidence for the diagnosis (Blackwell *et al.*, 1983). Gas–liquid chromatography can show the typical changes in fatty acid concentrations in the secretions—the normal lactate:succinate relationship, in which lactate predominates, is reversed so that the S:L ratio becomes >0.4 (Amsel *et al.*, 1983).

Metronidazole is the drug of choice for bacterial vaginosis. Although *G. vaginalis* is resistant *in vitro*, the anaerobes are highly susceptible and treatment eliminates the synergic mixture. A single 2 g dose is as effective as 400 mg twice daily for 5–7 days in producing immediate relief, although the relapse rate is a little higher (Jones *et al.*, 1985). The longer course may be used if cure is not achieved with the single dose. However, high doses of metronidazole cause an unpleasant taste and some patients find that it produces nausea; the interaction with alcohol may also discourage its use. Various alternative antimicrobials are also effective, e.g., tinidazole, ampicillin, amoxycillin, or metronidazole + tetracycline to inhibit mycoplasmas and other organisms concomitantly. As vaginosis-associated bacteria are part of the normal vaginal flora and endogenous infection develops when vaginal environmental changes allow multiplication, contact tracing and treatment of male partners would not appear to be appropriate. However, *G. vaginalis* and *Bacteroides* spp. may be isolated from the subpreputial sac and urethra in men, some of whom have balanitis–balanoposthitis (Kinghorn *et al.*, 1982b), and these men may represent a reservoir of reinfection for some women. Therefore, partners of women with repeated episodes of vaginosis should be investigated and, if necessary, treated. The aetiology, pathogenesis, diagnosis and management of bacterial vaginosis are reviewed in monographs edited by Mårdh (1982), Csángó (1983) and Mårdh and Taylor-Robinson (1984).

Infections related to pregnancy

During normal pregnancy, the uterus and its contents (placenta, membranes, amniotic fluid and fetus) are fairly resistant to ascending infection.

Anaerobic bacteria are present in the vagina, although in smaller numbers as pregnancy proceeds. They should not normally gain access to the uterine contents but Pearson and Anderson (1970a) found that bacteria from the normal vaginal flora caused amnionitis in about 10 per cent of deliveries. However, termination of the pregnancy, either by natural delivery around term, or by spontaneous or induced abortion earlier, leaves the uterus in a relatively unprotected state. The cervical os is dilated, giving easy access for ascending bacterial infection. The normal endometrial surface has not been regenerated, and damaged, necrotic tissue in the form of retained placental fragments or the products of conception provides ideal conditions for bacterial infection, and particularly for anaerobic infection.

Amnionitis

Ascending infection of the amniotic fluid and surrounding membranes causing clinical signs is rare while the membranes remain intact. Most of the cases of amnionitis reported by Pearson and Anderson (1970a) were apparently insignificant, although there seemed to be an association with perinatal mortality (Pearson and Anderson, 1967). It becomes a serious complication when there is early and prolonged rupture of the membranes. This allows access to the amniotic sac for organisms from the vagina, including anaerobes, particularly the vaginal *Bacteroides* spp. such as *B. bivius*, *B. disiens* and *B. melaninogenicus*. Gibbs *et al.* (1987) found antibodies to *B. bivius* in the serum of women with intra-amniotic infection. In clinically apparent amnionitis there is also infection of the endometrium (endometritis) which may spread to surrounding tissues producing parametritis, pelvic abscess and peritonitis. These infections produce fever and other general signs of infection, lower abdominal tenderness and a foul discharge through the cervix. The development of such an infection would lead quickly to spontaneous or medically-induced delivery of the fetus, leaving the uterine sepsis to be dealt with as a postpartum infection. Delivery by caesarian section would be preferred in many such cases to avoid the risks of further spread of infection implicit in a vaginal delivery, particularly if instrumental assistance were needed.

Postpartum (puerperal) infections

The appreciation of the role of anaerobes in postpartum infections has increased considerably in recent years. Up to the introduction of penicillin and other antimicrobial agents, the most important cause of postpartum infection was the β-haemolytic streptococcus of Lancefield's group A (*Streptococcus pyogenes*). This had been so common, and could be such a fulminant infection with a high mortality, that it overshadowed all else. The only anaerobes to be recognized as rare, but also very serious, causes of postpartum infection were *C. perfringens* and, even more rarely, *F. necrophorum*. Postpartum uterine gas gangrene caused by *C. perfringens* is associated with difficult vaginal deliveries, particularly those in which forceps have to be used and in which there is considerable trauma to the cervical and uterine tissues, leaving bruised and damaged tissue as an ideal focus for anaerobic infection. Postpartum gas gangrene is a rapidly progressive, fulminant disease with a high mortality, and although it is now rare in developed countries with high standards of obstetric hygiene (Dylewski *et al.*, 1989), it remains a serious problem in many parts of the world. Under suitable predisposing conditions the bacteria infect the endometrium and myometrium to cause necrosis and gas formation. The incubation period may be several days or as short as a few hours. There is abnormal vaginal bleeding, uterine tenderness and generalized systemic signs of infection. These may include intravascular haemolysis, which is rare in other forms of gas gangrene, and the destruction of leucocytes and platelets, leading to disseminated intravascular coagulation (Becker *et al.*, 1987; Willis and Phillips, 1988). Characteristic patterns of gas may be produced in the uterus and these can be detected by abdominal radiography, ultrasound examination or computerized tomography. The uterine infection may lead to clostridial septicaemia, which has a poor prognosis.

As with other forms of gas gangrene, immediate and effective treatment is essential. Penicillin or other agents effective against clostridia, e.g., metronidazole, should be given in high dosage. Hysterectomy may be necessary, sometimes at an early stage when infection is extensive, or when there is haemolysis, gas production or renal failure, to remove the necrotic focus of infection (Decker and Hall, 1966; Pritchard and Whalley, 1971). Early diagnosis and treatment are critical to the outcome.

F. necrophorum is another rare but well recognized cause of postpartum uterine infection which can spread rapidly to cause widespread systemic infection. The source of *F. necrophorum* in these infections has been a matter for debate. *F. necrophorum* is not part of the normal vaginal flora; it is primarily an organism of the upper respiratory tract (necrotic tonsillitis with bacteraemia and metastatic abscess formation—known as necrobacillosis—is the most common form of *F. necrophorum* infection; Alston, 1955) and this is probably the primary source of infection—either from the patient herself or from the obstetric attendants, i.e. an epidemiological situation similar to that for the group A β-haemolytic streptococcus.

Although the incidence of postpartum infections in developed countries with high standards of obstetric hygiene is much less than in the preantibiotic era, anaerobic bacteria have emerged as major causes. As has already been said, infection is particularly associated with difficult deliveries, instrumentation (forceps) and consequent tissue damage, and with retained products of conception, which provide conditions suitable for the proliferation of anaerobes; but infections are not confined to these high risk groups. Although the anaerobic flora of the vagina is reduced during pregnancy, after delivery there is rapid recolonization, often to higher than normal levels, in the immediate postpartum period. These organisms do not remain solely in the vagina and may be found in the uterine cavity in 70–80 per cent of women, even after uncomplicated delivery. Most of these women remain healthy but some develop endometritis, progressing in the more severe cases to pelvic abscess formation, peritonitis, pelvic septic thrombophlebitis and bacteraemia (Willis and Phillips, 1988). The common organisms are a mixture of those found in the normal vagina and some of faecal origin that probably reach the genital tract via perineal and introital contamination. Overall, bacteroides of the melaninogenicus–oralis group and gram-positive anaerobic cocci are the most common isolates from the uterus. In those cases which progress to clinical infection, *B. bivius*, *B. disiens* and *B. melaninogenicus*, together, account for about as many cases as those due to *B. fragilis* (Duerden, 1980b), with anaerobic cocci as part of a generally mixed flora.

Clinical features

The clinical features of uterine gas gangrene are described below (septic abortion) and in Chapter 18. The more common infections with gram-negative anaerobic bacilli usually develop over 1–7 days. There is uterine tenderness, the woman is generally unwell and has a mild to moderate fever, and the increasingly purulent lochia (the secretion from the uterus after childbirth) becomes brownish and is foul smelling, as in other anaerobic infections. When there is abscess formation and peritonitis there is often an associated bacteraemia, but anaerobic blood cultures may need to be incubated for several days before growth becomes evident because some of the organisms involved grow relatively slowly in standard blood culture broths.

Treatment

The diagnosis of anaerobic uterine infection often indicates the presence of retained products of conception as infected foci and removal of such material by dilatation and curettage is an essential part of the treatment; hysterectomy may be necessary in severe cases, as in gas gangrene. Metronidazole is the antibacterial agent of choice to be used in conjunction with surgical intervention because although many of the anaerobic species causing these infections are sensitive to penicillin, up to one-third of the vaginal *Bacteroides* strains may be resistant, and about half of the infections are caused by *B. fragilis*, which is generally resistant to most commonly used β-lactam agents.

Septic abortion

Uterine sepsis has always been a serious and feared complication of abortion, whether spontaneous or induced. In spontaneous abortion, there is always a considerable chance that products of conception will be retained and form a focus for infection with vaginal or perineal (faecal) anaerobes, principally *Bacteroides* spp. When there is any doubt about the completeness of a spontaneous abortion, early uterine evacuation is an important measure for the prevention of serious sepsis. As would be expected, the risks of infection following induced, instrumental abortion are far higher. Even in controlled medical termination of pregnancy, with all the aseptic precautions and careful antiseptic preparation of the vagina, instruments are introduced through the vagina into the uterus, creating damage and providing a ready route of infection. The greatest hazard, however, is with non-medical terminations—illegal, 'back-street', or 'folk-medicine' abortions. Here the introduction of non-sterile instruments or materials into the uterus to induce abortion causes a very high incidence of infection, particularly with anaerobic bacteria. The most feared organism in this context is *C. perfringens* causing postabortal gas gangrene (Decker and Hall, 1966; Pritchard and Whalley, 1971). This has a high mortality even with prompt diagnosis and treatment, but the very circumstances of the infection usually mean that there are delays in seeking medical help, resulting in even worse prognosis.

Non-clostridial anaerobic uterine infections are less dramatic but much more common, and are a major cause of postabortal morbidity. As with postpartum uterine infection, the clinical diagnosis is based upon uterine tenderness, systemic signs of infection, such as fever and a foul smelling (anaerobic) discharge from the cervix. Infections of the uterus have easy access to the bloodstream and bacteraemia is a common feature. Treatment with antibacterial agents, of which metronidazole is the first choice, may be effective in the early stages, but when there is widespread involvement of the uterine tissues, hysterectomy to remove the necrotic focus is essential.

Pelvic infections in non-pregnant women

Anaerobic bacteria make a major contribution to the group of deep pelvic infections, unrelated to pregnancy, that comprise infective endometritis, salpingitis, tubo-ovarian abscess and pelvic abscess. One of the problems in the diagnosis and investigation of these conditions is the difficulty of obtaining a precise diagnosis without recourse to invasive and potentially damaging investigations, and they tend to be grouped together as the general diagnostic category of pelvic inflammatory disease (PID). The anaerobic species generally implicated in this group of infections are the non-clostridial anaerobes—*Bacteroides* spp. and anaerobic gram-positive cocci

—from the normal vaginal flora. None of these are primary pathogens initiating disease in normal, healthy, undamaged tissues, but, as in many other clinical situations, anaerobes are important secondary invaders of damaged sites where, once given the opportunity to become established, they may then cause considerable damage.

Endometritis

The endometrial lining of the uterus, with its cyclical changes and regeneration, is normally sterile; the normal vaginal flora does not extend beyond the endocervical canal. Infective endometritis involving anaerobic bacteria develops only when there is some pathological change in the uterus, such as the presence of benign or malignant tumours growing in the uterine wall and the lumen and disturbing the normal function, or in the presence of a foreign body. A chronic, low grade endometritis is commonly associated with benign leiomyomas (fibroids) and there have even been reports of uterine gas gangrene and bacteraemia caused by *C. perfringens* related to these common tumours (Kaufman *et al.*, 1974). Serious infection is more common with endometrial carcinoma (Braverman *et al.*, 1987) and choriocarcinoma (Lacey *et al.*, 1976). Not only is the right environment created by the invasive and necrotic tumours in these malignant conditions, but the chemotherapy and radiotherapy used to treat the tumours both reduce the body's defence mechanisms and produce more necrotic, anaerobic tissue remnants suitable for anaerobic infection.

The most common foreign body in the uterus is an intrauterine contraceptive device (IUCD). The use of IUCDs is a very effective method of contraception and many women have used them without problems for years. However, it has been evident even from early reports on their use that there is a higher incidence of endometritis, salpingitis and PID in general in women who use IUCDs. The presence of the foreign body forms an ideal focus for bacterial colonization, and the need to leave the 'tails' of the devices passing through the cervical canal provides a ready route for bacteria from the vagina to reach the device in the body of the uterus. Some women who are particularly susceptible to these episodes of PID have to discontinue this method of contraception. Not only are the individual infections, with pelvic inflammation, pain and often more generalized systemic symptoms, distressing but the repeated infections and inflammatory episodes lead to tubal damage and pelvic fibrosis which may result in infertility. For this reason, IUCDs are more appropriate for women who have completed their families.

Most of these episodes of PID probably involve anaerobes (but see below for the problems of establishing a clear bacteriological diagnosis). However, in some reports one group of organisms has been associated specifically with uterine infection in IUCD users—*Actinomyces* spp. This association was first suspected when the filamentous structures of actinomyces were seen in cervical cytology preparations (Papanicolou smears) from women using IUCDs (Jones *et al.*, 1979). Many reports of actinomyces infection have been based solely upon such microscopic evidence, but some studies with more detailed bacteriological investigation, including culture of endocervical samples and removed IUCDs, have supported the association between IUCDs and actinomyces (Charnock and Chambers, 1979). The *Actinomyces* spp. isolated include *A. israelii*, but these infections do not result in typical actinomycotic abscesses; they may cause a chronic, low grade infection of the uterus and associated structures. However, not all studies have come to the same conclusion. Grice and Hafiz (1983) found no significant difference between the isolation rates of actinomyces from IUCD users (25 per cent) and from controls (19 per cent) and the significance of actinomyces filaments in patients with IUCDs remains in doubt.

Pyometra and parametritis

Although most uterine infections are confined to the lumen and endometrium (endometritis), untreated chronic infections or infections caused by particularly virulent organisms may spread to involve the body of the uterus (pyometra) and the surrounding tissues (parametritis).

Salpingitis

Infection of the fallopian tubes is probably the most common specific condition contributing to the overall diagnosis of PID and the most common cause of acquired infertility (Eschenbach, 1980). The rising incidence of acute salpingitis in the last 20 years has mirrored the rising incidence of sexually transmitted diseases, suggesting that the level of sexual activity and the acquisition of sexually transmitted pathogens is a key factor in the development of salpingitis (Cunningham, 1979). There is no evidence that anaerobes themselves are sexually transmitted pathogens. The gonococcus and *Chlamydia trachomatis* are probably the two most significant primary pathogens in acute salpingitis; both cause cervicitis and ascending infections that can be manifest as salpingitis and tubal abscesses; mycoplasmas may also contribute to this primary pathogenesis (Mårdh, 1980). However, many women who present with acute PID do not have active gonococcal or chlamydial infection, and the causes of these acute episodes may include anaerobic bacteria colonizing and causing infection in tubes already damaged by previous infections with the primary pathogens. The scarred mucosa, without active cilia, narrows the tubal lumen and causes obstruction, providing suitable conditions for ascending secondary infection with anaerobes. Bacteriological confirmation of the diagnosis is particularly difficult in salpingitis and PID (Kinghorn *et al.*, 1986).

Observation of the tubes at laparoscopy is an important investigation to confirm the presence of inflammation, i.e., to confirm the clinical diagnosis, but useful bacteriological specimens can be obtained in only a small proportion of cases. A swab can be passed through the laparoscope but organisms cannot be recovered from the outside of inflamed tubes, and only when there is pus or exudate emerging from the fimbriate end, or when there is free pus in the pelvis, can a reliable specimen be obtained. Such studies that have been done have shown that active gonococcal infection is important in PID, but also that non-sporing anaerobes of the vaginal types can be found in a proportion of cases. An indirect approach used in some studies has been to sample the endocervical flora in patients with PID confirmed by laparoscopy. Again, three groups of women have been delineated amongst those with PID or salpingitis—one group with active gonococcal infection, a second with *Ch. trachomatis* and a third group with a much higher than normal incidence of anaerobic bacteria in the endocervical flora, but this may merely reflect the known epidemiological association that salpingitis is more common in women with bacterial vaginosis (Kinghorn *et al.*, 1986).

Tubo-ovarian abscess

The development of a tubal infection (salpingitis) into a frank tubal or tubo-ovarian abscess is a matter of degree rather than a different condition. Their basic pathogenesis is the same—damage from a specific primary infection, e.g., gonococcal salpingitis, resulting in a narrowed, obstructed tube which is susceptible to secondary infection; infection of an obstructed tube leads to abscess formation. The pus from such abscesses obtained at operation generally contains a mixture of bacteria, including a high proportion of anaerobes.

Pelvic abscess

Once pelvic infection (PID) has progressed to the stage of producing a pelvic abscess, there is considerable evidence for the involvement of anaerobes. By definition, a collection of infected material is present and the management of pelvic abscess demands surgical intervention to drain the abscess. Unlike the situation with PID that falls short of abscess formation, a specimen, which is suitable for bacteriological investigation is now available and can provide clear proof of the contribution of different bacteria. As with other intra-abdominal abscesses, virtually all pelvic abscesses yield a mixture of bacteria, in which anaerobic bacteria form a major proportion of the total flora (Monif and Baer, 1982).

Postoperative infections in gynaecological surgery

Most gynaecological operations fall within the category of dirty, or at least significantly contaminated, operations. As in surgery of the gastrointestinal tract, although less obviously heavily contaminated than the large intestine, operations of the female genital tract involve a series of hollow viscera, lined with a mucosa and open to the body surface. For at least part of that tract the mucosa is host to a large and complex, predominantly anaerobic, normal bacterial flora and, as indicated above, in many of the pathological states that require surgery the anatomical or physiological changes necessitating surgery will have resulted in ascending colonization and, often, clinically evident infection.

Before the realization of the role in postoperative sepsis of organisms, and particularly the anaerobes, from the normal flora, and before the introduction of routine peroperative prophylaxis aimed primarily at anaerobes, infection rates for operations such as hysterectomy or vaginal repair were around 20 per cent and most of these were anaerobic infections. The infections were mostly wound infections, local abscesses and also pelvic abscesses. Bacteraemia, with the risk of spread of infection to cause metastatic abscesses, is an added complication of such infections. As would be expected, the most common organisms isolated from the postoperative infections are *Bacteroides* spp. of the melaninogenicus–oralis group, i.e., vaginal bacteroides; Willis et al. (1975) found that pigmented bacteroides ('*B. melaninogenicus*') were the most common. However, several groups have also found that the preponderance of these species is less than might have been expected, and that *B. fragilis* may be only a little less common, accounting for 30–40 per cent of isolates.

The introduction of peroperative prophylaxis with metronidazole for such gynaecological surgical procedures in the late 1970s made an immediate impact and postoperative sepsis became a relative rarity; with adequate prophylaxis, infection rates are < 5 per cent and generally around 1 per cent for uncomplicated operations where there is no evidence of pre-existing infection before surgery.

Anaerobic infections in men

The role of anaerobes in male genitourinary infections has received less general attention than their role in female infections. Similarly, convincing data on the contribution of anaerobes to the normal genital flora in men is scanty. It is clear that anaerobes do not form part of the normal flora of the external genitalia, including the subpreputial sac, but anaerobic cocci have been reported in the urethral flora by several groups (see Chapter 11) and recent studies do indicate that *B. ureolyticus*, but not other *Bacteroides* spp., may be isolated from the urethras of up to one-third of normal, healthy men with no signs or symptoms of urethritis (A. Eley and K.W. Bennett, personal communication). However, there is considerable evidence that anaerobes are involved in a group of superficial, necrotizing infections of the external genitalia that can be very severe and destructive. They range from relatively minor cases of balanitis or balanoposthitis—superficial inflammation, with or without ulceration, of the glans penis (balanitis) or the glans plus the prepuce (balanoposthitis)—to genital ulceration and superficial abscesses, and, at their most severe, to the forms of synergic scrotal and inguinal gangrene with widespread tissue destruction described by Meleny (1931, 1933) and Fournier (see Randall, 1920). None of these conditions is necessarily caused by a single, specific pathogen, and, as in genital ulcers in women (see above) the initial damage may be caused by trauma, neglect of personal hygiene or a virus infection such as genital herpes. However, in the established conditions there is a mixed bacterial flora, in which anaerobes predominate and which would lead to a reasonable description of these conditions as synergic bacterial infections. The contribution of anaerobes to other genitourinary infections in men —such as urethritis, prostatitis and epididymitis—is far less clear.

Balanitis and balanoposthitis

Inflammation and superficial erosion or ulceration of the glans penis and prepuce are common conditions in men attending genitourinary medicine clinics. They are clearly not specific bacterial infections and their incidence is related to sexual activity, to minor trauma and to lack of personal hygiene. In some cases infection with herpes simplex virus can be shown to be the primary initiating factor. However, whatever the initial cause, studies of the bacterial flora of men with balanitis–balanoposthitis have shown that anaerobes are present in the majority of cases whereas the normal flora at this site is predominantly aerobic. Masfari *et al.* (1983) found small numbers of anaerobes in only 21 per cent of 28 normal controls, whereas anaerobes were isolated from >75 per cent of patients with balanitis–balanoposthitis. Not only were anaerobes present in most cases, but when they were present they were the predominant microbial flora. The most common groups of anaerobes isolated were *Bacteroides* spp. —they were the predominant isolates found in 90 per cent of the patients (Table 14.3). *B. asaccharolyticus* was the most common single species, followed by *B. intermedius*, *B. ureolyticus* and *B. bivius*. The asaccharolytic bacteroides (*B. asaccharolyticus* and *B. ureolyticus*) are associated with a wide range of superficial, ulcerative conditions and in this series

Table 14.3 Anaerobes isolated from 98 patients with balanitis-balanoposthitis*

Species	Number of isolates
All *Bacteroides* spp.	92
Asaccharolytic group	(47)
B. asaccharolyticus	34
B. ureolyticus	13
Melaninogenicus–oralis group	(45)
B. melaninogenicus	7
B. intermedius	15
B. levii	2
B. bivius	11
B. disiens	5
B. oralis group	5
Anaerobic cocci	20

*From Masfari *et al.* (1983)

they were more commonly associated with the more severe forms of erosive balanoposthitis than with the milder, non-erosive cases; in particular, the combination of the two species was higher in erosive disease. The other *Bacteroides* spp. isolated from these conditions were mostly of the vaginal types (e.g., *B. bivius*) except for *B. intermedius*, which is a common oral species but much less common in the vagina. Gram-positive anaerobic cocci were also common isolates and the combination of *Bacteroides* spp. and anaerobic cocci may represent a synergic association related to continuing tissue damage. Although a direct causal relationship has yet to be established, the findings implicate the mixed, predominantly anaerobic flora of the lesions in the pathogenesis of these conditions (Duerden, 1987).

Genital ulcers

There is considerable overlap between the various categories of superficial, ulcerative conditions. Erosive balanitis–balanoposthitis described above could be considered to be one end of a spectrum of genital ulcers in which the more severe forms would be classified as synergic bacterial gangrene. Between the two extremes are a varied mixture of superficial, necrotizing ulcerative conditions affecting the penis (glans, prepuce and shaft) scrotum and inguinal region. As already described for ulcers in women and balanitis–balanoposthitis in men, the primary initiating factors are varied but herpes simplex virus is probably the primary cause in about half the ulcers seen in patients attending genitourinary medicine clinics. However, by the time most patients are seen, the ulcers have passed the typical herpetic stage of shallow, weeping, intensely painful ulcers and are deeper, with undermined edges and the base covered with a necrotic slough. Once established, the microbial flora of ulcers is fairly consistent, regardless of the primary cause. The flora of genital ulcers is predominantly anaerobic (Chapel et al., 1978; Masfari et al., 1985); anaerobes form the major part of the cultivable flora and *Bacteroides* spp., which are not found normally on the external genitalia, are the most common isolates. Masfari et al. (1985) found that in men, even more than in women, the asaccharolytic species *B. asaccharolyticus* and *B. ureolyticus* accounted for almost all the excess isolation rate of bacteroides in ulcer patients over controls (Table 14.4). The melaninogenicus–oralis group of vaginal bacteroides that were found in women, and whose presence could not be distinguished from simple local contamination, were present in much smaller numbers in men. Gram-positive anaerobic cocci were also found, but they were isolated much less frequently than the asaccharolytic bacteroides, which seem to have a particular association with superficial, necrotizing, ulcerative lesions of the genitalia, as they also do with similar lesions elsewhere on the body, e.g., decubitus and varicose ulcers, diabetic gangrene, etc.

Synergic gangrene

The most severe end of this spectrum of ulcerating and necrotizing conditions is represented by the highly destructive mixed bacterial infections known as synergic bacterial gangrene. The role of mixtures of anaerobic, microaerophilic and facultative organisms in these conditions has been recognized since the early descriptions by Meleney (1931, 1933) and Fournier (see Randall, 1920). This spreading gangrene can occur anywhere in the general abdominal or perineal region but it is particularly associated with the inguinoscrotal area, although it may spread rapidly and widely to involve large areas of the surrounding superficial tissues of the abdomen and thighs. It usually begins at a site of some relatively

Table 14.4 Anaerobes in men and women with genital ulcers*

Species	Number (and percentage) of subjects from whom the given species was isolated			
	Male controls (24)	Male patients (40)	Female patients (45)	All patients (91)
All anaerobes	5 (21)***	33 (72)***	37 (82)	70 (77)
Bacteroides spp.	3 (13)***	30 (65)***	35 (78)	65 (71)
Asaccharolytic group	2 (8)***	26 (57)***	21 (47)	47 (52)
B. asaccharolyticus	2 (8)***	23 (50)***	15 (33)	38 (42)
B. ureolyticus	0**	16 (35)**	15 (33)	31 (34)
Melaninogenicus–oralis group	1 (4)	10 (22)**	23 (51)**	33 (36)
B. melaninogenicus	1 (4)	0	3 (7)	3 (3)
B. intermedius	0	3 (7)	9 (20)	12 (13)
B. bivius	0	6 (13)	7 (16)	13 (14)
B. disiens	0	1 (2)	3 (7)	4 (4)
B. oralis group	0	0	1 (2)	1 (1)
B. fragilis	0	1 (2)	2 (4)	3 (3)
Anaerobic cocci	2 (8)	10 (22)	15 (33)	25 (27)

*From Masfari et al. (1985)
**$p\ 0.01$
***$p\ 0.001$

minor trauma, or a surgical wound, in the groin or scrotum; this becomes infected with a particularly virulent combination of bacteria which, acting together, produce a mixture of potent exoenzymes and aggressins that allow them to spread rapidly through the superficial tissues destroying the overlying skin and leaving large denuded areas covered with a necrotic slough. Fournier's gangrene is the specific description of a highly destructive, and fortunately uncommon, synergic anaerobic necrotizing fasciitis of the perineum. It is more common in men but can occur in women. Although it may arise apparently spontaneously on the scrotum or labia, it is usually a complication of some trauma or surgery, or of a perineal or perirectal abscess. As well as local destruction there is severe systemic toxaemia. Mortality may be up to 70 per cent. Treatment requires early, radical excision of affected tissue and high doses of suitable antimicrobials (Willis and Phillips, 1988). Appropriate treatment would be with metronidazole plus another antimicrobial, e.g., penicillin, depending upon the particular infecting combination. Initial reports described mixtures of aerobic and microaerophilic or anaerobic cocci in the lesions, but improved anaerobic techniques have shown that most of the lesions also contain gram-negative anaerobes. The asaccharolytic bacteroides have been isolated from several severe cases, usually in combination with anaerobic cocci, and this may represent a particularly virulent combination.

A very similar and equally, if not more, destructive clinical condition that may affect the external genitalia and surrounding superficial tissues is *noma*, a necrotizing or gangrenous state that is part of the fusospirochaetal complex of disease. It starts at mucocutaneous junctions and can spread rapidly to produce large areas of destruction. It occurs most commonly around the mouth and nose in poorly nourished and debilitated children, destroying large areas of the facial tissues in the condition known as *cancrum oris* (see Chapters 8 and 15), but it may also affect the vulva, prepuce or anus. The spreading necrosis is associated with a foul odour and is clearly an anaerobic infection. Predisposing factors include systemic disease, other chronic infections, malnutrition and poor hygiene. It is generally regarded as a fusospirochaetal synergic infection and both fusiform organisms and spirochaetes may be seen in stained preparations from the affected sites, but as with the more common fusospirochaetal infection of Vincent's gingivitis (acute ulcerative gingivitis) other anaerobes, particularly the pigmented *Bacteroides* spp., may have an important role in its pathogenesis (Kaufman *et al.*, 1972).

Urethritis

With the recognition that anaerobes had an important role in women in the condition that used to be known as non-specific vaginitis, similar attention was turned on the most common clinical condition in men attending genitourinary medicine clinics —urethritis. Two specific pathogens are clearly recognized as primary causes of urethritis—*N. gonorrhoeae* and *Ch. trachomatis*. *Ch. trachomatis* is more common, but the two account for only 50–60 per cent of cases with the clinical signs and symptoms of urethral infection. Anaerobic organisms are thought to be among other candidate organisms for a pathogenic role in the remaining cases of non-gonococcal non-chlamydial urethritis. However, there is still no clear proof of a specific pathogenic role. Several studies have shown that anaerobic gram-positive cocci are part of the normal urethral flora in men (Fontaine *et al.*, 1982) but that bacteroides are uncommon. However, in patients with urethritis the urethral flora becomes predominantly anaerobic and *Bacteroides* spp. can be isolated in high numbers, irrespective of the specific cause of urethritis (Masfari *et al.*, 1983). The findings on anaerobic culture are the same in patients with proven gonococcal or chlamydial urethritis, or with neither specific pathogen. Their presence indicates some abnormality but the association does not prove a causal relationship. The clearest evidence for a causative role for an anaerobe in urethritis came from the work of Fontaine and his colleagues (Fontaine *et al.*, 1982). They found a gram-negative anaerobic bacillus with a clear association with urethritis, and subsequently showed that patients responded to treatment with metronidazole (Hawkins *et al.*, 1988). In their hands it was a fastidious organism that required formate and fumarate from growth and highly reduced conditions for transport and primary isolation, but it had the general characteristics of *B. ureolyticus* and subsequent detailed studies confirmed that the urethral strains were typical of this species (Fontaine *et al.*, 1984, 1986; Taylor *et al.*, 1986, 1987). It was present in 50 per cent of 64 men with non-gonococcal urethritis but was rarely found in 30 control subjects. However, this has not been confirmed in other studies. Masfari *et al.* (1983) and subsequent studies in our laboratory (Duerden, 1987) confirmed that anaerobes were common in men with urethritis, but that *B. ureolyticus* represented only a small proportion of our isolates, most were *B. bivius*, *B. disiens*, *B. melaninogenicus* and *B. asaccharolyticus*. Further doubt has been cast on the pathogenicity of *B. ureolyticus* in urethritis by other studies in which *B. ureolyticus* has been isolated from the urethras of normal healthy controls as often, if not more often, than from patients with urethritis (A Eley and KW Bennett, personal communications). Therefore, the enigma of non-gonococcal non-chlamydial urethritis remains, and the isolation of bacteroides from the urethra is, at present, no more than an indication that the normal flora has been dsiturbed.

Prostatitis and epididymitis

The causes of infection and inflammation of the prostate and epididymis are often obscure. Sampling is difficult and material suitable for bacteriological investigation is rarely obtained. There have been many suggestions that anaerobes may contribute to these infections, but firm evidence is restricted to a small number of case reports in which anaerobic infection has been proved (e.g., Bartlett *et al.*, 1978).

Conclusion

Anaerobic bacteria are part of the normal bacterial flora of the genitourinary tract in men and women. They are also involved in a wide range of infective conditions from superficial abscesses and ulcers to deep abscesses and systemic sepsis. Anaerobes are the primary pathogens in few of these infections, but they are effective secondary invaders contributing considerably to tissue damage and necrosis.

References

Alston J M. (1955). Necrobacillosis in Great Britain. *British Medical Journal* 2: 1524–1528.

Amsel R, Totten P A, Spiegel C A, Chen K C S, Eschenbach D, Holmes K K. (1983). Nonspecific vaginitis: diagnostic criteria and microbial and epidemiologic considerations. *American Journal of Medicine* 74: 14.

Bartlett J G, Polk B F. (1984). Bacterial flora of the vagina: quantitative study. *Reviews of Infectious Diseases* 6 Suppl 1: S67–S72.

Bartlett J G, Weinstein W M, Gorbach S L. (1978). Prostatic abscess involving anaerobic bacteria. *Archives of Internal Medicine* 138: 1369–1371.

Becker R C, Giuliani M, Savage R A, Weick S K. (1987). Massive haemolysis in *Clostridium perfringens* infections. *Journal of Surgical Oncology* 35: 13–18.

Blackwell A L, Phillips I, Fox A R, Barlow D. (1983). Anaerobic vaginosis (nonspecific vaginitis): clinical, microbiological and therapeutic findings. *Lancet* 2: 1327–1328.

Bosu W K, Duerden B I, Bennett K W. (1987). Metabolic interactions of vaginosis-associated bacteria. *Journal of Medical Microbiology* 23: xii.

Bramley H M, Dixon R A, Jones B M. (1981). *Haemophilus vaginalis* (*Corynebacterium vaginale*, *Gardnerella vaginalis*) in a family planning clinic population. *British Journal of Venereal Diseases* 57: 62–66.

Braverman J *et al.* (1987). Spontaneous clostridial gas gangrene of uterus associated with endometrial malignancy. *American Journal of Obstetrics and Gynecology* 156: 1205–1207.

Chapel T, Brown W J, Jeffries C, Stewart J A. (1978). The microbiological flora of penile ulceration. *Journal of Infectious Diseases* 137: 50–56.

Charnock M, Chambers T J. (1979). Pelvic actinomycosis and intrauterine contraceptive devices. *Lancet* 1: 1239–1240.

Chen K C S, Amsel R, Eschenbach D A, Holmes K K. (1982). Biochemical diagnosis of vaginitis: determination of diamines in vaginal fluid. *Journal of Infectious Diseases* 145: 337–345.

Chen K C S, Forsyth P S, Buchanan T M, Holmes K K. (1979). Amine content of vaginal fluid from untreated and treated patients with nonspecific vaginitis. *Journal of Clinical Investigation* 63: 828–835.

Chow A W, Galpin J E, Guze L B. (1977). Clindamycin for treatment of sepsis caused by decubitus ulcers. *Journal of Infectious Diseases* 135 Suppl: S65–S68.

Csángó P A (ed). (1983). Proceedings of the first international conference on vaginosis (nonspecific vaginitis). *Scandinavian Journal of Infectious Diseases* Suppl 40.

Cunningham F G. (1979). The etiology and pathogenesis of acute pelvic inflammatory disease. *Sexually Transmitted Diseases* 6: 221–223.

Curtis A H. (1913). A motile curved anaerobic bacillus in uterine discharges. *Journal of Infectious Diseases* 12: 165–169.

Decker W H, Hall W. (1966). Treatment of abortions infected with *Clostridium welchii*. *American Journal of Obstetrics and Gynecology* 93: 394–399.

Duerden B I. (1980a). The isolation and identification of *Bacteroides* spp. from the normal human vaginal flora. *Journal of Medical Microbiology* **13**: 79–87.

Duerden B I. (1980b). The identification of gram-negative anaerobic bacilli isolated from clinical infections. *Journal of Hygiene* **84**: 301–313.

Duerden B I. (1987). Anaerobes in non-gonococcal urethritis, balanoposthitis and genital ulcers. In *Recent advances in anaerobic bacteriology*, pp. 231–241. Edited by Borriello S P, Hardie J M. Martinus Nijhoff, Dordrect.

Duerden B I, Bennett K W, Faulkner J. (1982). Isolation of *Bacteroides ureolyticus* (*B. corrodens*) from clinical infections. *Journal of Clinical Pathology* **35**: 309–312.

Duncan M O, Bilgeri Y R, Fehler H G, Ballard R C. (1981). The diagnosis of sexually acquired ulcerations in black patients in Johannesburg. *South African Journal of Sexually Transmitted Diseases* **1**: 20–23.

Dunkelberg W E, Skaggs R, Kellogg D S. (1970). Method for the isolation and identification of *Corynebacterium vaginale* (*Haemophilus vaginalis*). *Applied Microbiology* **19**: 47–52.

Dylewski J, Wiesenfeld H, Latow A. (1989). Postpartum uterine infection with *Clostridium perfringens*. *Reviews of Infectious Diseases* **11**: 470–473.

Eschenbach D A. (1980). Epidemiology and diagnosis of acute pelvic inflammatory disease. *Obstetrics and Gynecology* **55**: 142S–152S.

Fierer J, Daniel D, Davis C. (1979). The fetid foot: lower extremity infections in patients with diabetes mellitus. *Reviews of Infectious Diseases* **1**: 210–217.

Finegold S M. (1977). *Anaerobic bacteria in human disease*. Academic Press, New York.

Fontaine E A R, Borriello S P, Taylor-Robinson D, Davies H A. (1984). Characteristics of a gram-negative anaerobe isolated from men with non-gonococcal urethritis. *Journal of Medical Microbiology* **17**: 129–140.

Fontaine E A R, Bryant T N, Taylor-Robinson D, Borriello S P, Davies H A. (1986). A numerical taxonomic study of anaerobic gram-negative bacilli classified as *Bacteroides ureolyticus* isolated from patients with non-gonococcal urethritis. *Journal of General Microbiology* **132**: 3137–3146.

Fontaine E A R, Taylor-Robinson D, Hanna N F, Coufalik E D. (1982). Anaerobes in men with urethritis. *British Journal of Venereal Diseases* **58**: 321–326.

Gardner H L, Dukes C D. (1954). New etiologic agent in non-specific bacterial vaginitis. *Science* **120**: 853.

Gardner H L, Dukes C D. (1955). *Haemophilus vaginalis* vaginitis. *American Journal of Obstetrics and Gynecology* **69**: 962–976.

Gibbs R S, Forman J, St Clair P J, Baseman J B. (1987). Detection of serum antibody response to *Bacteroides bivius* by enzyme-linked immunosorbent assay in women with intraamniotic infections. *Obstetrics and Gynecology* **69**: 208–213.

Gorbach S L, Menda K B, Thadepalli H, Keith L. (1973). Anaerobic microflora of the cervix in healthy women. *American Journal of Obstetrics and Gynecology* **117**: 1053–1055.

Greenwood J R, Picket M J. (1980). Transfer of *Haemophilus vaginalis* to a new genus, *Gardnerella*: *G. vaginalis* (Gardner and Dukes) comb. nov. *International Journal of Systematic Bacteriology* **30**: 170–178.

Grice G C, Hafiz S. (1983). Actinomyces in the female genital tract. *British Journal of Venereal Diseases* **59**: 317–319.

Hafiz S, McEntegart M G, Morton R S, Waitkins S A. (1975). *Clostridium difficile* in the urogenital tract of males and females. *Lancet* **1**: 420–421.

Hammann R. (1982). A reassessment of the microbial flora of the female genital tract, with special reference to the occurrence of *Bacteroides* species. *Journal of Medical Microbiology* **15**: 293–302.

Hammond G W, Slutchuk M, Scatliff J, Sherman E, Wilt J C, Ronald A R. (1980). Epidemiological, clinical, laboratory and therapeutic features of an urban outbreak of chancroid in North America. *Reviews of Infectious Diseases* **2**: 862–879.

Hawkins D A, Fontaine E A R, Thomas B J, Boustouller Y L, Taylor-Robinson D. (1988). The enigma of non-gonococcal urethritis: role for *Bacteroides ureolyticus*. *Genitourinary Medicine* **64**: 10–13.

Holtz F, Mauch E W. (1962). Gas gangrene of the uterus. Survival following hysterectomy. *Obstetrics and Gynecology* **19**: 545–548.

Jones B M. (1983). *Gardnerella vaginalis*-associated vaginitis—a 'new' sexually transmitted disease. *Medical Laboratory Sciences* **40**: 53–57.

Jones B M, Geary I, Alawattagama A A, Kinghorn G R, Duerden B I. (1985). In-vitro and in-vivo activity of metronidazole against *Gardnerella vaginalis*, *Bacteroides* spp. and *Mobiluncus* spp. in bacterial vaginosis. *Journal of Antimicrobial Chemotherapy* **16**: 189–197.

Jones M C, Borschmann B D, Dowling E M, Pollock H M. (1979). The prevalence of actinomyces-like organisms found in cervico-vaginal smears of 300 IUD wearers. *Acta Cytologica* **23**: 282–283.

Kaufman B M, Cooper J M, Cookson P. (1974). *Clostridium perfringens* septicemia complicating degenerating uterine leiomyomas. *American Journal of Obstetrics and Gynecology* **118**: 877–878.

Kaufman E J et al. (1972). Fusobacterial infection: enhancement by cell free extracts of *Bacteroides melaninogenicus* possessing collagenolytic activity. *Archives of Oral Biology* **17**: 577–580.

Kinghorn G R, Duerden B I, Hafiz S. (1986). Clinical and microbiological investigation of women with acute salpingitis and their consorts. *British Journal of Obstetrics and Gynecology* **93**: 869–880.

Kinghorn G R, Hafiz S, McEntegart M G. (1982a). Pathogenic microbial flora of genital ulcers in Sheffield with particular reference to Herpes simplex virus and

Haemophilus ducreyi. *British Journal of Venereal Diseases* **58**: 377–380.

Kinghorn G R, Jones B M, Chowdhury F H, Geary I. (1982b). Balanoposthitis associated with *Gardnerella vaginalis* infection in men. *British Journal of Venereal Diseases* **58**: 127–129.

Lacey C G, Futoran R, Morrow C P. (1976). *Clostridium perfringens* infection complicating chemotherapy for choriocarcinoma. *Obstetrics and Gynecology* **47**: 337–341.

McCormack W M et al. (1977). Vaginal colonisation with *Corynebacterium vaginale*. *Journal of Infectious Diseases* **136**: 740–745.

Mårdh P A. (1980). An overview of infectious agents of salpingitis. *American Journal of Obstetrics and Gynecology* **138**: 933–951.

Mårdh P A. (1982). *Gardnerella vaginalis* and non-specific vaginitis. *European Journal of Clinical Microbiology* **3**: 283–325.

Mårdh P A, Taylor-Robinson D. (1984). *Bacterial vaginosis*. Almqvist and Wiksell International, Stockholm.

Masfari A N, Duerden B I, Kinghorn G R. (1986). Quantitative studies of vaginal bacteria. *Genitourinary Medicine* **62**: 256–263.

Masfari A N, Kinghorn G R, Duerden B I. (1983). Anaerobes in genitourinary infections in men. *British Journal of Venereal Diseases* **59**: 255–259.

Masfari A N, Kinghorn G R, Hafiz S, Barton I G, Duerden B I. (1985). Anaerobic bacteria and herpes simplex virus in genital ulceration. *Genitourinary Medicine* **61**: 109–113.

Meislin H W et al. (1977). Cutaneous abscesses: anaerobic bacteriology and out-patient management. *Annals of Internal Medicine* **87**: 145–149.

Meleney F L. (1931). Bacterial synergism in disease process with a confirmation of the synergistic bacterial aetiology of a certain type of progressive gangrene of the stomach wall. *Annals of Surgery* **19**: 961–968.

Meleny F L. (1933). A differential diagnosis between certain types of infectious gangrene of the skin: with particular reference to haemolytic streptococcus gangrene and bacterial synergistic gangrene. *Surgery, Gynecology and Obstetrics* **56**: 847–867.

Monif G R F, Baer H. (1982). Impact of diverging anaerobic technology on cul-de-sac isolates from patients with endometritis–salpingitis–peritonitis. *American Journal of Obstetrics and Gynecology* **142**: 896–900.

Nsanze H, Fast M V, D'Costa L J, Tukei P, Curran J, Ronald A R. (1981). Genital ulcers in Kenya: clinical and laboratory study. *British Journal of Venereal Diseases* **57**: 378–381.

O'Farrell S, Wilks M, Nash J Q, Tabaqchali S. (1984). A selective enrichment broth for the isolation of *Clostridium difficile*. *Journal of Clinical Pathology* **37**: 98–99.

Ohm M J, Galask R P. (1975). Bacterial flora of the cervix from 100 prehysterectomy patients. *American Journal of Obstetrics and Gynecology* **122**: 683–687.

Pearson H E, Anderson G V. (1967). Perinatal deaths associated with bacteroides infections. *Obstetrics and Gynecology* **30**: 486–492.

Pearson H E, Anderson G V. (1970a). Bacteroides infections and pregnancy. *Obstetrics and Gynecology* **35**: 31–36.

Pearson H E, Anderson G V. (1970b). Genital bacteroidal abscesses in women. *American Journal of Obstetrics and Gynecology* **107**: 1264–1265.

Phillips I, Taylor E. (1982). Anaerobic curved rods in vaginitis. *Lancet* **1**: 221.

Pritchard J A, Whalley P J. (1971). Abortion complicated by *Clostridium perfringens* infection. *American Journal of Obstetrics and Gynecology* **111**: 484–492.

Randall A. (1920). Idiopathic gangrene of the scrotum. *Journal of Urology* **4**: 219–235.

Sadhu K, Domingue P A G, Chow A W, Nelligan J, Cheng N, Costerton J W. (1989). *Gardnerella vaginalis* has a gram-positive cell-wall ultrastructure and lacks classical cell-wall lipopolysaccharide. *Journal of Medical Microbiology* **29**: 229–235.

Sanderson B E, White E, Balsdon M J. (1983). Amine content of vaginal fluid from patients with trichomoniasis and *Gardnerella*-associated nonspecific vaginitis. *British Journal of Venereal Diseases* **59**: 302–305.

Spiegel C A. (1987). New developments in the etiology and pathogensis of bacterial vaginosis. *Advances in Experimental Biology* **224**: 127–134.

Spiegel C A, Amsel R, Eschenbach D A, Schrenknecht F, Holmes K K. (1980). Anaerobic bacteria in nonspecific vaginitis. *New England Journal of Medicine* **303**: 601–607.

Spiegel C A, Eschenbach D A, Amsel R, Holmes K K. (1983). Curved anaerobic bacteria in bacterial vaginosis and their response to antimicrobial therapy. *Journal of Infectious Diseases* **148**: 817–822.

Spiegel C A, Roberts M. (1984). *Mobiluncus* gen. nov., *Mobiluncus curtisii* subsp. *curtisii* sp. nov., *Mobiluncus curtisii* subsp. *holmesii* sp. nov. and *Mobiluncus mulieris* sp. nov., curved rods from the human vagina. *International Journal of Systematic Bacteriology* **34**: 177–184.

Sprott M S, Pattman R S, Ingham H R, Short G R, Narang H K, Selkon J B. (1982). Anaerobic curved rods in vaginitis. *Lancet* **1**: 54–55.

Sprott M S et al. (1983). Characteristics of motile curved rods in vaginal secretions. *Journal of Medical Microbiology* **16**: 175–182.

Taylor A J, Costas M, Owen R J. (1987). Numerical analysis of electrophoretic protein patterns of *Bacteroides ureolyticus* clinical isolates. *Journal of Clinical Microbiology* **25**: 660–666.

Taylor A J, Dawson C A, Owen R J. (1986). The identification of *Bacteroides ureolyticus* from patients with non-gonococcal urethritis by conventional biochemical tests and by DNA and protein analysis. *Journal of Medical Microbiology* **21**: 109–116.

Wilks M, Tabaqchali S. (1987a). Quantitative bacteriology of the vaginal flora during the menstrual cycle. *Journal of Medical Microbiology* **24**: 241–245.

(1987b). The anaerobic bacterial flora of the vagina in health and disease. In *Recent advances in anaerobic bacteriology*, pp. 195–204. Edited by Borriello S P, Hardie J M. Martinus Nijhoff, Dordrecht.

Wilks M, Thin R N, Tabaqchali S. (1982). Quantitative methods for studies on vaginal flora. *Journal of Medical Microbiology* **15**: 141–147.

Wilks M, Thin R N, Tabaqchali S. (1984). Quantitative bacteriology of the vaginal flora in genital disease. *Journal of Medical Microbiology* **18**: 217–231.

Willis A T. (1977). *Anaerobic bacteriology: clinical and laboratory practice*. Butterworths, London.

Willis A T, Phillips K D. (1988). *Anaerobic infections: clinical and laboratory practice*. Public Health Laboratory Service, London.

Willis A T et al. (1975). An evaluation of metronidazole in the prophylaxis and treatment of anaerobic infections in surgical patients. Report by a study group. *Journal of Antimicrobial Chemotherapy* **1**: 393–401.

15

Dental and oral infection

J M Hardie

The mouth as a microbial habitat
The source of organisms in oral and dental infection
The pathogenesis of polymicrobial infection
Anaerobes and dental caries
   Roof canal infections
Dental abscesses
  Types of abscess in the mouth
  Recognition of the role of anaerobes
  Bacteriology of dental abscesses
  Quantitative studies
  Laboratory diagnosis and handling of specimens
  Clinical management of antimicrobial chemotherapy
Periodontal diseases
  Types of disease
  Anaerobes in periodontal disease
  Experimental gingivitis
  Juvenile periodontitis
  Acute ulcerative necrotizing gingivitis
  Adult periodontitis
  Indicators of periodontal diseases
  Pathogenic mechanisms in periodontal diseases
Other oral infections
  Perimandibular space infections and Ludwig's angina
  Actinomycosis
  Osteomyelitis
  Maxillary sinusitus
References

The mouth as a microbial habitat

At first glance it may seem strange that the mouth harbours large numbers of a wide variety of anaerobes (Hardie, 1974). The number of different organisms present in the human oral cavity cannot be stated exactly, but over 300 species have been found in samples from the human gingival crevice, many of them anaerobes (Moore, 1987). Several hitherto unclassified organisms of oral origin have

Table 15.1 Anaerobic genera found in the mouth

Gram-positive	Gram-negative
Peptococcus	Veillonella
Peptostreptococcus	
(Streptococcus)*	
(Actinomyces)*	Bacteroides
Arachnia	Centipeda
Bifidobacterium	Fusobacterium
Eubacterium	Leptotrichia
(Lactobacillus)*	Mitsuokella†
Propionibacterium	Porphyromonas‡
	Prevotella‡
	Selenomonas
	Wolinella
	Treponema

*Genera shown in brackets are usually facultatively anaerobic but include some species that are obligate anaerobes
†New genus name for some former *Bacteroides* species (Haapasalo *et al.*, 1986)
‡Proposed new genera for species formerly called *B. asaccharolyticus*, *B. gingivalis* and *B. endodontalis* (Shah and Collins, 1988) and *B. melaninogenicus*, *B. oralis* and several other species (Shah and Collins, 1990)

monly found in the mouth, in health or disease, are listed in Table 15.1. These include gram-positive and gram-negative cocci, rods and filaments. Spore-forming bacteria of the genus *Clostridium* do not appear to be a regular component of the oral microflora although they have been reported occasionally (Van Reenan and Coogan, 1970; Loesche *et al.*, 1972). In addition to the obligate anaerobes, several facultative, microaerophilic and capnophilic genera also colonize the oral activity, but these are not listed in the Table.

The microbial flora of the oral cavity varies according to the anatomical site, so that the mouth actually contains a series of different microenvironments or ecological niches (Hardie and Bowden, 1974). Thus, somewhat different habitats are provided by the keratinized and non-keratinized epithelial surfaces of the soft tissues; the hard, non-shedding surfaces of the teeth; and the space between the gingivae and the teeth (gingival crevice in health, periodontal pocket in periodontal disease). In particular, the dorsal surface of the tongue harbours a large flora which appears to resemble the microbial composition of pooled saliva, and has recently been shown to act as a nidus for anaerobic bacteria involved in periodontal diseases (Van der Velden *et al.*, 1986; Van Winkelhoff *et al.*, 1988b).

The oral tissues are bathed in saliva from the major and minor salivary glands. Secretions from these glands vary in flow rate, physical property and chemical composition, thus giving rise to further differences in the physiological characteristics of different parts of the mouth. Around the neck of the teeth there is a flow of gingival crevicular fluid which increases in volume with gingival inflammation, and

been isolated and there are probably many more that have yet to be successfully grown in the laboratory. Examination, by dark-field or phase-contrast microscopy, of dental plaque samples scraped from the teeth, particularly from below the gingival margin, usually reveals several morphological varieties of bacteria that are not recovered by conventional anaerobic culture techniques. In particular, spirochaetes of various sizes are frequently seen in such preparations, but these are generally ignored in cultural studies.

The strictly anaerobic genera which are com-

Table 15.2 Effect of oxygen tension and pocket depth on proportions of selected members of the subgingival flora*

Organism	pO_2†		Pocket depth‡	
	Low	High	Moderate	Deep
Intermediate spirochaete	++	(−)		+
Bacteroides intermedius	++			+
Fusobacterium nucleatum		+		+
Capnocytophaga spp.		++		
Actinomyces naeslundii		++	+	
Streptococcus mutans	++		+	

*Adapted from Loesche *et al.* (1983)
†Low $pO_2 \leq 15$ mmHg; high $pO_2 > 15$ mmHg
‡moderate pockets, 5–6 mm; deep pockets ≥ 7 mm

Table 15.3 Some factors which influence the composition of the oral microflora (from Hardie and Shah, 1982)

Anatomical	Pharmaceutical agents
site	antimicrobials
type of surface	other drugs
Salivary secretions	Oral and dental diseases
composition	other systemic
rate of flow	diseases
Crevicular fluid	dental treatment
Diet	and appliances
pO_2 and Eh	Oral hygiene procedures
Microbial nutrition	Hormonal factors
or metabolism	Genetic factors
Microbial adherence	Smoking

this exerts a particular influence on the environment of the gingival crevice.

Both the oxidation–reduction potential (Eh) and the level of oxygen available have been shown to vary at different sites within the oral cavity (Kenney and Ash, 1969; Eskow and Loesche, 1971) and this may be related to observed variations in the numbers and properties of anaerobes found (Loesche et al., 1983). As dental plaque develops on a tooth surface over a period of five to seven days, the Eh falls from an initial value of +244–294 mV to around −112–141 mV (Kenney and Ash, 1969).

A relationship between oxygen tension, pocket depth and types of bacteria present has been demonstrated (Loesche et al., 1983), the most anaerobic species being found where pO_2 values < 15 mmHg were recorded (Table 15.2).

The composition of the oral microflora may be influenced by several factors, as indicated in Table 15.3. However, the precise effects of most of these factors, particularly on the anaerobic component of the flora, are not well documented. Variations in diet, and particularly carbohydrate intake, affect the numbers and biochemical activities of saccharolytic and extracellular polysaccharide-producing organisms such as streptococci, but the influence of other dietary constituents has not been fully investigated. In sites such as the gingival crevice or periodontal pocket, where anaerobes comprise the major part of the flora, dietary factors may be of less significance than in other parts of the mouth (perhaps due to lack of penetration or access) although firm evidence for this view is lacking. In these situations the presence of crevicular fluid, or bleeding, may be more relevant, and it is notable that several of the strict anaerobes present in periodontal pockets are asaccharolytic, utilizing amino acids or peptides rather than carbohydrates.

The source of organisms in oral and dental infection

Oral and dental infections can broadly be classified as shown in Table 15.4. The majority of the common conditions, many of which involve anaerobes, are thought to be endogenous in origin since the bacteria isolated from these infections are commonly found as part of the normal oral flora. The reasons for the organisms involved causing disease

Table 15.4 Classification of bacterial infections of the mouth

Source of infection	Examples
Endogenous infections	
Dental plaque-related	Dental caries
	Periodontal diseases
Other conditions	Periapical abscess
	Periodontal abscess
	Pericoronitis
	Ludwig's angina
	Maxillary sinusitis
Exogenous infections*	
Primary infections of the mouth	Syphilis, gonorrhoea
Secondary manifestations of systemic infections	Tuberculosis, tetanus, syphilis

*Excluding viral, chlamydial, rickettsial, fungal and protozoal infections with oral manifestations

in these particular cases and not in others, and the route by which they reach the deeper tissues, are not always apparent. Changes in local conditions brought about by trauma or other factors, alterations in local or systemic host defence, effects of drug therapy, changes in the properties or behaviour of the micro-organisms themselves, or their translocation to a new site, are among the explanations which can be offered to account for the occurrence of these endogenous infections.

Exogenous infections of the mouth, whether primary or secondary manifestations of systemic infections, may be due to bacteria, viruses, fungi and other aetiological agents (MacFarlane, 1980). They are less common than the endogenous conditions and rarely involve anaerobes. Clostridial infections of the head and neck are unusual, although gas gangrene does occur occasionally following trauma to cervicofacial tissues. Spasm of the muscles of mastication (trismus or lockjaw) is a well recognized symptom of tetanus, although the site of infection with *Clostridium tetani* is usually distant from the mouth (Loescher, 1987). Sepsis in the orofacial region following traumatic injuries or surgery can be caused by almost any pathogenic micro-organism, including anaerobes.

The pathogenesis of polymicrobial infection

As described in greater detail below, most oral infections are polymicrobial in nature and it is quite unusual to find any that are clearly due to a single species. Mixed infections of varying complexity appear to be the norm, and anaerobic species are frequently present, usually associated with facultative bacteria such as streptococci.

The relative contribution or importance of different bacterial components of a mixed infection to the infectious process is difficult to determine. If the mixture includes organisms known to be highly virulent, such as *Staphylococus aureus* or *Streptococcus pyogenes*, these may be assumed to be particularly significant, but in most cases such organisms are not found in oral infections. Of the better-known anaerobic pathogens, neither *Clostridium* spp. nor *Bacteroides fragilis* are commonly associated with oral infections. Several species of black-pigmented *Bacteroides* are found and some of these may be of particular significance in the pathogenesis of oral infections (van Winkelhoff *et al.*, 1988a). *Actinomyces israelii*, a common component of dental plaque, is usually assumed to be significant if found in pus, particularly when actinomycosis is suspected clinically, although, once again, the flora is frequently polymicrobial.

Research on the pathogenesis of polymicrobial oral infections is hampered by the lack of a good model system for studying the relative contribution of different bacteria in a mixture. Early studies, using a guinea-pig skin model, indicated that black-pigmented *Bacteroides* spp. were of key importance in experimental mixed infections (MacDonald *et al.*, 1956; Socransky and Gibbons, 1965). Several other investigators have tested the pathogenicity of oral micro-organisms, either singly or in combinations, in various animal models, including mice, rats and guinea-pigs (Tazakoe and Nakamura, 1971; Sundqvist *et al.*, 1979; Kastelein *et al.*, 1981; Van Steenbergen *et al.*, 1982, 1984a; Brook *et al.*, 1983; Pancholi *et al.*, 1985; Brook and Walker, 1986). One recent study of the pathogenicity of 20 strains representing nine species from dental abscesses, using a mouse abscess model, demonstrated a range of tissue reactions (Lewis *et al.*, 1988). The most aggressive combinations (pairs) of species were those containing anaerobic gram-negative rods (*B. intermedius* or *Fusobacterium nucleatum*), indicating that these may be major pathogens in acute dento-alveolar abscesses.

Anaerobes and dental caries

Dental caries is initiated at specific sites by demineralization of the enamel of the crown of the tooth, or, less commonly, of cementum in root–surface caries. Although there have been several theories concerning the aetiology of caries over the centuries, the idea that this process is brought about by acids produced from carbohydrates by saccharolytic bacteria on the tooth surface is now generally accepted (Silverstone *et al.*, 1981). The bacteria believed to be chiefly responsible for caries initiation are streptococci and lactobacilli and one species in particular, *Streptococcus mutans*, is generally regarded as the principal aetiological agent (Loesche, 1986).

Caries is a destructive process and when the

surface of the dental hard tissues has been breached it may progress to form a cavity which involves the dentine and, if unchecked, may reach the pulp. There is no evidence to implicate obligate anaerobes in the initial stages of dental caries in enamel, but once the disease has spread to involve the dentine the situation is markedly different. The flora associated with such advanced lesions is more complex than that of the early lesion, which is dominated by acidogenic organisms, and contains several different anaerobes. A study of the prevalence of different bacterial groups in deep dentinal caries showed the presence of significant numbers of gram-positive rods (including *Actinomyces*, *Bifidobacterium*, *Eubacterium*, *Propionibacterium* and *Lactobacillus* spp.) together with lower numbers of streptococci, veillonellae, *Selenomonas* and *Bacteroides* spp. (Edwardsson, 1974). Gram-negative species were far less prevalent than gram-positive organisms in carious dentine, as shown previously (Loesche and Syed, 1973).

Root canal infections

When caries progresses to the extent that bacteria reach the pulp of the tooth, pulpitis ensues and, because of the anatomical constraints of the pulp chamber and root canal, this usually leads to necrosis of pulp tissue. At this stage the entire root canal becomes infected and, if untreated, the infection may spread through the apex of the tooth and into the surrounding tissues. Root canal infections also occur as a result of trauma, particularly when fractures of the tooth or operative procedures expose the pulp to bacterial contamination. When indicated clinically, conservative treatment of necrotic and infected teeth is possible by means of root canal therapy (endodontic treatment), the object of which is to clean out, disinfect and fill the entire pulp chamber and root canal.

A number of investigators have studied the bacterial flora associated with infected root canals over the years, using increasingly sophisticated techniques for isolation and characterization of the organisms present (Carlsson and Sundqvist, 1980). The importance of good sampling techniques, avoiding contamination from saliva or oral mucosa, when taking cultures from the root canal was stressed over 20 years ago by Möller (1966), who also demonstrated the presence of significant numbers of anaerobes. Since then several other reports have confirmed that root canal infections are normally polymicrobial, with a predominance of anaerobic species (Berg and Nord, 1973; Kantz and Henry, 1974; Wittgow and Sabiston, 1975; Sundqvist, 1976). Quantitative bacteriological investigation of ten endodontic samples showed a mean concentration of $10^{7.7}$ bacteria/g, with an average of five species/specimen, representing a wide range of aerobes and anaerobes (Zavistoski *et al.*, 1980).

A recent study of 62 necrotic root canal infections also showed the presence of a variety of obligate and facultative anaerobes (Haapasalo, 1986), including black-pigmented and non-pigmented *Bacteroides* spp. and fusobacteria (Table 15.5). A new species,

Table 15.5 Bacteriological findings in 62 necrotic root canal infections*

Organism	Percentage of all strains isolated
Anaerobes	
Non-pigmented *Bacteroides* spp.	16
Black-pigmented *Bacteroides* spp.	14
Fusobacterium spp.	12
Gram-negative motile rods	9
Gram-positive cocci and rods	33
Facultative anaerobes	
α-haemolytic streptococci	10
Eikenella corrodens	1
Capnocytophaga spp.	1
Actinobacillus actinomycetemcomitans	<1

*From Haapasalo (1986)

Mitsuokella dentalis, was described from the gram-negative anaerobes isolated in this study (Haapasalo *et al.*, 1986) and of the non-pigmented *Bacteroides* spp. found, *B. buccae* appeared to be particularly associated with symtomatic teeth (Haapasalo, 1986), as were the black-pigmented *B. gingivalis* and *B. endodontalis* (described by Van Steenbergen *et al.*, 1984; Haapasalo *et al.*, 1986). The relationship of the black-pigmented *Bacteroides* spp. to pain and other symptoms associated with root canal infections has been noted by other investigators, although it is not always possible to determine the exact species involved from earlier reports (Sundqvist, 1976; Griffee *et al.*, 1980).

Dental abscesses

Types of abscess in the mouth

Many dental abscesses (dentoalveolar abscesses) originate from necrotic and infected pulps via the dental root canal, but they may also occur as a result of a flare-up of infection within the gingival crevice (usually associated with local trauma) or, more likely, an existing periodontal pocket. The former type is variously referred to as a periapical, apical or endodontal abscess, while those which start in the periodontium are usually described as periodontal or paradontal abscesses.

Either type of abscess may present clinically in an acute form with pain and swelling. The amount and route of spread of infection depends upon the location of the affected tooth (i.e., upper or lower, anterior or posterior) as well as the virulence of the causative organisms (Piecuch, 1982). Some acute dentoalveolar abscesses discharge pus spontaneously, either externally through the skin or internally into the oral cavity. Periapical infections may become chronic, if not adequately treated, and form periapical granulomas, some of which subsequently become cystic. Acute exacerbations of such chronic lesions may then supervene, once again leading to the production of pus and the clinical signs of acute inflammation.

Recognition of role of anaerobes

It has been recognized for many years that dental abscesses are usually polymicrobial infections, and that a wide variety of facultative and obligately anaerobic bacteria have been isolated from them. A significant trend in recent years has been an increased appreciation of the frequency of occurrence and importance of strict anaerobes in oral pus samples. Some years ago, in a survey of 1000 consecutive specimens of pus from oral lesions (Sims, 1974), 'viridans' streptococci were found in 90 per cent of cases, whereas anaerobic gram-negative rods (*Bacteroides* and *Fusobacterium* spp.) were only reported in 2.6 per cent. However, the author pointed out that the isolation frequency of strict anaerobes, using routine clinical laboratory methods, was probably a gross underestimate which could be improved by the use of strict anaerobiosis at all stages in the handling of specimens. In fact, similar conclusions were drawn from an earlier study of 73 submucous oral abscesses by Feldman and Larje (1966). These authors also found viridans streptococci to be most common, but observed a higher isolation frequency of anaerobic bacteria than previously reported, this was attributed to the use of improved anaerobic isolation techniques.

Much of the earlier literature on the microbiology of dental pyogenic infections was reviewed by Sabiston and Grigsby (1977) and these authors emphasized the importance of proper sampling and attention to anaerobic techniques in order to obtain a true picture of the organisms present. Their conclusion that these infections are usually mixed anaerobic infections caused by indigenous bacteria is supported by several more recent studies.

The pathogenic potential of individual isolates and pairs of bacteria representing nine species recovered from dentoalveolar abscesses has been demonstrated experimentally in mice (Lewis *et al.* 1988b). These studies indicate a major role for anaerobic gram-negative rods in such infections.

Bacteriology of dental abscesses

Since the mid-1970s there have been several studies on the bacterial composition of pus from various types of dental abscess in which modern anaerobic isolation and identification methods have been employed (e.g., Moore and Russell, 1972; Sabiston *et al.*, 1976; Chow *et al.*, 1978; Greenberg *et al.*, 1979; Kannarigara *et al.*, 1980; Zavistoki *et al.*, 1980; Aderhold *et al.*, 1981; Brook *et al.*, 1981; Von Konow *et al.*, 1981; McKee *et al.*, 1982; Oguntebi *et al.*, 1982;

Table 15.6 Number of bacterial isolates from purulent oral infections

Authors	Number of samples	Mean number of isolates/sample	Range	Proportion of anaerobic isolates (%)
Feldman and Large (1966)	73	NR	1–6	High
Sabiston et al. (1976)	58	3.8	1–12	66
Aderhold et al. (1981)	50	3.7	1–7	73
Brook et al. (1981)	12	4.4	NR	90
Oguntebi et al. (1982)	10	2.5	2–4	48
McKee et al. (1982)	7	5.0	1–10	80
Williams et al. (1983)	10	4.5	1–8	70
Heimdahl et al. (1985)				
Mild infections	24	3.0	NR	90
Severe infections	31	3.5	NR	88
Lewis et al. (1986a)	50	3.3	1–8	75

NR, Not recorded

Heimdahl and Nord, 1983; Labriola et al., 1983; Williams et al., 1983; Morey et al., 1984; Van Winkelhoff et al., 1985; Haapasalo et al., 1986; Lewis et al., 1986a).

All investigations indicate that these infections are usually polymicrobial and that monoinfections are very uncommon. The number of isolates recovered per specimen varies from study to study; anything from one to twelve or more species may be found, although the mean number is usually between 2.5 and 5.0 (Table 15.6).

Most recent investigators have reported a high proportion of strictly anaerobic isolates from dental abscesses, the examples shown in Table 15.6 indicating that these often comprise 70–90 per cent of the bacteria recovered. The remaining organisms usually include one or more streptococcal species and, sometimes, other facultative bacteria.

Some idea of the range of strictly anaerobic genera and species that have been isolated is given in Table 15.7, which summarizes the findings in seven published papers. Direct comparisons between such studies are not easy because of differences in the selection of clinical cases, sample collection and transport, laboratory methods and taxonomic or nomenclatural criteria employed by different workers. The last of these variables applies particularly to the rapidly changing area of *Bacteroides* nomenclature and also to the anaerobic gram-positive cocci. Similarly, nomenclatural problems also exist with the streptococci which predominate among the aerobic and facultative component of these infections (Hardie, 1986a).

A few general points can be made about the anaerobic flora of dental abscesses. Anaerobic gram-positive cocci, various pigmented and non-pigmented *Bacteroides* spp., and *Fusobacterium nucleatum* feature regularly in most of the published reports. *Bacteroides fragilis* is usually not observed, although it has been reported in 29.5 per cent of one series of 61 cases of pyogenic dental infections (Kannangara et al., 1980) and in three patients with mandibular osteomyelites (Chow et al., 1978).

One isolate of *Clostridium clostridiiformis* was reported by Heimdahl et al. (1985) but most investigators have not isolated *Clostridium* spp. from purulent oral infections.

In the study by Heimdahl et al. (1985) the bacteriological data were analysed separately for patients with mild and severe oral infections, these categories being determined by body temperature (above or below 38°C) and extent of infection beyond the alveolar process. Anaerobic gram-negative rods were more frequently isolated from the severe cases and *F. nucleatum*, in particular, appeared to be associated with severity. Aerobes were isolated more often from patients with severe infections in this study, all but one of the strains being identified as *Str. milleri*. *Str. milleri* was found in 25 of the 50 abscesses examined by Lewis et al. (1986a)—a far greater frequency than any other streptococcal species. With the closely-related species *Str. intermedius* and *Str. constellatus*, which together comprise the '*Str. milleri*' group, (Hardie, 1986a), *Str. milleri* appears to be a particularly signi-

Table 15.7 Isolation of strict anaerobes from dental abscesses in seven published studies

Genera/species	Number of isolations in study no.							
	1	2	3	4	5	6		7
						mild cases	severe cases	
Peptostreptococcus spp.	22	9	0	3	6	8	10	14
Peptococcus spp.	2	26	2	0	0	0	0	32
Anaerobic streptococci	12	7	6	0	1	8	4	4
Unidentified gram-positive cocci	8	0	0	0	1	0	0	0
Propionibacterium acnes	0	2	0	0	0	0	0	1
Eubacterium lentum	1	6	0	0	1	5	6	1
Unidentified *Eubacterium* spp.	4	0	1	0	0	5	2	0
Bifidobacterium spp.	6	0	0	0	0	0	1	0
Actinomyces spp.	10	0	3	0	0	0	0	0
Unidentified gram-positive rods	1	6	0	0	1	0	0	0
Anaerobic lactobacilli	0	0	3	0	0	2	5	0
Clostridium clostridiiforme	0	0	0	0	0	0	1	0
Veillonella spp.	10	5	4	0	0	4	6	3
Unidentified *Bacteroides* spp.	3	2	3	0	8	8	10	0
B. asaccharolyticus	0	15	2	0	1	1	2	0
B. melaninogenicus	2	18	6	0	3	5	10	12
B. intermedius	8	1	1	2	0	0	0	5
B. gingivalis	0	0	0	0	0	0	0	14
B. oralis	3	1	3	0	3	1	4	20
B. ruminicola	8	1	0	0	2	3	14	6
B. distasonis	0	0	0	0	0	0	1	1
B. ureolyticus/B. corrodens	0	3	3	0	3	3	2	1
B. capillosus	0	0	0	0	0	0	2	1
B. uniformis	0	0	0	0	0	0	1	1
B. bivius	0	0	1	0	0	0	0	0
B. disiens	0	0	0	0	0	0	1	0
B. pneumosintes	0	0	0	0	0	1	0	0
Fusobacterium spp.	1	9	2	0	1	0	0	1
F. nucleatum	14	17	3	7	6	7	19	6
Number of cases examined	58	50	12	10	10	24	31	50

1, Sabiston *et al.* (1976); 2, Aderhold *et al.* (1981); 3, Brook *et al.* (1981); 4, Oguntebi *et al.* (1982); 5, Williams *et al.* (1983); 6, Heimdahl *et al.* (1985); 7, Lewis *et al.* (1986a)

ficant pathogen in purulent infections in other parts of the body as well as the mouth (Gossling, 1988). Since such a varied collection of bacteria, usually in mixed culture, is found in oral abscesses it is difficult to ascribe a high level of specificity or significance to any particular genus or species. Capsulate strains of *Bacteroides* and gram-positive anaerobic cocci occur more frequently in pus from orofacial abscesses, including periapical abscesses, than in strains from the normal oropharyngeal flora (Brook, 1986). This observation applied particularly to the *B. melaninogenicus* group, of which 84 per cent of abscess strains possessed capsules, in contrast to 23 per cent of control pharyngeal culture strains.

Quantitative studies

There have been few quantitative studies on the number of organisms present in dental abscesses. The range of cfu/ml pus for six specimens was found to be (3.3×10^6)–(1.2×10^9) (Williams *et al.*, 1983). Quantitative data from a more extensive study of 50 dental abscesses have recently been provided by

Table 15.8 Mean concentration of bacterial groups isolated from 50 acute dentoalveolar abscesses

Bacterial group	Mean concentration ($\log_{10}$ cfu/ml)
Facultative anaerobes	5.7 ± 0.2
Gram-positive cocci	6.1 ± 0.2
Gram-positive bacilli	4.6 ± 0.4
Gram-negative bacilli	5.6 ± 0.4
Strict anaerobes	6.2 ± 0.1
Gram-positive cocci	6.3 ± 0.2
Gram-positive bacilli	5.7 ± 0.8
Gram-negative cocci	5.9 ± 0.7
Gram-negative bacilli	6.1 ± 0.2

Lewis *et al.* (1986a) (Table 15.8). These figures indicate the relative numbers of the main groups of bacteria isolated and underline the preponderance of strict anaerobes. The mean viable bacterial load in this study was $10^{6.9}$ (range $10^{4.7}$–$10^{9.4}$) cfu/ml, similar to that reported previously by Williams *et al.* (1983). The identity and isolation frequency of the 166 strains recovered from 50 acute dental abscesses by Lewis *et al.* (1986a) are reproduced in Table 15.9.

It has been suggested that streptococci are involved in the early phase of abscess formation and that this prepares the environment for the subsequent growth of anaerobes (Aderhold *et al.*, 1981). Data from Lewis *et al.* (1986a) appear to support this observation, since these authors only found *Str. milleri* in pure culture when abscesses were sampled on the first day of clinical symptoms; after longer periods a mixed anaerobic flora was invariably found.

In summary, the flora of dental abscesses is usually polymicrobial, commonly yielding a mixture of aerobic or facultatively anaerobic species (predominantly α- or non-haemolytic streptococci) together with a preponderance of strictly anaerobic bacteria, which frequently include gram-positive anaerobic cocci, black-pigmented and non-pigmented *Bacteroides* spp. and *Fusobacterium* spp.

Laboratory diagnosis and handling of specimens

Not every dental abscess necessarily requires a bacteriological investigation; many patients respond quite satisfactorily to empirical treatment without the clinician knowing exactly which bacteria are

Table 15.9 Identity of 166 bacterial strains isolated from 50 acute dento-alveolar abscesses

Facultative anaerobes	Number of isolates	Strict anaerobes	Number of isolates
*Streptococcus milleri**	25	*Peptostreptococcus* spp.	14
Streptococcus mitior	3	*Peptococcus* spp.	32
Streptococcus sanguis	3	*Streptococcus intermedius**	3
Streptococcus mutans	1	*Streptococcus constellatus**	1
Lactobacillus fermentum	2	*Propionibacterium acnes*	1
Lactobacillus salivarius	1	*Eubacterium lentum*	1
Actinomyces odontolyticus	1	*Veillonella parvula*	3
Actinomyces naeslundii	1	*Bacteroides oralis*	20
Actinomyces meyeri	1	*Bacteroides gingivalis*	14
Arachnia propionica	1	*Bacteroides melaninogenicus*	12
Haemophilus parainfluenzae	2	*Bacteroides intermedius*	5
Capnocytophaga ochracea	1	*Bacteroides ruminicola*	6
Eikenella corrodens	1	*Bacteroides distasonis*	1
		Bacteroides ureolyticus	1
		Bacteroides capillosus	1
		Bacteroides uniformis	1
		Fusobacterium nucleatum	6
		Fusobacterium mortiferum	1
Total	43	Total	123

*These streptococcal species comprise part of the '*Streptococcus milleri*' group (Hardie, 1986a; Gossling, 1988)

present or their antibiotic sensitivity. However, for epidemiological purposes it is important that such data are collected periodically by some reference centre, in order to keep track of possible changes in the types of organisms responsible or in their antibiotic susceptibility. In addition, some patients require full bacteriological investigations for particular clinical reasons, such as failure to respond to conventional treatment or extensive and life-threatening spread of infection.

When cultures are indicated, great care must be taken during sampling to avoid contamination of the specimen with saliva or with bacteria present on the skin or oral mucosa. After appropriate surface disinfection, specimens should ideally be collected by aspiration with a sterile syringe and needle. Rapid transport to the laboratory, or maintenance in a reduced environment, will help to preserve the viability of obligate anaerobes present in the pus sample (Finegold and Edelstein, 1988).

Careful handling in the laboratory, with good anaerobic technique, is essential for optimum recovery of many of the species commonly found in oral specimens (Hardie, 1986b). Although the availability of an efficient anaerobic cabinet will undoubtedly facilitate the isolation of anaerobes, good recoveries can also be obtained by conventional anaerobic jar techniques, provided that freshly prepared or prereduced media are used, specimens are processed promptly and the anaerobic jar is properly maintained and put up. In one study (McKee et al., 1982) no significant differences were observed in the types of bacteria or their relative numbers when oral pus specimens were divided and processed in parallel in an anaerobic cabinet and in anaerobic jars.

Clinical management and antimicrobial chemotherapy

Surgical drainage is almost invariably necessary in the treatment of dental abscesses. Depending on the clinical situation this may be obtained through the pulp chamber and root canal (if it is intended to preserve the tooth), via the tooth socket after extracting the offending tooth or by incision, either through the oral mucosa or the skin. In uncomplicated, well localized cases antibiotic treatment may not be necessary, but when there is extensive spread of infection or systemic symptoms (such as fever) antimicrobial agents are usually prescribed.

The choice of a single chemotherapeutic agent to cover all possible combinations of bacteria that may be encountered in dental pyogenic infections is problematical. In most cases treatment needs to be initiated before bacteriological data are available and even when laboratory results have been obtained it is not easy to decide which organisms in a complex mixture are most significant. Primary antibiotic sensitivity testing on the original pus sample is sometimes helpful in determining which agents to select (Lewis et al., 1988a), although such tests may not be ideal for estimating the antibiotic sensitivity of the more slowly growing anaerobic species.

The results of antibiotic sensitivity testing on dental abscess isolates have been reported by several authors and much of the earlier data is reviewed by Sabiston and Grigsby (1977). Most of the streptococci and many of the anaerobic bacteria appear to be sensitive to penicillin, and this continues to be the usual drug of choice (Sims, 1974). However, treatment failures due to β-lactamase-producing *Bacteroides* strains have been recorded (Heimdahl et al., 1980). Erythromycin has often been considered as a good alternative for patients who are sensitive to penicillin, but resistance to erythromycin was found to be common among isolates in one recent study, particularly in strains of *F. nucleatum* (Heimdahl et al., 1985). In the same study, all isolates were susceptible to clindamycin, almost all to penicillin, while a few strains were resistant to doxycyline.

Metronidazole and other nitroimidazoles, such as tinidazole and ornidazole, are effective against all obligate anaerobes found in oral purulent infections, and have been used successfully in the treatment of acute dental infections (Ingham et al., 1977; Hood, 1978; Rood, 1980; Von Konow and Nord, 1983). From these reports it would appear that the nitroimidazoles are a good alternative to penicillin for orofacial infections, including other conditions such as acute pericoronitis (McGowan et al., 1977) and dry socket (Mitchell, 1984), even though they are ineffective against aerobic and facultative bacteria.

Recently, a clinical study has shown that a short course of high dose amoxycillin (two 3g sachets, taken 8 h apart) was as effective as a conventional five day course of oral penicillin (250 mg four times daily) in the treatment of acute dental abscesses (Lewis et al., 1986b). The simplicity of this treatment has many attractions to both general practitioners and patients.

Periodontal diseases

Types of disease

Inflammatory periodontal disease in man may be confined to the soft tissues of the gingiva (gingivitis), or extend to involve the deeper supporting structures of the periodontal membrane and alveolar bone (periodontitis). In periodontitis there is loss of attachment of the teeth, with the formation of periodontal pockets of varying depth, depending on the severity of the condition. Ultimately this may lead to loosening and loss of the affected teeth.

In recent years there has been an increasing awareness not only of the several different forms of human periodontal disease which can be recognized clinically, but also that these conditions may have different aetiologies and prognoses. Gingivitis can either be localized to one or more specific sites in the mouth, or can occur in a generalized form. It may also be acute or chronic, perhaps the most common presentation being chronic marginal gingivitis associated with poor oral hygiene and dental plaque accumulation. Particular types of gingivitis occur where there are underlying systemic conditions, such as pregnancy, insulin-dependent diabetes and Papillon–Lefevre syndrome, and in the relatively uncommon clinical condition of acute necrotizing ulcerative gingivitis (ANUG), also known as Vincent's gingivitis. In addition to naturally occurring gingivitis, there is a much studied experimental model of gingivitis in man which can be induced by withholding toothbrushing and other oral hygiene measures for 7–10 days (Loe et al., 1965).

Juvenile and adult forms of periodontitis are recognized and both forms may be localized or generalized. The severity and rate of progression of periodontitis varies and gives rise to qualifying descriptions of the clinical presentation, such as 'moderate', 'severe', rapidly-progressive' and 'destructive' periodontitis. Thus, by using a combination of type and extent of tissue involvement, rate of progress, age of patient and presence or absence of underlying systemic conditions, a wide variety of clinical types of periodontal disease may be recognized. Whether each of these has a distinctive microbial aetiology is far from clear at present, but, as described below, there is evidence of association between some types of periodontal disease and particular bacterial species.

It has recently been observed that periodontal diseases may progress episodically, with 'bursts' of destructive activity, rather than at a slow continuous rate as previously assumed (Goodson et al., 1982; Haffajee et al., 1986). As a result, interest in the possible microbiological changes or activities that may be found in sites where such bursts of activity occur has grown (Dzink et al., 1985).

Anaerobes in periodontal disease

The gingival crevice and periodontal pocket provide a reduced environment which is especially favourable to the growth of obligate anaerobes (Socransky et al., 1963; Kenney and Ash, 1969; Loesche et al., 1983). The use of improved sampling techniques and application of modern anaerobic isolation methods to subgingival samples in more recent studies has revealed high numbers of a very wide range of anaerobes which may be involved in periodontal disease (Aranki et al., 1969; Gordon et al., 1971; Newman et al., 1976; Slots, 1979; Tanner et al., 1979; Van Palenstein Helderman, 1981; Loesche et al., 1985; Moore, 1987). The complexity of the microflora found is well illustrated in the studies by the Department of Anaerobic Microbiology of the Virginia Polytechnic Institute (Moore et al., 1982a, b, 1983), in which over 325 species of bacteria were isolated, many of them hitherto undescribed or of uncertain taxonomic status (Moore, 1987).

One of the difficulties with research on periodontal disease has been to establish unequivocal evidence for cause and effect relationships with specific organisms (Socransky, 1979). However, several studies have indicated associations between putative pathogens and particular conditions, usually in cross-sectional studies. Possible virulence determinants have been examined in vitro, attempts have been made to eliminate or suppress these organisms therapeutically and further insights have been gained by investigating humoral and cellular responses in the host. Many of these studies tend to support a 'specific' rather than a 'non-specific' hypothesis for the aetiology of periodontal disease (Loesche, 1982).

Experimental gingivitis

The original studies on experimental gingivitis in humans showed that gingivitis occurs when the early, predominantly gram-positive plaque develops into a more complex flora containing many gram-negative, motile and spiral forms (Loe et al., 1965). More detailed bacteriological studies with the same experimental model have indicated that, with time, there is a shift towards an *Actinomyces*-dominated supragingival plaque, and that the onset of gingivitis is associated with *Actinomyces israelii* (Loesche and Syed, 1978; Syed and Loesche, 1978). The presence of gingival bleeding is associated with isolation of *A. viscosus* and non-saccharolytic, black-pigmented *Bacteroides* (probably *B. gingivalis*). In a later study on four healthy subjects, Moore et al. (1982a) reported that *A. naeslundii* (serotype III and serologically unreactive strains), *F. nucleatum*, *Lactobacillus* spp. (D-2), *Str. anginosus*, *Veillonella parvula* and *Treponema* spp. (A) appeared to be the most likely aetiological agents in this type of gingivitis. In contrast to the previously quoted study, these authors did not find evidence for the involvement of black-pigmented *Bacteroides* spp. in their experimental subjects. Although there are differences between these two studies, it does appear that experimental gingivitis occurs as a result of accumulation of increased numbers of some bacterial species that are normally present in supragingival plaque, and it seems unlikely that there is a single aetiological agent.

Juvenile periodontitis

This form of periodontal disease, previously known as periodontosis, causes extensive destruction of the periodontal tissues in young people. The condition is often localized to the first permanent molars and incisors, but a more generalized form is also recognized. In addition to any specific micro-organisms that may be involved, immunological factors are also of special significance in this condition (Genco and Slots, 1984). Studies by Newman et al. (1976) and Slots (1976) indicated that a variety of gram-negative rods could be isolated from the pockets of patients with juvenile periodontitis. In particular, these sites were found to harbour increased proportions of *Actinobacillus actinomycetemcomitans* (sometimes referred to as *Haemophilus actinomycetemcomitans*) frequently in association with *Capnocytophaga* spp. or *Eikenella corrodens* (Newman and Socransky, 1977; Mandell and Socransky, 1981; Eisenmann et al., 1983; Savitt and Socransky, 1984; Zambon et al., 1983a).

In a review of recent publications, Moore (1987) pointed out that the rate of isolation of *Actinobac. actinomycetemcomitans* from patients with juvenile periodontitis varies considerably from one study to another. In the cross-sectional studies analysed, the species was found to represent from 0 to 90 per cent of the flora at different sites; if it is assumed, arbitrarily, that 1 per cent of the flora is required for the organism to cause tissue destruction, Moore (1987) concludes that *Actinobac. actinomycetemcomitans* could have been the causative agent in approximately one-third of the sites examined. The reported concentration of this species was also found to vary in longitudinal studies, where it represented 0–1 per cent in 37 per cent of 43 active sites, 1–10 per cent in 60 per cent of sites, and 74 per cent in 2 per cent of sites (Mandell, 1984; Hafferjee et al., 1984, 1985; Dzink et al., 1985; Moore, 1987). Hafferjee et al. (1985) found that juvenile periodontitis patients with initially high levels of *Actinobac. actinomycetemcomitans*, *B. intermedius* and *Capnocytophaga ochracea* responded well to treatment with surgery and tetracycline, whereas those with high levels of *F. nucleatum*, *E. corrodens*, *B. intermedius* and *Cap. sputigena* responded poorly.

In a study of eight patients with juvenile periodontitis (Mandell et al., 1987) it was found that numbers of *E. corrodens* were significantly elevated in progressive sites which also harboured *Actinobac. actinomycetemcomitans*, and the possibility of synergy between these two species was suggested. All patients had elevated serum IgG antibodies to the same serotype of *Actinobac. actinomycetemcomitans* as that isolated from the periodontal pocket, but elevated responses to several other oral organisms were also noted.

Although there may still be some doubt as to whether *Actinobac. actinomycetemcomitans* is invariably associated with juvenile periodontitis, there is sufficient evidence to suggest that it is a significant pathogen in at least some cases of this disease, particularly the localized form. The ability of this species to produce a potent leucotoxin which affects human polymorphonuclear leucocytes (PMNL) is one virulence factor to attract particular attention (Taichman et al., 1982; Zambon et al., 1983b). The possibility that virulence of *Actinobac. actinomycetemcomitans* may be associated with bacteriophage infection has recently been suggested (Preus et al., 1987).

Acute ulcerative necrotizing gingivitis (ANUG)

In Europe and the USA this relatively uncommon type of gingivitis occurs mainly in young adults, in whom it often appears to be associated with stress (Goldhaber and Giddon, 1964).

A different picture has been reported in African countries, such as Nigeria, where ANUG is seen in children and may progress to cause massive destruction of orofacial tissues in the condition called *cancrum oris* or *noma* (Enwonwu, 1972). On the basis of microscopical examination of samples from affected patients, it has long been thought that a fusospirochaetal flora is associated with ANUG (Smith, 1932; Rosebury et al., 1950) although early cultural and experimental infection studies indicated an important role for *B. melaninogenicus* (MacDonald et al., 1956).

Antibiotic therapy is usually indicated as part of the treatment of ANUG, using either penicillin or metronidazole (Shinn, 1962; Duckworth et al., 1966). The early observation of the efficacy of metronidazole in the management of ANUG was especially significant in view of the subsequent recognition of its value in the treatment of other, life-threatening anaerobic infections.

In one of the few reported bacteriological studies on ANUG, Loesche et al. (1982) examined samples from 22 ulcerated sites in eight patients, using cultural and microscopical methods to determine the composition of the predominant flora present. On the basis of the microscopic counts, *Treponema* and *Selenomonas* spp. comprised mean levels of 32 per cent and 6 per cent, respectively, of the flora of diseases sites, while *B. intermedius* and *Fusobacterium* spp. averaged 24 per cent and 3 per cent of the viable counts. The constant presence of these four organisms was considered to be pathognomonic for ANUG. Significant reductions in the proportions of these bacteria were observed for two to three months following a one-week course of metronidazole, coinciding with a rapid resolution of clinical symptoms.

Adult periodontitis

The complexity of the periodontal microflora and the variety of study methods employed by different investigators have led to some confusion about the significance of specific organisms in adult periodontitis and other forms of periodontal disease (Moore, 1987). However, there can be little doubt that gram-negative bacteria play a major part in the pathogenesis of these conditions (Socransky, 1977; Slots, 1979; Duerden et al., 1987). Three species in particular, the anaerobes *B. gingivalis* and *B. intermedius*, and the facultative *Actinobac. actinomycetemcomitans*, are considered to be of special importance because they are frequently isolated from advancing progressive periodontitis lesions and are known to possess several potential virulence factors (Slots 1986a,b). Several other bacteria may also be regarded as possible periodontopathogens, including *Wolinella recta*, *B. forsythus*, *E. corrodens*, *Fusobacterium* and *Treponema* spp., and even some gram-positive spp., but the published data on these candidates pathogens is less extensive than on the three 'favourites'. Several recent reviews of the rapidly proliferating literature on this subject have been published (e.g., Slots 1986a,b; Newman, 1985; Zambon, 1985; Moore, 1987; Slots and Listgarten, 1988, Van Winkelhoff et al., 1988a).

Differences in the composition of the subgingival microflora between healthy and diseased sites have been demonstrated by examining wet preparations with dark-field or phase-contrast microscopy. In healthy sites, cocci and non-motile rods were found to comprise >90 per cent of the total flora, while motile rods and spirochaetes were rarely seen (Listgarten and Hellden, 1978). In contrast, samples from diseased periodontal pockets contained mean levels of motile rods and spirochaetes of 14 per cent and 38 per cent respectively, and the ratio of motile to non-motile organisms seen was approximately 1:1, compared to 1:49 in the healthy sites. In a later study the proportions of spirochaetes with or without motile rods were found to be good predictors of subsequent periodontal deterioration (Listgarten and Levin, 1981).

Several other investigators have also used microscopical counts of spirochaetes and motile rods as a means of assessing the periodontal flora and for monitoring treatment (Keyes *et al.*, 1978; Lindhe *et al.*, 1980; Armitage *et al.*, 1982; Macon *et al.*, 1982). A recent study on the prevalence of oral spirochaetes in prepubertal Tanzanian and Dutch children, using dark-field microscopy, has shown that small, medium and large spirochaetes can be found in the normal gingival crevice, indicating that inflammation is not an essential prerequisite, but increased proportions of spirochaetes were associated with gingival inflammation (Mikx *et al.*, 1986). Differences in the numbers of spirochaetes were observed between the two populations and higher levels appeared to be present at bleeding sites.

Most investigators have relied on assessment by microscopy of spirochaetes because of the difficulty of culturing these organisms, and only a few reports have included data on the isolation of spirochaetes from periodontal samples (Moore *et al.*, 1982a,b, 1983). Although there appears to be an association between the presence of high numbers of spirochaetes and human periodontitis, which may be a useful indicator of disease activity, the role of individual *Treponema* spp. in the aetiology of periodontal disease is still uncertain (Loesche and Laughon, 1982).

Since the mid-1970s, many reports of cultural studies on human periodontitis have been published (e.g., Williams *et al.*, 1976; Darwish *et al.*, 1978; Slots, 1979; Tanner *et al.*, 1979, 1984; Moore *et al.*, 1982b, 1983; Savitt and Socransky, 1984; Dzink *et al.*, 1985; Loesche *et al.*, 1985; Slots *et al.*, 1986). In a study of active destructive periodontal lesions in 19 patients (aged 11–60, mean age 19.5 ± 10.6) proportions of the predominant gram-negative species in 50 active disease sites were compared to those in 69 inactive sites of comparable pocket depth (Dzink *et al.*, 1985). The proportions of gram-negative rods were higher in active sites (46 per cent compared to 35 per cent in inactive sites) and the individual species *B. intermedius*, 'fusiform' *Bacteroides* (now called *B. forsythus*), *Actinobac. actinomycetemcomitans* and *W. recta* were found to be significantly elevated only in active sites. Other organisms, such as *F. nucleatum*, *Cap. gingivalis* and *E. corrodens* were found in significantly increased proportions in active sites of some subjects, but in inactive sites of others.

B. forsythus is a slow-growing, fusiform organism with a distinctive cell wall structure that has been isolated from advanced periodontal lesions (Dzink *et al.*, 1985; Tanner *et al.*, 1986) and which is known to produce a trypsin-like enzyme (Tanner *et al.*, 1985). Lai *et al.* (1987) have investigated the prevalence of *B. forsythus* in adults with varying degrees of periodontal disease, using an indirect immunofluorescence method for detecting the organism (which is difficult to isolate routinely). Higher proportions of *B. forsythus* were found in supragingival samples from patients with mild gingivitis (1.3 per cent), severe gingivitis (1.0 per cent) and periodontitis (0.9 per cent) than from those with healthy gingivae (0.2 per cent) and the levels were also significantly higher in subgingival samples (15.3 per cent) than in supragingival samples (0.9 per cent) in adult periodontitis patients. Previously-treated patients with moderate or severe periodontitis who were monitored for 12 months showed a significant increase in proportions of *B. forsythus* at sites where further breakdown occurred compared to stable sites. As the authors acknowledge, whether *B. forsythus* initiates periodontal destruction, or whether it is a secondary colonizer of diseased sites remains to be determined, but it may be a useful diagnostic indicator for monitoring patients under treatment.

One recent comparison of 15 active progressive sites with paired control sites in six subjects failed to find statistically significant evidence for the association of particular species with loss of attachment (Ranney *et al.*, 1987), even though there was a suggestion that *Eubacterium timidum*, *B. intermedius* and *A. israelii* levels may have differed. However, it is quite possible that there are different causes of disease progression in different instances, and that disease activity may be related to the presence of various combinations of species rather than individual taxa.

The relationship of *Actinobac. actinomycetemcomitans*, *B. gingivalis* and *B. intermedius* to progressive disease was investigated in 146 adults with advanced periodontitis, in whom 105 'non-progressing' and 130 'progressing' sites were examined (Slots *et al.*, 1986; Bragd *et al.*, 1987). One or more of these bacteria were recovered from 99 per cent of the progressive sites, compared to 4 per cent of the non-progressive sites, and the authors concluded that they were closely related to disease activity. The suggestion that these three periodontopathic species might be useful indicators or predictors of future periodontal disease activity has been taken up by other investigators. In one prospective study (Wennstrom et al., 1987), no clinically significant loss of attachment was observed in the absence of *Actinobac. actinomycetemcomitans* and *B. gingivalis* or the presence of less than 5 per cent *B. intermedius*. In those sites where the indicator species were present, only 5 out of 25 showed evidence of disease progression. Thus it appeared that the absence of these organisms was a better predictor of 'non-activity' than their presence was for future disease progression.

The need to consider the role of organisms other than the three main indicator species studied by some investigators is apparent from a recent study on subjects with refractory periodontal disease (Haffajee et al., 1988). The 13 patients who showed further disease progression following surgery and systemic tetracycline therapy were found to have three major microbial complexes in their subgingival flora: *B. forsythus*, *F. nucleatum* and *W. recta* (three subjects); *Str. intermedius*, *B. gingivalis* and *Peptostreptococcus micros* (three subjects); *Str. intermedius* and *F. nucleatum* with or without *B. gingivalis* (seven subjects). Combining antibiotic therapy with conventional periodontal treatment may have a significant effect on the flora of periodontal pockets, as has been demonstrated with metronidazole (Loesche et al., 1981; Gusberti et al., 1988) which can result in significant reductions in numbers of some gram-negative rods over a six-month period.

Indicators of periodontal diseases

It is difficult to summarize simply the many published studies and produce a clear indication of the significance of each potential periodontal pathogen. Some of the suggested associations between particular organisms and different forms of juvenile and adult periodontitis are shown in Table 15.10.

The quest for microbial and other indicators of periodontal disease activity has led to the development of alternative, more rapid methods for identifying and detecting key organisms. Tanner et al. (1987) have described the use of SDS-PAGE for rapid identification of 'activity-related gram-negative species', and other workers have used immunofluorescence (Bonta et al., 1985; Christersson et al., 1987a,b) or DNA probes (French et al., 1986; Kuritza et al., 1986; Roberts et al., 1987; Strzempko et al., 1987). Recently, the application of a DNA-RNA dot hybridization procedure using synthetic oligodeoxynucleotide probes for detection of *Haemophilus aphrophilus*, *Actinobac. actinomycetemcomitans*, *B. gingivalis*, *B. asaccharolyticus* and *B. intermedius* has been described (Chuba et al., 1988).

In addition to possible microbiological indicators of periodontal disease activity, the detection of bacterial products, both immunological and non-

Table 15.10 Bacteria associated with some types of human periodontitis*

Species	Clinical condition†			
	Localized JP	Generalized JP	Early onset AP	Severe AP
Actinobac. actinomycetemcomitans	++	+	?	++
B. gingivalis	−	++	+	++
B. intermedius	−	+	+	++
B. forsythus	?	?	+	+
Fusobacterium spp.	−	?	++	+
Capnocytophaga spp.	+	−	−	−
'Corroding organisms'‡	+	+	+	+
Eubacterium spp.	−	?	++	−
Treponema spp.	?	?	+	+

++, strong association; +, some evidence of association; ?, insufficient data available
*Adapted from Slots and Dahlen (1985)
†JP, Juvenile periodontitis; AP = adult periodontitis
‡ 'Corroding organisms' include *Wolinella* spp., *B. gracilis* and *E. corrodens*

immunological host reactions and products of tissue injury in epithelia, connective tissue or bone are alternative approaches which may be useful (Fine and Mandel, 1986).

Pathogenic mechanisms in periodontal diseases

The mechanisms by which periodontal disease is produced are not fully understood, but appear to be complex. As pointed out by Listgarten (1987), periodontal pathogens may cause tissue damage either by direct or indirect toxicity. A number of potentially toxic products or virulence factors have been described, including exotoxins, endotoxins (LPS), a variety of tissue-damaging enzymes and low molecular weight metabolic products (Slots and Genco, 1984; Slots and Dahlen, 1985; Listgarten, 1987). Indirect toxic effects may be exerted on host cells, for example, by neutralizing or modifying host defence mechanisms, or may occur as a result of bacterial interactions.

Environmental factors may alter the virulence of periodontal pathogens as has been shown, for example, with the effect of haemin on *B. gingivalis* (McKee *et al.*, 1986).

In addition to the bacterial factors, it is almost certain that some of the tissue destruction seen in periodontal diseases is host mediated (Genco and Slots, 1984; Taichman *et al.*, 1984). Both humoral and cell mediated immune responses have been implicated, as have complement activation and the release of various potentially damaging factors during the inflammatory process in the periodontal tissues (Listgarten, 1987).

Other oral infections

Perimandibular space infections and Ludwig's angina

Odontogenic infections may spread in several directions along fascial planes into the surrounding soft tissues, resulting in a variety of clinical presentations (Guralnick, 1984). Such conditions can be extremely serious and life-threatening, particularly in the submandibular region, where the airway may be obstructed, or when the infection spreads to other vital organs such as the brain (Ingham *et al.*, 1978).

A study of 21 perimandibular closed-space infections yielded an average of six species per specimen, about four of which were usually anaerobic (Bartlett and O'Keefe, 1979). Together with streptococci, the anaerobes most commonly found were anaerobic cocci, black-pigmented *Bacteroides* and *F. nucleatum*, but a wide range of other species were isolated on occasions, including several anaerobic gram-positive rods. As with studies of dental abscesses, these authors demonstrated that perimandibular space infections are almost invariably polymicrobial and they highlighted once again the important role of obligate anaerobes. Most strains were shown to be susceptible to penicillin and clindamycin, whereas tetracycline and erythromycin were less reliable (Bartlett and O'Keefe, 1979).

In the condition known as Ludwig's angina there is an acute, rapid, diffuse, bilateral cellulitis of the floor of the mouth and the neck, involving submental, submandibular and sublingual spaces. It is characterized by severe swelling and board-like ('woody') induration of the submandibular tissues and elevation of the tongue, which may cause serious obstruction of the airway (Marks *et al.*, 1974; Hought *et al.*, 1980). Further spread of infection to the chest, resulting in pleural effusion, empyema, pulmonary infiltration and pericarditis has been reported (Strauss *et al.*, 1980), and formation of subphrenic and mediastinal abscesses is another possible complication (Bounds, 1985).

As with other purulent orofacial infections, the bacterial flora associated with Ludwig's angina appears to be polymicrobial, usually involving one or more anaerobes, including *Bacteroides* spp., and streptococci (Marks *et al.*, 1974; Gross *et al.*, 1976; Hought *et al.*, 1980). Prompt and effective treatment is required to maintain the airway, establish drainage and initiate antimicrobial chemotherapy.

Actinomycosis

A. israelii and other *Actinomyces* spp. that are associated with human actinomycosis are all found as part of the commensal flora of the mouth, particularly in dental plaque. These organisms are facultative rather than obligate anaerobes but are usually isolated only after anaerobic incubation for several days (Bowden and Hardie, 1973). Cervicofacial actinomycosis is the most common form of the disease, although actinomyces infections also occur in the thorax, abdomen, female genital tract, brain and, occasionally, other body sites. In one series of 943 cases seen at the Institute of Hygiene in Cologne between 1970 and 1979, 98.8 per cent of the lesions were in the head and neck (Schaal, 1981). In this extensive and detailed study, *A. israelii* was isolated from 78 per cent of the cases, usually in a mixture with between one and eight other bacterial species (most commonly with two or three concomitant organisms).

Bacteria isolated alongside the *Actinomyces* spp. included streptococci, staphylococci, *Actinobac. actinomycetemcomitans*, *Bacteroides* spp. (pigmented and non-pigmented), fusobacteria and other anaerobes (Schaal, 1981). Cervicofacial actinomycosis usually presents as a soft-tissue infection, with extensive inflammation and swelling and sometimes one or more sinuses discharging through the skin (Richtsmeier *et al.*, 1979). Examination of pus for typical 'sulphur granules' and isolation of the causative *Actinomyces* spp. remains the basis of laboratory diagnosis (Slack and Gerencser, 1975). These infections may also affect bone and several cases of mandibular actinomycosis have been reported (e.g., Samuels *et al.*, 1974; Silbermann *et al.*, 1975; Chow *et al.*, 1978).

Osteomyelitis

Osteomyelitis of the jaws may occur by spread of infection from dentoalveolar abscesses or following fractures and other traumatic injuries. Almost any pathogenic bacteria, including *Staphylococcus aureus*, may be involved in these bone infections, often in mixtures including *Bacteroides* spp. and other anaerobes (Leake, 1972; Monaldo *et al.*, 1974; Sharpe *et al.* 1974).

Maxillary sinusitus

Infection of the maxillary sinus may occur by spread of bacteria from the nose, via the ostium, or from the mouth (Hardie, 1982). Oral bacteria can reach the sinus by direct spread of infection associated with the roots of the upper teeth (from the canine to the second molar, which may be in close proximity to the floor of the antrum), following surgical mishaps when attempting to extract these teeth, or as a result of fractures of the maxilla. Sinus infections of nasal origin, often following a respiratory virus infection, are commonly caused by *H. influenzae* or *Str. pneumonia*, but those originating from the mouth are usually mixed and may contain a variety of anaerobes, including anaerobic cocci, *Bacteroides* and *Fusobacterium* spp. (Lundberg *et al.*, 1979; Lundberg, 1980). The importance of using an appropriate sampling technique, such as antral aspiration or antral biopsy specimens, for studies on the maxillary sinus flora has been emphasized (Carenfelt *et al.*, 1978).

References

Aderhold L, Knothe H, Frenkel G. (1981). The bacteriology of dentogenous pyogenic infections. *Oral Surgery, Oral medicine, Oral Pathology* **52**: 583–587.

Aranki A, Syed S A, Kenney E B, Freter R. (1969). Isolation of anaerobic bacteria from human gingiva and mouse cecum by means of a simplified glove box procedure. *Applied Microbiology* **17**: 568–576.

Armitage C G, Dickinson W R, Jenderseck R S, Levine S M, Chambers D W. (1982). Relationship between the percentage of subgingival spirochaetes and the severity of periodontal disease. *Journal of Periodontology* **53**: 550–556.

Bartlett J G, O'Keefe P. (1979). The bacteriology of perimandibular space in Ludwigs angina: report of case. *Journal of Oral Surgery* **37**: 407–409.

Berg J D, Nord C E. (1973). A method for isolation of anaerobic bacteria from endodontic specimens. *Scandinavian Journal of Dental Research* **81**: 163–166.

Bonta Y, Zambon J J, Genco R J, Neiders M E. (1985). Rapid identification of periodontal pathogens in subgingival plaque: comparison of indirect immunofluorescence microscopy with bacterial culture for the detection of *Actinobacillus actinomycetemcomitans*. *Journal of Dental Research* **64**: 793–798.

Bounds G A. (1985). Subphrenic and mediastinal abscess formation: complication of Ludwig's angina. *British Journal of Oral and Maxillofacial Surgery* **23**: 313–321.

Bowden G H, Hardie J M. (1973). Commensal and pathogenic *Actinomyces* species in man. In *Actinomycetales: characteristics and practical importance*, pp. 277–299. Edited by Sykes G, Skinner F A. Academic Press, London.

Bragd L, Dahlen G, Wikstrom M, Slots J. (1987). The capability of *Actinobacillus actinomycetemcomitans*, *Bacteroides gingivalis* and *Bacteroides intermedius* to indicate progressive periodontitis; a retrospective study. *Journal of Clinical Periodontology* **14**: 95–99.

Brook I. (1986). Isolation of capsulate anaerobic bacteria from orofacial abscesses. *Journal of Medical Microbiology* **22**: 171–174.

Brook I, Gillmore J D, Coolbaugh J C, Walker R I. (1983).

Pathogenicity of encapsulated *Bacteroides melaninogenicus* group, *B. oralis* and *B. ruminicola* subsp. *brevis* in abscesses in mice. *Journal of Infection* 7: 218–226.

Brook I, Grimm S, Kielich R B. (1981). Bacteriology of acute periapical abscess in children. *Journal of Endodontics* 7: 378–380.

Brook I, Walker R I. (1986). The relationship between *Fusobacterium* species and other flora of mixed infection. *Journal of Medical Microbiology* 21: 93–100.

Carenfelt C, Lundberg C, Nord C E, Wretlind B. (1978). Bacteriology of maxillary sinusitis in relation to quality of the retained secretion. *Acta Otolaryngologica* 86: 289–302.

Carlsson J, Sundqvist G. (1980). Evaluation of methods of transport and cultivation of bacterial specimens from infected dental root canals. *Oral Surgery, Oral Medicine, Oral Pathology* 49: 451–454.

Chow A, Roser S, Brady F. (1978). Orofacial odontogenic infections. *Annals of Internal Medicine* 88: 392–402.

Christersson L A, Albini B, Zambon J J, Genco R J. (1987a). Tissue localization of *Actinobacillus actinomycetemcomitans* in human periodontitis. I. Light immunofluorescence and electron microscope studies *Journal of Periodontology* 58: 529–539.

Christersson L A, Wikesjo U M, Albini B, Zambon J J, Genco R J. (1987b). Tissue localization of *Actinobacillus actinomycetemcomitans* in human periodontitis. II. Correlation between immunofluorescence and culture techniques. *Journal of Periodontology* 58: 540–545.

Chuba P J, Pelz K, Krekeler G, De Isele T S, Gobel U. (1988). Synthetic oligodeoxynucleotide probes for the rapid detection of bacteria associated with human periodontitis. *Journal of General Microbiology* 134: 1931–1938.

Darwish S, Hyppa T, Socransky S S. (1978). Studies on the predominant cultivable microbiota of early periodontitis. *Journal of Periodontal Research* 13: 1–16.

Duckworth R, Waterhouse J P, Britton D E R, Nuki K, Sheiham A, Winter R. (1966). Acute ulcerative gingivitis. A double-blind controlled clinical trial of metronidazole. *British Dental Journal* 120: 599–602.

Duerden B I, Goodwin L, O'Neill T C A. (1987). Identification of *Bacteroides* species from adult periodontal disease. *Journal of Medical Microbiology* 24: 133–137.

Dzink J L, Tanner A C R, Haffajee A D Z, Socransky S S. (1985). Gram-negative species associated with active destructive periodontal lesions. *Journal of Clinical Periodontology* 12: 648–659.

Edwardsson S. (1974). Bacteriological studies on deep areas of carious dentine. *Odontologisk Revy* 25 Suppl 32.

Eisenmann A G C, Eisenmann R, Sousa O, Slots J. (1983). Microbiological study of localized juvenile periodontitis in Panama. *Journal of Periodontology* 54: 712–713.

Enwonwu C O. (1972). Epidemiological and biochemical studies of necrotizing ulcerative gingivitis and noma (cancrum oris) in Nigerian children. *Archives of Oral Biology* 17: 1357–1371.

Eskow R N, Loesche W J. (1971). Oxygen tensions in the human oral cavity. *Archives of Oral Biology* 16: 1127–1128.

Feldmann G, Larje O. (1966). The bacterial flora of submucous abscesses originating from chronic exacerbating osteitis. *Acta Odontologica Scandinavica* 24: 129–145.

Fine D H, Mandel I. (1986). Indicators of periodontal disease activity: an evaluation. *Journal of Clinical Periodontology* 13: 533–546.

Finegold S M, Edelstein M A C. (1988). Coping with anaerobes in the 80s. In *Anaerobes today*, pp. 1–10. Edited by Hardie J M, Borriello S P. John Wiley and Sons, Chichester.

French C K *et al.* (1986). DNA probe detection of periodontal pathogens. *Oral Microbiology and Immunology* 1: 58–62.

Genco R J, Slots J. (1984). Host responses in periodontal diseases. *Journal of Dental Research* 63: 441–451.

Goldhaber P, Giddon D B. (1964). Present concepts concerning the etiology and treatment of acute necrotizing ulcerative gingivitis. *International Dental Journal* 14: 468–496.

Goodson J M, Tanner A C R, Haffajee A D. (1982). Patterns of progression and regression of advanced destructive periodontal disease. *Journal of Clinical Periodontology* 9: 472–481.

Gordon D F, Stutman M, Loesche W J. (1971). Improved isolation of anaerobic bacteria from the gingival crevice area of man. *Applied Microbiology* 21: 1016–1021.

Gossling J. (1988). Occurrence and pathogencity of the *Streptococcus milleri* group. *Reviews of Infectious Diseases* 10: 257–285.

Greenberg R, James R, Marner R, Wood W, Sanders C, Kent J. (1979). Microbiological and antibiotic aspects of infections in the oral and maxillofacial region. *Journal of Oral Surgery* 37: 873–884.

Griffee M, Patterson S, Miller C, Kafrawy A, Newton C. (1980). The relationship of *Bacteroides melaninogenicus* to symptoms associated with pulpal necrosis. *Oral Surgery* 50: 457–461.

Gross B D, Roark D T, Meador R C, Cohen A M. (1976). Ludwig's angina due to bacteroides. *Journal of Oral Surgery* 34: 456–460.

Guralnick W. (1984). Odontogenic infections. *British Dental Journal* 156: 440–447.

Gusberti F A, Syed S A, Lang N P. (1988). Combined antibiotic (metronidazole) and mechanical treatment effects on the subgingival bacterial flora of sites with recurrent periodontal disease. *Journal of Clinical Periodontology* 15: 353–359.

Haapasalo M. (1986). *Bacteroides buccae* and related taxa in necrotic root canal infections. *Journal of Clinical Microbiology* 24: 940–944.

Haapasalo M, Ranta H, Shah H N, Ranta K, Lounatmaa K, Kroppenstedt R M. (1986). *Mitsuokella dentalis* sp.nov. from dental root canals. *International Journal of Systematic Bacteriology* 36: 566–568.

Haffajee A D, Socransky S S. (1986). Attachment level changes in destructive periodontal diseases. *Journal of Clinical Periodontology* 13: 461–472.

Haffajee A, Socransky S, Dzink J, Taubman M, Ebersole J. (1988). Clinical, microbiological and immunological features of subjects with refractory periodontal diseases. *Journal of Clinical Periodontology* 15: 390–398.

Haffajee A D, Socransky S S, Ebersole J L. (1985). Survival analysis of periodontal sites before and after periodontal therapy. *Journal of Clinical Periodontology* 12: 553–567.

Haffajee A D, Socranksy S S, Ebersole J L, Smith D L. (1984). Clinical, microbiological and immunological features associated with the treatment of active periodontitis lesions. *Journal of Clinical Periodontology* 11: 600–618.

Hardie J M. (1974). Anaerobes in the mouth. In *Infection with non-sporing anaerobic bacteria*, pp. 99–130. Edited by Phillips I, Sussmann M. Churchill Livingstone, Edinburgh.

Hardie J M. (1982). The bacteriology of chronic maxillary sinusitis, pp. 49–55. Proceedings of European Metronidazole Symposium, Academy Professional Information Services Inc, New York.

Hardie J M. (1986a). Oral streptococci. In *Bergey's Manual of systematic bacteriology* vol 2, pp. 1054–1063. Edited by Sneath P H A *et al.*, Williams and Wilkins, Baltimore.

Hardie J M. (1986b). Methods for the isolation and identification of anaerobes. In *Anaerobic bacteria in habitats other than man*, pp. 397–410. Edited by Barnes E M, Mead G C. Blackwell Scientific Publications, Oxford.

Hardie J M, Bowden G H. (1974). The normal microbial flora of the mouth. In *The normal microbial flora of man*, pp. 47–83. Edited by Skinner F A, Carr J G. Academic Press, London.

Hardie J M, Shah H N. (1982). Factors controlling the microbial flora of the mouth. *European Journal of Chemotherapy and Antibiotics* 2: 3–11.

Heimdahl A, Konow L Von, Nord C E. (1980). Isolation of β-lactamase-producing *Bacteroides* strains associated with clinical failures with penicillin treatment of human orofacial infections. *Archives of Oral Biology* 25: 689–692.

Heimdahl A, Konow Von K, Satoh, T and Nord C E. (1985). Clinical appearance of orofacial infections of odontogenic origin in relation to microbiological findings. *Journal of Clinical Microbiology* 22: 299–302.

Heimdahl A, Nord C E. (1983). Orofacial infections of odontogenic origin. *Scandinavian Journal of Infectious Diseases* Suppl 39: 86–91.

Hood F J C. (1978). The place of metronidazole therapy in the treatment of acute orofacial infection. *Journal of Antimicrobial Chemotherapy* 4: 71–73.

Hought R T, Fitzgerald B E, Latta J E, Zallen R D. (1980). Ludwig's angina: report of two cases and review of the literature from 1945 to January 1979. *Journal of Oral Surgery* 38: 849–855.

Ingham H R, High A S, Kalbag R M, Sengupta R P, Tharagonnet D, Selkon J B. (1978). Abscesses of the frontal lobe of the brain secondary to covert dental sepsis. *Lancet* 2: 497–499.

Ingham H R, Hood F J C, Bradnum P, Tharagonnet D, Selkon J B. (1977). Metronidazole compared with penicillin in the treatment of acute dental infections. *British Journal of Oral Surgery* 14: 264–269.

Kannangara, D W, Thadepalli H, McQuirter J. (1980). Bacteriology and treatment of dental infections. *Oral Surgery, Oral Medicine, Oral Pathology* 50: 103–109.

Kantz W E, Henry C A. (1974). Isolation and classification of anaerobic bacteria from intact chambers of nonvital teeth in man. *Archives of Oral Biology* 19: 91–96.

Kastelein P, Van Steenbergen T J M, Braas J M, De Graff J. (1981). An experimentally induced phlegmonous abscess by a strain of *Bacteroides gingivalis* in guinea pigs and mice. *Antonie van Leeuwenhoek* 47: 1–9.

Kenney E B, Ash M. (1969). Oxidation reduction potential of developing plaque, periodontal pockets and gingival sulci. *Journal of Periodontology* 40: 630–633.

Keyes P H, Wright W E, Howard S A. (1978). The use of phase-contrast microscopy and chemotherapy in the diagnosis and treatment of periodontal lesions—an initial report. *Quintessence International* 9: 51–56.

Kuritza A P, Getty C E, Shaughnessy P, Hesse R, Salyers A A. (1986). DNA probes for identification of clinically important *Bacteroides* species. *Journal of Clinical Microbiology* 23: 343–349.

Labriola J D, Mascaro J, Alpert B. (1983). The microbiological flora of orofacial abscesses. *Journal of Oral and Maxillofacial Surgery* 41: 711–714.

Lai C-H, Listgarten M A, Shirakawa M, Slots J. (1987). *Bacteroides forsythus* in adult gingivitis and periodontitis. *Oral Microbiology and Immunology* 2: 152–157.

Leake D L. (1972). Bacteroides osteomyelitis of the mandible. A report of two cases. *Oral Surgery, Oral Medicine, Oral Pathology* 34: 585–588.

Lewis M A O, MacFarlane T W, McGowan D A. (1986a). Quantitative bacteriology of acute dento-alveolar abscesses. *Journal of Medical Microbiology* 21: 101–104.

Lewis M A O, McGowan D A, MacFarlane T W. (1986b). Short-course high-dosage amoxycillin in the treatment of acute dento-alveolar abscess. *British Dental Journal* 161: 299–302.

Lewis M A O, MacFarlane T W, McGowan D A. (1988a). Reliability of sensitivity testing of primary culture of dentoalveolar abscess. *Oral Microbiology and Immunology* 3: 177–180.

Lewis M A O, MacFarlane T W, McGowan D A, MacDonald D G. (1988b). Assessment of the pathogenicity of bacterial species isolated from acute dentoalveolar abscesses. *Journal of Medical Microbiology* **27**: 109–116.

Lindhe J, Liljenberg B, Listgarten M A. (1980). Some microbiological and histopathological features of periodontal disease in man. *Journal of Periodontology* **51**: 264–269.

Listgarten M A. (1987). Nature of periodontal diseases: pathogenic mechanisms. *Journal of Periodontal Research* **22**: 172–178.

Listgarten M A, Hellden L. (1978). Relative distribution of bacteria at clinically healthy and periodontally diseased sites in humans. *Journal of Clinical Periodontology* **5**: 115–132.

Listgarten M A, Levin S. (1981). Positive correlation between the proportions subgingival spirochaetes and mobile bacteria and susceptibility of human subjects to periodontal deterioration. *Journal of Clinical Periodontology* **8**: 122–138.

Loe H, Theilade E, Jensen S B. (1965). Experimental gingivitis in man. *Journal of Periodontology* **36**: 177–178.

Loesche W J. (1982). The bacterial etiology of dental decay and periodontal disease: the specific plaque hypothesis. *Clinical Dentistry* **2**: 1–13.

Loesche W J. (1986). Role of *Streptococcus mutans* in human dental decay. *Microbiological Reviews* **50**: 353–380.

Loesche W J, Gusberti F, Mettraux G, Higgins T, Syed S A. (1983). Relationship between oxygen tension and subgingival bacterial flora in untreated human periodontal pockets. *Infection and Immunity* **42**: 659–667.

Loesche W J, Hockett R N, Syed S A. (1972). The predominant cultivable flora of tooth surface plaque removed from institutionalised subjects. *Archives of Oral Biology* **17**: 1311–1325.

Loesche W J, Laughon B E. (1982). Role of spirochaetes in periodontal disease. In *Host-parasite interactions in periodontal diseases*, pp. 62–75. Edited by Genco R J, Mergenhagen S E. American Society for Microbiology, Washington DC.

Loesche W J, Syed S A. (1973). The predominant cultivable flora of carious plaque and carious dentine. *Caries Research* **7**: 201–216.

Loesche W J, Syed S A. (1978). Bacteriology of human experimental gingivitis: effect of plaque and gingivitis score. *Infection and Immunity* **21**: 830–829.

Loesche W J, Syed S A, Laughon B E, Stoll J. (1982). The bacteriology of acute necrotizing ulcerative gingivitis. *Journal of Periodontology* **53**: 223–230.

Loesche W J, Syed S A, Morrison E C, Laughon B E, Grossman N S. (1981). Treatment of periodontal infections due to anaerobic infections due to anaerobic bacteria with short-term treatment with metronidazole. *Journal of Clinical Periodontology* **8**: 29–44.

Loesche W J, Syed S A, Schmidt E, Morrison E C. (1985). Bacterial profiles of subgingival plaques in periodontitis. *Journal of Periodontal Research* **56**: 447–456.

Loescher A. (1987). Tetanus: an unusual case of trismus. *British Dental Journal* **162**: 301–302.

Lundberg C. (1980). Dental sinusitis. *Swedish Dental Journal* **4**: 63–67.

Lundberg C, Carenfelt C, Engquist S, Nord C E. (1979). Anaerobic bacteria in maxillary sinusitus. *Scandinavian Journal of Infectious Diseases* Suppl 19: 74–76.

MacDonald A C R, Sutton R M, Knoll M L, Madlener E M, Grainger R M. (1956). The pathogenic components of an experimental mixed infection. *Journal of Infectious Diseases* **98**: 15–20.

MacFarlane T W. (1980). Systemic infections. In *Oral manifestations of systemic disease*, pp. 66–101. Edited by Jones J H, Mason D K. Saunders, London.

Mandell R L. (1984). A longitudinal microbiological investigation of *Actinobacillus actinomycetemcomitans* and *Eikenella corrodens* in juvenile periodontitis. *Infection and Immunity* **45**: 778–780.

Mandell R L, Ebersole J L, Socransky S S. (1987). Clinical immunological and microbiological features of active diseased sites in juvenile periodontitis. *Journal of Clinical Periodontology* **14**: 534–540.

Mandell R L, Socransky S S. (1981). A selective medium for *Actinobacillus actinomycetecomitans* and the incidence of the organism of juvenile periodontitis. *Journal of Periodontology* **52**: 593–598.

Marks R, Akin R, Walters P, Ellis D. (1974). Ludwigs angina: report of a case. *Journal of Oral Surgery* **32**: 462–464.

McGowan D, Murphy K, Sheiham A. (1977). Metronidazole in the treatment of severe acute pericoronitis: a clinical trial. *British Dental Journal* **142**: 221–223.

McKee A S, McDermid A S, Baskerville A, Dowsett A B, Ellwood D C, Marsh P D. (1986). Effect of hemin on the physiology and virulence of *Bacteroides gingivalis* W50. *Infection and Immunity* **52**: 349–355.

McKee A S, Pascal C, Aiken A, West L, Hardie J M. (1982). A preliminary microbiological study of dental abscesses. *Journal of Dental Research* **61**: 558 (abstract no. 206).

Mikx F H M, Matee M I, Schaeken M J M. (1986). The prevalence of spirochaetes in the subgingival microbiota of Tanzanian and Dutch children. *Journal of Clinical Periodontology* **13**: 289–293.

Mitchell L. (1984). Topical metronidazole in the treatment of 'dry socket'. *British Dental Journal* **156**: 132–134.

Möller A J R. (1966). Microbiological examination of root canals and periapical tissues of human teeth. *Odontologiste Tidskrift* (Special issue) **74**: 1–380.

Monaldo L, Bellome J, Zegarelli D, Ragaini V E. (1974). Bacteroides infection of the mandible with secondary spread to the neck. *Journal of Oral Surgery* **32**: 370–372.

Moore J, Russell C. (1972). Bacteriological investigation of dental abscesses. *The Dental Practitioner* 390–392.

Moore W E C. (1987). Microbiology of periodontal disease. *Journal of Periodontal Research* **22**: 335–341.

Moore W E C, Holdeman L V, Cato E P, Smibert R M, Burmeister J A, Ranney R R. (1983). Bacteriology of moderate (chronic) periodontitis in mature adult humans. Infection and Immunity **42**: 510–515.

Moore W E C et al. (1982a). Bacteriology of experimental gingivitis in young adult humans. *Infection and Immunity* **38**: 651–667.

Moore W E C, Holdeman L V, Smibert R M, Hash D E, Burmeister J A, Ranney R R. (1982b). Bacteriology of severe periodontitis in young adult humans. *Infections and Immunity* **38**: 1137–1148.

Morey E F, Moule A J, Higgins T J. (1984). Pyogenic dental infections—a retrospective analysis. *Australian Dental Journal* **29**: 150–153.

Newman M G. (1985). Current concepts of the pathogenesis of periodontal disease. Microbiology emphasis. *Journal of Periodontology* **56**: 734–739.

Newman M G, Socransky S S. (1977). Predominant cultivable microflora of periodontitis. *Journal of Periodontology* **47**: 373–379.

Newman M G, Socransky S S, Savitt E D, Propas D A, Crawford A. (1976). Studies on the microbiology of periodontosis. *Journal of Periodontology* **47**: 373–379.

Oguntebi B, Slee A M, Tanzer J M, Langeland K. (1982). Predominant microflora associated with human dental periapical abscesses. *Journal of Clinical Microbiology* **15**: 964–966.

Pancholi V, Ayyagari A S, Agarwal K C. (1985). Experimentally induced subcutaneous infection by black-pigmented *Bacteroides* species in rats. *Journal of Infection* **11**: 131–137.

Piecuch J. (1982). Odontogenic infections. *Dental Clinics of North America* **26**: 129–145.

Preus H R, Olsen I, Namork E. (1987). Association between bacteriophage-infected *Actinobacillus actinomycetemcomitans* and rapid periodontal destruction. *Journal of Clinical Periodontology* **14**: 245–247.

Ranney R R, Best A M, Breen T J, Moore W E C, Moore L V H. (1987). Bacterial flora of progressing periodontitis lesions. *Journal of Periodontal Research* **22**: 205–206.

Richtsmeier W J, Johns M E. (1979). Actinomycosis of the head and neck. *CRC Critical Reviews of Clinical Laboratory Science* **11**: 175–202.

Roberts M C, Moncla B, Kenny G E. (1987). Chromosomel DNA probes for identification of *Bacteroides* species. *Journal of General Microbiology* **133**: 1423–1430.

Rood J. (1980). The value of metronidazole in dental and oral surgery. *Dental Update* **7**: 293–300.

Rosebury T, MacDonald J B, Clark A R. (1950). A bacteriological survey of gingival scrapings from periodontal infections by direct examination, guinea-pig inoculation, and anaerobic cultivation. *Journal of Dental Research* **29**: 718–731.

Sabiston C B, Grigsby W R. (1977). The microbiology of dental pyogenic infections. *CRC Critical Reviews in Clinical Laboratory Science* **8**: 213–240.

Sabiston C B, Grigsby W R, Segerstrom N. (1976). Bacteriology of pyogenic infections of dental origin. *Oral Surgery, Oral Medicine, Oral Pathology* **41**: 430–435.

Samuels H S, Branham G B, Vogt P J, Peterson L J. (1974). Actinomycosis of the mandible. *Journal of Oral Surgery* **32**: 679–681.

Savitt E D, Socransky S S. (1984). Distribution of certain subgingival microbial species in selected periodontal conditions. *Journal of Periodontal Research* **19**: 218–230.

Schaal K P. (1981). Actinomycoses. *Revue de L'Institut Pasteur de Lyon* **14**: 279–288.

Shah H N, Collins M D. (1988). Proposal for reclassification of *Bacteroides asaccharolyticus*, *Bacteroides gingivalis* and, *Bacteroides endodontalis* in a new genus, *Porphyromonas*. *International Journal of Systematic Bacteriology* **38**: 128–131.

Shah H N, Collins M D. (1990). Prevotella, a new genus to include *Bacteroides melaninogeniais* and related species formerly classified in the genus *Bacteroides*. *International Journal of Systematic Bacteriology* **40**: 205–206.

Sharp P, Meador R, Martin R. (1974). A case of mixed anaerobic infection of the jaw. *Journal of Oral Surgery* **32**: 457–459.

Shinn D L S. (1962). Metronidazole in acute ulcerative gingivitis. *Lancet* **1**: 1191.

Silbermann M, Chiminello F J, Doku H C, Maloney P L. (1975). Mandibular actinomycosis: report of a case. *Journal of the American Dental Association* **90**: 162–165.

Silverstone L M, Johnson N W, Hardie J M, Williams R A D. (1981). *Dental Caries: aetiology, pathology and prevention*. MacMillan Press Limited, London.

Sims W. (1974). The clinical bacteriology of purulent oral infections. *British Journal of Oral Surgery* **12**: 1–12.

Slack J M, Gerencser M A. (1975). *Actinomyces, filamentous bacteria*. Burgess Publishing Company, Minneapolis.

Slots J. (1976). The predominant cultivable microflora in juvenile periodontitis. *Scandinavian Journal of Dental Research* **84**: 1–10.

Slots J. (1979). Subgingival microflora and periodontal disease. *Journal of Clinical Periodontology* **6**: 351–382.

Slots J. (1986a). Bacterial specificity in adult periodontitis. A summary of recent work. *Journal of Clinical Periodontology* **13**: 570–577.

Slots J. (1986b). Virulence factors of the bacteria that cause periodontal diseases. *Compendium of Continuing Education in Dentistry* **7**: 665–671.

Slots J, Bragd L, Wikstrom M, Dahlen G. (1986). The occurrence of *Actinobacillus actinomycetemcomitans*, *Bacteroides gingivalis* and *Bacteroides intermedius* in destructive periodontal disease in adults. *Journal of Clinical Periodontology* **13**: 570–577.

Slots J, Dahlen G. (1985). Subgingival micro-organisms and bacterial virulence factors in periodontitis. *Scandinavian Journal of Dental Research* **93**: 119–127.

Slots J, Genco R J. (1984). Black-pigmented *Bacteroides*

species, *Capnocytophaga* species and *Actinobacillus actinomycetemcomitans* in human periodontal diseases: virulence factors in colonization, survival and tissue destruction. *Journal of Dental Research* 63: 412–421.

Slots J, Listgarten M A. (1988). *Bacteroides gingivalis*, *Bacteroides intermedius* and *Actinobacillus actinomycetemcomitans* in human periodontal diseases. *Journal of Clinical Periodontology* 15: 85–93.

Smith D T. (1932). *Oral spirochaetes and related organisms in fuso-spirochaetal disease*. Williams and Wilkins Co, Baltimore.

Socransky S S. (1977). Microbiology of periodontal disease-present status and future considerations. *Journal of Periodontology* 48: 497–504.

Socransky S S. (1979). Criteria for the infectious agents in dental caries and periodontal disease. *Journal of Clinical Periodontology* 6: 16–21.

Socransky S S, Gibbons R J. (1965). Required role of *Bacteroides melaninogenicus* in mixed anaerobic infections. *Journal of Infectious Diseases* 115: 247–253.

Socransky S S, Gibbons R J, Dale A G C, Bortnick L, Rosenthal E, MacDonald J B. (1963). The microbiota of the gingival crevice area of man. I. Total microscopic and viable counts of specific organisms. *Archives of Oral Biology* 8: 275–280.

Strauss H R, Tilghman D M, Hankins J. (1980). Ludwig's angina: empyema, pulmonary infiltration, and pericarditis secondary to extraction of a tooth. *Journal of Oral Surgery* 38: 223–229.

Strzempko M N et al. (1987). A cross-reactivity study of whole genomic DNA probes for *Haemophilus actinomycetemcomitans*, *Bacteroides intermedius* and *Bacteroides gingivalis*. *Journal of Dental Research* 66: 1543–1546.

Sundqvist G. (1976). Bacteriological studies of necrotic dental pulps. Umea University Odontological Dissertations 49: 494–497.

Sundqvist G K, Eckerbom M I, Larson A P, Sjogren U T. (1979). Capacity of anaerobic bacteria from necrotic dental pulps to induce purulent infections. *Infection and Immunity* 25: 685–693.

Syed S A, Loesche W J. (1978). Bacteriology of human experimental gingivitis: effect of plaque age. *Infection and Immunity* 21: 821–829.

Taichman N S et al. (1982). Leukocidal mechanisms of *Actinobacillus actinomycetemcomitans*. In *Host-parasite interactions in periodontal diseases*, pp. 261–269. Edited by Genco R J, Mergenhagen S E. American Society for Microbiology, Washington DC.

Taichman N S et al. (1984). Neutrophil interactions with oral bacteria as a pathogenic mechanism in periodontal diseases. In *Advances in inflammation research*, vol 8. pp. 113–142. Edited by Weissman G. Raven Press, New York.

Takazoe I, Nakamura T. (1971). Experimental mixed infection by human gingival crevice material. *Bulletin of Tokyo Dental College* 12: 85–93.

Tanner A C R, Dzink J L, Socransky S S, Des Roches C L. (1987). Diagnosis of periodontal disease using rapid identification of activity-related gram-negative species. *Journal of Periodontal Research* 22: 207–208.

Tanner A C R, Haffer C, Bratthall G T, Visconti A, Socransky S S. (1979). A study of the bacteria associated with advancing periodontal disease in man. *Journal of Clinical Periodontology* 6: 278–307.

Tanner A C R, Listgarten M A, Ebersole J L, Strzempko M N. (1986). *Bacteroides forsythus* sp. nov a slow-growing fusiform *Bacteroides* sp. from the human oral cavity. *International Journal of Systematic Bacteriology* 36: 213–221.

Tanner A C R, Socransky S S, Goodson J M. (1984). Microbiota of periodontal pockets losing crestal alveolar bone. *Journal of Periodontal Research* 19: 279–291.

Tanner A C R, Strzempko M N, Belsky C A, McKinley G A. (1985). API-anadent reactions of fastidious gram-negative species. *Journal of Clinical Microbiology* 22: 333–335.

Van der Velden V, Van Winkehoff A J B, Abbas F, De Graaff J. (1986). The habitat of periodontopathic micro-organisms. *Journal of Clinical Periodontology* 13: 243–248.

Van Palenstein Helderman W H. (1981). Microbial etiology of periodontal disease *Journal of Clinical Periodontology* 8: 261–180.

Van Reenan J F, Coogan M M. (1970). Clostridia isolated from human mouths. *Archives of Oral Biology* 15: 845–848.

Van Steenbergen T J M, Kastelein P, Touw J J A, De Graaff J. (1982). Virulence of black-pigmented *Bacteroides* strains from periodontal pockets and other sites in experimentally induced skin lesions in mice. *Journal of Periodontal Research* 17: 41–49.

Van Steenbergen T J M, Van Winkelhoff A J, De Graaff J. (1984a). Pathogenic synergy: mixed infections in the oral cavity. *Antonie van Leeuwenhoek* 50: 789–798.

Van Steenbergen T J M, Van Winkelhoff A J, Mayrand D, Grenier D, De Graaff J. (1984b). *Bacteroides endodontalis* sp.nov., an asaccharolytic black-pigmented *Bacteroides* species from infected dental root canals. *International Journal of Systematic Bacteriology* 34: 118–120.

Van Winkelhoff A J, Carlee A W, De Graaff J. (1985). *Bacteroides edondontalis* and other black-pigmented *Bacteroides* species in odontogenic abscesses. 49: 494–49.

Van Winkelhoff A J, Van Steenbergen T J M, De Graaff J. (1988a). The role of black-pigmented *Bacteroides* in human oral infections. *Journal of Clinical Periodontology* 15: 145–155.

Van Winkelhoff A J, Van Der Velden V, Clement M, De Graaff J. (1988b). Intra-oral distribution of black-pigmented *Bacteroides* species in periodontitis patients. *Oral Microbiological Immunology* 3: 83–85.

Von Konow L, Nord C E, Nordenram A. (1981). Anaerobic

bacteria in dentoalveolar infections. *International Journal of Oral Surgery* **10**: 313–322.

Von Konow L, Nord C E. (1983). Ornidazole compared to phenoxymethyl penicillin in the treatment of orofocial infections. *Journal of Antimicrobial Chemotherapy* **11**: 207–215.

Wennstrom J L, Dahlen G, Svensson J, Nyman S. (1987). *Actinobacillus actynomycetemcomitans, Bacteroides gingivalis* and *Bacteroides intermedius*: predictors of attachment loss? *Oral Microbiology and Immunology* **2**: 158–163.

Williams B L, McCann G F, Schoenknecht F D. (1983). Bacteriology of dental abscesses of endodontic origin. *Journal of Clinical Microbiology* **18**: 770–774.

Williams B L, Pantalone R M, Sherris J C. (1976). Subgingival microflora and periodontitis. *Journal of Periodontal Research* **11**: 1–18.

Wittgow W C, Sabiston C B. (1975). Micro-organisms from pulpal chambers of intact teeth with necrotic pulps. *Journal of Endodontics* **1**: 168–171.

Zambon J J. (1985). *Actinobacillus actinomycetemcomitans* in human periodontal disease. *Journal of Clinical Periodontology* **12**: 1–20.

Zambon J J, Christersson L A, Slots J. (1983a). *Actinobacillus actinomycetemcomitans* in human periodontal disease: prevalence in patient groups and distribution of biotypes and serotypes within families. *Journal of Periodontology* **54**: 707–711.

Zambon J J, DeLuca C, Slots J, Genco R J. (1983b). Studies of leukotoxin from *Actinobacillus actinomycetemcomitans* using the promyelotic HL-60 cell line. *Infection and Immunity* **40**: 205–212.

Zavistoski J, Dzink J, Onderdonk A, Bartlett J. (1980). Quantitative bacteriology of endodontic infections. *Oral Surgery, Oral Medicine, Oral Pathology* **49**: 171–174.

16
Respiratory, ENT and CNS infections

H R Ingham and P R Sisson

Oropharyngeal infections
    Necrobacillosis
    Tonsillar infections
    Retropharyngeal and pharyngeal abscess
Otitis media
    Secretory otitis media
    Acute otitis media
    Chronic suppurative otitis media
    Therapy
Infections of the paranasal air sinuses
Intracranial infections
    Infections of intracranial venous sinuses
    Brain abscess
Anaerobic pleuropulmonary infection
    Aspiration
    Impaired drainage
    Metastatic spread
    Direct extension
    Aspiration pneumonia
    Lung abscess
    Empyema thoracis
    Therapy
References

Infections due to anaerobic bacteria in these regions of the body are almost always polymicrobial, reflecting the flora of the common primary source—the oropharynx—although in some infections, notably otogenic brain abscess and pulmonary infections originating from subdiaphragmatic sources, the obligate and facultative anaerobes isolated are more typical of the colonic microflora. Anaerobic bacteria commonly involved are fusobacteria, bacteroides, anaerobic gram-positive cocci, veillonellae and non-sporing gram-positive bacilli. *Fusobacterium* spp. encountered include *Fusobacterium nucleatum* (also known as *F. polymorphum*) and, less often, *F. necrophorum* and related organisms. *Bacteroides* spp. include *Bacteroides melaninogenicus*, *B. intermedius*, *B. gingivalis*, *B. oralis*, *B. ruminicola* and *B. ureolyticus* (formerly *B. corrodens*). *B. fragilis* is often present in chronic suppurative otitis media, otogenic cerebral abscess and certain pulmonary infections.

Oropharyngeal infections

Necrobacillosis

F. necrophorum, a commensal of the intestinal tract, vagina and oropharynx, may cause a form of acute suppurative tonsillitis, commonly referred to as necrobacillosis (Alston, 1955). This disease is characterized by severe necrotizing tonsillitis or peritonsillar abscess, local extension of which may cause pneumonia, lung abscess, empyema and thrombophlebitis of the internal jugular vein. A more generalized infection described by Lemierre (1936) is characterized by high fever, confusion, septicaemia and involvement of the joints, meninges, kidneys or liver, as well as the tonsils. Some isolates from patients with Lemierre's syndrome have been identified as *F. naviforme* (Hudson et al., 1984; Moore-Gillon et al., 1984; Wardle et al., 1984).

Necrobacillosis is now relatively rare, presumably due to the early use of antimicrobials in the treatment of tonsillitis. However, the diagnosis should be considered in previously healthy adolescents or young adults who develop systemic illness with pulmonary involvement after an initial sore throat. The organism may be isolated from blood and infected sites; because of the tendency to form spheroplasts and L-forms it may be very pleomorphic in stained smears. Penicillin is usually prescribed but the optimal treatment remains to be established; in several recent reports metronidazole has been effective (Hudson et al., 1984; Moore-Gillon et al., 1984; Wardle et al., 1984).

Tonsillar infections

Peritonsillar abscess

Peritonsillar abscess occurs most commonly in adolescents and adults as a complication of tonsillitis or glandular fever, when infection extends to adjacent tissue (Goldstein, 1978; Johnsen, 1981). If uncontrolled, the infection may spread to the parapharyngeal space, the great vessels of the neck, the mediastinum or within the skull. The condition may arise *de novo* or there may be an antecedent history of tonsillitis responding initially to antimicrobial therapy, with recurrence of the original symptoms, fever, trismus, dysarthria and dysphagia. In the more common, unilateral infection there is oedema and protrusion of the anterior palatine arch with displacement of the uvula; pus evacuated from the abscess is characteristically malodourous (Sugita et al., 1982).

Although *Streptococcus pyogenes* is the common pathogen in acute bacterial tonsillitis it is only present in about 25 per cent of peritonsillar abscesses (Table 16.1). In some instances failure to isolate this organism may be due to the effect of prior chemotherapy and it is noteworthy that 'sterile' pus or 'normal flora' were reported in 153 (46 per cent) patients in the series reviewed in Table 16.1. However, anaerobic bacteria figure prominently in some

Table 16.1 Bacteria isolated from peritonsillar abscesses

Author(s)	Number of patients	Number of strains isolated (%)			
		Group A streptococci	Anaerobes	Normal flora	No growth
Beeden and Evans (1970)	69	18 (26)	0 (0)	36 (52)	13 (19)
Flodstrom and Hallander (1976)	37	16 (43)	28 (76)	0 (0)	0 (0)
McCurdy (1977)	68	16* (24)	0 (0)	13 (19)	39 (57)
Muller (1978)	73	19 (26)	1 (1)	20 (27)	1 (1)
Brook (1981a)	16	4 (25)	16 (100)	0 (0)	0 (0)
Holt and Tinsley (1981)	33	7 (21)	4 (12)	12 (36)	10 (30)
Sugita et al. (1982)	30	12 (40)	23 (77)	3 (10)	0 (0)
Gray (1984)	8	1 (13)	1 (13)	6 (75)	0 (0)
Total	334	93 (28)	73 (22)	90 (27)	63 (19)

*β-haemolytic streptococci

of these studies (Flodstrom and Hallander, 1976; Brook, 1981a; Sugita et al., 1982) and it seems likely that these organisms are important in this condition.

Penicillin or erythromycin are commonly employed in the chemotherapy of these infections, which often require surgical drainage, sometimes with immediate tonsillectomy.

Tonsillitis

Anaerobic bacteria, predominantly of the *B. melaninogenicus* group, have been isolated from the tonsils of patients with acute tonsillitis (Brook et al., 1981b). Such strains of *B. melaninogenicus* were more often capsulate than those from healthy controls and consistently produced abscesses in experimental animals, observations supporting a possible pathogenic role in acute tonsillitis (Brook and Gober, 1983). Anaerobic bacteria other than *F. necrophorum* have also been isolated in large numbers from operative samples of tonsils from children and young adults with chronic tonsillar infections (Reilly and Willis, 1980; Brook et al., 1981b; Brook and Yocum, 1984); chronic tonsillitis is often associated with a fetid odour—an invariable hallmark of anaerobic bacteria. *Bacteroides* spp., *Fusobacterium* spp. and anaerobic gram-positive cocci predominate in these studies; Lancefield Group A β-haemolytic streptococci (GABHS) constitute only 10–30 per cent of isolates.

There is further evidence of the potential contribution of anaerobic bacteria to the maintenance of chronic tonsillar infection in reports of the isolation of β-lactamase-producing strains (Brook et al., 1981a; Brook and Yocum, 1984). The number and variety of β-lactamase-producing anaerobes in the tonsils increases in patients who have been previously exposed to penicillin (Heimdahl and Nord, 1979), a likely occurrence with recurrent tonsillitis. Treatment failure with penicillin in the presence of β-lactamase-producing bacteroides occurred in five patients with infections of the oropharynx (Heimdahl et al., 1980). β-lactamase production by such strains impairs the efficacy of penicillin in experimental abscesses in animals (Hackman and Wilkins, 1976) and such organisms may account for the failure of penicillin to eradicate GABHS in recurrent tonsillitis (Gastanaduy et al., 1980; Ingham et al., 1980; Brook et al., 1983a). In 45 patients with recurrent tonsillitis associated with GABHS, β-lactamase-producing aerobic and anaerobic bacteria were observed in 43 (96 per cent) (Brook and Hirokawa, 1985). Colonization with GABHS was eradicated in two out of 15 patients treated with penicillin, six out of 15 treated with erythromycin and 14 out of 15 treated with clindamycin. A year later, 12 out of 14 treated with penicillin, eight out of 14 treated with erythromycin and only one out of 15 treated with clindamycin continued to experience recurrent tonsillitis. These results confirm similar observations by Tuner and Nord (1982). While the superiority of clindamycin may result directly from activity against *Str. pyogenes* it may equally reflect killing of anaerobic bacteria.

These data suggest that the spectrum of activity of antimicrobial agents employed in the therapy of recurrent tonsillitis should include β-lactamase-producing anaerobic bacteria, especially when treatment with penicillin has failed (see above). Although metronidazole reduces the numbers of anaerobes in patients undergoing tonsillectomy (Reilly et al., 1981), it remains to be seen whether it has a role in the treatment of recurrent tonsillitis.

Retropharyngeal and parapharyngeal abscess

The retropharyngeal space (Fig. 16.1) is an anatomical plane, located between fascial layers extending from the base of the skull into the superior mediastinum. Posteriorly, the 'danger space' extends from the base of the skull into the posterior mediastinum and behind this is the prevertebral space. The latter runs from the base of the skull to the coccyx, allowing deep neck infections to spread caudally as far as the psoas sheath. Within the retropharyngeal space, on either side of the midline, are the retropharyngeal lymph nodes draining adjacent structures—the nose, paranasal sinuses, nasopharynx, pharynx, middle ears and eustachian tubes.

Abscesses in the prevertebral space usually stem from osteomyelitis of the cervical spine and are staphylococcal or tuberculous; fungi or anaerobic organisms are rarely implicated. By contrast, abscesses in the retropharyngeal space are usually secondary to local infection or trauma. Retropharyngeal abscess used to be most commonly seen in young children, secondary to infections of the ear, nose and throat. In older age groups the retropharyngeal lymph nodes become atrophic and retropharyngeal abscesses originating from the and β-haemolytic streptococci are usually cited as causative organisms but reports of sterile pus are not uncommon and obligate anaerobes, typical of the oral flora, are almost certainly the important pathogens (Bryan *et al.*, 1974; Sprinkle *et al.*, 1974; Barratt *et al.*, 1984). Chemotherapy must therefore also include an agent which is active against anaerobes, a suitable combination being penicillin or clindamycin plus metronidazole.

Otitis media

Otitis media is an inflammation of the middle ear which may or may not be infective in origin. Broadly it may be divided into three categories: secretory otitis media, acute otitis media and chronic suppurative otitis media.

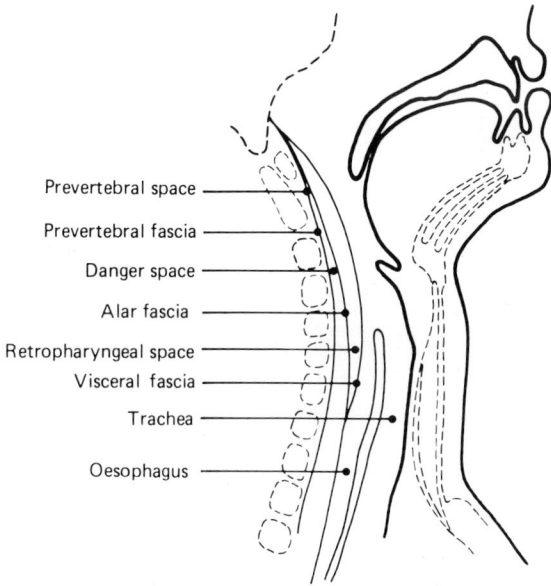

Fig. 16.1 Lateral view showing relationship of fascia to prevertebral space, 'danger space' and retropharyngeal space (Reproduced from *Laryngoscope* 1984, vol 95 p 456, by kind permission of the Editor)

orpharynx are more likely to be due to dental sepsis or trauma.

Sepsis in the orpharynx may extend widely via communication between the various anatomical spaces of the neck, the submental compartment, the peritonsillar, parapharyngeal, retropharyngeal, masticator, subhyoid and parotid spaces. Spread is almost certainly facilitated by the action of enzymes produced by oral anaerobes and ultimately may involve the mediastinum. Such peripharyngeal infections may result from trauma or may complicate medical procedures such as endotracheal intubation (Heath and Peirce, 1977) or laryngoscopy (Yeoh *et al.*, 1985); more often a dental source is implicated (Johnson *et al.*, 1984). Characteristically, previously fit individuals present with acute airway obstruction, inflammation or swelling of the neck, trismus and fever (Mukau, 1985). Lateral soft tissue X-ray may show gas spreading diffusely in tissue planes (Fig. 16.2). Early diagnosis and recognition of the likely primary source are essential to a favourable outcome.

Treatment consists of the establishment of an airway, surgical drainage and the administration of appropriate chemotherapy. *Staphylococcus aureus*

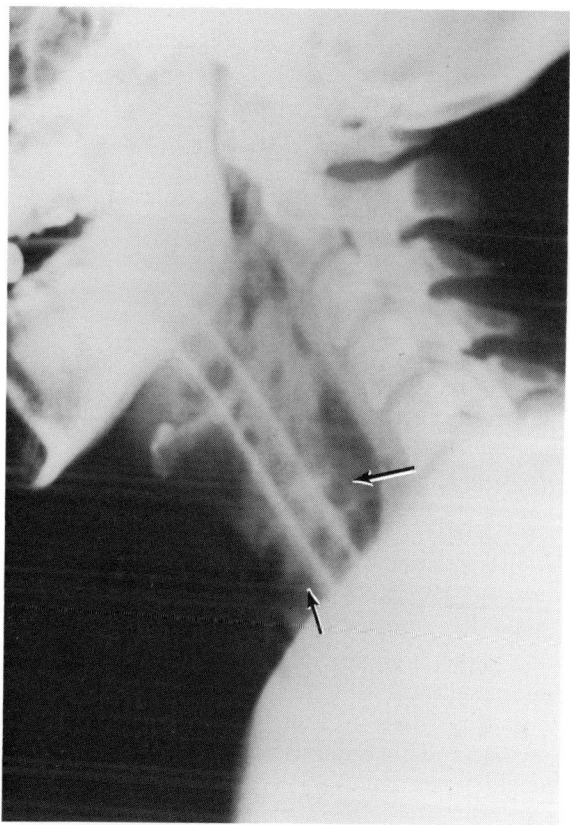

Fig. 16.2 Lateral radiograph of patient with peripharyngeal sepsis due to oral anaerobes, resulting from trauma during difficult intubation. Note collections of gas (arrowed)

Secretory otitis media

Non-suppurative or secretory otitis media (glue ear) is primarily a disease of children, resulting from obstruction of the eustachian tube due to dysfunction, infection, allergy, enlargement of the adenoids or, more rarely, tumour. Falling pressure within the tube causes retraction of the eardrum, impairment of hearing and in many instances there is an effusion. Although long considered to be sterile it is now clear that bacteria, including *Haemophilus influenzae*, *Str. pneumoniae* or *S. aureus*, may be isolated from up to 40 per cent of these effusions (Senturia *et al.*, 1958; Healy and Teele, 1977; Riding *et al.*, 1978; Brook *et al.*, 1983b); anaerobic bacteria have been isolated with less frequency (Fulghum *et al.*, 1977; Giebink *et al.*, 1979; Brook *et al.*, 1983b).

It is commonly believed that, if glue ear is left untreated, it may progress to cholesteatoma (see below) with the attendant danger of intracranial suppuration. The primary objective in the management of glue ear is the restoration of patency of the eustachian tube and ventilation of the middle ear. The latter may require surgical intervention such as the insertion of grommets. In the absence of acute infective exacerbations the role of antibiotics in this condition is unclear.

Acute otitis media

Acute suppurative otitis media may exacerbate glue ear, but more commonly complicates upper respiratory tract viral infections and tonsillitis. *Str. pneumoniae*, *H. influenzae*, *Branhamella catarrhalis* and, occasionally, group A β-haemolytic streptococci are the usual bacterial pathogens (Kamme *et al.*, 1971). Occasional anaerobes have been reported, but in one study were found in 27 per cent (17 out of 62) of patients (Brook *et al.*, 1978). The anaerobes present were *Propionibacterium acnes* and peptococci, the sole isolates in nine patients, and associated with facultative anaerobes in eight instances. There are scattered reports of acute otitis media and mastoiditis characterized by foul-smelling pus yielding *F. necrophorum*, *B. corrodens* (Reynolds *et al.*, 1985), and penicillin-resistant *Bacteroides* spp. (Swanston *et al.*, 1983). Anaerobic bacteria thus play a numerically small but clinically significant role in acute infections of the middle ear.

Chronic suppurative otitis media

There are two forms of chronic suppurative otitis media (CSOM). The so-called 'safe', or tubotympanic type of CSOM is associated with a persistent central perforation of the tympanic membrane secondary to acute suppurative otitis media. Discharge from this type is often profuse, and can cause secondary otitis externa. The 'dangerous type' of CSOM (tympano-mastoid or atticoantral) is characterized by malodourous discharge and is associated with the presence of cholesteatoma, a bag-like cystic structure lined with keratinizing stratified squamous epithelium resting on a fibrous stroma of variable thickness. Although there are several theories as to the origin of these structures they are held to arise most commonly as a result of invagination of the *pars flaccida* of the tympanic membrane due to the reduction in pressure in the eustachian tube which is associated with non-secretory otitis media. Squames and epithelial debris accumulate, causing the structure to enlarge progressively and a mixed bacterial flora develops (Table 16.2).

The organisms commonly present in CSOM are generally more characteristic of the faecal flora than that of the nasopharynx, and include *Escherichia coli*, *Proteus* spp., *Klebsiella* spp., *Str. faecalis*, *S. aureus* and *Pseudomonas aeruginosa*. The anaerobes normally present include anaerobic cocci, members of the *B. melaninogenicus* group, eubacteria and, significantly, *B. fragilis* and related species; fusobacteria are rarely reported (Jokipii *et al.*, 1977; Brook and Finegold, 1979). The nature of the organisms found in this condition strongly suggests introduction via the external auditory meatus rather than the eustachian tube.

Therapy

It will seldom be necessary to consider anaerobes in the treatment of non-suppurative and acute suppurative otitis media. However, in the rare instances where such organisms have been implicated, metronidazole has usually been included in the regimens employed (Swanston *et al.*, 1983; Reynolds *et al.*, 1985).

The treatment of CSOM is dictated by the type of disease. In the tubotympanic variety efforts are directed at maintaining a dry ear with local treatment which may include antimicrobials, an acute

Table 16.2 Bacteriology of cholesteatoma

Author(s)	Number of patients	Number of strains isolated (%)			
		Aerobes alone	Anaerobes alone	Aerobes and anaerobes	Sterile
Harker and Koontz (1978)	30	6 (20)	5 (17)	15 (50)	4 (13)
Brook (1981b)	24	8 (33)	4 (17)	12 (50)	0 (0)

exacerbation indicating a need for systemic chemotherapy. The presence of cholesteatoma in tympanomastoid CSOM is an absolute indication for surgical intervention because of the considerable danger of intracranial complications. It would seem wise to cover such operations with appropriate antimicrobials—a suitable regimen being ampicillin + metronidazole, with gentamicin as an additional option (see page 279; chemotherapy of intracranial sepsis).

Infections of the paranasal air sinuses

The contiguity of the sinuses with the nasopharyngeal cavity dictates the bacterial flora of these organs. Contrary to long held assumptions that healthy sinuses are sterile, recent evidence indicates colonization by bacteria representative of the normal mouth flora (Brook, 1981c). Sinusitis may thus result from overgrowth of sinus flora when drainage is impaired, spread of acute nasopharyngeal infections or a combination of both mechanisms. Obstruction to drainage of the sinuses most commonly results from upper respiratory tract infections but may be secondary to enlarged adenoids. Infection may also spread to the sinuses as a consequence of surgery, trauma, or secondary to apical dental disease.

The organisms usually reported in sinusitis are *H. influenzae*, *S. aureus*, *Str. pneumoniae* and *Str. milleri* (Carenfelt *et al*., 1978; Bridger, 1980; Blayney *et al*., 1984) but anaerobic bacteria are also often present. Indirect evidence to this effect comes from the increasing frequency with which anaerobic bacteria are isolated from intracranial complications of sinusitis, such as extradural and subdural empyema, meningitis and intracerebral abscesses, thrombosis of the cavernous and lateral sinuses, and orbital cellulitis (Brook *et al*., 1980; Sprott *et al*., 1981; Krohel *et al*., 1982; Weiss *et al*., 1983; Grace and Drake-Lee, 1984). The prevalence of anaerobic bacteria in sinusitis is almost certainly underestimated, as demonstrated by the fact that 95 of 547 (17 per cent) of the sinus cultures reported in eight studies were sterile; anaerobic bacteria were present in 128

Table 16.3 Bacteria isolated from infected paranasal air sinuses

Author(s)	Number of patients/ sinuses	Number yielding				
		Aerobes	Anaerobes	Aerobes and anaerobes	No growth	Other
Urdal and Berdal (1949)	81	53	6	2	20	0
Palva *et al*. (1962)	88	55	0*	0	33	0
Frederick and Braude (1974)	83	19	26	17	21	0
Evans *et al*. (1975)	24	14	1	2	7	3†
Carenfelt *et al*. (1978)	190	142	30	13	5	0
Karma *et al*. (1979)	61	35	2	15	9	0
Brook *et al*. (1980)	8	0	0	8	0	0
Skau *et al*. (1984)	12	3	6	0	0	3‡

*Anaerobic culture not carried out
†2 rhinovirus, 1 *Penicillium*
‡Not cultured

(23 per cent) (Table 16.3). The anaerobic bacteria reported in these studies were usually typical of the mouth flora; however *B. fragilis* and other β-lactamase-producing strains of *Bacteroides* have also been isolated, which may have implications for chemotherapy (Frederick and Braude, 1974; Brook, 1981c).

Intracranial infections

Intracranial sepsis results from local extension of infection in the head and neck, metastatic spread or as a complication of congenital cyanotic heart disease. Since the advent of antimicrobials, metastatic spread has become much less common except in certain groups of patients such as intravenous drug abusers. Brain abscess complicating congenital cyanotic heart disease may result from endocarditis but is more commonly due to shunting of blood from the right to the left side of the heart, allowing access of organisms to areas of brain substance which have been rendered hypoxic by highly viscous, polycythaemic blood. Although mouth commensals are the common anaerobes in the latter abscesses, colonic anaerobes, typified by *B. fragilis*, are sometimes present.

The spread of infection from local foci in the head and neck is responsible for the majority of intracranial infections due to anaerobic bacteria. Primary foci include the middle ear, the paranasal air sinuses, the teeth and pyogenic infections of the oropharynx. Less commonly anaerobic intracranial infections complicate local surgical procedures or trauma. Infections caused by spread from local foci include meningitis, epidural and subdural abscesses, cavernous and lateral sinus thrombosis and brain abscess. The bacteriology of these infections is often complex but certain general observations apply. Pyogenic organisms such as *S. aureus* and *Str. pyogenes*, which dominated in the preantimicrobial era, are now very uncommon, having been largely supplanted by organisms constituting the normal flora of the gingival crevice. An exception to this is seen in infections secondary to chronic suppurative otitis media in which faecal organisms abound.

Intracranial sepsis due to these opportunistic, synergic infections often develops insidiously, sometimes with absent or minimal signs in the primary focus, thus presenting considerable diagnostic problems. It is clear that the prevalence of anaerobes in such mixed infections is still underestimated, as seen by the number of series in which they have not figured and in which sterile abscesses were reported (Heineman and Braude, 1963; Shaw and Russell, 1975; McClelland *et al.*, 1978).

The complex bacteriology characterizing intracranial infections involving anaerobic bacteria dictates initial use in all such infections of appropriate broad-spectrum combinations of antimicrobials, the rationale for which is discussed in the section on the chemotherapy of brain abscess (page 279).

Meningitis

Anaerobic bacteria are infrequently encountered in meningitis. Rarely, they will gain access to the meninges by metastatic spread, as in Lemierre's syndrome, and less commonly in other septicaemic conditions. Anaerobic organisms may also be found in meningitis resulting from parameningeal sources. Fractures of the base of the skull and, more rarely, congenital abnormalities of the mastoid bone (Durham *et al.*, 1986) may first present with meningitis due to *Str. pneumoniae*, which may be difficult to differentiate from the classic meningitis due to this organism. The additional presence of anaerobic bacteria in this situation signals an alternative pathology and the need for further investigation, and usually surgical intervention.

Anaerobic meningitis may be encountered as a terminal feature of cavernous sinus thrombosis and as a complication of CSOM. In the latter there may be an initial good response to appropriate chemotherapy but brain abscess must always be carefully excluded. Initial computerized tomography (CT) may reveal only cerebritis, which usually proceeds to a fully developed abscess within 48 h, and which almost invariably requires surgical intervention.

Epidural and subdural empyema

Epidural and subdural empyemas are common sequelae of infections of the paranasal air sinuses. Dental foci and, more rarely, otitic sources, trauma, surgery or metastatic infections may be implicated.

Frontal sinusitis is the primary focus in most extradural abscesses and in the much more common subdural empyema. Spread of infection follows osteitis or infective thrombophlebitis of communicating veins. Infections involve the frontal poles and may extend posteriorly over the frontal lobe, the falx and the Sylvian fissure (Fig. 16.3). In the much less common otogenic subdural abscess, pus may be found in the posterior fossa or around the occipital lobe.

Concurrent involvement of the maxillary sinus, particularly when multiple mouth organisms are demonstrated in the pus, should direct attention to a primary focus in the roots of maxillary teeth. In acute sinusitis the most commonly reported organisms are *Str. milleri* and *S. aureus*. Obligate anaerobes such as fusobacteria, bacteroides and anaerobic cocci also feature, possibly indicating underlying chronic sinusitis or dental sources (Yoshikawa *et al.*, 1975; Sprott *et al.*, 1981; Brook and Friedman, 1982). Reduction in the mortality from subdural abscess, still as high as 40 per cent in some series, demands a high index of suspicion in patients presenting with fever, progressive loss of consciousness and a focal deficit. The differential diagnosis includes tuberculous or viral meningitis, herpes simplex and other forms of encephalitis. Lumbar puncture is contraindicated because of the danger of herniation of brain substance and diagnosis depends upon the use of CT scanning. Treatment consists of surgical drainage of the empyema and attention to the primary focus together with appropriate chemotherapy.

Infections of the intracranial venous sinuses

Cavernous sinus

Facial and nasal infections, causing cavernous sinus thrombosis via the facial and orbital veins, are now uncommon sources of infection, presumably due to early antimicrobial therapy of staphylococcal and streptococcal facial infections. However, the cavernous sinus also communicates with the venous drainage of the paranasal sinuses, oropharynx, middle and inner ear and the mastoid, all of which may be infective foci. Cavernous sinus thrombosis may present spontaneously but also as a direct sequel of operative procedures including dental treatment (Stern *et al.*, 1981; Goteiner *et al.*, 1982; Palmersheim and Hamilton, 1982; Harbour *et al.*, 1984). The diagnosis may be clouded by minimal signs of the primary infection and indolent presentation characteristic of retrograde spread (against the

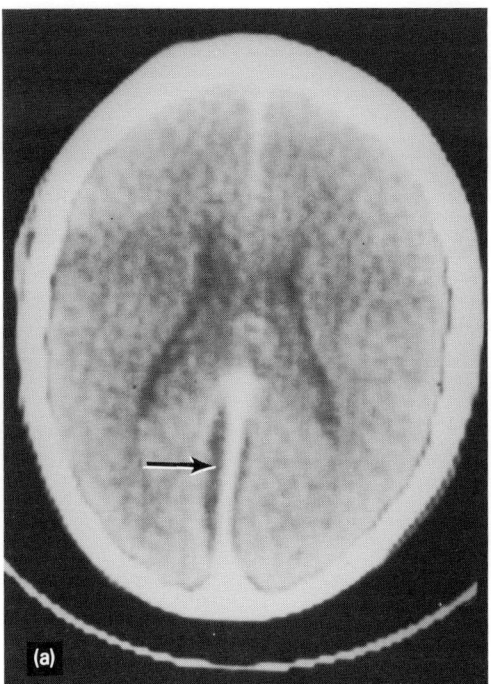

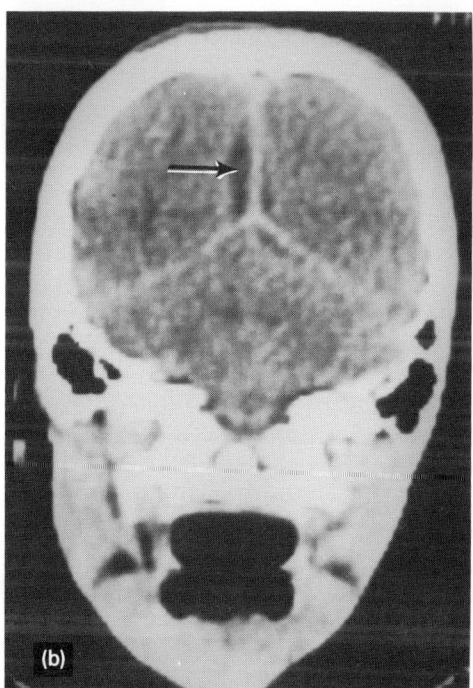

Fig. 16.3 Computerized tomograms of parafalcine subdural empyema complicating odontogenic maxillary and frontal sinusitis. Arrows indicate location of pus. (a) axial view; (b) coronal view

normal direction of venous flow) from sites such as the posterior mouth, pharynx, middle and inner ear.

Clinical features include headache, vomiting, high fever, cranial nerve palsies and exophthalmos, which may be bilateral; blood cultures may be positive. The condition carries a high mortality and successful outcome depends upon early diagnosis, identification and drainage of the primary focus and appropriate chemotherapy and anticoagulation.

Lateral sinus

Lateral sinus thrombosis is a rare sequel of extension of infection from the middle ear. As acute mastoiditis is now very uncommon, most examples of lateral sinus thrombosis are complications of cholesteatoma associated with CSOM (Courville, 1955; Hawkins, 1985). Common features include chills, intermittent fever, headaches, pulmonary emboli and discomfort in the neck; signs of the primary mastoid infection may or may not be apparent. As with infections of the cavernous sinus, blood cultures may be positive, the mortality is high and management requires appropriate chemotherapy, anticoagulation and surgical drainage of the primary focus.

The primary sources of infection in lateral and cavernous sinus thrombosis point to a complex bacteriological picture that is confirmed by polymicrobial infections involving anaerobic bacteria typical of those present in the oropharynx or chronic otitis media (Goteiner *et al.*, 1982; Palmersheim and Hamilton., 1982; Teichgraeber *et al.*, 1982; Pallares *et al.*, 1983; Harbour *et al.*, 1984; Hawkins 1985). Organisms classically associated with pyogenic infections, such as *S. aureus* and *Str. pyogenes*, are now rarely reported.

Brain abscess

Reports from the last 40 years show that the mortality from brain abscess has remained high, in some instances exceeding 50 per cent (Table 16.4). The role of obligate anaerobes in this field has been grossly underestimated, a fact graphically illustrated by Heineman and Braude's review (1963) of 17 papers on brain abscess. This demonstrated that there was a direct relationship between 'sterile'

Table 16.4 Mortality from brain abscess

Author(s)	Period	Mortality (%)
Tutton (1953)	1948–52	7/54 (13)
Pennybacker (1961)*	1938–50	18/50 (36)
	1950–60	2/35 (5.7)
Garfield (1969)	1951–57	37/100 (37)
	1962–67	43/100 (42)
Shaw and Russell (1975)†	1950–73	19/47 (41)
de Louvois *et al.* (1977)	1974–77	7/35 (20)
McClelland *et al.* (1978)	1946–76	61/113 (54)
Alderson *et al.* (1981)	1964–68	13/31 (42)
	1969–73	6/28 (21)
	1974–78	3/31 (9.7)
Bradley and Shaw (1983)	1950–59	88/177 (50)
	1960–69	32/82 (39)
	1970–79	19/61 (31)
Jadavji *et al.* (1985)	1960–84	8/74 (11)

*Otogenic
†Cerebellar

cultures and a poor yield of anaerobes; the reverse was also true.

The aetiology of brain abscess falls into seven main groups:
traumatic, either accidental or surgical;
metastatic;
congenital cyanotic heart disease;
odontogenic;
cryptogenic;
otogenic;
sinugenic.

The last two are the most common. Anaerobic bacteria may be found in any of these but are relatively rare in traumatic abscesses unless there is communication with a potentially infected source, e.g., paranasal sinus or oropharynx. The bacteriology of metastatic abscesses reflects that of the primary source; anaerobes may be found in those secondary to abdominal or pelvic sepsis and to lung infections due to aspiration or, more rarely, to intravenous drug abuse. Such abscesses tend to be multiple and may occur in any part of the brain but are most common in the distribution of the middle cerebral artery, i.e., the parietal and frontoparietal regions.

Anaerobes are often present in abscesses secondary to congenital cyanotic heart disease, with other organisms which mainly come from the oropharynx or the large bowel. The remaining groups of abscesses are associated with direct extension from infected foci anatomically adjacent to the brain. Spread occurs via communicating veins or an osteomyelitic process, and less commonly through congenital defects of bone or venous structures. The most common organism reported in sinugenic abscesses (typically located in the frontal lobe) is *Str. milleri*, usually Lancefield group F (de Louvois *et al.*, 1977). Obligate anaerobes have also been reported in these abscesses, originating from infective foci in either the paranasal air sinuses or the maxillary teeth (Heineman and Braude, 1963; Brook and Friedman, 1982; Grace and Drake-Lee, 1984).

Brain abscess may uncommonly complicate acute dental infections or dental procedures (Haymaker, 1945; Gold, 1949; Hollin *et al.*, 1967; Henig *et al.*, 1978). Frontal lobe abscesses occurring in the absence of sinusitis or other obvious primary foci, yielding organisms typical of the oral flora, have been considered to be secondary to covert dental sources (Ingham *et al.*, 1978; Baddour *et al.*, 1979). Direct spread from such foci probably occurs by retrograde spread through normal venous channels or via persistent fetal venous communications.

The bacterial flora of otogenic brain abscesses, found in the cerebellum or temporal lobe, is typically faecal. In addition to organisms such as *E. coli*, *Str. faecalis*, *Proteus* spp., anaerobic cocci, and *B. melaninogenicus*, *B. fragilis* is frequently present. This organism is rarely found in abscesses originating from the oropharynx or related structures. The pathogenesis of otogenic brain abscesses is closely associated with cholesteatoma (see page 272), which is usually regarded as a complication of glue ear. The faecal nature of the bacterial flora of cholesteatomas suggests anoaural transfer.

Improved bacteriological techniques allowing better characterization of flora have resulted in a more precise indication of primary sources with the consequence that fewer brain abscesses are attributed to cryptogenic foci.

Management

The successful management of brain abscess demands a high degree of clinical suspicion in any patient presenting with early signs of raised intracranial pressure—headache, nausea or vomiting and alteration in the level of consciousness. Suspicion will be heightened by the presence of a possible primary focus (see above), other indications of infection, such as fever and leucocytosis, are not consistently present. Among other signs which may be associated with cerebral abscess are convulsions, long tract involvement, papilloedema and, when the cerebellum is involved, nystagmus and ataxia. These may not be present initially but this must not defer serious consideration of the diagnosis.

A most dangerous situation arises when the primary focus is not obvious; under these circumstances attention may be diverted to alternative diagnoses such as encephalitis or meningitis and lead to the performance of a lumbar puncture. This procedure is absolutely contraindicated in brain abscess because of the danger of herniation of the brain substance through the foramen magnum; this has contributed significantly to the mortality in this condition (Garfield, 1969); CSF findings in patients with brain abscess may, in any case, be unremarkable, or at best misleading.

In the early stages CT scanning, the investigation of choice, may reveal only a low density lesion due to early cerebritis in the affected portion of the brain (Fig. 16.4). In some instances there is mid-line shift and narrowing of the anterior and posterior horns of the ipsilateral ventricles, less often there may be evidence of hydrocephalus. Repeat scanning, after 24–48 h, mandatory even in patients receiving optimal chemotherapy and showing clinical improvement, will almost invariably reveal development of the classic ring enhancing lesion, typical of brain abscess (Fig. 16.4). The importance of monitoring the development of these lesions cannot be stressed enough, if sudden, rapid deterioration in the patient's condition is to be avoided. The CT appearances of slowly-developing abscesses, particularly in so-called 'silent areas' such as the frontal lobe, can be very difficult to differentiate from brain tumour; and their true nature may only be revealed through surgical intervention. This clearly needs to be borne in mind because the clinical evolution of tumour is relatively slow, whereas dramatic deterioration in the patient's condition can occur with an abscess.

Surgery

Some neurosurgeons prescribe dexamethasone to aid reduction in the intracranial pressure but surgical decompression is an essential first step. This is

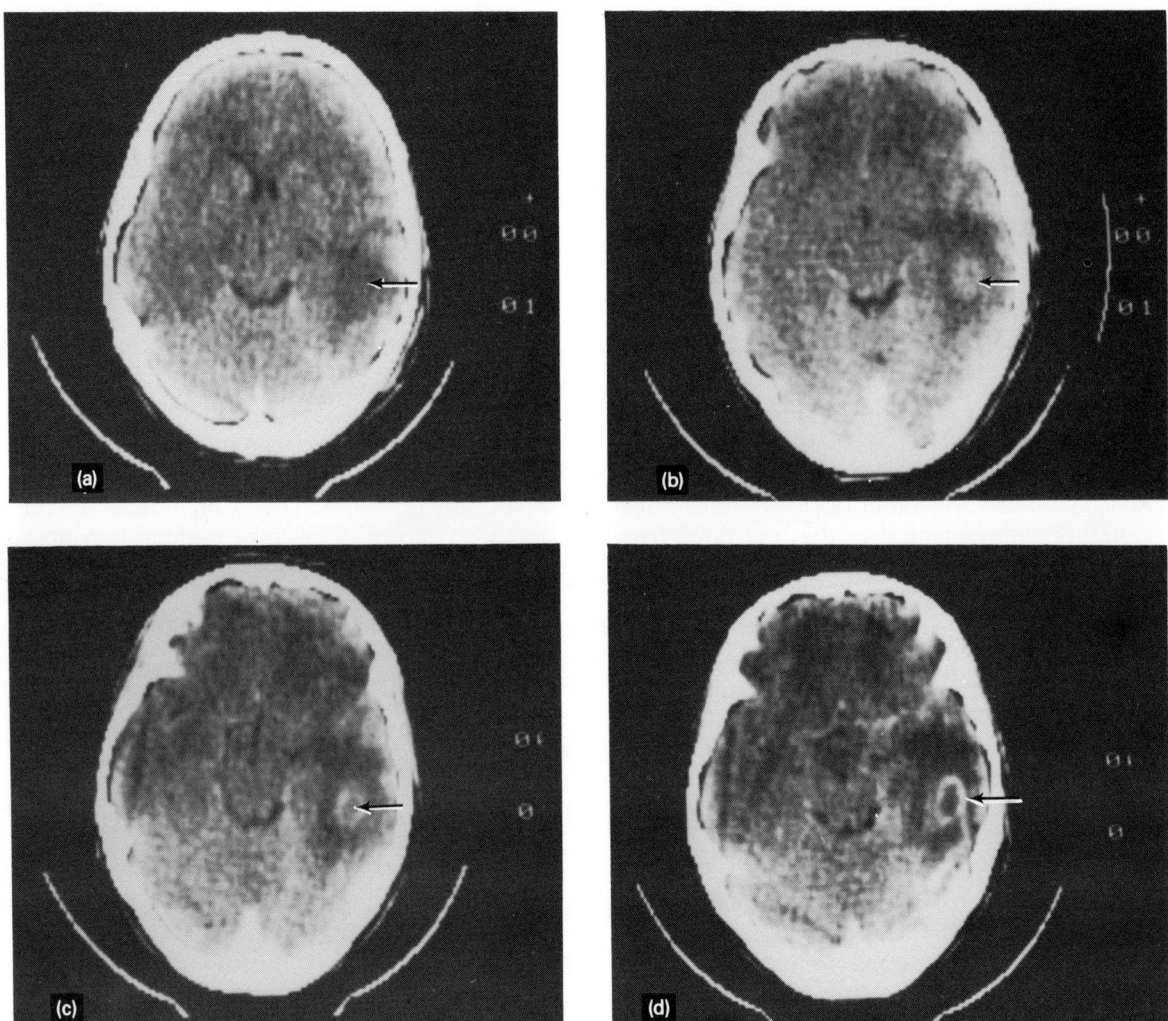

Fig. 16.4 Computerized tomograms, with contrast medium, of abscess of the temporal lobe of the brain associated with cholesteatoma. (a) initial scan when patient presented with extradural and subdural abscesses. These were drained and chemotherapy with ampicillin, gentamicin and metronidazole was commenced. Note low density lesion indicating cerebritis (arrowed); (b and c) repeat scans 4 and 6 days later, respectively. Low density lesion increased in size but clinical progress satisfactory; (d) scan after 13 days, patient asymptomatic, abscess considerably enlarged; note ring enhancement (arrowed). Burr hole aspiration carried out. Culture of the pus revealed *B. fragilis* and *Proteus* species; chemotherapy continued and abscess resolved.

often achieved by simply aspirating the abscess contents, via a burr hole in the skull, with a wide bore needle; in many instances, in combination with appropriate chemotherapy, this is all that is required. Reaspiration, which should seldom be necessary, is easily performed and in rare instances where it is deemed essential, secondary excision of the abscess is a viable option. It is generally acknowledged that primary excision is the treatment of choice for abscesses in the cerebellum. Although some advocate wider use of this procedure for all brain abscesses (Choudhury *et al.*, 1977) the more explicit knowledge of the bacteriology of brain abscess coupled with the availability of effective antimicrobial agents makes this approach much less tenable. When drainage of the abscess has been achieved immediate steps must be taken to identify and extirpate the primary focus of infection.

Aspiration may not be possible where multiple abscesses are present, as occurs with metastatic sources and in patients with congenital cyanotic heart disease, or when the lesion is deep seated and inaccessible. In these circumstances chemotherapy alone must be relied upon, and some success has been reported with this approach (Heineman et al., 1971; Barsoum et al., 1981; Weisberg, 1981. It must be stressed, however, that this is not the preferred treatment. Even in patients presenting only with CT evidence of cerebritis, the condition almost invariably progresses to abscess formation.

Chemotherapy

Treatment should be initiated immediately the diagnosis is suspected. The low incidence of brain abscess makes assessment of the efficacy of chemotherapy difficult, and in most series standard regimens have not been employed. The range of primary foci, and thus the nature of the likely infecting organisms, must be considered so that an appropriate initial regimen can be instituted pending results of culture. As Table 16.5 demonstrates, these considerations indicate a requirement for a broad-spectrum combination encompassing both obligate and facultative anaerobes. Such therapy is

Table 16.5 Source and bacterial isolates from 90 patients with brain abscess in Newcastle upon Tyne 1964–1978*

Bacteria	Source (number of patients)						
	Oto- genic (45)	Sinu- genic (4)	Metas- tatic† (15)	Trau- matic (10)	Den- tal (2)	Un- known (14)	Total
Anaerobes							
Anaerobic coccus	36	—	7	1	3	6	53
B. fragilis	28	—	3	1	—	1	33
B. melaninogenicus	17	—	—	—	3	—	20
Fusobacterium spp.	2	—	—	—	2	—	4
Actinomyces spp.	1	—	—	—	1	1	3
Clostridium spp.	2	—	—	—	—	—	2
Eubacterium spp.	11	—	—	—	—	1	12
Veillonella spp.	1	—	—	—	1	—	2
Propionibacterium spp.	—	—	—	—	1	—	2
							131
Aerobes							
Proteus spp.	20	—	1	1	—	—	22
E. coli	14	—	2	—	—	—	16
Klebsiella spp.	1	—	2	—	—	—	3
S. aureus	4	2	1	6	—	1	14
Non-haemolytic streptococci	12	1	—	2	—	1	16
Str. milleri	—	1	—	—	2	2	5
Str. pneumoniae	2	—	1	—	—	1	4
Enterococcus	1	—	—	—	—	—	1
Haemophilus spp.	3	—	—	—	—	2	5
Eikenella spp.	—	—	1	—	1	—	2
Acinetobacter spp.	—	—	1	—	—	—	1
Actinobacillus spp.	—	—	—	—	1	—	1
Lactobacillus spp.	1	—	—	—	—	—	1
							91

*Alderson et al. (1981)
† Includes abscesses due to congenital cyanotic heart disease

required not only to eradicate anaerobes *per se* but also to ameliorate the effects of β-lactamase production by organisms such as *B. fragilis* and some strains of oral anaerobes; such β-lactamase production has been shown to impair the clinical efficacy of penicillin both in animal models and in human infections (Hackman and Wilkins, 1976; Heimdahl *et al.*, 1980).

Antimicrobials are not infrequently instilled into abscess cavities. This practice undoubtedly reflects earlier attempts at control when systemic agents failed. The precise bacteriological data and effective antimicrobials now available allow satisfactory therapy by the systemic route. There is mounting evidence that metronidazole is the agent of choice for the anaerobic component of these infections (Ingham *et al.*, 1977; Alderson *et al.*, 1981; Walsh *et al.*, 1982; Mathisen *et al.*, 1984). The agents employed against facultative anaerobes vary but the spectrum must encompass *S. aureus*, streptococci, including *Str. faecalis*, and gram-negative facultative anaerobes. Some favour chloramphenicol, and we have found that gentamicin and ampicillin fulfil this purpose. It is believed that the use of this regimen reduced the mortality of brain abscess in this unit from 26 per cent during 1969–73 to 9.7 per cent in 1974–78, when metronidazole had replaced other agents such as clindamycin in the treatment of the anaerobic component (Alderson *et al.*, 1981); the mortality during 1979–83 was held at 8 per cent (unpublished data). The efficacy of chemotherapy should be monitored by clinical assessment and regular CT scanning; with an appropriate response, therapy rarely needs to exceed 3–4 weeks.

Anaerobic pleuropulmonary infection

Anaerobic pleuropulmonary infections, while commonly reported in the USA, do not figure prominently in European publications. These differences in prevalence are surprising in view of the pathogenetic factors common to most of these infections, which arise as a result of aspiration, obstruction or abnormal patency of the bronchi, metastatic spread, or direct extension from contiguous infective foci.

In the preantibiotic era it was well known that certain types of pleuropulmonary infections were of obscure aetiology and were not associated with the presence of the common bacterial pathogens such as *S. aureus*. A prime example of these 'non-specific' infections was lung abscess, from which a third of patients died, one-third recovered and the remainder were left with severe recurrent disease (Allen and Blackman, 1936). Although anaerobic bacteria were demonstrated in early operative samples from lung abscesses (Cohen, 1932) the catalyst to serious consideration of the pathogenic role of such organisms was the marked reduction in mortality effected by penicillin (Weiss and Flippin, 1967; Weiss and Cherniack, 1974). Subsequent studies employing transtracheal aspiration, which avoided contamination of samples with oral flora, and appropriate cultural techniques showed that the organisms involved in primary lung abscess were derived from the normal biota of the mouth, predominantly consisting of obligate anaerobes (Bartlett *et al.*, 1973; Bartlett *et al.*, 1974b).

Early animal studies (Smith, 1927) showed that the introduction of gingival flora into the bronchial tree resulted in fatal pneumonia or pulmonary abscesses, it was also known that aspiration of small amounts of material from the oropharynx occurs during sleep in normal individuals (Amberson, 1937). These findings indicated that the anaerobic bacteria responsible for certain types of pleuropulmonary sepsis are aspirated from the oral cavity.

Aspiration

Retention in the lungs of material aspirated from the oropharynx is likely to occur during loss of consciousness, as associated with epileptiform convulsions, general anaesthesia, head injury and drug or alcohol overdosage or as a consequence of neurological disorders interfering with normal deglutition or any kind of oesophageal obstruction. In healthy individuals, anaerobic bacteria are present in saliva in concentrations of approximately 10^8 cfu/ml, outnumbering facultative anaerobes by a ratio of 10:1 (Bartlett and Finegold, 1974). In certain acute or chronic oropharyngeal infections the numbers of anaerobic bacteria rise significantly, substantially increasing the aspirated inoculum. Thus, the pulmonary infection due to *F. necrophorum*, characteristic of Lemierre's syndrome (Lemierre, 1936), is preceded by a tonsillar infection; another important element of other episodes of anaerobic pleuropulmonary infections is the presence of dental sepsis.

It is likely that aspiration also plays a role in

anaerobic infections complicating accidental trauma or, more rarely, thoracic surgery.

Impaired drainage

Obstruction, abnormal patency or alteration in the quality of secretions of the bronchial tree interfere with the normal drainage of small quantities of oropharyngeal secretions which intermittently gain access to the lungs. Obstruction may result from tumour, inhalation of foreign bodies, blood or particulate matter or may be caused by atelectasis complicating pneumothorax or trauma. In bronchiectasis the airways are abnormally large and the clearance mechanisms mediated by the ciliary mucosa are impaired. In cystic fibrosis the viscid secretions also interfere with clearance of aspirated material but anaerobic infections are less commonly reported (Lester et al., 1983).

Metastatic spread

Anaerobic bacteria may spread to the lungs from any septic focus, although most commonly from foci in the abdomen or pelvis; a rare source is cavernous or lateral sinus thrombosis. Haematogenous spread of anaerobic bacteria to the lungs may also be a sequel to intravenous drug abuse (Finegold, 1977).

Direct extension

Anaerobic infections may reach the lungs and pleural cavity from contiguous foci, e.g., subphrenic abscess secondary to abdominal sepsis or mediastinitis resulting from deep infections of the neck or oesophageal perforation.

The types of anaerobic bacteria responsible for pleuropulmonary infections generally reflect the primary sources and pathogenesis. Infections secondary to aspiration, obstruction or impaired drainage are characterized by anaerobes typical of the gingival flora, although *B. fragilis* is sometimes reported (Bartlett et al., 1974a; Lorber and Swenson, 1974). Where infection is metastatic, blood borne or from contiguous foci the faecal anaerobes usually reflect abdominal or pelvic sources. Oral anaerobes are again implicated in the rarer forms of spread from peripharyngeal foci, thrombosed cavernous and lateral sinuses or from infections of the mediastinum.

The clinical syndromes typifying anaerobic pleuropulmonary infections due to aspiration reflect a spectrum of pathological processes ranging from acute pneumonitis to lung abscess and empyema thoracis. Either lung may be affected but due to the anatomy of the bronchial tree material aspirated in the recumbent position more commonly lodges in the right lung (Brock, 1954). The sites implicated in both lungs are the posterior segment of the upper lobes and superior segments of the lower lobes. Aspiration in the upright position results in basilar involvement.

Aspiration pneumonia

Within hours, aspiration of gastric contents may result in an acute chemical pneumonitis due to the action of gastric acid on the lung parenchyma (Mendelson, 1946). The infectious consequences of aspiration are not usually immediately apparent, but evidenced by the subsequent development of either pneumonia or lung abscess. Pneumonia presents within seven days with fever, leucocytosis, non-productive cough and non-cavitating infiltrates on radiographs. Aspiration of large numbers of organisms may result in necrotizing pneumonitis, characterized by multiple small abscesses and a high mortality. Failure of resolution of aspiration pneumonia may likewise cause death or progression to lung abscess with or without empyemata.

Chronic destructive pneumonia (CDP) may develop in some patients with repeated episodes of aspiration, particularly where there is poor oral hygiene (Le Roux and Dodds, 1968) (Fig. 16.5). In the active phase of CDP the pathology is consolidation, with pus formation and cavities filled with necrotic material together with purulent bronchitis. During relative quiescence, fibrosis occurs with granulating cavities and bronchiectasis. The disease, which is chronic, may result in exsanguinating haemoptyses, empyema or metastatic spread.

Lung abscess

Primary lung abscess usually presents within two weeks of the initiating episode, and most patients will have had symptoms for more than one week (Bartlett and Finegold, 1974). One or more cavitating pulmonary lesions are evident on radiographs. There is fever and leucocytosis and frequently

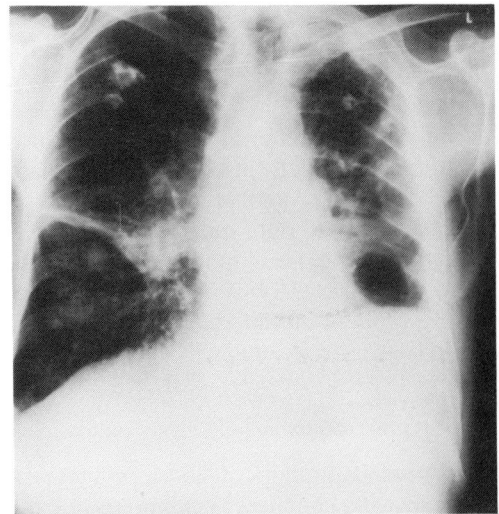

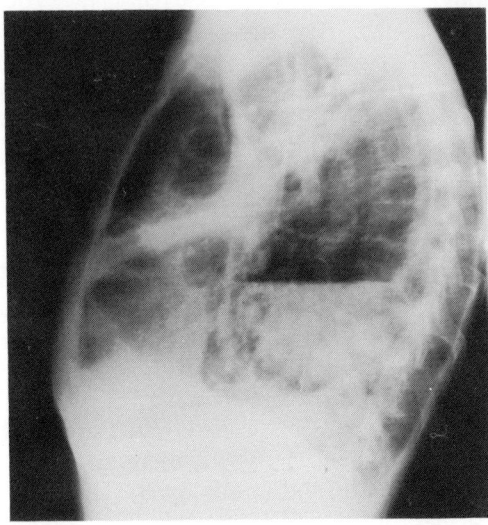

Fig. 16.5 Chest radiograph demonstrating multiple primary lung abscesses due to repeated aspiration of oral anaerobes in a patient with chronic periodontal sepsis

weight loss, anaemia and cough, which may produce foul-smelling sputum; in approximately one-third of patients there is an associated empyema.

Empyema thoracis

This condition used to be a sequel to pneumonia caused by *Str. pneumoniae* or *S. aureus*. Presumably due to the widespread exhibition of antibiotics it is now a rare complication of acute bacterial pneumonia; when encountered it is usually secondary to pulmonary abscess or operations on the oesophagus or lung. The organisms now seen are anaerobic bacteria typical of the oral flora and, especially in postoperative patients, gram-negative facultative organisms. As the condition often complicates pre-existing infections it does not usually respond satisfactorily to chemotherapy alone and some type of surgical drainage is frequently required.

Therapy

Aspiration pneumonia in non-hospitalized patients is typically caused by mixed oral organisms; in in-patients the condition may be complicated by the presence of gram-negative facultative anaerobes or *Ps. aeruginosa* (Bartlett *et al.*, 1974a; Bartlett *et al.*, 1986) reflecting the change in oral flora which accompanies hospitalization (McNamara *et al.*, 1967; Johanson *et al.*, 1969). Despite this, penicillin is highly effective in the treatment of aspiration pneumonia and primary lung abscess (Bartlett *et al.*, 1974a). Optimal response of primary lung abscess to chemotherapy, as judged by cavity closure, requires prolonged high dosage regimens, by parenteral or oral route; a small proportion of patients may require some form of surgical intervention. The presence of *B. fragilis* in aspiration penumonia and primary lung abscess in about 15 per cent of patients (Bartlett *et al.*, 1974a; Bartlett and Gorbach, 1975; Levison *et al.*, 1983) has stimulated interest in agents other than penicillin; clindamycin appears to be an effective alternative (Sen *et al.*, 1974; Bartlett and Gorbach, 1975; Levison *et al.*, 1983). Metronidazole, with either penicillin or a cephalosporin, has greatly improved the outcome of chronic destructive pneumonia (Cameron, 1978). Despite doubts over the efficacy of metronidazole as the sole agent in the treatment of other pleuropulmonary infections (Sanders *et al.*, 1979; Perlino, 1981) it seems likely that it will establish a niche in the chemotherapy of the anaerobic component of such conditions.

References

Alderson D, Strong A J, Ingham H R, Selkon J B. (1981). Fifteen-year review of the mortality of brain abscess. *Neurosurgery* **8**: 1–6.

Allen C I, Blackman J F. (1936). Treatment of lung abscess with a report of 100 consecutive cases. *Journal of Thoracic Surgery* **6**: 156–172.

Alston J M. (1955). Necrobacillosis in Great Britain. *British Medical Journal* 2: 1524–1528.

Amberson J B. (1937). Aspiration bronchopneumonia. *International Clinics* 3: 126–138.

Baddour H M, Durst N L, Tilson H B. (1979). Frontal lobe abscess of dental origin. *Oral Surgery, Oral Medicine, Oral Pathology* 47: 303–306.

Barratt G E, Koopmann C F, Coulthard S W. (1984). Retropharyngeal abscess—a ten-year experience. *Laryngoscope* 94: 455–463.

Barsoum A H, Lewis H C, Cannillo K L. (1981). Nonoperative treatment of multiple brain abscesses. *Surgical Neurology* 16: 283–287.

Bartlett J G, Finegold S M. (1974). Anaerobic infections of the lung and pleural space. *American Review of Respiratory Disease* 110: 56–77.

Bartlett J G, Gorbach S L. (1975). Treatment of aspiration pneumonia and primary lung abscess. *Journal of the American Medical Association* 234: 935–937.

Bartlett J G, Gorbach S L, Finegold S M. (1974a). The bacteriology of aspiration pneumonia. *American Journal of Medicine* 56: 202–207.

Bartlett J G, Gorbach S L, Tally F P, Finegold S M. (1974b). Bacteriology and treatment of primary lung abscess. *American Review of Respiratory Disease* 109: 510–518.

Bartlett J G, O'Keefe P, Tally F P, Louie T J, Gorbach S L. (1986). Bacteriology of hospital-acquired pneumonia. *Archives of Internal Medicine* 146: 868–71.

Bartlett J G, Rosenblatt J E, Finegold S M. (1973). Percutaneous transtracheal aspiration in the diagnosis of anaerobic pulmonary infection. *Annals of Internal Medicine* 79: 535–540.

Beeden A G, Evans J N G. (1970). Quinsy tonsillectomy—a further report. *Journal of Laryngology and Otology* 84: 443–448.

Blayney A W, Frootko N J, Mitchell R G. (1984). Complications of sinusitis caused by *Streptococcus milleri*. *Journal of Laryngology and Otology* 98: 895–899.

Bradley P J, Shaw M D M. (1983). Three decades of brain abscess in Merseyside. *Journal of the Royal College of Surgeons of Edinburgh* 28: 223–228.

Bridger R C. (1980). Sinusitis: an improved regime of investigation for the clinical laboratory. *Journal of Clinical Pathology* 33: 276–281.

Brock R C. (1954). In *The anatomy of the bronchial tree*, 2nd edn, pp. 13–20. Edited by Cumberlidge G. Oxford University Press, London.

Brook I. (1981a). Aerobic and anaerobic bacteriology of peritonsillar abscess in children. *Acta Paediatrica Scandanavica* 70: 831–835.

—— (1981b). Aerobic and anaerobic bacteriology of cholesteatoma. *Laryngoscope* 91: 250–253.

—— (1981c). Aerobic and anaerobic bacterial flora of normal maxillary sinuses. *Laryngoscope* 91: 372–376.

Brook I, Anthony B F, Finegold S M. (1978). Aerobic and anaerobic bacteriology of acute otitis media in children. *Journal of Pediatrics* 92: 13–16.

Brook I, Finegold S M. (1979). Bacteriology of chronic otitis media. *Journal of the American Medical Association* 241: 487–488.

Brook I, Friedman E M. (1982). Intracranial complications of sinusitis in children a sequela of periapical abscess. *Annals of Otology* 91: 41–43.

Brook I, Friedman E M, Rodriguez W J, Controni G. (1980). Complications of sinusitis in children. *Pediatrics* 66: 568–572.

Brook I, Gober A E. (1983). *Bacteroides melaninogenicus*: its recovery from tonsils of children with acute tonsillitis. *Archives of Otolaryngology* 109: 818–820.

Brook I, Hirokawa R. (1985). Treatment of patients with a history of recurrent tonsillitis due to group A β-haemolytic streptococci. *Clinical Pediatrics* 24: 331–336.

Brook I, Pazzaglia G, Coolbaugh J C, Walker R I. (1983a) In-vivo protection of group A β-haemolytic streptococci from penicillin by β-lactamase-producing *Bacteroides* species. *Journal of Antimicrobial Chemotherapy* 12: 599–606.

Brook I, Yocum P. (1984). Bacteriology of chronic tonsillitis in young adults. *Archives of Otolaryngology* 110: 803–805.

Brook I, Yocum P, Calhoun L. (1981a). β-Lactamase producing *Bacteroides* species recovered from children: possible clue to failure of penicillin treatment. *Lancet* i: 332.

Brook I, Yocum P, Friedman E M. (1981b). Aerobic and anaerobic bacteria in tonsils of children with recurrent tonsillitis. *Annals of Otology, Rhinology and Laryngology* 90: 261–263.

Brook I, Yocum P, Shah K, Feldman B, Epstein S. (1983b). Aerobic and anaerobic bacteriologic features of serous otitis media in children. *American Journal of Otolaryngology* 4: 389–392.

Bryan C S, King B G, Bryant R E. (1974). Retropharyngeal infection in adults. *Archives of Internal Medicine* 134: 127–130.

Cameron E W J. (1978). Treatment of chronic destructive pneumonia with cephalosporins, penicillin and metronidazole. *South African Medical Journal* 54: 57–60.

Carenfelt C, Lundberg C, Nord C E, Wretlind B. (1978). Bacteriology of maxillary sinusitis in relation to quality of the retained secretion. *Acta Otolaryngology* 84: 298–302.

Choudhury A R, Taylor J C, Whitaker R. (1977). Primary excision of brain abscess. *British Medical Journal* 2: 1119–1121.

Cohen J. (1932). The bacteriology of abscess of the lung and methods for its study. *Archives of Surgery* 24: 171–88.

Courville C B. (1955). Intracranial complications of otitis media and mastoiditis in the antibiotic era. *Laryngoscope* 65: 31–46.

Durham L, MacKenzie I J, Foy P. Bowden A. (1986).

Recurring meningitis: beware the normal looking ear. *British Medical Journal* **293**: 1230.

Evans F O *et al.* (1975). Sinusitis of the maxillary antrum. *New England Journal of Medicine* **293**: 735–739.

Finegold S M. (1977). *Anaerobic bacteria in human disease*, p. 202. Academic Press, New York.

Flodstrom A, Hallander H O. (1976). Microbiological aspects on peritonsillar abscesses. *Scandinavian Journal of Infectious Diseases* **8**: 157–160.

Frederick J, Braude A I. (1974). Anaerobic infection of the paranasal sinuses. *New England Journal of Medicine* **290**: 135–137.

Fulghum R S, Daniel H J, Yarborough J G. (1977). Anaerobic bacteria in otitis media. *Annals of Otology* **86**: 197–203.

Garfield J. (1969). Management of supratentorial intracranial abscess: a review of 200 cases. *British Medical Journal* **1**: 7–11.

Gastanaduy A S, Kaplan E L, Huwe B B, McKay C, Wannamaker L W. (1980). Failure of penicillin to eradicate group A streptococci during an outbreak of pharyngitis. *Lancet* ii: 498–502.

Giebink G S *et al.* (1979). The microbiology of serous and mucoid otitis media. *Pediatrics* **63**: 915–919.

Gold L. (1949). Brain abscess secondary to dental infection. Report and discussion of a case. *Oral Surgery* **2**: 1107–1117.

Goldstein J C. (1978). Epstein–Barr virus the ravager. *Annals of Otology, Rhinology and Laryngology* **87**: 729–735.

Goteiner D, Sonis S T, Fasciano R. (1982). Cavernous sinus thrombosis and brain abscess initiated and maintained by periodontally involved teeth. *Journal of Oral Medicine* **37**: 80–94.

Grace A, Drake-Lee A. (1984). Role of anaerobes in cerebral abscesses of sinus origin. *British Medical Journal* **288**: 758–759.

Gray W C. (1984). Throat culture in impending peritonsillar abscess. *Southern Medical Journal* **77**: 1545–1547.

Hackman A S, Wilkins T D. (1976). Influence of penicillinase production by strains of *Bacteroides melaninogenicus* and *Bacteroides oralis* on penicillin therapy of an experimental mixed anaerobic infection in mice. *Archives of Oral Biology* **21**: 385–389.

Harbour R C, Trobe J D, Ballinger W E. (1984). Septic cavernous sinus thrombosis associated with gingivitis and parapharyngeal abscess. *Archives of Ophthalmology* **102**: 94–97.

Harker L A, Koontz F P. (1978). The bacteriology of cholesteatoma. *1st International Conference on Cholesteatoma*, pp. 264–267. McCabe and Abramson Aesculepius, Alabama.

Hawkins D B. (1985). Lateral sinus thrombosis: a sometimes unexpected diagnosis. *Laryngoscope* **95**: 674–677.

Haymaker W. (1945). Fatal infections of the central nervous system and meninges after tooth extraction. *American Journal of Orthodontics and Oral Surgery* **31**: 117–188.

Healy G B, Teele D W. (1977). The microbiology of chronic middle ear effusions in children. *Laryngoscope* **87**: 1472–1478.

Heath L K, Peirce T H. (1977). Retropharynngeal abscess following endotracheal intubation. *Chest* **72**: 776–777.

Heimdahl A, Nord C E. (1979). Effect of phenoxymethyl-penicillin and clindamycin on the oral, throat and faecal microflora of man. *Scandanavian Journal of Infectious Diseases* **11**: 233–242.

Heimdahl A, von Konow L, Nord C E. (1980). Isolation of β-lactamase-producing *Bacteroides* strains associated with clinical failures with penicillin treatment of human orofacial infections. *Archives of Oral Biology* **25**: 689–692.

Heineman H S, Braude A I. (1963). Anaerobic infection of the brain. *American Journal of Medicine* **35**: 682–697.

Heineman H S, Braude A I, Osterholm J L. (1971). Intracranial suppurative disease. Early presumptive diagnosis and successful treatment without surgery. *Journal of the American Medical Association* **218**: 1542–1547.

Henig E F, Derschowitz T, Shalit M, Toledoe E. (1978). Brain abscess following dental infection. *Oral Surgery* **45**: 955–958.

Hollin S A, Hayashi H, Gross S W. (1967). Intracranial abscesses of odontogenic origin. *Oral Surgery, Oral Medicine and Oral Pathology* **23**: 277–293.

Holt G R, Tinsley P P. (1981). Peritonsillar abscesses in children. *Laryngoscope* **91**: 1226–1230.

Hudson S, Maddocks A C, Stacey A. (1984). Necrobacillosis. *British Medical Journal* **288**: 1915–1916.

Ingham H R, High A S, Kalbag R M, Sengupta R P, Tharagonnet D, Selkon J B. (1978). Abscesses of the frontal lobe of the brain secondary to covert dental sepsis. *Lancet* ii: 497–499.

Ingham H R, Selkon J B, Roxby C M. (1977). Bacteriological study of otogenic cerebral abscesses; chemotherapeutic role of metronidazole. *British Medical Journal* **2**: 991–993.

Ingham HR, Sprott MS, Selkon JB. (1980). β-lactamase-producing anaerobes. *Lancet* ii: 748.

Jadavji T, Humphreys R P, Prober C G. (1985). Brain abscesses in infants and children. *Pediatric Infectious Disease* **4**: 394–398.

Johansen W G, Pierce A K, Sanford J P. (1969). Changing pharyngeal bacterial flora of hospitalized patients. *New England Journal of Medicine* **281**: 1137–1140.

Johnsen T. (1981). Infectious mononucleosis and peritonsillar abscess. *Journal of Laryngology and Otology* **95**: 873–876.

Johnson A P, Gray R F, Smelt G J C. (1984). Life-threatening peripharyngeal suppuration. *Journal of Laryngology and Otology* **98**: 429–435.

Jokipii A M M, Karma P, Ojala K, Jokipii L. (1977). Anaerobic bacteria in chronic otitis media. *Archives of Otolaryngology* **103**: 278–280.

Kamme C, Lundgren K, Mårdh P A. (1971). The aetiology of acute otitis media in children. *Scandinavian Journal of Infectious Diseases* 3: 217–223.

Karma P, Jokipii L, Sipila P, Luotonen J, Jokipii A M M. (1979). Bacteria in chronic maxillary sinusitis. *Archives of Otolaryngology* 105: 386–390.

Krohel G B, Krauss H R, Winnick J. (1982). Orbital abscess. Presentation, diagnosis, therapy and sequelae. *Ophthalmology* 89: 492–98.

Lemierre A. (1936). On certain septicaemias due to anaerobic organisms. *Lancet* i: 701–703.

Le Roux B T, Dodds T C. (1968). *A second portfolio of chest radiographs*, pp. 86–87. Churchill Livingstone, Edinburgh.

Lester L A, Egge A, Hubbard V S, Di Sant'Agnese P A. (1983). Aspiration and lung abscess in cystic fibrosis. *American Review of Respiratory Disease* 127: 786–787.

Levison M E *et al.* (1983). Clindamycin compared with penicillin for the treatment of anaerobic lung abscess. *Annals of Internal Medicine* 98: 466–471.

Lorber B, Swenson R M. (1974). Bacteriology of aspiration pneumonia. A prospective study of community- and hospital-acquired cases. *Annals of Internal Medicine* 81: 329–331.

de Louvois J, Gortvai P, Hurley R. (1977). Bacteriology of abscesses of the central nervous system: a multicentre prospective study. *British Medical Journal* 2: 981–984.

McClelland C J, Craig B F, Crockard H A. (1978). Brain abscesses in Northern Ireland: a 30 year community review. *Journal of Neurology, Neurosurgery and Psychiatry* 41: 1043–1047.

McCurdy J A. (1977). Peritonsillar abscess: A comparison of treatment by immediate tonsillectomy and interval tonsillectomy. *Archives of Otolaryngology* 103: 414–415.

McNamara M J, Hill M C, Balows A, Tucker E B. (1967). A study of the bacteriologic patterns of hospital infections. *Annals of Internal Medicine* 66: 480–488.

Mathisen G E, Meyer R D, George W L, Citron D M, Finegold S M. (1984). Brain abscess and cerebritis. *Reviews of Infectious Diseases* 6: S101–S106.

Mendelson C L. (1946). The aspiration of stomach contents into the lungs during obstetric anaesthesia. *American Journal of Obstetrics and Gynecology* 52: 191–205.

Moore-Gillon J, Lee T H, Eykyn S J, Phillips I. (1984). Necrobacillosis: a forgotten disease. *British Medical Journal* 288: 1526–1527.

Mukau L. (1985). Dissecting retropharyngeal abscess due to *Fusobacterium necrophorum* in an adult. *Southern Medical Journal* 78: 476–478.

Muller S P. (1978). Peritonsillar abscess: a prospective study of pathogens, treatment, and morbidity. *Ear, Nose and Throat Journal* 57: 439–444.

Pallares R, Santamaria J, Ariza X, Gudiol F. (1983). Polymicrobial anaerobic septicemia due to lateral sinus thrombophlebitis. *Archives of Internal Medicine* 143: 164–165.

Palmersheim L A, Hamilton M K. (1982). Fatal cavernous sinus thrombosis secondary to third molar removal. *Journal of Oral and Maxillofacial Surgery* 40: 371–376.

Palva T, Gronroos J A, Palva A. (1962). Bacteriology and pathology of chronic maxillary sinusitis. *Acta Otolaryngology* 54: 159–175.

Pennybacker J. (1961). Discussion on intracranial complications of otogenic origin. *Proceedings of the Royal Society of Medicine* 54: 309–320.

Perlino C A. (1981). Metronidazole vs clindamycin treatment of anaerobic pulmonary infection. *Archives of Internal Medicine* 141: 1424–1427.

Reilly S, Willis A T. (1980). β-lactamase-producing anaerobes. *Lancet* ii: 970–971.

Reilly S, Timmis P, Beeden A G, Willis A T. (1981). Possible role of the anaerobe in tonsillitis. *Journal of Clinical Pathology* 34: 542–547.

Reynolds M A, Hart C A, Harris F, Taitz L S. (1985). Anaerobes in acute otitis media. *Journal of Infection* 10: 262–264.

Riding K H, Bluestone C D, Michaels R H, Cantekin E I, Doyle W J, Poziviak C S. (1978). Microbiology of recurrent and chronic otitis media with effusion. *Journal of Pediatrics* 93: 739–743.

Sanders C V, Hanna B J, Lewis A C. (1979). Metronidazole in the treatment of anaerobic infections. *American Review of Respiratory Disease* 120: 337–343.

Sen P, Tecson F, Kapila R, Louria D B. (1974). Clindamycin in the oral treatment of putative anaerobic pneumonias. *Archives of Internal Medicine* 134: 73–77.

Senturia B H, Gessert C F, Carr C D, Baumann E S. (1958). Studies concerned with tubotympanitis. *Annals of Otology, Rhinology and Laryngology* 67: 440–467.

Shaw M D M, Russell J A. (1975). Cerebellar abscess. A review of 47 cases. *Journal of Neurology, Neurosurgery and Psychiatry* 38: 429–435.

Skau N K, Nielsen K O, Osgaard O, Molgaard I L, Peitersen E. (1984). Intracranial and orbital complications of frontal and ethmoidal sinusitis. *Acta Otolaryngology (Stockholm)*, Suppl 412: 91–94.

Smith D T. (1927). Experimental aspiratory abscess. *Archives of Surgery* 14: 231–238.

Sprinkle P M, Veltri R W, Kantor L M. (1974). Abscesses of the head and neck. *Laryngoscope* 84: 1142–1148.

Sprott M S, Newman P K, Hall K, Welbury R R, Ingham H R. (1981). Subdural abscess secondary to covert dental sepsis. *Postgraduate Medical Journal* 57: 39–41.

Stern N S, Shensa D R, Trop R C. (1981). Cavernous sinus thrombosis: a complication of maxillary surgery. *Journal of Oral Surgery* 39: 436–438.

Sugita R, Kawamura S, Icikawa G, Fujimaki Y, Oguri T, Deguchi K. (1982). Microorganisms isolated from peritonsillar abscess and indicated chemotherapy. *Archives of Otolaryngology* 108: 655–658.

Swanston R, Grace A R H, Drake-Lee A B, Moffat D A. (1983). Anaerobic infection in acute mastoiditis. *Journal of Laryngology and Otology* 97: 633–634.

Teichgraeber J F, Per-Lee J H, Turner J S. (1982). Lateral sinus thrombosis: a modern perspective. *Laryngoscope* **92**: 744–751.

Tuner K, Nord C E. (1982). β-lactamase-producing anaerobic bacteria in recurrent tonsillitis. *Journal of Antimicrobial Chemotherapy* **10** Suppl A: 153–156.

Tutton G K. (1953). Cerebral abscess—the present position. *Annals of the Royal College of Surgeons of England* **13**: 281–311.

Urdal K, Berdal P. (1949). The microbial flora in 81 cases of maxillary sinusitis. *Acta Otolaryngology (Stockholm)* **37**: 20–25.

Walsh T J, Weinstein R A, Malinoff H, Breyer M D, Berelowitz B A. (1982). Meningorectal fistula as a cause of polymicrobial anaerobic meningitis. *American Journal of Clinical Pathology* **78**: 127–130.

Wardle J K, Connolly M J, Ingham H R, Hudgson P, Snow M H. (1984). Necrobacillosis. *British Medical Journal* **288**: 1916.

Weisberg L A. (1981). Nonsurgical management of focal intracranial infection. *Neurology (New York)* **31**: 575–580.

Weiss A, Friendly D, Eglin K, Chang M, Gold B. (1983). Bacterial periorbital and orbital cellulitis in childhood. *Ophthalmology* **90**: 195–201.

Weiss W, Cherniack N S. (1974). Acute nonspecific lung abscess: a controlled study comparing orally and parenterally administered penicillin G. *Chest* **66**: 348–351.

Weiss W, Flippin H F. (1967). Treatment of acute nonspecific primary lung abscess. *Archives of Internal Medicine* **120**: 8–11.

Yeoh L H, Singh S D, Rogers J H. (1985). Retropharyngeal abscess in a children's hospital. *Journal of Laryngology and Otology* **99**: 555–566.

Yoshikawa T T, Chow A W, Guze L B. (1975). Role of anaerobic bacteria in subdural empyema. *The American Journal of Medicine* **58**: 99–104.

17

Superficial ulceration and necrosis

B Adriaans, B S Drasar and B I Duerden

The anatomy of the skin
The normal bacterial flora of the skin
    Control of skin flora
Infections of the skin and associated structures
    Pustules and abscesses
    Ulcerative lesions and superficial gangrene
References

The skin of a normal, healthy person is generally highly resistant to invasion by the bacteria to which it is constantly exposed. However, local trauma that disturbs the normal defences, e.g., a wound or abrasion, the introduction of a foreign body, or a systemic debilitating disease such as malnutrition or diabetes mellitus, that reduces both the body's systemic defences and the quality of the skin envelope, opens the way for superficial infections. The more common acute skin infections are caused by staphylococci and streptococci, but many of the more subacute or chronic infections, and especially those in patients with debilitating underlying diseases, are caused by anaerobes or by synergic combinations of anaerobes and facultative species.

The way in which skin infections develop is determined to a large extent by the structure of skin and by the activity of its normal bacterial flora.

The anatomy of the skin

The skin is a vitally important organ with many diverse functions; it consists of three structurally different layers.

The epidermis

This is the outermost layer, and is in direct contact with the environment. It consists of fragile living cells, the keratinocytes, which are in a state of steady biological flux. They divide regularly to form the stratified epithelial tissue, and as they pass from the basal layer to the stratum corneum they mature to form the tough, resistant outer barrier. Numerous intercellular connections between keratinocytes help to maintain this barrier, which provides protection against infection (Breathnach, 1985). The epidermis is fully developed at birth, and underdeveloped in the premature neonate, which is thus predisposed to superficial infections.

Several non-keratinocyte cells are present in the epidermis: Langerhans cells, Merkel cells and melanocytes. The Langerhans cells, which originate in the bone marrow, are dendritic cells which circulate through the dermis. They are found predominantly in a suprabasal position in the epidermis and constitute about 3 per cent of the epidermal cell population. Their major function is in antigen presentation; they are associated with the initiation of the inflammatory response when viruses invade the epidermis, functioning as the most peripheral part of the immune system (Czernielewski et al., 1985). The Merkel cells appear to have mechanoreceptor and chemoreceptor functions. The melanocytes are pigment-producing cells.

The dermis

The dermis is the middle layer of the skin, derived mainly from mesodermal elements and composed of fibroelastic material, principally collagen and procollagen, interspersed with networks of vascular and lymphatic channels.

The subcutaneous tissue

This is also of mesodermal origin, and consists mainly of fat cells surrounded by a fibrous stroma.

The normal bacterial flora of the skin

The skin is sterile at birth but is colonized rapidly in the first 24 h. Organisms capable of multiplication and survival are termed 'resident' flora, while those deposited from the environment often do not multiply on the skin and are termed 'transient'. The 'normal' skin flora varies with age and it has been suggested that the low level of fatty acids on the skin of children may account for their more varied bacterial flora. (Kligman et al., 1976; Leyden et al., 1975). The flora is predominantly gram-positive and the most common anaerobic species is *Propionibacterium acnes* (see Chapter 11).

Control of skin flora

Humidity

This is an important factor in determining the level of bacterial colonization of the skin. Quantitative studies at various sites show more bacteria in the moist, intertriginous areas of the body, Bacteria rapidly decrease in number on areas of dry skin. Under prolonged occlusion the relative numbers of micro-organisms also change; coagulase-negative staphylococci increase initially, peak at about four days and then decrease in number, whereas lipophilic coryneforms, which are virtually undetectable before occlusion, increase considerably afterwards (Aly et al., 1978).

Secretions and skin lipids

Eccrine sweat contains factors which are inhibitory to the growth of some bacteria, such as *Streptococcus pyogenes*, *Corynebacterium diphtheriae* and certain strains of *Staphylococcus aureus*. Yet for some surface bacteria the skin lipids may be stimulatory rather than inhibitory. Puhvel and Reisner (1970) found that saturated fatty acids inhibited *P. acnes* whereas the unsaturated fatty acid, oleic acid, was stimulatory. Washing the skin with acetone allows greater persistence of artificially applied bacteria. Gram-negative aerobic and facultative bacteria such as *Escherichia coli* and *Pseudomonas aeruginosa* are not sensitive to the skin lipids. Various other antibacterial substances produced in sebaceous secretions may also eliminate some bacteria from the skin, yet allow other bacteria to survive and multiply. Those micro-organisms which survive contribute to the 'normal cutaneous flora' (Kearney et al., 1984).

Sites

There is considerable variation in the composition of the skin microbial flora at different sites of the body—the anaerobic species *P. acnes* is associated with sebaceous follicles; aerobic diphtheroids are found mainly in the axillae and on the interdigital skin of the feet; *S. epidermidis* is found over a significant area of the body. Anaerobic gram-positive cocci, and facultative gram-negative bacilli such as *Proteus* spp., *E. coli* and *Enterobacter* spp., are usually present in the moist areas of the body. Exposed areas tend to have higher colony counts than non-exposed areas.

Age

The bacterial flora varies with age. The number of *P. acnes* of the skin is high in infants and young children and decreases in children aged between five and ten years. At puberty the number of organisms increases, with significantly higher numbers occurring in late adolescence. The levels increase even further in young adults. The changes seem to correlate with sebum production. After the age of about 25 years levels remain fairly constant until about 70 years, when they decrease (Leyden et al., 1975).

Sex

The normal flora is similar in males and females under the age of 20 years. Males over 20 carry significantly more *P. acnes* than females.

Microbial antagonism

Most of the resident flora are not pathogenic and often it is not possible to distinguish between the 'resident' and 'transient' groups of micro-organisms. One function of the resident flora is protection, i.e., preventing overgrowth of harmful micro-organisms by microbial antagonism. The mechanisms for this regulation are complex and incompletely understood but one that has been investigated is the role of bacteriocins, produced by staphylococci, streptococci and *Bacillus* spp. The resident flora produces other substances that may inhibit the growth of other organisms, including H_2O_2, lactic and propionic acids, toxins, haemolysin and other enzymes. When the balance is upset, disease may result.

Infections of the skin and associated structures

Breaches of the protective mechanisms of the skin can result in three main types of infection: pustules (or, if larger, abscesses), cellulitis, and ulceration.

Pustules and abscesses

Most of the small pustules and many of the larger abscesses (e.g. carbuncles) are caused by infection of hair follicles or sweat glands with *S. aureus*. However, some of the larger abscesses have an important anaerobic component. The predominant infecting organisms in sebaceous cysts and pilonidal sinuses (two closely related lesions developing in blocked and damaged sebaceous glands) are anaerobes (Sandusky *et al.*, 1942; Stokes, 1958), principally gram-positive cocci. *Bacteroides* spp., mostly of the asaccharolytic group, are also common (Bornstein *et al.*, 1964; Pien *et al.*, 1972) and there may be synergy between the cocci and the *Bacteroides* spp. The particular *Bacteroides* spp. associated with the anaerobic cocci in these necrotic lesions are the pigmented *Bacteroides asaccharolyticus* and the non-pigmented, corroding species *B. ureolyticus* (Duerden, 1980; Duerden *et al.*, 1982, 1989). The foul smell of the pus from these lesions reflects the role of these putrefactive anaerobes in the infections. A similar mixture of anaerobic bacteria is found in a particular type of breast abscess that is not related to lactation. Most abscesses of the lactating breast are staphylococcal, but in the less common abscesses of the non-lactating breast, anaerobic cocci and *Bacteroides* spp. play a major role (Pearson, 1967). *B. ureolyticus* has been associated with some of these abscesses (Duerden *et al.*, 1982).

The involvement of anaerobes in superficial infections is often overlooked. This is probably the case in paronychia, in which anaerobic cocci and *Bacteroides* spp., usually of the asaccharolytic group, are involved more commonly than is usually suspected (Sandusky *et al.*, 1942). Brook (1981) found anaerobes, principally *Bacteroides* spp. *Fusobacterium nucleatum* and gram-positive anaerobic cocci, in 26 out of 33 children with paronychia.

A common and problematic pustular lesion is acne. The role of bacteria in acne remains debatable but there is clear evidence that, although the underlying pathogenesis is metabolic, the degradation of sebaceous secretions by *P. acnes* is important in the development of the inflammation and irritation around the pustule. *P. acnes* produces lipase which cleaves triglycerides, liberating free fatty acids (Reisner *et al.*, 1968). These are irritants and may initiate the characteristic inflammatory response. This was thought to be the basis for the beneficial effects of tetracyclines in the treatment of acne—not only do they inhibit *P. acnes*, but they were also thought to inhibit the lipase activity (Weber *et al.*, 1971) thus reducing the liberation of fatty acids (Freinkel *et al.*, 1965). However, subsequent work has produced a more complex hypothesis of a role of *P. acnes*. The amount of free fatty acid needed to induce inflammation is greater than is normally produced in comedones but other products may be more potent. With excessive sebaceous secretion due to hormonal changes, follicles become blocked by keratin. Seretion continues and *P. acnes* grows in the blocked follicles, producing closed comedones which are not inflamed. The *P. acnes* may then produce protease, hyaluronidase and other enzymes which cause the follicle to become 'leaky' to the body's defence system. The cell wall of the *P. acnes* activates complement and this induces leucocyte accumulation and inflammation (Noble, 1984).

Ulcerative lesions and superficial gangrene

Superficial ulceration may occur as a result of infection, lack of adequate nutrition, neoplasia or other causes; often, it results from a combination of these factors. Anaerobic infections may cause ulceration as a primary phenomenon, predominantly by the release of a toxin, or toxins, or they may invade an already ulcerated area as a secondary event and play a part in the subsequent prolonged duration of the ulcer.

Decubitus ulcers

The term 'decubitus' is used synonymously with pressure sore. The latter is probably the more correct term, because decubitus comes from the Latin word for 'lying down'. Though many decubitus ulcers occur while lying down, many occur in the sitting position too. They are common in patients with spinal cord injuries and in some reported series the incidence of these ulcers ranges from 25 to 85 per cent of patients admitted to institutions for long term care. Despite the large number of underlying conditions that predispose to these ulcers, there are several characteristics common to all these patients. They are usually elderly, frequently incontinent and unable to move unaided.

Aetiology and pathogenesis

Decubitus ulcers result from prolonged pressure to a relatively small area of skin. This compresses the vasculature in the area, causing tissue hypoxia and damage. Bacterial infection may occur in this devitalized area, causing further local damage, and infection may spread by the haematogenous route and result in secondary metastatic abscesses. Decubitus ulcers may follow an abrasion or other trauma to the skin, as occurs in an area of diminished sensation. The deeper layers may then become infected, with spreading ulceration.

Trauma which produces a haematoma may also contribute to impaired perfusion and further tissue hypoxia. Reuler and Cooney (1981) suggested that four factors contribute to the development of a decubitus ulcer—pressure, friction, shearing forces and moisture.

Pressure is usually unevenly distributed while lying down, being concentrated over the bony prominences. The total tissue pressure increases with pressure from the outside, leading to increased arteriolar pressure-filtration of fluid from capillaries, oedema and autolysis. The impaired blood flow causes tissue hypoxia and subsequent necrosis. Infection often develops in this devitalized area and may spread to other sites by the haematogenous route. A further effect of pressure is compression of the lymphatics, with accumulation of metabolic waste products which may themselves lead to tissue necrosis.

Shearing forces come into play when the patient lies on a bed with the head elevated and the posterior sacral skin fixed. The subcutaneous fat is also very vulnerable to mechanical forces which accentuate the shearing forces.

Friction, which arises when the patient is moved across the bed, causes loss of the stratum corneum, perhaps also predisposing to ulceration.

As mentioned previously, moisture facilitates the growth of many bacteria and these may thus colonize a superficial ulcer.

The prevalence of decubitus ulcers is affected by many factors; e.g., genetic constitution, nutritional status, environmental conditions and occupation. Women are affected more often than men, some reports quoting a ratio of 2:1, others 4:1. The skin ulceration usually occurs on the lower half of the body, over the bony prominences or over the malleoli or heels. Ulcers are often precipitated by trauma, which may be physical or chemical in nature.

Clinical manifestations

Early ulcers are generally irregular in shape, whereas chronic ulcers usually have a well defined edge. Around the area of ulceration the skin is often oedematous due to the damage to surrounding superficial lymphatics. Obliteration of the lymphatics may cause gross swelling and surrounding hyperkeratosis. Recurrent infection leads to scarring in the long term, with further distortion of the lymphatics. Periosteitis of the underlying bone may occur in long standing, chronic ulcers.

Diagnosis

The clinical setting of a patient with a history of peripheral ischaemia, or who is immobile as the result of a stroke or paraplegia, leads to the suspicion of a decubitus ulcer if the ulcer is at a pressure site. If it is in an area of diminished sensation, this can be confirmed by examination. Decubitus ulcers should be differentiated from vascular ulcers, which may present at a point of pressure, e.g., in a patient immobile with rheumatoid arthritis.

Bacteriology

Samples from these lesions frequently yield a very mixed growth. This is not surprising because the ulcers are often contaminated with faecal and urinary micro-organisms as a result of incontinence or immobility. Anaerobes have been isolated from these lesions, usually from the depths of the ulcers below the superficial debris. These contribute to the prolonged duration of the ulcers, if they are not treated appropriately. Galpin et al. (1976) showed the presence of anaerobes in 50 per cent of their samples. When anaerobes have been sought specifically, *Bacteroides* spp. have been the major isolates from decubitus ulcers (Finegold, 1977) and their significance has been most apparent when the ulcers have been the sources of bacteroides bacteraemia (Felner and Dowell, 1971; Schoutens et al., 1973; Chapter 19). *B. fragilis* has been the species most commonly named in these reports but most isolates have been listed as *Bacteroides* spp. When more detailed identification has been made, *B. asaccharolyticus* and *B. ureolyticus* have been found to be particularly important (Peromet et al., 1973; Duerden et al., 1982, 1989). The asaccharolytic *Bacteroides* spp. are often associated with anaerobic gram-positive cocci in these mixed infections. Anaerobes are always isolated in conjunction with aerobes or facultative anaerobes, but are probably the more damaging component of the mixed infections.

Complications

Septicaemia has been reported from infected decubitus ulcers; the morbidity and mortality resulting from this are considerable. Extension of the infection to underlying bone may occur, with resultant osteomyelitis, pyoarthrosis or joint disarticulation. Ulcers may even extend into viscera such as the bladder, or sinus tracts may reach the bowel. Long standing infections may lead to amyloidosis.

Treatment (see Reuler and Cooney, 1981)

Preventing prolonged pressure on a small area of soft tissue over a bony prominence should reduce the incidence of these lesions. This calls for regular and dedicated nursing, which may place a strain on limited resources, but failure to prevent pressure will only lead to more time consuming nursing once patients develop ulcers. The aim is to spread the body's weight over a larger area to avoid pressure being concentrated in one part. If the patients are mentally alert they can be trained to inspect their skin regularly with a long handled mirror. Proper relief of pressure in areas of erythema and early treatment of superficial skin necrosis will prevent the subsequent formation of deeper ulcers. Patients should be turned at least every 2 h to prevent pressure necrosis and they should not be in bed with the top end elevated. A turning frame or a Circo electric bed may be useful. A good sitting position should be adopted while in a wheelchair and thick foam rubber cushions with declivities cut in the area of the ischial tuberosities should be used. Even with such a device patients should be regularly lifted out of these chairs to prevent the skin breaking down over the pressure areas. Moisture should be kept to a minimum. This involves frequent inspection of the skin. Sheepskins have a very high capacity to absorb water vapour and are useful in the prevention of decubitus ulcers.

Lesions confined to the upper dermis will usually respond only to topical therapy. This involves regular cleansing, followed by dressing of the wound with a semipermeable dressing which may be left in place for up to two days. The commercially available polyurethane and hydrocolloid dressings keep normal and periulcer skin relatively dry and do not cause maceration of the skin. Lesions extending deeper into the fascia may require surgical intervention and debridement. Sinography may be indicated to assess the extent of the depth of the lesions. Systemic metronidazole therapy may improve the rate of healing of deeper ulcers.

Ischaemic ulcers

A second type of superficial ulcer resulting from poor tissue perfusion and superficial injury is the ischaemic ulcer. This usually affects the lower leg in patients with poor vascular supply and results in tissue ischaemia. Anaerobic cocci are commonly found in ischaemic ulcers (Sandusky et al., 1942; Pieri et al., 1972). *Bacteroides* spp. are common in the depths of the ulcers, beneath the undermined edges and at the advancing edge where damage is continuing and, as in the decubitus ulcers, the pigmented strains (probably *B. asaccharolyticus*) and *B. urolyticus* seem to have a special role (Myers et al., 1969; Duerden et al., 1982).

Diabetic ulcers

Diabetic ulcers result from a combination of abnormalities of the healing and repair mechanisms which particularly affect the vascular system. These include the following:
1. Abnormal leukocyte function
   – diminished chemotaxis of polymorphs,
   – diminished phagocytosis,
   – diminished intracellular killing,
   these improve with improvement in diabetic control;
2. Diminished leucocyte migration through the capillary wall;
3. Impaired repair after minor trauma because of damaged dermal vasculature. This facilitates access for pathogenic bacteria;
4. Impaired cell-mediated immunity;
5. Impaired opsonization; serum-dependent uptake of radiolabelled bacteria such as *S. aureus* and *E. coli* is reduced;
6. Atherosclerosis often accompanies diabetes and impairs wound healing, thus providing favourable conditions for secondary infection. Reduced blood flow also decreases oxygen supply to the tissues thereby enhancing the growth of microaerophilic and anaerobic organisms;
7. Malnutrition and dehydration; frequent swings in blood glucose may impair clearance of endotoxin and other microbial products, leading to altered host defences and increasing the risk of infection.

A combination of these factors is usually found, contributing to an increased incidence of infections in the diabetic patient. These factors may act synergically to augment the severity of an established infection.

Pathogenesis

Intimal hyperplasia is the most common vascular abnormality noted in the blood vessels in diabetes mellitus. This results from modulation and proliferation of smooth muscle cells originating in the subendothelium. The basement membrane in the muscle capillaries is also thickened and a change is found in the skin capillaries. Diabetics may also have a functional abnormality of their capillaries, which may explain their predisposition to vascular disease. Certainly, the outflux of albumen from vessels is increased. Experimental cultures of endothelial cells have more receptors for insulin than aortic endothelial cells and this may contribute to the ulcerative lesions in diabetics (Logerfo and Coffman, 1984).

Neuropathy, predominantly of the sensory type, is also found regularly in diabetics. This may contribute to ulceration, especially of the feet, by predisposing to trauma from poorly fitting shoes or from unnoticed objects. Motor neuropathy is less common but it may lead to an abnormal gait, with excessive pressure on the metatarsal heads which may produce callosities, infection occurs when the skin splits.

Diagnosis

Patients should be educated to examine their feet regularly and the feet should also be examined when the patient is seen at clinic visits. Even small abrasions should not be ignored. Other causes of ulceration should be excluded, e.g., vascular disease as a primary event, or hypertension. After removing surface debris, swabs should be taken from the depths of the ulcer for aerobic and anaerobic culture. If an abscess develops, pus should be carefully collected for bacteriological examination.

Bacteriology of diabetic infections

Infections in diabetes are generally polymicrobial; results of studies on the microbiology of diabetic foot infections have shown that approximately 70 per cent of infections are polymicrobial (Louie et al., 1976; Wheat et al., 1986). *S. aureus* and *S. epidermidis* were isolated with the same frequency. Coryneform organisms, which used to be regarded as harmless commensals, were isolated from 20 per cent of cases and may be important in foot infections.

Of the anaerobic organisms isolated, anaerobic cocci, predominantly *Peptostreptococcus* spp., were seen most frequently. *Bacteroides* spp. were isolated less frequently and *Clostridium* spp. were found only rarely. Anaerobes were more frequent in necrotizing infections spreading to involve bone and other deep tissues than in superficial infections. Obligate anaerobes, particularly non-sporing anaerobes, were never isolated in pure culture from swabs taken from ulcers, abscesses or deep tissue infections. They were always isolated with either facultative anaerobes or aerobes.

Infections

Diabetics have always been thought to be more prone to infection than non-diabetics, although this statement has not been adequately confirmed (Larkin and Ireland, 1985). Certain infections are common in diabetics but rarely seen in normal individuals. One such infection is necrotizing cellulitis, a severe infection involving the deep tissues such as the fascia. However, less extensive foot infections are common in diabetics and frequently cause ketoacidosis and are thus the cause of admission to hospital because of loss of diabetic control. Infection in the diabetic correlates well with poor control of the diabetes and the aim should be for good diabetic control (Rayfield et al., 1982). Infections range from superficial ulcers to deep abscesses. When the infection spreads deeper, subcutaneous soft tissue necrosis follows. Other deep infections include osteomyelitis, necrotizing fasciitis or septicaemia.

Treatment

Good control of diabetes is absolutely essential. Most of the evidence suggests that infections are better controlled and occur less commonly when the blood sugar level is normal. Cigarette smoking should not be allowed. Feet should be kept warm, dry and comfortable with well fitting shoes to minimize friction, trauma and thus ulceration. Patients should be thoroughly examined for vascular disease. If considerable vascular disease exists, reconstructive surgery, if feasible, should be carried out to improve the blood supply.

Antibiotics are usually given after the results of the cultures are available; their use depends on the clinical features of the culture. Chronic lesions with minimal inflammation or cellulitis settle with local treatment and minimum systemic antibiotic therapy. Indications for antimicrobial therapy include extensive involvement of soft tissues, extension to bone or evidence of systemic infection. If antibiotic therapy is given empirically a broad-spectrum β-lactam antibiotic provides cover against the facultative gram-positive cocci and some of the anaerobes present. When anaerobes have been thought to be a major factor in the ulcer, systemic metronidazole has been found useful in controlling infection and promoting healing.

Tropical ulcer

Aetiology and pathogenesis

Tropical ulcer follows trauma to the skin (since organisms cannot enter the intact skin); the trauma may be minor, such as an insect bite or a skin abrasion. Evidence for an infective aetiology was provided by MacAdam (1966) who found that it was possible to transmit infections by inoculating pus from tropical ulcers into normal healthy subjects. One prerequisite for the development of infection was a skin abrasion, which could be as minor as that resulting from smallpox vaccination. Following the introduction of the infecting organisms there is damage to the collagen, as seen by electron microscopy (Adriaans et al., 1987b) but the body's defence mechanisms clear the organisms fairly quickly; few organisms are visible on histological examination of gram-stained sections of ulcers that are six (or more) weeks old. The disease remains localized to the skin, with no evidence of systemic spread of the infection. Malnutrition was thought to be a predisposing factor in the susceptibility to tropical ulcer; but a recent study has failed to confirm this (Robinson et al., 1988).

Histopathological examination shows evidence of epidermal and dermal ulceration with a mixed inflammatory infiltrate in the dermis. Along the margins of the ulcer there is considerable acanthosis and spongiosis. The blood vessels are not primarily affected.

Clinical manifestations

Tropical ulcer is an area of acute or chronic skin necrosis. The disease occurs predominantly in (but is not confined to) the tropics. Lesions are most common in children between the ages of 5 and 15 years, although older children may also be affected. The ulcers are usually seen on the lower legs and are less common on the upper legs or the arms; the trunk is usually spared. The initial lesion consists of a non-specific papule 2–3 mm in diameter; this breaks down to form the ulcer which may be a few centimeters in diameter. Early lesions are painful but in chronic ulcers pain is not a major symptom. Ulcers tend to remain localized and may be single or multiple, affecting one or more limbs at a time. Similar lesions may recur, as one infection does not confer immunity from recurrence. Healing may take place spontaneously, over many weeks or months.

Ulcers generally extend from the epidermis to the upper dermis, but may also extend to the reticular dermis and, rarely, to the subcutaneous tissues. Scarring is an inevitable complication because of the dermal involvement. The scars are thin and measure several centimeters in diameter.

Diagnosis

The diagnosis is made on clinical grounds. When it occurs in an endemic area a history of a painful papule on the lower leg, which is followed a few days later by the development of an ulcer with a foul smelling purulent slough, is highly suggestive of a tropical ulcer.

Tropical ulcer should be differentiated from Buruli ulcers, which are painless and extend to the subcutaneous tissues, Buruli ulcers are caused by *Mycobacterium ulcerans* and the hallmark of the disease is the subcutaneous spread of the disease with the formation of a characteristic undermined edge.

Tropical ulcer should also be differentiated from ulceration of the lower legs, which is associated with the hereditary haemolytic anaemias. Sickle-cell anaemia and the sickle-cell trait are common in certain areas where tropical ulcer occurs. However, the lesions caused by the haemolytic diseases affect a wider age range of patients and are localized to the lower third of the leg, often on the medial or lateral side of the ankle. Superficial hypertensive ulcers, which usually start on the anterolateral aspect of the leg at the junction of the middle and lower third, are associated with ischaemia. There are no associated venous abnormalities, patients are older, and the ulcers tend to be superficial.

Bacteriological findings

Several authors have reported on the microbiology of tropical ulcers. Most of the reports were based on the microscopical examination of smears taken from ulcers, rather than on cultures. The results of previous reports relating to the bacterial flora found in tropical ulcers are listed in Table 17.1. Despite a clinical suspicion that anaerobes were involved, no anaerobes had been cultured before Adriaans (1986). This is probably a result of inadequate trans-

Table 17.1 Bacteria detected in tropical ulcers

Author	Organism	Method of detection
Adriaans et al. (1986)	*F. ulcerans* Anaerobic cocci Coliforms Spirochaetes	Culture Smears
Ampofo and Findlay (1951)	Fusobacteria Spirochaetes Cocci	Smears
Lasbrey (1952)	*S. aureus* Streptococci Vincent's organisms Fusobacteria	Smears
Lindner and Adeniyi-Jones (1968)	Vincent's organisms	Smears

portation of the samples together with poor facilities for anaerobic culture. The isolation of anaerobes from any sample depends largely on adequate care being taken of the sample from the time it is taken to the time it is cultured.

Swabs from tropical ulcers, transported in an anaerobic transport medium, may allow adequate isolation of anaerobes in culture several weeks after the collection of the samples. A recent survey of patients with tropical ulcers reported by Adriaans *et al.* (1986) demonstrated the presence of various anaerobes, including fusobacteria and anaerobic cocci. Fusobacteria of a previously unreported species, *Fusobacterium ulcerans* (Adriaans and Drasar, 1987), were isolated. These organisms were thought to have been introduced into a minor wound from mud, which also contained fusobacteria indistinguishable from the ulcer isolates. *Bacteroides* spp., particularly *B. fragilis*, were also isolated, but less frequently. Anaerobes were always found in conjunction with either aerobes or facultative organisms (Adriaans *et al.*, 1987a) and were detected predominantly in ulcers less than one month old. Spirochaetes were seen on dark-ground microscopy in about 25 per cent of the samples. Gram-stained sections of skin biopsies from the edge of ulcers demonstrated the presence of many organisms in the epidermis and dermis and a Dieterle stain showed spirochaetes in the papillary dermis. These findings suggest that tropical ulcer is most likely to be a polymicrobial infection of the skin, in which *F. ulcerans* plays a significant part.

Treatment

Specific therapy should be aimed at eliminating the infection. Most of the bacteria present are sensitive to penicillin and this should be given early, at the onset of the disease. Antibiotics given late in the course of the disease have little effect on the subsequent course of the ulceration once it is established as a chronic ulcer. Tetracycline and metronidazole should be administered if a patient is known to be hypersensitive to penicillin. Patients should be give a 10-day course of oral antibiotics to ensure adequate therapy; inadequate treatment may lead to incomplete resolution of the ulcer and the development of a chronic lesion. Intramuscular antibiotics should be given if there are any doubts about compliance. Topical treatment should aim to provide adequate cleansing without the risks of sensitization. Povidone iodine is a good antiseptic, effective against many gram-positive and gram-negative organisms, but may cause sensitization, although not commonly. It does not persist on the skin for very long and thus does not provide residual action. Chlorhexidine is another antiseptic with good antibacterial properties and prolonged action and is generally not irritant to the skin; it rarely causes sensitization. Topical antibiotics (penicillin, bacitracin or aminoglycosides) have no place in the treatment of these ulcers as they invariably cause sensitization, which limits their overall usefulness.

Ulcers may heal spontaneously over many months. Failure to heal and persistent ulceration may result in a squamous carcinoma of the skin several years after the onset of the ulcer.

Noma (gangrenous stomatitis, cancrum oris)

Noma is a destructive but painless lesion of the skin and usually affects the face in the region of the mouth. Fusobacteria, bacteroides and spirochaetes are present in large numbers in the lesion. It is not a primary disease, but occurs in conjunction with some other underlying disease. Malnutrition is inevitably present, mainly in the form of Kwashiokor. Infants and children are affected most often although more recently the disease has been recognized in adults. The disease occurs predominantly in the tropics. Some authors regard Noma as a general term to embrace all midline facial granulomas, but this leads to confusion and does not reflect the pathogenesis of the various groups of diseases.

The early lesions of Noma start with an area of ulcerative stomatitis with tender, firm swelling of the upper gum and some swelling of the overlying part of the face. The teeth then loosen and inflammation spreads to the underlying bone, with the formation of an osteitis and further sequestrum formation. The cheek usually ulcerates, causing cavitation into the mouth. Constitutional features are uncommon, but as the disease spreads fever and general apathy occur, together with weakness. A heavy, purulent foul-smelling exudate develops, in which anaerobes predominate. Lesions may extend from the cheek to the nose, and across to the other side of the face. Other sites involved include the perineum and vulva. There is rapid spread, leading ultimately to death if left untreated.

Neonatal Noma has been reported in low birthweight babies. *Ps. aeruginosa* was isolated from the majority of these patients and was thought to be the aetiological agent in this subsection of Noma patients. The infants affected were aged between 6 and 23 days and were all of low birth weight. The clinical features were otherwise identical to Noma (the anaerobic type) found in older children.

Diagnosis

The diagnosis is again a clinical one. The clinical picture is characteristic when the face is involved, particularly around the mouth. The presence of underlying malnutrition, malaria or measles points to the secondary nature of the disease. The diagnosis is less clear when the ulceration is at a site other than the face. A skin biopsy of the lesion shows thrombosis of the vessels in the area affected, this accounts for the lack of haemorrhage at the site of the lesion.

The condition should be distinguished from other ulcerative lesions. Zinc deficiency causes Acrodermatitis enteropathica, with similar ulceration around the gingival and genital areas. Zinc deficiency may occur as a genetic defect or as a secondary phenomenon following chronic malnutrition, chronic inflammatory bowel disease or parental nutrition without adequate supplementation. Alopecia and an eczematous eruption in the genital area are associated with zinc deficiency. These additional clinical findings should help to separate the two conditions.

Bacteriological findings

Smears from the ulcers show a predominance of fusiform bacilli and spirochaetes. Hence the disease is thought to be a fusospirochaetal infection, although adequate microbiological investigation of a good series of cases has not yet been reported. *B. melaninogenicus* has been cultured from these lesions in large numbers, and may play an important role in the pathogenesis of the condition (Emslie, 1963; Kaufman *et al.*, 1972).

Treatment

Specific treatment is aimed at the presumed pathogenic organisms and these usually respond to benzyl penicillin. If antibiotics are not administered early the area becomes demarcated and sloughs, leaving a large denuded area which requires extensive skin grafting and reconstruction. Unfortunately, facilities for these procedures are not often available in the areas where Noma occurs. The underlying diseases which predispose to this condition, particularly malnutrition, should also be treated. Measles frequently precedes Noma and vaccination campaigns should reduce the morbidity from this disease. Poor oral hygiene also requires attention.

Progressive synergic gangrene (Meleney's gangrene)

This destructive ulceration of the skin is caused by a mixed infection, in which anaerobes are a significant component. Initial reports attributed the disease to infection with microaerophilic streptococci and *S. aureus*, sometimes accompanied by a gram-negative bacillus (Meleney, 1933); Meleney was able, experimentally, to produce a similar lesion in dogs. Clinically, the lesion starts as a local area of redness, usually one to two weeks after abdominal or thoracic surgery, although lesions have been reported as soon as one day, and as late as seven weeks, after surgery. In the past, the lesion was thought to be associated with through-and-through sutures. However, it may occur without prior surgery. The initial lesion starts around the margins of the incision or at the suture areas. Thereafter, the affected area develops into an enormous ulcer with a shaggy centre and a gangrenous rim. As the lesion spreads outwards the inner margin of the gangrenous zone becomes undermined and necrotic. The necrosis is accompanied by the formation of slough, which may extend from the dermis into the subcutaneous tissues, or even into the muscle layer. There are relatively few systemic manifestations. If it is left untreated the ulcer may result in extensive tissue destruction. It is extremely painful and the tissues are very tender. The condition is more common in males than females.

Diagnosis

Synergic gangrene should be suspected if, after abdominal or thoracic surgery, a patient develops an ulcer which fails to heal and which continues to extend. Specimens should be taken for aerobic and anaerobic culture. In descriptions of the classic form of the disease microaerophilic streptococci were said to be cultured mainly from the extending area, and *S. aureus* mainly from the central gangrenous area (Archer *et al.*, 1984), although group A β-haemolytic streptococci and *S. aureus* have been reported to cause similar lesions (Colebunders *et al.*, 1984). However, in some studies the streptococci have been found to be obligate anaerobes and they have been isolated with a wide range of other bacteria, of which *Proteus* is one of the most common (Hutchinson *et al.*, 1976). In more recent reports, the role of *Bacteroides* spp. in association with anaerobic cocci has been more widely recognized (Duerden *et al.*, 1982).

The form of synergic gangrene that affects the perineum and vulva, scrotum or penis (Fournier's gangrene, or necrotizing dermogenital syndrome) is a spreading anaerobic cellulitis and necrotizing fasciitis which can spread to destroy large areas of the perineum, abdomen, buttocks and thighs. Mixtures of anaerobes, which include gram-positive cocci and *Bacteroides* spp. (often the asaccharolytic group) are usually found (Saksena *et al.*, 1968).

Treatment

Antibiotics given systemically can be life-saving. The choice of antibiotic depends on the organisms cultured and their sensitivities. Wide excision used to be thought to be mandatory, but there have been reports of cures with antibiotics alone. Debridement has the advantage of removing the necrotic skin and soft tissues but has the disadvantage that normal skin will also be removed. Healing is slow because the amount of tissue to be replaced is large. Skin grafting to cover the defect increases the rate of healing, since this would otherwise take place by granulation tissue filling the defect before re-epithelialization. Occlusive dressings promote healing and allow the defects to fill in without the deep depression of scar tissue which follows healing by conventional methods. Patients are often in a state of shock and may need fluid and electrolyte replacement. Any underlying predisposing cause such as diabetes mellitus should be excluded.

References

Adriaans B, Drasar B S. (1987). The isolation of fusobacteria from tropical ulcers. *Epidemiology and Infection* 99: 361–372.

Adriaans B, Hay R J, Drasar B S., Robinson D C. (1986). Anaerobic bacteria in tropical ulcer—the application of a new transport system for their isolation. *Transactions of the Royal Society of Tropical Medicine and Hygiene* 80: 793–794.

(1987a). The infectious aetiology of tropical ulcer—a study of the role of anaerobic bacteria. *British Journal of Dermatology* 116: 31–37.

Adriaans B, Hay R, Lucas S, Robinson D C. (1987b). Light and electron microscopic features of tropical ulcers. *Journal of Clinical Pathology* 40: 1231–1234.

Aly R, Shirley C, Cunico B, Maibach H. (1978). Effect of prolonged occlusion on the microbial flora, pH, carbon dioxide, and transepidermal water loss of human skin. *Journal of Investigative Dermatology* 71: 378–381.

Ampofo O, Findlay G M. (1951). The treatment of tropical ulcers with aureomycin ointment. *Transactions of the Royal Society of Tropical Medicine and Hygiene* 45: 650–652.

Archer C B, Rosenberg W M C, Scott G W, MacDonald D M. (1984). Progressive bacterial synergistic gangrene in a patient with diabetes mellitus. *Journal of the Royal Society of Medicine* 77 Suppl 4: 1.

Breathnach A. (1975). Aspects of epidermal ultrastructure *Journal of Investigative Dermatology* 65: 2–15.

Bornstein D L, Weinberg A N, Schwartz M N, Kurz L. (1964). Anaerobic infections—a review of current experience. *Medicine, Baltimore* 43: 207.

Brook I. (1981). Bacteriology of paronychia in children. *American Journal of Surgery* 141: 703–705.

Colebunders R, Matthys R, Neujens F, Verselder R. (1984). Synergistic bacterial gangrene caused by a group A β-haemolytic streptococcus and a *Staphylococcus aureus*. *Dermatologica* 168: 150–151.

Czernielewski J, Vaigot P, Prunieras M. (1985). Epidermal Langerhans cells—a cycling population. *Journal of Investigative Dermatology* 84: 424–426.

Duerden B I. (1980). The identification of gram-negative anaerobic bacilli isolated from clinical infections. *Journal of Hygiene* 84: 301–313.

Duerden B I, Bennett K W, Faulkner J. (1982). The isolation of *Bacteroides ureolyticus* from clinical infections. *Journal of Clinical Pathology* 35: 309–312.

Duerden B I, Eley A, Goodwin L, Magee J T, Hindmarch J M, Bennett K W. (1989). A comparison of *Bacteroides ureolyticus* isolates from different clinical sources. *Journal of Medical Microbiology* 29: 63–73.

Emslie R D. (1963). Cancrum oris. *Dental Practice* 13: 481.

Felner J M, Dowel V R. (1971). Bacteroides bacteraemia. *American Journal of Medicine* 50: 787–796.

Finegold S M. (1977). *Anaerobic bacteria in human disease*, pp. 392–393. Academic Press Inc, New York.

Freinkel R K, Strauss J S, Pochi R E. (1965). The effect of tetracyline on the composition of sebum in acne vulgaris. *New England Journal of Medicine* 273: 850–854.

Galpin J E, Chow A W, Bayer A S, Guze L B. (1976). Sepsis

associated with decubitus ulcers. *American Journal of Medicine* **61**: 346–350.

Hutchinson P E, Summerly R, Lawson L J. (1976). Post-operative progressive gangrene: a reminder. *British Journal of Dermatology* **94**: 89–91.

Israil A. (1984). Recent data on the molecular biology of bacteriocins. *Archives Roumaines de Pathologie Expérimentale et de Microbiologie* **43**: 5–30.

Kaufman E J, Mashims P A, Aausman E, Hanks C T, Ellison S A. (1972). Fusobacterial infection: enhancement by cell-free extracts of *Bacteroides melaninogenicus* possessing collagenase activity. *Archives of Oral Biology* **17**: 577.

Kearney J N, Ingham E, Cunliffe W, Holland K. (1984). Correlations between human skin bacteria and skin lipids. *British Journal of Dermatology* **110**: 593–599.

Kligman A, Leyden J, McGinley K. (1976). Bacteriology. *Journal of Investigative Dermatology* **67**: 160–167.

Larkin J G, Ireland J T. (1985). Diabetes and infection. *Postgraduate Medical Journal* **61**: 233–237.

Lasbrey A. (1952). Tropical ulcers of the leg and nail bed. *South African Medical Journal* **26**: 66–69.

Leyden J, McGinley K, Mills M, Kligman A. (1975). Age related changes in the resident bacterial flora of the human face. *Journal of Investigative Dermatology* **65**: 379–381.

Lindner R R, Adeniyi-Jones C. (1968). The effect of metronidazole on tropical ulcers. *Transaction of the Royal Society of Tropical Medicine and Hygiene* **62**: 712–714.

Logerfo F W, Coffman J D. (1984). Vascular and microvascular disease of the foot in diabetes. *New England Journal of Medicine* **311**: 1615–1619.

Louie T J, Bartlett J G, Tally F P et al. (1976). Aerobic and anaerobic bacteria in diabetic foot ulcers. *Annals of Internal Medicine* **85**: 461–463.

MacAdam I. (1966). Tropical phagedermic ulcers in Uganda. Report of an investigation. *Journal of the Royal College of Surgeons of Edinburgh* **11**: 196–205.

Meleney F L. (1933). A differential diagnosis between certain types of infectious gangrene of the skin: with particular reference to haemolytic streptococcus gangrene and bacterial synergistic gangrene. *Surgery Gynecology and Obstetrics* **56**: 847–867.

Myers M B, Cherry G, Bommsiele B B, Bommsiele G T. (1969). Ultraviolet red fluorescence of *Bacteroides melaninogenicus Applied Microbiology* **17**: 760.

Noble W C. (1984). Skin microbiology: coming of age. *Journal of Medical Microbiology* **17**: 1–12.

Pearson H E. (1967). *Bacteroides* in areolar breast abscess. *Surgery, Gynecology and Obstetrics* **125**: 800.

Peromet M, Labbe M, Yourassowsky E, Schoutens E. (1973). Etude des germes anaerobes isoles des escarres de decubitus. *Acta Clinica Belgica* **28**: 117.

Pien F D, Thompson R L, Martin W J. (1972). Clinical and bacteriologic studies of anaerobic gram-positive cocci. *Mayo Clinic Proceedings* **47**: 251.

Puhvel M, Reisner R M. (1970). Effect of fatty acids on the growth of *Corynebacterium acnes in vitro. Journal of Investigative Dermatology* **54**: 48–52.

Rayfeld E J, Auit M J, Kensch G T. (1982). Infection and diabetes; the case for control. *American Journal of Medicine* **72**: 439–449.

Reisner R M, Silver D Z, Puhvel M, Sternberg T H. (1968). Lipolytic activity of *Corynebacterium acnes. Journal of Investigative Dermatology* **51**: 190.

Reuler J, Cooney T. (1981). The pressure sore: pathophysiology, and principles of management. *Annals of Internal Medicine* **94**: 661–666.

Robinson D C, Adriaans B, Hay R J, Yesudian P. (1988). The clinical and epidemiologic features of tropical ulcer (tropical phagedenic ulcer). *International Journal of Dermatology* **27**: 49–53.

Saksena D S, Block M A, Mettenry M C, Truant J P. (1968). Bacteroidaceae: anaerobic organisms encountered in surgical infections. *Surgery* **63**: 261.

Sandusky W R, Pulaski E J, Johnson B A, Meleney F L. (1942). The anaerobic non-haemolytic streptococci in surgical infections on a general service. *Surgery, Gynecology and Obstetrics* **75**: 145.

Schoutens E, Labbe M, Yourassowsky E. (1973). '*Bacteroides fragilis*' septicaemias: incidence and sensitivity of the strain to antibodies. *Pathological Biology* **21**: 349.

Stokes E J. (1958). Anaerobes in routine diagnostic cultures. *Lancet* **1**: 668.

Weber K, Freedman R, Endy W W. (1971). Tetracycline inhibition of a lipase from *Corynebacterium acnes. Applied Microbiology* **21**: 639.

Wheat L J, Allen S D, Henry M et al. (1986). Diabetic foot infections. Bacteriologic analysis. *Archives of Internal Medicine* **146**: 1935–1940.

18

Gas gangrene and clostridial cellulitis

A. Trevor Willis

Bacteriology
    Contamination and colonization
    Infection
The genesis of gas gangrene
    Host determinants
    Bacterial determinants
Clinical considerations
    Clostridial cellulitis
    Gas gangrene (clostridial myonecrosis)
Laboratory investigations
    Radiology
    Microbiology
Management
    Prophylaxis
    Treatment
References

Gas gangrene has been variously referred to in the past by such terms as malignant oedema and gas infection, although MacLennan (1943b–d) considered the terms anaerobic myositis and clostridial myositis to provide a more appropriate alternative nomenclature. However, since the host reaction in gas gangrenous muscle is negligible, and since most of the local pathological changes seen in *Clostridium perfringens* infections can be produced *in vitro* in sections of normal muscle by exposure to sterile culture filtrates, the condition is properly regarded as clostridial myonecrosis (Power, 1945; Robb-Smith, 1945a; Oakley, 1954).

Descriptions of gas gangrene are to be found scattered diffusely through the early medical literature. Millar (1932), in his brief historical survey, provides the following account which is attributable to Hippocrates (460–355 BC): 'Citron of Thasa commenced to experience pain in his foot, in his great toe... He went to bed the same day. He had a slight chill, some nausea, and then a little fever; he became delirious during the night. On the second day there was swelling of the entire foot and over the whole ankle, which was a little red and tender; there were present tiny black blebs and he had a great fever. The sick one was completely out of his head. There were frequent evacuations of bilious matter. He died the second day after the onset of the illness...' Also among the Epidemics of Hippocrates, Sussman (1958) identified a clearly recognizable clinical description of *C. histolyticum* gas gangrene.

The disease was rarely encountered during the Napoleonic Wars (1803–14), the Crimean War (1854–56) and the Franco–Prussian War (1870) and, curiously, was not seen at all during the American Civil War (1861–65) (Keen, 1915). In his review of British military surgery in the time of Hunter (1728–93) and in World War I, Bowlby (1919) declared that 'gas gangrene is so striking and terrible a malady that it could not possibly have been overlooked if it were at all frequent. Yet I find no description of it in Hunter's work or in those of any of the early writers on war surgery.' It is also clear that the commonly encountered hospital gangrene of the pre-Listerian surgical era was not *gas* gangrene, but probably anaerobic necrotizing fasciitis.

It is thus clear that gas gangrene did not come into any great prominence until World War I, when its common occurrence among battle casualties engendered a prodigious literature on the subject. The devastating frequency of gas gangrene among the wounded on the fields of Flanders led to its being regarded as essentially a disease of war, in which extensive, heavily contaminated wounds are likely to abound. In civilian practice, serious wounds of the dimensions and variety of those seen in battle casualties are much less frequent, and for those comparable wounds that do occur, early and adequate surgical treatment, and hence prophylaxis against clostridial infection, is much more rapidly and readily available.

Apart from gas gangrene and related infections of accidental wounds, clostridia may also be implicated in such conditions as postoperative sepsis (the most common form of gas gangrene seen in modern communities), postinjectional gas gangrene, and postabortal and puerperal sepsis. These, and other conditions, will be referred to in the following pages.

Bacteriology

Three types of clostridial wound infections are distinguished (MacLennan, 1943b, 1962):
1. Simple contamination in which one or more clostridia are present, but from which subsequent invasion of the underlying tissues does not necessarily occur;
2. Clostridial cellulitis, a condition characterized by the invasion of fascial planes by the organisms, but without invasion of muscle tissue, and with minimal toxin production;
3. Clostridial myonecrosis, in which there is invasion of healthy muscle tissue, with abundant formation of exotoxins.

Contamination and colonization

Simple contamination of wounds with pathogenic clostridia is not uncommon, and many such wounds heal by first intention without special treatment, and without sequelae. 'The presence of *C. welchii* or of other toxigenic clostridia in a wound is not necessarily an indication of gas gangrene. These organisms are often present in the wounds of patients not suffering from that disease' (Medical Research Council, 1943); and MacLennan (1943b) commented that 'their absence from war wounds is a matter of surprise rather than satisfaction; their pre-

Table 18.1 The incidence of clostridial contamination in accidental wounds*

Authors	Number of wounds	Percentage showing presence of		
		C. perfringens	Other clostridia	Subsequent gas gangrene
Stoddard (1918)	137	23	—	2.9
Pulaski *et al.* (1941)	200	12	11	0
Altemeier and Gibbs (1944)	99	39.4	6	—
Cutler and Sandusky (1944)	110	?	24	—
Greenberg (1945)	87	2.3	6.9	0
Feeney (1947)	101	6	21	0
Roy *et al.* (1954)	409	8	10.2	0
Lowbury and Lilly (1958)	454	34.8	—	0

*From Willis (1969)

sence, for resignation rather than alarm.' The findings of some studies of the incidence of clostridial contamination of war wounds and civilian injuries are summarized in Table 18.1. The relatively common occurrence of clostridia in accidental wounds in the absence of anaerobic infection is largely attributable to the fact that conditions in the lesions are unsuitable for the multiplication of the contaminating organisms or for toxin production by them. A high oxidation–reduction potential (Eh) provided by adequately perfused and healthy surrounding tissues is the chief factor preventing active invasion of the tissues. In the absence of treatment, clostridial cellulitis or myonecrosis may develop from simple contamination, so that the three types of 'infection' may be regarded as three ascending grades of severity.

Infection

Several pathogenic and related clostridia are likely to be encountered in wounds that are the seat of gas gangrenous infection. The most important pathogens are *C. perfringens* type A, *C. septicum*, *C. novyi* type A and *C. histolyticum*; *C. sordellii*, *C. fallax* and *C. carnis* are of less importance. Of the non-pathogenic species, *C. sporogenes*, *C. bifermentans* and *C. tertium* are among those most frequently isolated.

Polymicrobial participation

Although monomicrobic infections do occasionally occur (Cooke et al., 1945), it is usual to find more than one species of *Clostridium* participating in the infection. Indeed, as with most anaerobic infections polymicrobial participation is the rule in gas gangrene, and may involve not only multiple species of clostridia and non-sporing anaerobes such as anaerobic cocci and *Bacteroides* species (MacLennan, 1943a–d, 1944) but it frequently involves facultative organisms such as *Escherichia coli*, *Proteus* species and staphylococci. The bacterial flora of a gas gangrenous wound of a battle casualty in the Korean War is illustrative (Strawitz et al., 1955); the anaerobes isolated from a patient who had received shell fragment wounds were *C. perfringens*, *C. novyi*, *C. bifermentans*, *C. sporogenes* and *C. paraputrificum*, while aerobic isolates consisted of *Streptococcus pyogenes*, *Str. faecalis*, viridans streptococci, *Sarcina lutea*, *E. coli* and a *Bacillus* sp.

Wound botulism

C. botulinum sometimes occurs as a wound contaminant from which botulinum neurointoxication may occur, although wound botulism is rare. About 30 cases of this unusual syndrome appear in the world literature, most of which occurred in the United States; they were due to *C. botulinum* type A (the most common type) or type B. Nine of these cases were reviewed by Merson and Dowell (1973). The rarity of botulism following a true wound infection is presumably due not only to the fact that the organism is sparsely distributed in the environment, but also because its spores fail to germinate readily in the tissues (Keppie, 1951). Major and trivial wounds have been implicated, including an infected hypodermic needle wound in a cocaine addict (Morbidity and Mortality Weekly Report, 1982). The development of the disease is similar to tetanus in that there is an incubation period of 4–14 days between wounding and the onset of neurological symptoms. Wound botulism is fully considered in Chapter 23.

Classic gas gangrene

The results of surveys of the anaerobic flora of gas gangrenous wounds by various workers show a remarkable unanimity. Thus, among the pathogenic clostridia *C. perfringens* has been the organism most frequently encountered, *C. novyi* and *C. septicum* have rivalled one another for second place while *C. histolyticum* has been reported infrequently; the most common non-pathogens are *C. sporogenes*, *C. tertium* and *C. bifermentans*. The detailed findings of six surveys of clostridia present in gas gangrenous wounds are summarized in Table 18.2.

In general, the incidence of pathogenic and related clostridia in gas gangrene is a reflection of the incidence of these organisms in soil (Keen, 1915; MacLennan, 1943c, 1962). It is not surprising, therefore, that the incidence of gas gangrene among battle casualties should vary from one theatre of war to another. Thus, in the desert fighting during World War II, gas gangrene was comparatively rare, a finding which can be related to the sparse anaerobic flora of the Western Desert (MacLennan, 1943c). The

Table 18.2 Clostridia isolated from gas gangrenous wounds by different workers*

Organism	Percentage of cases					
	MacLennan (1943c) 146 cases	Hamilton (1944–45) 25 cases	Cooke et al. (1945) 72 cases	Smith and George (1946)	Stock (1947) 30 cases	Dhayagude and Purandare (1949) 25 cases
C. perfringens	56	68	48	39	80	52
C. novyi	37	20	0	32	50	0
C. septicum	19	0	0	0	3	36
C. histolyticum	6	0	0	0	0	16
C. sporogenes	37	20	11	54	63	0
C. bifermentans	4	28	2	54	23	0
C. tetani	13	4	0	4	10	0
C. tertium	30	4	1	3	13	0
C. butyricum	13	0	2	3	7	0
C. cadaveris	5	0	3	3	0	0
C. fallax	1	0	6	3	3	0
C. cochlearium	9	0	0	2	3	0
C. lentoputrescens†	19	0	0	2	0	0
C. sphenoides	3	0	0	2	0	0
C. tetanomorphum†	2	0	0	0	0	0
C. hastiforme	3	0	0	0	0	0
C. regulare†	0	0	0	2	0	0
C. multifermentans†	0	0	0	5	0	0
C. carnis	0	8	0	0	0	0
Unidentified clostridia	16	12	3	53	13	0

† Species no longer recognized
*From Willis (1969)

abundance of the organisms in the cultivated soil of Flanders, on the other hand, accounted for the high incidence of the infection on the Western Front during World War I. The anaerobic flora of man's immediate environment—his clothing—is not dissimilar to that of non-arid soil, and is doubtless the result of daily endogenous contamination with faecal material, and of exogenous contamination with dust. In a study of 110 wounds sustained during aerial combat, Cutler and Sandusky (1944) noted that most of them contained small particles of flying-suit, many of the particles being embedded in deep recesses of the wound and lying in intimate contact with traumatized tissues. They isolated clostridia from 24 per cent of these wounds, three of which subsequently became gas gangrenous.

In cases of gas gangrene that follow accidental trauma the organisms most commonly contaminate the tissues at the time of wounding, and it is clear that almost any foreign material that enters the tissues may carry the offending pathogens. The implantation of soil, clothing or skin are obvious examples. In the case of bullet wounds, it has been shown that they can become infected with clostridial spores present on the missile (it is a popular misconception that a bullet is sterilized by firing) and that contaminated clothing, whether covering the entrance or exit wound, can lead to infection as a result of organisms being sucked into the wound by the cavitational process (Thoresby, 1966; Thoresby and Darlow, 1967; Thoresby and Watts, 1967).

Clostridial cellulitis

Anaerobic (clostridial) cellulitis (MacLennan, 1943a–d) is a comparatively uncommon condition characterized by a foul, seropurulent infection of the depths and crevices of a wound, often with local extension along fascial planes, but without progressive involvement of healthy muscle and without marked toxaemia. The predominant organisms are proteolytic and non-toxigenic clostridia (Table

18.3). *C. perfringens* is often present, but other pathogens are infrequent.

Postoperative and postinjectional gas gangrene

While gas gangrenous infections of wounds are characteristically associated with accidental trauma including severe thermal burns (Monafo *et al.*, 1966; Davis *et al.*, 1979), they sometimes follow clean and elective surgical procedures (Bornstein *et al.*, 1964; British Medical Journal, 1964; Aldrete and Judd, 1965; Parker, 1967; Braithwaite *et al.*, 1982; Gledhill, 1982) and have followed the injection of adrenaline and other substances (Copper, 1946; Marshall and Sims, 1960; Koons and Boyden, 1961; Berggren *et al.*, 1964; Harvey and Purnell, 1967; Maguire and Langley, 1967; British Medical Journal, 1968; Nahir *et al.*, 1978; Seradge and Anderson, 1980; Yangco *et al.*, 1982; Teo and Balasubramanian, 1983). In most cases, the offending organism is *C. perfringens*. This ubiquity of *C. perfringens* is highlighted by its frequent presence in the air in hospitals and operating theatres (Lowbury and Lilly, 1958; Gye *et al.*, 1961). Moreover, its endogenous contamination of the skin from the large bowel, especially at such injectional sites as the buttocks and thighs, is normal.

Postoperative gas gangrene after limb surgery

In clostridial infections following clean limb surgery the infecting organisms may be derived from exogenous sources as a result of some breakdown in theatre ventilation or sterility. But, more commonly, they are endogenously derived from the patient's skin (Roberts *et al.*, 1933) and are implanted into the underlying tissues at the time of surgery. Although upper limb surgery is not exempt (Braithwaite *et al.*, 1982) most cases of gas gangrene that complicate 'clean' surgery follow operations on the lower limb, the skin of which is more constantly and more heavily contaminated with clostridia than the skin in other areas, especially in older patients. Mid-thigh amputation for vascular disease, especially in diabetes, and orthopaedic operations involving the insertion of foreign bodies, e.g., pinning the neck of the femur, are procedures that carry a special risk of postoperative gas gangrene (Gye *et al.*, 1961; Heineman and Braude, 1961; Parker, 1967; British Medical Journal, 1967; Knutsdon, 1983; Laszlow and Elo, 1983). Consequently, careful attention to preoperative skin preparation is of the utmost importance in patients who are to be submitted to lower limb surgery (Lowbury *et al.*, 1964). The hazard of postoperative gas gangrene dictates the need for antimicrobial prophylaxis in patients undergoing lower limb orthopaedic surgery in general, and especially for amputation and for operations involving the hip (Taylor, 1960). The drugs of choice are metronidazole and penicillin.

Postoperative gas gangrene of the abdominal wall

Fortunately, gas gangrene of the abdominal wall is a relatively infrequent complication of abdominal surgery. The infection is of endogenous origin, the organisms being derived from the intestinal or biliary tracts and contaminating the abdominal wound at the time of operation. As might be expected, it occurs less often after operations on the stomach and duodenum than after those involving the colon, and is almost always due to *C. perfringens*.

Table 18.3 The clostridia isolated from 17 cases of anaerobic cellulitis*

Organism	Percentage of cases	Organism	Percentage of cases
C. sporogenes	70	*C. cochlearium*	12
C. perfringens	41	*C. bifermentans*	6
C. tertium	35	*C. novyi*	6
C. lentoputrescens†	18	*C. sphenoides*	6
C. butyricum	12	*C. tetani*	6
C. cadaveris	12	*C. tetanomorphum*†	6

† Species no longer recognized
* Reproduced from MacLennan (1943) by courtesy of the editor of *Lancet*

Postinjectional gas gangrene

In postinjectional gas gangrene, skin contamination by clostridia derived from the bowel is also causal; skin disinfection before injection is usually very perfunctory and the majority of cases of postinjectional gas gangrene have occurred after injections into the hip, thigh and buttock. It has been suggested that this hazard can be minimized by avoiding the administration of injections below the waist (Harney, 1939; Brewster, 1962; British Medical Journal, 1964). The other possibility, that gas gangrene bacilli may contaminate parenteral solutions, needles, syringes or antiseptic solutions used for skin preparation at the time of injection, must never be overlooked (Bowie, 1956; Rubbo and Gardner, 1968). However, in these days of modern quality control the chance of infection derived from these sources is remote.

Uterine gas gangrenous infections

Most clostridial infections of the uterus are due to *C. perfringens*, the organism being derived mainly from endogenous sources; *C. perfringens* is a transient inhabitant of the lower vagina. The high isolation rates reported in earlier studies (Butler, 1941; Sadusk and Manahan, 1941; Salm, 1944; Ramsay, 1949; Slotnick *et al.*, 1963) are in sharp contrast to more recent experience. The numerous contemporary studies of the microbial flora of the vagina have been singularly consistent in the infrequency with which clostridia have been identified. The incidence has varied from nil (de Louvois *et al.*, 1975; Lindner *et al.*, 1978), through 1–2 per cent (Corbishley, 1977; Thadepalli *et al.*, 1978) to an exceptional 16 per cent (Mead, 1978; Hammerschlag *et al.*, 1978). Interestingly, and in contrast, Tabaqchali *et al.* (1984) isolated *C. difficile* from the vaginas of 11 per cent of 82 consecutive pregnant women attending an antinatal clinic. *C. perfringens* doubtless gains access to the introitus and vagina from the perianal area, where it occurs frequently and abundantly. The organism seems to be most commonly present following abortion (Bartizal *et al.*, 1974; Kowen *et al.*, 1979) but there are few figures available concerning its incidence in the vaginas of healthy, non-pregnant women (Bartlett *et al.*, 1977; Corbishley, 1977; Lindner *et al.*, 1978). Its inconstant presence postpartum, and the fact that it is present more frequently following deliveries in which there has been manual or instrumental interference (Salm, 1944) suggests that the organism is derived in these cases from the perineum. That most clostridial uterine infections follow 'self-induced' abortion, and that *C. perfringens* is rarely present in the uterine contents of puerperal patients (Peckham, 1936) also suggests that endogenous infections are derived in the first instance from the patient's own faecal flora (Magarey *et al.*, 1927).

In a comparison of the frequencies of puerperal and abortal infections due to *C. perfringens* over the 20-year period from 1940 to 1959, Hill (1964) noted that of 83 fatal *C. perfringens* uterine infections, 99 per cent were postabortal and only one per cent was puerperal.

Puerperal infections are partly the result of unskilful management of labour, and are more frequent when difficulties arise during delivery. Thus, introduction of endogenous organisms into the uterus is favoured by prolonged and difficult labour, by instrumentation and by repeated vaginal examination and manipulation. In contrast to infections complicating 'self-induced' abortion, exogenous contamination of the puerperal uterus by unsterile instruments and non-surgical techniques is a rare event. In modern obstetrics, the avoidance of prolonged and traumatic deliveries and the use of antimicrobial prophylaxis accounts for the extreme rarity of this form of puerperal sepsis. The manner of entry of clostridia into the uterine cavity in cases of 'self-induced' abortion is not difficult to understand, exogenous and endogenous contamination occurring as the result of unskilful manipulations, and the use of unsterile and unclean 'instruments' and abortifacients.

The genesis of gas gangrene

The most important single factor which enables anaerobes to flourish in a wound is low oxygen tension. The Eh of healthy tissues (at around pH 7.5) is well above the levels necessary for the initiation of anaerobic growth, so that anaerobes are unable to multiply in normal tissues and are therefore unable to produce disease. This was demonstrated as long ago as 1892 by Vaillard and Rouget, who showed that the atraumatic inoculation of guinea-pigs with toxin-free spores of *C. tetani* did not give rise to the disease. However, most wounds are not atraumatic, and it is the trauma of wounding that usually initiates the chain of events leading to colonization of the tissues by contaminating anaerobes.

Host determinants

Trauma of wounding

The type of wound produced by a high velocity missile provides ideal conditions for the proliferation of anaerobes. On entry, the missile produces a relatively small wound in the skin compared with the extensive damage beneath. Shock wave pressures cause tearing of soft tissues and blood vessels, extensive cavity formation and fracture of long bones at a distance from the site of wounding. Since many large muscles receive their main blood supply from only one or two main arteries, damage to these vessels may cut part or all of the blood supply to the muscle (Power, 1945a). Not only does this result in immediate tissue anoxia, but necrosis of muscle is also likely to occur, since the colateral circulation is usually inadequate and revascularization is too long delayed. Indeed, vascular damage is the most important predisposing factor in anaerobic infections. North (1947) reported that 72 per cent of cases of gas gangrene had injuries to the vessels supplying the affected muscle, while Lowry and Curtis (1947) recorded that 87 per cent of cases submitted to amputation for gas gangrene showed evidence of vascular damage.

Cavitation, which is the process by which a high velocity missile causes tissue destruction, was the subject of review by Thoresby (1966). In essence, cavitation is due to the transfer of part of the missile's kinetic energy to the tissues, the missile causing 'crushing and attrition of the tissues directly and indirectly: directly by the immediate action of the bullet itself, and indirectly by the communication of a part of its energy to the solid and liquid particles which it displaces' (Stevenson, 1910). If the 'missile' is a hypodermic needle then the energy involved is small and its time of action long, so that the subsequent restoration of the displaced tissue is virtually complete. If, however, the missile is a high velocity bullet that transfers most of its strike energy to the wound then the sudden, intense and widespread distortion of tissues will result in extensive tissue damage. In general, the amount of tissue damage is roughly proportional to the amount of energy absorbed by the tissues from the time of strike to the time of exit in the case of a perforating wound (or complete retardation in the case of a retained missile); only those missiles that rapidly transfer a high proportion of their kinetic energy to the tissues will cause severe wounds. As might be expected, severe cavitational wounds are likely to result when a high velocity missile is rapidly retarded in the tissues by striking a bone.

Host reaction to trauma

Consequent upon the tissue damage inflicted by the missile, extensive haemorrhages occur in the damaged tissues, local tissue perfusion ceases and large areas of muscle become anoxic. The Eh and pH of the damaged tissues fall and these changes, together with the breakdown of some protein to amino acids, produce conditions highly favourable to the growth of anaerobic organisms. Anaerobiosis is also favoured by the presence of foreign bodies, such as missile fragments or pieces of clothing that have been carried into the tissues during penetration of the missile. Contamination of the wound with soil also encourages the development of anaerobiosis, due to the presence of calcium chloride, which is itself a necrotizing agent (Bullock and Cramer, 1919; Altemeier and Furste, 1949). Anoxia may be further intensified by an increasing pressure in the wounded area due to the accumulation of extravasated blood and tissue fluids that are unable to escape because of the relatively small entry and exit wounds in the skin. Additionally, poor perfusion of the tissues is aggravated by traumatic shock and blood loss. Finally, a series of chemical changes that occurs within the damaged and anoxic tissues provides an ideal environment for the development of clostridia (Oakley, 1954).

Local biochemical changes

Breakdown of the carbohydrate in anoxic muscle continues and the oxidation–reduction potential falls. Extravasated haemoglobin and myohaemoglobin are reduced and cease to act as oxygen carriers. As a result, aerobic oxidation is halted and anaerobic reduction of pyruvate to lactate results in a further reduction in Eh. The alkali reserve is depleted and the pH falls because of the accumulation of lactate in the muscle. As a consequence of the lowered Eh and pH muscle proteinases are activated, and these hydrolyse protein with a consequent increase of amino acids and a further fall in pH. This increase in the concentration of amino acids not only provides contaminating clostridia with nutrients for growth, but also enables the organisms to multiply at a higher oxidation–reduction potential.

Once bacterial growth is established there is a further fall in Eh, and concomitant production of bacterial toxins and other products of metabolism. These factors promote the injury and invasion of adjacent healthy muscle so that gas gangrene is established and progresses by the 'domino' effect. Anaerobic infection is further aided by the fact that the natural defences of the body, both cellular and humoral, are hampered; neither phagocytic cells nor antibodies can enter the necrotic area. Moreover, parenterally administered antimicrobial agents, even in large doses, are unlikely to succeed as a prophylactic measure because they also fail to enter the necrotic target tissues.

It has been noted that gas gangrene can complicate much less traumatic lesions than those produced by high velocity missiles. However, in all cases the underlying precipitating host factor is reduced tissue perfusion and tissue anoxia. Thus, postoperative gas gangrene is most common in patients with arterial insufficiency (as in diabetes mellitus), arterial thrombosis, senile gangrene and frostbite. Limb surgery involving the use of tourniquets or bloodless field techniques favours colonization of the tissues by anaerobes, as does any procedure in which there is imperfect haemostasis, excessive heat coagulation of tissues, prolonged and traumatic use of retractors and the application of tight plasters or dressings (Pearson et al., 1980).

In the setting of postinjectional gas gangrene, the most dangerous substance has been adrenaline, its notoriety in this respect being attributable to its vasoconstrictor effect. Evans et al. (1948) showed that the ischaemia induced by 2 µg of adrenaline injected into the tissues of the guinea-pig caused reduction of the Eh and inhibition of the inflammatory migration of leucocytes and of the exudation of blood fluids. These effects, which lasted for about 2 h, rendered the tissues much more susceptible to anaerobic infection; with *C. perfringens* and *C. septicum* the enhancement of infection was about 100 000-fold.

The progression of events in uterine gas gangrene follows much the same pattern as outlined for gas gangrene at other sites, except that dangerous bacteraemia is much more likely to occur. Once the interior of the puerperal or postabortal uterus has been contaminated by a minimal infective number of clostridia, fragments of blood clot and necrotic tissue (fetal, placental or maternal) and damaged maternal tissue provide conditions favourable for the multiplication of contaminating anaerobes. For the development of uterine gas gangrene, Wrigley (1930) concluded that the organism must be introduced into the uterus from without, that the fetus must be dead or must remain *in situ* for a sufficient incubation period, and that damage to maternal tissues must have occurred. Although these postulates are in keeping with the clinical findings in most cases (Russell and Roach, 1939) it is clear that the presence of a dead fetus is not a prerequesite for sepsis (Toombs, 1932; Hill, 1936).

Bacterial determinants

Gas gangrene is considered to have commenced 'not when a wound has become infected with the pathogenic anaerobes, but from the moment when a group of these bacteria have been enabled to surround themselves with a toxin sufficiently concentrated to abolish the local defences of the tissues' (Committee upon Anaerobic Bacteria and Infections, 1919). Further multiplication and toxin production leads to rapid invasion of healthy tissue and the development of clinical gas gangrene. While there is much evidence to show that the exotoxins produced by the clostridia are largely responsible for the nature of the lesion and the systemic manifestations, the modes of action of the toxins and the individual importance of each are complex and uncertain, and are understood in only the most general terms.

Lecithinases C, such as *C. perfringens* α-toxin and *C. novyi* γ-toxin, damage or destroy cell membranes. The effect of these enzymes on capillaries may render them freely permeable to fluid and protein, leading to increased tension in affected muscles with a resulting increased anoxia. Lysis of erythrocytes by α-toxin is the cause of the intravascular haemolysis and haemoglobinuria sometimes associated with *C.*

perfringens bacteraemia. In *C. perfringens* type A infections the lecithinase is the most important lethal factor (Evans, 1945a,b; Kass *et al.*, 1945). The collagenases of *C. perfringens* (κ-toxin) and *C. histolyticum* (β-toxin) destroy collagen barriers in the tissues, which might otherwise tend to localize the infection, while destruction of reticulin around capillaries leads to haemorrhage and thrombosis. In addition to producing these damaging effects, proteolysis also serves to provide additional amino acids and peptides for bacterial growth (Smith, 1949). The hyaluronidase of *C. perfringens* (μ-toxin) greatly facilitates the spread of the organism through the tissues, an effect which is probably due both to the removal of a physical barrier and to the liberation of fermentable carbohydrates following breakdown of hyaluronic acid.

Plausible roles for the activities of other soluble antigens of the pathogenic clostridia can be suggested, but proof of their involvement in the disease process is difficult; deoxyribonucleases (*C. perfringens* ν-antigen, *C. septicum* β-toxin) lipases (*C. novyi* ε-antigen), proteinases (*C. histolyticum* β-, γ- and δ-toxins) and fibrinolysins have activities that are almost certainly exerted on appropriate tissue elements *in vivo*. However, it is much more difficult to define the extent of the role of the lethal toxins (the α-toxins of *C. perfringens*, *C. novyi*, *C. septicum* and *C. histolyticum*, and *C. perfringens* θ-toxin) in producing the classical clinical syndrome of gas gangrene. The lethal toxin about which most is known is the α-toxin of *C. perfringens*, the lethal activity of which is thought to be due to its specific lecithinase C activity. Because lecithin is an important component of many diverse tissues, slow systemic absorption of the α-toxin is likely to lead to many serious chemical changes in vital organs.

The toxicology of the histotoxic clostridia was reviewed in some detail by Willis (1969).

Among other factors that may contribute to the systemic events in clostridial myonecrosis are acidosis (MacLennan, 1962), central nervous system effects (Nora *et al.*, 1966) and destruction of energy-yielding enzyme systems (Kielley and Meyerhof, 1950; Macfarlane and Datta, 1954). These matters have been discussed by Evans (1945b), Smith (1949), Macfarlane (1955), MacLennan (1962), Hauschild (1971) and Ispolatovskaya (1971).

Clinical considerations

There is a recent history and clinical evidence of wounding, except in spontaneous gas gangrene, and although the preceding trauma is usually extensive and severe, anaerobic myonecrosis may follow minor injuries, e.g., hypodermic injections. Like most other anaerobic infections gas gangrene is not transmissible from person to person, so that theatre and hospital cross-infection does not occur. There is no more need to isolate a patient with gas gangrene than there is to isolate a patient because he has clostridia in the stool (Williams *et al.*, 1960).

Clostridial cellulitis

Excellent descriptions of clostridial cellulitis have been published by MacLennan (1943a–c, 1944, 1962). Unlike clostridial myonecrosis, anaerobic cellulitis is not a life threatening infection. Typically, poorly toxigenic and non-pathogenic clostridia of limited invasive power are restricted to the depths and crevices of the infected wound and to the interconnecting tissue spaces; muscle tissue is not involved. The infection may vary in extent from a limited gas abscess to extensive involvement of a whole limb. The onset of the infection is more gradual than gas gangrene, the incubation period averaging 3–4 days or longer and there is usually no associated systemic toxaemia or shock. Locally, the dirty wound exudes a brownish seropurulent malodorous discharge. Gas is a constant and prominent feature 'extending diffusely between the muscle groups, crackling in the subcutaneous tissues, and bubbling up through the clot and discharge of the wound, but never, it should be emphasized, to be found intramuscularly. The presence of this gas must be regarded as of importance in the differential diagnosis; if it is abundant, extensive and easily demonstrable, and there is no pronounced toxaemia, the condition is almost certainly not gas gangrene' (MacLennan, 1943b). The skin is rarely discoloured and there is little or no oedema (Table 18.4). In civilian practice anaerobic cellulitis complicating accidental trauma is rare.

Gas gangrene (clostridial myonecrosis)

There are many excellent accounts of gas gangrene in the literature. Good general descriptions have

Table 18.4 The clinical differences between clostridial cellulitis, clostridial myonecrosis and anaerobic streptococcal myositis*

Feature	Clostridial cellulitis	Clostridial myonecrosis	Anaerobic streptococcal myositis
Incubation period	More than 3 days	Less than 3 days	3 days
Onset	Gradual	Acute	Subacute
Toxaemia	Nil or slight	Constant; severe	Mild; varies with temperature
Temperature	99–100°F	100–102°F	High
Pulse	Slightly fast	Rapid; poor quality	High; proportionate to temperature
Blood pressure	Normal	Low	Normal
Anaemia	Absent	Usually present	Not characteristic
Pain	Absent	Often present; severe	Often present; severe
Wound:			
discharge	Nil or slight	Slightly watery to profuse brown	Wet; oedematous; profuse; seropurulent blood-stained;
odour	Foul	Inconstant	Foul
gas	Abundant	Rarely abundant	Slight
crepitation	Characteristic	Inconstant	Slight; late
muscle	Not involved	Always involved; diffuse	Usually involved; focal

*Reproduced from Lowry and Curtis (1947) by courtesy of the Editor of the *American Journal of Surgery*

been published by the Medical Research Council (1943), Cooke *et al.* (1945) and MacLennan (1962).

The incubation period of gas gangrene after injury may be as short as 7 h, but in most cases the disease develops within 7 days of wounding. The average incubation periods of the three main aetiological types are: 10–48 h with *C. perfringens*, 2–3 days with *C. septicum*, and 5–6 days with *C. novyi*.

One of the earliest prodromal signs is the development of pain in the region of the wound. This may be of quite sudden onset but it usually develops gradually, steadily increasing in intensity. There is progressive swelling and oedema of the affected part. Concomitant with increasing pain is a steep rise in the pulse rate, and sometimes a rise in temperature, although pyrexia is not a marked feature and is often absent.

Locally, there is obvious oedema about the wound and the area is extremely tender. Early in the infection there is a thin watery discharge, but as the disease progresses there is a profuse serous or serosanguinous discharge which may be so copious as to cause haemoconcentration. The area of oedema rapidly increases, bubbles of gas may be seen escaping from the wound and the tissues may become crepitant. With increasing swelling and tension the skin becomes white and marbled. In untreated cases the infective process extends rapidly and inexorably, not ceasing even with the patient's death. Local clinical signs in cases of postinjectional gas gangrene may be relatively covert, belying the gross progressive pathology of the deep tissues. However, systemic illness is always profound.

Gas may not be an obvious feature, either because it is not produced in any considerable volume, as in infections due to *C. novyi* and *C. histolyticum*, or because its presence is masked by its deep intramuscular location and by the intense local oedema. In any event, overt gas formation in cases of clostridial myonecrosis is usually a late manifestation. Moreover, the presence of gas in wounds, including infected wounds, does not necessarily imply clostridial infection (McDonald, 1947; Culbertson, 1958; Nichols and Smith, 1975; Van Beek *et al.*, 1974). Thus, infections due to other organisms, especially *E. coli*, are sometimes associated with copious gas formation (Kemp and Vollum, 1946; Lewis *et al.*, 1978; Fitzpatrick *et al.*, 1979). Bacteriogenic gas may also be due to bacterial colonization of devitalized tissues apart from trauma, as occurs commonly in diabetic gangrene of the foot. Air inclusions commonly result from air being forced into the tissues at the time of wounding by a high velocity missile due to the cavitational effects, and in industrial accidents with compressed air (Stammers, 1945; Desmond, 1947; Saleh and Bollen, 1984). A 'gas abscess' may develop in association with a foreign body in a wound, which resolves when wound toilet is performed and the foreign body removed, while irrigation of wounds with hydrogen peroxide may produce fairly diffuse collections of gas.

The characteristic muscle changes are seen to best advantage at operation. In the early stages of infection there may be little apart from oedema and pallor. Later, however, involved muscle is slate-blue, brick-red or purplish in colour. It is non-contractile and poorly or non-bleeding, and some gas may be present. Still later, with progression to diffuse myonecrosis, muscle masses become friable and dark purple or black in colour. The appearance of the tissues shows considerable variation from case to case, depending on the nature of the infecting organisms (Table 18.5). In infections due to *C. novyi*, for example, oedema is the most conspicuous feature, while in those due to *C. perfringens* oedema is less pronounced but gas production is likely to be marked and globules of free fat are characteristically present in the affected muscle and oedema fluid. In pure *C. histolyticum* infections, which are very rare, there is rapid and extensive digestion and liquefaction of all the soft tissues, while in wounds infected with *C. histolyticum* along with other pathogenic clostridia, blackening and digestion of muscle tissue may be the most prominent features.

By far the most serious effect of gas gangrene is the profound systemic intoxication, about which Macfarlane and MacLennan (1945) published the following classic account: 'The patient lies collapsed and obviously desperately ill. He has a livid pallor, the extremities are cold, and sometimes the veins cannot be filled sufficiently to make venepuncture possible. The pulse is often impalpable; it is feeble and irregular, and we have noticed it to be markedly dicrotic in some cases. The blood-pressure, particularly the diastolic pressure, is low. In some cases there is a very large pulse-pressure, with systolic readings of about 100 mm of mercury, while the diastolic pressure is too low to be recorded. Mentally, the patient is usually alert and clear, anxious, even terrified, and apparently fully aware of his danger. Sometimes he lapses into coma or delirium before death, but more often he dies suddenly, particularly during some disturbance such as being moved or anaesthetised. Death appears to be due to circulatory failure'. Sudden, unexpected death has been commented upon by a number of workers (Keen, 1915); 'We have seen a patient die while engaged in completing a football forecast' (Rains and Ritchie, 1977).

It is most unusual for gas gangrenous wound infections to be complicated by intravascular haemolysis. This is much more commonly associated with clostridial (usually *C. perfringens*) infections of the uterus, and of the biliary tract following surgery. The salient features of the four main bacterial types of gas gangrene are summarized in Table 18.5.

Postoperative gas gangrene

Postoperative gas gangrene of the abdominal wall, although rare, is of special importance not merely because it is often fatal but because its development is usually unexpected and its recognition delayed; it may complicate colonic and biliary tract surgery. Reports of the condition have been published by Quinn *et al.* (1942), Elliot-Smith and Ellis (1957), Spann and McGill (1957), Shapiro *et al.* (1963),

Table 18.5 Chief clinical features of the four main varieties of gas gangrene*

Feature	C. perfringens	C. novyi	C. septicum	C. histolyticum
Incubation period	10–48 hours	5–6 days	2–3 days	2–3 days
Onset	Sudden	Sudden	Sudden	?
Course	Rapid	Rapid	Very rapid	Slow
Toxaemia	Moderate	Severe	Severe	Moderate
Swelling, oedema	Moderate	Marked	Moderate	Slight
Skin reaction	Bronzing	Pallor	Erythema	Digestion
Gas	Moderate	Slight	Marked	None
Odour	Sweetish	None	Sweetish	Foul
Exudate	Moderate; haemorrhagic	Copious; yellow	Moderate; haemorrhagic	Moderate; brown
Muscle	Slate-blue	Pale	Red	Black

*Reproduced from MacLennan (1962) by courtesy of the Editor of *Bacteriological Reviews*

McNally and Crile (1964), Isenberg (1966) and Gledhill (1982) and they emphasize the need for a high degree of clinical awareness of the condition and for early and vigorous treatment.

The onset of infection usually occurs 36–72 h after operation, but may be delayed for as long as 4 days. The first evidence of infection is usually an abrupt rise in the pulse rate, which is disproportionate to the temperature, and hypotension; there is increasing general toxicity, pain in the region of the wound and a rapidly developing secondary anaemia. There is an early yellowish-brown discoloration of the skin at the margins of the wound, from which a brownish serosanguinous exudate may be expressed. A firm diagnosis is easily established at this stage by the demonstration of gram-positive bacilli in stained films of the exudate. Later, the skin around the wound takes on a bronzy colour, which rapidly darkens, and the abdominal wall becomes frankly gangrenous. The mortality is between 50 per cent and 60 per cent (Quinn et al., 1942; Hitchcock et al., 1975). Gas gangrene of the abdominal wall must be distinguished from Meleney's progressive synergic gangrene and related infections (see Chapter 17).

Gas gangrenous infections of the uterus

Toombs and Michelson (1928) reviewed and analysed the reports of 41 cases of uterine gas gangrene that had appeared in the literature up to that time and Toombs (1932) described a further example of the condition and referred to an additional 13 case reports published during the intervening period. Since then many reports of postabortal infections, and a few of puerperal gas gangrenous infections of the uterus, have appeared in the world literature. Reports from the United Kingdom include those of Wrigley (1930), Dawbarn and Williams (1938), Baker (1939), Ramsay et al. (1948), Ramsay (1949, 1950), Lee et al. (1966), Hanson et al. (1966), Ruenberg et al. (1967), Stephen (1977), Symonds and Robertson (1978) and Bradbury (1985). The overwhelming proportion of clostridial infections of the uterus follow abortion, while puerperal infections are rare (Hill and Butler, 1948). Infection has been reported as a complication of caesarean section (Mariona and Ismail, 1980) and amniocentesis (Fray et al., 1984). Of extreme rarity are gas gangrenous infections occurring in the non-gravid uterus (Gabriel and Kingsbury, 1922; Spencer, 1922; Rutherford, 1938).

Hill (1936, 1950, 1964) divided severe uterine infections into five main clinical varieties:

Infection with rapidly progressive haemolytic jaundice

The patient with severe postabortal infection due to *C. perfringens* often presents with a characteristic picture. Following manipulation there is the onset of fever, chills and lower abdominal pain, often associated with vomiting and occasionally with diarrhoea. The patient is usually alert and well orientated and the pulse is disproportionately high in relation to the degree of fever. When bacteraemia and intravascular haemolysis occur the patient develops jaundice, which in some cases rapidly deepens to a curious mahogany colour due to concurrent vascular collapse and cyanosis. A profound anaemia may develop very rapidly due to haemolysis and excretion of haemoglobin gives the urine a burgundy-type of colour. The blood plasma is also discoloured and there is a marked leucocytosis. It is noteworthy that infection with jaundice may proceed to circulatory failure without signs of haemolysis in the blood or urine, and that patients with haemolysis whose infection is controlled are likely to require renal dialysis.

Infection with collapse

Collapse as a primary feature of the disease can usually be suspected early, for it is commonly preceded for some hours by measurable hypotension. The syndrome may develop so rapidly that the patient goes into peripheral circulatory collapse, with cardiac failure and pulmonary oedema within a matter of hours.

True uterine gas gangrene

True gas gangrene develops more slowly and generally takes 48 h to present signs of peritonitis localizing over those areas of the uterus in which gangrene has reached its peritoneal surface. With gas formation, the uterus may distend or become emphysematous, developments that may be demonstrated radiographically (Poppel and Silverman, 1941; Doehner et al., 1960; Lacey et al., 1976). There is usually exquisite uterine tenderness with general prostration.

Metastatic gas gangrene

Metastatic gas gangrene is rare but overwhelmingly rapid in its development. There is dissemination of organisms to produce foci of infection in remote parts of the body, with or without other clinical signs of bacteraemia. The development of severe pain in a skeletal muscle group, such as the glutei, is characteristic of this form of the infection.

Mild bacteraemia

Bacteraemia sometimes develops, in which poorly virulent strains of the organism may be isolated from the blood. There are no clinical signs of bacteraemia other than a relatively high fever lasting for 2–3 days, from which recovery is the rule.

In all of Hill's five clinical varieties of uterine infection, except that of simple mild bacteraemia, unless the infection is quickly controlled, the terminal picture is one of peripheral circulatory failure, the patient usually maintaining clear consciousness to the end.

Bacteraemia

Although the demonstration of pathogenic clostridia in the bloodstream of patients with gas gangrenous infections is often of grave prognostic significance, early or transient bacteraemias, which are probably produced by mechanical causes rather than by active invasion, are of little pathogenic or prognostic consequence (Baugher, 1914; Hill, 1964). On occasion, *C. perfringens* or other species of *Clostridium* may be recovered from the blood of patients who show no evidence of gas gangrene, who are not acutely ill and who subsequently make an uneventful recovery (Butler, 1937; Gorbach and Thadepalli, 1975). In some such cases isolation of the organism is due to blood culture contamination from the skin (Ahmad and Darrell, 1976). By the same token, however, the skin may provide the source of organisms from which overt clinical *C. perfringens* bacteraemia is derived (Rose, 1979).

Significant invasion of the bloodstream by clostridia may occur rarely as a late manifestation of anaerobic myonecrosis, and is usually due to *C. perfringens* or *C. septicum*. Clostridial bacteraemias, especially those due to *C. perfringens*, are much more commonly associated with gas gangrenous postabortal and puerperal infections of the uterus, and with *C. perfringens* infections that occur as a complication of biliary tract surgery. *C. septicum* bacteraemia is a well recognized form of so-called spontaneous gas gangrenous infection (see below).

Postcholecystectomy bacteraemia

C. perfringens bacteraemia following elective cholecystectomy is a rare but devastating event. The condition differs in no essential way from clostridial invasion of the bloodstream that may complicate anaerobic infections at other sites, being characterized by its dramatic onset and overwhelmingly rapid course, with extensive intravascular haemolysis, haemoglobinuria, haemoglobinaemia and rapidly deepening jaundice. The patient is collapsed and has pyrexia, a rapid pulse, a falling blood pressure and increasing cyanosis and jaundice. The organism is not only recoverable from the bloodstream by culture, but can sometimes also be seen in direct films of the blood stained by Gram's method. As long ago as 1948, Brown and Milch considered that '*Cl. welchii* infection is a considerable factor in mortality in biliary tract surgery, occurring more frequently than has been generally recognized'. Case reports are to be found in the publications of Pyrtek and Bartus (1962), Plimpton (1964), Turner (1964) and Yudis and Zucker (1967).

Spontaneous (non-traumatic) gas gangrene

Non-traumatic clostridial infections may present as a bacteraemia, as a metastatic localized lesion or as myonecrosis due to contiguous extension of intra-abdominal sepsis. Most of these infections occur in elderly, debilitated patients, who are often diabetics or are immunocompromised and who have some underlying ulcerating lesion of the gastrointestinal, biliary or genitourinary tract.

C. septicum bacteraemia

The relationship of clostridial bacteraemia to neoplastic disease is now increasingly acknowledged, and the syndrome of *C. septicum* (and sometimes *C. perfringens*) bacteraemia associated particularly with malignant disease of the large bowel or haematological abnormality is not infrequently encountered (Valentine, 1956–57; Kapusta et al., 1972; Proetz et al., 1974; Collier et al., 1983; Bekassy et al., 1984; Pelfrey et al., 1984; Narula and Khatib, 1985). In their valuable review of *C. septicum* bacteraemia, Koransky et al. (1979) noted that 71 per cent of cases had malignancies; half of these patients had haematological malignancies and half had solid tumors of the colon. Thus, while bacteraemia may be the first declaration of some deep seated neoplastic pathology, the trilogy of pyrexia, abdominal symptoms and hypotension in a neutropenic patient must raise the suspicion of endogenous clostridial infection. This syndrome is fully considered in Chapter 19.

Metastatic gas gangrene

Purely metastatic gas gangrene is an exceedingly rare complication that may occur in patients with an existing gas gangrenous lesion or in those with a soiled wound or with intra-abdominal lesions containing pathogenic clostridia, but showing no evidence of active infection (Bourns, 1946; Cairns et al., 1947; Hill, 1964; Gazzaniga, 1967; Marty and Filler, 1969; Engeset et al., 1973). The sudden onset of severe pain in an affected muscle group is characteristic of this condition, which usually runs a rapidly fulminating and fatal course. A very unusual example of metastatic infection was described by Warren and Mason (1970) in which the presenting feature of an occult colonic carcinoma was a *C. septicum* thyroditis.

Endogenous gas gangrene of intra-abdominal origin

Another well documented form of endogenous gas gangrene commences as an intra-abdominal visceral clostridial cellulitis complicating some pre-existing pathology with subsequent spread to adjacent muscles. There may be direct involvement of the abdominal wall, or spread of infection may extend along the retroperitoneal tissue planes including the psoas muscles and into the lower limb. In either event, there is usually a massive myonecrosis of the abdominal wall, the pelvic wall or the thigh, often with a fatal outcome. *C. perfringens* is the most usual pathogen (Gordon, 1936–37; Wyman, 1949; Whyland and Levin, 1960; Nichols, 1961; Mzabi et al., 1975) but other intestinal clostridia such as *C. sporogenes*, *C. septicum* and *C. sordelli* have also been implicated (Jones et al., 1960; Thys et al., 1980; Lee et al., 1981; Gatt, 1985). In the case reported by Gordon, an appendiceal abscess tracked down over the iliacus muscle; the peritoneum and fascia overlying the muscle were eroded and invasion of the iliacus by *C. perfringens* and extension of this infection into the thigh followed. In the reports of both Wyman and of Nichols, endogenous contamination originated from the colon in patients with carcinoma and diverticular disease respectively. In both cases, a pericolic abscess tracked down into the thigh along the line of the psoas muscle, leading to massive gas gangrene of the lower extremity.

Gas gangrene of the scrotum and perineum

Perineal gas gangrenous infections, ascribed especially to *C. perfringens* and developing as a complication of a relatively mild perianal abscess or fistula, are occasionally encountered (Berkow and Tolk, 1923; Rutherford, 1938; Thomas, 1945–46; Sterling et al., 1954; Himal and Duff, 1967). It seems unlikely that most of these were cases of true gas gangrene, but were more probably examples of Fournier's gangrene—anaerobic necrotizing fasciitis of the perineum—due to non-sporing anaerobes (see Chapters 14 and 17).

Laboratory investigations

As with tetanus, the diagnosis of gas gangrene is reached entirely on clinical grounds, which may be aided by the radiological demonstration of the presence of gas in the tissues. The term 'gas gangrene' is to some extent misleading, because the presence of gas may not be an obvious feature of the disease. Moreover, gas in a wound or other infected tissue, as has been noted, may be due to various causes other than clostridial infection (Filler et al., 1968; Chopra and Mukherjee, 1970; Brightmore and Greenwood, 1974; Bessman and Wagner, 1975).

Radiology

Radiological examination has an essential place both in establishing an early diagnosis of gas gangrene and in determining the extent of the infection. Small air inclusions in the immediate vicinity of the wound are commonly seen in open fractures and lacerated wounds of the limbs; they spontaneously disappear within a few days. In clostridial cellulitis gas translucencies are confined to the subcutaneous tissues and intermuscular planes, so that there is no outlining of muscle fibres. In gas gangrene, on the other hand, there is additional infiltration of gas within intramuscular planes, this results in delineation of individual muscle fibres, producing the classic 'pectinate appearance' described by Kemp (1945).

Microbiology

Bacteriological investigations provide confirmation of the clinical diagnosis, and subsequent identification and enumeration of the infecting organisms. However, it is important to recognize that the presence of pathogenic clostridia, especially *C. perfringens*, in a wound is of no diagnostic significance in itself. In cases of anaerobic streptococcal myositis, a rare condition which, in its early stages, may be clinically indistinguishable from gas gangrene, bacteriological examination is of the utmost importance. Distinction between the two conditions is important, since radical surgery may be contraindicated in cases of anaerobic streptococcal myositis (MacLennan, 1943a; Hayward and Pilcher, 1945; Anderson *et al.*, 1972) (Table 18.4, page 308).

Microscopic examination of pathological material

Tentative confirmation of the clinical diagnosis may be aided by microscopic examination of pathological material. Thus, the presence of regularly-shaped gram-positive bacilli without spores is strongly suggestive of a *C. perfringens* infection, and Butler (1945) drew attention to the correlation between the severity of *C. perfringens* infections on the one hand, and the degree of capsulation of the organism and the extent of leucocyte damage on the other (Stratford, 1973).

The presence of *C. perfringens* in the genital canal more often than not has no clinical significance (Butler, 1941; Ramsay, 1949). However, in patients who develop signs of postabortal or puerperal sepsis the detection of *C. perfringens* in the genital tract, especially from intracervical swabs, may be of the greatest importance. The simplest method for the rapid detection of *C. perfringens* is the demonstration of typical gram-positive capsulate rods in direct smears from the cervical os or canal. Leucocytes are plentiful in the early stages of infection and most of them are intact, but in the later stages extensive destruction of leucocytes is evident. At no stage does phagocytosis of bacilli occur. The detection of non-capsulate strains of *C. perfringens* in the uterine contents or the vagina is usually of no significance (Butler, 1943, 1945; Hill, 1964).

Culture from pathological material

The most important step in the identification of the infecting organisms is culture, and this should include cultures for both aerobes and anaerobes. The object is to identify the bacterial species present in the wound, and to assess their relative numbers and significance (Lindsey, 1959). For this reason, primary direct plating of the pathological material is essential. Preliminary enrichment without recourse to direct plating may give a completely false impression of the relative importance of the anaerobes ultimately isolated. Full accounts of the methods of isolation and identification of the anaerobes have been published by Willis (1977) and Willis and Phillips (1983, 1986); see also Chapter 12. Clostridial systematics are considered by Willis (1969; 1983).

The following method of examination of wounds for clostridia is recommended (Fig. 18.1); it is based on that of Hayward (1945). This procedure aims at obtaining all the relevant organisms in pure culture. However, its full implementation is not always necessary, and *must never delay the issue of preliminary reports, which can usually be made with confidence (and by telephone) in less than 24 h.*

1. Inoculate exudate or tissue on to: (a) aerobic and aerobic horse-blood agar; (b) anaerobic egg-yolk agar, with and without added neomycin; (c) anaerobic egg-yolk agar. Lactose–egg-yolk–milk agar cultures give much more definitive information than those on plain egg-yolk plates but require more experience for their interpretation. The complex egg medium is included in fig. 18.1;

2. Inoculate material into four universal containers of cooked meat medium (or other suitable anaerobic broth) and heat to 80°C for 5, 10, 15 and 20 min. The primary object here is to isolate *C. novyi* and *C. septicum*, the enriched cultures being subcultured after incubation for 24 and 48 h to aerobic and anaerobic fresh blood agar and to

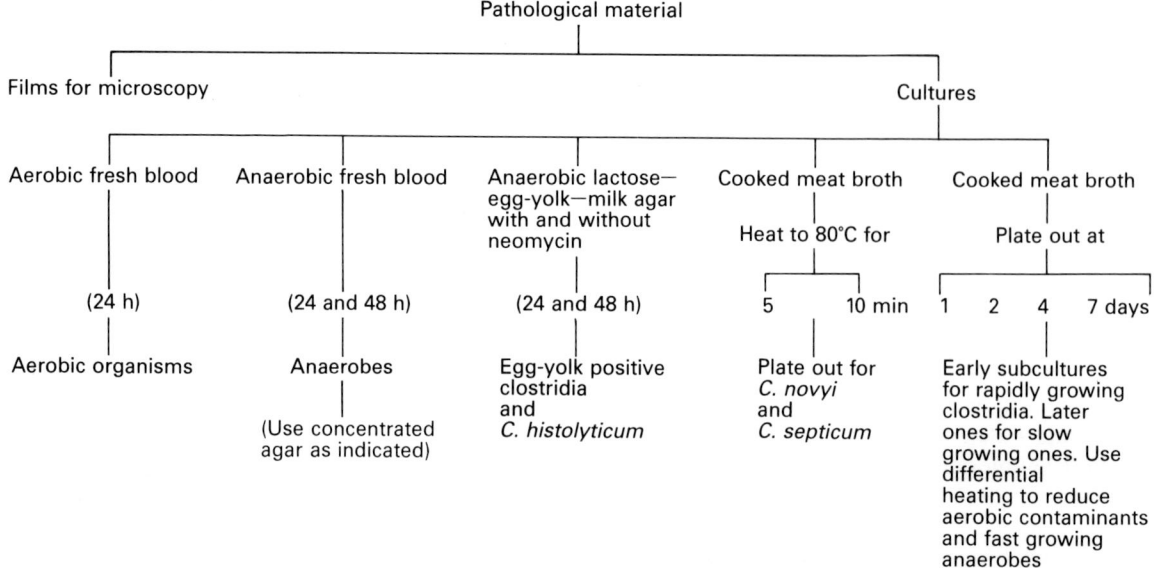

Fig. 18.1 Scheme for the isolation of clostridia from wounds (adapted from Willis, 1977)

aerobic and anaerobic egg-yolk agar plates, neomycin being included in the latter medium for anaerobic incubation;

3. Inoculate material into cooked meat broth (or other suitable anaerobic broth) for anaerobic incubation. This enrichment broth is subcultured to plates, as in 1, above, at 1, 2, 4 and 7 days. It is useful to make these subcultures in duplicate, the inoculum in one case being heated to 80°C for 10 min before transfer.

Pasteurization methods may be supplemented or even replaced by ethanol treatment of pathological material or of cultures;

Note: To assist in distinguishing between the growth of anaerobes and facultative organisms in the non-selective anaerobic plate cultures, paired discs of gentamicin (10 μg) and metronidazole (5 μg) are placed about 15 mm apart in the area of primary inoculation before incubation.

It is often helpful to use concentrated agar plates for anaerobic culture to prevent swarming growth; these plates are best used in addition to those mentioned above. The procedures noted in 1 and 2 are carried out to isolate particular species. The egg-yolk agar plates indicate the presence of *C. perfringens*, *C. bifermentans*, *C. sordellii*, *C. novyi* and *C. sporogenes*. An incubation period of 48 h is required for full development of the cultural characters of the last two clostridia. If an anaerobic cabinet is not available a separate anaerobic jar should be used for examining cultures at 24 h, since brief exposure to air at this time may prevent further development of tiny, and as yet unrecognizable, colonies of *C. novyi*. If the amount of specimen submitted for examination is limited, step 2 may be omitted from the above scheme. Clearly, growth on the aerobic control plates will provide essential information about the associated facultative flora. However, it is less well recognized that some 'facultatively aerobic' clostridia, notably *C. histolyticum*, *C. tertium* and *C. carnis* produce small surface colonies after incubation in aerobic conditions.

Inoculation of pathological material directly into animals is often successful in separating pathogenic from non-pathogenic clostridia, although it is considered that purification is best accomplished by *in vitro* culture methods, and that animal inoculation experiments are most usefully reserved for the demonstration of pathogenicity or toxigenicity of pure cultures.

Blood cultures

Blood cultures are often positive, especially in cases of *C. perfringens* and *C. septicum* gas gangrene, although *C. perfringens* bacteraemia can occur in the absence of overt infection (Butler, 1937; Rathbun, 1968; Ellner and O'Donnell, 1969). It is therefore important that blood cultures should be made, suitable media being cooked meat broth, glucose broth and Fastidious Anaerobe Broth (Lab M). These are incubated anaerobically and are dealt with in the usual way.

Other laboratory diagnostic methods

In an attempt to expedite the early diagnosis of clostridial wound infections, McClean and his colleagues (McClean et al., 1943; McClean and Rogers, 1943, 1944) sought to demonstrate the presence of bacterial toxins in the oedema fluid and infected tissues of guinea-pigs with experimental clostridial myonecrosis. They were regularly able to detect hyaluronidase in the material from animals infected with *C. perfringens*, *C. septicum* and with some strains of *C. novyi*. Lecithinase C activity could usually be demonstrated in *C. perfringens* infections, but rarely in those due to *C. novyi*, while *C. septicum* haemolysin was not demonstrable. Although these findings were not substantiated by the subsequent studies of MacLennan and Macfarlane (1945) for gas gangrenous infections in man, Noyes and Easterling (1967) reported successful detection of clostridial haemolysins in the wound exudate from goats experimentally infected with *C. perfringens*, *C. novyi* and *C. septicum*.

Unlike its use in non-clostridial anaerobic sepsis, direct gas–liquid chromatographic analysis of pathological material has scarcely been explored in gas gangrenous infections. Indeed, no planned study seems to have been carried out in human infections, doubtless a reflection of the scarcity of relevant clinical material. References made to the likely value of chromatography in this setting are encouraging, especially in the early examination of blood cultures for clostridial bacteraemia (Phillips et al., 1976; Sondag et al., 1980; Reig et al., 1981; Edson et al., 1982). Unpublished findings from my laboratory on two cases of clostridial infection are relevant. Direct chromatography of exudate from a patient with fatal postinjectional *C. perfringens* gas gangrene gave a tracing typical of the organism, with an acetic and a large butyric peak; exudate from the mid-thigh amputation stump of a diabetic patient, which was colonized with *C. bifermentans*, gave an identifiable *bifermentans/sordellii* tracing.

Management

Prophylaxis

Surgery

Surgery to the at-risk wound is by far the most important prophylactic measure. Radical excision is imperative; all damaged muscle is removed leaving only well-vascularized tissue and eliminating most of the infecting or contaminating organisms. Irrigation of the wound assists in removing blood clot and fragments of necrotic tissue and foreign material, and repair of arterial injuries re-establishes an adequate circulation to the part. Closure of the wound is delayed for 5–6 days until it is clear that the wound is free from infection. Not only must surgery be adequate, but it must be undertaken as early as possible, consistent with management of traumatic shock. Delays in surgical intervention greatly increase the risk of a contaminated wound becoming the seat of active infection, while the later surgery is applied to patients with an established infection the greater are the risks that the infection will terminate fatally.

The efficacy of early surgery is substantiated by experience during modern military operations. During the Korean War, Howard and Inui (1954) showed that severe clostridial infections in battle casualties were rare when early and adequate surgical treatment was available; Latta (1951) reported only three cases among 1850 wounded. During the military operations in Borneo (1963–65) there was no reported gas gangrene among 119 wounded (Wheatley, 1967) and in Vietnam, Moffat (1967) recorded only two cases among 60 wounded personnel. During the Falklands conflict, gas gangrene did not occur at all among British forces fighting ashore, although Argentine personnel were afflicted (Shouler, 1983; Lancet, 1984). This difference is attributable to the primary suture of entry and exit wounds, and administration of oral tetracycline practiced on Argentine troops, compared with the British practice of wide excision of penetrating wounds, delayed closure, and systemic β-lactam antibiotic therapy.

Clearly, the extent of prophylactic surgical excision of a wound must be tempered by a regard to the future function of the part, so that in some regions, such as the thigh, complete debridement may not be possible if important structures are to be preserved. This consideration may explain, at least in part, the significant differences in the incidence of gas gangrene in different anatomical sites. Jeffrey and Thomas (1944) concluded that the anatomical situation of a wound exerts its effect by determining the extent to which excision is practicable; wounds of the buttock, for example, which usually permit reasonably easy excision, are far less likely to become the seat of a fatal or serious infection than wounds of the thigh, where radical excision is far more difficult.

Antimicrobial prophylaxis

Surgical debridement is combined with local and parenteral antimicrobial therapy. One appropriate prophylactic regimen utilizing benzylpenicillin would be 2 million units daily given intramuscularly in divided doses (Garrod et al., 1973). Metronidazole is the prophylactic agent of choice, and is commenced preoperatively with an intravenous dose of 500 mg; this dose is repeated 8-hourly for 24 h, and the schedule is then reduced to a 12-hourly one. At any convenient time intravenous administration of metronidazole may be replaced by the rectal route (1 g, 12-hourly) or the oral route (400 mg 6-hourly). Whatever antimicrobial schedule is chosen, it should commence before surgical toilet is undertaken. There appear to be few indications for the use of prophylactic gas gangrene antitoxin; indeed, gas gangrene antitoxin is no longer commercially available in the United Kingdom.

It is pertinent to recall here that patients submitted to elective lower limb surgery are at risk from endogenously derived gas gangrene, especially if there is obliterative arterial disease. Prophylaxis in these patients is provided by thorough preoperative skin preparation (Ayliffe and Lowbury, 1969), and by the use of peroperative antimicrobial prophylaxis with metronidazole or benzylpenicillin; prophylaxis for 48 h is usually sufficient.

Treatment

In the treatment of established gas gangrene, and in the absence of a hyperbaric oxygen facility, less consideration can be given to the preservation of 'life *and* limb', the condition demanding surgical intervention that is sufficiently radical to save life. 'The surgical measures demanded in true gas gangrene consist of an uncompromising excision of gangrenous and infected muscle tissue. Where one muscle, or one group of muscles, is involved, that muscle, or muscle-group, must be extirpated from origin to insertion ... In cases of segmental gangrene, where the whole limb or a segment of a limb is involved, and in patients with fulminating gangrene, amputation holds out the best hope of saving life' (Medical Research Council, 1943).

Hyperbaric oxygen

The treatment of established gas gangrene has been revolutionized in recent years by the introduction of hyperbaric oxygen therapy. The early pioneer reports of its value in the management of gas gangrene by Brummelkamp and his colleagues (1961, 1963) and Smith *et al.* (1962) were fully substantiated by subsequent experience (Irvin and Smith, 1968; Johnson *et al.*, 1969; Slack *et al.*, 1969; Rodin *et al.*, 1972; Jackson and Waddell, 1973; Schweigel and Shim, 1973; Holland *et al.*, 1975; British Medical Journal, 1978; Tonjum *et al.*, 1980; Ellis and Mandal, 1983; Unsworth and Sharp, 1984). The rationale of hyperbaric oxygen therapy is simply that it aims to increase the oxygen concentration in the immediate vicinity of an active clostridial infection, so that anaerobic growth and toxin production are prevented. The effects of hyperbaric oxygen on micro-organisms and the factors affecting the response of clostridial infection to hyperbaric oxygen therapy were reviewed by Gottlieb (1971).

Surgery and antimicrobials

The modern management of gas gangrene is along the following lines. When the patient is admitted to the unit, any sutures are removed from the wound, which is widely opened; no surgical excision is performed at this time. Antimicrobial therapy is commenced after collection of pathological specimens; Slack *et al.* (1969) advocated a combination of ampicillin 500 mg intramuscularly 6-hourly and benzylpenicillin 1 million units intramuscularly 6-hourly for at least 5 days. These days the preferred alternative to benzylpenicillin is metronidazole 500 mg intravenously 8-hourly for 24 h, then 12-hourly for 4 days. Treatment sessions with hyperbaric oxygen are then commenced. Since hyperbaric oxygen drenching does not rejuvenate necrotic tissue, surgical toilet of the wound is performed as soon as the patient's general condition has improved. There is no place for the use of gas gangrene antitoxin in this regimen.

Uterine gas gangrene

The treatment of uterine gas gangrene involves the immediate application of a number of measures —parenteral metronidazole therapy, treatment of shock, uterine curettage and management of renal failure. Although immediate curettage is desirable, it should not be attempted until the patient has recovered sufficiently from shock, and restoration of blood loss and intravenous metronidazole therapy have begun. Hysterectomy may be necessary. As with gas gangrene at other sites, hyperbaric oxygen drenching is of first importance in the treatment of uterine infections (Hanson *et al.*, 1966).

References

Ahmad F J, Darrell J H. (1976). Significance of the isolation of *Clostridium welchii* from routine blood cultures. *Journal of Clinical Pathology* 29: 185–186.
Aldrete J S, Judd E S. (1965). Gas gangrene: a complication of elective abdominal surgery. *Archives of Surgery* 90: 745–754.
Altemeier W A. Furste W L. (1949). Studies in virulence of *Clostridium welchii*. *Surgery* 25: 12–19.
Altemeier W A, Gibbs E W. (1944). Bacterial flora of fresh accidental wounds. *Surgery, Gynecology and Obstetrics* 78: 164–168.
Anderson C B, Marr J M, Jaffe B M. (1972). Anaerobic streptococcal infections simulating gas gangrene. *Archives of Surgery* 104: 186–189.
Ayliffe G A J, Lowbury E J L. (1969). Sources of gas gangrene in hospital. *British Medical Journal* 2: 333–337.
Baker J K. (1939). *Clostridium welchii* septicaemia and peritonitis in the puerperium. *Lancet* 2: 646–647.
Bartizal F J, Pacheco J C, Malkasian G D, Washington J A. (1974). Microbial flora found in the products of conception in spontaneous abortions. *Obstetrics and Gynecology* 43: 109–112.
Bartlett J G et al. (1977). Quantitative bacteriology of the vaginal flora. *Journal of Infectious Diseases* 136: 271–277.
Baugher A H. (1914). The *Bacillus aerogenes capsulatus* in blood-cultures with recoveries. *Journal of the American Medical Association* 62: 1153–1155.

Bekassy A N, Garwicz S, Wiebe T. (1984). Fulminating clostridial septicemia in children treated for lymphoproliferative disorders. *Scandinavian Journal of Infectious Diseases* 16: 157–159.
Berggren R B, Batterton T D, McArdle G, Erb, W H. (1964). Clostridial myositis after parenteral injections. *Journal of the American Medical Association* 188: 1044–1048.
Berkow S G. Tolk N R. (1923). Ischiorectal abscess followed by gas gangrene; gas gangrene following trauma. Report of two cases. *Journal of the American Medical Association* 80: 1689–1691.
Bessman A N, Wagner W. (1975). Nonclostridial gas gangrene: report of 48 cases and review of the literature. *Journal of the American Medical Association* 233: 958–963.
Bornstein D L, Weinberg A N, Swartz M N, Kunz L J. (1964). Anaerobic infections—review of current experience. *Medicine* 43: 207–232.
Bourns H K. (1946). A case of metastatic gas gangrene. *British Medical Journal* 1: 570–571.
Bowie J H. (1956). Gas gangrene after injection. *Lancet* 2: 997–998.
Bowlby A. (1919). British military surgery in the time of Hunter and in the Great War. *Lancet* 1: 285–293.
Bradbury S M. (1985). Incontinence pads and clostridium infection. *Journal of Hospital Infection* 6: 115.
Braithwaite P A, Challis D C, McCartney P W. (1982). Clostridial myonecrosis after resection of skin tumours in an immunosuppressed patient. *Medical Journal of Australia* 1: 515, 518–519.
Brewster E S. (1962). Gas gangrene following an intramuscular injection of concentrated liver extract. Report of case with fatal termination by fulminating *Clostridium perfringens* bacillus septicaemia. *Journal of the American Medical Association* 181: 901–903.
Brightmore T, Greenwood T W. (1974). The significance of tissue gas and clostridial organisms in the differential diagnosis of gas gangrene. *British Journal of Clinical Practice* 28: 43–50.
British Medical Journal. (1964). Clostridial infection: 2: 1150–1151.
(1967). Postoperative gas gangrene. 2: 68–69.
(1968). Gas gangrene from adrenaline. 1: 721.
(1978). Hyperbaric oxygen. 1: 1012.
Brown R K, Milch E. (1948). Gallbladder gas gangrene. *Gastroenterology* 10: 626–633.
Brummelkamp W H, Boerema I, Hoogendyk L. (1963). Treatment of clostridial infections with hyperbaric oxygen drenching. A report on 26 cases. *Lancet* 1: 235–238.
Brummelkamp W H, Hogendijk J, Boerema I. (1961). Treatment of anaerobic infections (clostridial myositis) by drenching the tissues with oxygen under high atmospheric pressure. *Surgery* 49: 299–302.
Bullock W E, Cramer W. (1919). On the mechanism of bacterial infection, with reference to gas gangrene. *6th Report of the Imperial Cancer Research Fund* 23–69.

Butler H M. (1937). *Blood cultures and their significance.* Churchill, London.

—— (1941). The bacteriological diagnosis of severe *Clostridium welchii* infection following abortion. *Medical Journal of Australia,* 1: 33–38.

—— (1943). The pathogenicity of washed *Clostridium welchii* and the mode of development of *Clostridium welchii* infections in man. *Medical Journal of Australia* 2: 224–227.

—— (1945). Bacteriological studies of *Clostridium welchii* infections in man. *Surgery, Gynecology and Obstetrics* 81: 475–486.

Cairns H, Calvert C A, Daniel P, Northcroft G B. (1947). Complications of head wounds, with special reference to infection. *British Journal of Surgery,* War Surgery Suppl I, 198–243.

Chopra I B, Mukherjee P. (1970). Non-clostridial crepitant cellulitis. *Journal of the Indian Medical Association* 54: 177–179.

Collier P E, Diamond D L, Young J C. (1983). Nontraumatic *Clostridium septicum* gangrenous myonecrosis. *Diseases of the Colon and Rectum* 26: 703–704.

Committee upon Anaerobic Bacteria and Infections. (1919). Special Report Series of the Medical Research Committee, No. 39. London, HMSO.

Cooke W T et al. (1945). Clostridial infections in war wounds. *Lancet* 1: 487–493.

Cooper E V. (1946). Gas gangrene following injection of adrenaline. *Lancet* 1: 459–461.

Corbishely C M. (1977). Microbial flora of the vagina and cervix. *Journal of Clinical Pathology* 30: 745–748.

Culbertson W R. (1958). Acute nonclostridial crepitant cellulitis. *Archives of Surgery, Chicago* 77: 462–468.

Cutler E C. Sandusky W R. (1944). Treatment of clostridial infections with penicillin. *British Journal of Surgery* 32: 168–176.

Davis D M. et al. (1979). Survival after major burn complicated by gas gangrene, acute renal failure, and toxic myocarditis. *British Medical Journal* 1: 718–719.

Dawbarn R Y, Williams B. (1938). Three cases of *Cl. welchii* infection following abortion. *British Medical Journal* 2: 279–281.

Desmond A M. (1947). Surgical emphysema due to compressed air. *British Medical Journal* 1: 842–843.

Dhayagude R G, Purandare N M. (1949). Studies on anaerobic wound infection. *Indian Journal of Medical Research* 37: 283–292.

Doehner G A, Klinges K G, Pisani B J. (1960). The X-ray diagnosis of gas gangrene of the uterus. *American Journal of Obstetrics and Gynecology* 79: 542–544.

Edson R S, Rosenblatt J E, Washington J A, Stewart J B. (1982). Gas–liquid chromatography of positive blood cultures for rapid presumptive diagnosis of anaerobic bacteremia. *Journal of Clinical Microbiology* 15: 1059–1061.

Elliot-Smith A, Ellis H. (1957). *Clostridium welchii* infection following cholecystectomy. *Lancet* 2: 723.

Ellis M E, Mandal B K. (1983). Hyperbaric oxygen treatment: 10 years' experience of a Regional Infectious Diseases Unit. *Journal of Infection* 6: 17–28.

Ellner P D, O'Donnell E D. (1969). Non-fatal *Clostridium perfringens* bacteraemia. *Journal of the American Geriatrics Society* 17: 644–647.

Engeset J, MacIntyre J, Smith G, Welch G. (1973). *Clostridium welchii* infection: an unusual case. *British Medical Journal* 1: 91–92.

Evans D G. (1945a). The *in-vitro* production of alpha-toxin, theta-haemolysin and hyaluronidase by strains of *Cl. welchii* type A, and the relationship of *in-vitro* properties to virulence for guinea-pigs. *Journal of Pathology and Bacteriology* 57: 75–85.

—— (1945b). Gas gangrene. *Lancet* 2: 478.

Evans D G, Miles A A, Niven J F. (1948). The enhancement of bacterial infections by adrenaline. *British Journal of Experimental Pathology* 29: 20–39.

Feeney R E. (1947). Wound bacteriology in the Philippines. *Military Surgeon* 100: 514–518.

Filler R M, Griscom N T, Pappas A. (1968). Post-traumatic crepitation falsely suggesting gas gangrene. *New England Journal of Medicine* 278: 758–761.

Fitzpatrick D J, Turnbull P C B, Keane C T, English L F. (1979). Two gas-gangrene-like infections due to *Bacillus cereus*. *British Journal of Surgery* 66: 577–579.

Fray R E, Davis T P, Brown E A. (1984). *Clostridium welchii* infection after amniocentesis. *British Medical Journal* 1: 901–902.

Gabriel W B, Kingsbury A N. (1922). A case of acute anaerobic (*B. welchii*) infection of uterine fibroids, with reference to the aetiology of necrobiosis. *Lancet* 1: 172–175.

Garrod L P, Lambert H P, O'Grady F, Waterworth P M. (1973). *Antibiotics and chemotherapy,* 4th edn. Churchill Livingstone, Edinburgh.

Gatt D. (1985). Non-traumatic metastatic clostridial myonecrosis in a seventeen-year-old. *British Journal of Surgery* 72: 240–241.

Gazzaniga A B. (1967). Non-traumatic, clostridial, gas gangrene of the right arm and adenocarcinoma of the cecum. Report of a case. *Diseases of the Colon and Rectum* 10: 298–300.

Gledhill T. (1982). Gas gangrene following cholecystectomy. *British Journal of Clinical Practice* 36: 23–25.

Gorbach S L, Thadepalli H. (1975). Isolation of *Clostridium* in human infections: evaluation of 114 cases. *Journal of Infectious Diseases,* 131 Suppl: S81–S85.

Gordon S. (1936–37). *B. welchii* infection complicating conservatively treated appendicial abscess. *British Journal of Surgery* 24: 399–401.

Gottlieb S F. (1971). Effect of hyperbaric oxygen on microorganisms. *Annual Review of Microbiology* 25: 111–152.

Greenberg L. (1945). A bacteriological study of war wounds in Italy. *Canadian Medical Services Journal* 2: 133–141.

Gye R, Rountree P M, Lowenthal J. (1961). Infection of surgical wounds with *Clostridium welchii. Medical Journal of Australia* **1**: 761–764.

Hamilton J. (1944–45). Anaerobic infection in war wounds. *Canadian Medical Services Journal* **2**: 387–407.

Hammerschlag M R, et al. (1978). Anaerobic microflora of the vagina in children. *American Journal of Obstetrics and Gynecology* **131**: 853–856.

Hanson G C, Slack W K, Chew H E R, Thomas D A. (1966). Clostridial infection of the uterus—a review of treatment with hyperbaric oxygen. *Postgraduate Medical Journal* **42**: 499–505.

Harney C H. (1939). Gas gangrene following therapeutic injections. *Annals of Surgery* **109**: 304–308.

Harvey P W, Purnell G V. (1967). Fatal case of gas gangrene associated with intramuscular injections. *British Medical Journal* **1**: 744–746.

Hauschild A H W. (1971). *Clostridium perfringens* toxins types B, C, D, and E. In *Microbial toxins*, vol 2A, p. 159. Edited by Kadis S et al. Academic Press, New York.

Hayward N J. (1945). The examination of wounds for clostridia. *Proceedings of the Association of Clinical Pathologists* **1**: 5–17.

Hayward N J, Pilcher R. (1945). Anaerobic streptococcal myositis. *Lancet* **2**: 560–562.

Heineman H S, Braude A I. (1961). Shock in infectious diseases. *Disease-a-Month* October: p. 24.

Hill A M. (1936). Post-abortal and puerperal gas gangrene. *Journal of Obstetrics and Gynaecology of the British Empire* **43**: 201–251.

(1950). The changing face of obstetric infection. *Medical Journal of Australia* **2**: 782–787.

(1964). Why be morbid? Paths of progress in the control of obstetric infection, 1931 to 1960. *Medical Journal of Australia* **1**: 101–111.

Hill A M, Butler H M. (1948). The diagnosis, prevention and treatment of puerperal infection. *Medical Journal of Australia* **1**: 227–235.

Himal H S, Duff J H. (1967). Endogenous gas gangrene: a report of three cases. *Canadian Medical Association Journal* **97**: 1541–1543.

Hitchcock C R, Demello F J, Haglin J J. (1975). Gangrene infection. New approaches to an old disease. *Surgical Clinics of North America* **55**: 1403–1410.

Holland J A, Hill G B, Wolfe W G, Osterhout S, Saltzman H A, Brown I W. (1975). Experimental and clinical experience with hyperbaric oxygen in the treatment of clostridial myonecrosis. *Surgery, St Louis* **77**: 75–85.

Howard J M, Inui K K. (1954). Clostridial myositis—gas gangrene. Observations of battle casualties in Korea. *Surgery* **36**: 1115–1118.

Irvin T T, Smith G. (1968). Treatment of bacterial infections with hyperbaric oxygen. *Surgery* **63**: 363–367.

Isenberg A N. (1966). *Clostridium welchii* infection. A clinical evaluation. *Archives of Surgery* **92**: 727–731.

Ispolatovskaya M V. (1971). Type A *Clostridium perfringens* toxin. In *Microbial toxins*, vol 2A. p. 109. Edited by Kadis S et al. Academic Press, New York.

Jackson R W, Waddell J P. (1973). Hyperbaric oxygen in the management of clostridial myonecrosis (gas gangrene). In *Clinical orthopedics and related research*, No. 96. p. 271. Edited by Urist M R. Lippincott, Philadelphia.

Jeffrey J S, Thomson S. (1944). Gas gangrene in Italy. A study of 33 cases treated with penicillin. *British Journal of Surgery* **32**: 159–167.

Johnson J T, Gillespie T E, Cole J R, Markowitz H A. (1969). Hyperbaric oxygen therapy for gas gangrene in war wounds. *American Journal of Surgery* **118**: 839–843.

Jones L E, Wirth W A, Farrow C C. (1960). Clostridial gas gangrene and septicemia complicating leukemia. *Southern Medical Journal* **53**: 863–866.

Kapusta M A, Mendelson J, Niloff P. (1972). Non-traumatic gas gangrene: report of a case with long term survival. *Canadian Medical Association Journal* **106**: 863, 866.

Kass E H, Lichstein H C, Waisbren B A. (1945). Occurrence of hyaluronidase and lecithinase in relation to virulence in *Clostridium welchii. Proceedings of the Society for Experimental Biology and Medicine* **58**: 172–175.

Keen W W. (1915). The contrast between the surgery of the Civil War and that of the present war. *New York Medical Journal* **150**: 817–824.

Kemp F H. (1945). X-rays in diagnosis and location of gas-gangrene. *Lancet* **1**: 332–336.

Kemp F H, Vollum R L. (1946). Anaerobic cellulitis due to actinomyces, associated with gas production. *British Journal of Radiology* **19**: 248–249.

Keppie J. (1951). The pathogenicity of the spores of *Clostridium botulinum. Journal of Hygiene* **49**: 36–45.

Kielley W W, Meyerhof O. (1950). Studies on adenosinetriphosphatase of muscle. III. The lipoprotein nature of the magnesium-activated adenosinetriphosphatase. *Journal of Biological Chemistry* **183**: 391–401.

Knutson L. (1983). Postoperative gas gangrene in abdomen and in extremity. *Acta Chirurgica Scandinavica* **149**: 567–571.

Koons T A, Boyden G M. (1961). Gas gangrene from parenteral injection. *Journal of the American Medical Association* **175**: 46–47.

Koransky J R, Stargel M D, Dowell V R. (1979). *Clostridium septicum* bacteremia. Its clinical significance. *American Journal of Medicine* **66**: 63–66.

Kowen D E, Hanslo D H, Botha P L, Davey D A. (1979). Incidence of aerobic and anaerobic infection in patients with incomplete abortion. *South African Medical Journal* **55**: 129–132.

Lacey C G, Futoran R, Morrow C P. (1976). *Clostridium perfringens* infection complicating chemotherapy for choriocarcinoma. *Obstetrics and Gynecology* **47**: 337–341.

Lancet. (1984). Remembering gas gangrene **2**: 851–852.

Laszlow G, Elo G. (1983). The risk of malignant edema following lower limb amputation for ischemic gangrene. *Archives of Orthopedic and Traumatic Surgery* **102**: 49–50.

Latta R M. (1951). Management of battle casualties from Korea. *Lancet* **1**: 228–231.

Lee A B, Waffle C M, Trebbin W M, Solomon R J. (1981). Clostridial myonecrosis. Origin from an obturator hernia in a dialysis patient. *Journal of the American Medical Association* **246**: 1232–1233.

Lee H A, Hills E A, Brundenell J M. (1966). Management of abortion complicated by *Clostridium welchii* infection and acute renal failure. *British Journal of Clinical Practice* **20**: 169–175.

Lewis V L, Myers M B, Griffith B H. (1978). Early diagnosis of crepitant gangrene caused by *Bacteroides melaninogenicus*. *Plastic and Reconstructive Surgery* **62**: 276–279.

Lindner J G E M, Plantema F H F, Hoogkamp-Korstanje J A A. (1978). Quantitative studies of the vaginal flora of healthy women and of obstetric and gynaecological patients. *Journal of Medical Microbiology* **11**: 233–241.

Lindsey D. (1959). Gas gangrene. Clinical inferences from experimental data. *American Journal of Surgery* **97**: 582–592.

de Louvois J, Harley R, Stanley V C. (1975). Microbial flora of the lower genital tract during pregnancy: relationship to morbidity. *Journal of Clinical Pathology* **28**: 731–735.

Lowbury E J L, Lilly H A. (1958). The sources of hospital infection of wounds with *Clostridium welchii*. *Journal of Hygiene*, **56**: 169–182.

Lowbury E J L, Lilly H A, Bull J P. (1964). Methods for disinfection of hands and operation sites. *British Medical Journal* **2**: 531–536.

Lowry K F, Curtis G M. (1947). Diagnosis of clostridial myositis. *American Journal of Surgery* **74**: 752–757.

Macfarlane M G. (1955). On the biochemical mechanism of action of gas-gangrene toxins. In *Mechanisms of microbial pathogenicity*, 5th Symposium of the Society for General Microbiology, p. 57. Edited by Howie J W, O'Hea A J. Cambridge University Press, Cambridge.

Macfarlane M G, Datta N. (1954). Observations on the immunological and biochemical properties of liver mitochondria with reference to the action of *Clostridium welchii* toxin. *British Journal of Experimental Pathology* **35**: 191–202.

Macfarlane R G, MacLennan J D. (1945). The toxaemia of gas-gangrene. *Lancet* **2**: 328–331.

MacLennan J D. (1943a). Streptococcal infection of muscle. *Lancet* **1**: 582–584.

(1943b,c,d). Anaerobic infections of war wounds in the Middle East. *Lancet* **2**: 63–66, 94–99, 123–126.

(1944). Anaerobic infections in Tripolitania and Tunisia. *Lancet* **1**: 203–207.

(1962). The histotoxic clostridial infections of man. *Bacteriological Reviews* **26**: 177–276.

MacLennan J D, Macfarlane R G. (1945). Toxin and antitoxin studies of gas gangrene in man. *Lancet* **2**: 301–305.

Magarey R, Cleland J B, Sleeman J G. (1927). Gas infections of the uterus with jaundice due to *Bacillus welchii* following abortions. *Medical Journal of Australia* **1**: 787–789.

Maguire W B, Langley N F. (1967). Gas gangrene following an adrenaline-in-oil injection into the left thigh with survival. *Medical Journal of Australia* **1**: 973–975.

Mariona F G, Ismail M A. (1980). *Clostridium perfringens* septicemia following cesarean section. *Obstetrics and Gynecology* **56**: 518–521.

Marshall V, Sims P. (1960). Gas gangrene after the injection of adrenaline-in-oil, with a report of three cases. *Medical Journal of Australia* **2**: 653–656.

Marty A T, Filler R M. (1969). Recovery from non-traumatic localized gas gangrene and clostridial septicaemia. *Lancet* **2**: 79–81.

McClean D, Rogers H J. (1943). Early diagnosis of wound infection with special reference to mixed infections. *Lancet* **1**: 707–708.

(1944). Detection of bacterial enzymes in infected tissues. *Lancet* **2**: 434–435.

McClean D, Rogers H J, Williams B W, Hale C W. (1943). Early diagnosis of wound infections with special reference to gas gangrene. *Lancet* **1**: 355–360.

McDonald E J. (1947). The clinical significance of gas shadows in X-ray examinations of compound wounds. *Medical Annals of the District of Columbia* **16**: 595–598, 648.

McNally M J, Crile G. (1964). Diagnosis and treatment of gas gangrene of the abdominal wall. *Surgery, Gynecology and Obstetrics* **118**: 1046–1050.

Mead P B. (1978). Cervical-vaginal flora of women with invasive cervical cancer. *Obstetrics and Gynecology* **52**: 601–604.

Medical Research Council. (1943). Notes on gas gangrene prevention, diagnosis and treatment. *War Memorandum*. No. 2, 2nd edn London, HMSO.

Merson M H, Dowell V R. (1973). Epidemiologic, clinical and laboratory aspects of wound botulism. *New England Journal of Medicine* **289**: 1005–1010.

Millar W M. (1932). Gas gangrene in civil life. *Surgery, Gynecology and Obstetrics* **54**: 232–238.

Moffat W C. (1967). Recent experience in the management of battle casualties. *Journal of the Royal Army Medical Corps* **113**: 25–30.

Monafo W W, Brentano L, Gravens D L, Kempson R, Moyer C A. (1966). Gas gangrene and mixed clostridial infections of muscle complicating deep thermal burns. *Archives of Surgery* **92**: 212–221.

Morbidity and Mortality Weekly Report. (1982). Wound botulism associated with parenteral cocaine abuse—New York City **31**: 87–88.

Mzabi R, Himal H S, MacLean L D. (1975). Gas gangrene of the extremity: the presenting clinical picture in perforating carcinoma of the caecum. *British Journal of Surgery* **62**: 373–374.

Nahir A M, Hashmonai M, Merzbach D, Scharf J. (1978).

Gas gangrene following intramuscular injection of aqueous solution. *New York State Journal of Medicine* **78**: 1948–1949.

Narula A, Khatib R. (1985). Characteristic manifestations of *Clostridium* induced spontaneous gangrenous myositis. *Scandinavian Journal of Infectious Diseases* **17**: 291–294.

Nichols J B. (1961). A case of massive gas gangrene. *British Medical Journal* **2**: 940.

Nichols R L, Smith J W. (1975). Gas in the wound: what does it mean? *Surgical Clinics of North America* **55**: 1289–1296.

Nora P F, Mousavipour M, Mittelpunkt A, Resenberg M, Laufman H. (1966). Brain as target organ in *Clostridium perfringens* exotoxin toxicity. *Archives of Surgery* **92**: 243–246.

North J P. (1947). Clostridial wound infections and gas gangrene. Arterial damage as a modifying factor *Surgery* **21**: 364–372.

Noyes H E, Easterling R. (1967). Detection of clostridial haemolysin formed *in vivo*. *Journal of Bacteriology* **93**: 1254–1261.

Oakley C L. (1954). Gas gangrene. *British Medical Bulletin* **10**: 52–58.

Parker M T. (1967). Clostridial sepsis. *British Medical Journal* **1**: 698.

Pearson R D, Valenti W M, Steigbigel R T. (1980). *Clostridium perfringens* wound infection associated with elastic bandages. *Journal of the American Medical Association* **244**: 1128–1130.

Peckham C H. (1936). Statistical studies on puerperal infection. II. An analysis of 545 cases of puerperal infection (including a comparison between them and a similar group of cases with normal puerperia). *American Journal of Obstetrics and Gynecology* **31**: 582–597.

Pelfrey T M, Turk R P, Peoples J B, Elliott D W. (1984). Surgical aspects of *Clostridium septicum* septicemia. *Archives of Surgery* **119**: 546–550.

Phillips K D, Tearle P V, Willis A T. (1976). Rapid diagnosis of anaerobic infections by gas–liquid chromatography of clinical material. *Journal of Clinical Pathology* **29**: 428–432.

Plimpton N C. (1964). *Clostridium perfringens* infection. *Archives of Surgery* **89**: 499–507.

Poppel M H, Silverman M. (1941). Gas gangrene of the uterus. *Radiology* **37**: 491–492.

Power R W. (1945a). Gas gangrene with special reference to vascularization of muscles. *British Medical Journal* **1**: 656–658.

(1945b). Gas gangrene. *Lancet* **2**: 477–478.

Proetz D M, Wood L, Park C. (1974). Adenocarcinoma of the colon presenting as *Clostridium septicum* cellulitis of the left thigh. *Southern Medical Journal* **67**: 862–864.

Pulaski E J, Meleney F L, Speath W L C. (1941). Bacterial flora of acute traumatic wounds. *Surgery, Gynecology and Obstetrics* **72**: 982–988.

Pyrtek L J, Bartus S H. (1962). *Clostridium welchii* infection complicating biliary-tract surgery. *New England Journal of Medicine* **266**: 689–693.

Quinn W C, Lord J W, Wade L J. (1942). Gas gangrene of the abdominal wall. *Surgery* **11**: 233–243.

Rains A J H, Ritchie H D. (1977). *Bailey and Love's short practice of surgery*, 17th edn. Lewis, London.

Ramsay A M. (1949). The significance of *Clostridium welchii* in the cervical swab and blood stream in postpartum and postabortal sepsis. *Journal of Obstetrics and Gynaecology of the British Empire* **56**: 247–258.

(1950). The problem of the infected abortion. *Clinical Journal* **79**: 27–37.

Ramsay A M, Bishop I R, Burnett C W F, Stallworthy J: (1948). Discussion on treatment of septic abortion. *Proceedings of the Royal Society of Medicine* **41**: 317–322.

Rathbun H K. (1968). Clostridial bacteremia without haemolysis. *Archives of Internal Medicine* **122**: 496–501.

Reig M, Molina D, Loza E, Ledesma M A, Meseguer M A. (1981). Gas–liquid chromatography in routine processing of blood cultures for detecting anaerobic bacteraemia. *Journal of Clinical Pathology* **34**: 189–193.

Robb-Smith A H T. (1945). Tissue changes induced by *Cl. welchii* type A filtrates. *Lancet* **2**: 362–368.

Roberts K, Johnson W W, Bruckner H S. (1933). The aseptic peritoneal cavity—a misnomer. *Surgery, Gynecology and Obstetrics* **57**: 752–761.

Rodin B, Groeneveld P H A, Boerema I. (1972). Ten years experience in the treatment of gas gangrene with hyperbaric oxygen. *Surgery, Gynecology and Obstetrics* **134**: 579–585.

Rose H D. (1979). Gas gangrene and *Clostridium perfringens* septicemia associated with the use of an indwelling radial artery catheter. *Canadian Medical Association Journal* **121**: 1595–1596.

Roy T E, Hamilton J D, Greenberg L. (1954). Wound contamination and wound infection. *Journal of the Royal Army Medical Corps* **100**: 276–295.

Rubbo S D, Gardner J F. (1968). Intramuscular injections and gas gangrene. *British Medical Journal* **1**: 241–242.

Ruenberg M L, Baker L R I, McBride J A, Sevitt L H, Brain M C. (1967). Intravascular coagulation in a case of *Clostridium perfringens* septicaemia. Treatment by exchange transfusion and heparin. *British Medical Journal* **2**: 271–274.

Russel P B, Roach M J. (1939). *B. welchii* infections in pregnancy: with a review of the literature and a report of seventeen cases. *American Journal of Obstetrics and Gynecology* **38**: 437–447.

Rutherford R. (1938). Abortion associated with broad ligament cysts and *B. welchii* infection. *British Medical Journal* **1**: 173.

Sadusk J F, Manahan C P. (1941). Observations on the occurrence of *Clostridium welchii* in the vagina of pregnant women. *American Journal of Obstetrics and Gynecology* **41**: 856–861.

Saleh M, Bollen S R. (1984). Sucking wound of the knee: not gas gangrene. *British Medical Journal* **2**: 1348.

Salm R. (1944). The occurrence and significance of *Clostridium welchii* in the female genital tract. *Journal of Obstetrics and Gynaecology of the British Empire* **51**: 121–126.

Schweigel J F, Shim S S. (1973). A comparison of the treatment of gas gangrene with and without hyperbaric oxygen. *Surgery, Gynecology and Obstetrics* **136**: 969–974.

Seradge H, Anderson M G. (1980). Clostridial myonecrosis following intra-articular steroid injection. *Clinical Orthopedics* **147**: 207–209.

Shapiro B, Rohman M, Cooper P. (1963). Clostridial infection following abdominal surgery. *Annals of Surgery* **158**: 27–30.

Shouler P J. (1983). The management of missile injuries. *Journal of the Royal Naval Medical Service* **69**: 80–84.

Slack W K, Hanson G C, Chew H E R. (1969). Hyperbaric oxygen in the treatment of gas gangrene and clostridial infection. A report of 40 patients treated in a single person hyperbaric oxygen chamber. *British Journal of Surgery* **56**: 505–510.

Slotnick I J, Stelluto M, Prystowsky H. (1963). Microbiology of the female genital tract. III. Comparative investigation of the cervical flora of parturients receiving either rectal or vaginal examinations. *American Journal of Obstetrics and Gynecology* **85**: 519–526.

Smith G, Sillar W, Norman J N, Ledingham I McA, Bates E H, Scott A C. (1962). Inhalation of oxygen of 2 atmospheres for *Clostridium welchii* infections. *Lancet* **2**: 756–757.

Smith L D S. (1949). Clostridia in gas gangrene. *Bacteriological Reviews* **13**: 233–254.

Smith L D S, George R L. (1946). The anaerobic bacterial flora of clostridial myositis. *Journal of Bacteriology* **51**: 271–279.

Sondag J E, Mariam A, Murray P R. (1980). Rapid presumptive identification of anaerobes in blood cultures by gas-liquid chromatography. *Journal of Clinical Microbiology* **11**: 274–277.

Spann J L, McGill R A. (1957). Clostridial myositis following resection of bowel. Case report. *Annals of Surgery* **146**: 98–104.

Spencer H R. (1922). Acute anaerobic (*B. welchii*) infection of fibroids. *Lancet* **1**: 250–251.

Stammers F A R. (1945). Discussion on the toxaemia of gas gangrene. *Proceedings of the Royal Society of Medicine* **39**: 291–293.

Stephen G M. (1977). Fulminating obstetric gas gangrene. *Journal of the Royal Army Medical Corps* **123**: 45–48.

Sterling J A, Ong T-G, Klaus I G, Goldsmith R. (1954). Gas gangrene of the perineum. *American Journal of Surgery* **87**: 874–876.

Stevenson W F. (1910). *Wounds in war. The mechanism of their production and their treatment*, 3rd edn. Longmans, London.

Stock A H. (1947). Clostridia in gas gangrene and local anaerobic infections during the Italian campaign. *Journal of Bacteriology* **54**: 169–174.

Stoddard J L. (1918). The occurrence and significance of *B. welchii* in certain wounds. *Journal of the American Medical Association* **71**: 1400–1402.

Stratford B C. (1973). Gas gangrene. *Medical Journal of Australia* **2**: 47.

Strawitz J G, Wetzler T F, Marshall J D, Lindberg R B, Howard J M, Artz C P. (1955). The bacterial flora of healing wounds. A study of the Korean battle casualty. *Surgery* **37**: 400–408.

Sussman M. (1958). A description of *Clostridium histolyticum* gas gangrene in the epidemics of Hippocrates. *Medical History* **2**: 226.

Symonds R P, Robertson A G. (1978). *Clostridium welchii* septicaemia after intrauterine caesium insertion. *British Medical Journal* **1**: 754–755.

Tabaqchali S, O'Farrell S, Nash J Q, Wilks M. (1984). Vaginal carriage and neonatal acquisition of *Clostridium difficile*. *Journal of Medical Microbiology* **18**: 47–53.

Taylor G W. (1960). Preventive use of antibiotics in surgery. *British Medical Bulletin* **16**: 51–54.

Teo W S, Balasubramanian P. (1983). Gas gangrene after intramuscular injection of adrenaline. *Clinical Orthopedics* **174**: 206–207.

Thadepalli H, Chan W H, Maidman J E, Davidson E C. (1978). Microflora of the cervix during normal labor and the puerperium. *Journal of Infectious Diseases* **137**: 568–572.

Thomas I J. (1945–46). Fulminating gas gangrene infection following an ischiorectal abscess. *British Journal of Surgery* **33**: 292–295.

Thoresby F P. (1966). 'Cavitation.' The wounding process of the high velocity missile, a review. *Journal of the Royal Army Medical Corps* **112**: 89–99.

Thoresby F P, Darlow H M. (1967). The mechanisms of primary infection of bullet wounds. *British Journal of Surgery* **54**: 359–361.

Thoresby F P, Watts J C. (1967). Gas gangrene in the high-velocity missile wound. *British Journal of Surgery* **54**: 25–29.

Thys J P, Ectors P, Noel P. (1980). Non-traumatic clostridial myositis: an unusual feature of brain death. *Postgraduate Medical Journal* **56**: 501–503.

Tonjum S, Digranes A, Alho A, Gjengsto H, Eidsvik S. (1980). Hyperbaric oxygen treatment in gas-producing infections. *Acta Chirurgica Scandinavica* **146**: 235–241.

Toombs P W. (1932. Report of an additional case of puerperal septicemia due to infection by *Clostridium welchii*. *American Journal of Obstetrics and Gynecology* **24**: 415–421.

Toombs P W, Michelson I D. (1928). *Clostridium welchii*

septicemia complicating prolonged labor due to obstructing myoma of uterus, with report of case. *American Journal of Obstetrics and Gynecology* **15**: 379–389.

Turner F P. (1964). Fatal *Clostridium welchii* septicemia following cholecystectomy. *American Journal of Surgery* **108**: 3–7.

Unsworth I P, Sharp P A. (1984). Gas gangrene. An 11-year review of 73 cases managed with hyperbaric oxygen. *Medical Journal of Australia* **1**: 256–260.

Vaillard L, Rouget J. (1892). Contribution a l'etude du tetanus etiologie. *Annales de l'Institut Pasteur* **6**: 385–435.

Valentine J C. (1956–57). Gas gangrene septicaemia due to carcinoma of the caecum and muscular trauma. *British Journal of Surgery* **44**: 630–632.

Van Beek A, Zook E, Yaw P, Gardner R, Smith R, Glover J L. (1974). Nonclostridial gas-forming infections. *Archives of Surgery* **108**: 552–557.

Warren C P W, Mason B J. (1970). *Clostridium septicum* infection of the thyroid gland. *Postgraduate Medical Journal* **46**: 586–588.

Wheatley P R. (1967). Research on missile wounds—the Borneo operations January 1962—June 1965. *Journal of the Royal Army Medical Corps* **113**: 18–24.

Whyland W A, Levin M N. (1960). Gas gangrene without visible portal of entry. *Annals of Surgery* **99**: 77–79.

Williams R E O, Blowers R, Garrod L P, Shooter R A. (1960). *Hospital infection*. Lloyd Luke, London.

Willis A T. (1969). *Clostridia of wound infection*. Butterworths, London.

(1977). *Anaerobic bacteriology, clinical and laboratory practice*. 3rd edn. Butterworths, London.

(1983). *Clostridium*: the spore-bearing anaerobes. In *Topley and Wilson's Principles of bacteriology, virology and immunity*. 7th edn, vol. 2. p. 442. Edited by Parker M T. Edward Arnold, London.

Willis A T, Phillips K D. (1983). *Anaerobic infections*. Public Health Laboratory Service Monograph Series No 3, 2nd edn. HMSO, London.

(1988). *Anaerobic infections: clinical and laboratory practice*. Public Health Laboratory Service, London.

Wrigley A J. (1930). Puerperal infection by pathogenic anaerobic bacteria. *Proceedings of the Royal Society of Medicine* **23**: 1645–1654.

Wyman A L. (1949). Endogenous gas gangrene complicating carcinoma of colon. Report of a case. *British Medical Journal* **1**: 266–267.

Yangco B G, Germain B F, Deresinski S C. (1982). Fatal gas gangrene following intra-articular steroid injection. *American Journal of the Medical Sciences* **283**: 94–98.

Yudis M, Zucker S. (1967). *Clostridium welchii* bacteraemia: A case report with survival and review of the literature. *Postgraduate Medical Journal* **43**: 487–490.

19

Anaerobic bacteraemia

R C Spencer

Incidence
   Bacteroides
   Clostridia
   Peptococcaceae
   Fusobacteria
   Capnocytophaga
   Bifidobacteria and Eubacteria
The role of mixed bacteraemia
Endocarditis
Clinical features
Isolation techniques
   Media
    Incubation methods
    New techniques
Treatment
Acknowledgements
References

Incidence

Anaerobic bacteraemia is often a complication of pre-existing anaerobic infection. Its true incidence is very difficult to determine, because the optimal techniques for anaerobic isolation are not always used by laboratories and there is a lack of uniformity in anaerobic blood culture technique (Finegold, 1977). While reports on the incidence of anaerobic bacteraemia surfaced in the 1930s (Thompson and Beaver, 1932; Lemierre, 1936; Lemierre and Reilly, 1938; Rosenow and Brown, 1938) and occasionally in the 1950s (Gunn, 1956) it was only after 1960 that the increased incidence and importance of anaerobic septicaemia, with especial reference to the *Bacteroides* spp., was recognized (Bornstein *et al.*, 1964; Gelb and Seligman, 1970; Alpern and Dowell, 1971; Sandford, 1972; Washington, 1971, 1972; Washington and Martin, 1973; Finegold, 1977). Rosenow and Brown (1938) reported six cases of bacteroides bacteraemia, representing 4 per cent of a total of 144 clinical cases of septicaemia. Later studies, from several centres, indicated that anaerobes accounted for 8–11 per cent of all bacteraemias (Washington, 1971; Wilson *et al.*, 1972; Sonnenwirth, 1974; Chow and Guze, 1974; Paisley *et al.* 1978). In the study by Wilson *et al.* (1972), 151 of 360 anaerobic isolates were *Propionibacterium acnes*, only one of which was clinically significant. Some studies by Sandford (1972) and Washington and Martin (1973) reported much higher isolation rates of 27 per cent and 28.8 per cent respectively. However, these studies involved selected patient populations rather than general hospital populations. More recent surveys report an average of 5–13 per cent of bacteraemias as having an anaerobic component (Table 19.1).

Much of the increase 20 years ago was accounted for by the rise in the rate of isolation of *Bacteroides* spp., especially *B. fragilis*. Even in more recent studies, *B. fragilis* accounts for over 50 per cent of anaerobic gram-negative isolates. In many of the earlier studies, organisms were not fully classified, leading authors to label all such cases as 'bacteroides septicaemia'. When full identification is performed, 4–20 per cent are actually *Fusobacterium* spp.; these were especially prevalent in the preantibiotic era when the upper respiratory tract was a common portal of entry (Felner and Dowell, 1971; Washington, 1971; Wilson *et al.*, 1972). The reported incidences of clostridia and anaerobic cocci are much lower, and the frequency has not changed appreciably. Bifidobacteria and eubacteria are rarely encountered.

Fig. 19.1 shows the incidence of anaerobic bacteraemia in England and Wales, as determined by laboratory returns of the PHLS Communicable Disease Surveillance Centre. Out of a return of 125 094 patients with bacteraemia, reported by laboratories from 1975 to 1985, 6928 reports (5.5 per cent) were due to anaerobic organisms. The proportion of

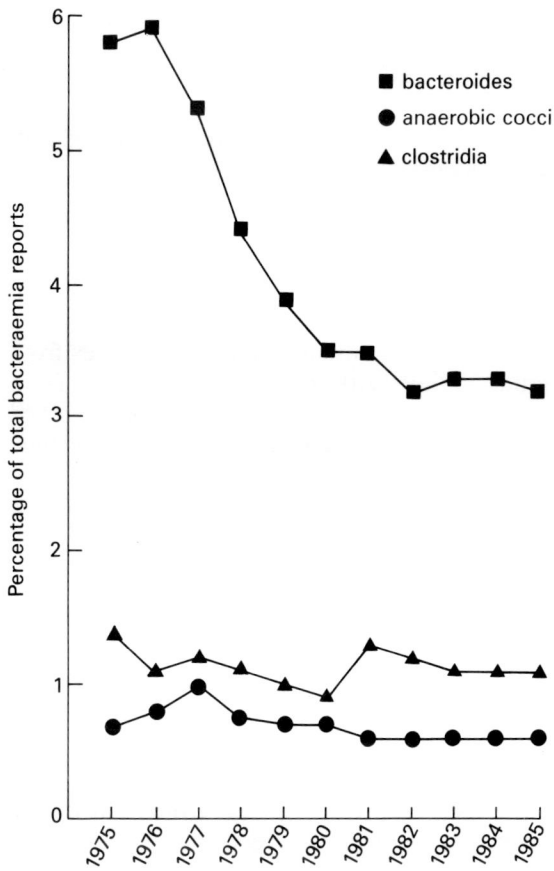

Fig. 19.1 Bacteraemia due to anaerobes (6928 reports, 5.5 per cent of total bacteraemias)

Table 19.1 Recent surveys of anaerobic isolates from blood cultures

Reference	Country	Total number of blood cultures	Number (%) containing anaerobes		Species isolated				
					Bacteroides	*Clostridia*	*Cocci*	*Fusobacteria*	*Others*
Roberts (1980)	USA	639	60	(9)	38	14	8	—	—
Michel and Priem (1981)	Netherlands	1289	49	(4)	43	6	—	—	—
Weinstein *et al.* (1983)	USA	623	81	(13)	35	14	16	5	11
Bouza *et al.* (1985)*	Spain	2454	212	(9)	96	88	25	10	13
Spencer (1986) (unpublished)	UK	1884	97	(5)	63	20	9	5	—

*232 anaerobic strains isolated

infections with *Bacteroides* spp. has decreased considerably from 5.8 per cent in 1975 to 3.2 per cent in 1985. Clostridial infections also fell as a proportion of reports from 1.4 to 1.1 per cent, while anaerobic streptococcal septicaemia has remained constant, with incidences of 0.7 and 0.6 per cent in 1975 and 1985 respectively. On the submitted report forms there was no information concerning perioperative antibiotic prophylaxis during gastrointestinal surgery, but it is probable that the falls in incidence were due in part to this.

The increase in the isolation of anaerobes from blood in the 1970s reflected improvements in anaerobic methodology and the use of improved media. It should be mentioned here that at one time it was not routine for blood–culture broths to be subcultured anaerobically (Bornstein et al., 1964). By 1971, however, a survey by Bartlett (1973) of blood-culture techniques in 21 clinical laboratories demonstrated that only one laboratory did not routinely use some anaerobic procedure. In my own survey into blood culture techniques in the United Kingdom, all 73 laboratories questioned used anaerobic subculture (Spencer, 1984 unpublished). The other major predisposing factor to the increased isolation of anaerobes was the increase in complex surgical procedures on the gastrointestinal tract, especially in debilitated patients with serious underlying conditions such as malignancy or hepatic disease. Anaerobic septicaemia is often a complication of pre-existing anaerobic infection and usually occurs in patients with underlying disease such as malignancy, haematological neoplasia, diabetes mellitus, hepatic and biliary disorders and surgical wound infections. Many of the same species of anaerobes that are present in the intestinal tract are also present in human infections. Of approximately 40 different species of anaerobes isolated from infected human tissue, including blood, all but three or four were found in the normal intestinal tract (Moore et al., 1969). The major portals of entry are the gastrointestinal and female genital tracts, especially for bacteroides and clostridia. For fusobacteria, the respiratory tract is the usual portal of entry. Anaerobic bacteraemia is seldom associated with the urinary tract, in contrast with infection by facultative bacteria. The majority of cases occur in the older age groups (over 60 years). Anaerobic septicaemia is seldom seen in children (Brook et al., 1980). Over a 5-year period at two different hospitals, 29 anaerobes were isolated from 28 paediatric patients. Organisms isolated were *Bacteroides* spp., 14 (*B. fragilis* 11); clostridia, 4; anaerobic cocci, 4; *P. acnes*, 4; and fusobacteria 3. No other organisms were isolated with them. The main portals of entry were the gastrointestinal tract (13) and upper respiratory tract (7). Five children died and antibiotic treatment was used, on average, for 20 days.

Bacteroides

B. fragilis is the most common clinically significant anaerobe isolated from blood cultures. It accounts for 70–90 per cent of the anaerobic gram-negative bacilli isolated (Marcoux et al., 1970; Wilson et al., 1972; Nobles, 1973; Chow and Guze, 1974). In an examination of the occurrence of the five subspecies of *B. fragilis* (now species within the fragilis group) in bacteraemia, Sonnenwirth (1974) found 80–90 per cent were ss. *fragilis*, whereas ss. *thetaiotaomicron* accounted for 10–15 per cent. Unfortunately, when bacteroides septicaemia was common the majority of strains were not classified (Tynes and Frommeyer, 1962).

In 1936 Lemierre described the classical clinical syndrome of bacteroides septicaemia originating in the oropharynx of young adult patients.

Tonsillitis → Abscess → Cervical
Peritonsillitis   adenitis

→ Distant metastases
especially to the lungs

Jaundice was a common accompanying feature. A world review of the literature by Gunn (1956) covered the period 1899 to 1954 and included 173 cases of bacteroides bacteraemia. When the cases were subdivided into those occurring before and after the introduction of sulphonamides, mortality in the first group was 81 per cent. In the 'post sulphonamide' series, of those treated by sulphonamides alone 64 per cent were fatal; combined therapy with penicillin, streptomycin and sulphonamides reduced the mortality to 50 per cent. It is interesting to note that in the 148 cases for which the site of the primary lesion was recorded, nasopharyngeal infections were the most frequent source, with 54 patients. The sequence was as described by Lemierre (1936), with thrombophlebitis of the internal jugular vein a serious terminal complication. For a time, ligation of the vessel was recommended as treatment, but in Kissling's series (1929), ten out of twelve patients died in spite of this procedure. In Gunn's report (1956), lesions of the abdominal tract were mentioned in only 26 cases, appendicitis being the principal predisposing cause in such patients. Nowadays the most common portals of entry for cases of bacteroides bacteraemia are the gastrointestinal and female genital tracts. Abdominal or pelvic surgery, gynaecological instrumentation, malignancies and other causes of breaches in the mucosal surfaces of the abdominal and genital tract are common underlying conditions (Tynes and Frommeyer, 1962; Gelb and Seligman, 1970; Kagnoff et al., 1972; Olsen, 1976; Fry et al., 1979). Skin infections and the oropharynx are less common portals of entry (Roberts, 1980). Due to its paucity of anaerobes as pathogens, the urinary tract is seldom the initial focus of infection (Segura et al., 1972). The analysis of cases supports the importance of intra-abdominal sepsis. Marcoux et al. (1970) found that gastrointestinal lesions, of which the most common underlying problem was malignancy, were the portal of entry in 62 per cent of 123 patients. Bodner et al. (1970) found that 24 of 39 cases of bacteroides bacteraemia originated in the gastrointestinal tract. Similar findings were reported by Kuklinca and Gavan (1971) in their survey of post-mortem blood cultures.

Chow and Guze (1974) reported their clinical experience of 112 patients seen between 1969 and 1972 in a comprehensive review of the subject which includes 211 references. Their cases accounted for 9.3 per cent of all septicaemic patients, a larger percentage than the low frequencies seen in earlier reports before 1968 (Dalton and Allison, 1967; Watt and Okubadejo, 1967). This increase after 1968 was most likely due to the greater awareness of bacteroides as pathogens and also to improved culture techniques. The gastrointestinal tract was again the most common portal of entry in 48 cases, followed by the female genital tract, 32 and decubitus ulcers, 12. Appropriate antimicrobial therapy resulted in a 16 per cent mortality compared with a mortality of 60 per cent when inappropriate treatment was used. At the time, chloramphenicol was the most commonly used appropriate antibiotic. The authors emphasized the importance of surgical drainage. Shock was seen in 31 patients, of whom seven had evidence of disseminated intravascular coagulation (DIC). Septic shock due to bacteroides appears to involve endotoxin, as in aerobic or facultative gram-negative bacillary sepsis. However its lipid-A lacks 2-keto-3-deoxyoctonate (KDO) and is less likely to cause DIC, although in severe infections this can occur (Yoshukawa et al., 1974).

Clostridia

Bloodstream invasion by *Clostridium perfringens* rarely occurs in anaerobic myonecrosis (Weinstein and Barza, 1973), it is much more common in postabortal and puerperal infections of the uterus (Willis, 1969). The classic clinical picture was described by Hill (1936) with acute toxaemia, hypotension, rapid pulse, jaundice and port-wine coloured urine. Women with septic abortions have positive clostridial blood cultures in 18–27 per cent of cases (Ramsey, 1949; Decker and Hall, 1966; Pritchard and Whalley, 1971; Smith et al., 1971). In these reports, however, the full syndrome of uterine gas gangrene was rare and in the majority of the women the infection ran a benign, self-limiting course; this was also noted in Ramsay's series, reported before the general availability of antibiotics. Surgical curettage appears to be just as important as antibiotic treatment. Septicaemia has also been reported in patients with various underlying diseases and conditions such as malignancy (Wynne and Armstrong, 1972) and decubitus ulcers (Rathbun, 1968). In Rathbun's report of 20 cases, mostly due to *C. perfringens*, none showed evidence of intravascular haemolysis.

In a series of 14 patients with neoplastic disease reported by Wynne and Armstrong (1972), *C. perfringens* accounted for 11 episodes and 13 patients had extensive neoplastic disease. Fourteen of the 15 anaerobic episodes reported appeared to originate from the gastrointestinal tract. Haemolysis was the exception rather than the rule and 12 of the patients died, all within 24 h. Cabrera et al. (1965) made the observation that clostridial infections in cancer patients tend to be of bloodstream invasion rather than localized skin or soft tissue infections. Fry et al. (1981) reviewed 47 patients with a mean age of 62; overall, 27 died. These were predominantly elderly individuals, with underlying disease and 35 did not have a primary focus of infection. *C. perfringens* was again the most common isolate in 21 cases. The authors were able to identify three patient groups:

elderly and debilitated with multiple associated illnesses;

alcoholics with systemic complications of alcoholic disease;

patients with a surgically treatable primary focus.

Antibiotic treatment did not affect the survival rate. Perhaps ineffective hepatic-reticuloendothelial function and portal bacteraemia may be significant factors in those patients without a primary focus of infection.

Species other than *C. perfringens* occasionally cause septicaemia (Alpern and Dowell, 1969; Spencer *et al.*, 1984). An association between *C. septicum* bacteraemia and the presence of underlying malignancy has been reported (Alpern and Dowell, 1969; Koransky *et al.*, 1979). Malignancies were present in 75 per cent (65 out of 86) of patients from whom *C. septicum* was isolated, and 53 (62 per cent) of these died. Fig. 19.2 shows such a patient, a 64-year-old man who was admitted '*in-extremis*' with blister formation on arms, shoulders and the upper part of the trunk. *C. septicum* was isolated from blister fluid and blood. Autopsy showed carcinoma of the ascending colon.

During an outbreak of necrotizing enterocolitis, Howard *et al.* (1977) isolated *C. butyricum* from the blood of seven babies. However, other workers (Gothefors and Blenkharn, 1978; Smith *et al.*, 1980) have criticized these findings; their own results showing that *C. butyricum* cannot be implicated as the sole aetiological agent in such cases. Neutropenic enterocolitis can be associated with septicaemia due to clostridial species, especially *C. septicum*. It is thought that mucosal damage due to disease or treatment predisposes to invasion of the mucosal wall and hence septicaemia (King *et al.*, 1984).

In 1975, Gorbach and Thadepalli reported on blood cultures which yielded 65 strains of clostridia from 49 patients, representing 2.6 per cent of 2168 positive blood cultures. *C. perfringens* was the most common species, accounting for 57 per cent of the

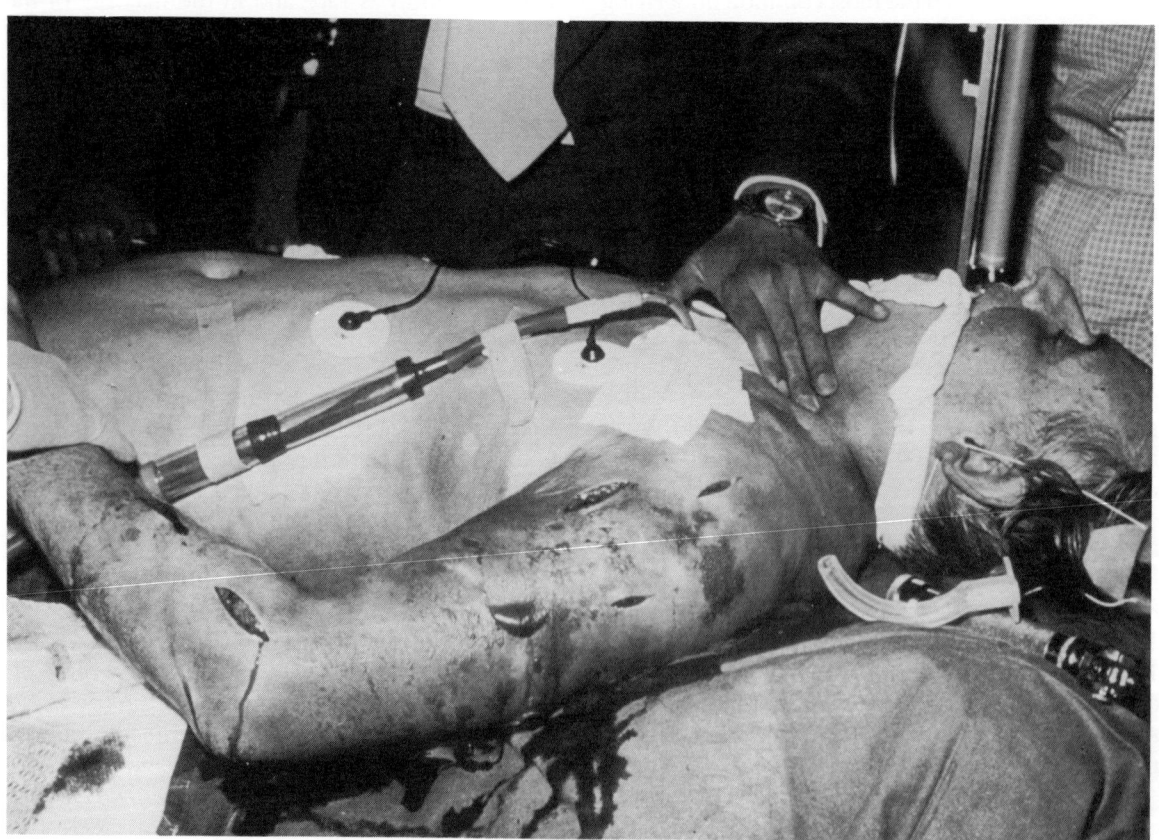

Fig. 19.2 Septicaemia due to *Clostridium septicum* with underlying malignancy

isolates. Septicaemia was often unrelated to clinical conditions, the study included patients with lobar pneumonia, meningococcaemia, pulmonary tuberculosis and even frostbite, albeit in a chronic alcoholic. Only twelve patients had concurrent soft-tissue infection with the same organism. The authors emphasized that clostridial isolation from blood should be interpreted with caution. A similar experience was noted at Cook County Hospital, Chicago, where Gorbach and Bartlett (1974) isolated 65 strains of clostridia, including eleven different species, over a 14-month period. Most of the patients had no apparent portal of entry and their reasons for admission to hospital were unrelated. The authors concluded that although clostridia isolated from a blood culture cannot be regarded legitimately as a contaminant, it may occur as a benign event in a patient with serious underlying disease. However, a study by Ahmad and Darrell (1976) of 100 patients showed that 21 per cent had antecubital fossa carriage of *C. perfringens*. The majority of carriers were long stay elderly female patients. Investigations highlighted the ineffective nature of skin preparation before venepuncture. The authors concluded that finding *C. perfringens* in blood cultures was not always clinically significant.

Peptococcaceae

Bacteraemia due to anaerobic cocci is relatively uncommon, although transient bacteraemia with these oganisms can occur following dental manipulation (De Vries *et al.*, 1972; Baltch *et al.*, 1982), bronchoscopy (Burman ,1960) and gastrointestinal procedures (Adami *et al.*, 1981); these are rarely of clinical significance. The incidence has varied from 1 per cent of anaerobic bacteraemias (Roberts, 1980; Topiel and Simon, 1986) to ≥14 per cent (Washington, 1971; Martin, 1974). Some studies have indicated that very few patients (<1 per cent) with infections due to anaerobic gram-positive cocci have bacteraemic sequelae (Pien *et al.*, 1972; Bourgault *et al.*, 1980). These two reports may, however, be biased because obstetric patients, who are the group at greatest risk of developing bacteraemia, were not included. Ramsay (1949) reported that anaerobic cocci accounted for 88 per cent of all anaerobic infections in postpartum patients. Smith *et al.* (1970) found peptostreptococci in 50 per cent of 76 patients with postseptic abortion septiceamia, results similar to those in studies by Rotherham and Schick (1969) and Pritchard and Whalley (1971). Dizerega *et al.* (1980) found that peptococci were the organisms most frequently isolated from bacteraemic patients with postcaesarian section endometritis. Of 41 patients with bacteraemia, 12 had peptococci in their blood. In their review of peptococcal bacteraemia, Topiel and Simon (1986) reported on 12 cases over a 20-month period, of whom seven had postpartum endometritis. Identification of the organisms revealed *Peptostreptococcus micros*, 5; *Pstr. asaccharolyticus*, 5; *Pstr. magnus*, 1 and *Peptococcus* spp., 1.

There is a low mortality with septicaemia due to peptococcaceae (Rotherham and Schick, 1969; Smith *et al.*, 1970; Wilson *et al.*, 1972; Topiel and Simon, 1986). This could be related to a young age group, absence of serious underlying disease, no associated hypotension and the lack of virulence factors such as the polysaccharide capsule of *B. fragilis*, which helps resist phagocytosis by neutrophils (Simon *et al.*, 1982). However, suppurative thrombophlebitis and metastatic infection can occur following septicaemia. Anaerobic cocci (and bacteroides) are known to induce septic thrombophlebitis more readily than aerobic bacteria (Schwartz and Dieckmann, 1927; Tynes and Frommeyer, 1962). Thus, septic thrombophlebitis originating in the uterine sinusoids may extend to the large pelvic veins and become resistant to therpay. (Collins *et al.*, 1951). Since many anaerobes are strongly proteolytic, digestion of venous clots results in the showering of the blood with microemboli containing bacteria. Descriptions of the metastatic abscesses that result from this septicaemia are found in reports dealing with anaerobic infection (Bornstein *et al.*, 1964).

Fusobacteria

The genus *Fucobacterium* was first proposed by Knorr in 1923 for non-sporulating, gram-negative, rod-shaped anaerobic organisms. *Fusobacterium necrophorum* is a virulent organism that causes a serious suppurative illness, 'necrobacillosis', a phrase coined by Alston (1955), in man and animals which, before the advent of antibiotics was often fatal. Today, necrobacillosis is rarely diagnosed, leading some authors to describe it as a 'forgotten disease' (Moore-Gillon *et al.*, 1984). This, despite the characteristic presenting features of postanginal septicaemia which were described in detail by Lemierre (1936) who stated that the septicaemias were always accompanied by the formation of distant metastatic abscesses. In the preantibiotic era, necrotic tonsillitis caused by *F. necrophorum* often resulted in bacteraemia. Tynes and Utz (1960) found 53 cases of presumptive bacteraemia in the world literature, of which 20 were confirmed by culture of blood or metastatic abscesses. Felner and Dowell (1971) found 55 patients from whom *Fusobacterium* spp. were isolated from the blood. A more recent study described 26 patients seen over a 5-year period at the Boston City Hospital (Henry *et al.*, 1983b). Bacteraemia occurred primarily in young adults or patients over the age of 60. Death occurred in three patients, all over 60 years, and metastatic infection occurred in only one patient, causing osteomyelitis. Postpartum fusobacteraemia occurred in six patients and was uniformly benign. The most common primary foci in this series were the female genital tract, upper and lower respiratory tract and the oral cavity. In a recent survey form 1979 to 1986, Dr S Eykyn at St Thomas' Hospital, London (private communication) has collected 29 cases of septicaemia caused by fusobacteria—*F. necrophorum*, 18; *F. nucleatum*, 8; *F. varium*, 2, and *F. necrogenes*, 1. In these cases septicaemia occurred following necrotic tonsillitis, liver abscesses, gastrointestinal disease, ear infections and meningitis. Other small series have been reported by several authors (O'Grady and Ralph, 1976; Seidenfeld *et al.*, 1982; Adams *et al.*, 1983; Prout and Glymph, 1985). Fusobacteria are considered to be difficult to recognize in the routine diagnostic laboratory because only *F. nucleatum* is consistently fusiform and many isolates are indistinguishable from strains of *Bacteroides* (Bennett and Duerden, 1985).

Capnocytophaga

Capnophilic gram-negative bacilli were originally described by Prévot in 1955 (Prévot *et al.*, 1956). They were subsequently designated as *Fusiformis nucleatus* var. *ochraceus*, *Ristella ochracea*, *B. ochraceus* var. *elongatus*, *B. ochraceus* and CDF biogroup DF-1 and were included among the 'anaerobes' as members of the family Bacteroidaceae. In 1979 two independent reports established the validity of the genus designated *Capnocytophaga* (Newman *et al.*, 1979; Williams *et al.*, 1979). Recent interest has centred on the role of this organism as an opportunist pathogen causing septicaemia in the immunocompromised host (Fortenza *et al.*, 1980; Hawkey *et al.*, 1984; Mycock and Azadian, 1985). The source of the organism is the mouth, where it forms part of the normal flora and is implicated in periodontitis (Newman *et al.*, 1976). On subculture some authors have found that *Cap. ochracea* will grow in air (Kristiansen *et al.*, 1984; Mycock and Azadian, 1985) while others consider carbon dioxide to be an absolute requirement (Hawkey *et al.*, 1984). Antimicrobial susceptibility tests suggest that penicillin, ampicillin, erythromycin and clindamycin are agents of choice. Metronidazole, tetracycline and chloramphenicol are only moderately active (Fortenza *et al.*, 1981).

Bifidobacteria and eubacteria

Despite their prevalence in the faecal flora of man, with 10^4–10^{11} organisms/g of faeces, these genera are rarely isolated from blood cultures, or even from other clinical material (Moore *et al.*, 1969). In the two series reported by Washington (1971, 1972) of 3950 positive blood cultures, six yielded bifidobacteria (0.15 per cent) and seven eubacteria (0.17 per cent).

The role of mixed bacteraemia

The incidence of polymicrobial bacteraemia seems to be increasing and it is associated with higher than expected case fatality rates (Saravolatz *et al.*, 1978; Kiani *et al.*, 1979; Kreger *et al.*, 1980; Ing *et al.*, 1981). The condition is often seen in patients with underlying malignancies (Hermans and Washington, 1970; Kiani *et al.*, 1979; Elting *et al.*, 1986; Spencer and Nicol, 1986). The report by Sandford (1972) showed that in 188 bacteraemic patients, 23 (12.2 per cent) had polymicrobial bacteraemia, and in 57 per cent of these, one (or more) anaerobic species was incriminated as a causative organism. In the review by Wilson *et al.* (1972), polymicrobial bacteraemia with other anaerobes or facultative organisms was present in 31 per cent of patients with bacteroides bacteraemia and Henry *et al.* (1983) reported that 38 per cent of fusobacterial bacteraemias were polymicrobial. In a survey of polymicrobial septicaemia

The role of mixed bacteraemia 331

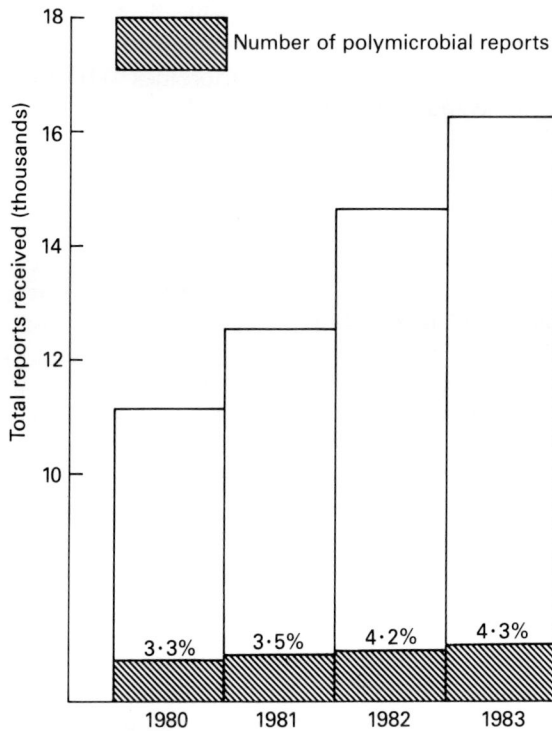

Fig. 19.3 Total number of bacteraemia reports received by Communicable Disease Surveillance Centre 1980–1983

in cancer patients over a 10 year period (Elting et al., 1986) anaerobes were implicated in 21 per cent of 507 episodes, with *Bacteroides* spp. the most common isolates.

In a survey of polymicrobial bacteraemia from returns to the PHLS Communicable Disease Surveillance Centre, Colindale, 2113 patient reports were submitted from 1980 to 1983, which represented 3.3–4.3 per cent (annually) of total bacteraemia reports (Fig. 19.3). Of the total of 4443 isolates reported, 591 (13 per cent) were anaerobic bacteria —*Bacteroides* spp., 334; *Clostridium* spp., 150; anaerobic cocci, 83; others, 24. Of 1852 patients from whom two organisms were isolated there were 307 (17 per cent) in whom at least one was an anaerobe (Table 19.2). Of 261 patients with three or more organisms, 100 (38 per cent) had at least one anaerobe. The main groups of polymicrobial patients and their association with anaerobic bacteria are shown in Table 19.3.

In many instances Enterobacteriaceae and

Table 19.2 Combinations of two organisms in 307 patients in whom an anaerobe was cultured

Additional isolate	Number
Enterobacteriaceae	133
Anaerobe	80
Streptococcus	65
Staphylococcus	20
Others	9
Total	307

anaerobes are a common mixture. Some *in vitro* investigations (Spencer, unpublished) have shown that the presence of enterobacteria can reduce the Eh in blood culture media, enabling small numbers of anaerobes, if present, to increase to sufficient numbers to be detected by subculture on to solid media. A report by von Graevenitz and Sabella (1971) indicated that many mixtures of gram-negative facultative and obligately anaerobic bacilli will be missed unless subcultures are performed on selective anaerobic media containing antibiotics. One method to improve the rate of polymicrobial isolation may be to reincubate positive blood cultures and then to repeat subcultures (Hanse and Hetmanski, 1983; Weinstein et al., 1983; Spencer et al., 1984) although the value of this approach has been disputed (Harkness et al., 1975; Stetz and Martin, 1985; Towne and Gay, 1985). In a prospective 2-year study, repeated subcultures were performed on 572 blood cultures that had yielded a single

Table 19.3 Groups of patients and their association with polymicrobial bacteraemia and anaerobic organisms

Underlying disease	Total number	Percentage with anaerobes
Hepatic	403	18
Urinary trace	367	6
Gastrointestinal (excluding neoplasia)	337	27
Neoplasms	252	25
Haematological	237	11
Dermatology	183	32
Respiratory infection	147	11
Intravenous lines	133	<1
Diabetes	100	23
Obstetrics/gynaecology	22	50

significant organism. This yielded 82 additional isolates from 70 patients of which 14 (17 per cent) were anaerobes—six being *B. fragilis* (Spencer and Nicol, 1985, 1986). However, it was found that additional antibiotics were only considered necessary in six (8.5 per cent) patients. So, although repeat subculture did increase polymicrobial isolation, such a procedure, while increasing the time and cost of processing blood culture samples, did not benefit the majority of patients.

Endocarditis

As we shall see, anaerobes are uncommon but important causes of infective endocarditis, and are associated with a high mortality (Nord, 1982). In his review of endocarditis, Felner (1974) reported that most cases were due to *P. acnes*, *B. fragilis*, and fusobacteria. The predominance of the anaerobic cocci seen in early reports has not been repeated in subsequent studies. It is probable that some earlier isolates were CO_2-dependent rather than truly obligate anaerobes. As many cases of infective endocarditis are culture negative, Nord (1982) suggested that these may have an anaerobic aetiology. It is, therefore, possible that with improved isolation techniques, the incidence of anaerobic endocarditis may rise.

Transient anaerobic bacteraemia occurs following dental manipulation in patients with periodontal disease (Rogosa *et al.*, 1960; De Vries *et al.*, 1972) and may also result from instrumentation of gastrointestinal and genitourinary tracts (Le Frock *et al.*, 1973, 1975; Chow and Guze, 1974). However, endocarditis due to anaerobic organisms remains rare (Schnurr *et al.*, 1977). In their review of 33 patients with anaerobic endocarditis, Felner and Dowell (1970) found 18 cases due to bacteroides and fusobacteria. Five cases were caused by *P. acnes*. Masri and Grieco (1972) reviewed 27 cases of whom 60 per cent had pre-existing heart disease; the mortality rate was 36 per cent. Felner (1974) reported similar findings in his review of 22 cases. *Bacteroides* spp. were the predominant isolates in both series. Nastro and Finegold (1973) reported a case of endocarditis due to *B. fragilis* which developed after appendicitis. Therapy with several agents over a 2-year period failed to eradicate the organism and the patient died. Unfortunately, metronidazole, to which the organism was sensitive, was not available for clinical evaluation at the time. The authors reviewed a further 37 cases and found bacteroides to be more invasive and destructive than viridans endocarditis. Overall, mortality again was high at 46 per cent. Among 100 cases of endocarditis studied by Lerner and Weinstein (1966), 13 were caused by microaerophilic streptococci and three by anaerobic cocci. Dormer (1958) studied 82 cases over a 12-year period and found one due to '*B. fusiformis*' and eight due to microaerophilic streptococci. In 408 cases, Cates and Christie (1951) found only six due to microaerophilic streptococci and Werner *et al.* (1967) found one case caused by a peptostreptococcus and one due to a bacteroides out of 206 patients. In a recent survey, Baylis *et al.* (1983) reported 544 episodes of infective endocarditis, in which one case was due to *B. fragilis* and one due to *F. necrophorum*. Two reports accounting for 133 patients found no cases of anaerobic endocarditis (Lowes *et al.*, 1980; Moulsdale *et al.*, 1980).

Major embolic episodes are reported to be characteristic prominent complications of bacteroides endocarditis (Nastro and Finegold, 1973). Heparinase production by the *Bacteroides* spp. may account for the increased incidence of thrombophlebitis and embolization seen in cases of endocarditis (as well as septicaemia) caused by these organisms (Gesner and Jenkin, 1961). *Bacteroides* spp. also account for most of the fatalities seen in anaerobic endocarditis (Sapico and Sarma, 1982).

C. perfringens was isolated from 11 of 100 samples of street heroin and 31 of 100 samples of injection material (Tuazon *et al.*, 1974). However, in a recent review over one year of 74 instances of endocarditis in drug addicts, anaerobes were implicated in only two polymicrobial cases, neither of which yielded clostridia. (Levine *et al.*, 1986).

Head *et al.* (1984) studied the effects of penicillin V, or metronidazole, 2 g given 1 h before dental procedures, on the elimination of anaerobes from postextraction bacteraemias. Penicillin V reduced the occurrence of anaerobes to a greater degree, although anaerobic gram-negative bacilli were still detected in four out of 25 patients. No anaerobic gram-negative bacilli were isolated in the metronidazole-treated group. Penicillin V and metronidazole might therefore be an effective combination for use in prophylaxis. A report by Joseffson *et al.* (1985) reported the incidence of anaerobic bacteraemia during dental procedures in three groups of patients —control (untreated), penicillin V 2 g, and erythromycin 500 mg given prophylactically. Ten minutes after operation the incidences of bacteraemia in the two treatment groups were significantly reduced. These two observations are of importance, as dental and oral disease or manipulation are not an uncommon precursor of endocarditis.

Clinical features

The clinical features of anaerobic bacteraemia are similar in most respects to those associated with aerobes—fever, systemic toxicity and leucocytosis. In septicaemia due to anaerobic gram-negative bacilli, septic shock has been reported to occur in 16–37 per cent of patients (Gelb and Seligman, 1970; Marcoux et al., 1970; Wilson et al., 1972; Nobles, 1973; Chow and Guze, 1974). However, some distinguishing features can occur, such as an increased incidence of jaundice, septic thrombophlebitis and distant metastatic spread (Gorbach and Bartlett, 1974). As stated previously, the usual portals of entry are the gastrointestinal and female genital tracts, though for *Fusobacterium* spp. the oropharynx remains the most common. The urinary tract is seldom implicated. Most published reports indicate that anaerobic septicaemia occurs more frequently in adults, particularly in elderly patients. In a series reported by Kalager and Solberg (1985), the condition was not found in anyone under 30 years old. Chow and Guze (1974) reported an incidence of 26 per cent of all septicaemias in a nursery, but others have been unable to substantiate this finding (Escheverria and Smith, 1978; Brook et al., 1980). Reported mortality figures are around 30 per cent but important differences occur within individual patient populations. Severe or ultimately fatal underlying conditions are associated with a poor prognosis. In the series by Kalager and Solberg (1985) the mortality was 43 per cent, which can be explained by the elderly age group and the high proportion of underlying malignancies or preceding complicated abdominal surgery. Pearson and Anderson (1967, 1970) reported that all obstetric patients with anaerobic septicaemia survived; however, infant mortality in this group was 90 per cent.

In cases of clostridial bacteraemia, clinicians frequently interpret such a finding with the presence of devitalized tissue, haemolysis, jaundice, renal failure and shock. Although this is recognized as a well established syndrome, patients with clostridial bacteraemia often have a benign disease. In Ramsay's series (1949), 17 out of 28 gynaecological patients with *C. perfringens* in the blood had localized infection with no clinical evidence of systemic toxicity.

Isolation techniques

Due to the low number of cfu/ml of blood, the cultivation of 20–30 ml of blood per culture from adults is essential for the optimal isolation of bacteria, including anaerobes (Ilstrup and Washington, 1983). Neonates and infants have higher levels of circulating bacteria, thus permitting less blood to be sampled for a single culture. Positive blood cultures are often taken late in disease and bacteriological reports reach the clinician too late for appropriate antibiotic therapy (Okubadejo et al., 1973). Empiric appropriate antimicrobial therapy should always be given to those patients suspected of anaerobic septicaemia.

Media

A routine blood culture set usually includes at least two different broths, one suitable for aerobes and the other for anaerobes. However, the necessity for a specific culture medium for the isolation of anaerobes may be questionable because unvented bottles maintain sufficiently low redox potentials (−175 mV) to allow anaerobic growth. Blood culture media vary considerably in their ability to support the early growth of strict anaerobes (Shanson, 1974). In the 1970s, cooked meat media were associated with the late isolation of *Bacteroides* spp. and other non-sporing organisms, although their viability was good once growth had occurred (Forgan-Smith and Darrell, 1974; Shanson, 1974). Other media, such as thioglycollate, allowed detection at an early stage of incubation although the bacteria quickly died after growth had commenced. This was a problem with some of the commercial thioglycollate media available at the time (Szarwatkowski, 1976). Washington and Martin (1973) compared brain–heart infusion (BHI), tryptic soy and thioglycollate broths and failed to demonstrate any significant difference in isolation rates of anaerobic bacteria from blood. During the last few years there has been much improvement in the range of blood culture media available, such as fresh homemade cooked meat mixed with BHI medium (Collee *et al.*, 1977). Although cooked meat medium is cheap and gives good results, it is time consuming to prepare. Other preparations include Fastidious Anaerobe Broth (Ganguli *et al*, 1982); thioglycollate plus Panmede (Hunt and Price, 1982) and BHI plus cysteine (Shanson and Singh, 1981). Recently it has been shown that BHI supplemented with both cysteine and Panmede is excellent for the growth of non-fastidious anaerobes in simulated blood cultures (Shanson and Pratt, 1983). Whatever liquid medium is used, it should be noted that the anaerobic performance deteriorates with storage (Barr, 1980). Liquoid is not recommended for inclusion in broth used primarily for the isolation of strict anaerobes, because occasional strains of anaerobic cocci are inhibited (Hoare, 1939; Graves *et al.*, 1974). Sodium amylosulphate is less inhibitory than liquoid for anaerobic cocci, notably peptostreptococci, but it is less efficient in its ability to neutralize the antibacterial effects of human blood (Kocka *et al.*, 1974).

Incubation methods

All conventional systems should include a vented bottle for the recovery of strict aerobes—pseudomonads, yeasts—and an unvented bottle for the recovery of strict anaerobes (Gantz *et al.*, 1974; Blazevic *et al.*, 1975). Although instead of being unvented, the bottle can be incubated, with its cap loosened and plugged, in an anaerobic atmosphere including CO_2 10 per cent (Watt, 1973). Whatever the size of the bottle, it is important that the addition of blood leaves little or no gas space above the liquid. A study by Martin *et al.* (1984) showed that 9 per cent of anaerobic isolates from patients were recovered only from vented bottles. This study emphasized the need to check vented bottles for anaerobes too, as failure to do this would mean that a significant number of these bacteria would be missed.

All cultures should be routinely incubated for up to seven days (Cumitech, 1982; Martin *et al.*, 1984; Murray, 1985). Prolonged incubation is unnecessary for the routine detection of anaerobes. Most laboratories subculture aerobically at 24 h and anaerobically at 48 h, and then after seven days, if cultures *appear* negative. In a survey to evaluate the necessity for the final routine subculture, Campbell and Washington (1980) subcultured 2780 previously negative blood cultures. Of four bottles found to be positive, three yielded the same organism previously isolated and one was considered to be a contaminant. The expense and time of routine terminal subcultures appears, therefore, not to be justified. Also in question is the routine anaerobic subculturing of macroscopically negative blood-culture bottles. Two studies using tryptic soy broth and thiol broth (Murray and Sondag, 1978) and tryptic soy broth alone (Paisley *et al.*, 1978) found that blind subculture of macroscopically negative cultures did not significantly increase the yield of anaerobes, nor did the rare isolation improve the clinical management of the patient. However, these two groups of workers assumed that anaerobic growth can be easily detected macroscopically. With their media this may be so, but many laboratories now use fastidious anaerobe broth which is initially turbid. As it is not always possible to detect macroscopic growth, therefore, blind subculture for anaerobes is still to be recommended. Broths under vacuum are best subcultured with syringe and needle to help maintain the vacuum.

The isolation and identification of anaerobic bacteria can be time consuming and, with bacteroides septicaemia especially, there is a substantial mortality rate. Early detection is important. Studies have shown the effectiveness of gas–liquid chromatographic analysis of the characteristic volatile fatty acids produced by anaerobes during glucose metabolism in blood–culture media (Sondag and Murray, 1980; Edson *et al.*, 1982). In the latter study *iso*valeric and butyric acids, alone or together, were found in 39 (90 per cent) of 43 blood cultures that contained anaerobes, but were absent in all 44 cultures that contained facultative or aerobic bacteria.

New techniques

Biphasic blood culture systems were originally developed in 1947 by Castaneda for the isolation of brucella. Modifications and improvements have been made and the most recent version is a biphasic slide culture system which incorporates a slide unit on to a traditional blood culture bottle. Evaluations of this system have demonstrated that it is not readily adaptable for anaerobic culturing (Bryan, 1981). Although some workers do not consider this to be a major disadvantage of the system (in view of the data that suggests routine anaerobic subculturing is unnecessary; Pfaller et al., 1982) others feel that this new system should be used in combination with an unvented bottle for anaerobic isolation (Henry et al., 1984).

The gas capture system is a single blood culture system which detects gas produced by organisms from substrates in a specially formulated medium. Its sensitivity was found to be comparable to that of a radiometric method, although the number of anaerobes studied was small (King et al., 1986). Experiments with a stock culture of C. perfringens failed to produce any increase in pressure, thereby failing to signal growth, although the broth was obviously turbid. A drawback appears to be the small amount of blood cultured—5 ml/bottle. At least four bottles will be necessary to culture an adequate amount of blood, so increasing cost.

Deland and Wagner (1969) described a radiometric method for the detection of bacteria in blood. This is now available as a commercial product in which C^{14} CO_2 produced from C^{14} labelled substrates in the culture medium is measured and taken to be an indicator of bacterial growth. Identification of the organism is performed as with conventional broth cultures. In a Danish study (Arpi et al., 1985) the radiometric method had a shorter average detection time than the conventional broth system (1 vs 1.7 days). However, a quarter of all the positive blood cultures was due to contamination, primarily with coagulase-negative staphylococci. Detection of anaerobes has been disappointing, but the introduction of specific anaerobic media such as prereduced tryptic soy broth may improve matters. It has been suggested that as many as 17 per cent of all clinically significant isolates will be missed unless four bottles are used instead of the two routinely recommended (Plorde et al., 1985). This will naturally produce a substantial increase in cost.

Automated screening of blood cultures could be considered desirable because conventional methods are time consuming, tedious and expensive in materials and labour. A system of screening blood cultures by measuring electrical conductance has been investigated (Brown et al., 1984). This system was only able to detect seven Bacteroides and Clostridium spp. while conventional BHI broth detected 12 species. The incorporation of dextran into the broth may eliminate the effect of sedimented blood cells which can mask early signals from bacteria (Curtis et al., 1985).

The determination of bacterial adenosine triphosphate by bioluminescence has been adapted to a blood culture system so as to make possible the early detection of bacterial growth (Beckurs and Lang, 1983). This technique is in its infancy and much work remains to be done.

Two alternative methods are available for concentrating organisms and thereby eliminating the need for broth culture—lysis centrifugation (Dorn and Smith, 1978) and lysis filtration (Lamberg et al., 1983). The lysis centrifugation technique utilizes a double-stoppered evacuated tube containing haemolysing/anticoagulant agents and a stabilized high density hyrophobic cushion in which the organisms are concentrated. Samples of the hydrophobic cushion are cultured on conventional agar media. There is early recovery of microorganisms and increased numbers are recovered in this system compared with conventional blood cultures. However, in a study by Henry et al. (1983a) this system was not the optimal method for isolating anaerobic bacteria, although the reasons for this are unclear. In lysis filtration, bacteria are concentrated by filtration through a membrane. A Swedish study (Heimdahl et al., 1985) showed that this technique was far superior to a conventional one of BHI plus peptone meat broth from the detection of anaerobic bacteraemia.

Treatment

The introduction of metronidazole has revolutionized the treatment of anaerobic bacteraemia, with especial regard to those caused by Bacteroides spp. It has excellent antimicrobial activity against the major groups of anaerobes—bacteroides, clostridia, fusobacteria and most obligate anaerobic cocci. Penicillin remains the treatment of choice where Clostridium spp. are the pathogens. If a collection of pus is the source of bacteraemia then surgical drainage and debridement remain important. Antibacterial therapy should be administered as early as possible to prevent metastatic suppurative complications. Because of the inability to distinguish initially between aerobic and anaerobic septicaemia, metronidazole should be given intravenously, together with an aminoglycoside such as gentamicin or a broad-spectrum cephalosporin such as cefuroxime. Administration is usually as a buffered isotonic 0.5 per cent w/v aqueous solution in 100 ml bottles or packs containing 500 mg of metronidazole infused over 20 min every 8 h. Other effective antianaerobe agents include clindamycin and cefoxitin, but each has its drawbacks. Clindamycin has been associated with pseudomembranous colitis (Cohen et al., 1973) and cefoxitin is recognized as an excellent inducer of chromosomal β-lactamases. Cefoxitin resistance has been found in strains of the B. fragilis group (Dornbusch et al., 1980) with sporadic occurrences of clindamycin resistance in the same group of organisms (Tally et al., 1979). Data have been presented that suggest that patients with recent exposure to clindamycin may have clindamycin-resistant anaerobic bacteria in a current infection which may prevent the infection from responding to the clindamycin treatment (Ohm-Smith et al., 1986). However, reports of strictly anaerobic bacteria showing resistance to metronidazole are still rare (Ingham et al., 1978).

Before metronidazole was introduced chloramphenicol was advocated as suitable treatment but there have been reports of therapeutic failures including persistent bacteroides septicaemia in the face of adequate therapy (Mitre and Rotherham, 1974; Thadepalli et al., 1978). Several other antibiotics were also used—tetracycline, pencillin, erythromycin and cephalothin (Marcoux et al., 1970). Metronidazole was first used successfully in the treatment of anaerobic septicaemia in the early 1970s (Mitre and Rotherham, 1974; Baron et al., 1975). Cases of septicaemia which had failed to respond to treatment with parenteral clindamycin or lincomycin were successfully treated with oral metronidazole (Sharp et al., 1977; Galgiani et al., 1978).

The common occurrence of anaerobic sepsis, both local and systemic, as a complication of abdominal and gynaecological surgery has been a continuing problem for many years. Over this time efforts by microbiologists and surgeons to reduce the incidence of infection have led to the employment of various agents (Watt, 1973). However, the prophylactic use of metronidazole in such circumstances has been conspicuously successful in the prevention of systemic postoperative anaerobic infection (Willis, 1978), including bacteraemia.

Endocarditis due to anaerobic gram-negative bacilli is a serious infection with a high mortality. In 37 cases, Nastro and Finegold (1973) reported a death rate of 46 per cent. The use of agents without good bactericidal activity usually results in failure. Metronidazole offers the possibility of a more effective therapy. However, due to the relative scarcity of this condition, well controlled studies are not available. Clindamycin and metronidazole have been advocated as initial therapy, although I believe metroinidazole alone is just as effective. Galgiani et al. (1978) reported two patients with B. fragilis bacteraemia, one of whom had endocarditis. This patient remained febrile and bacteraemic on clindamycin; addition of metronidazole resulted in clinical cure without surgical intervention. A rabbit model has been used to investigate the effect of antibiotics on the prevention of experimental endocarditis due to B. fragilis (Goldman et al., 1978). Clindamycin was the most effective prophylactic agent, while metronidazole was found to be ineffective. Other intravascular infections, such as septic thrombophlebitis, have again been shown to be highly susceptible to metronidazole therapy when other antibiotic treatment has been ineffective (Mitre and Rotherham, 1974).

Acknowledgements

I am grateful to the Communicable Disease Surveillance Centre of the Public Health Laboratory Service for supplying the data relevant to Figs. 1 and 2; and to Dr S. Eykyn of St Thomas' Hospital London for information regarding fusobacteria.

References

Adami, B, Eckardt V F, Suermann R B, Karbach V, Eine K. (1981). Bacteremia after proctoscopy and hemorrhoidal injection sclerotherapy. *Diseases of the Colon and Rectum* 24: 373–374.

Adams J, Capistrant T, Crossley K, Johanssen R, Liston S. (1983). *Fusobacterium necrophorum* septicemia. *Journal of the American Medical Association* 250: 35–36.

Ahmad F J, Darrell J H. (1976). Significance of the isolation of *Clostridium welchii* from routine blood cultures. *Journal of Clinical Pathology* 29: 185–186.

Alpern R J, Dowell V R. (1969). *Clostridium septicum* infections and malignancy. *Journal of the American Medical Association* 209: 385–388.

(1971). Non-histotoxic clostridial bacteremia. *American Journal of Clinical Pathology* 55: 717–722.

Alston J M. (1955). Necrobacillosis in Great Britain. *British Medical Journal* ii: 1524–1528.

Arpi M, Lester A, Frederiksen W. (1985). A comparison of a conventional and radiometric examination of clinical blood cultures with respect to recovery rate and detection time of microorganisms. *Acta Pathologica Microbiologica et Immunologica Scandinavica* 93B: 263–271.

Baltch A L, et al. (1982). Bacteremia following dental cleaning in patients with and without penicillin prophylaxis. *American Heart Journal* 104: 1335–1339.

Baron D, Drugeon H, Nicolas F. Courteiu A L. (1975). Interêt du metronidazole dans les septicemias a *Bacteroides fragilis*. Deux observations. *La Nouvelle Presse Medicale* 4: 667.

Barr J G. (1980). A cooked-meat blood culture medium: shelf-life and clinical evaluation compared with other systems. *Journal of Infection* 2: 247–258.

Bartlett R C. (1973). Contemporary blood culture techniques. In *Bacteremia—laboratory and clinical aspects*, p. 15–35. Edited by Sonnerwith A C. Thomas, Springfield.

Bayliss R, Clarke C, Oakley C M, Somerville W, Whitfield A G W, Young S E J. (1983). The microbiology and pathogenesis of infective endocarditis. *British Heart Journal* 51: 339–345.

Beckurs B, Lang H R M. (1983). Rapid diagnosis of bacteriaemia by measuring bacterial adenosine triphosphate in blood culture bottles using bioluminescence. *Medical Microbiology and Immunology* 172: 117–122.

Bennett K W, Duerden B I. (1985). Identification of fusobacteria in a routine diagnostic laboratory. *Journal of Applied Bacteriology* 59: 171–181.

Blazevic D J, Stemper J E, Matsen J M. (1975). Effect of aerobic and anaerobic atmospheres on isolation of organisms from blood cultures. *Journal of Clinical Microbiology* 1: 154–156.

Bodner S J, Koenig G, Goodman J S. (1970). Bacteremic bacteroides infections. *Annals of Internal Medicine* 73: 537–544.

Bornstein D L, Weinberg A N, Swartz M N, Kunz L J. (1964). Anaerobic infection. Review of current experience. *Medicine* 43: 207–232.

Bourgault A M, Rosenblatt J E, Fitzgerald R H. (1980). *Peptococcus magnus*: a significant human pathogen. *Annals of Internal Medicine* 93: 244–248.

Bouza E et al. (1985). Retrospective analysis of two hundred and twelve cases of bacteremia due to anaerobic microorganisms. *European Journal of Clinical Microbiology* 4: 262–267.

Brook I, Controni G, Rodriquez W J, Martin W J. (1980). Anaerobic bacteremia in children. *American Journal of Diseases of Children* 134: 1052–1056.

Brown D F J, Warner M, Taylor C E D, Warren R E. (1984). Automated detection of micro-organisms in blood cultures by means of the Malthus Microbiological Growth Analyser. *Journal of Clinical Pathology* 37: 65–69.

Bryan L E. (1981). Comparison of a slide blood culture system with a supplemented peptone broth culture method. *Journal of Clinical Microbiology* 14: 389–392.

Burman S O. (1960). Bronchoscopy and bacteremia. *Journal of Thoracic and Cardiovascular Surgery* 40: 635–639.

Cabrera A, Tsukada Y, Pickren J W. (1965). Clostridial gas gangrene and septicemia in malignant disease. *Cancer* 18: 800–806.

Campbell J, Washington J A. (1980). Evaluation of the necessity for routine terminal subcultures of previously negative blood cultures. *Journal of Clinical Microbiology* 12: 576–578.

Castaneda M R. (1947). A practical method for routine blood cultures for brucellosis. *Proceedings of the Society for Experimental Biology and Medicine* 64: 114–115.

Cates J E, Christie R V. (1951). Subacute bacterial endocarditis. A review of 442 patients treated in 14 centres appointed by the Penicillin Trials Committee of the Medical Research Council. *Quarterly Journal of Medicine* 20: 93–130.

Chow A W, Guze L B. (1974). Bacteroidaceae bacteremia: clinical experience with 112 patients. *Medicine (Baltimore)* 53: 93–126.

Cohen L E, McNeill C J, Wells R F. (1973). Clindamycin-associated colitis. *Journal of the American Medical Association* 233: 1379–1380.

Collee J G, Duerden B I, Brown R. (1977). Recovery of anaerobic bacteria from small inocula: a model for blood culture studies. *Journal of Clinical Pathology* 30: 609–614.

Collins C G, MacCullum E A, Nelson E W, Weinstein B B, Collins J H. (1951). Suppurative pelvic thrombophlebitis; incidence, pathology and etiology; study of seventy patients treated by ligation of inferior vena cava and ovarian vessels. *Surgery* 30: 298–328.

Cumitech I A. (1982). *Blood cultures II* American Society for Microbiology, Washington DC.

Curtis G D W, Thomas C D, Johnston H H. (1985). A note on the use of dextran in blood cultures monitored by conductance methods. *Journal of Applied Bacteriology* 58: 571–575.

Dalton H P, Allison M J. (1967). Etiology of bacteremia. *Applied Microbiology* 15: 808–814.

Decker W H, Hall W. (1966). Treatment of abortions infected with *Clostridium welchii*. *American Journal of Obstetrics and Gynecology* 95: 394–399.

Deland F H, Wagner H N. (1969). Early detection of bacterial growth, with carbon-14-labelled glucose. *Radiology* **92**: 154–155.

De Vries J, Francis L E, Lang D. (1972). Control of post-extraction bacteremias in the penicillin hypersensitive patient. *Journal of the Canadian Dental Association* **38**: 63–66.

Dizerega G S, Yonekura M L, Keegan K, Roy S, Nakamura R, Ledger W. (1980). Bacteremia in post-cesarean section endomyometritis: differential response to therapy. *Obstetrics and Gynecology* **55**: 587–590.

Dormer A E. (1958). Bacterial endocarditis. Survey of patients treated between 1945 and 1956. *British Medical Journal* i: 63–69.

Dorn G L, Smith K. (1978). New centrifugation blood culture device. *Journal of Clinical Microbiology* **7**: 52–54.

Dornbusch K, Olsson-Liljeqvist B, Nord C E. (1980). Antibacterial activity of new β-lactam antibiotics on cefoxitin-resistant strains of *Bacteroides fragilis*. *Journal of Antimicrobial Chemotherapy* **6**: 207–216.

Edson R S, Rosenblatt J E, Washington J A, Stewart J B. (1982). Gas–liquid chromatography of positive blood cultures for rapid presumptive diagnosis of anaerobic bacteremia. *Journal of Clinical Microbiology* **15**: 1059–1061.

Elting L S, Bodey G P, Fainstein V. (1986). Polymicrobial septicaemia in the cancer patient. *Medicine* **65**: 218–225.

Escheverria P, Smith A L. (1978). Anaerobic bacteremia as observed in a children's hospital. *Clinical Pediatrics* **53**: 93–126.

Felner J M. (1974). Infective endocarditis caused by anaerobic bacteria. In *Anaerobic bacteria: role in disease*, pp. 345–358. Edited by Balows A *et al*. Thomas, Springfield.

Felner J M, Dowell V R. (1970). Anaerobic bacterial endocarditis. *New England Journal of Medicine* **283**: 1188–1192.

Felner J M, Dowell V R. (1971). 'Bacteroides' bacteremia. *American Journal of Medicine* **50**: 787–796.

Finegold S M. (1977). *Anaerobic bacteria in human disease*. Academic Press, New York.

Forgan-Smith W R, Darrell J H. (1974). A comparison of media used in-vitro to isolate non-sporing gram-negative anaerobes from blood. *Journal of Clinical Pathology* **27**: 280–283.

Fortenza S W, Newman M G, Lipsey A I, Siegal S E, Blachman U. (1980). Capnocytophaga sepsis: a newly recognised clinical entity in granulocytopenic patients. *Lancet* i: 567–568.

Fortenza S W, Newman M G, Horikoshi A L, Blachman U. (1981). Antimicrobial susceptibility of Capnocytophaga. *Antimicrobial Agents and Chemotherapy* **19**: 144–146.

Fry D E, Garrison R N, Polk H C. (1979). Clinical implications in bacteroides bacteremia. *Surgery, Gynecology and Obstetrics* **149**: 189–192.

Fry D E, Klamer T W, Garrison R N, Polk H C. (1981). Atypical clostridial bacteremia. *Surgery, Gynecology and Obstetrics* **153**: 28–30.

Galgiani J N, Busch D F, Brass C, Reunaus L W, Mangels J I, Stevens D A. (1978). *Bacteroides fragilis* endocarditis, bacteremia and other infections treated with oral or intravenous metronidazole. *American Journal of Medicine* **65**: 284–289.

Ganguli L A, Turton L G, Tillotson G S. (1982). Evaluation of Fastidious Anaerobe Broth as a blood culture medium. *Journal of Clinical Pathology* **35**: 458–461.

Gantz N M, Medeiros A A, Swain J L, O'Brien T F. (1974). Vacuum blood-culture bottles. Inhibiting growth of candida and fostering growth of bacteroides. *Lancet* ii: 1174–1176.

Gelb A F, Seligman S J. (1970). Bacteroidaceae bacteremia —effect of age and focus of infection upon clinical course. *Journal of the American Medical Association* **212**: 1038–1041.

Gesner B M, Jenkin C R. (1961). Production of heparinase by bacteroides. *Journal of Bacteriology* **81**: 595–604.

Goldman P L, Durack D T, Petersdorf R G. (1978). Effect of antibiotics on the prevention of experimental *Bacteroides fragilis* endocarditis. *Antimicrobial Agents and Chemotherapy* **14**: 755–760.

Gorbach S L, Bartlett J G. (1974). Anaerobic infections. *New England Journal of Medicine* **290**: 1237–1245.

Gorbach S L, Thadepalli H. (1975). Isolation of *Clostridium* in human infections: evaluation of 114 cases. *Journal of Infectious Diseases* **131**: S81–S85.

Gothefors L, Blenkharn I. (1978). *Clostridium butyricum* and necrotising entercolitis. *Lancet* i: 52–53.

Graves M H, Morello J A, Kocka F E. (1974). Sodium polyanethol sulfonate sensitivity of anaerobic cocci. *Applied Microbiology* **27**: 1131–1133.

Gunn A A. (1956). Bacteroides septicaemia. *Journal of the Royal College of Surgeons, Edinburgh* **2**: 41–50.

Hansen S L, Hetmanski J. (1983). Enhanced detection of polymicrobial bacteremia by repeat sub culture of previously positive blood cultures. *Journal of Clinical Microbiology* **18**: 208–210.

Harkness J L, Hall M, Ilstrup D M, Washington J A. (1975). Effects of atmosphere of incubation and of routine subcultures on detection of bacteremia in vacuum blood culture bottles. *Journal of Clinical Microbiology* **2**: 296–299.

Hawkey P M, Malnick H, Glover S, Cook N, Watts J A. (1984). *Capnocytophaga ochracea* infection. *Journal of Clinical Pathology* **37**: 1066–1070.

Head T W, Bentley K C, Millar E P, de Vries J A. (1984). A comparative study of the effectiveness of metronidazole and penicillin V in eliminating anaerobes from post-extraction bacteremias. *Oral Surgery, Oral Medicine and Oral Pathology* **58**: 152–155.

Heimdahl A, Josefsson K, von Konow L, Nord C E. (1985). Detection of anaerobic bacteria in blood cultures by lysis filtration. *European Journal of Clinical Microbiology* 4: 404–407.

Henry N K, Grewell C M, McLimans C A, Washington J A. (1984). Comparison of the Roche Septi-Check blood culture bottle with a Brain Heart Infusion biphasic medium bottle and with a Tryptic Soy Broth bottle. *Journal of Clinical Microbiology* 19: 315–317.

Henry N K, McLimans C A, Wright A J, Thompson R L, Wilson W R, Washington J A. (1983a). Microbiological and clinical evaluation of the Isolator lysis—centrifugation blood culture tube. *Journal of Clinical Microbiology* 17: 864–869.

Henry S, De Maria A, McCabe W R. (1983b). Bacteremia due to *Fusobacterium* species. *American Journal of Medicine* 75: 225–231.

Hermans P E, Washington J A. (1970). Polymicrobial bacteremia. *Annals of Internal Medicine* 73: 387–392.

Hill A M. (1936). Post-abortal and puerperal gas gangrene; report of 30 cases. *Journal of Obstetrics and Gynaecology of the British Empire* 43: 201–251.

Hoare R D. (1939). The suitability of 'liquoid' for use in blood culture media, with particular reference to anaerobic streptococci. *Journal of Pathology and Bacteriology* 48: 573–577.

Howard F M, Flynn D M, Bradley J M, Noone P, Szawatkowski M. (1977). Outbreak of nectrotising enterocolitis caused by *Clostridium butyricum*. *Lancet* ii: 1099–1102.

Hunt G H, Price E H. (1982). Comparison of a homemade blood culture containing a papain digest of liver, with four commercially available media for the isolation of anaerobes from simulated paediatric blood cultures. *Journal of Clinical Pathology* 35: 1142–1149.

Ilstrup D M, Washington J A. (1983). The importance of volume of blood cultured in the detection of bacteremia and fungemia. *Diagnostic Microbiology and Infectious Diseases* 1: 107–110.

Ing A F M, McLean P H, Meakins J L. (1981). Multiple organism bacteremia in the surgical intensive care unit: a sign of intraperitoneal sepsis. *Surgery* 90: 779–786.

Ingham H R, Eaton S, Venables C W, Adams P C. (1978). *Bacteroides fragilis* resistant to metronidazole after long term therapy. *Lancet* i: 214.

Joseffson K, Heimdahl A, von Konow L, Nord C E. (1985). Effect of phenoxymethyl penicillin and erythromycin prophylaxis on anaerobic bacteraemia after oral surgery. *Journal of Antimicrobial Chemotherapy* 16: 243–252.

Kagnoff M F, Armstrong D, Bleuins A. (1972). Bacteroides bacteremia. *Cancer* 29: 245–251.

Kalager T, Solberg C O. (1985). Treatment of anaerobic septicaemia. *Scandinavian Journal of Infectious Diseases* Suppl 46: 96–100.

Kiani D, Quinn E L, Burch K H, Madhavan T, Saravolatz L D, Neblett T R. (1979). The increasing importance of polymicrobial bacteremia. *Journal of the American Medical Association* 242: 1044–1047.

King A, Rampling A, Wright D G D, Warren R E. (1984). Neutropenic enterocolitis due to *Clostridium septicum* infection. *Journal of Clinical Pathology* 37: 335–343.

King A, Bone G, Phillips I. (1986). Comparison of radiometric and gas capture system for blood cultures. *Journal of Clinical Pathology* 39: 661–665.

Kissling K. (1929). Ueber postanginose Sepsis. *Munchener medizinische Wochenschrift* 76: 1163–1168.

Knorr M. (1923). Uber die fusisipillare Symbiose, die Gattung *Fusobacterium*. *Zentralblatt fur Bakteriologie, Parasitenkunde und Infectionskrankheiten* 89: 4–22.

Kocka F E, Arthur E J, Searcy R L. (1974). Comparative effects of two sulfated polyanions used in blood culture on anaerobic cocci. *American Journal of Clinical Pathology* 61: 25–27.

Koransky J R, Stargel M D, Dowell V R. (1979). *Clostridium septicum* bacteremia. *American Journal of Medicine* 66: 63–66.

Kreger B E, Craven D E, Carling P C, McCabe W R. (1980). Gram-negative bacteremia. III. Reassessment of etiology, epidemiology and ecology in 612 patients. *American Journal of Medicine* 68: 332–343.

Kristiansen J E, Bremmelgaard A, Busk H E, Aeltberg O, Frederiksen W, Jutesen T. (1984). Rapid identification of *Capnocytophaga* isolated from septicaemic patients. *European Journal of Clinical Microbiology* 3: 326–240.

Kuklinca A G, Gavan T L. (1971). Anaerobic bacteria in postmortem blood cultures. Correlation with lesions of the gastro-intestinal tract. *Cleveland Clinic Quarterly* 38: 5–11.

Lamberg R E, Schell R F, LeFrock J L. (1983). Detection and quantitation of simulated anaerobic bacteremia by centrifugation and filtration. *Journal of Clinical Microbiology* 17: 856–859.

LeFrock J L, Ellis C A, Turchik J B, Weinstein L. (1973). Transient bacteremia associated with sigmoidoscopy. *New England Journal of Medicine* 289: 467–469.

LeFrock J L, Ellis C A, Klainer A S, Weinstein L. (1975). Transient bacteremia associated with barium enema. *Archives of Internal Medicine* 135: 835–837.

Lemierre A. (1936). On certain septicaemias due to anaerobic organisms. *Lancet* i: 701–703.

Lemierre A, Reilly J. (1938). Les bacteriemies a '*Bacillus rumosus*'. *La Presse Medicale* 46: 385–387.

Lerner P I, Weinstein L. (1966). Infective endocarditis in the antibiotic era. *New England Journal of Medicine* 274: 199–206.

Levine D P, Crane L R, Zervos M J. (1986). Bacteremia in narcotic addicts at the Detroit medical center. II Infectious endocarditis: a prospective comparative study. *Reviews of Infectious Diseases* 8: 374–396.

Lowes J A *et al.* (1980). 10 years of infective endocarditis at St Bartholomew's Hospital: analysis of clinical features and treatment in relation to prognosis and mortality. *Lancet* i: 133–136.

Marcoux J A, Zabransky R A, Washington J A, Wellman W E,

Martin W J. (1970). Bacteroides bacteremia: a review of 123 cases. *Minnesota Medicine* 53: 1169–1176.

Martin W J. (1974). Isolation and identification of anaerobic bacteria in the clinical laboratory: a 2-year experience. *Mayo Clinic Proceedings* 49: 300–308.

Martin W J, Wilhelm P A, Bruckner D. (1984). Recovery of anaerobic bacteria from vented blood-culture bottles. *Reviews of Infectious Diseases* 6: S59–S61.

Masri A F, Grieco M H. (1972). Bacteroides endocarditis. Report of a case. *American Journal of the Medical Sciences* 263: 357–367.

Michel M F, Priem C C. (1981). Positive blood cultures in a University Hospital in the Netherlands. *Infection* 9: 283–289.

Mitre R J, Rotherham E B. (1974). Anaerobic septicemia from thrombophlebitis of the internal jugular vein. *Journal of the American Medical Association* 230: 1168–1169.

Moore W E C, Cato E P, Holdeman L V. (1969). Anaerobic bacteria of the gastrointestinal flora and their occurrence in clinical infectious. *Journal of Infectious Diseases* 119: 641–649.

Moore-Gillon J, Lee T H, Eykyn S J, Phillips I. (1984). Necrobacillosis: a forgotten disease. *British Medical Journal* 288: 1526–1527.

Moulsdale M T, Eykyn S J, Phillips I. (1980). Infective endocarditis 1970–79: a study of culture positive cases in St. Thomas's Hospital. *Quarterly Journal of Medicine* 49: 315–328.

Murray P R. (1985). Determination of the optimum incubation period of blood culture broths for the detection of clinically significant septicemia. *Journal of Clinical Microbiology* 27: 481–485.

Murray P R, Sondag J E. (1978). Evaluation of routine subcultures of macroscopically negative blood cultures for detection of anaerobes. *Journal of Clinical Microbiology* 8: 427–430.

Mycock G, Azadian B S. (1985). *Capnocytophaga ochracea*: an unusual opportunistic pathogen. *Journal of Clinical Pathology* 38: 1081–1082.

Nastro L J, Finegold S M. (1973). Endocarditis due to anaerobic gram-negative bacilli. *American Journal of Medicine* 54: 482–296.

Newman M G, Socransky S S, Savitt E D, Propas D, Crawford A. (1976). Studies on the microbiology of periodontosis. *Journal of Periodontology* 47: 373–379.

Newman M G et al. (1979). Detection, identification and comparison of *Capnocytophaga, Bacteroides ochraceus* and DF-1. *Journal of Clinical Microbiology* 10: 557–562.

Nobles E R. (1973). Bacteroides infections. *Annals of Surgery* 177: 601–606.

Nord C E. (1982). Anaerobic bacteria in septicaemia and endocarditis. *Scandinavian Journal of Infectious Diseases* 31 Suppl: 95–104.

O'Grady L R, Ralph E D. (1976). Anaerobic meningitis and bacteremia caused by *Fusobacterium* species. *American Journal of Diseases in Children* 130: 871–873.

Ohm-Smith M J, Sweet R L, Hadley W K. (1986). Occurrence of clindamycin-resistant anaerobic bacteria isolated from cultures taken following clindamycin therapy. *Antimicrobial Agents and Chemotherapy* 30: 11–14.

Okubadejo O A, Green P J, Payne D J H. (1973). Bacteroides infections among hospital patients. *British Medical Journal* 2: 212–214.

Olsen H. (1976). Bacteroides bacteraemia. A clinical and bacteriological analysis of 51 patients. *Scandinavian Journal of Infectious Diseases* 8: 107–111.

Paisley J W, Rosenblatt J E, Hall M, Washington J A. (1978). Evaluation of a routine anaerobic subculture of blood cultures for detection of anaerobic bacteremia. *Journal of Clinical Microbiology* 8: 764–766.

Pearson H E, Anderson G V. (1967). Perinatal deaths associated with bacteroides infections. *Obstetrics and Gynecology* 30: 486–492.

(1970). Bacteroides infections and pregnancy. *Obstetrics and Gynecology* 35: 31–36.

Pfaller M A, Sibley T K, Westfall L M, Hoppe-Bauer J E, Keating M A, Murray P R. (1982). Clinical comparison of a slide blood culture system with a conventional broth system. *Journal of Clinical Microbiology* 16: 525–530.

Pien F D, Thompson R L, Martin W J. (1972). Clinical and bacteriologic studies of anaerobic Gram-positive cocci. *Mayo Clinic Proceedings* 47: 251–257.

Plorde J J, Tenover F C, Carlson L G. (1985). Specimen volume versus yield in the BACTEC blood culture system. *Journal of Clinical Microbiology* 22: 292–295.

Prévot A R, Tardieux P, Joubert L, de Cadore F. (1956). Reserches sur *Fusiformis nucleatus* (Knorr) et son pouvoir pathogene pour l'homme et les animaux. *Annals de l'Institut Pasteur* 91: 788–798.

Pritchard J A, Whalley P J. (1971). Abortion complicated by *Clostridium perfringens* infection. *American Journal of Obstetrics and Gynecology* 111: 484–492.

Prout J, Glymph R. (1985). *Fusobacterium mortiferum* septicemia. *Clinical Microbiology Newsletter* 7: 29–30.

Ramsay A M. (1949). The significance of *Clostridium welchii* in the cervical swab and blood-stream in post-partum and postabortum sepsis. *Journal of Obstetrics and Gynaecology of the British Commonwealth* 56: 247–258.

Rathbun H K. (1968). Clostridial bacteremia without hemolysis. *Archives of Internal Medicine* 122: 496–501.

Roberts F J. (1980). A review of positive blood cultures: identification and source of microorganisms and patterns of sensitivity to antibiotics. *Reviews of Infectious Diseases* 2: 329–339.

Rogosa M, Hampp E G, Nevin T A, Wagner H N, Discoll E J, Baer P N. (1960). Blood sampling and cultural studies in the detection of postoperative bacteremias. *Journal of the American Dental Association* 60: 171–180.

Rosenow E C, Brown A E. (1938). Septicaemia: A review of cases, 1934–1936 inclusive. *Proceedings of the Staff Meetings at the Mayo Clinic* **13**: 89–94.

Rotherham E B, Schick S F. (1969). Non-clostridial anaerobic bacteria in septic abortion. *American Journal of Medicine* **46**: 80–89.

Sandford J P. (1972). Generalised gram-negative bacillary infections. *Journal of the Royal College of Physicians* **6**: 189–193.

Sapico F L, Sarma R J. (1982). Infective endocarditis due to anaerobic and microaerophilic bacteria. *Western Journal of Medicine* **137**: 18–23.

Saravolatz L D, Burch K H, Quinn E L, Cox F, Maghavan T, Fisher E. (1978). Polymicrobial infective endocarditis: an increasing clinical entity. *American Heart Journal* **95**: 163–168.

Schnurr L P, Ball A P, Geddes A M, Gray J, McGhie D. (1977). Bacterial endocarditis in England in the 1970's: a review of 70 patients. *Quarterly Journal of Medicine* **46**: 499–512.

Schwartz O H, Dieckmann W J. (1927). Puerperal infection due to anaerobic streptococci. *American Journal of Obstetrics and Gynecology* **13**: 467–485.

Segura J W, Kelalis P P, Martin W J, Smith L H. (1972). Anaerobic bacteria in the urinary tract. *Mayo Clinic Proceedings* **47**: 30–33.

Seidenfeld S M, Sutker W L, Luby J P. (1982). *Fusobacterium necrophorum* septicemia following oropharyngeal infection. *Journal of the American Medical Association* **348**: 1348–1350.

Shanson D C. (1974). An experimental assessment of different anaerobic blood culture methods. *Journal of Clinical Pathology* **27**: 273–279.

Shanson D C, Singh J. (1981). Effect of adding cysteine to brain–heart infusion broth on the isolation of *Bacteroides fragilis* from experimental blood cultures. *Journal of Clinical Pathology* **34**: 221–223.

Shanson D C, Pratt J. (1983). The effect of adding a papain digest of ox liver to brain–heart infusion cysteine broth on the recovery of non-sporing anaerobes from simulated blood cultures. *Journal of Clinical Pathology* **36**: 678–682.

Sharp D J, Corringham R E T, Nye E B, Sagor G P, Noone P. (1977). Successful treatment of *Bacteroides* bacteraemia with metronidazole after failure with clindamycin and lincomycin. *Journal of Antimicrobial Chemotherapy* **3**: 233–237.

Simon G L, Klempner M S, Kasper D L, Gorbach S L. (1982). Alterations in opsonophagocytic killing by neutrophils of *Bacteroides fragilis* associated with animal and laboratory passage: effect of capsular polysaccharide. *Journal of Infectious Diseases* **145**: 72–77.

Smith J W, Southern P M, Lehmann J D. (1970). Bacteremia in septic abortion: complications and treatment. *Obstetrics and Gynecology* **35**: 704–708.

Smith L P, McLean A P, Maughan G B. (1971). *Clostridium welchii* septitoxemia. *American Journal of Obstetrics and Gynecology* **110**: 135–149.

Smith M F, Borriello S P, Clayden G S, Casewell M W. (1980). Clinical and bacteriological findings in necrotising enterocolitis: a controlled study. *Journal of Infection* **2**: 23–31.

Sondag J E, Murray P R. (1980). Rapid presumptive identification of anaerobes in blood cultures by gas–liquid chromatography. *Journal of Clinical Microbiology* **11**: 274–277.

Sonnenwirth A C, Yin E T, Sarnuents E M, Wessler S. (1972). Bacteroidaceae endotoxin detection by Limulus assay. *American Journal of Clinical Nutrition* **25**: 1452–1454.

Sonnenwirth A C. (1974). Incidence of intestinal anaerobes in blood cultures. In *Anaerobic bacteria: role in disease*, pp. 157–171. Edited by Balows A *et al.* Thomas, Springfield.

Spencer R C, Courtney S P, Nicol C D. (1984). Polymicrobial septicaemia due to *Clostridium difficile* and *Bacteroides fragilis*. *British Medical Journal* **289**: 531–532.

Spencer R C, Nicol C D. (1985). Polymicrobial septicaemia. *Lancet* **i**: 1210.

(1986). Increased detection of polymicrobial septicaemia by repeat subculture. *Journal of Medical Microbiology* **22**: 85–87.

Stetz E M, Martin W J. (1985). Repeat subculture of known positive blood cultures: costly and ineffective in detecting polymicrobial bacteraemias. *Diagnostic Microbiology and Infectious Diseases* **3**: 113–118.

Szarwatkowski M V. (1976). A comparison of three readily available types of anaerobic blood culture media. *Medical Laboratory Sciences* **23**: 5–12.

Tally F P, Snydman D R, Gorbach S L, Malamy M H. (1979). Plasmid-mediated, transferable resistance to clindamycin and erythromycin in *Bacteroides fragilis*. *Journal of Infectious Diseases* **139**: 83–88.

Thadepalli H, Gorbach S L, Bartlett J G. (1978). Apparent failure of chloramphenicol in anaerobic infections. *Obstetrics, Gynecology and Surgery* **33**: 334–335.

Thompson L, Beaver D C. (1932). Bacteremia due to anaerobic gram-negative organisms of the genus *Bacteroides*. *Medical Clinics of North America* **15**: 1611–1626.

Topiel M S, Simon G L. (1986). *Peptococcaceae* bacteremia. *Diagnostic Microbiology and Infectious Disease* **4**: 109–117.

Towne A R, Gay R M. (1985). Evaluation of the efficacy of reincubation and subsequent subculture of initially positive blood cultures in the detection of additional clinically significant isolates. *Journal of Clinical Microbiology* **21**: 155–157.

Tuazon C U, Hill R, Sheagren J N. (1974). Microbiological study of street heroin and injection paraphernalia. *Journal of Infectious Diseases* **129**: 327–329.

Tynes B S, Frommeyer W B. (1962). Bacteroides septicaemia: cultural, clinical and therapeutic features in a

series of twenty-five patients. *Annals of Internal Medicine* **56**: 12–26.
Tynes B S, Utz J P. (1960). *Fusobacterium* septicemia. *American Journal of Medicine* **29**: 879–887.
Von Graevenitz A, Sabella W. (1971). Unmasking additional bacilli in gram-negative rod bacteremia. *Journal of Medicine (Basel)* **2**: 185–191.
Washington J A. (1971). Comparison of two commercially available media for detection of bacteremia. *Applied Microbiology* **22**: 604–607.
(1972). Evaluation of two commercially available media for detection of bacteremia. *Applied Microbiology* **23**: 956–959.
Washington J A, Martin W J. (1973). Comparison of three blood culture media for recovery of anaerobic bacteria. *Applied Microbiology* **25**: 70–71.
Watt B. (1973). The influence of carbon dioxide on the growth of obligate anaerobes on solid media. *Journal of Medical Microbiology* **6**: 307–313.
Watt P J, Okubadejo O K. (1967). Changes in incidence and aetiology of bacteraemia arising in hospital practice. *British Medical Journal* **1**: 210–211.
Weinstein L, Barza M A. (1973). Gas gangrene. *New England Journal of Medicine* **289**: 1129–1131.
Weinstein M P, Reller L B, Murphy J R, Lichtenstein K A. (1983). The clinical significance of positive blood cultures: a comprehensive analysis of 500 episodes of bacteremia and fungemia in adults. I. Laboratory and epidemiologic observations. *Review of Infectious Diseases* **5**: 35–53.
Werner A S, Cobbs C G, Kaye D, Hook E W. (1967). Studies on the bacteremia of bacterial endocarditis. *Journal of the American Medical Association* **202**: 199–203.
Williams B L, Hollis D, Holdeman L V. (1979). Synonomy of strains of Center for Disease Control group DF-1 with species of *Capnocytophaga*. *Journal of Clinical Microbiology* **10**: 550–556.
Willis A T. (1969). *Clostridia of wound infections.* Butterworths, London.
(1978). The treatment of anaerobic bacterial infections. *British Journal of Hospital Medicine* **20**: 579–585.
Wilson W R, Martin W J, Wilkowske C J, Washington J A. (1972). Anaerobic bacteremia. *Mayo Clinic Proceedings* **47**: 639–646.
Wynne J W, Armstrong D. (1972). Clostridial septicemia. *Cancer* **29**: 215–221.
Yoshikawa T T, Chow A W, Guze L B. (1974). Bacteroidaceae bacteremia with disseminated intravascular coagulation. *American Journal of Medicine* **56**: 725–728.

20

Anaerobes in intestinal homeostasis

S P Borriello and F E Barclay

Mechanisms of colonization resistance
    Direct mechanisms of colonization resistance
    Indirect mechanisms of colonization resistance
    Protective effect of anaerobes in the gut
    Colonization resistance with respect to *Clostridium difficile*
Conclusions
References

The term 'colonization resistance' was first coined by van der Waaij (1971) who used it to describe the resistance of the healthy gastrointestinal tract to colonization by exogenous organisms. This resistance is provided chiefly by a stable indigenous microflora which acts as a barrier to help prevent potential pathogens from establishing themselves and causing disease. This effect is achieved in a number of ways, both directly and indirectly. The most familiar example of the importance of colonization resistance is probably in infection with *Clostridium difficile*. The normal colon is resistant to infection but following antibiotic administration, which disrupts the normal gut flora and its associated barrier effect, *C. difficile* can cause severe disease. Another example is that of infant botulism, in which a not yet fully developed normal flora permits *C. botulinum* to become established in the gut where it releases toxin and causes paralytic disease.

In this chapter we will describe the mechanisms that could account for colonization resistance (giving some well established examples), highlight the role of anaerobes in this process and finally discuss in some detail colonization resistance to *C. difficile* and the application of this concept to prophylaxis and therapy in preventing or treating *C. difficile* gut disease.

Mechanisms of colonization resistance

Direct mechanisms of colonization resistance

Direct resistance may be achieved by ecological interaction, i.e., the ecological niches (such as specific mucosal receptor sites) required by the invading pathogen in order to grow and multiply have already been utilized by other bacteria. Unable to establish itself in the new environment the invader is excluded from the body by peristalsis. However, disruption of the resisting flora may effectively remove resident bacteria from these niches and allow potential pathogens to utilize them. Experiments with germ-free animals have shown them to be very susceptible to colonization by pathogenic organisms. However, if a 'normal' flora is established, by direct oral administration of bacterial isolates, by faecal or caecal material from conventional animals or by housing with conventional animals, then colonization resistance is restored (Syed *et al.*,

1970; Freter and Abrams, 1972; Freter et al., 1983).

The intestinal flora itself may be viewed as a complex organ and interactions between individual members play a part in colonization resistance. Metabolites produced by some of the organisms present in the gut can often be utilized by other bacteria, either directly or indirectly, as nutrients or growth factors. They may also be detrimental, due to their toxicity or as a result of changes in pH creating an acidic or alkaline environment. The phenomenon by which one bacterial species is able to inhibit or kill another is known as 'bacterial interference' and this is discussed further below. Lastly, competition for available nutrients is likely to be an important factor in deciding which species dominate the intestinal flora. The most successful species, i.e., those present in the highest numbers, will be those that can most readily utilize the nutrients available.

Indirect mechanisms of colonization resistance

The intestinal microflora is thought to play an important, although indirect part in normal gastrointestinal function. Abrams and Bishop (1966) studied intestinal motility in conventional and germ-free mice using a radioactive tracer and observed that a far greater proportion of an administered feed remained in the small intestine of the germ-free animal than in that of its conventional counterpart. This suggests that the presence of a resident flora stimulates peristaltic emptying. If fed a dose of viable bacteria, the small intestine of the germ-free host retained a significant portion of that dose for many hours. Retention of bacteria within the intestine for long periods can result in a build-up of large intraluminal populations to which the mucosa is exposed. This may explain, in part, the findings of Sprinz et al. (1961) and Formal et al. (1963) that the germ-free guinea-pig, unlike its conventional counterpart, is extremely susceptible to infection of the small intestine with *Shigella flexneri*. Events resulting from cessation of peristalsis in animal models can be studied by ligating the bowel to form a closed tube or stagnant loop. Abrams and Bishop (1966) ligated the terminal ileum of conventional and germ-free mice prior to intraduodenal challenge with *Salmonella typhimurium* and found that growth of the organism within the bowel was greater than in the unligated mice. Moreover, the numbers of *S. typhimurium* attained from the conventional and germ-free groups of mice with ligated intestines were the same.

Bacterial interference

Many organisms are capable of preventing the growth of other species *in vitro* and it is thought that antagonism *in vivo* may be an important factor in colonization resistance. Investigators have been aware of this phenomenon of bacterial interference, and its potential uses, for over 100 years. In 1885, Cantani used a culture of a non-pathogenic organism in an attempt to replace *Mycobacterium tuberculosis* in the human lung. Two years later, Pasteur and Joubert (1887) noted inhibition of growth of *Bacillus anthracis* in urine contaminated with 'common' micro-organisms and speculated that the contaminants might be useful as therapy against anthrax. Emmerich (1887) extended this work, showing that rabbits were protected from death if streptococci were added to the inoculum of anthrax bacilli. Over the ensuing years several reports have appeared describing bacterial interference *in vitro*, most often between commensal bacteria and known pathogens (Hsu and Wiseman, 1967; Sanders, 1969; Hentges and Maier, 1970; Mitchell and Kenworthy, 1976; Wilhelm et al., 1977; El Sanousi et al., 1978; Sanders and Sanders, 1982; James and Tagg, 1988). Interest in these observations in the early years was stimulated by the discovery of antibiotics, when it was thought that these antibacterial substances produced by micro-organisms may be of therapeutic use. Emphasis was put upon their isolation and purification, but in the search for broad-spectrum antibiotics it was found that they were of limited use. With modern diagnostic facilities which allow most pathogens to be detected or isolated and identified rapidly, the use of broad-spectrum antibiotics may be less desirable, particularly in the light of the currently held belief that such antibiotics have the greatest detrimental effect on colonization resistance. An antibiotic with a specific action which has minimal effect on the commensal flora while still being effective against the pathogen would be preferable. However, no generalizations can be made, especially as the components of the gut flora involved in colonization resistance to different organisms may differ. Also, an effect on a narrow range of micro-organisms may in turn cause other more significant changes in the complex ecosystem exposed to a particular antibiotic. These factors, coupled with the pharmokinetics of the antibiotic, its route of administration, its rate of inactivation and its combined effects with coadministration of other antimicrobials, all serve to make it difficult to predict the impact of a compound on a complex microbial ecosystem from *in vitro* activity against pure isolates. From a colonization resistance point of view, spectrum of activity should be defined as the impact of a compound on a particular microbial ecosystem. This view is supported by the work of Nord and colleagues (1988).

Some bacteria produce bactericidal proteins which are active against other bacterial strains but not against the producing organism. These proteins have the general name 'bacteriocins', but are usually named after the producing organism, i.e., 'perfringocin' from *C. perfringens* and 'colicin' from *Escherichia coli*. It is thought that bacteriocinogeny is genetically determined by plasmids, a fact well established for the Enterobacteriaceae with which most work has been performed. Most of the work on bacteriocins has been performed *in vitro* and the relevance of bacteriocin production *in vivo* has yet to be established. There is some evidence that they are, in fact, of very little ecological significance in the gut. Ikari *et al.* (1969) inoculated germ-free mice with both a colicin-producing and a colicin-sensitive strain of *E. coli*. Examination of faecal samples revealed no suppression of the colicin-sensitive *E. coli* which, in fact, outnumbered the colicin-producing strain. Similarly, Booth *et al.* (1977) found three bacteriocin-producing strains of *Bacteroides* co-existing with a larger population of sensitive strains in the faeces of an individual. Interestingly, Tompkins and Tagg (1989) produced evidence to show that production of active bacteriocins *in situ* selected a relatively bacteriocin-resistant accompanying microbiota in the mouth.

Other factors involved in colonization resistance

Antibacterial substances secreted into the gastrointestinal tract form an important defence mechanism against exogenous pathogens. These substances include bile acid, lysozyme, proteases and gastric acid. Relatively high concentrations of bile acids in the small intestine may account for the limited number of bacteria at this site because many species are known to be inhibited by bile salts. The so-called gastric acid barrier is also an important factor in controlling the species of exogenous bacteria that come into contact with the small and large bowel. Gastric secretions create an acidic environment that is not conducive to bacterial multiplication. The gut is also well serviced with immunological defence mechanisms which include the immunoglobulins G, M and E, secretory IgA, and cell-mediated immunity. The role of these mechanisms has not yet been fully elucidated, although there is evidence that secretory IgA and cell-mediated immunity play an active role (Williams and Gibbons, 1972; Futura and Freter, 1973, Müller-Schoop and Good, 1975).

There are mechanisms other than the production of antibacterial substances by which antagonism of one or more species by another may occur. Inhibition may be due to the depletion of essential nutrients from the growth medium. Not all bacteria grow at the same rate and those with the shortest generation time will have a distinct advantage in the competition for nutrients. The large number of nutrients available, which derive directly from the diet, from the host or as bacterial breakdown products, may account for the diversity of the microflora, which consists of many different species of bacteria with different energy requirements. It is thought that the competition for nutrients is a major mechanism by which the number and type of organisms colonizing the gut are controlled. End products of bacterial metabolism may also result in a change in pH, thus affecting those organisms which are particularly sensitive to acid or alkali. Metabolism itself may also create a restrictive environment by changing the redox potential.

Almost all species of bacteria are susceptible to infection by bacteriophages. These bacterial viruses can only reproduce within the host bacterial cell, eventually causing its lysis and death. They exhibit a high specificity, each being able to infect only one group of closely related organisms, usually a single species or occasionally only a few strains of a single species.

All these factors have a direct influence on the ability of the intestinal flora to counter invasion by both exogenous and endogenous 'opportunist' pathogens.

Protective effect of anaerobes in the gut

It is generally accepted that anaerobes play a major

role in colonization resistance in the gastrointestinal tract. Non-sporing anaerobes, principally bacteroides, eubacteria, bifidobacteria and propionibacteria predominate in the normal intestinal flora (see Chapter 11). These outnumber the facultative anaerobes (streptococci, *E. coli* and lactobacilli) by 100–1000 to 1. To show the importance of the anaerobic flora in colonization resistance, Van der Waaij *et al*. (1971) treated conventional mice with antibiotics so that only the anaerobic component remained. Germ-free mice were then colonized with this anaerobic flora and their offspring challenged with *E. coli*. These mice, which contained an exclusively anaerobic flora, showed a high resistance to colonization by *E. coli* and were, therefore, called colonization resistance factor (CRF) mice.

Wensinck and Ruseler-van Embden (1971) infected CRF mice with graded doses of *E. coli* and other gram-negative organisms and found that they were unable to colonize. On examination, the protective flora consisted of at least five species of anaerobe, although no attempt was made to identify them. Welling *et al*. (1980) went further; they isolated different strains of anaerobic bacteria from CRF mice and gradually associated germ-free mice with an increasing number of the strains while monitoring a number of different parameters including colonization resistance to *E. coli*. They found that the levels of colonizing *E. coli* dropped as the mice became associated with more strains of anaerobic bacteria. The lowest level of colonization was reached after association with 46 strains. Interestingly, this level of colonization with *E. coli* was considerably lower than that reached in the CRF control mice. No *E. coli* could be detected in the conventional (SPF) mice. Using an *in vitro* method to demonstrate colonization resistance in man, Boriello and Barclay (1986) showed that faecal material which had been aerated to remove non-sporing anaerobes supported the growth of *C. difficile*, whereas untreated material inhibited and actively killed the organism (Borriello and Barclay, unpublished data). They also added the antibiotic aztreonam, which is active only against aerobic bacteria, to faecal material and found that it was still able to inhibit *C. difficile* (unpublished data), an observation also made with hamster caecal material (Borriello *et al*., 1988a). Aztreonam has been shown not to predispose to infection with *C. difficile* *in vivo* in the hamster model of *C. difficile* enterocolitis (Borriello *et al*., 1988a).

It has been suggested that the large amounts of volatile fatty acids produced by many anaerobic bacteria as metabolic by-products may inhibit the growth of other bacteria due to the creation of an acidic environment. Such inhibition has been demonstrated many times *in vitro*. Barclay (1985), investigating the apparent inhibition of *C. difficile* by a strain of *C. beijerinckii* on a solid medium, found that inhibition also occurred in broth. The broth cultures containing *C. beijerinckii*, which produces large amounts of acetic and butyric acids, were found to have pH values as low as 5.5. It was demonstrated that *C. difficile* was inhibited when inoculated into a fresh broth adjusted to pH 5.5 and that its cytotoxin was inactivated between pH 5.0 and 5.5. *C. beijerinckii* was also found to inhibit a selection of other organisms, both aerobic and anaerobic and, interestingly, all of them gram-positive (Barclay and Borriello, 1982).

The role of acetic and butyric acids in the control of gram-negative facultative bacilli in the gastrointestinal tract has been studied in infants (Bullen *et al*., 1976) and mice (Lee and Gemmell, 1972; Byrne and Dankert, 1979; Pongpech and Hentges, 1989). In essence, these studies showed that low levels of volatile fatty acids were associated with high levels of Enterobacteriaceae and vice-versa, measurements being performed on faecal and caecal samples. Lee and Gemmell (1972) found that elimination, with penicillin, of anaerobic fusiform bacteria from the intestines of young mice resulted in a significant drop in the level of butyric acid and a million-fold increase in the numbers of coliform bacilli.

Hentges and Maier (1970) added *Sh. flexneri* to an established culture of *Bacteroides fragilis* grown in a glucose-containing medium and found that it failed to multiply. The cultures were found to be of low pH (5.45–5.70). Without glucose in the growth medium the *B. fragilis* cultures remained above pH 7.0 and *Sh. flexneri* was able to grow. Antagonism between these two organisms was also investigated *in vivo* (Maier and Hentges, 1972). It was found that the levels of *Sh. flexneri* were the same in germ-free mice associated with *B. fragilis* as they were in those mice not associated with *B. fragilis*.

Thus it would appear that the antagonisms between different organisms, which are so often noted *in vitro*, may not necessarily operate *in vivo*.

Colonization resistance is multifactorial and, while apparently not achieving resistance by themselves, single factors such as antagonism due to acid production and pH sensitivity probably play their part in the whole.

Although many lines of evidence indicate that anaerobes are important in colonization resistance to various pathogens the evidence still falls short of proving that they are the only important component. Indeed, Gorbach et al. (1988) questioned the hypothesis that colonization resistance in man is related to the anaerobic flora. However, it must be borne in mind that this work suffers just as much from broad generalizations based on relatively superficial microbiological investigation as the work supporting the contention that anaerobes are involved in this process in man.

Colonization resistance with respect to *Clostridium difficile*

It is generally accepted that the majority of cases of *C. difficile*-mediated disease occur as a direct result of antibiotic therapy. The administration of antibiotics, particularly over a long period or as a 'cocktail' or mixture of drugs, effectively disrupts some of the normal defence mechanisms that operate to exclude *C. difficile* from the gut and, as a consequence, render the patient susceptible to infection with this organism. Although the normal bacterial flora of the gut is a major defence mechanism against *C. difficile* it is known that *C. difficile* can become intimately associated with the gut mucosa in man (Borriello, 1979) and in hamsters with *C. difficile*-mediated ileocaecitis (Borriello, 1984; Borriello et al., 1988b) and it may be that the organism must associate with the gut wall before it can fully express its potential virulence (Borriello et al., 1988b). It seems probable that, in order to prevent colonization by *C. difficile*, the organisms must be denied access to specific mucosal receptors and that this is accomplished by the presence of a normal bacterial flora. Based on these propositions, a number of regimes involving recolonization with members of the normal gut flora have been tried for prophylaxis or treatment in both man and animals. Some workers have attempted to simulate the protection afforded by the full normal flora by using a limited number of species derived from it, or even species not normally found in the gut. In this way it is hoped to identify individual species or well defined mixtures of microorganisms that could be used in prevention or treatment. As this topic has been extensively reviewed elsewhere (Borriello, 1989) only the experiments in man will be discussed here.

Various approaches have been used to treat *C. difficile*-associated gastrointestinal disease in man, such as the use of a non-toxigenic avirulent strain of *C. difficile*, the use of a 'complete' flora or a simplified flora in the form of a faecal enema, or a strain of lactobacillus from a human source *Lactobacillus* GG. Although these approaches have been generally successful, in all cases experience is limited and no controlled trial data are available. However, the experience with *Lactobacillus* GG has been greatly expanded. In addition *Saccharomyces boulardii* has been used to prevent antibiotic-associated diarrhoea.

In the late 1950s, when staphylococcal pseudomembranous colitis was becoming a problem, particularly when it followed tetracycline therapy, Eiseman et al. (1958) noted a marked improvement in four patients following administration of faecal enemas. Bowden and colleagues (1981) used a similar approach in the treatment of 16 patients with pseudomembranous colitis over an 18-year period. Of these only one was sufficiently recent to have been examined for the presence of *C. difficile* cytotoxin, and the result of the test was positive. Thirteen of the patients, including the *C. difficile* cytotoxin-positive patient, responded dramatically. In total, 23 patients have been treated with faecal enemas and 19 have improved. However, only four of these patients are known to have been *C. difficile*-positive, the best documented cases being those of Schwan and colleagues (1984) and Tvede and Rask-Madsen (1989). One of these cases was one of the four that failed to respond. In an attempt to simplify the flora used, Tvede and Rask-Madsen (1989) attempted to emulate the success of rectal instillation of homologous faeces with a simplified mixture of faecal bacteria. The mixture consisted of *Str. faecalis, C. innocuum, C. ramosum, C. bifermentans, Peptostreptococcus productus, B. ovatus, B. vulgatus, B. thetaiotaomicron* and two strains of *E. coli*. However, the consistent finding that *Bacteroides* spp. could not be isolated from the faeces of the patients in relapse and that, of the mixture of bacteria used bacteriotherapeutically, recovery was associated with establishment of the bacteroides, implies that the *Bacteroides* spp. play an important role in protection.

An alternative approach has been to use a single species of the faecal flora. A human *Lactobacillus* strain has been used to treat five patients with relapsing *C. difficile* diarrhoea. Four of these patients responded immediately to this treatment, but one patient relapsed (Gorbach *et al.*, 1987). This work has now been extended to at least 15 similar cases with a high rate of success.

The use of non-gut-derived micro-organisms has also been tried. Although, in this case, the micro-organism, *S. boulardii*, was being evaluated for its general ability to prevent antibiotic-associated diarrhoea (Surawicz *et al.*, 1989), it was possible to analyse its effect on *C. difficile* carriage. Of 48 *C. difficile*-positive specimens, five out of 16 on placebo had diarrhoea (31 per cent), compared with only three out of 32 (9.4 per cent) taking *S. boulardii*.

A different approach was used by Seal and co-workers (1987) based on the hamster animal work of Wilson and Sheagren (1983) and Borriello and Barclay (1985) demonstrating the protective effects of non-toxigenic strains of *C. difficile*. Over a period of three days they gave a well characterized non-toxigenic strain of *C. difficile* suspended in milk to two elderly patients who had suffered continual relapses of *C. difficile* diarrhoea. At the time of administration both patients had just completed a course of antibiotics to clear the toxigenic *C. difficile* but one patient was found still to be colonized. Daily faecal samples showed that the non-toxigenic strain colonized both patients and eventually neither toxigenic nor non-toxigenic strains were found. The diarrhoea resolved and neither patient relapsed. The results of this study indicate that non-toxigenic *C. difficile* may be capable of replacing toxigenic strains in the human gut and that, while they prevent these strains from re-establishing, they allow time for the normal flora to be reinstated and colonization resistance to be restored.

As relapse after antimicrobial therapy directed against *C. difficile* can occur in as many as 20 per cent of cases (Bartlett *et al.*, 1980; Fekety *et al.*, 1981) bacteriotherapy based on the approaches outlined above will provide a useful alternative to further antimicrobial therapy in those cases.

Conclusions

It is evident that an individual's stable normal flora provides protection against colonization by potential pathogens, this being especially true in the gut. There is also evidence to show that for many potential pathogens the anaerobic component of the flora contributes to this effect. However, this may not be true in all cases, and for others the evidence is relatively weak. The work with *C. difficile* highlights some of the problems as there is good evidence to show that anaerobes are important in excluding *C. difficile* from the gut, but that this can also be achieved by lactobacilli or *S. boulardii*. It would seem that the normal mechanism of exclusion is achieved by anaerobes in the gut (colonization resistance) but that other organisms can be used to achieve similar results when the gut flora has been compromised (bacteriotherapy or prophylaxis). During interpretation of experimentation one must distinguish between what contributes to colonization resistance under normal conditions and what can be used to achieve similar effects under experimental conditions, as these are frequently not one and the same.

References

Abrams G D, Bishop J E. (1966). Effect of the normal microbial flora on the resistance of the small intestine to infection. *Journal of Bacteriology* **92**: 1604–1608.

Barclay F E. (1985). Colonization resistance of the digestive tract with respect to infection with *Clostridium difficile*. FIMLS Thesis.

Barclay F E, Borriello S P. (1982) *In vitro* inhibition of *C. difficile*. *European Journal of Chemotherapy and Antibiotics* **2**: 155.

Bartlett J G, Tedesco F J, Shull S, Lowe B, Chang T-W. (1980). Symptomatic relapse after oral vancomycin therapy of antibiotic-associated pseudomembranous colitis. *Gastroenterology* **78**: 431–434.

Booth S J, Johnson J L, Wilkins T D. (1977). Bacteriocin production by strains of bacteroides isolated from human faeces and the role of these strains isolated in the bacterial ecology of the colon. *Antimicrobial Agents and Chemotherapy* **11**: 718–724.

Borriello S P. (1979). *Clostridium difficile* and its toxin in the gastrointestinal tract in health and disease. *Research and Clinical Forums* **1**: 33–35.

Borriello S P. (1984). *Clostridium difficile* and gut disease. In *Microbes and infections of the gut*. p. 327. Edited by Goodwin C S. Blackwell Scientific Publications, Oxford.

Borriello S P. (1989). Influence of the normal flora of the gut on *Clostridium difficile*. In *The regulatory and protective role of the normal microflora*, p. 239. Edited by Grubb R *et al*. Macmillan Press.

Borriello S P, Barclay F E. (1985). Protection of hamsters against *Clostridium difficile* ileocaecitis by prior colonization with non-pathogenic strains. *Journal of Medical Microbiology* 19: 339–350.

Borriello S P, Barclay F E. (1986). An *in vitro* model of colonization resistance to *Clostridium difficile* infection. *Journal of Medical Microbiology* 21: 299–309.

Borriello S P, Barclay F E, Welch A R. (1988a). Evaluation of the predictive capability of an in-vitro model of colonization resistance to *Clostridium difficile* infection. *Microbial Ecology in Health and Disease* 1: 61–64.

Borriello S P, Welch A R, Barclay F E, Davies H E. (1988b). Mucosal association by *Clostridium difficile* in the hamster gastrointestinal tract. *Journal of Medical Microbiology* 25: 191–196.

Bowden T A, Mansberger A R, Lykins L E. (1981). Pseudomembranous enterocolitis: mechanism of restoring flora homeostasis. *American Surgeon* 47: 178–183.

Bullen C L, Tearle P V, Willis A T. (1976). Bifidobacteria in the intestinal tract of infants: an *in-vivo* study. *Journal of Medical Microbiology* 9: 325–333.

Byrne B M, Dankert Y. (1979). Volatile fatty acids and aerobic flora in the gastrointestinal tract of mice under various conditions. *Infection and Immunity* 23: 559–563.

Cantani A. (1885). Un tentativo di batterioterapia. *Giornale Internazionale delle Scienze Mediche, Napoli* 7: 493–496.

Eiseman B, Silem W, Bascomb W S, Kanvor A J. (1958). Fecal enema as an adjunct in the treatment of pseudomembranous enterocolitis. *Surgery* 44: 854–858.

El Sanousi S M, El Azhari G, Adlan A M. (1978). The inhibition of *Clostridium chauvoei* by *Bacillus cereus* metabolites. *Australian Veterinary Journal* 54: 455–456.

Emmerich R. (1887). Die Heilung des Milzbrandes. *Archives of Hygiene (Berlin)* 6: 442–501.

Fekety R *et al*. (1981). Treatment of antibiotic-associated enterocolitis with vancomycin. *Reviews of Infectious Diseases* 3 Suppl: S273–S281.

Freter R, Abrams G D. (1972). Function of various intestinal bacteria in converting germfree mice to the normal state. *Infection and Immunity* 6: 119–126.

Freter R, Brickner H, Botney M, Cleven D, Aranki A. (1983). Mechanisms which control bacterial populations in continuous flow culture models of mouse large intestinal flora. *Infection and Immunity* 39: 676–685.

Formal S B, Abrams G D, Schneider H, Sprinz H. (1963). Experimental shigella infections. VI. Role of the small intestine in an experimental infection in guinea-pigs. *Journal of Bacteriology* 85: 119–125.

Futura E S, Freter R. (1973). Protection against enteric-bacterial infections by SIgA. *Journal of Immunology* 111: 395–403.

Gorbach S L, Barza M, Giuliano M, Jacobus N V. (1988). Colonisation resistance of the human intestinal microflora: testing the hypothesis in normal volunteers. *European Journal of Clinical Microbiology and Infectious Diseases* 7: 98–102.

Gorbach S L, Chang T-W, Goldin B. (1987). Successful treatment of relapsing *Clostridium difficile* colitis with lactobacillus GG. *Lancet* 2: 1519.

Hentges D J, Maier B R. (1970). Inhibition of *Shigella flexneri* by the normal intestinal flora. III Interactions with *Bacteroides fragilis* strains *in vitro*. *Infection and Immunity* 2: 364–370.

Hsu C-Y, Wiseman G M. (1967). Antibacterial substances from staphylococci. *Canadian Journal of Microbiology* 13: 947–955.

Ikari N S, Kenton D M, Young V M. (1969). Interaction in the germfree mouse intestine of colicinogenic and colicin-sensitive microorganisms. *Proceedings of the Society of Experimental Biology and Medicine* 130: 1280–1284.

James S M, Tagg J R. (1988). A search within the genera *Streptococcus*, *Enterococcus* and *Lactobacillus* for organisms inhibitory to mutans streptococci. *Microbial Ecology in Health and Disease* 1: 153–162.

Lee A, Gemmell E. (1972). Changes in the mouse intestinal flora during weaning: role of volatile fatty acids. *Infection and Immunity* 5: 1–7.

Maier B R, Hentges D J. (1972). Experimental shigella infections in laboratory animals. I. Antagonism by normal flora components in gnotobiotic mice. *Infection and Immunity* 6: 168–173.

Mitchell D C G, Kenworthy R. (1976). Investigations on a metabolite from *Lactobacillus bulgaricus* which neutralizes the effect of enterotoxin from *Escherichia coli* pathogenic for pigs. *Journal of Applied Bacteriology* 41: 163–174.

Müller-Schoop J W, Good R A. (1975). Functional studies of Peyer's patches: evidence for their participation in intestinal immune responses. *Journal of Immunology* 114: 1757.

Nord C E, Kager L, Malmborg A S. (1988). Effects of antimicrobial prophylaxis on colonisation resistance. *Journal of Hospital Infection* 11 Suppl A: 259–264.

Pasteur L. Joubert J F. (1877). Carbon et septicémie. *Comptes Rendus Academie des Sciences (Paris)* 85: 101.

Pongpech P, Hentges D J. (1989). Inhibition of *Shigella sonnei* and enterotoxigenic *Escherichia coli* by volatile fatty acids in mice. *Microbial Ecology in Health and Disease* 2: 153–161.

Sanders E. (1969). Bacterial interference. 1. Its occurrence among the respiratory tract flora and characterization of inhibition of Group A streptococci by Viridan's streptococci. *Journal of Infectious Diseases* 120: 698–707.

Sanders C C, Sanders W E. (1982). Enocin: an antibiotic

produced by *Streptococcus salivarius* that may contribute to protection against infections due to group A streptococci. *Journal of Infectious Diseases* **146**: 683–690.

Schwan A, Sjolin S, Trottestam U, Aronsson B. (1984). Relapsing *Clostridium difficile* enterocolitis cured by rectal infusion of normal faeces. *Scandinavian Journal of Infectious Diseases* **16**: 211–215.

Seal D, Borriello S P, Barclay F E, Welch A, Piper M, Bonnycastle M. (1987). Treatment of relapsing *Clostridiium difficile* diarrhoea by administration of a non-toxigenic strain. *European Journal of Clinical Microbiology* **6**: 51–53.

Sprinz H, Kundel D W, Dammin G J, Horowitz H, Schneider H, Formal S B. (1961). The response of the germ-free guinea-pig to oral bacterial challenge with *Escherichia coli* and *Shigella flexneri*. *American Journal of Pathology* **39**: 681–695.

Surawicz C M, Elmer G W, Speelman P, McFarland L V, Chinn J, Van Belle G. (1989). A clinical trial to test the ability of *Saccharomyces boulardii* to prevent antibiotic associated diarrhoea. *Microecology and Therapy* **17**: (in press).

Syed S A, Abrams G D, Freter R. (1970). Efficiency of various intestinal bacteria in assuming normal functions of enteric flora after association with germfree mice. *Infection and Immunity* **2**: 376–386.

Tompkins G R, Tagg J R. (1989). The ecology of bacteriocin-producing strains of *Streptococcus salivarius*. *Microbial Ecology in Health and Disease* **2**: 19–28.

Tvede M, Rask-Madsen J. (1989). Bacteriotherapy for chronic relapsing *Clostridium difficile* diarrhoea in six patients. *Lancet* **i**: 1156–1160.

Van der Waaij D, Berghius-de Vries J M, Lekkerkerk-van der Wees J E C. (1971). Colonization resistance of the digestive tract in conventional and antibiotic-treated mice. *Journal of Hygiene* **69**: 405–411.

Welling G W, Groen G, Tuinte J H M, Koopman J P, Kennis H M. (1980). Biochemical effects on germ-free mice of association with several strains of anaerobic bacteria. *Journal of General Microbiology* **117**: 57–63.

Wensinck F, Ruseler-van Embden J G H. (1971). The intestinal flora of colonization-resistant mice. *Journal of Hygiene* **69**: 413–421.

Wilhelm M P, Lee D T, Rosenblatt J E. (1987). Bacterial interference by anaerobic species isolated from human feces. *European Journal of Clinical Microbiology* **6**: 266–270.

Williams R C, Gibbons R J. (1972). Inhibition of bacterial adherence by secretory immunoglobulin A: a mechanism of antigen disposal. *Science* **177**: 697–701.

Wilson K H, Sheagren J N. (1983). Antagonism of toxigenic *Clostridium difficile* by non-toxigenic *C. difficile*. *Journal of Infectious Diseases* **147**: 733–736.

21

Anaerobic bacteria in diarrhoea and colitis

S P Borriello

Diarrhoea and colitis due to *Clostridium difficile*
Non-food poisoning *Clostridium perfringens* enterotoxin-associated diarrhoea
Clostridium perfringens type C and enteritis necroticans (Pig-Bel)
Clostridium septicum and neutropenic enterocolitis
   Other clostridial causes of diarrhoea and colitis
Enterotoxigenic *Bacteroides fragilis* and diarrhoea
Anaerobiospirillum and diarrhoea
Evidence implicating anaerobes in neonatal necrotizing enterocolitis
Evidence implicating anaerobes in inflammatory bowel disease
References

The clostridia are the most important of the anaerobes known to be involved in causing diarrhoea or colitis. Consequently most of this chapter is devoted to them. However, other anaerobes are implicated in some cases, and this chapter will review the evidence implicating *Bacteroides vulgatus* and *B. fragilis* in enteric disease and the possible role of *Anaerobiospirillum* in diarrhoea. The possible role of spirochaetes in diarrhoea is reviewed in Chapter 8.

Diarrhoea and colitis due to *Clostridium difficile*

Next to *Clostridium perfringens*, which causes food poisoning (Chapter 22) *C. difficile* ranks as the most common identifiable anaerobe causing diarrhoea, and is the most common identifiable bacterial cause of nosocomial diarrhoea. Willis (1988) stated that 'Among all the many advances that have been made in our understanding of anaerobic bacteria and anaerobic bacterial disease during the last 40 years, elucidation of the syndrome of antibiotic-associated pseudomembranous colitis ranks as a major triumph'. These observations highlight the importance of *C. difficile* as an enteric pathogen. Of all the anaerobes involved in gastrointestinal disease this is the one clinicians are most likely to encounter.

C. difficile was confirmed as the aetiological agent of pseudomembranous colitis (PMC) in 1978 (Bartlett *et al.*, 1978; George *et al.*, 1978; Larson *et al.*, 1978). Since then *C. difficile* has also been implicated in many cases of antibiotic-associated diarrhoea (AAD) (Burden, 1984). The major virulence factors are two heat labile protein toxins designated toxins A and B, the main characteristics of which are presented in Table 21.1. Before describing the role of these toxins in the disease it is pertinent to discuss the association of PMC and AAD with antibiotics and the factors that permit colonization by *C. difficile*.

The normal stable gut flora serves as an efficient barrier against the establishment of pathogens. This barrier effect has been termed colonization resistance (CR). Though effective against many enteric pathogens, it appears to be most effective against *C.*

Table 21.1 Main characteristics of *C. difficile* toxins A and B

Characteristic	Toxin A	Toxin B
Mol. wt (10^3)	308	c. 250
Heat lability (60°C)	+	+
pH lability (pH 4 and 10)	−	+
Isoelectric point	c. 5.2	c. 4.1
Inactivation with:		
trypsin	−	+
chymotrypsin	+	+
pronase	+	+
oxidizing agents	+	+
Induction of fluid		
accumulation	+	−
Cytotoxicity	+	+
Increase in vascular		
permeability	+	+
Lethal following intra-		
peritoneal injection	+	+
Identified receptors	+*	−

*Galα1-3Galβ1-4GlcNAc

difficile since, in the majority of cases, this organism cannot become established in the gut unless the normal flora and its associated CR is first disrupted, e.g., by antibiotics. A variety of antibiotics has been reported to predispose to *C. difficile* infection (Tedesco, 1984) notable exceptions being parenterally administered aminoglycosides. Further interest in CR to *C. difficile* has been stimulated by the high incidence of relapse following treatment. It is hoped that a fuller understanding of the operative mechanisms will lead to effective and safe bacteriotherapeutic or bacterioprophylactic approaches to the control of the disease. (An account of CR with special reference to *C. difficile* is given in Chapter 20.)

Most cases of *C. difficile*-mediated gut disease result from nosocomial infections. This is apparent from the many reports of outbreaks of disease (e.g., Delmee *et al.*, 1985) and the fact that *C. difficile* is not a normal component of the faecal flora of healthy adults. Many typing schemes have been developed to study the epidemiology of this disease (Mulligan, 1988); these include serotyping, immunoblot typing, protein profiling, plasmid typing and restriction endonuclease DNA fragment profiles. In most cases the typing schemes support the evidence of cross-infection. In general it is recommended that patients with *C. difficile* diarrhoea should be isolated (in keeping with practise for any infectious diarrhoea) and if an outbreak occurs the group of patients infected should be 'barrier nursed'. The standard regimen for treatment is oral vancomycin (0.5–2.0 g/day) for 7–14 days, although metronidazole has also been used to good effect (1.0–1.5 g/day) for about 5 days.

Laboratory diagnosis consists of detection of the organism or its toxins (Gerding *et al.*, 1988). Many methods have been described for the indirect and direct detection of *C. difficile* in stool, as well as its isolation on selective media or by selective procedures (Borriello and Honour, 1984). However, of these methods, isolation on a selective medium incorporating cycloserine (250 μg/ml) and cefoxitin (10 μg/ml) has proved to be the best. Culture does not give information on whether the isolate is toxigenic or, if it is, whether or not toxin was produced *in vivo*. Because of this many laboratories rely on detection of faecal toxin for diagnosis. Various methods are available for this—immunological, such as ELISA, or biological, such as cytotoxicity for cell cultures (Borriello and Welch, 1984). The most commonly employed method is the detection of cytotoxicity in tissue culture coupled to its neutralization with the cross-reacting *C. sordellii* antitoxin. Problems in laboratory diagnosis relate to analysis of faecal specimens from infants who can harbour *C. difficile* and excrete toxins A and B and yet remain asymptomatic. Therefore, analysis of material from this age group is of little value in this respect.

It has been suggested that toxin A is the more important of the two toxins in the disease process. Evidence for this comes from the observation that toxin A is responsible for the fluid accumulation and tissue damage in ligated intestinal loops (Mitchell *et al.*, 1986; Lima *et al.*, 1988). The mechanism of action is not fully understood, but it does seem clear that the fluid accumulation is the result of leakage secondary to tissue damage and not due to activation of a secretory mechanism (Mitchell *et al.*, 1987). The mechanism of action of both toxins has recently been reviewed (Donta, 1988; Borriello, 1990) the major points being that the cytotoxicity of both is due to disruption of the cytoskeleton and that this cytotoxicity, and fluid accumulation due to toxin A, involves calcium.

A reflection of the interest in this organism is the publication of two books on the subject (Borriello,

1984; Rolfe and Finegold, 1988), to which the interested reader is guided.

Non-food poisoning *Clostridium perfringens* enterotoxin-associated diarrhoea

It was shown recently that enterotoxigenic strains of *C. perfringens* could cause diarrhoea by a mechanism other than that of classic food poisoning (Borriello *et al.*, 1984; Jackson *et al.*, 1986). The main differences between food poisoning and the non-food poisoning disease appears to be the lack of apparent involvement of contaminated foods, and the fact that many of the cases that occur in institutions are sporadic. Most importantly, the non-food-poisoning disease seems to be more severe than food poisoning, with more profuse diarrhoea. This may continue for many weeks, with the frequent presence of both blood and mucus (Borriello *et al.*, 1987; Larson and Borriello, 1988). Histological examination of rectal mucosa reveals only mild oedema. However, the frequent presence of blood indicates that there are likely to be more severe lesions in proximal regions of the gut. Additional features are the relatively frequent occurrence of abdominal pain, in some cases this is severe, and the infrequency of vomiting (Borriello *et al.*, 1987; Larson and Borriello, 1988).

The disease occurs almost exclusively in geriatric patients; the vast majority of cases occur in those over 60 years old, and the disease is antibiotic-associated in about 60 per cent of cases (Borriello *et al.*, 1987; Larson and Borriello, 1988). Although this association with antibiotics could be coincidental, in a recent study comparing a cluster of cases of patients with *C. perfringens* diarrhoea with patients who were on the same geriatric unit, but who were without diarrhoea, the association was shown to be highly significant ($P = 0.0001$) (Williams *et al.*, 1985).

It is difficult to explain why the disease should be restricted almost exclusively to old people. It may be because this age group is more susceptible to colonization with *C. perfringens*, and the findings of Yamagishi and colleagues (1976) and Stringer (1981) showing persistent carriage of large numbers of *C. perfringens* in healthy institutionalized geriatric patients would support this. Although this would explain the higher incidence of non-AAD, it does not explain why all age groups should not be equally susceptible following antibiotics. Because of the frequent carriage of high numbers of *C. perfringens* in the population group most commonly affected, little reliance can be placed on the presence of high faecal levels of *C. perfringens* spores as a diagnostic criterion, although in the study by Jackson *et al.* (1986) 95 per cent of the enterotoxin-positive specimens had *C. perfringens* spore levels of greater than 10^6/g of faeces, and in the series reported by Larson and Borriello (1988) neither diarrhoea nor *C. perfringens* enterotoxin were present in subjects with less than 10^7 *C. perfringens* spores/g of faeces. The most reliable method of laboratory diagnosis for this condition is detection of the enterotoxin in faeces by tissue culture (Borriello *et al.*, 1984; Larson and Borriello, 1988) or by ELISA (Jackson *et al.*, 1986). A reverse, passive latex agglutination test developed for the detection of faecal enterotoxin in cases of *C. perfringens* food poisoning has been evaluated and shown to be as sensitive and specific as an ELISA method (Berry *et al.*, 1986). This latex agglutination test, which is available from Oxoid, would no doubt also be useful in diagnosing *C. perfringens* diarrhoea which is unrelated to food poisoning.

The management of *C. perfringens* diarrhoea shares many of the features of *C. difficile* diarrhoea in that metronidazole (Borriello and Williams, 1985) or vancomycin can be used in treatment, patient relapse (unpublished data) and cross-infection is a potential problem (Borriello *et al.*, 1985). A detailed epidemiological investigation of a cluser of cases (Borriello *et al.*, 1985) demonstrated wide-spread contamination of the hospital environment by the incriminated serotype of *C. perfringens*, which was found in 59 per cent of areas around patients during the outbreak, in 27 per cent of similar areas once the diarrhoea had resolved but in only 9 per cent of areas where there was no history of disease. This particular strain was also isolated from the hands of patients and nursing personnel. In the study of different serotypes by Jackson *et al.* (1986) carriage appeared to be ward-related. The serotypes associated with 46 cases from various centres in the UK are presented in Table 21.2. Of these, only serotype 41 is considered a common outbreak strain, while serotypes 27 and 34 have been found in a hospital outbreak (Stringer, 1985). Interestingly, plasmid analysis of some of the strains described in the

Table 21.2 *C. perfringens* serotypes isolated from 46 patients with *C. perfringens* enterotoxin-associated diarrhoea

Serotype	Number of isolates*	Number of centres contributing samples
28	7	3
41	5	1
27	3	2
24	2	1
34	2	2
36	2	1
48	2	2
53	2	1
6, 16, 18, 29, 30, 31, 38, 47, 51, 59, 62, 68, 72, 80, 83, 86	1	6
Not typable	10	

*Total exceeds 46 because more than one serotype was isolated from a few patients

epidemiological study of Borriello *et al.* (1985) demonstrated a promising additional differentiating capability when used to supplement serotyping in the epidemiological study of this condition (Mahony *et al.*, 1987).

Diarrhoea is a common problem on geriatric units and it is evident that *C. perfringens* is an important cause.

Clostridium perfringens type C and enteritis necroticans (Pig-Bel)

Enteritis necroticans, a necrotizing haemorrhagic jejunitis caused by *C. perfringens* type C, is most commonly recognized in the young adult population of the Highlands of Papua New Guinea. The disease is not restricted to Papua New Guinea, but also occurs in other parts of Southeast Asia. It is also likely that 'Darmbrand' (burnt bowel) disease, which occurred in Germany and Norway shortly after World War II was in fact enteritis necroticans.

The disease, which is commonly referred to as Pig-Bel due to its association with pig feasts, can occur in two forms:

1. Acute Pig-Bel, which is either a fulminating, often fatal disease with death occurring within 24 h, or is a subacute form which is difficult to distinguish from gastroenteritis and from which the patient usually recovers;
2. Chronic Pig-Bel, which is rarer, and in which patients present some time after an acute episode with gastrointestinal symptoms such as subacute intestinal obstruction.

The primary site of damage is usually the jejunum, although the ileum can also be affected. Symptoms usually consist of abdominal pain with distention, vomiting and blood in the faeces. Unlike most of the other clostridial gut diseases which cause diarrhoea, constipation can ensue within 48 h. Macroscopic changes can vary from mild oedema to necrotic patches, over which a membrane normally forms, and in late stages there may be gangrenous segments. The affected areas frequently appear as 'skip lesions' in segmental fashion with diseased and apparently normal tissues lying adjacent to each other. Microscopic examination reveals thrombi in the smaller blood vessels, oedema and polymorph infiltration in the submucosa and large gas cysts in any layer of the bowel wall. In addition many bacteria, including gram-positive rods, can be seen in the necrotic regions.

C. perfringens type C also causes several gastrointestinal diseases in animals, including necrotic enteritis in piglets, calves, chickens, ducks and sheep, which have been reviewed elsewhere (Borriello and Carman 1985, 1988). The most analogous to Pig-Bel is 'struck' in sheep. In late winter or early spring sheep moved to certain pastures or given a new diet, present with a fatal enterotoxaemia. It is presumed that it is this dietary change which triggers infection. Features of the diseased small intestine range from an acute to a mild congestion with ulceration of one or more sections of the jejunal mucous membranes. The site of the primary necrotic lesion is the superficial surface of the mucosa. These lesions eventually extend down into the mucosal lining in which adherent bacteria can be detected.

In addition to these natural diseases there are some well documented animal models, especially in the guinea-pig, that have proved useful in delineating the pathogenesis of Pig-Bel (Lawrence and Cooke, 1980).

C. perfringens type C is present in the jejunal

contents of diseased patients, but is also widespread in the environment and is part of the normal faecal flora in the population of Papua New Guinea. The conditions that permit this common potential pathogen to express its virulence are interesting. Disease usually follows feasting on a high protein diet such as pork, although other high protein diets are also associated with the disease. This periodic high protein load, which is frequently contaminated with *C. perfringens* type C spores, is ingested at a time when the host chymotrypsin levels are depressed due to malnutrition. In addition, sweet potato, which traditionally accompanies the meal, contains trypsin inhibitors, and round worms (*Ascaris lumbricoides*) which commonly infect this population secrete protease inhibitors. These events result in the ingestion of *C. perfringens* type C with nutrients that facilitate rapid growth in subjects with depressed or inactivated levels of proteases, preventing the enzymatic destruction of the released β-toxin, which is the toxin important in the disease process.

An understanding of the pathogenesis of this disease has enabled the development of prophylaxis based on immunization with *C. perfringens* type C toxoid. Field trials have shown this to be extremely effective for at least two years. However, protection is lost after four years, necessitating reimmunization (Lawrence *et al.*, 1979; Walker, 1985). In this respect it is interesting to compare Darmbrand, which spontaneously resolved and has not reappeared, with Pig-Bel, which has persisted. The difference is most probably a reflection of nutrition and hygiene and implies that the eventual eradication of Pig-Bel will depend on improved nutrition and education with respect to basic hygiene, especially in relation to traditional practices during ceremonial feasts. The interested reader is directed to a comprehensive review of Pig-Bel by Walker (1985).

Clostridium septicum and neutropenic enterocolitis

The gastrointestinal symptoms, characterized by necrosis, that may occur in patients with leukaemia probably represent a number of distinct clinical entities, i.e., PMC due to *C. difficile*, ischaemic colitis and neutropenic colitis.

Patients with neutropenic enterocolitis often present with symptoms resembling appendicitis. The lesion, which shows necrosis, haemorrhage, pronounced oedema and wall thickening, is usually restricted to the caecum. The aetiology of this disease was unknown until the recent work implicating *C. septicum* (Rifkin, 1980; King *et al.*, 1984). Septicaemia due to *C. septicum* occurs relatively frequently in patients with neutropenia, leukaemia and large bowel malignancy (Alpern and Dowell, 1969; Koransky *et al.*, 1979; Katlic *et al.*, 1981) sometimes in association with ileocaecal disease (Bignold and Harvey, 1979; Rifkin, 1980; Hopkins and Kushner, 1983; King *et al.*, 1984). It was this association that led Rifkin (1980) to propose that *C. septicum* was responsible for the formation of the enterocolitic lesion. Support for this hypothesis was to be later provided by King and colleagues (1984) who examined bowel resection material taken early in the course of the illness from three patients and showed the presence of *C. septicum* in the bowel wall by specific immunofluorescence staining techniques.

Although the evidence is far from conclusive and it is possible to argue that the association between *C. septicum* and neutropenic enterocolitis is a consequence of the disease, it is difficult to believe that this toxigenic micro-organism does not play an aetiological role. It is likely that the source of the organism is endogenous, for although it is not a common component of the faecal flora of healthy Western subjects it appears to be isolated more frequently from patients with large bowel malignancies (George and Finegold, 1985) and has been reported to be present frequently in the normal appendix (Lanz and Travel, 1904) although this latter work is dated and may not reflect the present situation. However, *C. septicum* is an organism of worldwide distribution, whose natural habitats are soil and the intestinal tract of a number of animals. Despite this rather widespread occurrence it is only known to cause one other intestinal infection, 'braxy', a disease of the rumen of lambs during their first year. *C. septicum* can be isolated in pure culture from regions of haemorrhagic inflammation in the abomasum of affected sheep, and immunization against *C septicum* prevents disease. This raises the possibility of active immunization for at-risk patients. Treatment has been primarily restricted to surgery (Ikard, 1981), although administration of broad-spectrum antibiotics coupled to bowel rest and fluid therapy has been successful (Shaked *et al.*, 1983). These treatment regimes preceded evidence that *C. septicum* was the cause of the gastrointestinal disease and it may now be possible to make a more rational approach. However, prevention is preferable and although no vaccine is available, or is likely to be for such a relatively rare disease, it may be possible to reduce the risk of disease by varying the antimicrobials received by these patients. It is possible that components of the normal gut flora serve to exclude or repress *C. septicum* in the gut (this concept of 'colonization resistance' is discussed in Chapter 20) and that the varying incidence of the disease between centres reflects antibiotic policies which differently affect CR.

There is a report of an association between *C. tertium* and neutropenic enterocolitis, although no good evidence has been presented to implicate it in the disease process (Yates *et al.*, 1988).

Other clostridial causes of diarrhoea and colitis

There are several isolated reports of diarrhoea and colitis due to other clostridia (Table 21.3). There is also a case report of gastrointestinal infection with *C. perfringens* type C, which causes Pig-Bel, giving rise to antibiotic-associated pseudomembranous colitis (Schwartz *et al.*, 1980). As all of these reports are single case studies it is difficult to determine the importance of the findings. A detailed appraisal of all of these cases is to be found elsewhere (Borriello and Carman, 1988). Due to the wide range of toxins produced by the clostridia and the known potential for natural transfer of the capability for toxin production between species, it is likely that many different clostridia are the cause of isolated cases of gastrointestinal disease.

Enterotoxigenic *Bacteroides fragilis* and diarrhoea

B. fragilis is a gram-negative obligate anaerobe present as a component of the normal faecal flora. Although well established as a cause of extraintestinal infections e.g., abscesses, it is only recently that evidence has been provided to prove that it can behave as an enteric pathogen. The first observations in this respect were made in lambs (Myers *et al.*, 1984), calves (Myers *et al.*, 1985) and piglets (Myers and Shoop, 1987). Calves and lambs yielded strains of *B. fragilis* that elaborated, *in vitro*, an enterotoxin that could be detected in calf and lamb ligated ileal loop assays following concentration. One of three porcine enterotoxigenic isolates injected intraileally caused severe enteric disease in adult rabbits with ligated caeca (Myers and Shoop, 1987). Since this work, the same group (Myers *et al.*, 1987) has isolated, from the stool specimens of eight patients with diarrhoea, strains of *B. fragilis* which are enterotoxigenic in the lamb ligated ileal loop assay. Seven of the eight yielded no recognized enteric pathogens. Four of these seven patients were three years old or younger. The individuals with enterotoxigenic *B. fragilis* had watery diarrhoea which usually lasted for 1–4 weeks, often with mild to moderate intestinal cramping. The infants had hyperthermia, vomiting and blood in the stools. In all cases the diarrhoea was self-limiting. The enterotoxin was heat labile (70°C for 30 min), induced the formation of a pseudomembrane (fibrin and leucocytes) in lamb ligated ileal loops and caused enteric disease with diarrhoea and death in rabbits with ligated caeca. Non-enterotoxigenic strains of *B. fragilis* had no activity. Until these observations are repeated by other workers and a simple method to readily dis-

Table 21.3 Other clostridial causes of diarrhoea and colitis

Clostridium sp. implicated	Disease	Reference
C. perfringens type C	Antibiotic-associated pseudomembranous colitis	Schwartz *et al.* (1980)
C. sphenoides	Diarrhoea	Sullivan *et al.* (1980)
C. spiroforme	Diarrhoea	Babudieri *et al.* (1986)
Clostridium spp.	Pseudomembranous colitis	Chiu and Abraham, (1982)

tinguish enterotoxigenic and non-enterotoxigenic strains of *B. fragilis* is developed, the importance of the observation and the frequency with which these organisms cause disease in man will remain unknown.

Anaerobiospirillum and diarrhoea

Anaerobiospirillum spp. are anaerobic, bipolar, spiral-shaped, flagellate gram-negative bacteria which have been isolated from the faeces of some patients with diarrhoea (Malnick et al., 1983; Malnick, 1990). The most recent figures for the UK are that *A. succiniproducens* and *Anaerobiospirillum* spp. have been isolated from the faeces of 17 patients (age range 6 months—64 years) with diarrhoea. Interestingly, all of these strains were isolated on Skirrow's selective medium for campylobacters after incubation for 48 h under microaerobic conditions. Recently, a selective medium based on the incorporation of vancomycin, sulphamethoxazole, polymyxin and Victoria blue into Fastidious Anaerobe Agar (Lab M) with lysed horse blood, has been developed (Malnick, 1990). This should help determine the frequency with which these organisms are isolated from faeces and their true association with diarrhoea. At present, in the absence of an animal model or of any putative mechanism for causing diarrhoea, its role in diarrhoea must remain speculative. Although similar problems are associated with explaining how campylobacters cause diarrhoea, their association with disease is much more clear cut.

Evidence implicating anaerobes in neonatal necrotizing enterocolitis

The aetiology of neonatal necrotizing enterocolitis (NNEC) remains unknown. Therefore, before it is possible to consider the possible role of anaerobes, it is necessary to review the evidence implicating an infectious aetiology. The strongest evidence is the temporal and geographical clustering of cases and the effectiveness of antimicrobial agents and infection control measures in controlling NNEC outbreaks (Kliegman, 1985). Additional evidence is the high incidence of positive blood cultures, appearance of septic-like symptoms, and the frequent presence of pneumatosis cystoides intestinalis (PCI) where the gas (hydrogen) is of microbial origin (Yale and Balish, 1985). However, all of these later features could be a consequence, and not the cause, of the disease and as such should be interpreted with caution. The favoured explanation at present is that NNEC is the outcome of an interaction between predisposing factors, such as ischaemia, leading to devitalized segments of bowel allowing gut flora opportunist pathogens to exhibit their virulence leading to the necrosis and pneumatosis that are features of the disease.

If bacteria play a role in the disease, the ones that have received the most attention are the clostridia, in particular *C. perfringens* and *C. butyricum*, and to a lesser extent *C. difficile*. Three recent reviews fully discuss the evidence implicating clostridia in NNEC (Borriello and Stephens, 1984; Kliegman, 1985; Borriello and Carman, 1988). Therefore, only the salient points and the most recent work will be discussed here.

The major problem faced by researchers with respect to NNEC is that of implicating a normal component of the infant gut flora in the disease, especially as the three organisms that have received most attention are frequently present at the same time (Smith et al., 1980). In addition, both major toxins of *C. difficile* can be present in high levels in healthy infants and neonates. It has been proposed that *C. perfringens* does not in itself initiate the lesion but takes advantage of an initial lesion induced, for example, by ischaemia. There is evidence for this from the work of Sibbons et al. (1990) who investigated the aetiology and pathogenesis of NNEC in a low birthweight (LBW) neonatal piglet model. They clearly demonstrated that the full histopathological spectrum of human NNEC could be induced in LBW piglets by using clinically associated experimental compromising insults. The initiating insult in this model was local circulatory breakdown in the vasculature of the distal ileum. However, for complete NNEC to develop, arterial and lymphatic occlusion were necessary, as was the presence of gas-producing organisms in the ileum. Of interest here is that *C. perfringens*, which has been shown to be capable of producing PCI in an experimental rat model (Yale and Balish, 1985), could serve as one of the gas-producing micro-organisms.

There is no doubt that *C. perfringens* can be found in necrotic tissue from cases of necrotizing enterocolitis, but this is far removed from saying that the organism *causes* the disease. However the work described above furnishes some experimental evidence of its potential to be a contributory factor.

The problems of trying to implicate *C. butyricum* in NNEC are the same as the problems of implicating the other clostridia, i.e., it is frequently present as a commensal and not all infants with necrotizing enterocolitis harbour *C. butyricum*; conversely, a high proportion of asymptomatic infants do. Furthermore, there are no veterinary precedents for assuming a potential role in gastrointestinal disease and, until recently, there were no known potential virulence factors. The first people to raise the possibility that *C. butyricum* may be involved in the disease were Howard and colleagues (1977), who isolated the organism from seven of eleven blood cultures and two of ten faecal specimens from patients with NNEC. Two isolates tested for cytotoxigenicity gave negative results. A report describing the isolation of a cytotoxigenic strain of *C. butyricum* from a case of NNEC followed three years later (Sturm *et al.*, 1980). The organism was isolated from blood and cerebrospinal fluid (CSF). The CSF isolate was cytotoxigenic to Walker cells but not to human fibroblast (foreskin) cells, whereas the blood isolate was toxic to both cell lines. Popoff (1990) found that all 55 strains of *C. butyricum* from NNEC cases produced a low titre cytotoxic effect on Vero cells, but that this was most likely due to production of butyric acid.

An apparently successful attempt to develop an animal model in germ-free rats infected with *C. butyricum* (Lawrence *et al.*, 1982) could not be repeated (Lawrence and Bates, 1983). However Popoff and colleagues have recently described the induction of experimental lesions in ligated intestinal loops in guinea-pigs (Popoff and Ravisse, 1985) and caecitis in gnotoxenic chickens (Popoff *et al.*, 1985) and quails (Bousseboua, 1989). The effects seen in guinea-pigs consisted of vascular alterations such as congestion, patchy haemorrhage and inflammatory infiltration of the gut wall. However, the other lesions of necrotizing enterocolitis, extensive necrosis and pneumatosis, were absent, although they were present in experimental caecitis induced in axenic chickens and quails monoassociated with *C. butyricum*.

The only clinically adverse effect noted initially was a slow gain in body weight. There was no diarrhoea or mortality. However, post-mortem examination revealed caecal lesions after 3-4 weeks in chickens and about one week earlier in quails; the quails also developed crop lesions. Interestingly the presence of dietary lactose (4 per cent) was a prerequisite for the development of caecitis.

An additional finding was that there was no difference in the ability to induce disease in the avian models between strains of *C. butyricum* isolated from cases of NNEC and those from healthy neonates. If the clostridia do play a primary role in the aetiology of NNEC, it is most likely that there are one or more initiating events that allow these commensal clostridia to express their pathogenic potential. In this respect the pathogenesis of necrotizing jejunitis (Pig-Bel) is of particular interest (see pages 354-355) as it clearly demonstrates how a commensal *C. perfringens* type C can cause severe gastrointestinal disease under the appropriate conditions.

Of the models described above, the LBW piglet model would appear to be the most appropriate, with its potential for combining relevant clinically associated insults with a bacterial interaction. Although the aetiology of NNEC may be multifactorial one should be mindful of the fact that the list of proposed aetiologies for PMC, before the role of *C. difficile* was discovered, was very similar to the current list for NNEC.

Evidence implicating anaerobes in inflammatory bowel disease

The causes of ulcerative colitis and Crohn's disease remain unknown and these inflammatory gut diseases have attracted the attention of microbiologists. Despite this interest it has not been possible to define a bacterial aetiology. The improvement in Crohn's disease in response to metronidazole (Krook *et al.*, 1981) implies some role for the gut flora, most likely the anaerobic component. Also, it is possible that the beneficial effects of the sulphasalazine in ulcerative colitis is due in part to the antibacterial activity of the sulphonamide component.

Changes related to disease occur in the faecal flora, e.g., elevated levels of bacteroides, fusobacteria, eubacteria and peptostreptococci (Wensinck *et al.*, 1981). An autoimmune response to a bacterial antigen of gut origin has also been proposed. In this context elevated levels of circulating antibodies to bacteroides, *Escherichia coli*, eubacterial and peptostreptococcal antigens have been demonstrated in the sera of subjects with Crohn's disease (Tvede *et al.*, 1983; Wensinck *et al.*, 1983).

Some interesting observations that provide experimental evidence of an antibody response to a particular anaerobe in the pathogenesis of the disease have been made in the carrageenan-induced colitis guinea-pig model (Breeling *et al.* 1988). It has been shown that *B. vulgatus* recolonization of gnotobiotic guinea-pigs provokes colonic ulceration and that this can be prevented with metronidazole. In addition, immunization with *B. vulgatus* prior to feeding carrageenan results in a more rapid intestinal ulceration response. The most recent findings from this group are that the antigens on the outer membrane of *B. vulgatus* (mol. wt (10^3) 100, 57, 34 and 21) were capable of enhancing the inflammatory response (Breeling *et al.*, 1988).

Although there is some evidence for a role for anaerobes in the pathogenesis of inflammatory bowel disease, a great deal more work needs to be done to confirm this suspicion.

References

Alpern R J, Dowell V R. (1969). *Clostridium septicum* infections and malignancy. *Journal of the American Medical Association* **209**: 385–388.

Babudieri S, Borriello S P, Pantosti A, Luzzi I, Testore G P, Panichi G. (1986). Diarrhoea associated with toxigenic *Clostridium spiroforme*. *Journal of Infection* **12**: 278–279.

Bartlett J G, Chang T W, Gurwith M, Gorbach S L, Onderdonk A B. (1978). Antibiotic-associated pseudomembranous colitis due to toxin-producing clostridia. *New England Journal of Medicine* **298**: 531–534.

Berry P R, Stringer M F, Uemura T. (1986). Comparison of latex agglutination and ELISA for the detection of *Clostridium perfringens* type A enterotoxin in faeces. *Letters in Applied Microbiology* **2**: 101–102.

Bignold L P, Harvey H P B. (1979). Necrotizing enterocolitis associated with invasion by *Clostridium septicum* complicating cyclic neutropenia. *Australian and New Zealand Journal of Medicine* **9**: 426–429.

Borriello S P. (ed.). (1984). *Antibiotic associated diarrhoea and colitis: the role of Clostridium difficile in gastrointestinal disorders*. Martinus Nijhoff, Boston.

Borriello S P. (1990). Mechanism of action of *Clostridium difficile* toxins A and B. In *Bacterial protein toxins IV*. Edited by Rappuoli R. Springer Verlag, Berlin (in Press).

Borriello S P *et al.* (1985). Epidemiology of diarrhoea caused by enterotoxigenic *Clostridium perfringens*. *Journal of Medical Microbiology* **20**: 363–372.

Borriello S P, Carman R J. (1985). Clostridial diseases of the gastrointestinal tract in animals. In *Clostridia in gastrointestinal disease,* p. 195. Edited by Borriello S P. CRC Press, Boca Raton.

Borriello S P, Carman R J. (1988). Other clostridial causes of diarrhoea and colitis in man and animals. In *Clostridium difficile: its role in intestinal disease*, p. 66. Edited by Rolfe R D, Finegold S M. Academic Press, San Diego.

Borriello S P, Honour P. (1984). Detection, isolation and identification of *Clostridium difficile*. In *Antibiotic associated diarrhoea and colitis: the role of Clostridium difficile in gastrointestinal disorders*, p. 37. Edited by Borriello S P. Martinus Nijhoff, Boston.

Borriello S P, Larson H E, Barclay F E, Welch A R. (1987). *Clostridium perfringens* enterotoxin-associated diarrhoea. In *Recent advances in anaerobic bacteriology*, pp. 33–42. Edited by Borriello S P, Hardie J M. Martinus Nijhoff, Boston.

Borriello S P, Larson H E, Welch A R, Barclay F, Stringer M F, Bartholomew B A. (1984). Enterotoxigenic *Clostridium perfringens*: a possible cause of antibiotic-associated diarrhoea. *Lancet* **i**: 305–307.

Borriello S P, Stephens S. (1984). The development of the infant gut flora: and the medical microbiology of infant botulism and necrotizing enterocolits. In *Microbes and infections of the gut*, p. 1. Edited by Goodwin C S. Blackwell Scientific Publications, Melbourne.

Borriello S P, Welch A R. (1984). Detection of *Clostridium difficile* toxins. In *Antibiotic associated diarrhoea and colitis: the role of Clostridium difficile in gastrointestinal disorders*, p. 49. Edited by Borriello S P. Martinus Nijhoff, Boston.

Borriello S P, Williams R K T. (1985). Treatment of *Clostridium perfringens* enterotoxin-associated diarrhoea with metronidazole. *Journal of Infection* **10**: 65–67.

Bousseboua H *et al.* (1989). Experimental cecitis in gnotobiotic quails monoassociated with *Clostridium butyricum* strains isolated from patients with neonatal necrotizing enterocolitis and from healthy newborns. *Infection and Immunity* **57**: 932–936.

Breeling J L, Onderdonk A B, Cisneros R L, Kasper D L. (1988). *Bacteroides vulgatus* outer membrane antigens associated with carrageenan-induced colitis in guinea pigs. *Infection and Immunity* **56**: 1754–1759.

Burdon D W. (1984). Spectrum of disease. In *Antibiotic associated diarrhoea and colitis: the role of Clostridium difficile in gastrointestinal disorders*, p. 9. Edited by Borriello S P. Martinus Nijhoff, Boston.

Chiu A O, Abraham A A. (1982). Pseudomembranous colitis associated with an unidentified species of *Clostridium*. *American Journal of Clinical Pathology* **78**: 398–402.

Delmee M, Homel M, Wauters G. (1985). Serogrouping of *Clostridium difficile* strains by slide agglutination. *Journal of Clinical Microbiology* **21**: 323–327.

Donta S T. (1988). Mechanism of action of *Clostridium difficile* toxins. In *Clostridium difficile: its role in intestinal disease*, p. 169. Edited by Rolfe R D, Finegold S M. Academic Press, San Diego.

Gerding D N, Gebhard R L, Sumner H W, Peterson L R. (1988). Pathology and diagnosis of *Clostridium difficile* disease. In *Clostridium difficile: its role in intestinal disease*, p. 260. Edited by Rolfe R D, Finegold S M. Academic Press, San Diego.

George R H *et al.* (1978). Identification of *Clostridium difficile* as a cause of pseudomembranous colitis. *British Medical Journal* 1: 695.

George W L, Finegold S M. (1985). Clostridia in the human gastrointestinal flora. In *Clostridia in gastrointestinal disease*, p. 1. Edited by Borriello S P. CRC Press, Boca Raton.

Hopkins D G, Kushner J P. (1983). Clostridial species in the pathogenesis of necrotizing enterocolitis in patients with neutropenia. *American Journal of Hematology* 14: 289–294.

Howard F M, Flynn D M, Bradley J M, Noone P, Szawatkowski M. (1977). Outbreak of necrotising enterocolitis caused by *Clostridium butyricum*. *Lancet* ii: 1099–1102.

Ikard R W. (1981). Neutropenic typhlitis in adults. *Archives of Surgery* 116: 943–945.

Jackson S G, Yip-chuck D A, Clark J B, Brodsky M H. (1986). Diagnostic importance of *Clostridium perfringens* enterotoxin analysis in recurring enteritis among elderly, chronic care psychiatric patients. *Journal of Clinical Microbiology* 23: 748–751.

Katlic M R, Derkac W M, Coleman W S. (1981). *Clostridium septicum* infections and malignancy. *Annals of Surgery* 193: 361–364.

King A, Rampling A, Wight D G D, Warren R E. (1984). Neutropenic enterocolitis due to *Clostridium septicum* infection. *Journal of Clinical Pathology* 37: 335–343.

Kliegman R M. (1985). Role of clostridia in the pathogenesis of neonatal enterocolitis. In *Clostridia in gastrointestinal disease*, p. 67. Edited by Borriello S P. CRC Press, Boca Raton.

Koransky J R, Stargel M D, Dowell V R. (1979). *Clostridium septicum* bacteraemia; its clinical significance. *American Journal of Medicine* 66: 63–66.

Krook A, Jarnerot G, Danielsson D. (1981). Clinical effect of metronidazole and sulfasalazine on Crohn's disease in relation to changes in the fecal flora. *Scandinavian Journal of Gastroenterology* 16: 569–575.

Lanz O, Tavel E. (1904). Bacteriologie de l'appendicite. *Revue de Chirurgie (Paris)* 43: 215.

Larson H E, Borriello S P. (1988). Infectious diarrhoea due to *Clostridium perfringens*. *Journal of Infectious Diseases* 157: 390–391.

Larson H E, Price A B, Honour P, Borriello S P. (1978). *Clostridium difficile* and the aetiology of pseudomembranous colitis. *Lancet* i: 1063–1066.

Lawrence G W, Bates J. (1983). Pathogenesis of neonatal necrotizing enterocolitis. *Lancet* i: 540.

Lawrence G W, Bates J, Gaul. (1982). Pathogenesis of neonatal necrotizing enterocolitis. *Lancet* i: 137–139.

Lawrence G, Cooke R. (1980). Experimental pigbel: the production and pathology of necrotising enteritis due to *Clostridium welchii* type C in the guinea-pig. *British Journal of Experimental Pathology* 61: 261–271.

Lawrence G, Walker P D, Garap J, Avusi M. (1979). The occurrence of *Clostridium welchii* type C in Papua New Guinea. *Papua New Guinea Medical Journal* 22: 69.

Lima A A M, Lyerly D M, Wilkins T D, Innes D J, Guerrant R L. (1988). Effects of *Clostridium difficile* toxins A and B in rabbit small and large intestine *in vivo* and on cultured cells *in vitro*. *Infection and Immunity* 56: 582–588.

Mahony D R, Stringer M F, Borriello S P, Mader J A. (1987). Plasmid analysis as a means of strain differentiation in *Clostridium perfringens*. *Journal of Clinical Microbiology* 25: 1333–1335.

Malnick H. (1990). A medium for the isolation of *Anaerobiospirillum* species from faeces. In *Clinical and molecular aspects of anaerobes*, p. 301. Edited by Borriello S P. Wrightson Biomedical, London (in press).

Malnick H, Thomas M E M, Lotay H, Robbins M. (1983). *Anaerobiospirillum* species isolated from humans with diarrhoea. *Journal of Clinical Pathology* 36: 1097–1101.

Mitchell T J, Ketley J M, Burdon D W, Candy D C A, Stephen J. (1987). Biological mode of action of *Clostridium difficile* toxin A: a novel enterotoxin. *Journal of Medical Microbiology* 23: 211–219.

Mitchell T J *et al.* (1986). Effect of toxin A and B of *Clostridium difficile* on rabbit ileum and colon. *Gut* 27: 78–85.

Mulligan M E. (1988). General epidemiology, potential reservoirs, and typing procedures. In *Clostridium difficile: its role in intestinal disease*, p. 230. Edited by Rolfe R D, Finegold S M. Academic Press, San Diego.

Myers L L, Firehammer B D, Schoop D S, Border M M. (1984). *Bacteroides fragilis*: a possible cause of acute diarrhoeal disease in newborn lambs. *Infection and Immunity* 44: 241–244.

Myers L L, Schoop D S. (1987). Association of enterotoxigenic *Bacteroides fragilis* with diarrheal disease in young pigs. *American Journal of Veterinary Research* 48: 774–775.

Myers L L, Schoop D S, Firehammer B D, Border M M. (1985). Association of enterotoxigenic *Bacteroides fragilis* with diarrheal disease in calves. *Journal of Infectious Diseases* 152: 1344–1347.

Myers L L *et al.* (1987). Isolation of enterotoxigenic *Bacteroides fragilis* from humans with diarrhoea. *Journal of Clinical Microbiology* 25: 2330–2333.

Popoff M R. (1990). Are anaerobes involved in neonatal necrotizing enterocolitis? In *Clinical and molecular aspects of anaerobes*, p. 49. Edited by Borriello S P. Wrightson Biomedical, London. (in press).

Popoff M R, Ravisse P. (1985). Lesions produced by *Clostridium butyricum* strain CB1002 in ligated intestinal loops in guinea pigs. *Journal of Medical Microbiology* 19: 351–357.

Popoff M R, Szylit O, Ravisse P, Dabard J, Ohayon H. (1985). Experimental cecitis in gnotoxenic chickens monoassociated with *Clostridium butyricum* strains isolated from patients with neonatal necrotising enterocolitis. *Infection and Immunity* 47: 697–703.

Rifkin G D. (1980). Neutropenic enterocolitis and *Clostridium septicum* infections in patients with agranulocytosis. *Archives of Internal Medicine* 140: 834–835.

Rolfe R D, Finegold S M. (eds). (1988). *Clostridium difficile: its role in intestinal disease*. Academic Press, San Diego.

Schwartz J N et al. (1980). Ampicillin-induced enterocolitis: implication of toxigenic *Clostridium perfringens* type C. *Journal of Pediatrics* 97: 661–663.

Shaked A, Shinar E, Freund H. (1983). Neutropenic typhlitis: a plea for conservation. *Diseases of the Colon and Rectum* 26: 351–352.

Sibbons P, Spitz L, van Velzen D, Bullock G. (1990). Investigations in the aetiology and pathogenesis of neonatal necrotizing enterocolitis using a low-birth-weight neonatal piglet model. In *Clinical and molecular aspects of anaerobes*, p. 59. Edited by Borriello S P. Wrightson Biomedical, London.

Smith M E, Borriello S P, Clayden G S, Casewell M W. (1980). Clinical and bacteriological findings in necrotizing enterocolitis: a controlled study. *Journal of Infection* 2: 23–31.

Stringer M F. (1981). Studies on *Clostridium perfringens* food poisoning PhD Thesis, University of London.

Stringer M F. (1985). *Clostridium perfringens* type A food poisoning. In *Clostridia in gastrointestinal disease*, p. 177. Edited by Borriello S P. CRC Press, Boca Raton.

Sturm R, Staneck J L, Stauffer L R, Neblett W W. (1980). Neonatal necrotizing enterocolitis associated with penicillin-resistant, toxigenic *Clostridium butyricum*. *Pediatrics* 66: 928–930.

Sullivan S N, Darwich R J, Schieven B C. (1980). Severe diarrhoea due to *Clostridium sphenoides*: a case report. *Canadian Medical Association Journal* 123: 398.

Tedesco F J. (1984). Antibiotics associated with *Clostridium difficile* mediated diarrhoea and/or colitis. In *Antibiotic associated diarrhoea and colitis: the role of Clostridium difficile in gastrointestinal disorders*, p. 3. Edited by Borriello S P. Martinus Nijhoff, Boston.

Tvede M, Bondesen S, Haagen Nielsen O, Rasmussen N. (1983). Serum antibodies to *Bacteroides* species in chronic inflammatory bowel disease. *Scandinavian Journal of Gastroenterology* 18: 783–789.

Walker P D. (1985). Pig-Bel. In *Clostridia in gastrointestinal disease*, p. 93. Edited by Borriello S P. CRC Press, Boca Raton.

Wensinck F, Custers-van Lieshout L M C, Poppelaars-Kustermans P A J, Schroder A M. (1981). The faecal flora of patients with Crohn's disease. *Journal of Hygiene* 87: 1–12.

Wensinck F, van de Merwe J P, Mayberry J F. (1983). An international study of agglutinins to *Eubacterium*, *Peptostreptococcus* and *Coprococcus* species in Crohn's disease, ulcerative colitis and control subjects. *Digestion* 27: 63–69.

Williams R et al. (1985). Diarrhoea due to enterotoxigenic *Clostridium perfringens*: clinical features and management of a cluster of ten cases. *Age and Ageing* 14: 296–302.

Willis A T. (1988). Historical aspects. In *Clostridium difficile: its role in intestinal disease*, p. 15. Edited by Rolfe R D, Finegold S M. Academic Press, San Diego.

Yale C E, Balish E. (1985). Evidence for clostridial involvement in pneumatosis cystoides intestinalis. In *Clostridia in gastrointestinal disease*, p. 59. Edited by Borriello S P. CRC Press, Boca Raton.

Yamagishi T, Serikawa T, Morita R, Nakamura S, Nishida S. (1976). Persistent high numbers of *Clostridium perfringens* in the intestines of Japanese aged adults. *Japanese Journal of Microbiology* 20: 397–403.

Yates P, MacGowen A P, Potter M, White H, Slade R R. (1988). Clostridia and neutropenic enterocolitis (Letter) *Lancet* i: 185.

22

Clostridium perfringens type A food poisoning

P R Berry and R J Gilbert

History
Clinical features
Epidemiology
   Isolation and identification
   Epidemiological typing
Enterotoxin
   Sporulation and enterotoxin production *in vitro*
Mode of action
Methods of detection
   Biological assays
   Immunological assays
Control and prevention
Acknowledgement
References

History

That food poisoning could be caused by *Clostridium perfringens* was established only comparatively recently, yet early observations had pointed to the importance of this organism before the turn of the century. Klein (1895) attributed an outbreak of diarrhoea in St Bartholomew's Hospital, London, to *C. perfringens* and Andrewes (1899) recorded the presence of the organism both in the faeces of ill patients and in a rice-pudding dish in a subsequent outbreak at the same hospital. However, the fact that *C. perfringens* was commonly found in faeces, food and the environment hampered any serious thoughts about its role in food poisoning, until Knox and MacDonald (1943) and McClung (1945) reported several outbreaks of food poisoning in which *C. perfringens* was the only significant organism isolated. The importance of the organism in producing the typical symptoms of *C. perfringens* food poisoning was shown by Cravitz and Gillmore (1946), who reproduced the illness in both man and animals following oral administration of a culture isolated in the outbreak described by McClung (1945). Osterling (1952) found similar results using isolates from food incriminated in food poisoning outbreaks in Sweden.

Clinical features

C. perfringens food poisoning is typically characterized by diarrhoea and abdominal pain 8–24 h after the ingestion of food containing large numbers of the vegetative cell. Nausea is common, although vomiting, shivering, headaches and pyrexia are only rarely encountered. The duration of the illness is short and symptoms usually disappear within 12–24 h. Fatalities occur occasionally among debilitated persons, particularly geriatric patients. Between 1976 and 1985, there were 14 deaths among the 16 384 reported cases of *C. perfringens* food poisoning in England and Wales (Table 22.1). The clinical symptoms are caused by an enterotoxin, which is produced during sporulation of the organism in the small intestine. The toxin is not excreted from the sporulating cell, but is liberated from the sporangium on lysis and is distinct from the major and minor lethal toxins. The role of the enterotoxin in causing the typical symptoms was confirmed by oral administration of pure enterotoxin to human volunteers (Skjelkvale and Uemura, 1977).

Table 22.1 Incidence of outbreaks of *C. perfringens* food poisoning in England and Wales, 1976–1985

Year	General outbreaks	Number of Family outbreaks	Cases	Deaths
1976	84	3	2924	
1977	78	5	2576	5
1978	38	4	1042	
1979	53	3	1607	6
1980	53	2	1056	0
1981	48	2	918	0
1982	65	4	1455	0
1983	63	5	1624	0
1984	64	4	1716	0
1985	58	6	1466	3

Epidemiology

C. perfringens may be the most ubiquitous of all pathogenic bacteria (Smith, 1975). The organism is commonly found in dust, soil, sewage and the intestinal tract of man and animals (Taylor and Gordon, 1940; Hobbs *et al.*, 1953) and can be readily isolated from many common foodstuffs such as meat, poultry, fish, vegetables and spices (Hall and Angelotti, 1965; Sutton and Hobbs, 1969; Taniguti and Zenitani, 1969; Gibbs, 1971).

Although *C. perfringens* can be isolated from the faeces of almost 100 per cent of people by enrichment culture and repeated examination, food poisoning outbreaks are not associated with human excretors. The organism normally originates from the food itself, although postcooking contamination from dust or by handling has been suggested as a cause of some outbreaks (McKillop, 1959; Hauschild, 1973).

The vegetative cells of *C. perfringens* are destroyed by normal cooking procedures, but the spores are able to survive. Heat-resistant strains, i.e., those producing spores capable of surviving boiling for longer than 1 h, were originally thought to be responsible for food poisoning outbreaks (Hobbs *et al.*, 1953). However, it is now known that heat-sensitive strains account for a considerable proportion of the outbreaks. Sutton and Hobbs (1965) suggested that heat-sensitive strains were responsible for more outbreaks than had originally been thought because the use of isolation methods favoured heat-resistant strains.

Most outbreaks occur in hospitals, canteens, schools and other institutions where large-scale catering occurs (Table 22.2). Meat based meals such as roasts, stews, pies, poultry and soups are most often implicated. The factors contributing to the problem are inadequate cooling and refrigeration of cooked meat products, a delay of several hours between preparation and consumption, and inadequate reheating.

Table 22.2 Location of general and family outbreaks of *C. perfringens* food poisoning in England and Wales, 1969–1986

Location	Number of outbreaks
Hospitals	227
Restaurants and public houses	205
Institutions	171
Schools and canteens	250
Private houses	59
Others or Unknown	69

Table 22.3 Counts of *C. perfringens* in foods incriminated in 146 outbreaks of food poisoning in the UK

Colony count/g of food	Number of outbreaks
$<10^4$	19
10^4–9.9×10^4	24
10^5–9.9×10^5	31
10^6–9.9×10^6	28
10^7–9.9×10^7	25
10^8–9.9×10^8	17
10^9–9.9×10^9	2
$>10^{10}$	0

Median count/g = 7×10^5

Isolation and identification

Foods incriminated in food poisoning outbreaks often have *C. perfringens* counts exceeding 10^5/g (Table 22.3) and since the organism sporulates very poorly in most foods these counts are also a measure of the numbers of vegetative cells. However, this level is not an absolute value, since the viability of the organism declines fairly rapidly in left-over foods stored at low temperatures, particularly if these are frozen (Harmon and Duncan, 1984).

Spores and vegetative cells are found in numbers exceeding 10^6/g in faeces from patients involved in *C. perfringens* food poisoning outbreaks. Since counts in the faeces of people not involved in a food poisoning incident normally do not exceed 10^4/g, this high count can be used as a criterion for the laboratory confirmation of an outbreak (Sutton *et al.*, 1971). However, it must be noted that some long stay geriatric in-patients can carry *C. perfringens* in numbers exceeding 10^6/g of faeces (Stringer *et al.*, 1985), and so the isolation of large numbers of the organism from this group of patients is not necessarily significant.

Many different media have been formulated for the isolation of *C. perfringens*; these have been reviewed by Labbe (1983) and Mead (1985). One commonly used medium is that of Harmon *et al.* (1971), a tryptose–sulphite–cycloserine (TSC) agar, or its egg-yolk-free variant (Hauschild and Hilsheimer, 1974). These media were found to be best in an evaluation carried out by Hauschild *et al.* (1977; 1979) for the International Commission on Microbiological Specifications for Foods.

In the UK, blood agar which contains neomycin is recommended for the isolation of *C. perfringens* from food poisoning outbreaks, since it is readily available in hospital and Public Health Laboratories (Sutton *et al.*, 1971). However, some strains of heat-resistant *C. perfringens* may, very occasionally, be inhibited by neomycin (Spencer, 1969).

Confirmation of an isolate as *C. perfringens* is usually done by demonstrating the Nagler reaction, i.e., neutralization of the lecithinase reaction by type A lecithinase antitoxin. Very occasionally, a *C. perfringens* food poisoning outbreak is caused by a lecithinase-negative strain (Pinegar and Stringer, 1977). In these cases, confirmation can be achieved by testing for non-motility, nitrate reduction, lactose fermentation and gelatin liquifaction.

Epidemiological typing

Since counts alone are not always of value for the laboratory confirmation of an outbreak, *C. perfringens* isolates can be typed for epidemiological confirmation. Several typing methods have been described (Mahony, 1979; Watson, 1985; Mahony *et al.*, 1987) but, at the present time, only serological typing of the capsular polysaccharides has proved sufficiently discriminating to be of particular use. Serotyping is a valuable tool for confirming a *C. perfringens* outbreak, because it is important to establish that:

1. Strains isolated from different patients are of the same type;
2. Strains isolated from the incriminated food and faecal specimens are of the same type;
3. A particular food is responsible when several foods have been eaten.

Serotyping is performed at the Food Hygiene Laboratory, Central Public Health Laboratory, Colindale, London, which acts as an international reference centre for *C. perfringens* food poisoning.

The antisera are raised in rabbits against whole cells according to the method devised by Henderson (1940). The value of serological typing was first recognized in 1953, when Hobbs *et al.* described its use for the typing of non-haemolytic heat-resistant strains associated with outbreaks of food poisoning. The range of antisera was then increased over the following years until a total of 75 English serotypes was established (Sutton, 1969; Hughes *et al.*, 1976; Stringer *et al.*, 1980). Thirty-four American and 34

Japanese strains were added to these English strains following a meeting of the Tenth International Congress of Microbiology in Mexico, where it was decided that an international serotyping scheme, based in the UK, would be a worthwhile development of, and an aid to, the investigation of food-borne disease caused by *C. perfringens* (Stringer et al., 1980). The development and use of serological typing has been reviewed by Stringer et al. (1982).

Since carriage of *C. perfringens* in the intestine of man is normal, it is important to test more than one isolate from each patient by serotyping to avoid picking the patients' own serotype, which would confuse the results. Serotyping may also be of limited value when testing isolates from hospitalized long stay geriatric patients who can carry the same serotype for extended periods (Stringer et al., 1985).

Loss of capsular polysaccharide results in the appearance of 'rough' colonies which will be non-typable or auto-agglutinable. 'Smooth' colonies can occasionally be regenerated from the rough forms by rapid (i.e., every 4 h) subculture several times on blood agar.

Enterotoxin

Sporulation and enterotoxin production *in vitro*

It has been suggested that enterotoxin production occurs during sporulation, and that the toxin is released during lysis of the sporangium (Duncan et al., 1972). The enterotoxin was thought to be a sporulation-specific gene product (Duncan et al., 1972) and a structural component of the spore coat (Frieben and Duncan, 1973). The relationship between sporulation and enterotoxin production is also supported by genetic and biochemical evidence (Labbe, 1980). However, some studies (Walker et al., 1975; Roper et al., 1976) suggested that there is no direct relationship between enterotoxin and spore-coat proteins. Goldner et al. (1986) reported that a non-sporulating mutant strain produced enterotoxin in the absence of sporulation. Studies have shown that high levels of sporulation occur with enterotoxin-negative strains and that enterotoxin can be produced when the frequency of sporulation is very low (Uemura et al., 1973; Skjelkvale et al., 1979; Craven et al., 1981). It is clear that the yield of toxin per sporulating cell varies considerably between strains. It seems likely that the enterotoxin is not a spore-specific protein, and it is possible that all strains of *C. perfringens* are enterotoxin-positive when the protein can be detected at the 0.1 ng level (Granum et al., 1984). The association of enterotoxin with cell components is not restricted to spores; Loffler and Labbe (1985) showed that inclusion bodies synthesized during sporulation were serologically related to enterotoxin. The inclusion bodies may form intracellularly following over-production of spore components (Somerville et al., 1968).

Several workers have shown that heat activation of the cultures prior to incubation results in increased sporulation and toxin production (Uemura et al., 1973; Tsai et al., 1974). The exact mechanism of heat activation is unknown, although heat may select for those cells with an increased inherent ability to produce spores.

All *C. perfringens* type A isolates produce the same serologically related enterotoxin, except for the mutant strain 8-6, which produces a spore without a coat, and an enterotoxin with only a slight similarity to the classic *C. perfringens* enterotoxin (Lindsay et al., 1985). Antiserum to the classic type A toxin did not neutralize the type A (8-6) enterotoxin in mouse or erythema tests.

Enterotoxin has also been detected in a few strains of *C. perfringens* types C and D (Skjelkvale and Duncan, 1975; Uemura and Skjelkvale, 1976). The proteins in these cases appear very similar to those of the classic toxin from type A strains.

There are many different formulations of media available for the enhancement of sporulation of *C. perfringens*, but no single medium appears suitable for all strains (Ellner, 1956; Duncan and Strong, 1968; Tsai et al., 1974). Improvements of these media are still continuing (Harmon and Kautter, 1986a; Ushijima et al., 1987) using liquid or semi-solid recipes. Phillips (1986) has developed a solid medium for the sporulation of *C. perfringens* which incorporates bile, bicarbonate and quinoline into a blood agar base. This medium produces excellent sporulation in all strains tested. Preliminary work shows that food poisoning strains do produce enterotoxin directly on this medium, whereas environmental strains, even with good sporulation, do not (Berry, unpublished results).

Mode of action

An enterotoxin was first considered to be the cause of illness when Duncan and Strong (1969) and Hauschild et al. (1970) reported that cell-free extracts of C. perfringens caused fluid accumulation in ligated ileal loops of rabbits and lambs. McDonel and Duncan (1977) showed that, in the rabbit, most damage was caused to the terminal ileum, with the jejunum and duodenum less affected. The colon appears unaffected by the enterotoxin (McDonel and Demers, 1982).

McDonel (1974) and McDonel and Duncan (1975) investigated the physiological effects of enterotoxin in perfused ileal loops treated with the toxin. They observed an increased net secretion of fluid, sodium and chloride ions, and a reduction in glucose absorption. The leakage of ^{51}Cr, nucleotides and lactate dehydrogenase, but not RNA, from cells treated with enterotoxin indicated that the main site of action was the cell membrane, and not cellular processes (McClane and McDonel, 1980; Skjelkvale et al., 1980). This is supported by the fact that osmotic stabilisers such as sucrose, polyethylene glycol and bovine serum albumin protect against toxin-induced changes (McClane and McDonel, 1981; McClane, 1984).

The exact nature of the enterotoxin receptor on the cell membrane has not been fully elucidated. Wnek and McClane (1983) have isolated a protein of mol. wt 50 000 from rabbit intestine brush-border membranes which binds the toxin. This protein was found to inhibit the biological activity of the enterotoxin on Vero cells.

From these studies, it would appear that the primary action of the enterotoxin involves alteration of membrane permeability, causing leakage of intracellular contents and eventual cell death. Horiguchi et al. (1985; 1986a) have shown that the action of the enterotoxin includes a temperature independent binding step, a temperature dependent ion transport alteration step and a calcium ion dependent step leading to the release of intracellular macromolecules.

Methods of detection

Some of the assay systems used for C. perfringens enterotoxin detection are summarized in Table 22.4.

Biological assays

The earliest biological assay developed was the rabbit ileal loop procedure (Duncan et al., 1968),

Table 22.4 Sensitivities of some assay methods for *C. perfringens* enterotoxin

Method	Minimum concentration of enterotoxin detected (μg/ml)	Reference
Biological		
Erythemal activity	1.25–2.50	Genigeorgis et al. (1973)
Rabbit ileal loop (standard)	9.6	Genigeorgis et al. (1973)
Rabbit ileal loop (90 min)	2.5	Hauschild et al. (1971)
Vero cell	0.001	McDonel and McClane (1981)
Immunological		
Single gel diffusion	0.9	Genigeorgis et al. (1973)
Double gel diffusion	0.5	Stringer et al. (1980)
Counterimmunoelectrophoresis	0.2	Naik and Duncan (1977)
Reversed passive haemagglutination	0.001	Genigeorgis et al. (1973)
Radioimmunoassay	0.001	Stelma et al. (1983)
ELISA	0.0001	Olsvik et al. (1982)
	0.001	Notermans et al. (1984)
	0.002	Bartholomew et al. (1985)
Reversed passive latex agglutination	0.003	Harmon and Kautter, (1986b)

which had been used previously for the detection of cholera toxin. A modified version of this method (Hauschild et al., 1971) reduced the assay time from 6 h to 90 min. Yamamoto et al. (1979) developed a more sensitive and economical procedure using the intestinal loops of mice. However, all these tests, which require surgery on the animal, are time consuming, relatively insensitive and expensive.

The skin erythema test in guinea-pigs or rabbits is a more sensitive and adaptable alternative to the ligated loop (Hauschild, 1970; Stark and Duncan, 1971). Animals are inoculated intradermally with the test sample; a positive result is indicated by the development of an area of erythema at the site of injection. The test can be made more rapid, sensitive and semiquantitative by intravenous injection of Evans Blue Dye 10–20 min after injection of the enterotoxin. The diameter of the blue zone that develops due to increased capillary permeability is proportional to the concentration of toxin injected (Stark and Duncan, 1972).

Tissue culture assays with various mammalian and human cell lines have been investigated for C. perfringens enterotoxin detection. Giugliano et al. (1983) found that enterotoxin produced a morphological change in 10 out of 12 cell lines tested, particularly Vero (African Green Monkey kidney) and MDCK (dog kidney). McDonel and McClane (1981) used Vero cells in a technique based on the ability of toxin to inhibit the plating efficiency of Vero cells grown in culture. The authors quoted a sensitivity for detection of toxin of 1 ng/ml, although this level is likely to be lower when testing faeces because of interfering factors. Although tissue culture assays are more sensitive than other biological procedures, Berry et al. (1988) found Vero cell assay less sensitive than enzyme linked immunosorbent assay (ELISA) or reversed passive latex agglutination (RPLA) which could, in part, be due to the biological activity of the enterotoxin being less stable than the serological activity (McDonel and McClane, 1981).

Immunological assays

The requirements for speed, sensitivity, reproducibility and convenience led to the development of serological assays for the enterotoxin. Several of the earlier methods, such as single and double gel diffusion, counter immunoelectrophoresis and reversed passive haemagglutination are given in Table 22.4. These methods have been used for many years with good effect, although they now have been generally replaced by the more sensitive and rapid methods of ELISA and RPLA.

Several ELISA procedures have been developed recently. Their sensitivity equals or exceeds that of radioimmunoassay and they do not require the use of labelled isotopes. Sensitivites of the tests are usually around the 1–2 ng/ml level (Table 22.4) and since the majority of faeces from outbreaks contain toxin initially in excess of 1 µg/g (Bartholomew et al., 1985) the toxin can be detected even when specimens are collected several days after the onset of symptoms. A modification of the ELISA using a 'dot immunoblot' procedure has been developed for detecting enterotoxin directly from colonies (Stelma et al., 1985).

The most recently developed test procedure for detection of C. perfringens enterotoxin is the RPLA test, marketed in the UK by Oxoid for Denka-Seiken Co. Ltd of Japan. Comparison of RPLA with ELISA has shown that RPLA is as sensitive (Berry et al., 1986) or slightly less sensitive (Harmon and Kautter, 1986b) than ELISA. The advantage of RPLA is that it is simple to perform and does not require specialized equipment. However, Berry et al. (1988) have found that approximately 1 per cent of faecal specimens can give false positive reactions, although the agglutination titres are very low and interpretation is unlikely to be a problem in clinical use.

Enterotoxin detection may be the only definitive way of confirming C. perfringens as the causative agent of diarrhoea in long stay geriatric in-patients. Carriage of large numbers of the same serotype is common in these patients (Stringer et al., 1985) and, therefore, cannot be used as a criterion for laboratory confirmation. Jackson et al. (1986) used an ELISA test to detect toxin in faeces and to show that C. perfringens was a likely cause of gastroenteritis in a population of elderly chronic care patients. Whether this type of patient excretes enterotoxin regularly without overt symptoms awaits confirmation.

Monoclonal antibodies (MAbs) to C. perfringens enterotoxin have been produced and characterized (Wnek et al., 1985; Horiguchi et al., 1986b), although use in immunological detection systems has not yet been fully investigated. The MAbs are likely to lead to a better understanding of the structure and function of the toxin. Wnek et al. (1985) produced two MAbs that neutralized the biological activity of the enterotoxin and also blocked the binding of enterotoxin to intestinal brush borders.

Control and prevention

Bryan (1978) and Roberts (1982) have shown that the principle reason for the occurrence of *C. perfringens* food poisoning outbreaks is the failure to refrigerate cooked foods, particularly when prepared in large quantities. The incidence is also related to inadequate reheating of the food before consumption, since the vegetative cells would be destroyed if the internal temperature of the reheated food reached 70°C. Cooked foods should be kept above 60°C or below 10°C (Labbe, 1988). Reheating by microwave cooking may have an influence on incidence in the future; Wright-Rudolph et al. (1986) showed that reductions in counts (from 10^5/g) of vegetative cells in patties during microwave cooking were only in the range $\log_{10}$ 0.75–1.48/g, compared to reductions of $\log_{10}$ 3.5–$\log_{10}$ 8.06/g for conventional cooking.

Thomas et al. (1977), following a large outbreak of *C. perfringens* food poisoning in a hospital, recommended:
1. That portions of meat for cooking should not exceed 3 kg;
2. The rapid consumption, or rapid cooling and refrigeration, of cooked food;
3. That reheated cooked foods should be subjected to a temperature of 100°C throughout their bulk to ensure killing of all vegetative cells;
4. Adequate staff training in hygiene and catering practice.

Acknowledgement

We are grateful to our colleague, Joanna Rodhouse, for her help in the production of this chapter.

References

Andrewes F W. (1899). On an outbreak of diarrhoea in the wards of St Bartholomew's Hospital, probably caused by infection of rice pudding with *Bacillus enteritidis sporogenes*. *Lancet* i: 8–9.

Bartholomew B A, Stringer M F, Watson G N, Gilbert R J. (1985). Development and application of an enzyme-linked immunosorbent assay for *Clostridium perfringens* type A enterotoxin. *Journal of Clinical Pathology* 38: 222–228.

Berry P R, Stringer M F, Uemura T. (1986). Comparison of latex agglutination and ELISA for the detection of *Clostridium perfringens* type A enterotoxin in faeces. *Letters in Applied Microbiology* 2: 101–102.

Berry P R, Rodhouse J C, Hughes S, Bartholomew B A, Gilbert R J. (1988). An evaluation of ELISA, RPLA and Vero cell assay for the detection of *C. perfringens* enterotoxin in faecal specimens. *Journal of Clinical Pathology* 41: 458–461.

Bryan F L. (1978). Factors that contribute to outbreaks of food-borne disease. *Journal of Food Protection* 41: 816–827.

Craven S E, Blankenship L C, McDonel J L. (1981). Relationship of sporulation, enterotoxin formation and spoilage during growth of *Clostridium perfringens* type A in cooked chicken. *Applied and Environmental Microbiology* 41: 1184–1191.

Cravitz L, Gillmore J D. (1946). The role of *Clostridium perfringens* in human food poisoning. Project X-756. Report No. 2. Naval Medical Research Institute, Bethesda. Cited by Hobbs B C. (1986). *Clostridium perfringens* food poisoning. In *Progress in food safety*, pp. 73–78. Edited by Cliver D O, Cochrane B A. Food Research Institute, University of Wisconsin–Madison.

Duncan C L, Strong D H. (1968). Improved medium for sporulation of *Clostridium perfringens*. *Applied Microbiology* 16: 82–89.

(1969). Ileal loop fluid accumulation and production of diarrhoea in rabbits by cell-free products of *Clostridium perfringens*. *Journal of Bacteriology* 100: 86–94.

Duncan C L, Sugiyama H, Strong D H. (1968). Rabbit ileal loop response to strains of *Clostridium perfringens*. *Journal of Bacteriology* 95: 1560–1566.

Duncan C L, Strong D H, Sebald M. (1972). Sporulation and enterotoxin production by mutants of *Clostridium perfringens*. *Journal of Bacteriology* 110: 378–391.

Ellner P. (1956). A medium promoting rapid quantitative sporulation in *Clostridium perfringens*. *Journal of Bacteriology* 71: 495–496.

Frieben W R, Duncan, C L. (1973). Homology between enterotoxin protein and spore structural protein in *Clostridium perfringens*. *European Journal of Biochemistry* 39: 393–401.

Genigeorgis C, Sakaguchi G, Riemann H. (1973). Assay methods for *Clostridium perfringens* type A enterotoxin. *Applied Microbiology* 26: 111–115.

Gibbs P A. (1971). The incidence of clostridia in poultry carcasses and poultry processing plants. *British Poultry Science* 12: 101–111.

Giugliano L G, Stringer M F, Drasar B S. (1983). Detection of *Clostridium perfringens* enterotoxin by tissue culture and double-gel diffusion methods. *Journal of Medical Microbiology* 16: 233–237.

Goldner S B, Solberg M, Jones S, Post L S. (1986). Enterotoxin synthesis by nonsporulating cultures of *Clostridium perfringens*. *Applied and Environmental Microbiology* **52**: 407–412.

Granum P E, Olsvik O, Staun A. (1984). Enterotoxin formation by *Clostridium perfringens* during sporulation and vegetative growth. In *Bacterial protein toxins*, pp. 121–122. Edited by Alouf J E. *et al.* Academic Press, London.

Hall H, Angelotti R. (1965). *Clostridium perfringens* in meat and meat products. *Applied Microbiology* **13**: 352–356.

Harmon S M, Duncan C L. (1984). *Clostridium perfringens*. In *Compendium of methods for the microbiological examination of foods* pp. 483–495. Edited by Speck M L. American Public Health Association.

Harmon S M, Kautter D A. (1986a). Improved media for sporulation and enterotoxin production by *Clostridium perfringens*. *Journal of Food Protection* **49**: 706–711.
(1986b). Evaluation of a reversed passive latex agglutination test kit for *Clostridium perfringens* enterotoxin. *Journal of Food Protection* **49**: 523–525.

Harmon S M, Kautter D A, Peeler J T. (1971). Improved medium for enumeration of *Clostridium perfringens*. *Applied Microbiology* **22**, 688–692.

Hauschild A H W. (1970). Erythemal activity of the cellular enteropathogenic factor of *Clostridium perfringens* type A. *Canadian Journal of Microbiology* **16**: 651–654.
(1973). Food poisoning by *Clostridium perfringens*. *Canadian Institute of Food Science and Technology Journal* **6**: 106–110.

Hauschild A H W, Desmarchelier P, Gilbert R J, Harmon S M, Vahlefeld R. (1979). ICMSF Methods Studies. XII. Comparative study for the enumeration of *Clostridium perfringens* in faeces. *Canadian Journal of Microbiology* **25**: 953–963.

Hauschild A H W, Gilbert R J, Harmon S M, O'Keefe M F, Vahlefeld R. (1977). ICMSF Methods Studies. VIII. Comparative study for the enumeration of *Clostridium perfringens* in foods. *Canadian Journal of Microbiology* **23**: 884–892.

Hauschild A H W, Hilsheimer R. (1974). Evaluation and modifications of media for enumeration of *Clostridium perfringens*. *Applied Microbiology* **27**: 78–82.

Hauschild A H W, Hilsheimer R, Rogers G G. (1971). Rapid detection of *Clostridium perfringens* enterotoxin by a modified ligated intestinal loop technique in rabbits. *Canadian Journal of Microbiology* **17**: 1475–1476.

Hauschild A H W, Niilo L, Dorward W J. (1970). Response of ligated intestinal loops in lambs to an enteropathogenic factor of *Clostridium perfringens* type A. *Canadian Journal of Microbiology* **16**: 339–343.

Henderson D W. (1940). The somatic antigen of the *Cl. welchii* group of organisms. *Journal of Hygiene* **40**: 501–502.

Hobbs B C, Smith M E, Oakley C L, Warrack G H, Cruickshank J C. (1953). *Clostridium welchii* food poisoning. *Journal of Hygiene* **51**: 75–101.

Horiguchi Y, Uemura T, Kozaki S, Sakaguchi G. (1985). The relationship between cytotoxic effect and binding to mammalian cultured cells of *Clostridium perfringens* enterotoxin. *FEMS Microbiology Letters* **28**: 131–135.
(1986a). Effects of Ca^{2+} and other cations on the action of *Clostridium perfringens* enterotoxin. *Biochimica et Biophysica Acta* **889**: 65–71.

Horiguchi Y, Uemura T, Kamata Y, Kozaki S, Sakaguchi G. (1986b). Production and characterisation of monoclonal antibodies to *Clostridium perfringens* enterotoxin. *Infection and Immunity* **52**: 31–35.

Hughes J A, Turnbull P C B, Stringer M F. (1976). A serotyping system for *Clostridium welchii* (*Cl. perfringens*) Type A, and studies on the type-specific antigens. *Journal of Medical Microbiology* **9**: 475–485.

Jackson S G, Yip-Chuck D, Clark J B, Brodsky M H. (1986). Diagnostic importance of *Clostridium perfringens* enterotoxin analysis in recurring enteritis among elderly chronic care psychiatric patients. *Journal of Clinical Microbiology* **23**: 748–751.

Klein E. (1895). Ueber einen pathogenen anaeroben Darmbacillus, *Bacillus enteritidis sporogenes*. *Zentralblatt fur Bakteriologie und Parasitenkunde* **18**: 737–743.

Knox R, MacDonald E K. (1943). Outbreaks of food poisoning in certain Leicester institutions. *Medical Officer* **69**: 21–22.

Labbe R G. (1980). Relationship between sporulation and enterotoxin production in *Clostridium perfringens*. *Food Technology* **34**: 88–90.
(1983). Enumeration and confirmation of *Clostridium perfringens*. *Journal of Food Protection* **46**: 68–73.
(1989). *Clostridium perfringens*. In *Foodborne bacterial pathogens* pp 191–234. Edited by Doyle M P. Maral Dekker, Inc., New York.

Lindsay J A, Sleigh R W, Ghitgas C, Davenport J B. (1985). Purification and properties of an enterotoxin from a coatless spore mutant of *Clostridium perfringens* type A. *European Journal of Biochemistry* **149**: 287–293.

Loffler A, Labbe R G. (1985). Isolation of an inclusion body from sporulating, enterotoxin-positive *Clostridium perfringens*. *FEMS Microbiology Letters* **27**: 143–147.

McClane B A. (1984). Osmotic stabilizers differentially inhibit permeability alterations induced in Vero cells by *Clostridium perfringens* enterotoxin. *Biochimica et Biophysica Acta* **777**: 99–106.

McClane B A, McDonel J L. (1980). Characterisation of membrane permeability alterations induced in Vero cells by *Clostridium perfringens*. *Biochimica et Biophysica Acta* **600**: 974–985.
(1981). Protective effect of osmotic stabilisers on morphological and permeability alterations induced in Vero cells by *Clostridium perfringens*. *Biochimica et Biophysica Acta* **641**: 401–411.

McClung L S. (1945). Human food poisoning due to growth of *Cl. perfringens* (*Cl. welchii*) in freshly cooked chicken: preliminary note. *Journal of Bacteriology* 50: 229–231.

McDonel J L. (1974). In vivo effects of *Clostridium perfringens* enteropathogenic factors on the rat ileum. *Infection and Immunity* 10: 1156–1162.

McDonel J L, Demers G W. (1982). In vivo effects of enterotoxin from *Clostridium perfringens* type A in the rabbit colon: binding vs. biologic activity. *Journal of Infectious Diseases* 145: 490–494.

McDonel J L, Duncan C L. (1975). Effects of *Clostridium perfringens* enterotoxin on metabolic indexes of everted ileal sacs. *Infection and Immunity* 12: 274–280. (1977). Regional localisation of activity of *Clostridium perfringens* type A enterotoxin in the rabbit ileum, jejunum and duodenum. *Journal of Infectious Diseases* 136: 661–666.

McDonel J L, McClane B A. (1981). Highly sensitive assay for *Clostridium perfringens* enterotoxin that uses inhibition of plating efficiency of Vero cells grown in culture. *Journal of Clinical Microbiology* 13: 940–946.

McKillop E J. (1959). Bacterial contamination of hospital food with special reference to *Clostridium welchii* food poisoning. *Journal of Hygiene*, 57: 31–46.

Mahony D E. (1979). Bacteriocin, bacteriophage and other epidemiological typing methods for the genus *Clostridium*. In *Methods in microbiology*, vol 13. pp. 1–30. Edited by Bergen T, Norris J R. Academic Press, London.

Mahony D E, Stringer M F, Borriello S P, Mader J A. (1987). Plasmid analysis as a means of strain differentiation in *Clostridium perfringens*. *Journal of Clinical Microbiology* 25: 1333–1335.

Mead G S. (1985). Selective and differential media for *Clostridium perfringens*. *International Journal of Food Microbiology* 2: 89–98.

Naik H S, Duncan C L. (1977). Rapid detection and quantitation of *Clostridium perfringens* enterotoxin by counterimmunoelectrophoresis. *Applied and Environmental Microbiology* 34: 125–128.

Notermans S, Heuvelman C, Beckers H, Uemura T. (1984). Evaluation of the ELISA as a tool in diagnosing *Clostridium perfringens* enterotoxin. *Zentralblatt fur Bakteriologie, Mikrobiologie und Hygiene*. I. *Abteilung Originale B.* 179: 225–234.

Olsvik O, Granum P E, Berdal B P. (1982). Detection of *Clostridium perfringens* type A enterotoxin by ELISA. *Acta Pathologica Microbiologica et Immunologica Scandinavica. Section B.* 90: 445–447

Osterling S. (1952). Food poisoning caused by *Clostridium perfringens* (*Cl. welchii*) in freshly cooked chicken: preliminary note. Cited by Stringer M F S. (1985). *Clostridium perfringens* type A food poisoning. In *Clostridia in gastrointestinal disease*, pp. 117–143. Edited by Borriello S P. CRC Press, Boca Raton.

Phillips K D. (1986). A sporulation medium for *Clostridium perfringens*. *Letters in Applied Microbiology* 3: 77–79.

Pinegar J A, Stringer M F. (1977). Outbreaks of food poisoning attributed to lecithinase-negative *Clostridium welchii*. *Journal of Clinical Pathology* 30: 491.

Roberts D. (1982). Factors contributing to outbreaks of food poisoning in England and Wales, 1970–1979. *Journal of Hygiene* 89: 491–498.

Roper G, Short J A, Walker P D. (1976). The ultrastructure of *Clostridium perfringens* spores. In *Spore research, 1976*, pp. 279–296. Edited by Baker A M *et al*. Academic Press, London.

Skjelkvale R, Duncan C L. (1975). Enterotoxin formation by different toxigenic types of *Clostridium perfringens*. *Infection and Immunity* 11: 563–575.

Skjelkvale R, Uemura T. (1977). Experimental diarrhoea in human volunteers following oral administration of *Clostridium perfringens* enterotoxin. *Journal of Applied Bacteriology* 43: 281–286.

Skjelkvale R, Stringer M F, Smart J L. (1979). Enterotoxin production by lecithinase-positive and lecithinase-negative *Clostridium perfringens* isolated from food poisoning outbreaks and other sources. *Journal of Applied Bacteriology* 47: 329–339.

Skjelkvale R, Granum P E. Jarmund T. (1980). *Mechanism of action of* Clostridium perfringens *enterotoxin*, pp. 980–984. Proceedings of the World Congress on Foodborne Infections and Intoxications.

Smith, L D S. (1975). *The pathogenic anaerobic bacteria*, pp. 115–176. Thomas, Springfield.

Somerville H J, Delafield F P, Rittenberg S C. (1968). Biochemical homology between crystal and spore protein of *Bacillus thuringiensis*. *Journal of Bacteriology* 96: 721–726.

Spencer R. (1969). Neomycin-containing media in the isolation of *Clostridium botulinum* and food-poisoning strains of *Clostridium welchii*. *Journal of Applied Bacteriology* 32: 170–174.

Stark R L, Duncan C L. (1971). Biological characteristics of *Clostridium perfringens* type A enterotoxin. *Infection and Immunity* 4: 89–96.
(1972). Transient increase in capillary permeability induced by *Clostridium perfringens* type A enterotoxin. *Infection and Immunity* 5: 147–150.

Stelma G N, Wimsatt J C, Kauffman P E, Shah D B. (1983). Radioimmunoassay for *Clostridium perfringens* enterotoxin and its use in screening isolates implicated in food poisoning outbreaks. *Journal of Food Protection* 46: 1069–1073.

Stelma G N, Johnson C H, Shah D B. (1985). Detection of enterotoxin in colonies of *Clostridium perfringens* by a solid phase enzyme-linked immunosorbent assay. *Journal of Food Protection* 48: 227–231.

Stringer M F, Turnbull P C B, Gilbert R J. (1980). Application of serological typing to the investigation of outbreaks of *Clostridium perfringens* food poisoning, 1970–78. *Journal of Hygiene* 84: 443–456.

Stringer M F, Watson G N, Gilbert R J. (1982). *Clostridium perfringens* type A: serological typing and methods for the detection of enterotoxin. In *Isolation and identification methods for food poisoning organisms*, pp. 111–135. Edited by Corry J E L *et al.* Society for Applied Bacteriology, Technical Series 17. Academic Press, London.

Stringer M F, *et al.* (1985). Faecal carriage of *Clostridium perfringens*. *Journal of Hygiene* **95**: 277–288.

Sutton R G A. (1969). The pathogensis and epidemiology of *Clostridium welchii* food poisoning. Ph.D. Thesis, University of London.

Sutton R G A, Hobbs B C. (1965). Food poisoning caused by heat-sensitive *Clostridium welchii*. A report of five recent outbreaks. *Journal of Hygiene* **66**: 135–146.

(1969). Enumeration of *Clostridium perfringens* in dried foods and feeding stuffs. In *The microbiology of dried foods*, pp. 243–249. Edited by Kampelmacher E H *et al.* Proceedings of the 6th International Symposium on Food Microbiology. Grafische Industrie, Haarlem.

Sutton R G A, Ghosh A C, Hobbs B C. (1971). Isolation and enumeration of *Clostridium welchii* from food and faeces. In *Isolation of anaerobes*, pp. 39–47. Edited by Shapton D, Board R. Society for Applied Bacteriology, Technical Series 5. Academic Press, London.

Taniguti T, Zenitani B. (1969). Incidence of *Clostridium perfringens* in fish. I. On the application of LAS medium to detection of *Clostridium perfringens*. *Journal of the Food Hygiene Society of Japan* **10**: 199–203.

Taylor A W, Gordon W S. (1940). A survey of the types of *Clostridium welchii* present in soil and in the intestinal content of animals and man. *Journal of Pathology and Bacteriology* **50**: 271–277.

Thomas M E *et al.* (1977). Hospital outbreak of *Clostridium perfringens* food poisoning. *Lancet* **i**: 1046–1048.

Tsai C C, Torres-Anjel M J, Riemann H P. (1974). Improved culture techniques and sporulation medium for enterotoxin production by *Clostridium perfringens* type A. *Journal of the Formosan Medical Association* **73**: 404–409.

Uemura T, Skjelkvale R. (1976). An enterotoxin produced by *Clostridium perfringens* type D. Purification by affinity chromatography. *Acta Pathologica, Microbiologica et Immunologica Scandinavica Section B*. **84**: 414–420.

Uemura T, Sakaguchi G, Riemann H P. (1973). In vitro production of *Clostridium perfringens* enterotoxin and its detection by reversed passive haemagglutination. *Applied Microbiology* **26**: 381–385.

Ushijima T, Sugitani A, Ozaki Y. (1987). A pair of semisolid media facilitate detection of spore and enterotoxin of *Clostridium perfringens*. *Journal of Microbiological Methods* **6**: 145–152.

Walker P D, Short J A, Roper G. (1975). Location of antigens on ultrathin sections of sporeforming bacteria. In *Spores VI*, pp. 572–579. Edited by Gerhardt P *et al.* American Society for Microbiology, Washington DC.

Watson G N. (1985). The assessment and application of a bacteriocin typing scheme for *Clostridium perfringens*. *Journal of Hygiene* **94**: 69–79.

Wnek A P, McClane B A. (1983). Identification of a 50 000 Mr protein from rabbit brush border membranes that binds *Clostridium perfringens* enterotoxin. *Biochemical and Biophysical Research Communications* **112**: 1099–1105.

Wnek A P, Strouse R J, McClane B A. (1985). Production and characterisation of monoclonal antibodies against *Clostridium perfringens* type A enterotoxin. *Infection and Immunity* **50**: 442–448.

Wright-Rudolph L, Walker H W, Parrish F C. (1986). Survival of *Clostridium perfringens* and aerobic bacteria in ground beef patties during microwave and conventional cookery. *Journal of Food Protection* **49**: 203–206.

Yamamoto K, Ohishi I, Sakaguchi G. (1979). Fluid accumulation in mouse ligated intestine inoculated with *Clostridium perfringens* enterotoxin. *Applied and Environmental Microbiology* **37**: 181–186.

23

Systemic toxigenic diseases (tetanus, botulism)

J G Collee and S van Heyningen

Tetanus
    Causative agent
    Occurrence of *C. tetani*
    Culture
    Tetanospasmin
    Toxin–antitoxin neutralization test
    Patients at risk
    Epidemiology
    Clinical diagnosis
    Treatment
    Prevention
    Neonatal tetanus
Botulism
    Causative agent
    Botulinum toxins
    Botulism in man
    Clinical features
    Action to be taken when botulism occurs
    Laboratory confirmation
    Clinical management
    Prevention
    Infant botulism
    Wound botulism
References

Tetanus and botulism have several features in common as they are both systemic toxic diseases caused by specific neurotoxins produced by sporing anaerobic bacilli. Much can be done to prevent these illnesses but management of the established diseases poses problems. Despite our advancing knowledge of the molecular biology of the toxins responsible for these diseases, and our awareness of some parallels in the actions of the toxins in the central nervous system, the mechanisms whereby the toxins produce such widely different syndromes are not understood.

Tetanus

Causative agent

Clostridium tetani is generally described as a slender, gram-positive rod with a typically round terminal spore that gives the organism a drumstick appearance (Fig. 23.1). This classic appearance is the exception rather than the rule; the tetanus bacillus is not particularly slender. It is usually seen as a straight rod of about 4–5 µm by 0.5–0.8 µm with rounded ends, but pleomorphic shorter forms and filaments may occur (Fig. 23.2). It has numerous peritrichous flagella (Fig. 23.3) and gentle motility can be seen in anaerobic wet preparations of broth cultures. Very young cultures may show gram-positive forms, but the organism's early tendency to lose its 'proper' gram-staining reaction ensures that its general appearance in smears is gram-negative.

Spores

Early spores, or forespores, are oval and subterminal. A small terminal cap of the vegetative cell is

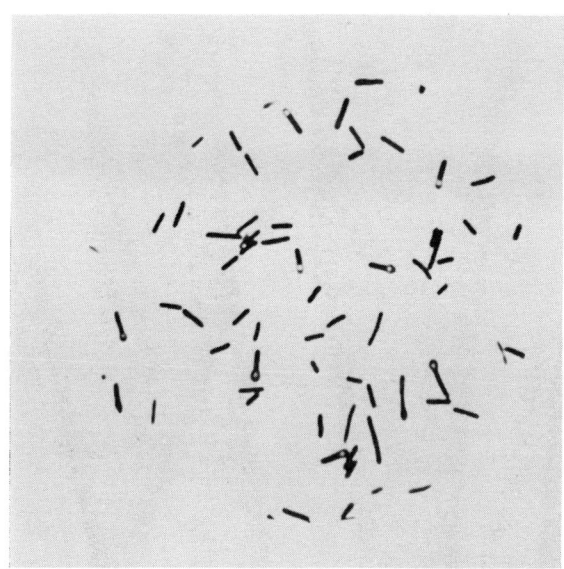

Fig. 23.2 Smear of a broth culture of *Clostridium tetani* showing a range of forms. Gram's stain × 1000

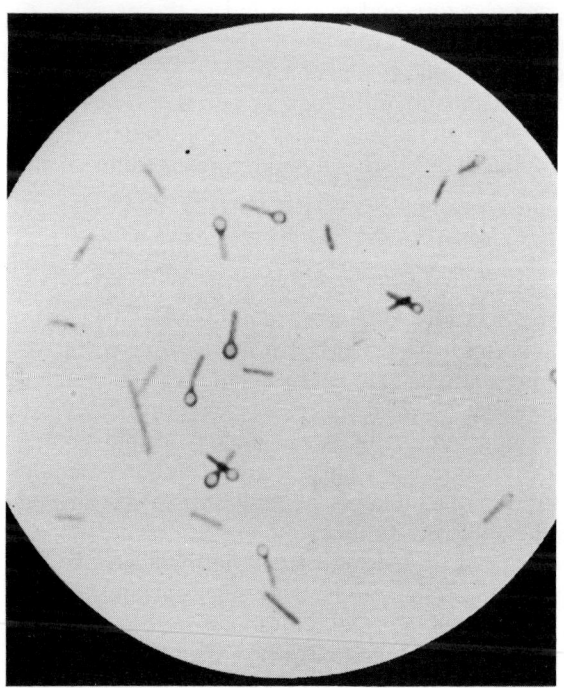

Fig. 23.1 Smear of a sporing culture of *Clostridium tetani*. Carbolfuchsin stain × 1000

often evident; mature spores are rounder. The spores are highly resistant to a wide range of deleterious agents. They ensure the organism's survival against desiccation, aeration, marked changes in pH, exposure to most household antiseptics and disinfectants and considerable extremes of temperature. The spores can survive boiling in water at neutral pH for some minutes. More resistant strains produce spores that may survive boiling for some hours, and may even survive autoclaving at 120°C for 15–20 min (see Finegold *et al.*, 1985). More work is needed to confirm this. We regard about 110°C as the upper limit for survival of tetanus spores in wet heat for 15–20 min; standard hospital autoclave practice generally accepts 121°C for 15–20 min, or higher temperatures for shorter times, as essentially safe in this context.

Occurrence of *C. tetani*

C. tetani occurs widely in soil, dust, air, water and in human and animal faeces and some animal products (Willis, 1969). The spores tend to be more abundant in manured soil and some evidence suggests that natives in tropical or subtropical climates are at greater risk—but the hazard is worldwide; reliable, comparative data on spore counts in soil are

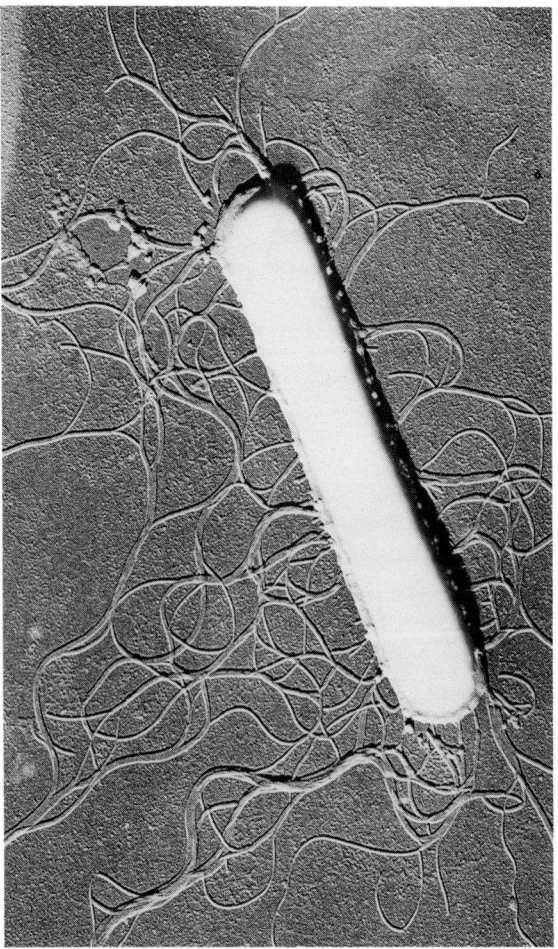

Fig. 23.3 Electron micrograph of a tetanus bacillus showing numerous peritrichous flagella. EM × 20 000 (prepared by Dr B Watt)

not available (Sanada and Nishida, 1965). Data on the occurrence of tetanus spores in human faeces show considerable variation, at least some of which may reflect differing standards of technical competence, differing habits of the population groups studied and laboratory developments during the period when the observations were made. Thus, Tenbroeck and Bauer (1922) reported that almost 35 per cent of the faeces samples examined from 78 subjects in Peking yielded *C. tetani*. Kerrin (1929) was unable to isolate *C. tetani* from more than 300 specimens of human faeces examined in Scotland in 1928–29, although Tulloch (1919–20) found *C. tetani* in 5 of 31 specimens of human faeces examined in Scotland. The human carrier rate seems to lie in the range 0–40 per cent (Lowbury and Lilly, 1958) or 10–25 per cent (Finegold *et al.*, 1985).

Kerrin's work on animal faeces (Kerrin, 1929) confirmed the presence of *C. tetani* in faeces of a wide range of animals, including horses, cows, sheep, dogs and poultry.

Culture

The organism is a strict anaerobe and will not grow well in culture unless a nutritious medium is poised at a suitably low Eh (Willis, 1969). Robertson's cooked meat broth medium, Brain–Heart Infusion broth or Brewer's medium enriched with Fildes' peptic digest all yield good growth. Broth cultures have a faintly rancid (butyric) odour. Culture in semi-solid glucose broth encourages sporulation of some laboratory strains (Collee *et al.*, 1987). Surface cultures of *C. tetani* on solid media such as blood agar with a good quality base show an early tendency to spread. This can be exploited in the laboratory, although many likely contaminant species such as *Proteus* spp., *Pseudomonas* spp. and other spreading clostridia (*C. septicum*, *C. sporogenes*) can outpace *C. tetani* and may invalidate the exercise. *C. tetani* spreads as very fine feathery projections at the edge of a delicate film of advancing growth; the film may be missed by the inexperienced observer. Non-flagellate strains do not spread. Spreading can be inhibited by increasing the concentration of the agar.

Laboratory identification

The definitive laboratory identification of *C. tetani* rests upon observations of its strict anaerobic requirement, its colonial morphology and spreading habit, its typical microscopic morphology, its specific immunofluorescence after suitable staining, its limited biochemical activity and the demonstration of its toxin in neutralization tests with specific antitoxin in mice. *C. tetani* is essentially non-saccharolytic, but some strains ferment glucose. It is weakly proteolytic, hydrolysing gelatin but not attacking other proteins in laboratory tests, although the meat in cooked meat broth cultures after prolonged incubation shows some blackening. Strains vary in their gelatin liquefying properties; some taking 2–4 days, others taking much longer. *C. tetani* is lecithinase negative and lipase negative. A neuraminidase is produced. Many strains produce indole; in our experience, tests for indole production are often negative after one day but positive after two days. Hydrogen sulphide is not produced. Gas–liquid chromatographic procedures show that acetic, propionic and butyric acids are major products. A detailed account of the laboratory identification of *C. tetani* is given by Collee *et al.* (1987), see also Willis (1969). On agar enriched with blood, a haemolytic effect is produced which can be neutralized by tetanus antitoxin (as a result of antitetanolysin antibodies present in the serum). This is the basis of the 'bench mouse' procedure advocated by Lowbury and Lilly (1958). Tetanolysin, like other bacterial oxygen-labile haemolysins, can be non-specifically inhibited and Willis (1969) has explained why this test is unreliable.

Tetanospasmin

Tetanus toxin is synthesized by *C. tetani* in amounts that vary greatly with strain and culture conditions; the toxin yield can reach more than 5 per cent of the dry weight of the organism. Watt and Brown (1975) showed that strains of *C. tetani* could be grown and could produce toxin on a defined medium. This is likely to be of great practical value in assessing factors needed for toxin production.

The yield of toxin is low during the growth phase and maximum yields are demonstrable in the stationary phase. Very little toxin is released into the growth medium before the cells are lysed.

Toxigenic strains sporulate poorly. When the heat resistance of tetanus spores is exploited to recover *C. tetani* from soil samples, better recovery rates and more highly toxigenic strains are obtained if the heat treatment is mild rather than severe (Sanada and Nishida, 1965).

The toxin is a simple protein, originally synthesized on the ribosome as a single chain with a mol. wt of about 150 000. (For more detailed reviews of tetanus toxin, see Mellanby and Green, 1981; Habermann and Dryer, 1986; Simpson, 1986; van Heyningen, 1986.) This single chain is sometimes called the 'intracellular' form of the toxin. It is cleaved, by proteolytic enzymes produced by the bacterium, into the 'extracellular' form, which has two chains linked by a disulphide bond: the heavy chain (100 000 mol. wt) and the light chain (50 000 mol. wt). The gene for the toxin (like that of many bacterial toxins) is on a plasmid which has been sequenced by two different laboratories (Eisel *et al.*, 1986; Fairweather and Lyness, 1986) so that a deduced amino acid sequence is now available. This sequence shows some similarities with what is known of the sequence of botulinum toxin (see page 387).

Tetanus toxin is highly toxic: the minimum lethal dose (MLD) for man is about 130 ng. Many different animal species are susceptible and there is a wide variation in the degree of susceptibility among species. If an effective dose of toxin is injected intramuscularly into the hind limb of a mouse, just at the base of the tail and lateral to it, a typical sequence of signs develops. What happens after a particular time depends critically and reproducibly on the dose of toxin; experienced workers can 'score' the symptoms produced by sublethal doses and thus estimate the dose remarkably accurately. Within a day or so after injection of a lethal dose, there is a

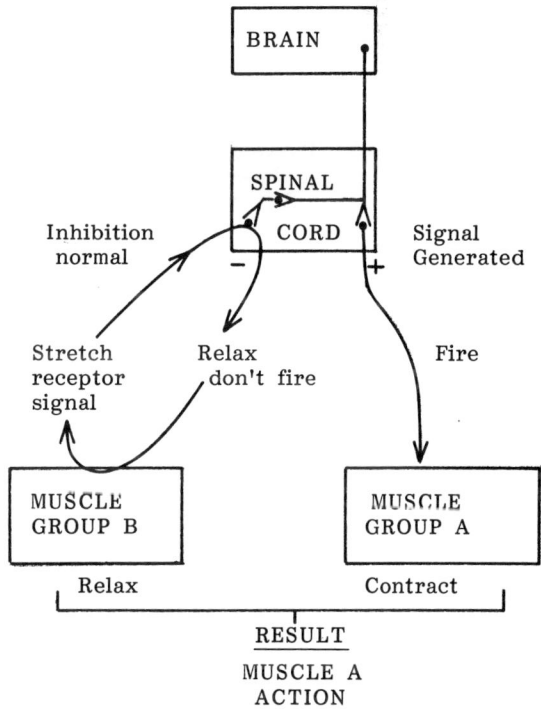

Fig. 23.4 A simplistic outline of the negative feedback control system in the spinal cord that allows a motor signal to operate (muscle group A) with relaxation of an opposing group (muscle group B)

slight stiffness in the limb on the side of the injection. The mouse then develops a limp but retains the use of the limb. Next, the limb loses its function but is still passively moveable. Within two or three days, the animal's tail is curved towards the side of the injection and the limb becomes fixed in spastic paralysis (Fig. 23.4). The disease progresses to a phase when the animal has generalized convulsions. If the animal is allowed to survive during this phase (and it should not be) death occurs, usually within a few days of the initial challenge. In the right circumstances, the effects of sublethal doses of toxin can persist for several weeks or even longer.

Although tetanus toxin has been studied for nearly a century, disappointingly little is known about how it produces these dramatic effects; certainly much less than is known about some other protein exotoxins such as those of cholera, diphtheria or pertussis. This may reflect the fact that we are dealing with aspects of neurophysiology and neurochemistry that

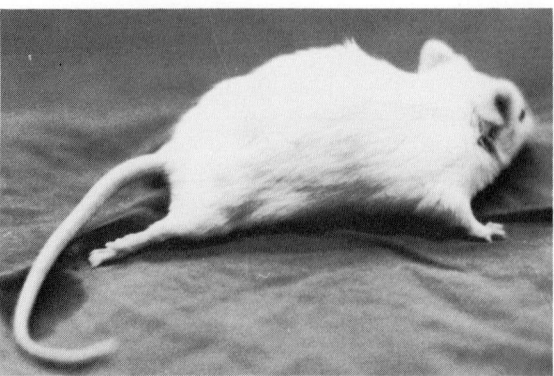

Fig. 23.6 The early effect of tetanus toxin in a mouse; 1–2 days after injection of culture supernate at the base of the tail on the right side, the tail curves to the right and the right hand limb becomes fixed in spastic paralysis

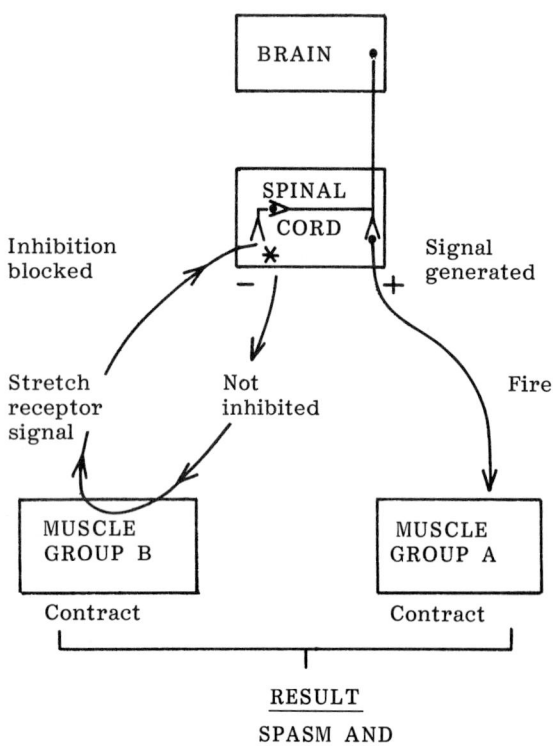

Fig. 23.5 The tetanus effect that occurs when the toxin produces a presynaptic block of the inhibitory pathway in the spinal cord (adapted from van Heyningen, 1968)

are still mysterious. It is fairly clear that the spastic symptoms are produced by a blocking action of the toxin on the release of inhibitory transmitters normally acting at the motor neurones (glycine and gamma-amino butyric acid, GABA). This presynaptic block causes the loss of the normal moderating effects of the upper motor neurone system over the lower motor neurone impulses in the anterior horn (Figs 23.5 and 23.6, and van Heyningen, 1968). The spastic paralysis and convulsions result when the feedback controls between agonist and antagonist muscle groups fail and both groups are activated. There is an initial phase of hypertonic contraction of affected muscles that may last for some days in human patients. Thereafter, the second phase of 'myostatic contracture' evolves and the muscle may remain contracted for many days or for some weeks.

It is clearly established that when the toxin is produced *in vivo* (Smith and MacIver, 1974) it is transported exclusively along alpha (not gamma) motor neurones from peripheral nerve endings. The transport of the toxin is retrograde and intra-axonal, such that its rate of spread within the nervous system seems to be many millimetres per hour.

Two clinical forms of tetanus have been distinguished: local tetanus, in which the symptoms are found only close to the site of infection, and the more common general tetanus, affecting the whole body as described above. In local tetanus, the transport of toxin is along only the nerves supplying the affected muscles; there is not enough toxin for a more general effect. In general tetanus, toxin originally enters the CNS from the nerves near the site of injection, but later gets to motor nerve terminals throughout the body via the lymphatics, which carry it from the initial site of its production into the bloodstream. It is doubtful whether blood-borne toxin can then get directly to the CNS via the floor of the fourth ventricle of the brain; it seems more likely that it passes to the CNS via motor nerve pathways and that the short pathways of the cranial nerves are thus affected early by blood-borne toxin. Hence, spasm of the masseter muscles ('lockjaw') is an early sign of tetanus and has given the disease its common name.

It has been known for some years that tetanus toxin binds rather specifically to ganglioside receptors: gangliosides GD1b and GT1 (with two and three sialic acid residues per molecule, respectively). Gangliosides are complex lipids which are dissolved in the outer membrane of virtually all animal cells, but this particular pair are essentially unique to neuronal cells—a possible reason for the specific action of the toxin. (Compare cholera toxin, which has no cell specificity because it binds to a different ganglioside, GM1, which is practically ubiquitous.) There is good evidence that this binding is significant *in vivo* and so has some role in the action of the toxin, but results of recent experiments suggest that there are also unidentified protein receptors for the toxin on the surfaces of cells (Pierce *et al.*, 1986). It may be that both ganglioside and protein are involved (Montecucco, 1986). The term 'receptor' is widely used in this context but may not be pharmacologically correct, as binding to the receptor alone is not enough to produce any effect. It has been normally assumed that the toxin needs to enter the cell before anything else happens, and there is clear evidence for such entry following binding to the cell surface. However, although we consider it virtually certain that this entry is a prerequisite for activity, there is very little direct experimental evidence that it is necessary.

The binding of the toxin to ganglioside is entirely through the heavy chain; the light chain does not bind. This observation has made the postulation that tetanus toxin has the two-component strategy of many other toxins, such as diphtheria and cholera (van Heyningen, 1982), almost irresistible. Binding of one part of the toxin molecule to the cell leads to entry into the cell of another part of the molecule, which catalyses a reaction producing the toxin effect. Perhaps the light chain of tetanus will turn out to have some important intracellular activity. It is an attractive hypothesis, likely to be correct, but supported at present by analogy rather than by experiment.

Very little is known about the molecular mechanism by which the toxin inhibits neurotransmitter secretion. Some experiments have shown an inhibitory effect of the toxin on the release of acetylcholine and GABA from nerve cell cultures and other preparations. Recently these observations have been extended to non-neuronal cells. For example, tetanus toxin was shown to inhibit the release of lysosomal contents from macrophages (Ho and Klempner, 1985). Furthermore, when added to a primary culture of bovine adrenal cells, botulinum toxin (for which the molecular mechanism is likely to be very similar) inhibited the release of catecholamines from these cells, which is normally evoked by treatment of the cells with nicotine or potassium ions (Knight *et al.*, 1985). Tetanus toxin directly microinjected into the same cells had a similar effect (Penner *et al.*, 1986). These experiments suggest that the action of the toxin is a general one on secretion from cells and its physiological restriction to neuronal cells may reflect the distribution of its ganglioside 'receptors'.

Toxin–antitoxin neutralization test

An inoculation of 0.2 ml of the fluid supernate of a 2–4 day culture in cooked meat broth should be made into the tissues to the right of the base of the tail of a mouse. A control animal is protected by giving 500–1000 units of tetanus antitoxin subcutaneously or intraperitoneally at least 1 h before the challenge. The test animal challenged with material from a toxigenic strain in this way will develop signs of tetanus, whereas the control animal will remain well.

Patients at risk

Any wounded tissue is less aerobic than normal tissue; this is true for some days after an accidental or a surgical wound. Obvious contributory factors to poor oxygenation are the interference with the normal blood supply, the accumulation of extravasated fluid with poor drainage and the presence of a foreign body or of soil, debris and necrotic tissue. The insertion of tight sutures may combine the disadvantages of a foreign body with the added disadvantage of tissue constriction. Similarly, the application of tight plaster of Paris or other splints or dressings might not make allowance for the tissue swelling that follows injury. Heavy bacterial contamination with aerobes and facultative anaerobes may overwhelm host defences so that bacterial proliferation occurs in the wound and this further decreases the available oxygen supply. The growth of many bacterial species under these circumstances adds further enzymic or toxic damage to host tissues. Moreover, a pyogenic response brings phagocytes to the site; these in turn need oxygen for their effective phagocytic and bactericidal functions. Paradoxically, this may be a significant component of the various systems that now operate in the wound to lower the Eh to the point at which the germination and outgrowth of anaerobic spore-forming bacteria is encouraged. If the phagocytes die at the site their hydrolytic enzymes are released and contribute to tissue damage.

The above account has obvious application to a severe injury with marked contamination, but neglected minor injuries with small retained foreign bodies, such as splinters or thorns, allow the same dangerous sequence of events in much less dramatically obvious lesions.

Other host factors at this stage of the early pathogenesis of tetanus are also important. For various and different reasons host defences are less adequate at the extremes of age. In the case of the neonate in an underprivileged society who is challenged by an application of animal faeces or contaminated ligatures to the umbilical stump, tetanus spores are implanted into tissues that are destined to be necrotic and highly favourable for the growth of anaerobic bacteria (see neonatal tetanus, page 383). In the case of an elderly patient with a wound in tissues served by a marginally adequate blood supply, various factors may operate to allow anaerobes to grow. When tetanus spores contaminate injections or injection sites used by drug addicts a potentially severe challenge may be delivered directly into the tissues of severely compromised individuals.

When an injury occurs as a result of a gunshot or road accident, contaminated clothing, skin and soil may be driven deep into tissues that are severely damaged and are on the borderline between necrosis and viability for some time. If there is delay in attending to such a severe wound, peripheral vasoconstriction and a sustained hypotension with shock add further threats to the critical level of tissue oxygenation. Prompt and adequate primary attention to wounds is accordingly a major factor in the prevention of infection generally, and in the prevention of dangerous anaerobic sequelae in particular (see Chapter 18).

When a contaminated thorn or splinter is driven into the bare foot, or hand, of a gardener, or into the auditory meatus of a child in the course of unduly enthusiastic ear cleaning, and if the splinter is not removed promptly, conditions may arise in a relatively trivial wound that may allow tetanus to develop. Agricultural workers, who often tread on nails or sometimes put the prong of a fork through a foot, and athletes who suffer injuries from shoe spikes and boot studs, are also candidates for tetanus. The successful control of tetanus in children in the UK can be attributed to the efficacy of the active immunization programme (Report, 1982). The evidence of recent surveys shows that injuries on playing fields and minor gardening accidents may be factors related respectively to the risk of tetanus in males aged 15–44 years and to older people, aged 65 and over (Atrakchi and Wilson, 1977; Galbraith *et al.* 1981; Report, 1986).

Cryptogenic tetanus

It is sometimes impossible to discover the initiating lesion, or at least to be sure of the site. For example, in barefoot native communities minor injuries and lesions of the feet and lower limbs are common, and tetanus spores abound in the environment. In some cases, tetanus bacilli can be isolated from the external auditory meatus. If there is an associated lesion in the ear canal, otogenic tetanus may be suspected. One of us has seen and investigated several examples in India and there is no doubt that toxigenic tetanus bacilli can be recovered in such circumstances, but similarly toxigenic strains were also isolated from the ears of control subjects in the same environment who did not have tetanus.

Epidemiology

Tetanus occurs throughout the world, but it is particularly recognized as a common and dangerous disease with high mortality in underprivileged tropical and subtropical countries, where neonatal tetanus is also a significant killer. Tetanus ranks within the leading four most common infectious causes of death in many such countries. The epidemiology of tetanus in man was reviewed by Bytchenko (1967) and other co-authors (see Eckmann, 1967). There must be much under-reporting of the disease from communities in which medical services are lacking or inadequate. Available data suggest that there may be 100 000–200 000 non-neonatal deaths attributable to tetanus annually in the world and almost 1 million deaths due to neonatal tetanus. Mortality rates range from 46 to 0.03 per 100 000 population for tetanus in the general population (Rubbo, 1966). For cases of established tetanus, case fatality rates of 15–75 per cent have been recorded, and 80–90 per cent or worse for neonatal tetanus. In 1966, Rubbo had Haiti, Malaya, Panama and Nigeria at the top of his tetanus mortality table and USA, Finland, England and Canada at the bottom. With intensive care and modern treatment, the case fatality rate for non-neonatal tetanus can be reduced to about 10 per cent (see below).

Clinical diagnosis

The diagnosis of tetanus is essentially clinical. All aspects of the organism and the disease have been most carefully considered by Willis (1969), Adams et al. (1969) and J W G Smith (1985). In the early stages of the disease, when trismus is a common presenting sign, the differential diagnosis may include a range of conditions affecting the jaw, such as restriction of movement associated with an unerupted wisdom tooth, a dental abscess, or a dental or maxillofacial tumour (Stoddart, 1979; Ranabir Bose et al., 1985). The unilateral features of these conditions may help to distinguish them from early tetanus, but so-called cephalic tetanus in patients with wounds of the head or neck may present with cranial nerve palsies that may initially be predominantly unilateral (Edmondson and Flowers, 1979). The patient with tetanus may complain of difficulty in swallowing or stiffness of the jaw. Bilateral stiffness of the temporomandibular joint caused by trauma or as an early feature of a more generalized arthralgic syndrome may mimic early tetanus. Some patients with tetanus complain of back pain early in the infection. The anterior abdominal muscles are rigid but not tender. The patient with tetanus is usually afebrile, rational and anxious. There is increasing muscle tone in the masseters, the abdominal muscles and the paraspinal muscles. When a spasm occurs, there is increased trismus and the ominous appearance of risus sardonicus is seen in some patients (Lancet, 1980). In a possibly encephalitic child with features very suggestive of tetanus, and who was seen by one of the authors, a definitive diagnosis was never made. In a severely injured patient with signs very suggestive of tetanus, we subsequently excluded tetanus and attributed the spasms to the effects of other drugs given in high dosage to control a difficult series of clinical problems.

Various systems have been devised to assess the severity of the disease and to provide an index of a patient's chances of survival (Phillips, 1967). Early referral of a patient to a centre that has special equipment and personnel with special experience can be life-saving. However, the achievements of Sanders et al. (1977) and Rennie (1979) should encourage workers in less privileged areas to believe that care and attention to detail in applying basic principles and dedicated nursing in the management of tetanus are the first essentials.

Treatment

Until the action of tetanus toxin is understood in sufficient detail to allow effective pharmacological interruption of the mechanism the clinical management of the disease will remain essentially passive, but it is essential to take the following steps promptly and positively:
1. Early administration of antitoxin to neutralize any residual unbound toxin;
2. Attention to any obvious wound that might be the site of infection and the initial source of the toxin;
3. Powerful sedation and support of the patient for as long as is necessary until the toxin has decayed and its effect subsided.

The use of muscle relaxants has an important place in the treatment of cases of moderate or marked severity. A combination of neuromuscular blockade with a curare-like drug to produce total paralysis, tracheostomy and intermittent positive pressure respiration in suitably equipped specialist units has achieved very considerable success. However, this calls for highly sophisticated intensive care with continuous monitoring. These facilities are unfortunately too costly in terms of finance, expertise and equipment to be generally available in the countries where tetanus is most common, but much can be done when resources are limited (Sanders et al., 1977; Rennie, 1979).

The clinical management of tetanus has been well documented by Edmondson and Flowers (1979) and Flowers and Edmondson (1980), who reviewed 100 cases (75 male, 25 female) treated during the period 1961–77. As severe cases were referred from other hospitals, this series was perhaps biased to the severe side of the spectrum. Seven mild cases did not require tracheostomy; three moderately ill patients received tracheostomy but not paralysis and 90 patients required tracheostomy with full paralysis for 5–44 days (average period of paralysis 21 days). The disease was graded as very severe in 21 of the 90 patients. The mortality rate in the whole series was 10 per cent. Treatment consisted of early and adequate excision of possibly contaminated tissue when a wound was identified as the likely source of the trouble, with delayed closure after packing with suitable antibacterial dressings. As a routine, benzyl penicillin 10^6 units 6-hourly was given intramuscularly for one week. Equine antitetanus serum or locally prepared human antitetanus serum (10 000–150 000 units) was given when the series began, but commercially prepared human antitetanus immunoglobulin (Humotet, Wellcome; 2000–4000 units intramuscularly) was given to the later patients. Diazepam and opiates were the mainstays for patients with mild tetanus. (Some workers positively avoid opiates.) Moderately severe cases required tracheostomy and sedation. In addition, the severe and very severe grades received paralysis with curare and intermittent positive pressure respiration. Patients with evidence of sympathetic overactivity required additional special management.

The modern therapy of tetanus continues to evolve. J W G Smith (1985) carefully evaluated the therapeutic usefulness of antitoxin and gave good reasons for favouring the intravenous route. On balance, and more evidence is still needed, we strongly agree that 10 000 units of antitoxin, preferably human tetanus immunoglobulin (HTIG) should be given intravenously as soon as tetanus is suspected. The makers of commercially available human tetanus immunoglobulin (Humotet, Wellcome) advise 30–300 units per kg body weight; hence the dose range for a 70 kg man would be 2100–21 000 units. This is recommended for intramuscular administration and the intravenous route is said to be contraindicated. The volume of 2100–21 000 units would be about 8–80 ml. This, and other considerations, encourage the view that the therapeutic dose of HTIG for an adult should be 10 000 units, given by slow intravenous injection. Perhaps the intraperitoneal route merits more study. Evidence favouring the early intrathecal administration of antitoxin continues to accumulate, although controlled observations are understandably scarce (Sanders et al., 1977; Rennie, 1979; J W G Smith, 1985).

A few cases of tetanus have been recorded in which patients apparently responding to initial management, which included antitoxin, have had recurrences of spasms within a few days and have accordingly been given more antitoxin. The persuasive reports of Crawford et al. (1980) and Passen and Anderson (1986) suggest that a further dose of antitoxin can be of positive value in such cases and also indicate the usefulness of early assays of a patient's levels of tetanus IgM and IgG before and during therapy.

Chemotherapy

It is reasonable to include antimicrobial therapy in the early management of tetanus to inhibit the growth and further toxin production of tetanus bacilli at or near the site of infection. As there is often a mixed infection present, including penicillinase-producing organisms, the use of penicillin alone can be justifiably criticized, but 10^6 units of crystalline penicillin 6-hourly by intramuscular injection or by intravenous infusion is often used as a routine. Ahmadsyah and Salim (1985) favoured metronidazole 500 mg orally 6-hourly or 1 g rectally by suppository 8-hourly but they used procaine penicillin in their comparative study. The antianaerobe activity of metronidazole and its favourable pharmacokinetics commend it for use in this situation. There is good justification for a 7-day course of metronidazole (with crystalline penicillin or an equivalent if penicillinase-producing organisms are detected) for any patient with tetanus; the antibiotics should be started at the earliest opportunity. There are good reasons for instituting early antitoxic and antimicrobial therapy before surgical excision of the wound.

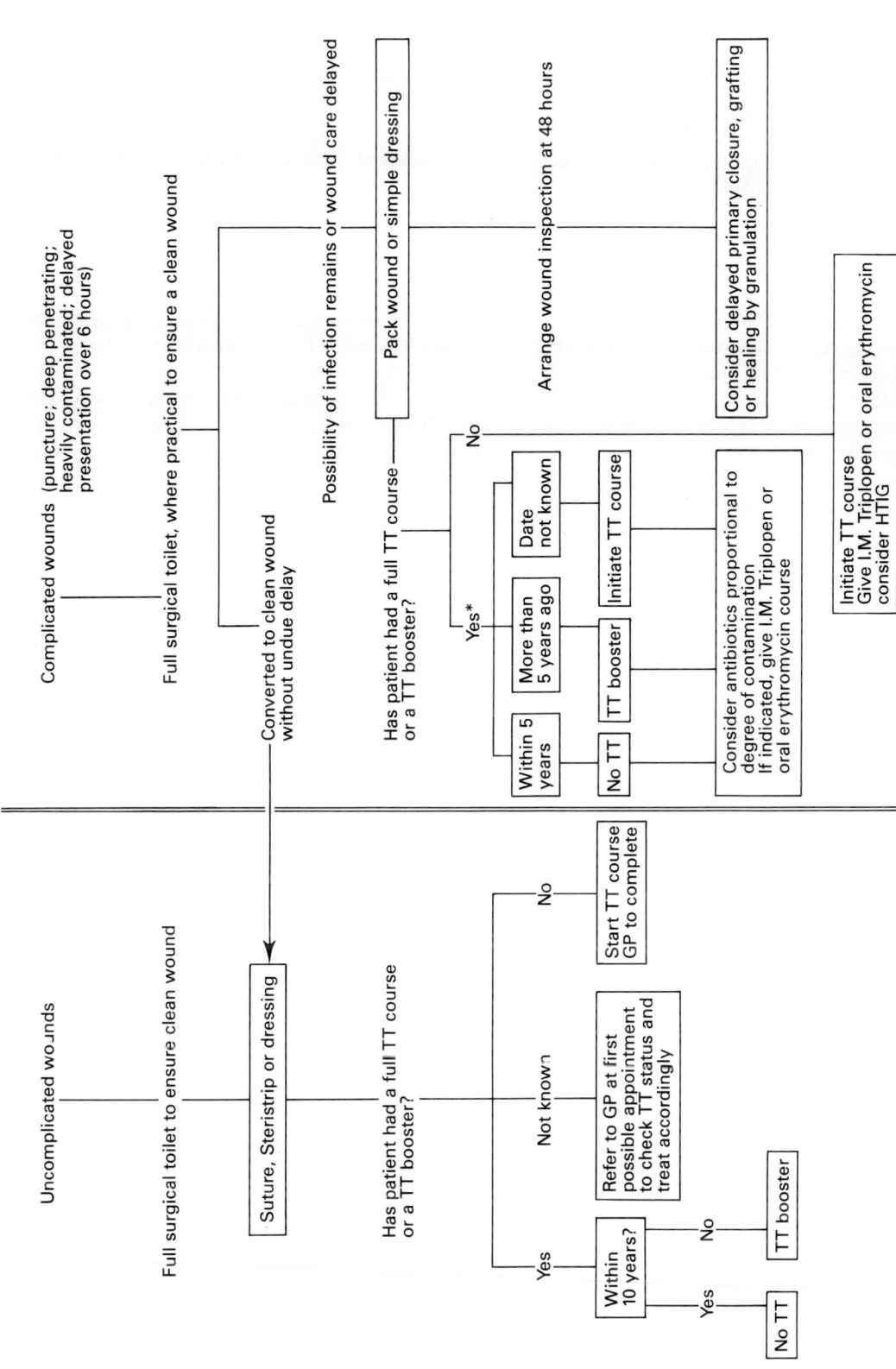

Fig. 23.7 A schedule of tetanus prophylaxis in relation to wound management

Prevention

The prevention of tetanus depends upon prompt and effective wound management, continuing delivery of an effective active immunization programme and socioeconomic improvements in underprivileged societies that are particularly vulnerable to the disease. These expensive ideals range from the availability of adequate footwear to the provision and availability of health education and health services. They include avoidance of practices associated with special risks of tetanus, such as primitive traditional native applications or ligatures to wounds and to umbilical stumps.

As tetanus prophylaxis cannot be divorced from wound management it is important to have combined guidelines. Major contributions have been made in this context by Rubbo (1966) and Smith et al. (1975). The wound management and tetanus prophylaxis schedule cited here (Fig. 23.7) (see also Collee et al., 1987) was developed in consultation with the Accident and Emergency Advisory Subcommittee of the Lothian Health Board in 1982. We are particularly indebted to Mr D A D Macleod, Dr K Little and other colleagues for their help in producing this schedule and gratefully acknowledge the Lothian Health Board's permission to publish it here.

Wound management & tetanus prophylaxis schedule

1. Even the most trivial wound can cause tetanus.
2. *Prompt and thorough surgical wound toilet is of paramount importance* Foreign materials and devitalized tissues must be removed if possible, and wound edges should be excised if they are not viable.
3. Protective immunity depends upon a full course of tetanus toxoid (TT) completed or boosted within the previous five to ten years. Even this does not give an absolute guarantee of protection. If it is 10–20 years since the patient last had a full course of tetanus toxoid without a booster, the full course of tetanus toxoid injections should be arranged.
4. *Active immunization* The patient may be uncertain, or unable to reply when questioned about previous active immunization with tetanus toxoid (TT). If the patient was born in the United Kingdom after 1955, it is reasonable to assume that he or she will have received immunization as a preschool child, with boosters at school entry. If doubt persists, patients should be given a letter and advised to seek clarification from their general practitioners at the first available appointment. Unnecessary TT injections can lead to painful local reactions; if a booster toxoid injection was given during the previous year, another TT injection should not be given.
5. *Passive immunization* Human tetanus immunoglobulin (HTIG) should be reserved for protecting unimmunized patients with wounds that are considered to be tetanus-prone. The prophylactic dose is 250 units intramuscularly. TT injections can safely be given at the same time as HTIG.
6. Ensure that a Tetanus Advice form is completed and distributed as instructed. The responsibility to complete a course of TT injections is placed on the patient and the general practitioner.
7. *Tetanus-prone wounds* Wounds complicated by a delay in treatment of more than 6 h, all stab wounds and other puncture wounds, all animal or human bite wounds, penetrating wounds and heavily contaminated wounds (especially if contaminated with agricultural or horticultural materials or soil) require particular care in their assessment and wound toilet. Special attention should be given to excision and exploration as necessary, and local antisepsis with an iodophor such as Betadine merits consideration. Such wounds should not be sutured as a routine, but Steristrips may be applied. Alternatively, packing with delayed primary closure or grafting should be considered, with follow-up wound inspection at 48–72 h.
   Note: The assessment of *puncture wounds* is difficult, and adequate excision and exploration is not always practical. All such wounds should continue to be managed as potentially contaminated (see Fig. 23.7).
8. *Antibiotic prophylaxis* Antibiotics are not a substitute for adequate surgical toilet, but they may be a necessary adjunct in certain cases. When penicillin is indicated, give one vial of long-acting penicillin (Triplopen) preferably just before the wound toilet and exploration takes place, if this is practical. If the patient is allergic to penicillin, or a more prolonged course of antibiotic is con-

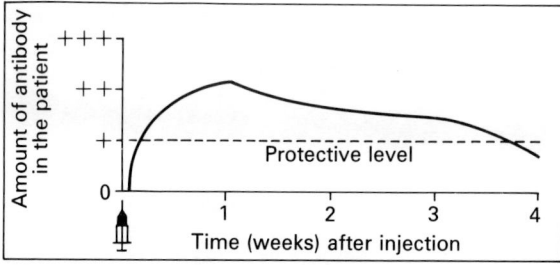

Fig. 23.8 An injection of human tetanus immunoglobulin (250 units) gives likely protection for 3–4 weeks (Courtesy of Update and Wellcome Medical Division)

sidered to be necessary, use erythromycin 500 mg, b.d., for 5 d, giving the first tablet just before wound toilet if practical.

9. These recommendations are not a substitute for applying the basic surgical principles of primary wound care and follow-up. In the management of some serious wounds, decisions are difficult and prompt senior advice should be taken.

Immunization

Passive protection

An intramuscular injection of human tetanus immunoglobulin (HTIG) 250 units (see British National Formulary, 1986b) provides a protective level of antitoxin for 3–4 weeks (Fig. 23.8).

Active immunization

Three doses of adsorbed tetanus toxoid given at intervals of 6 weeks and 6 months confer immunity

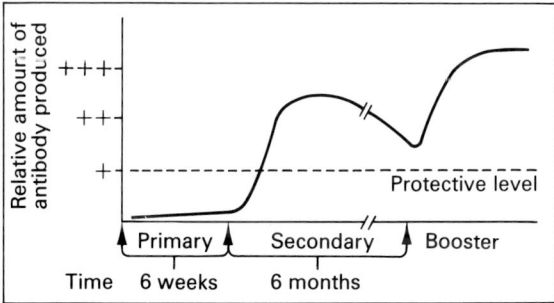

Fig. 23.9 Active immunization with a course of three injections of adsorbed tetanus toxoid given at intervals of 6 weeks and 6 months is likely to confer protection for 10 years

for 10 years. As indicated in fig. 23.9 the primary response is not protective but the secondary response is, and a booster injection confers durable protection.

Combined active and passive immunization may be part of a prophylactic approach to the care of the injured person (see Fig. 23.7) or may be given in the course of treatment of tetanus. The antitoxin and the adsorbed toxoid should be injected at different sites from separate syringes. Tetanus is not an immunizing disease; active immunization of a person surviving tetanus should be followed up with a second and third dose of adsorbed toxoid at the usual intervals.

Neonatal tetanus

It is estimated that a million neonatal deaths a year in the world are attributable to neonatal tetanus resulting from contamination of the umbilical stump wound with tetanus spores or bacilli. In some underprivileged communities in the Third World, neonatal tetanus heads the list of the causes of neonatal death. Stanfield and Galazka (1985) regarded neonatal tetanus mortality rates as inverse indices of the quality and delivery of maternal health services. They cited death rates ranging from 23 per cent to 72 per cent of all live births, with 67 per 1000 live births dying within the neonatal period in rural areas of Uttar Pradesh, India. Woodruff *et al.* (1984) reported that tetanus killed almost 1 per cent of the infants born in a district of Southern Sudan in 1984 in which there were 3595 live births per annum. These workers identified the fine string-like roots used for tying the umbilical cord as one of the vehicles of infection and showed that *C. tetani* could be cultured from root samples.

The disease in infants can be prevented by active immunization of mothers before or during pregnancy and by aseptic and antiseptic care and attention to the umbilical cord and the stump during healing. However, immunization of remote communities is difficult and programmes to achieve this and to teach hygienic principles of obstetric practice and neonatal care merit more support. Woodruff *et al.* (1984) advocated issuing a simple sterile kit, containing a blade, a ligature, adhesive dressings and swabs, to midwives and pregnant women in areas where neonatal tetanus is common.

Neonatal tetanus has a high death rate in countries that lack intensive care facilities. About 80–90 per cent of the affected children die, and more than 90 per cent of these deaths occur within two weeks of birth. Smythe *et al.* (1974) and Adams *et al.* (1979) have reported how the modern management of tetanus neonatorum can dramatically save lives, but a significant change in the world statistics must await effective prevention.

Botulism

Causative agent

Clostridium botulinum, the cause of botulism, is a relatively large, usually straight, spore-bearing, gram-positive, rod-shaped bacillus (4–6 µm × 0.9–1.2 µm) with rounded ends. Longer, shorter and gently curved forms also occur (Fig. 23.10). In smears of representative strains that we have examined, type A and B strains tend to be shorter and less stout than type E strains. The spores are oval, subterminal and distend the vegetative cell. When the cell is short this gives a navicular appearance. The organism is motile with 4–8 peritrichous flagella. Gram-negative forms are commonly seen, especially in older cultures and in sporulating cultures. The organism is strictly anaerobic. It grows on a range of nutritious culture media; it grows well in cooked meat broth or on blood agar. The optimum temperature for growth is 35–37°C for most types, but 30°C for type E strains. The growth of some strains can occur at relatively low temperatures (e.g., 6°C) with toxin production and this is important in food hygiene (see Chapter 22). There is some evidence that toxin production by type E strains is reduced or inhibited at 37°C.

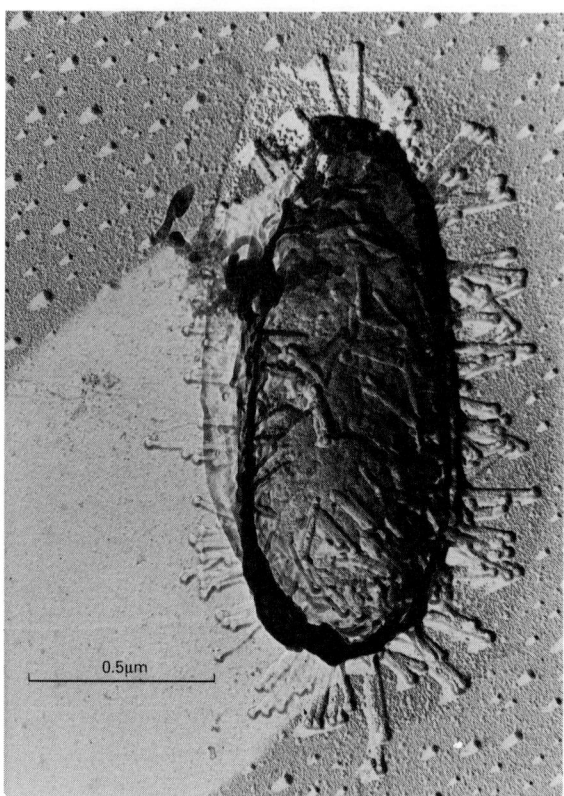

Fig. 23.11 Carbon replica EM preparation of a spore of *Clostridium botulinum* type E showing tubular appendages (With thanks to W Hodgkiss, Z J Ordal and the *Journal of Bacteriology*)

Spores

The spores are large, oval, thick-walled structures. Hodgkiss and Ordal (1966) demonstrated tubular appendages radiating from the surface of spores of type E strains (Fig. 23.11). The spore coat in these strains showed many surface irregularities. It seems

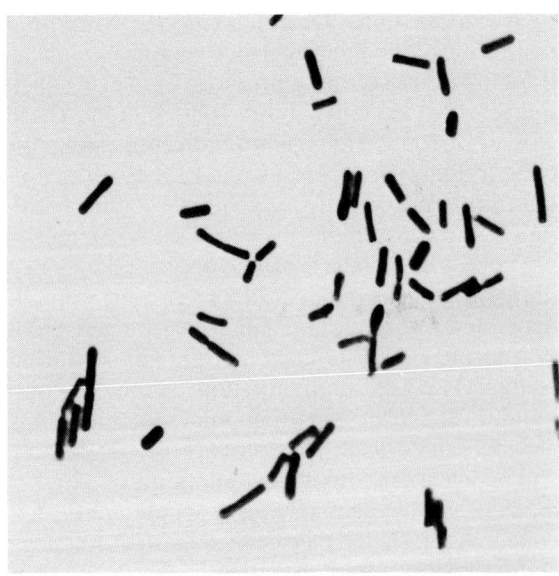

Fig. 23.10 *Clostridium botulinum* type E. Smear of a broth culture showing a range of forms. Gram's stain × 1250

that these unusual structures have not been observed in spores of other types of *C. botulinum*.

Sporulating forms of Group 1 strains (see below) are more distended than those of the other groups.

The heat resistance of *C. botulinum* spores has served as a yardstick for the thermal processing of canned foodstuffs. Spores of types A and B are markedly resistant to heat, surviving boiling at neutral pH for hours and even autoclaving at 115°C for 10–40 min. Autoclaving at 121°C for 15–20 min inactivates them. Decimal reduction (D_{121}) values for spores of types A and B at 121°C are within the range 0.1–0.21 min. Spores of types C and D are of intermediate heat resistance and type E spores are said to be relatively heat sensitive. The expressed resistance of type E strains may be moderated by their special germination requirements; the point is not merely academic and it is of practical importance.

The resistance of *C. botulinum* spores to radiation is of special relevance to food processing and consideration of food safety (see Chapter 22, and Genigeorgis and Riemann, 1979). The radiation dose that reduces counts of viable *C. botulinum* spores by a factor of 10^{12} was calculated for a highly resistant type B strain in phosphate buffer at 20°C as 3.5 Mrads.

The spores are killed by exposure to the usual sporicidal chemicals such as formaldehyde, glutaraldehyde and ethylene oxide.

Serotypes

Seven types of *C. botulinum* (A–G) are recognized on the basis of the serological specificities of their biologically similar neurotoxins. Each of these toxins is neutralized by its type-specific antitoxin, but the situation is not straightforward; types C and D produce multiple toxins. Type Cα produces toxins C1, C2 and D. Type Cβ produces C2 only. Type D produces C1 and D toxins.

Groups of *C. botulinum*

The types can be grouped according to their biochemical activities. Table 23.1 shows the relationship of the groups to the biochemical reactions of the types and the toxins produced.

Group I

Contains all type A strains and those of type B and F that are strongly saccharolytic and proteolytic and produce H_2S. They ferment glucose and fructose, not mannose or lactose, and give variable reactions with maltose and sucrose.

Group II

Contains all type E strains and those of types B and F that are saccharolytic but non-proteolytic. They do not produce H_2S. They ferment glucose, fructose, mannose and sucrose but not lactose and they give variable reactions with maltose.

Group III

Contains all strains of types C and D. They are saccharolytic. Although they are non-proteolytic, they are gelatinolytic. They do not produce H_2S. They ferment glucose, mannose and maltose, but not sucrose or lactose.

Group IV

Accommodates one rather unusual type (G) which is proteolytic but non-saccharolytic. It produces H_2S.

With the exception of the asaccharolytic Group IV, strains in the other groups ferment other sugars not listed here. All strains of *C. botulinum* produce gelatinase; all are indole negative and all are unable to reduce nitrates. Strains of Groups I, II and III produce lipase; some type C strains produce lecithinase.

Colonial characteristics

Colonies of Group I strains are moderately large (up to 8 mm) round, with a raised opaque centre and with a rhizoid edge, often spreading. Colonies of Group II and III strains are smaller (about 3 mm) irregular and translucent or semiopaque and matt. Colonies of Group IV are small (about 1 mm in diameter) raised, grey and smooth. The translucent colonies of toxigenic strains (TOX colonies) may revert to non-toxigenic opaque sporing (OS) colonial mutants (Dolman and Murakami, 1961).

Table 23.1 The relationship of the groups of C. botulinum to the biochemical reactions of the types and the toxins produced

Test	Characteristics of group			
	I	II	III	IV
Potent proteinase	+	−	−	+
Gelatinase	+	+	+	+
Fermentation of sugars:*				
glucose	+	+	+	−
mannose	−	+	+	−
maltose	v	v	+	−
sucrose	v	+	−	−
lactose	−	−	−	−
Major fermentation products†	A B iV	A B	A P B	
Production of:				
H_2S	+	−	−	+
Indole	−	−	−	−
Lipase	+	+	+	−
Lecithinase	−	−	−/+	−
Serotype	A BF	BF E	Cα Cβ D	G
Toxin(s) produced	A BF	BF E	C1 C1 C2 C2 D D	G

v, Variable
*Ammonia production may mask Group I fermentation reactions
†, A acetic, P propionic, B butyric, iV iso-valeric acid

Taxonomic problems

The C. botulinum complex embraces a considerable range of organisms with different biochemical activities and various cross-reactive antigens (Bom et al., 1986). In cultural and biochemical characteristics, excluding toxin production, Group I strains are indistinguishable from C. sporogenes, which may be considered as a non-toxigenic variant (Iida et al., 1981; Poxton, 1984). Phage mediated toxigenesis of type C strains can be lost and interconversion of types and interspecies conversion by phage is possible from C. botulinum to C. novyi type A. Poxton and Byrne (1984) showed significant cross-reactions between C. sporogenes antiserum and strains of C. botulinum of type A, proteolytic type B and type E strains, and between C. novyi type A antiserum and C. botulinum types C and D. Most strains of C. botulinum types A and proteolytic B share antigens with C. sporogenes, whereas types C and D share antigens with C. novyi. Analyses of somatic antigens and their inter-relationships are further confused by the occurrence of cross-reacting spore antigens in apparently non-sporing culture preparations.

Hoffman et al. (1982) and Hall et al. (1985) attributed a case of infant botulism to a toxigenic (type F) organism resembling C. barati. Aureli et al. (1986) attributed two cases to toxigenic (type E) strains of C. butyricum. Thus, the emerging evidence suggests that the gene responsible for toxin production in the C. botulinum complex may be on a plasmid that might be transferable. Alternatively, this may be another illustrative example of the pitfalls that lie in the path of medical microbiologists who attach undue taxonomic importance to the production of a toxin. We shall have to come to terms with the fact that the C. botulinum species (or genus) is already untidily heterogeneous as, on various grounds, it might now be said to embrace the non-toxigenic C. sporogenes, some members of C. novyi, the various members of C. botulinum, the old C. parabotulinum classification and toxin-producing organisms resembling C. barati and C. butyricum. The taxonomy of the C. botulinum complex is in urgent need of repair.

Botulinum toxins

Closely related neurotoxic exoproteins are produced by all serological types of the organism in serologically specific forms and, in some cases, in combinations of these forms (see Table 23.1 and Lewis, 1981). They are all basically similar in structure and action, except for type C2 toxin, which is distinct from all the others and should probably not be regarded as one of the botulinum neurotoxins. This point is further discussed below. For general reviews of these toxins see Simpson (1986), Sakaguchi (1986) and Habermann and Dryer (1986).

These toxins are among the most potent toxins known. They are active at various points, chiefly in the peripheral nervous system (Fig. 23.12); the possibility of their action in the central nervous system is debated. They are remarkable in that, despite being protein, they can be very active when delivered (as preformed toxin) by the oral route into the digestive tract. The lethal oral dose is certainly much more than the lethal parenteral (intravenous or intraperitoneal) dose; the ratio varies widely from 100 000 to 10, depending upon the test animal and the test toxin (L DS Smith, 1975). It is clear that there are marked differences in susceptibility to the different types of toxin and there are differences in the abilities of different types of toxin to pass in active form across the intestinal wall.

Botulism and tetanus are very different diseases: tetanus is a disease of spasticity with convulsions; botulism is a general flaccid paralysis. However, it is becoming increasingly clear that botulinum and tetanus toxins are very similar both in structure and in action at the molecular level (Mellanby, 1984; Simpson, 1986). There are several marked similarities in amino acid sequence between the two toxins. It is not yet possible to be specific about the degree of homology; although the entire sequence of tetanus toxin is known, that of botulinum toxin is not (Eisel *et al.*, 1986). Both the toxins act by blocking evoked and spontaneous release of neurotransmitters; the differences between the two diseases probably reflect binding and delivery of the toxins to different receptors and cells. Furthermore, the dose of botulinum toxin required to block neuromuscular action is about 500 times less than that of tetanus toxin. This is one likely reason for the fact that flaccid paralysis is produced only by botulinum toxin: tetanus toxin in sufficient quantities to give the same effect in peripheral tissues will already have killed the animal via the central nervous sytem. Sellin (1981) discusses the trophic effect of the nerve supply to a muscle and the interference produced by botulinum toxin in this system. Patients who survive poisoning by botulism may have long-term neuromuscular dysfunction and weakness related to these effects.

Botulinum toxins have the same two-chain structure as tetanus toxin (see above): a heavy chain (mol. wt of about 100 000) and a light chain (mol. wt of about 50 000) joined by a disulphide bond. However, they are generally first produced as protoxins that have to be activated by endogenous or exogenous proteases. There is evidence that they bind to similar receptors (including gangliosides) through their heavy chains: a binding fragment of tetanus toxin can antagonize the action of botulinum toxin at neuromuscular junctions. Much research has been

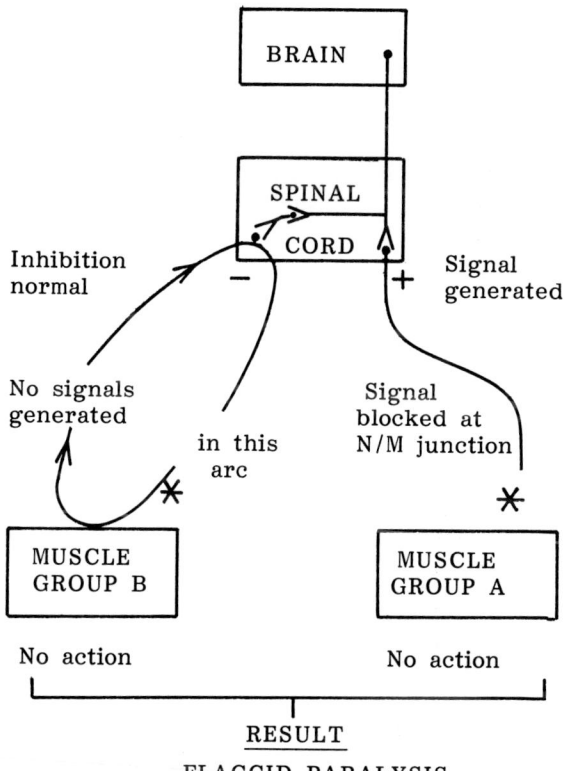

Fig. 23.12 The botulism effect that results when the toxin blocks transmission at neuromuscular junctions (Adapted from van Heyningen, 1968)

done on the binding and uptake of botulinum toxin at nerve terminals (Black and Dolly, 1986).

The two component strategy, which has been put forward for tetanus toxin (see page 375), has also been suggested for botulinum toxin and it is known that botulinum toxin binds to cells and inhibits secretion in the same way (e.g., Knight *et al.*, 1985; Montecucco, 1986). There is also evidence that the type D toxin is an enzyme and that it can catalyse a reaction in cell membranes (Ohashi and Narumiya, 1987). Almost all of the molecular activities shown for tetanus toxin have also been shown for botulinum toxin and vice-versa, although there are some reports of differences in the site of action.

The C2 toxin is different. It is made up of two quite distinct proteins (mol. wt of about 100 000 and 50 000) not joined by any covalent bond. Both components working together are needed for any lethal action and there is evidence that the larger protein binds to cells, while the smaller catalyses the ADP-ribosylation of non-muscle actin (Aktories *et al.*, 1986; Ohishi and Tsuyama, 1986). This multifactorial system also has analogies with other bacterial toxins, e.g., anthrax, that are made up of separate, interacting components.

Botulism in man

Human botulism is usually caused by the ingestion of food in which *C. botulinum* has grown and produced toxin. There is evidence that toxin production can occur as a result of germination and outgrowth of spores *in vivo* and that this may be a component of the pathogenesis of the disease in some cases. This must be true in wound botulism and in infant botulism (see page 391), but most of the toxin or protoxin causing adult human food-associated botulism is preformed in the food.

Human botulism is mainly associated with types A, B and E, whereas animal botulism is mainly associated with the type C complex (Table 23.2). Type A and B strains are widely distributed in cultivated and virgin soil. Type E strains are particularly associated with marine environments, fish and seafood (Hobbs, 1976). These associations are not exclusive. In recent surveys of coastal sediments and inland waterways in Britain, G. R. Smith *et al.* (1978) confirmed that type B occurs regularly, but that type A seems to be uncommon. Type A strains are common in soil in the western states of the USA, whereas type B strains

Table 23.2 Botulism in man and some animals

Species (and disease)	Type of C. botulinum associated with disease*	
Man (botulism)	A, E, B	(C, D, F, G)
Cows (lamsiekte), sheep	D, C	(B)
Horses (forage poisoning)	C	
Mink‡	C†	(A, B, E)
Chickens (limberneck)	C	(A)
Pheasants‡	C†	
Waterfowl§ (western duck sickness, alkali disease)	Cα	(Cβ)
Fish‡	E	

*Less common associations indicated in brackets
† the sub-group of C is not invariably recorded
‡ farmed and intensively reared
§ wildfowl

predominate in the soils of north-east and central states (CDC Handbook, 1979). Sakaguchi (1979) reviewed the distribution of spores and pointed out that the relative heat sensitivity of the spores of type E strains may have excluded them from some surveys in which a heat resistance step was used for selective isolation. The point is valid, although the apparent heat sensitivity of spores of type E strains may be partly related to their germination requirements.

G R Smith (1985) noted that *C. botulinum* sometimes occurs in the faeces of pigs, cattle, horses and waterfowl, but types A, B and E are not commonly found—and certainly not in human faeces.

L DS Smith (1977) reviewed the considerable occurrence of human botulism in the world and noted its apparently curious distribution—i.e. restricted predominantly to countries of the northern hemisphere; Poland, Germany, USSR, USA and Japan headed the list. He related the data to faulty domiciliary methods of food preparation or to the types of preservation traditionally practised in these countries—ranging from home curing of pork products in Poland (type B) to the preparation of uncooked fermented fish (izushi) in Japan (type E). He extended the link to include home preserved uncooked ham or sausage in France (type B) and conceded that there may be considerable underreporting of botulism, so that figures for some countries such as China and Hungary may be unreliable.

It is difficult to believe that botulism has not occurred more extensively in other countries. G R Smith (1985) reviewed data from several workers and produced tabulations of the number of episodes for different periods in the present century. In terms of the numbers of cases per recorded year, and with the reservations that must apply to analyses of such data, France, USSR, USA, Germany and Japan were again prominent. A recent report indicates that significant numbers of cases of botulism have occurred in Argentina (Ciccarelli and Gimenez, 1981). But until the medical services of many countries are properly developed, with adequate laboratory support, the true world prevalence of botulism will remain unknown. If advances in food processing proceed in parallel with the progress of medical services, we shall not know whether botulism occurred more extensively than the available records show—but the continuing evidence of man's carelessness in many approaches to modern food processing is a sobering consideration in this context.

Food-associated botulism

A wide range of food has been incriminated in outbreaks of human botulism. The food has almost always been improperly processed, or has been stored or preserved under conditions that allow the survival and subsequent outgrowth of resistant *C. botulinum* spores. Records show that items such as uncooked fermented whale meat, decomposing seal flipper and putrefying salmon in the diet of Eskimos and northern maritime Canadian or Alaskan people are not without hazard. Home canning and other amateur forms of home preservation of foods are also potentially dangerous; here the incriminated foods range from canned string beans to home cured meat. The commercial canning and food preservation industry enjoys a good reputation, but it must be mindful of the cost of occasional lapses. In Britain, the Loch Maree tragedy of 1923 (Leighton, 1923) is a classic example and the Birmingham outbreak of 1978 (Ball and Farrell, 1979) received wide publicity. While experienced home bottlers of acid fruit are unlikely to incur a risk of botulism, amateur attempts to preserve non-acid vegetables, mushrooms or meat pastes are much more hazardous. The proper preservation of food calls for a knowledge of inactivating temperatures, heat penetration times, the reduced microbicidal and sporicidal effects of dry heat, the water requirements for microbial growth, the water activity (a_w) in the environment, the ability of many organisms to grow under anaerobic conditions, the effect of pH on microbial growth and on microbial heat inactivation times, factors influencing the germination and outgrowth of spores and the usefulness or potential toxicity of preservative chemicals. Thus, the public's taste for milder curing processes and a perhaps undue concern about the potentially carcinogenic effects of nitrates or nitrites in the diet, might drive our producers to consider narrower margins of safety in preservative processes that relate to the inhibition of germination and outgrowth of *C. botulinum* spores.

Clinical features

The incubation period ranges from a few hours to more than a week, usually 1–2 days, depending on various circumstances. The careful report of the eight patients affected in the Loch Maree outbreak (Leighton, 1923) records times to death of 1–7 days from ingestion of the toxin, with symptoms and signs first appearing within 1–2 days.

Some patients have initial nausea and vomiting, which may be relatively non-specific features. Dizziness and double vision are early symptoms; the muscles controlling movement of the eyes and the eyelids are affected early in the disease. Leighton records that a patient would lift his eyelids with his hands to see. As the disease progresses, the eyeballs cannot move; the pupils may be dilated with loss of the light reflex and loss of accommodation; there may be blurring of vision.

There is a progressive descending symmetrical flaccid paralysis of the motor system with no sensory loss and with no paraesthesiae. This is important in the differential diagnosis. The lip muscles and laryngeal muscles are affected, with consequent difficulties in speech and later aphonia. There is general muscle weakness but no fever and no headache. There is dryness of the tongue and inability to protrude the tongue. The patient complains of thirst and a dry, sore throat. Difficulties in swallowing and breathing may be associated with bouts of restlessness and despair. Consciousness is not impaired until late in the disease, though the patient appears weak and sleepy between periods of restlessness as the disease progresses. Abdominal pain and tenderness or fullness sometimes occur, especially with type E botulism. Death is usually due to respiratory failure or cardiac failure.

Action to be taken when botulism occurs

When a suspected case of botulism occurs, it is most important:
1. To deal promptly and effectively with the patient;
2. To take immediate steps to detect other possible cases or subjects at risk;
3. To determine the suspected food and ensure that it and any likely ingredients or related batches are impounded pending laboratory tests;
4. To consider whether other foods may have been contaminated during or after preparation in the premises.

Prompt and reliable communications must be established with clinical colleagues, the local Medical Officer for Environmental Health and the local Environmental Health Department (or the equivalent of these agencies in other countries). Local and national microbiological expertise must be assured. Supplies of specific antitoxin must be made available without delay (Communicable Diseases Report, 1979). The CDC Handbook on Botulism (1979) gives much useful information.

Laboratory confirmation

The presence of botulinum toxin in the patient's serum is determined by toxin–antitoxin neutralization tests in mice. For this, 20–30 ml of clotted venous blood should be taken before therapeutic antitoxin is given. Specimens of the patient's faeces and vomit may also yield evidence of botulism and should be submitted, but the tests on the patient's serum are of primary and urgent diagnostic importance. Laboratory tests to detect botulinum toxin or organisms in suspected items of food by mouse tests and by culture for *C. botulinum* are also important.

C. botulinum is classed in containment category 2(V) of the Advisory Committee on Dangerous Pathogens (UK) 1984 as 'an organism that may cause human disease and might be a hazard to laboratory workers but is unlikely to spread in the community'.

Recommendations for work at containment level 2 apply. Similarly, health workers handling or transmitting specimens that may contain *C. botulinum* or its toxins must take special care to avoid accidental ingestion, inhalation or inoculation of traces of botulinum toxin or contamination of themselves or others with the organism or its spores or toxins.

The relevant laboratory should be advised by telephone that specimens are being urgently forwarded. Laboratory workers should distinguish clearly between the relatively routine referral of a clostridium to a reference laboratory for definitive identification as *C. botulinum*, when no associated disease has been reported, and the urgent referral of isolates for special confirmatory identification when suspected botulism has occurred. Laboratory workers who are likely to handle *C. botulinum* as a routine should be actively immunized with toxoid.

Clinical management

The management of a person thought to have ingested botulinum toxin calls for the following immediate steps:
1. Avoidance of further absorption of toxin by vomiting and purgation, and by gastric lavage. Special considerations apply (CDC Handbook, 1979).
2. Neutralization of unfixed toxin by polyvalent antitoxin.

The British National Formulary (1986a) advises that trivalent botulism antitoxin (anti-A 1000 i.u., anti-B 700 i.u. and anti-E 100 i.u./ml) is available for post-exposure prophylaxis and for treatment. There are cautions about possible hypersensitivity reactions and specific assurances should be obtained about a patient's possible allergy and past history of receiving heterologous sera. A prior sensitivity test should be done with 0.2 ml of the antitoxin. For prophylaxis for a patient thought to be at risk, 20 ml of the serum is given intramuscularly as soon as possible. For treatment, 20 ml is given intravenously with subsequent 10 ml doses at 2–4 h and thereafter at intervals of 12–24 h. The total amount of antitoxin given per patient in a series reviewed by Morris (1981) in the USA did not exceed 40 ml but it must be noted that the Canadian (Connaught) equine trivalent antitoxin used in the USA is more potent than the equivalent preparation generally available in the UK (Ball and Farrell, 1979). The Connaught serum contains anti-A 7500 i.u., anti-B 5500 i.u., and anti-E 8500 i.u./ml. The use of more refined heterologous preparations and the development of homologous antitoxic sera are clearly to be encouraged.

The record of antitoxin therapy in botulism is most encouraging in type E cases (L DS Smith, 1977; Morris, 1981) and less impressive in cases caused by type A and type B strains.

The success of type E antitoxin and major developments in intensive care of the paralysed patient, particularly in assisted respiration and the control of respiratory complications, have been responsible for significant improvements in the clinical management of botulism and for reductions in mortality rates. The contribution of antitoxin A and B to therapy is still debated, but there are good prima facie reasons for prompt and adequate antitoxic therapy in all cases. For special cases, polyvalent antitoxin with activity against A, B, C, D, E and F can be obtained from Denmark.

While antitoxin should be given as promptly as possible, some evidence indicates that it is still worthwhile to give it to a patient if the diagnosis has been delayed, especially in the case of type E botulism.

Some workers believe that patients with botulism should be given antibiotics to control any continuing challenge from toxin that might be being generated *in vivo* as a result of the outgrowth of ingested spores in the patient's gut. The evidence that this occurs is debated. The use of an aminoglycoside to this end is illogical, as anaerobes are resistant to aminoglycosides; moreover, these drugs have actions at neuromuscular junctions that should be avoided in botulism. If the *in-vivo* production of toxin is considered probable it would not be unreasonable to give the patient metronidazole to control the growth of anaerobes in the gut—but the achievement of high concentrations in the alimentary tract of a paralysed but conscious patient poses further problems of clinical management.

Prevention

The prevention of botulism rests upon the avoidance of foods that have been dangerously processed, the avoidance of amateur canning or bottling or home curing of a range of potentially dangerous items and the insistence on care and attention to rigid standards in professional canning, curing and food processing.

Active immunization with specific toxoid preparations is practised for the protection of farmed animals such as mink or cattle. A course of active immunization is recommended for the protection of laboratory workers in specialist laboratories where *C. botulinum* and its toxins may be processed.

Special precautions apply to the prevention of infant botulism and wound botulism (see below).

The importance of awareness, early diagnosis and prompt and positive management must apply to all forms of botulism. It is so much easier to diagnose the second case.

Infant botulism

Although it has been known for many years that botulism can kill patients of any age, infant botulism was not defined and recognized as a special disease until the 1970s (Arnon *et al.*, 1977). Adult botulism generally occurs as a food-borne intoxication with a common source affecting several victims, but infant botulism is an infectious disease affecting one patient at a time.

Type A or type B organisms are usually incriminated. The affected infant has a flaccid paralysis of the muscles of the head, face, neck and throat; this extends to the body and limbs with the development of the floppy child syndrome. However, the severity of the disorder ranges from a mild failure to thrive, extending over weeks, to a sudden catastrophe presenting as an unexpected cot death.

The disease results from absorption of toxin from *C. botulinum* growing in the gut of the affected infant, who is usually aged less than 6 months when the colonization resistance of the intestine for *C. botulinum* is assumed to be low. There is evidence that *C. botulinum* spores contaminate honey used by some mothers to coax babies to suck or to supplement their feeds. Arnon (1986) suggested that the minimum infecting dose of *C. botulinum* spores in these circumstances might be as low as 10–100 spores.

Wound botulism

A few cases are on record in which the symptoms and signs of botulism have developed as a result of the growth of *C. botulinum* in a wound (Center for Disease Control Handbook, 1979; Merson and Dowell, 1973). In the CDC series, the 18 patients were predominantly male and the wounds were mainly compound fractures or severe lacerations. Incubation periods ranged from 4 to 14 days. Type A botulism was diagnosed in eight cases and type B in three; the diagnosis was made on clinical grounds in the other seven cases with negative bacteriology and toxin tests. There were four deaths.

References

Adams E B, Laurence D R, Smith J W G. (1969). *Tetanus*. Blackwell Scientific Publications, Oxford.
Adams J M, Kennedy J D, Rudolph A J. (1979). Modern management of tetanus neonatorum. *Pediatrics*, 64: 472–477.
Ahmadsyah I, Salim A. (1985). Treatment of tetanus: an open study to compare the efficacy of procaine penicillin and metronidazole. *British Medical Journal* 291: 648–650.
Aktories K, Barman M, Ohishii I, Tsuyama S, Jakobs K H, Habermann E. (1986). Botulinum C2 toxin ADP-ribosylates actin. *Nature* 322: 390–392.
Arnon S S. (1986). Infant botulism: anticipating the second decade. *Journal of Infectious Diseases* 154: 201–206.
Arnon S S et al. (1977). Infant botulism. Epidemiological, clinical and laboratory aspects. *Journal of the American Medical Association* 237: 1946–1951.
Atrakchi S A, Wilson D H. (1977). Who is likely to get tetanus? *British Medical Journal* 1: 179.
Aureli P, Fenicia L, Pasolini B, Gianfranceschi M, McCroskey L M, Hatheway C L. (1986). Two cases of Type E infant botulism caused by neurotoxigenic *Clostridium butyricum* in Italy. *Journal of Infectious Diseases* 154: 207–211.
Ball A P, Farrell I D. (1979). Problems in human botulism. *Journal of Infection* 1: 121–125.
Black J D, Dolly J O. (1986). Interaction of ^{125}I-labelled botulinum neurotoxins with nerve terminals. I. Ultrastructural autoradiographic localization and quantitation of distinct membrane acceptors for types A and B on motor nerves. *Journal of Cell Biology* 103: 521–534.
Bom I J, Smelt J P P N, Kersters K, Verrits C T. (1986). Identification and grouping of *Clostridium botulinum* strains by numerical analysis of their electrophoretic protein patterns. *Journal of Applied Bacteriology* 60: 483–490.
British National Formulary (1986a). No. 11 Botulism 14.4.19. Botulism anti-toxin p. 395. British Medical Association and The Pharmaceutical Society of Great Britain, London.
(1986b). No. 12 Tetanus 14.4.14, p. 395. 14.5.2 Specific immunoglobulins, pp. 398–399. British Medical Association and The Pharmaceutical Society of Great Britain, London.
Bytchenko B. (1967). Tetanus as a world problem. In *Principles of tetanus*, pp. 21–41. Edited by Eckman L, et al. Shuber, Bern.
Center for Disease Control (CDC) Handbook (1979). See Handbook.
Communicable Diseases Report 79/22 of the Communicable Diseases Surveillance Centre. (1979). PHLS of England, Wales and Ireland.
Ciccarelli A S, Gimenez D F. (1981). Clinical and epidemiological aspects of botulism in Argentina. In *Biomedical aspects of botulism*, pp. 291–301. Edited by Lewis G E. Academic Press, New York.
Collee J G, Brown R, Poxton I R, Fraser A G. (1989). Clostridia of wound infection. In *Practical medical microbiology*, 13th edn, Chapter 37. Edited by Collee J G et al. Churchill Livingstone, Edinburgh.
Collee J G, Macleod D A D, Little K. (1987). Wound management including tetanus prophylaxis schedule. *Communicable Diseases in Scotland, Weekly Report*. CDS Unit, Ruchill Hospital, Glasgow.
Crawford R J, Dow B C, Mitchell R. (1980). Active and passive immunity to tetanus. *Communicable Diseases Scotland* 80/26, ix–x.
Dolman C E, Murakami L. (1961). The translucent colonies of toxigenic strains (Tox colonies) may revert to non-toxigenic opaque sporing (OS) colonial mutants. *Journal of Infectious Diseases* 109: 107–128.
Eckman L. (ed). (1967). Principles on tetanus. Proceedings of the International Conference on Tetanus, Bern, July 15–19, 1966. Shuber, Bern.
Edmondson R S, Flowers M W. (1979). Intensive care in tetanus: management, complications, and mortality in 100 cases. *British Medical Journal* 1: 1401–1404.
Eisel U et al. (1986). Tetanus toxin: primary structure, expression in *E. coli*, and homology with botulinum toxins. *The EMBO Journal* 5: 2495–2502.
Fairweather N F, Lyness V A. (1986). The complete nucleotide sequence of tetanus toxin. *Nucleic Acid Research* 14: 7809–7812.
Finegold S M, George W L, Mulligan M E. (1985). Tetanus. In *Disease-a-month* 31: Anaerobic Infections Part II, pp. 82–88. Yearbook Medical Publishers Inc, Chicago.
Flowers M W, Edmondson R S. (1980). Long-term recovery from tetanus: a study of 50 survivors. *British Medical Journal* 1: 303–305.
Galbraith N S, Forbes P, Tillett H. (1981). National surveillance of tetanus in England and Wales 1930–1979. *Journal of Infection* 3: 181–191.
Genigeorgis C, Riemann H. (1979). Food processing and hygiene. In *Foodborne infections and toxications*, 2nd edn., pp. 613–713. Edited by Riemann H, Bryan F L. Academic Press, New York.
Habermann E, Dryer F. (1986). Clostridial neurotoxins: handling and action at the molecular and cellular level. *Current Topics in Microbiology and Immunology* 129: 93–179.
Hall J D, McCroskey L M, Pincomb B J, Hatheway C L. (1985). Isolation of an organism resembling *Clostri-*

dium barati which produces type F botulinal toxin from an infant with botulism. *Journal of Clinical Microbiology* **21**: 654–655.

Handbook for Epidemiologists, Clinicians and Laboratory Workers (1979). *Botulism in the United States 1899 to 1977*. US Department of Health, Education and Welfare, Public Health Service, Center for Disease Control.

Ho J L, Klempner M S. (1985). Tetanus toxin inhibits secretion of lysosomal contents from human macrophages. *Journal of Infectious Diseases* **152**: 922–929.

Hobbs G. (1976). *Clostridium botulinum* and its importance in fishery products. *Advanced Food Research* **22**: 135.

Hodgkiss W, Ordal Z J. (1966). Morphology of the spore of some strains of *Clostridium botulinum* type E. *Journal of Bacteriology* **91**: 2031–2036.

Hoffman R E, Pincomb B J, Skeels M R, Burkhart M J. (1982). Type F botulism. *American Journal of Diseases of Childhood* **136**: 270–271.

Iida H, Oguma K, Inoue K. (1981). Toxin production and phage in *Clostridium botulinum* types C and D. In *Biomedical aspects of botulism*, p. 119. Edited by Lewis, G E. Academic Press, New York.

Kerrin J C. (1929). Distribution of *B. tetani* in the intestines of animals. *British Journal of Experimental Pathology* **10**: 370.

Knight D E, Tonge D A, Baker P F. (1985). Inhibition of exocytosis in bovine adrenal medullary cells by botulinum toxin type D. *Nature* **317**: 719–721.

Lancet. (1980). The diagnosis of tetanus. *Lancet* 1066.

Leighton G R. (1923). Report of the circumstances attending the deaths of eight persons from botulism at Loch Maree, Ross-shire. HMSO, Edinburgh.

Lewis G E, (ed). (1981). *Biomedical aspects of botulism*. Academic Press, New York.

Lowbury E J L, Lilly H A. (1958). Contamination of operating-theatre air with *Cl. tetani*. *British Medical Journal* **2**: 1334–1336.

Mellanby J. (1984). Comparative activities of tetanus and botulinum toxins. *Neuroscience* **11**: 29–34.

Mellanby J, Green J. (1981). How does tetanus toxin act? *Neuroscience* **6**: 281–300.

Merson M H, Dowell V R. (1973). Epidemiologic, clinical and laboratory aspects of wound botulism. *New England Journal of Medicine* **289**: 1005–1010.

Montecucco C. (1986). How do tetanus and botulinum toxins bind to neuronal membranes? *Trends in Biochemical Sciences* **11**: 314–317.

Morris J G. (1981). Current trends in therapy of botulism in the United States. In *Biomedical aspects of botulism*, pp. 317–326. Edited by Lewis G E. Academic Press, New York.

Ohashi Y, Narumiya S. (1987). ADP-ribosylation of a Mr 21 000 membrane protein by type D botulinum toxin. *Journal of Biological Chemistry* **262**: 1430–1433.

Ohishi I, Tsuyama S. (1986). ADP-ribosylation of non-muscle actin with component 1 of C2 toxin. *Biochemical and Biophysical Research Communication* **136**: 802–806.

Passen E L, Andersen B R. (1986). Clinical tetanus despite a 'protective' level of toxin-neutralizing antibody. *Journal of the American Medical Association* **255**: 1171–1173.

Penner R, Neher E, Dreyer F. (1986). Intracellularly injected tetanus toxin inhibits exocytosis in bovine adrenal chromaffin cells. *Nature* **324**: 76–78.

Phillips L A. (1967). A classification of tetanus. *Lancet* **1**: 1216–1217.

Pierce E J, Davison M D, Parton R G, Habig W H, Crichley D R. (1986). Characterization of tetanus toxin binding to rat brain membranes. *Journal of Biochemistry* **236**: 845–852.

Poxton I R P. (1984). Demonstration of the common antigens of *Clostridium botulinum*, *C. sporogenes* and *C. novyi* by an enzyme linked immunosorbent assay and electroblot transfer. *Journal of General Microbiology* **130**: 975–981.

Poxton I R P, Byrne M D. (1984). Demonstration of shared antigens in the genus *Clostridium* by an enzyme-linked immunosorbent assay. *Journal of Medical Microbiology* **17**: 171–176.

Ranabir Bose, *et al*. (1985). Diagnosis of tetanus by a simple test. *British Medical Journal* **290**: 30.

Rennie J. (1979). Management and severity of tetanus (Correspondence). *British Medical Journal* **1**: 1631.

Report (1982). Tetanus surveillance and prophylaxis. *British Medical Journal* **284**: 1715–1716.

Report (1986). Report from the PHLS Communicable Disease Surveillance Centre. *British Medical Journal* **293**: 680–682.

Rubbo S D. (1966). New approaches to tetanus prophylaxis. *Lancet* **2**: 449–453.

Sakaguchi G. (1979). In *Food-borne infections and intoxications*, 2nd edn., pp. 405–410. Edited by Riemann H, Bryan F L. Academic Press, New York.

(1986). *Clostridium botulinum* toxins. In *International encyclopedia of pharmacology and therapeutics*, Section 119, pp. 519–548. Edited by Dorner F, Drews J. Pergamon Press, Oxford.

Sanada I, Nishida S. (1965). Isolation of *Clostridium tetani* from soil. *Journal of Bacteriology* **89**: 626–629.

Sanders R K, Joseph R, Martyn B, Peacock M L. (1977). Intrathecal antitetanus serum (horse) in the treatment of tetanus. *Lancet* **1**: 974–977.

Sellin L C. (1981). Postsynaptic effects of botulinum toxin at the neuromuscular junction. In *Biomedical aspects of botulism*, pp. 81–92. Edited by Lewis G E. Academic Press, New York.

Simpson L L. (1981). Pharmacological studies on the cellular and sub-cellular effects of botulinum toxin. In *Biomedical aspects of botulism*, pp. 35–46. Edited by Lewis G E. Academic Press, New York.

(1986). Molecular pharmacology of botulinum toxin

and tetanus toxin. *Annual Review of Pharmacological Toxicology* **26**: 427–453.

Smith G R. (1985). Botulism. In *Topley and Wilson's Principles of bacteriology, virology and immunity*, 7th edn. Chapter 72, pp. 496–514. Edward Arnold, London.

Smith G R, Milligan R A, Moryson C J. (1978). *Clostridium botulinum* in aquatic environments in Great Britain and Ireland. *Journal of Hygiene* **80**: 431.

Smith J W G. (1985). Tetanus. In *Topley and Wilson's Principles of bacteriology, virology and immunity*, 7th edn. Chapter 64, pp. 345–368. Edward Arnold, London.

Smith J W G, Laurence D R, Evans D G. (1975). Prevention of tetanus in the wounded. *British Medical Journal* **3**: 453–455.

Smith J W G, MacIver A G. (1974). Growth and toxin production of tetanus bacilli *in vivo*. *Journal of Medical Microbiology* **7**: 497–504.

Smith L DS. (1975). *Clostridium botulinum*. In *The pathogenic anaerobic bacteria* 2nd edn., pp. 203–229. Thomas, Springfield.

(1977). Botulism—the organism, its toxins, the disease. Thomas Springfield.

Smythe P M, Bowie M D, Voss T J V. (1974). Treatment of tetanus neonatorum with muscle relaxants and intermittent positive-pressure ventilation. *British Medical Journal* **1**: 223–226.

Stanfield J P, Galazka A. (1985). Neonatal tetanus—an under-reported scourge. *World Health Forum* **6**: 127–129.

Stoddart J. (1979). Pseudotetanus. *Anaesthesia* **34**: 877–81.

Tenbroeck C, Bauer J H. (1922). The tetanus bacillus as an intestinal saprophyte in man. *Journal of Experimental Medicine* **36**: 261–271.

Tulloch W J. (1919–20). Report of bacteriological investigation of tetanus carried out on behalf of the War Office Committee for the study of tetanus. *Journal of Hygiene* **18**: 103–202.

van Heyningen S. (1982). Similarities in the action of different toxins. In *Molecular action of toxins and viruses*, pp. 169–190. Edited by Cohen P, van Heyningen S. Elsevier, Amsterdam.

van Heyningen S. (1986). Tetanus toxin. In *International encyclopaedia of pharmacology and therapeutics*, Section 119, pp. 549–569. Edited by Dorner F, Drews J. Pergamon Press, Oxford.

van Heyningen W E. (1968). Tetanus. *Scientific American* 69–77.

Watt B, Brown R. (1975). A defined medium for the growth of *Clostridium tetani* and other anaerobes of clinical interest. *Journal of Medical Microbiology* **8**: 167–172.

Willis A T. (1969). Clostridium tetani. In *Clostridia of wound infection*, pp. 385–461. Butterworths, London.

Woodruff A W, Grant J, El-Bashir E A, Baya E I, Yugusuk A Z, El-Suni A. (1984). Neonatal tetanus: mode of infection, prevalence and prevention in Southern Sudan. *Lancet* **1**: 378–379.

24

Metabolic and carcinogenic effects of anaerobes

M J Hill

Production of carcinogens/mutagens/promoters by anaerobic bacteria
    Products of glycoside hydrolysis
    Products of nitrogen metabolism
    Products of steroid metabolism
Bacterial metabolism and human cancer
    Cancer of the stomach
    Cancer of the biliary tract
    Cancer of the large bowel
    Cancer of the urinary bladder
Conclusions
References

It is generally accepted that the vast majority of cancers have an environmental aetiology. All external surfaces of the body are colonized by bacteria, the microflora being dominated by anaerobic bacteria at most sites. This bacterial flora is uniquely placed to mediate in the interaction between the host and the environment and may either detoxify or activate carcinogens and modulators of carcinogenesis in the environment.

A major component of the environment is the diet, and the interaction between the diet and the microflora of the digestive tract is becoming increasingly well documented. In this chapter I will begin by discussing the production of modulators of carcinogenesis (initiation, promotion or inhibition) by anaerobic bacteria. I will then describe hypotheses of the causation of cancer at various sites.

Production of carcinogens/mutagens/promoters by anaerobic bacteria

A wide range of compounds of potential importance in human carcinogenesis is produced by anaerobic bacteria. Some of these have been tested *in vivo* in animals (or have been studied epidemiologically in man) and have been shown to be carcinogens, cocarcinogens or tumour promoters. Others have been tested only *in vitro* in mammalian cell or microbial mutagenesis assays and so can only be described as mutagens or comutagens. Whereas the vast majority of carcinogens have been shown to be mutagens, the reverse is far from true.

In this section the carcinogens/mutagens/promoters to be discussed will be those released from inactive glycosides, those produced by metabolism of nitrogen compounds, and those produced by metabolism of steroids.

Products of glycoside hydrolysis

Two types of glycosidase are important in carcinogen release, β-glucosidase and β-glucuronidase. Plants produce a wide range of β-glucosides, many of which are of toxicological importance (e.g., the cardioactive glycosides, the cathartic glucosides, the cyanogenic glucosides, etc.). In most cases it is the aglycone produced by the action of bacterial glucosidase that is toxic. The metabolism of plant glucosides has been discussed in detail elsewhere (Hill, 1986a).

Cycasin, methylazoxymethanol-β-glucoside, is present in cycad nuts and is a potent colon carcinogen when given orally to rodents. However, cycasin is inactive when given to rats intravenously or when given by any route to germ-free rodents. Methylazoxymethanol is a potent carcinogen and is released from its β-glucoside by bacterial β-glucosidase. The colonic mucosa has β-glucosidase activity but the mucosal enzyme is highly substrate specific and has very little activity on plant glucosides. In contrast, the bacterial enzyme has a very wide substrate specificity and acts readily on the plant glucosides, including cycasin. For many years it was not known whether cycasin was a unique example of a carcinogenic glucoside; it is now known that it is one of a family of such compounds and, in the salmonella mutagenesis assay, it is now recommended that plant compounds be tested before and after treatment with β-glucosidase.

A wide range of lipophilic carcinogens, including the polycyclic aromatic hydrocarbons (PAHs) are ingested and are then excreted via the bile as their glucuronide conjugates. During passage through the intestine these conjugates may be hydrolysed by bacterial β-glucuronidase. Kinoshita and Gelboin (1978) demonstrated that during such hydrolysis a high energy intermediate is released which binds to DNA and is, therefore, potentially mutagenic.

Both β-glucosidase and β-glucuronidase are produced by most genera of gut bacteria (Table 24.1) but because of their numerical dominance in the colon the anaerobes are responsible for a high proportion of the enzyme activity in the intestine.

Products of nitrogen metabolism

The products of nitrogen metabolism of interest here are some metabolites of amino acids and the N-nitroso compounds produced by the action of nitrite on a suitable nitrogen compound (such as a secondary amine, an alkylurea, an amide, etc.).

Amino acid metabolites

Methionine is metabolized to its S-ethyl analogue ethionine by many species of bacteria (including *Escherichia coli*) when grown in a mineral salts medium supplemented with methionine, sulphate and glucose. Ethionine is a potent carcinogen in rodents but its activity in man has not been investigated. The production of ethionine by anaerobes has not been studied.

Tyrosine is metabolized in the mammalian colon to a range of volatile phenols, principally *p*-cresol and phenol; these are absorbed from the colon, conjugated in the liver as glucuronides and sulphates and excreted in the urine as the urinary volatile phenols (UVP). The amount of UVP is increased by a high protein diet and by small bowel overgrowth and is decreased by actions which decrease the number or activity of the colonic flora (e.g., colectomy, colonic washout, antibiotics). Interestingly, whereas total colectomy or right hemi-

Table 24.1 Production of glycosidases by various genera of gut bacteria

	Enzyme activity (moles substrate degraded/hour/10^8 cells		
	β-galactosidase	β-glucosidase	β-glucuronidase
E. coli	42.4 ± 3.2	5.8 ± 2.5	24.7 ± 2.1
Str. faecalis	53.8 ± 6.0	192.7 ± 19.5	2.9 ± 0.6
Lactobacillus spp.	90.6 ± 10.7	26.0 ± 7.4	1.6 ± 0.2
Bifidobacterium spp.	39.1 ± 4.7	29.3 ± 6.0	1.9 ± 0.8
Bacteroides spp.	50.7 ± 4.9	35.1 ± 4.8	6.0 ± 3.5
Clostridium spp.	13.7 ± 2.7	22.1 ± 5.0	11.3 ± 2.3

Table 24.2 Metabolites of tryptophan that have been shown to be mutagens or cocarcinogens

Test system	Indication	Positive metabolites
2-AAF treated rats	Cocarcinogenicity	Tryptophan, indole
Bladder implantation test	Cocarcinogenicity	3-hydroxykynurenine
		3-hydroxyanthranilic acid
		8-hydroxyquinaldic acid
		xanthenuric acid
		quinaldic acid
		kynurenine
Cultured mammalian cells	Mutagenesis	3-hydroxykynurenine
		3-hydroxyanthranilic acid

colectomy results in a massive decrease in UVP, left hemicolectomy has no effect, suggesting that the site of UVP production is the proximal colon. Many phenols, including p-cresol and phenol, have been shown to be promoters of skin carcinogenesis (Boutwell and Bosch, 1959) and it has been suggested by Bakke and Midtvedt (1970) that UVPs may be important aetiological agents in hepatomas in rats. The incidence of such hepatomas is decreased greatly in germ-free animals and is related to the amount of dietary protein.

Tryptophan is metabolized to a very wide range of metabolites in the mammalian colon. The principal metabolites are indole (produced as a result of the action of tryptophanase), tryptamine (resulting from decarboxylation) and indoleacetic acid (resulting from deamination and side chain shortening) in addition there are metabolites on the kynurenine pathway. These metabolites are absorbed from the colon and either undergo further hepatic metabolism or are excreted in the urine. Indole is hydroxylated and excreted in the urine as indoxyl sulphate (indican); the amount of urinary indican can be decreased by colonic surgery, antibiotic action or colonic washout and increased by increased dietary protein and by small bowel overgrowth. The other urinary tryptophan metabolites behave similarly to indican (and to UVP). Many tryptophan metabolites, principally those on the kynurenine pathway, have been claimed to be carcinogens, mutagens or tumour promoters (Table 24.2) and it has been suggested by Bryan (1971), amongst others, that the

Table 24.3 Evidence that bacteria are able to catalyse the N-nitrosation reaction

Reference*	Observation
Sander (1968)	First demonstrated N-nitrosation of a secondary amine at neutral pH in the presence of *E. coli* but not in their absence
Hawksworth and Hill (1971)	Confirmed the above; extended the range of organisms studied. Needed growing cultures
Klubes et al. (1972)	Repeated the above; showed that part of the activity was destroyed by autoclaving (and so could be enzymic) but part of the activity was resistant (and so was non-enzymic)
Brooks et al. (1972)	N-nitrosation by *Proteus* spp. *in vivo* in urine
Coloe and Heywood (1976)	Wide variation in activity of *E. coli* strains, studied the kinetics of the reaction
Tannenbaum et al. (1978)	Studied N-nitrosation in saliva *in vitro* and showed that live bacteria were needed
Ruddell et al. (1978)	Demonstrated N-nitrosation by bacteria in gastric juice
Kunisaka and Hayashi (1979)	Studied N-nitrosation by resting cells of *E. coli* and concluded that the reaction was enzymic
Leach et al. (1985)	Confirmed that the reaction is enzymic; denitrifiers are very much more active than non-denitrifiers; pH optimum between 7 and 8

*for details of references see Hill, 1986b

urinary tryptophan metabolites might be responsible for those bladder cancers not associated with industrial exposures to bladder carcinogens. In support of this, there is a large body of published work on the bladder carcinogenicity of tryptophan metabolites, principally using the mouse bladder implantation test. In this test a pellet of cholesterol mixed with the test compound is implanted subcutaneously in the bladder wall and the test compound allowed to leach out over a prolonged period of time. It is not clear whether the metabolites are acting as carcinogens, as promoters of cholesterol carcinogenesis or as promoters of the carcinogenesis caused by some other urinary metabolite.

Production of N-nitroso compounds

There is a large body of evidence that bacteria are able to catalyse the N-nitrosation of secondary amines (Table 24.3). The N-nitroso compounds form a family of potent carcinogens that cause cancer in every animal species in which they have been tested. Thus, although there is no proof that they are carcinogenic in man, there is no good reason to suppose that man is uniquely resistant to their action. A range of nitrogen compounds can form carcinogenic N-nitroso compounds: those from alkylureas or amides give local-acting N-nitroso products while those from secondary amines give products that are target organ specific. The carcinogenicity of the N-nitroso compounds has been well reviewed by Magee and Barnes (1967). N-nitrosation has been studied mainly as an acid catalysed reaction but bacteria are able to catalyse the reaction at neutral pH; the characteristics of the acid catalysed and the bacterially catalysed reactions are compared in Table 24.4. Production of N-nitroso compounds by bacteria has been incriminated in the causation of cancer of the stomach, urinary bladder and the colon, and this will be discussed in detail later.

The best substrates for bacterial N-nitrosation are the secondary amines. Those present in the body in the greatest amounts are dimethylamine, piperidine and pyrrolidine, which are produced by bacterial action in the colon. Dimethylamine is produced by N-dealkylation of choline, which itself is the product of the action of bacterial lecithinase on dietary or endogenous lecithin. Pyrrolidine is produced from ornithine by decarboxylation of the amino acid

Table 24.4 Comparison of the kinetic characteristics of the acid catalysed and the bacterially catalysed reaction

	Acid catalysis	Bacterial catalysis
Optimum pH	2–4	6–8
Kinetics:		
secondary amine	First order	First order
nitrite	Second order	First order

followed by cyclization of the resultant diamine. Piperidine is formed by a similar series of reactions on lysine. Each day 20–40 mg dimethylamine, 500–1000 µg piperidine and 100–500 µg pyrrolidine are produced in the human body and excreted in the urine; they are also transferred to most body secretions, as is nitrate, so that N-nitrosamines could, in principle, be formed at any site in the body where there is bacterial colonization. Most of the organisms that can catalyse the N-nitrosation reaction are either aerobic or facultative organisms growing anaerobically (and using nitrate or nitrite as the terminal electron acceptor). However, the anaerobic flora of the gut is undoubtedly of crucial importance in the production of the secondary amines described above.

Products of steroid metabolism

The products of steroid metabolism that have been implicated in human cancer are those of cholesterol and bile acids.

Cholesterol metabolites

Cholesterol is metabolized by the normal colonic bacterial flora to yield coprostanol (by reduction of the 5–6 double bond) and coprostanone (by oxidoreduction of the 3-hydroxyl group). In most people in most populations, 60–80 per cent of cholesterol is metabolized during transit through the colon (Table 24.5) principally by the anaerobic bacterial flora. This is of interest because of the claims, summarized by Heiger (1949), that cholesterol is carcinogenic. There has been a great deal of controversy about this work because of the inherent unlikelihood of a compound so widely distributed in the human body being carcinogenic. Some explanations attempted to implicate a con-

Table 24.5 The extent of cholesterol metabolism during colonic transit in various populations*

Population	Number of analyses	Percentage of cholesterol metabolized during colonic transit
England	71	71
Scotland	18	71
USA—White	42	68
—Black	12	72
South Africa—White	13	60
— Black	23	44
Uganda	11	55
India	18	62

*From Drasar and Hill (1974)

taminant in the cholesterol, while others offered included solid state carcinogenesis. In his excellent review of the subject, Bischoff (1969) favoured the latter explanation, with the cholesterol precipitating from the supersaturated oily solution at the injection site and acting as a non-specific solid. However, if cholesterol is a carcinogen then bacterial metabolism of the steroid should protect the host. Black African populations (see Table 24.5) have a relatively high proportion of undegraded cholesterol but a very low risk of colon cancer.

There are a number of situations where lack of bacterial cholesterol metabolism has been incriminated in human colorectal cancer. In adenomatosis coli, juvenile polyposis coli and cancer family syndrome (three diseases determined by autosomal dominant genes and carrying an excess risk of colorectal cancer) patients carrying the gene are characterized by a low percentage cholesterol degradation in faeces (Table 24.6). However, in adenomatosis coli (the best documented of these diseases) there is a small proportion of patients who clearly carry the gene but who have normal cholesterol metabolism; this suggests that the lack of cholesterol metabolism cannot be causally related to the cancer risk (although in these patients it is a good marker of risk).

In a study of normal individuals, Wilkins and Hackman (1974) claimed that 25–30 per cent of people were 'low metabolisers' of cholesterol, while the rest were normal metabolisers. They went on to suggest that the low metabolisers were at greater risk of subsequent colorectal cancer than the normal metabolisers. There is no evidence to support this hypothesis at all.

Bile acid metabolites

The human liver synthesises two bile acids—cholic acid and chenodeoxycholic acid. These are conjugated with the amino acids glycine or taurine before secretion in the bile. The conjugated bile acids undergo enterohepatic circulation, are recovered from the terminal ileum by an active transport system that is 95–98 per cent efficient, are returned to the liver via the portal blood system and are resecreted in the bile. The bile acid pool is circulated 5–10 times/day (depending on the amount of fat in the diet) so that in Western individuals 300–800 mg of bile acid escapes to the colon each day. In the colon the bile acids are vigorously degraded by the

Table 24.6 The metabolism of cholesterol during colonic transit in various patient groups

Patient group	Percentage of cholesterol as metabolites in faeces
Normal persons	60–90
Adenomatosis coli	0–15
Juvenile polyposis	0–15
Peutz–Jeghers syndrome	60–90
Cancer family syndrome	0–30

gut bacteria, although up to 30 per cent are recovered (by passive diffusion) to be returned to the bile pool; the ease of recovery appears to be decreased as the bile acids are more extensively degraded. Consequently, an individual on a Western diet would show a faecal loss of bile acids of 200–500 mg/day, depending on the amount of dietary fat. The enterohepatic circulation of bile acids has been reviewed by Hill (1983) and in more detail by Hoffmann (1977).

The major metabolic reactions carried out by gut bacteria on the bile acids are:
deconjugation (via the cholanoyl amide hydrolase);
dehydroxylation (via the bile acid 7α-dehydroxylase);
hydroxyloxidation to an oxo group or inversion to a β configuration.
In addition, there are minor reactions such as:
desulphation reactions;
nuclear dehydrogenations.

The deconjugation of bile salts is the most widely distributed of the bile degrading enzyme reactions and, because a high proportion of strains of anaerobic genera in the gut (Table 24.7) produces a highly active enzyme, almost all the bile salts are deconjugated rapidly in the proximal colon. The next most common metabolic reaction involving bile salts in the colon is the 7α-dehydroxylation, which yields deoxycholic acid from cholic acid and lithocholic acid from chenodeoxycholic acid. This reaction is carried out by a high proportion of strictly anaerobic gut species (Table 24.7) including the *Bacteroides fragilis* group of organisms, *Bifidobacterium* spp., *Eubacterium* spp., *Clostridium* spp. and *Veillonella* spp. Of the aerotolerant organisms, only the faecal streptococci carry out the reaction. In addition to being produced mainly by strictly anaerobic organisms, the 7-dehydroxylase enzyme is only produced when the pH is above 6, and it requires strictly anaerobic conditions for its activity. In the normal colon, 7-dehydroxylation of the bile acids is virtually complete and occurs in the proximal colon. Although the reaction is greatly impaired in people with diarrhoea, this is more likely to be due to an unfavourable oxygen tension in the caecum than to an insufficient transit time. From Table 24.7, and studies *in vitro*, the activity of the hydroxy steroid oxidoreductases might appear to be greater than that of the dehydroxylase. The enzyme is produced by a greater proportion of gut organisms (Table 24.7) and its activity under aerobic conditions is very much higher than that of the dehydroxylase. However the gut is strictly anaerobic and conditions are unfavourable to the dehydrogenases but extremely favourable to the dehydroxylase. Since the latter reaction is also irreversible, it is favoured very strongly in the gut.

Bacterial metabolism and human cancer

Bacterial metabolites have been implicated in carcinogenesis at a number of sites in the human body including the stomach, biliary tract, large bowel and bladder. For each of these sites, the role of bacterial metabolites will be put into the context of the epidemiology and, where understood, the histopathology of the disease.

Table 24.7 The production of bile-degrading enzymes by gut bacteria

Organisms	Number tested	Deconjugation (%)	Dehydroxylation (%)	Dehydrogenation		
				3α	7α	12α
E. coli	267	0	0	35	65	10
Str. faecalis	252	91	11	22	72	10
Str. salivarius	82	0	0	0	0	0
Lactobacillus spp.	48	0	0	0	0	0
Bacillus spp.	81	32	0	5	16	0
Clostridium spp.	130	95	34	45	88	10
Bacteroides spp.	54	79	44	24	60	10
Bifidobacterium spp.	192	82	40	19	52	10
Veillonella spp.	92	48		25	60	10
Anaerobic sarcinae	12	67		0	58	0

Table 24.8 The incidence of cancer of the stomach, biliary tract, colon and bladder in various populations (from Waterhouse *et al.*, 1982)

Population	Cancer incidence/100 000/annum, age-adjusted							
	Stomach		Biliary tract		Colon/rectum		Bladder	
	M	F	M	F	M	F	M	F
Africa								
Nigeria	2.6	1.7	0	0	0.4	0.4	0.7	0.6
Senegal	3.7	2.0	0	0.2	0.6	1.5	3.0	1.7
Zimbabwe	2.7	0.6	0.6	0.3	1.9	0.8	5.0	2.3
Asia								
Japan	88.0	42.0	6.4	9.4	8.3	9.2	5.3	1.6
India	9.7	5.9	0.5	0.7	3.5	4.5	3.5	0.9
China	55.5	21.0	1.1	1.5	6.7	9.0	7.5	1.4
Singapore–Chinese	43.7	17.6	1.5	1.8	14.9	14.2	7.0	2.0
South and Central America								
Brazil	45.7	19.0	2.1	4.1	11.4	7.9	15.4	3.4
Colombia	46.3	27.3	3.3	7.2	4.5	3.4	9.8	2.6
Puerto Rico	21.1	9.2	1.7	2.5	7.6	6.2	8.4	2.8
North America								
Canada	13.5	6.3	1.8	2.1	21.5	14.9	19.2	5.6
USA—White	10.5	4.5	1.7	1.8	25.6	14.6	19.6	6.1
—Black	18.5	7.8	1.8	1.2	28.4	6.7	12.0	4.5
Europe								
UK—Birmingham	22.1	10.1	1.3	1.7	16.3	16.7	17.1	4.5
—E Scotland	22.2	10.9	1.3	1.7	23.1	14.6	15.8	5.2
Denmark	17.1	8.9	2.1	3.3	19.0	17.0	22.0	5.6
W Germany	26.7	12.6	2.7	4.7	16.5	12.9	13.7	3.1
E Germany	30.3	14.3	3.6	7.7	10.7	13.3	10.1	1.8
Hungary	37.4	17.1	1.1	5.3	9.7	16.1	9.7	1.3
Australasia								
Australia—NSW	13.7	6.7	1.6	2.0	21.5	12.8	14.7	4.1
NZ—White	16.3	7.0	1.9	2.0	25.5	16.1	12.8	3.4
—Maori	41.7	20.3	1.2	3.3	9.0	9.8	5.0	3.8

Cancer of the stomach

Epidemiology

Gastric cancer is very common in eastern Asia, the Andean countries of South and Central America and in Eastern Europe; it is relatively uncommon in North America, Australasia and Western Europe (Table 24.8). Within Europe the disease is more common in the south and the east than in the west; within the British Isles the position is reversed, the incidence being higher in the north and west than in the south and east.

Within a population gastric cancer is more common in men than in women, and more common in urban than in rural areas. The incidence is inversely related to socioeconomic standards. There are a number of recognized predisposing states including pernicious anaemia, gastric surgery, hypogammaglobulinaemia, gastric atrophy and intestinal metaplasia of the gastric mucosa (Table 24.9). Although first order relatives of gastric cancer cases have an excess risk of the disease (in comparison to the general population) until the last 10 years the role of genetic factors has been thought to be weak. However, Lauren (1965) classified gastric cancer into the so-called diffuse, intestinal and intermediate types; Lehtola (1978) demonstrated that whereas in gastric

Table 24.9 Disease states predisposing to gastric cancer

Disease	Risk of gastric cancer relative to the normal population
Gastric atrophy	Increased
Chronic atrophic gastritis	Increased
Intestinal metaplasia	Increased
Pernicious anaemia	4–6-fold
Hypogammaglobulinaemia	50-fold
Gastric surgery:	
Polya partial gastrectomy	4–6-fold
vagotomy	6–10-fold

cancer cases *in toto* the excess risk of the disease in close relatives was small, in diffuse cases the excess risk was more than 7-fold, compared with almost no excess in relatives of intestinal-type cancer cases. In contrast, in numerous studies, summarized by Correa (1982), environmental factors play an important part in determining the risk of gastric cancer of the intestinal type but not of the diffuse type.

Histopathology

In 1975 Correa *et al.* postulated a histopathological sequence in gastric carcinogenesis. The first stage in this is chronic atrophic gastritis (CAG) which progresses to intestinal metaplasia (IM), followed by increasingly severe dysplasia and finally carcinoma. Although dysplasia may arise in CAG, or even in normal mucosa, it arises much more commonly in areas of IM. Similarly, although IM can arise in previously normal tissue, it is much more likely to arise in CAG. Dysplasia can be graded as mild, moderate or severe and whereas severe dysplasia appears to be irreversible and to progress to carcinoma, mild or moderate dysplasia appear to be reversible conditions. This histopathogenic sequence is a model only for the intestinal type of cancer and is not a model for the diffuse type.

Aetiology

Correa *et al.* (1975) postulated that, as a result of CAG, the gastric acid secretion was insufficient to be bactericidal. Consequently, the presence of a resident bacterial flora that reduces dietary nitrate to nitrite and then catalyses the N-nitrosation of endogenous or dietary nitrogen compounds is established. On the Correa hypothesis, therefore, it is necessary

to determine the causation of CAG;
to show that CAG is a precancerous lesion and is associated with an increased bacterial flora, gastric juice nitrite and N-nitroso compounds;
to show that the gastric juice analyses are correlated with progression of the lesion through the histopathological sequence.

The causation of CAG

There is very little information on the causation of CAG, but there is a growing body of data showing that gastritis, accompanied by loss of gastric acid secretion is associated with *Helicobacter (Campylobacter) pylori* infection (Marshall *et al.*, 1985). It has still to be proven that the association is causal but a mechanism has been hypothesized for such a relationship. *H. pylori* was first described as a campylobacter-like organism (CLO) by Marshall *et al.* (1985) and is isolated only from the gastric mucosa or from areas of gastric metaplasia in the duodenum. The organism is usually found beneath the mucin layer and adjacent to the mucosa. Although it is acid-sensitive *in vitro* the organism can be cultured from the mucosa of patients with an acid gastric lumen, suggesting that the mucin layer confers essential protection to *H. pylori* from the hostile acidic luminal environment.

H. pylori produces an extremely active urease and it has been hypothesized that this is important in the causation of gastritis (Hazell and Lee, 1986). Gastric juice and saliva both contain urea; it has been suggested that *H. pylori* urease hydrolyses the urea, releasing high local concentrations of ammonia which cause back diffusion of the hydrogen ions secreted by the parietal cells. The failure of the stimulated parietal cells to secrete hydrogen ions is suggested to be the cause of the tissue damage. Although there is no direct evidence in support of this hypothesis there is epidemiological evidence in its favour. There have been anecdotal reports of longitudinal investigations of outbreaks of infectious achlorhydria which showed that loss of acid secretion correlated with an increase in serum antibodies to *H. pylori*. This increase in serum antibodies was thought to discriminate between mucosal surface colonization and invasion. In numerous studies it has been found that serum antibodies were present in about 10 per cent of people with no gastric symptoms, 80–90 per cent of people with superficial gastritis and almost 100 per cent of people with chronic atrophic gastritis.

The major problem with the hypothesis relating *H. pylori* infection to CAG is the original source of the infection. Since the organism is acid-sensitive it must enter the stomach at a time when the gastric lumen is at a non-bactericidal pH (perhaps during a meal?) and must colonize the mucosa at a time when it is non-secreting. The proposed mechanism can only explain the inhibition of recovery of the gastric acid secretion, not the genesis of that condition. There have been numerous reports (e.g., Ramsey *et al.*, 1979; Gledhill *et al.*, 1982) associating virus infection with transient achlorhydria. A possible aetiology of CAG could begin with such a viral infection. If, during the course of the infection the host was not infected by *H. pylori*, the mucosa could recover and normal gastric acid secretion could resume. However, if, during the transient achlorhydria associated with the virus infection, the host had a secondary infection with *H. pylori*, the infected area would not be allowed to recover and progression to CAG would ensue. On this hypothesis, gastritis would occur initially in focal patches which could then grow in size and severity. This is supported by clinical observation of atrophic gastritis.

Analysis of luminal contents in CAG

There have been few studies of the relation between gastric juice analyses and the presence or absence of CAG, but there have been many studies of the relation between pH and gastric juice analyses and between pH and gastric mucosal disease.

In the normal acid stomach the gastric luminal contents are at a pH below 4 for most of the day and below 2.5 for much of the day (Thomas *et al.*, 1987). The pH increases during meals as a result of the buffering effect of food but the acid secretion in response to food rapidly re-establishes acidic conditions. The normal stomach is also free of nitrate-reducing bacteria (NRB) for most of the day, these organisms are present only during meals. These were, presumably, swallowed with saliva and re-establishment of an acidic environment effectively removes the NRB from the flora.

Table 24.10 lists some recent studies of the relationship between gastric pH and gastric juice analyses. In all cases the presence of high resting gastric pH was accompanied by a profuse bacterial flora, rich in NRB and with a high nitrite concentration. Where it has been measured reliably, a high pH is also accompanied by an increased concentration of N-nitroso compounds.

Acute gastritis is usually accompanied by an acidic environment in the gastric lumen, the initial lesions being foci of non-secreting tissue surrounded by normally secreting mucosa. As the lesion develops and the gastritis becomes chronic the atrophic area increases until, inevitably, a stage is reached where the secreted acid is insufficient to reduce the pH to bactericidal levels. Before this stage is reached the gastric flora is salivary in origin; when the conditions cease to be bactericidal for prolonged periods organisms are able to take up residence in the stomach and a flora suited to the gastric milieu is established. This flora tends to be rich in NRB.

Table 24.10 Studies of the relationship between increased gastric juice pH and gastric juice analysis

Patient group	Effect of increased pH on			Reference
	Bacterial flora	$[NO_2]$	$[NNO]$	
Assorted	Increased	Increased	?	Ruddell *et al.* (1976)
Partial gastrectomy		Increased		Jones *et al.* (1978)
Partial gastrectomy	Increased	Increased	Increased	Schlag *et al.* (1980)
Assorted	Increased	Increased	Increased	Reed *et al.* (1981)
Penicious anemia	Increased	Increased	Increased	Stockbrugger *et al.* (1984)
Various	Increased	Increased	No relation	Hall *et al.* (1984)
Healthy on cimetidine	Increased	Increased	Decreased	Milton-Thompson *et al.* (1982)
Various		Increased		Pignatelli *et al.* (1987)

Table 24.11 Some studies relating gastric juice analyses to the progression from CAG to gastric cancer

Patient group	Observation	Reference
Gastric surgery	Dysplasia correlated with NO_2 concentration	Jones et al. (1978)
Various groups	Dysplasia correlated with NO_2 and with the bacterial flora	Watt et al. (1984)
Gastric surgery	Dysplasia correlated with NO_2 concentration	Mortensen et al. (1984)
Various groups	Dysplasia correlated with NO_2 and with the bacterial flora	Hall et al. (1984)

To demonstrate why CAG is a precancerous lesion it is merely necessary to note the excess risk of gastric cancer associated with gastric hypoacidity of any aetiology (see Table 24.9, page 402). There are no disease states associated with chronic hypochlorhydria that do not predispose to gastric cancer. Clearly then, the precancerous nature of CAG is associated with the accompanying achlorhydria (and its sequelae) rather than to any other aspect of the CAG state.

Progression from CAG to cancer

Table 24.11 lists some recent reports where the gastric juice analyses have been related to the progression either to IM or to increasingly severe dysplasia. No relation could be demonstrated with either the numbers or the types of bacteria isolated; this was not surprising, since quantitative bacteriology is still far from an exact science. However, in a number of studies the progression of the lesion from CAG has been correlated with the nitrite concentration, as well as with the bile acid concentration.

Although N-nitrosamides, N-nitrosoureas, etc., are locally acting carcinogens, the N-nitrosamines are target organ specific agents. Gastric juice is rich in nitrosatable amines and it is likely that if locally acting N-nitroso compounds were formed in association with loss of gastric acidity, then the distally acting analogues would be likely to be produced as well. Thus, if the locally acting N-nitroso compounds are formed and are responsible for the excess gastric cancer risk in these patients we would expect to see an excess of cancers at distant sites as a result of the formation of the target organ specific analogues. This was looked for by Caygill et al. (1987) in a study of 5018 patients treated surgically for peptic ulcer. In addition to the excess risk of gastric cancer after a 20-year latency (Table 24.12) as reported by others, they also noted an excess risk of cancer of the biliary tract, colon and a number of other sites. The excess risk of cancer of the stomach, colon and biliary tract was observed by Caygill et al. (1984) in a pilot study of gastric surgery patients and in a study of 1000 patients with pernicious anaemia.

A problem with validating the Correa hypothesis, relating bacterially synthesized N-nitroso compounds to the causation of the progression from CAG to gastric cancer, has been the dearth of data on concentrations of N-nitroso compounds. The analysis of N-nitroso compounds has been a subject of much controversy and the analytical difficulties should not be underestimated. The results of analyses of N-nitroso compounds in gastric juice by various groups of workers are summarized in Table 24.10, page 403. Although Schlag et al. (1980) and Reed et al. (1981) showed a strong correlation between gastric pH, gastric cancer risk and N-nitroso compound concentration (in accordance with the Correa hypothesis) in studies carried out by Bavin et al. the reverse was observed (Hall et al., 1984; Kieghley et al., 1984). The analytical methods were critically examined by Pignatelli et al. (1987) who obtained results that were consistent with those of Reed et al.; their report contains criticisms of the method used by Bavin et al. that offer an explanation of the discrepancies in Table 24.10.

Table 24.12 The risk of cancer at various sites following gastric surgery for peptic ulcer*

Site	Excess risk of cancer after surgery	
	0–20 years	>20 years
Stomach	1.2	4.5
Large bowel	0.7	1.6
Biliary tract	1.9	8.6
Breast	0.5	4.0
Pancreas	0.7	3.8
Bladder	0.6	2.4

*From Caygill et al. (1987)

In conclusion:
1. All clinical states characterized by gastric hypochlorhydria carry an increased risk of gastric cancer. Hypochlorhydria is associated with infection by *H. pylori* and a mechanism has been proposed for the inhibition of gastric acid secretion (and the consequent mucosal damage) by this organism;
2. Hypochlorhydria is associated with a profuse resident bacterial flora in the stomach; this flora is rich in organisms which produce nitrate reductase and which can catalyse the N-nitrosation of secondary amines;
3. The extent of progression along the histopathological sequence proposed by Correa *et al.* is correlated with the gastric nitrite concentration and with the gastric juice N-nitroso compound formation. This supports the hypothesis that the excess risk of gastric cancer in patients with hypochlorhydria is caused by the local production of N-nitroso compounds by bacterial action;
4. If this is so, the cancers at distant sites should also be associated with gastric hypochlohydria; this has been observed.

Cancer of the biliary tract

Epidemiology

There have been several good reviews of the epidemiology of cancer of the biliary tract (e.g., Fraumeni, 1975; Mack and Menck, 1982) and some on the gallbladder alone (e.g., Devor, 1982). A major difficulty in interpreting the incidence data for gallbladder cancer is the high prevalence of cholecystectomy in some Western countries (approximately 10 per cent in the USA); such an operation, which is usually on patient groups at increased risk of malignancy, effectively removes the person from the at risk group and distorts the subsequent incidence data.

The incidence of cancer of the biliary tract is low in Africa, Asia (except Israel and Japan), North America and Australasia and is relatively high in the countries of Eastern Europe, parts of South America and in the hispanic peoples of California (Table 24.8, page 401). The highest national incidences are seen in urban Poland, Japan, Colombia and the German Democratic Republic; however the highest incidences of all are seen in the American Indians (Creagan and Fraumeni, 1972). Biliary tract cancers account for only 1 per cent of all cancers in Western populations; within racial groups (e.g., Caucasian, Mongoloid, Negroid) biliary tract cancer is correlated with gastric cancer incidence (Caygill *et al.*, 1983). The disease is very much more common in women than in men (in contrast to gastric cancer) and is a disease of the oldest age groups. In women the incidence of the disease is inversely related to social class, but this is not seen in men.

High risk disease states

A number of disease states has been shown to predispose to cancer of the biliary tract, the excess risk often being limited to certain subsites in the biliary tract (Table 24.13).

The best documented of the high risk groups is that of patients with gallstones. This group has been reviewed by Devor (1982) who tabulated 69 reports on gallbladder cancer, of which 59 gave information on gallstones. In those reports, 77 per cent of cases of gallbladder cancer were associated with gallstones; the association is very much stronger in women (80–90 per cent) than in men (about 60 per cent) and is very much stronger in Caucasian than Negroid individuals. There was no association with cancer of the bile duct.

Caygill *et al.* (1984, 1987) reported a 10-fold excess risk of cancer of the biliary tract in patients with Polya partial gastrectomy; the tumours had a latency of 20 years. Further analysis (Caygill *et al.*, 1987) showed no excess risk for cancer of the bile ducts but a 15-fold excess risk for gallbladder cancer. In this study there was no information on the prevalence of gallstones.

Table 24.13 Disease states which predispose to a high risk of cancer of the biliary tract

Disease state	Magnitude of excess risk	Subsite affected
Gallstones	High	Gallbladder
Polya partial gastrectomy	15-fold	Gallbladder
Choledochal cysts	80-fold	Bile ducts
Ulcerative colitis	10-fold	Bile ducts
Parasitic infection	>1-fold	Bile ducts
Typhoid carriage	6-fold	?

Three diseases predispose to cancer of the bile ducts—ulcerative colitis (Edwards and Truelove, 1964), parasitic infection of the biliary tract (Gibson, 1971) and choledochal cysts (Lowenfels, 1978).

Finally, Welton et al. (1979) showed that in known typhoid carriers in New York there was an excess risk (6-fold) of subsequent cancer of the biliary tract. This report has yet to be confirmed and it gave no indication of the latency period between infection and cancer diagnosis and no information on the subsite of the cancer.

Aetiology

A common feature of all of the high risk patient groups is biliary stasis and a high prevalence of gallbladder infection. This is best documented for gallstones where, when proper care is taken in sample collection, transport and culture, a very high proportion of samples are found to be colonized (e.g., England and Rosenblatt, 1977) often with anaerobic organisms such as *B. fragilis*, *Fusobacterium* spp. and *Clostridium* spp. in addition to coliforms and enterococci (Table 24.14).

Table 24.14 The bacterial flora of the infected biliary tract. Data are expressed as the percentage of samples from which the organism was isolated*

Organisms	Percentage isolation
Anaerobes only	2
Aerobes only	60
Mixed anaerobes and aerobes	39
Anaerobes	
B. fragilis	18
Fusobacterium spp.	15
Clostridium spp.	15
Non-sporing gram-positive rods	3
Enterobacteria	24
E. coli	17
Enterobacter spp.	4
Proteus spp.	3
Streptococci	
Enterococci	19
Viridans group	7
Staphylococci	
S. aureus	3
S. epidermidis	3

*From England and Rosenblatt (1977)

Lowenfels (1978) has postulated that the gallbladder flora is able to metabolize the bile acids to release tumour promoters such as deoxycholic or lithocholic acids. This would require only the deconjugating enzyme (which is produced at high activity by a high proportion of strains of *B. fragilis*, *Clostridium* spp. and *Str. faecalis*). The amounts would be increased if the cholanoyl 7-dehydroxylase were also produced; this enzyme is produced under favourable circumstances by strains of *B. fragilis* and *Clostridium* spp. It has yet to be demonstrated that these enzymes are produced or are active under the conditions prevalent in the bile.

A more likely route to carcinogenesis might be via the hydrolysis of the glucuronides of PAHs; the enterobacteria (including *Salmonella* spp.), enterococci and the anaerobes are good producers of β-glucuronidase. This would also be consistent with the observations by Welton et al. (1979) that chronic carriage of *S. typhi* predisposes to an increased subsequent risk of biliary tract cancer.

It is possible that the excess risk of gallbladder cancer in patients with Pólya's partial gastrectomy after a latency of 20 years is caused by N-nitroso compounds synthesized by bacterial action in the colonized stomach and with a target organ specificity for the gallbladder.

Cancer of the large bowel

Epidemiology

The epidemiology of cancer of the large bowel mirrors that of cancer of the stomach. The disease is common in North America, North and Western Europe, Australasia and the River Plate area of South America and is relatively rare in Asia, Africa, the Andean countries of South and Central America and parts of eastern Europe (Table 24.8, page 401). Within Europe the disease is more common in the north and west than in the south and east, this pattern is repeated within the British Isles.

Colon cancer has a higher incidence in the women within a population than in the men; the reverse is true for cancer of the rectum. The incidence of colon cancer is positively correlated with socioeconomic status (although this correlation is weak in the high incidence countries) and is much more common in urban than in rural areas. There are a number of predisposing or precursor diseases including familial adenomatous polyposis (FAP), cancer family syndrome (CFS), discrete adenomas of the large bowel, inflammatory bowel disease (IBD), gallbladder surgery and gastric surgery.

There is a growing body of evidence that first order relatives of colorectal cancer cases have an excess risk of developing the disease, the magnitude of the excess being estimated usually as 3–5-fold. However, studies of migrant populations either between or within countries, show that the risk of colorectal cancer is related to the current residence rather than to the country or race of origin. This suggests that the risk of the disease in a population is hardly affected by genetic factors and is determined almost entirely by environmental factors. These latter can be divided into two broad categories—the physical and the cultural environment.

The physical (or shared) environment includes those factors that are common to an area or population, such as latitude, altitude, climate, air pollution or water pollution.

The cultural (or chosen) environment includes those factors which are at the discretion of the individual such as diet, smoking or use of other drugs, personal hygiene habits, factors associated with religious beliefs, etc.

Comparison of subpopulations living in a particular location who have a shared physical environment but a distinct cultural environment (e.g., the racial groups in Johannesburg, the religious groups in Bombay or California, the subgroups of Chinese living in Singapore, etc.) show that the cultural factors are of overwhelming importance, and that the physical environment plays little part in determining the risk of the disease. This evidence, and that from the studies of relatives of colorectal cancer cases, can be rationalized if the genetic factors predisposing to the disease are distributed in a relatively uniform manner in all populations. The risk of the disease in a population (i.e., the proportion of the at risk subgroup which actually develops the disease) would then be determined by environmental factors, while the identity of those developing the disease in a relatively uniformly exposed population would be determined by genetics.

The cultural factor most strongly associated with the risk of colorectal cancer is diet and most epidemiologists are convinced that diet is important in the causation of the disease. Until recently there was little agreement on the nature of the causal dietary agent, most components having been implicated by somebody at some time (Table 24.15). In recent years a consensus has formed that the risk of bowel cancer in a population is related to the amount of dietary fat or meat but that, in populations with a high fat intake, the risk of the disease is inversely related to the intake of cereal fibre. However, there is no such consensus from case control studies and some possible reasons for this will be discussed later. There has been one major prospective study, carried out in Japan by Hirayama (1985). This has shown a positive correlation with the intake

Table 24.15 The relation between diet and colorectal cancer from epidemiological studies

Type of study*	Dietary items most strongly correlated with colorectal cancer risk
Population studies	
Gregor et al. (1969)	Animal protein (no other items reported)
Drasar and Irving (1973)	Animal protein and animal fat
Armstrong and Doll (1975)	Meat, animal fat and animal protein
Enstrom (1977)	Beer with rectal cancer
Liu et al. (1979)	Meat, animal fat and cholesterol
Bingham et al. (1979)	Inverse correlation with cereal fibre (no other items studied)
Eyssen and Brightsee (1984)	Fat and meat
Case-control studies	
Pernu (1960)	Cases consume more fat
Higginson (1966)	None
Wynder and Shigematsu (1967)	Fat
Wynder et al. (1969)	Fat
Haenszel et al. (1973)	Meat and string beans
Bjelke (1974)	Protective effect of vitamins A and C
Modan et al. (1975)	Protective effect of crude fibre
Haenszel et al. (1980)	None
Jain et al. (1982)	Fat
Prospective study	
Hirayama (1985)	Meat; strong protective effect of yellow-green leafy vegetables

*For details of references see Hill (1986b)

of fat and meat and a stronger protective effect of yellow–green leafy vegetables. Most studies of micronutrients have found an inverse correlation with vitamins A and C and a number of groups have found a positive correlation with beer intake, especially for rectal carcinoma.

Other cultural factors found to be related to the risk of colorectal cancer are parity (in women), obesity and intensity of physical exercise.

Histopathology

Most histopathologists who have studied colorectal tumours accept that the vast majority of colorectal cancers arise in preformed adenomas. Thus there is an adenoma–carcinoma sequence that was proposed by Morson (1974) and has been further modified and refined to the dysplasia–carcinoma sequence (Morson et al., 1983).

Benign adenomas are very common in individuals from Western countries; they may be detected in 50 per cent of males and 30 per cent of females aged over 70 years. The epidemiology of colorectal adenomas is summarized in Table 24.16. They are evenly distributed throughout the large bowel and have a high prevalence in Western populations and a relatively low prevalence in populations living in Africa, Asia and South America. Most adenomas remain small in all populations, but the risk of an adenoma growing to a large size (i.e., more than 20 mm diameter) is very much greater in the left than in the right colon and is also very much greater in Western populations than in those with a low risk of colorectal cancer.

Since, in the West, the prevalence of adenomas in people aged over 70 years is 30–50 per cent, and since the lifetime risk of colorectal cancer is 2–4 per cent, it is clear that the vast majority of adenomas in the large bowel remain benign. Morson and his colleagues have studied the malignant potential of adenomas. They have showed that the risk of malignant change increases greatly with increased adenoma size, with increasing severity of epithelial dysplasia and with increasing villousness of the tumour.

In summary, there are likely to be a number of stages in the causation of colorectal carcinogenesis, beginning with adenoma causation, followed by adenoma growth, followed by the development of increasingly severe epithelial dysplasia and finally the development of carcinoma.

Aetiology

Aries et al. (1969) have postulated that colorectal cancer is caused by a metabolite produced *in situ* in the lumen of the colon from some benign substrate. Evidence in support of this concept comes from the observation of tumour regression following diversion of the faecal stream and, in animal studies, the protective effect of germ-free status or of antibiotic treatment during the promotion stage. Almost all of the substrates listed in the earlier section are potential candidates in this hypothesis, including the volatile phenols, the tryptophan metabolites, cholesterol, bile acid metabolites and the products of PAH glucuronide hydrolysis.

Although they have not been studied intensively, there is much circumstantial evidence to suggest that the UVP are *not* important in the overall process of colorectal carcinogenesis, although they may, of course, be implicated in one of the stages. The mean UVP in populations is unrelated to the risk of colorectal cancer (Hill et al., 1982). The tryptophan metabolites have been incriminated in bladder cancer by Bryan (1971) and have been suggested to have

Table 24.16 The epidemiology of colorectal adenomas

Geographical distribution	High prevalence in populations living in Western Europe, North America, Australia; low prevalence in populations living in Africa, Asia, S America
Subsite distribution	Evenly distributed along the colorectum
Size	Vast majority are and remain small; proportion growing to >20 mm diameter greater in left colon than in right; greater in population with high incidence of colorectal cancer than in those with a low incidence
Sex ratio	Prevalence much higher in men than in women; by 50–150% in different populations
Role of genetic factors and environmental factors	Role for both genetic and environmental factors is well established

a role in colon carcinogenesis (Chung et al., 1975) but again, there is no relation between the concentration or amounts of faecal tryptophan metabolites in populations and the risk of colorectal cancer (Hill, 1986b). The products of PAH glucuronide hydrolysis have been implicated by Renwick and Drasar (1976) while cholesterol carcinogenesis has been proposed by Cruse et al. (1979) and by Wilkins and Hackman (1974).

The best evidence is in favour of the bile acid metabolites, in part because they have been the subject of the most work. The evidence in favour of a role for the bile acids was reviewed by Hill (1986b); in summary:

1. The faecal bile acid (FBA) concentration is increased by dietary fat and decreased by dietary cereal fibre. This is consistent with the epidemiology relating diet to the risk of colorectal cancer;
2. In the many studies comparing populations, the incidence of colorectal cancer is correlated with the mean FBA concentration;
3. There is a mass of data from animal and *in vitro* studies showing that the principal FBAs, deoxycholic acid (DC) and lithocholic acids (LA), are potent tumour promoters in the rodent colon and comutagens in the *in vitro* mutagenesis assays;
4. DC and LA are products of 7-dehydroxylation of the primary bile acids synthesized by the liver. The activity of the 7-dehydroxylase is much higher in colorectal cancer cases than in healthy controls;
5. Bile acid receptors are found in a high proportion of colorectal cancers but in only a small proportion of samples of control tissue;
6. Bile acid kinetics in patients with the precursor adenoma are very different from those in healthy controls.

This has led to further studies of two types. In one direction, workers have attempted to find an improved discriminant bile acid, while in the other, they have attempted to determine the point in the adenoma–carcinoma sequence at which the bile acids act.

A discriminant bile acid

Owen et al. (1984) developed a method that permitted the analysis of the major individual FBAs on large numbers of persons. In particular they assayed DC and LA to see whether one of these, or the sum of the two, provided a better discriminant than the total FBA; in fact, the ratio of LA:DC provided the best discriminant and this has since been rationalized on the basis of *in vitro* results (Wilpart et al., 1983; Owen et al., 1987a,b). The ratio is a good marker for the at risk individuals within a population (and so yields good results in case control studies); since even in high risk countries 95 per cent of the population will *not* develop colorectal cancer, it provides only weak discrimination in population studies.

The site of bile acid action

To determine whether the bile acids act at the adenoma formation or the adenoma growth stage in the adenoma–carcinoma sequence, Hill et al. (1978) studied 152 patients with colorectal polyps, and made comparisons with healthy controls. They showed that:

the faecal bile acids were not involved in adenoma formation;
the FBA concentration was strongly correlated with adenoma size;
although there was a correlation between the severity of epithelial dysplasia and FBA concentration, this could have been secondary to the strong correlation between dysplasia and size.

In a study of patients with chronic pancolitis (Hill et al., 1987) there was a good correlation between the FBA concentration and the severity of epithelial dysplasia. All of this data on adenomas and colitis was reviewed and described in detail by Hill (1986b).

In addition to the evidence incriminating the 7-dehydroxylase there is a body of evidence from case control and population studies implicating a bile acid 4-dehydrogenase. This enzyme produces steroids with a 4-en-3-one structure from 5β-3-oxo substrates; it is a dehydrogenase which, in contrast to the hydroxysteroid dehydrogenases, acts on the steroid nucleus and is therefore referred to as a nuclear dehydrogenase. It is only produced by a subgroup of lecithinase negative clostridia which produce butyric acid as an end product of fermentation; these clostridia are called nuclear dehydrogenating clostridia or NDC. These were first described by Aries et al. (1971) and characterized in more detail by Goddard et al. (1975). For the NDC to produce unsaturated bile acids they need a hydrogen aceptor, shown by Fernandez and Hill (1978) to be menaquinones of the electron transport chain. The mean source of faecal menaquinone is the *Bacteroides* spp.; there is a correlation between the number of *Bacteroides* spp./g of faeces and the risk of colorectal cancer in populations.

In summary, bile acid metabolites produced by bacterial action in the gut are a major component of the aetiology of colorectal cancer. The major organisms incriminated are the non-sporing, strictly anaerobic rods which are the main source of the steroid 7-dehydroxylase (which produces the potent tumour promoters deoxycholic and lithocholic acids from the primary bile acids synthesized by the liver), the NDC (which produce unsaturated bile acids and which are correlated with the risk of bowel cancer) and the bacteroides (which produce the menaquinones necessary for the action of the NDC on bile acids).

Cancer of the urinary bladder

Epidemiology

The epidemiology of cancer of the urinary bladder suggests that it is a disease of industrialization, since it is common in all heavily industrialized areas with the exception of Japan (see Table 24.8, page 401) and is relatively rare in the less industrialized countries of Africa, Asia and South America. The disease within a population is much more common in urban than rural areas and is more common in men than women; both of these aspects are consistent with the hypothesis that a major cause of bladder cancer comes from industrial sources. A number of chemicals have been identified as causal agents in bladder cancer and exposure to large amounts of benzidine, β-naphthylamine, α-toluidine etc. has been shown to be associated with a high risk of subsequent bladder cancer. However, it is clear that there must also be many non-industrial causes of the disease since, although there is an association between bladder cancer and heavy industry, the highest national incidences are reported from Switzerland.

Predisposing diseases

The disease states that predispose to bladder cancer are chronic urinary tract infection, schistosomiasis, paraplegia and urethral strictures.

Chronic urinary tract infection

Bacterial infection of the human urinary bladder is very common and is normally asymptomatic; these infections are very much more common in women than in men and the incidence increases with age. Patients with a history of urinary tract infection have an above average risk of bladder cancer in comparison with the general population (Radomski *et al.*, 1978).

Paraplegia

In paraplegia and in quadriplegia there is a high incidence of secondary chronic urinary tract infection, usually complex and presumably the cumulative result of repeated superinfection. These patients have an excess risk of bladder cancer (Davies, 1977).

Schistosomiasis

Patients with schistosomal infection of the bladder usually have a secondary bacterial infection, which is presumably potentiated by the foreign-body effect of schistosomal cysts. Because the infecting organism is able to invade the bladder wall at the site of the cyst implantation, repeated superinfections result in a bacterial flora similar in complexity to that of faeces (Hicks *et al.*, 1977) and including strictly anaerobic genera such as *Bacteroides* spp., *Propionibacterium* spp., *Fusobacterium* spp. etc. The risk of bladder cancer in schistosomal patients is greatly in excess of that in the general population.

Urethral strictures

In many parts of Africa there is a high incidence of bladder cancer associated not with schistosomal infection, but with urethral structures. These lesions might be expected to cause retention of urine and a consequently increased incidence of symptomatic bladder infection.

Aetiology

The common factor in all of the associated diseases listed is chronic bacterial infection of the bladder. This could result in the formation or release of a range of carcinogens or tumour promoters. The principal metabolites cited in the causation of bladder cancer are tryptophan metabolites and N-nitroso compounds.

Bryan (1971) has reviewed the data implicating tryptophan metabolites in the causation of bladder cancer of non-industrial aetiology. There is a large literature on this stemming from the original observations by Dunning et al. (1950) that dietary tryptophan increased the number of tumours in rats treated with 2-acetylaminofluorene. A number of epidemiological case control studies have related the urinary concentration of tryptophan metabolites, particularly those on the quinoline pathway, to the risk of non-industrial bladder cancer. However, the hypothesis has remained controversial, with many claims that the increased concentration of tryptophan metabolites is caused by the presence of the tumour, rather than vice versa (e.g., Teulings et al., 1978).

Urine is the main route of excretion of secondary amines, alkylureas, polyamines etc. (all of which are N-nitrosatable) as well as of nitrate. Thus, in patients with urinary tract infection the conditions are ideal for the bacterial catalysis of N-nitrosation; the formation of N-nitroso compounds in the infected urinary bladder has been demonstrated by many groups and has been incriminated in the causation of the bladder cancer associated with chronic urinary tract infection (Radomski et al., 1978) and with the infections secondary to schistosomiasis of the bladder (Hicks et al., 1977). In the latter study the urinary concentration of N-nitroso compounds was very much higher than that observed in the urine of patients with simple bladder infections; the risk of bladder cancer was also very much higher.

The bladder was one of the sites noted by Caygill et al. (1987) to have an increased cancer risk associated with gastric surgery (see Table 24.12, page 404). It has been suggested that these cancers may be caused by N-nitroso compounds formed in the infected gastric remnant and target organ specific for, in this case, the urinary bladder.

Conclusions

In this review I have attempted to describe some of the evidence for the role of bacteria, particularly the anaerobes, in human carcinogenesis. Such a role is virtually self-evident in principle, what is open to question is the relative importance of bacterially induced cancers in comparison with other causes. It is important to determine this because such cancers, as with other bacterial diseases, are preventable.

References

Aries V C, Crowther J S, Drasar B S, Hill M J. (1969). Degradation of bile salts by human gut bacteria. *Gut* **10**: 575–577.

Aries V C, Goddard P, Hill M J. (1971). Degradation of steroids by intestinal bacteria III 3-oxo-5β-steroid Δ^1-dehydrogenase and 3-oxo-5β-steroid Δ^4-dehydrogenase *Biochimica Biophysica Acta*. **248**: 482–488.

Armstrong B, Doll R. (1975). Environmental factors and the incidence and mortality from cancer in different countries with specific reference to dietary practice. *International Journal of Cancer* **15**: 617–631.

Bakke O M, Midtvedt T. (1970). Influence of germ-free status on the excretion of simple phenols of possible significance in tumour promotion. *Experientia* **26**: 519.

Bavin P M, Darkin D W, Viney N J. (1982). Total nitroso compounds in gastric juice. In *N nitroso compounds. occurrence and biological effects*, pp. 337–344. Edited by Bartsch H et al. IARC, Lyon.

Bingham S, Williams D R, Cole T J, James W P T. (1979). Dietary fibre and regional large bowel cancer mortality in Britain. *British Journal of Cancer* **40**: 456–463.

Bischoff F. (1969). Carcinogenic effects of steroids. *Advances in Lipid Research* **7**: 165–244.

Bjelke E. (1974). Epidemiological studies of cancer of the stomach, colon and rectum. *Scandinavian Journal of Gastroenterology* **9** Suppl 31.

Boutwell R K, Bosch D K. (1959). The tumour-promoting action of phenol and related compounds on the mouse skin. *Cancer Research* **19**: 413–427.

Brooks J B, Cherry W B, Thacker L, Alley C C. (1972). Analysis by gas chromatography of amino and nitrosamines produced *in vivo* and *in vitro* by *Proteus mirabilis*. *Journal of Infectious Diseases* **126**: 143–153.

Bryan G T. (1971). The role of urinary tryptophan metabolites in the etiology of bladder cancer. *American Journal of Clinical Nutrition* **24**: 841–847.

Caygill C, Hall R, Hill M J. (1983). Association between biliary tract cancer and gastric cancer. *Lancet* **ii**: 1204–1205.

Caygill C, Craven J, Hall R, Hill M J, Miller C. (1984). The relevance of gastric achlorhydria to human carcinogenesis. In: *N-nitroso compounds: occurrence, biological effects and relevance to human cancer*, pp. 895–900. IARC Scientific Publication No. 57. IARC, Lyon.

Caygill C P J, Hill M J, Kirkham J S, Northfield T C. (1986). Mortality from gastric cancer following gastric surgery for peptic ulcer. *Lancet* **i**: 929–931.

Caygill C P J, Hill M J, Hall C N, Kirkham J S, Northfield T C. (1987). Increased risk of cancer at multiple sites after gastric surgery for peptic ulcer. *Gut* **28**: 924–928.

Chung K T, Falk G E, Slein M W. (1975). Tryptophanase of faecal flora as a possible factor in the aetiology of colon cancer. *Journal of the National Cancer Institute* **54**: 1073–1078.

Coloe P J, Heywood N J. (1976). The importance of prolonged incubation for the synthesis of dimethylnitrosamine by enterobacteria. *Journal of Medical Microbiology* **9**: 211–224.

Correa P. (1982). Epidemiology of gastric cancer and its precurser lesions. In *Gastrointestinal cancer*, pp. 119–130. Edited by DeCosse J, Sherlock P. Martinus Nijhoff, Amsterdam.

Correa P, Haenszel W, Cuello C, Tannenbaum S, Archer M. (1975). A model for gastric cancer epidemiology. *Lancet* **ii**: 58–60.

Creagan E T, Fraumeni J F. (1972). Cancer mortality among american indians 1950–1967. *Journal of the National Cancer Institute* **49**: 959–967.

Cruse P, Lewin M, Clark C G. (1979). Cholesterol and colon cancer. *Lancet* **ii**: 43–44.

Davies J M. (1977). Two aspects of the epidemiology of bladder cancer in England and Wales. *Proceedings of the Royal Society of Medicine* **70**: 411–413.

Devor E J. (1982). Ethnographic patterns in gallbladder cancer. In *Epidemiology of cancer of the digestive tract*, pp. 197–226. Edited by Correa P, Haenszel W. Martinus Nijhoff, The Hague.

Drasar B S, Hill M J. (1974). *Human intestinal flora*. Academic Press, London.

Drasar B S, Irving D. (1973). Environmental factors and cancer of the colon and breast. *British Journal of Cancer* **27**: 167–172.

Dunning W F, Curtis M R, Maun M E. (1950). The effect of added dietary tryptophane on the occurrence of 2-acetylaminofluorene-induced liver and bladder cancers in rats. *Cancer Research* **10**: 454.

Edwards F C, Truelove S C. (1964). The course and prognosis of ulcerative colitis. IV, Carcinoma of the colon. *Gut* **5**: 15–22.

England D M, Rosenblatt J E. (1977). Anaerobes in human biliary tracts. *Journal of Clinical Microbiology* **6**: 494–570.

Enstrom J E. (1977). Colorectal cancer and beer drinking. *British Journal of Cancer* **35**: 674–683.

Eyssen G, Bright-See E. (1984). Dietary factors in colon cancer: international relationships. *Nutrition and Cancer* **6**: 160–170.

Fernandez F, Hill M J. (1978). A faecal hydrogen acceptor for clostridial 3-oxo-steroid Δ^4-dehydrogenase. *Transactions of the Biochemical Society* **6**: 376–377.

Fraumeni J F. (1975). Cancers of the pancreas and biliary tract: epidemiological considerations. *Cancer Research* **35**: 3437–3446.

Gibson J B. (1971). In *Liver cancer*, IARC Scientific Publication No. 1 IARC, Lyon.

Gledhill T, Buck M, Paull A, Hunt R H. (1982). Epidemic hypochlorhydria. *Gut* **23**: A888.

Goddard P, *et al*. (1975). The nuclear dehydrogenation of steroids by intestinal bacteria. *Journal of Medical Microbiology* **8**: 429–435.

Gregor O, Toman R, Prusova F. (1969). Gastrointestinal cancer and nutrition. *Gut* **1**: 1031–1034.

Haenszel W, Berg J, Segi M, Kurihara M, Locke F. (1973). Large bowel cancer in Hawaiian Japanese. *Journal of the National Cancer Institute* **51**: 1765–1779.

Haenszel W, Locke F, Segi M. (1980). A case-control study of large bowel cancer in Japan. *Journal of the National Cancer Institute* **64**: 17–21.

Hall C N, *et al*. (1984). Evaluation of the nitrosamine hypothesis of gastric carcinogenesis in man. *Clinical Science* **66**: 34P–35P.

Hawksworth G M, Hill M J. (1971). Bacteria and N-nitrosation of secondary amines. *British Journal of Cancer* **25**: 520–526.

Hazell S, Lee A. (1986). *Campylobacter pyloridis*, urease, hydrogen ion back diffusion and gastric ulcers. *Lancet* **ii**: 15–17.

Heiger I. (1949). Carcinogenicity of lipoid substances. *British Journal of Cancer* **3**: 123–128.

Hicks R M *et al*. (1977). Demonstration of nitrosamines in human urine: preliminary observations on a possible etiology for bladder cancer in association with chronic urinary tract infections. *Proceedings of the Royal Society of Medicine* **70**: 413–6.

Higginson J. (1966). Etiological factors in gastrointestinal cancer in man. *Journal of the National Cancer Institute* **37**: 527–545.

Hill M J. (1983). The enterohepatic circulation. In *Colon: structure and function*, pp. 223–251. Edited by Bustos Fernandez L. Plenum, New York.

(1986a) Metabolism of carbohydrates and glycosides. In *Microbial metabolism in the digestive tract*. Edited by M J Hill. CRC Press, Boca Raton.

(1986b) *Microbes and human carcinogenesis*. Edward Arnold, London.

Hill M J, Melville D, Lennard-Jones J, Neale K, Ritchie J K. (1987). Faecal bile acids, dysplasia and carcinoma in ulcerative colitis. *Lancet* **ii**: 185–186.

Hill M J, Morson B C, Bussey H J R. (1978). Etiology of adenoma—carcinoma sequence in large bowel *Lancet* **1**: 245–247.

Hill M J, Taylor A J, Thompson M H, Wait R. (1982). Faecal steroids and urinary volatile phenols in four Scandinavian populations. *Nutrition and Cancer* **4**: 5–19.

Hirayama T. (1985). Diet and cancer: feasibility and importance of prospective cohort study. In *Diet and human carcinogenesis*, pp. 191–197. Edited by Joossens *et al*. Excerpta Medica, Amsterdam.

Hofmann A F. (1977). The enterohepatic circulation of bile acids in man. *Clinical Gastroenterology* **6**: 3–24.

Jain M G, Harrison L, Howe G R, Miller A B. (1962). Evaluation of a self-administered dietary questionnaire for use in a cohort study. *American Journal of Clinical Nutrition* **36**: 931–935.

Jones S M, Davies P W, Savage A. (1978). Gastric juice nitrite and gastric cancer. *Lancet* i: 1355.

Kieghley M, Youngs D, Poxon V. (1984). Intragastric N-nitrosation is unlikely to be responsible for gastric carcinoma developing after operation for duodenal ulcer. *Gut* 25: 238–245.

Kinoshita, N. and Gelboin, H. (1978). β-glucuronidase catalysed hydrolysis of benzo-[a]-pyrene-3-glucuronide and binding to DNA. *Science* 199: 307–309.

Klubes P, Cernal I, Rabinowitz A D. (1972). Factors affecting the dimethylnitrosamine formation from simple precursors by rat intestinal bacteria. *Food and Cosmetic Toxicology* 10: 757–767.

Kunisaka N, Hayashi M. (1979). Formation of N-nitrosamines from secondary amines and nitrite by resting cells of *Escherichia coli* B. *Applied and Environmental Microbiology* 37: 279–282.

Lauren P. (1965). The two histological main types of gastric carcinoma: diffuse and so-called intestinal type. *Acta Pathologica et Microbiologica Scandinavica* 64: 31–49.

Leach S, Challis B, Cook A R, Hill M, Thompson M. (1985). Bacterial catalysis of the N-nitrosation of secondary amines. *Transactions of the Biochemical Society* 13: 380–381.

Lehtola J. (1978). Family study of gastric carcinoma; with special reference to histological types. *Scandinavian Journal of Gastroenterology* 13 Suppl. 50: 1–54.

Liu K, Moss M, Persky V, Stamler J, Garside D, Soltero I. (1979). Dietary cholesterol, fat and fibre and colon cancer mortality. *Lancet* ii: 782–785.

Lowenfels A. (1978). Does bile promote extra-colonic cancer? *Lancet* ii: 239–41.

Mack T M, Menck H R. (1982). Epidemiology of cancer of the gallbladder and extrahepatic biliary passages. In *Epidemiology of cancer of the digestive tract*, pp. 227–242. Edited by Correa P, Haenszel W. Martinus Nijhoff, The Hague.

Magee P, Barnes J M. (1967). Carcinogenic N-nitroso compounds. *Advances in Cancer Research* 10: 163–246.

Marshall B J, *et al.* (1985). Attempts to fulfil Koch's postulates for pyloric campylobacter. *Medical Journal of Australia* 142: 436–479.

Milton-Thompson G, *et al.* (1982). Intragastric acidity bacteria, nitrite and N-nitroso compounds before, during and after cimetidine treatment. *Lancet* 1: 1091–1094.

Modan B, Barell V, Lubin F, Modan M, Greenberg R, Graham S. (1975). Low fibre intake as an etiologic factor in cancer of the colon. *Journal of the National Cancer Institute* 55: 15–18.

Morson B C M. (1974). The polyp–cancer sequence in the large bowel. *Proceedings of the Royal Society of Medicine* 67: 457–457.

Morson B C, Bussey H J, Day D W, Hill M J. (1983). Adenomas of the large bowel. *Cancer Surveys* 2: 451–478.

Mortensen N J, *et al.* (1984). Relationship of duodenogastric reflux, gastric pH and nitrite levels to histological changes in the gastric mucosa. *British Journal of Surgery* 71: 363.

Newell D G. (1981). Identification of the outer membrane proteins of *Campylobacter pyloridis* antigenic cross reactivity between *C. pyloridis* and *C. jejuni*. *Journal of General Microbiology* 133: 163–170.

Owen R W, Dodo M, Thompson M, Hill M. (1987a). Fecal steroids and colorectal cancer. *Nutrition and Cancer* 9: 73–80.

Owen R W, Thompson M H, Hill M J. (1984). Analysis of metabolic profiles of steroids in faeces of healthy subjects undergoing chenodeocycholic acid treatment by liquid–gel chromatography and gas–liquid chromatography–mass spectrometry. *Journal of Steroid Biochemistry* 21: 593–600.

Owen R W, Thompson M H, Hill M J, Wilpart M, Mainguet, P, Roberfroid M. (1987b). The importance of the ratio of lithocholic to deoxycholic acid in large bowel carcinogenesis. *Nutrition and Cancer* 9: 67–72.

Pernu J. (1960). An epidemiological study on cancer of the digestive organs and respiratory system. *Annals of International Medicine, Finland* 49: 1–117.

Pignatelli S, Richard I, Bourgade M, Bartsch H. (1987). Improved group determination of total N-nitroso compounds in human gastric juice by chemical denitrosation and thermal energy analysis. *The Analyst* 112: 945–950.

Radomski J L, Greenwald D, Hearn W L, Block N L, Woods F M. (1978). Nitrosamine formation in bladder infections and its role in the etiology of bladder cancer. *Journal of Urology* 120: 48–56.

Ramsey E J, Carey K V, Peterson W L. *et al.* (1979). Epidemic gastritis with hypochlorhydria. *Gastroenterology* 76: 1449–1457.

Reed P I, Smith P L R, Haines K. *et al.* (1981). Gastric juice N-nitrosamines in health and gastroduodenal disease. *Lancet* 2: 550–552.

Renwick A G, Drasar B S. (1976). Environmental carcinogens and large bowel cancer. *Nature* 263: 234–235.

Ruddell W S J, Bone E S, Hill M J, Blendis L M, Walters C L. (1976). Gastric juice nitrite: a risk factor for cancer in the hypochlorhydric stomach. *Lancet* ii: 1037–1039.

Ruddell W S, Bone E S, Hill M J, Walters C L. (1978). Pathogenesis of gastric cancer in pernicious anemia. *Lancet* 1: 521–523.

Sander J. (1968). Nitrosaminsynthese durch Bakterien. *Zeiting fur Physiologische Chemi* 349: 429–432.

Schlag P, Bockler R, Ulrich H, Peter M, Merkle P, Herfath C H. (1980). Are nitrite and N-nitroso compounds in gastric juice risk factors for carcinoma in the operated stomach? *Lancet* i: 727–729.

Stockbrugger R. *et al.* (1984). Pernicious anemia intragastric bacterial overgrowth and possible consequences. *Scandinavian Journal of Gastroenterology* 19: 355–364.

Tannenbaum S R, Archer M C, Wishnok J S, Bishop W W.

(1978). Nitrosamine formation in human saliva. *Journal of the National Cancer Institute* **60**: 251.

Teulings F, Peters H, Hop W, Fokkens W, Haije W, Portengen H, Van der Werf-Messing B. (1978). A new aspect of the urinary excretion of tryptophan metabolites in patients with cancer of the bladder. *International Journal of Cancer* **21**: 140–146.

Thomas J M *et al.* (1987). Effect of one years treatment with ranitidine and of truncal vagotomy on gastric contents. *Gut* **28**: 726–738.

Waterhouse J, Muir C, Shanmuguratnam K. (1982). *Cancer incidence in five continents*, vol 4. IARC, Lyon.

Watt P C, *et al.* (1984). Late mortality after vagotomy and drainage for duodenal ulcer. *British Medical Journal* **288**: 1335–1338.

Welton J C, Marr J S, Friedman S M. (1979). An association between hepatobiliary cancer and the typhoid carriage state. *Lancet* **i**: 791–794.

Wilkins T G, Hackman A S. (1974). Two patterns of neutral steroid conversion in the faeces of normal North Americans. *Cancer Research* **34**: 2250–2254.

Wilpart M, Mainguet P, Maskens A, Roberfroid M. (1983). Structure activity relationship amongst biliary acids showing co-mutagenicity towards 1,2 dimethylhydrazine. *Carcinogenesis* **4**: 1239–1241.

Wynder E L, Kajitani T, Ishikawa S, Dodo H, Takano A. (1969). Environmental factors of cancer of the colon and rectum. *Cancer* **23**: 1210–1220.

Wynder E L, Shigematsu T. (1967). Environmental factors of cancer of the colon and rectum. *Cancer* **20**: 1520–1561.

25

Antibiotics and anaerobes

D Greenwood, B Watt and B I Duerden

Antimicrobial agents
   β-lactam antibiotics
   Inhibitors of bacterial protein synthesis
   Quinolones
   Nitroimidazoles
   Other antimicrobial agents
Resistance to antimicrobial agents among anaerobes
Sensitivity testing of anaerobes
   Variables affecting sensitivity testing of anaerobes
   Relationship of susceptibility testing to therapy
Treatment of infections due to anaerobes
   Treatment of specific anaerobic infections
   Treatment based upon site of infections
   Prophylactic use of antimicrobial agents
References

Antimicrobial agents

Anaerobes that inhabit the human body, and which occasionally cause disease, live in a variety of highly specialized environments, sometimes perilously close to oxygenated areas like the skin, or as part of a complex flora of the mouth, gut or vagina. The organisms themselves are consequently very diverse and well adapted to life in a hostile environment.

It might be expected, therefore, that anaerobes would exhibit markedly dissimilar patterns of response to antimicrobial agents, but in the event it turns out that the susceptibility of anaerobic bacteria to antimicrobial agents is, in general, remarkably uniform. Indeed, from the point of view of chemotherapy, anaerobes are generally much less troublesome than their aerobic relatives. They are usually susceptible to a wide variety of antimicrobial compounds, intrinsic resistance is often predictable and acquired resistance among susceptible species is ordinarily neither common nor easy to induce.

Not surprisingly, most of our knowledge of the mechanism of action of antimicrobial agents comes from investigations of aerobic bacteria that are easily handled in the laboratory. Except for the 5-nitroimidazoles, which have no useful activity against aerobes, little is known specifically about the ways antibacterial drugs achieve their effects in anaerobic organisms. It is naturally assumed that results obtained in aerobes are applicable to anaerobes as well, and in many cases there is sufficient circumstantial evidence to suggest that this is a reasonable assumption.

Agents with therapeutically useful activity against a wide spectrum of anaerobic species include most

Table 25.1 Summary of anti-anaerobe activity of the major groups of antimicrobial agents

Generally good	Poor or unreliable
Penicillins*	Aminoglycosides
Cephalosporins*	Mecillinam
Cephamycins	Monobactams
Latamoxef	Tetracyclines
Imipenem	Trimethoprim
Lincosamides	Sulphonamides
Macrolides	Polymyxins
Chloramphenicol	Fosfomycin
Fusidic acid	Quinolones
5-nitroimidazoles	Glycopeptides†
Rifampicin	
Nitrofurantoin	

*Poor activity against *B. fragilis* group;
† Good activity against gram-positive anaerobes

penicillins, certain other β-lactam antibiotics, a number of inhibitors of bacterial protein synthesis and compounds of the 5-nitroimidazole group (Table 25.1). Minimum inhibitory concentrations (MICs) of a selection of these compounds for medically important anaerobes are shown in Table 25.2. It should be noted that anaerobic bacteria are not the easiest of organisms to test and an unusually broad spread of MIC values is evident in reports from different laboratories; the values given represent a consensus view culled from various reports in the literature.

β-lactam antibiotics

Most anaerobic bacteria, including many *Bacteroides* strains (other than those belonging to the *B. fragilis* group), clostridia, anaerobic cocci and spirochaetes are highly susceptible to benzylpenicillin. This situation has not changed dramatically in more than 40 years of penicillin use, although there are reports from some centres, notably in the USA, of increasing resistance to penicillin among the non-*B. fragilis*-group *Bacteroides* strains (Edson *et al.*, 1982) and *Clostridium perfringens* (Marrie *et al.*, 1981).

Other β-lactam agents, with the exception of mecillinam, temocillan and the monobactams, also display broad-spectrum anti-anaerobe activity, but none (except perhaps the new carbapenem derivatives, imipenem and meropenem) is more active than benzylpenicillin (Sutter and Finegold, 1976; Appelbaum and Chatterton, 1978; Rolfe and Finegold, 1981) and there seems little reason to prefer one of the more expensive modern compounds in many anaerobic infections.

A serious defect in the anaerobic spectrum of benzylpenicillin and many other β-lactam agents exists among organisms of the *B. fragilis* group. The resistance is associated with the constitutive β-lactamase activity exhibited by most *B. fragilis*-group strains (Nord and Olsson-Liljequist, 1981) and full susceptibility to benzylpenicillin and other β-lactam agents can be restored by β-lactamase inhibitors such as clavulanic acid or sulbactam (Wise *et al.*, 1980; Lamothe *et al.*, 1984; Wexler *et al.*, 1985). These strains are also usually susceptible to imipenem (Martin *et al.*, 1982; Goldstein and Citron, 1985). Alternatively, cephalosporins of the cephamycin type, which carry a stabilizing methoxy group on the β-lactam ring (e.g., cefoxitin, cefotetan, latamoxef and cefmetazole) retain activity against *B. fragilis*-group strains, although they are not as active as clavulanate-potentiated penicillins or imipenem (Eley and Greenwood, 1986a,b), perhaps because of differential permeability through the cell wall outer membrane (Cuchural *et al.*, 1988a).

Susceptibility to cephamycins is not uniform among species within the *B. fragilis* group. In general, *B. fragilis sensu stricto*, *B. vulgatus* and *B. distasonis* are more susceptible than *B. ovatus* or *B. thetaiotaomicron* (Jenkins *et al.*, 1982). Moreover, although latamoxef usually appears considerably more active than cefoxitin as judged by conventional MIC titrations, the differential activity is much less striking if the compounds are compared on the basis of bactericidal or bacteriolytic activity (Eley and Greenwood, 1984).

Inhibitors of bacterial protein synthesis

All the commonly used inhibitors of bacterial protein synthesis, with the notable exception of the aminoglycosides, exhibit good activity against anaerobes (Sutter and Finegold, 1976; Appelbaum and Chatterton, 1978). Indeed, the lincosamide antibiotic, clindamycin, is among the most active of all currently available agents against *B. fragilis* and many other anaerobes (Sutter, 1983). This antibiotic, and its close relative, lincomycin, also possesses

Table 25.2 Minimum inhibitory concentrations of selected agents for medically-important anaerobic species (data from various sources)

Species	Minimum inhibitory concentration (mg/l)							
	Benzyl-penicillin*	Cefoxitin	Latamoxef	Imipenem	Erythro-mycin	Clinda-mycin	Chloram-phenicol	Metro-nidazole
B. fragilis group	16†	2–32	0.25–8	0.25	8	2	8	2
B. melaninogenicus group	0.5	0.5	4	0.12	0.5	0.1	1	0.5
Other *Bacteroides* spp.	0.5–8	0.5–8	2–32	0.5	1	0.5	4	1
Fusobacterium spp	0.1	0.5	2	0.5	8	0.1	0.5	0.5
Peptococcus spp.	0.1	0.5	2	0.06	2	0.5	2	1
Peptostreptococcus spp.	0.5	0.5	4	0.06	0.5	0.5	1	1
Veillonella spp.	1	1	64	0.12	4	0.06	2	1
C. perfringens	0.5	2	1	0.25	0.5	1	4	1
C. difficile	1	128	64	4	1	16	2	0.5
Other clostridia	1	2–32	4	0.5	0.5	4	8	1
Actinomyces spp.	0.5	0.5	0.5	0.25	0.5	0.1	2	128
Other gram-positive bacilli	0.1–1	0.25–4	0.5–128	0.06–1	0.1–1	0.1–1	0.25–4	0.5–>256

*Ampicillin, amoxycillin, acylureidopenicillins and carboxypenicillins exhibit similar activity; β-lactamase inhibitors, such as clavulanic acid and sulbactam, reduce the MICs of penicillins for most strains of *B. fragilis* group to around 0.5 mg/l
†

good anti-staphylococcal and anti-streptococcal activity—a potentially valuable property in certain blind treatment situations. However, enthusiasm for these drugs has been tempered by the severe diarrhoea, sometimes progressing to fatal pseudomembranous colitis, which has been associated with their use. These side effects are not, of course, restricted to lincosamides, but the incidence of both diarrhoea and pseudomembranous colitis appears to be more frequent with clindamycin than with other common culprits, including ampicillin (Gurthwith et al., 1977; Lusk et al., 1977).

A rare, but fatal side effect, aplastic anaemia, has also severely restricted the use of chloramphenicol, an antibiotic that otherwise possesses excellent activity against anaerobes and many other bacteria. Indeed, there is now no justification for use of this agent in anaerobic infection, with the possible exception of brain abscess, in which mixed infections involving anaerobes may be encountered. The related compound, thiamphenicol, which is said to lack the irreversible side effects of chloramphenicol, also exhibits good anti-anaerobe activity (Finegold, 1984) which might be exploited in those countries in which the drug is marketed.

Antibiotics of the macrolide group, of which erythromycin is the most familiar, are considerably less active than lincosamides against *Bacteroides* spp. and most other anaerobes. However, most strains of *Bacteroides*, *Fusobacterium*, *Clostridium* and anaerobic gram-positive cocci are inhibited by concentrations of erythromycin achievable in serum, at least by parenteral therapy (Harvey et al., 1981).

As in aerobes, relative resistance to macrolides among anaerobes is thought to be due to difficulty in penetration through the cell wall outer membrane. Accumulation of macrolides in *B. fragilis* is related to hydrophobicity (Muto et al., 1989) so that the possibility of designing new derivatives with improved anti-anaerobic activity cannot be discounted. However, among the new derivatives of erythromycin under development (Wise, 1989) none appears to be distinguished by enhanced activity against anaerobes; even azithromycin, which displays substantially better activity than erythromycin against Enterobacteriaceae, exhibits only modest activity against anaerobes (Retsema et al., 1987).

Assessment of the true clinical potential of erythromycin and other macrolides is made even more difficult because of the great influence of methodological variation on the results of susceptibility testing with these agents. This is exemplified by the effect of CO_2 in the incubation atmosphere (Ingham et al., 1970) and of surface pH changes during growth (Watt and Brown, 1975; 1985). The activity of erythromycin, in particular, is dramatically reduced at the acid pH values produced by growth of anaerobes in most test systems, and results of *in vitro* tests may not correlate with clinical response.

Among tetracyclines, minocycline and doxycycline exhibit the best anti-anaerobic activity (Chow et al. 1975; Robbins et al., 1987) but many strains are resistant to concentrations of tetracyclines achieved therapeutically and these antibiotics are no longer regarded as reliable anti-anaerobic agents.

It is not widely recognized that the anti-staphylococcal steroid antibiotic, fusidic acid, possesses quite good activity, at least against *Bacteroides* strains (Steinkraus and McCarthy, 1979). Stirling and Goodwin (1977) have reported good clinical results in a small number of patients with *B. fragilis* infections who were treated with fusidic acid; this agent perhaps deserves more attention as a potential anti-anaerobe agent.

Anaerobes constitute an important gap in the antibacterial spectrum of the aminoglycoside antibiotics. The deficiency also extends to streptococci, bacteria which generally grow well under aerobic or microaerophilic conditions, but which display an anaerobic type of metabolism. The basis of the resistance (which also extends to aerobes grown under anaerobic conditions) has been pinpointed to the absence of respiratory quinones involved in the transport of aminoglycosides across the bacterial cell membrane (Bryan and Kwan, 1981).

Quinolones

Nalidixic acid and its early congeners, including oxolinic acid, cinoxacin, pipemidic acid and flumequine, display no notable activity against anaerobes. Indeed, nalidixic acid is a useful selective agent for the isolation of anaerobes. However, the newer fluorinated, piperazine-substituted compounds (which include ciprofloxacin, norfloxacin, enoxacin, ofloxacin and pefloxacin) do possess some activity against anaerobic organisms (Goldstein and Citron, 1985; Phillips et al., 1988). The activity is not sufficient to raise hopes that these agents might be useful in anaerobic infection, but some attention has been paid to the effect that these incompletely absorbed agents might have on the gut flora. It turns out that, although the aerobic flora of the gut is severely depleted by the newer quinolones, the anaerobic component escapes virtually unscathed (Reeves, 1986). Fluoroquinolones that exhibit enhanced activity against anaerobes are under development (Phillips et al., 1988).

Nitroimidazoles

It was the observation of an alert patient, whose ulcerative gingivitis was simultaneously cured during treatment for trichomonal vaginitis, that first indicated that the 5-nitroimidazole, metronidazole, might exhibit antibacterial activity (Shinn, 1962). Metronidazole had originally been marketed solely for the treatment of trichomoniasis, although this was later extended to amoebiasis and giardiasis —conditions for which 5-nitroimidazoles remain the agents of first choice. The antibacterial activity of metronidazole was found to be restricted to anaerobes and the factor that linked these organisms with certain protozoa turned out to be anaerobic metabolism.

The reason for the restricted spectrum of metronidazole and other 5-nitroimidazoles lies in the fact that reduction of the nitro group is required for antimicrobial activity; this occurs only at low Eh values, in which conditions the compound acts as an electron sink to siphon off electrons from ferredoxin and other electron carriers. The reduced intermediate is generated intracellularly in anaerobes and it is this compound that is believed to act by causing strand breakage in DNA (Edwards, 1979).

The activity of metronidazole against anaerobic bacteria is remarkably consistent and very reliable; primary resistance in clinical isolates appears to be very rare. Oxygen-tolerant or microaerophilic species, including *Actinomyces* spp. and *Propionibacterium* spp., are commonly resistant to nitroimidazoles, but surprisingly these agents may display useful activity against some strains of *Campylobacter jejuni* (Vanhoof et al., 1978) and *Gardnerella vaginalis* (Jones et al., 1985). The 2-hydroxymethyl derivative of metronidazole, which is a major metabolite of the drug, exhibits even better activity than the parent compound against *G. vaginalis* (Easmon et al., 1982; Jones et al., 1985).

Other 5-nitroimidazoles, including tinidazole, nimorazole and ornidazole share the activity and spectrum of metronidazole (Reynolds et al., 1975). None appears to have any great advantage over metronidazole, although tinidazole may exhibit marginally better intrinsic activity (Jokipii and Jokipii, 1988) and pharmacokinetic properties (Welling and Munro, 1972).

Other antimicrobial agents

Among other antimicrobial agents, vancomycin and the related glycopeptide, teicoplanin, deserve special mention. Although these antibiotics display no useful activity against most gram-negative anaerobes, they possess good activity against clostridia and anaerobic streptococci (Glupczynski et al., 1984; Greenwood and Palfreyman, 1987). Some pigmented strains of bacteroides are unusually susceptible to teicoplanin (Greenwood et al., 1988) and to vancomycin (Watt, unpublished observation). Vancomycin has established a useful place in the management of enterocolitis caused by *C. difficile* toxins (Fekety et al., 1984; Burdon, 1984).

Sulphonamides and trimethoprim are, at best, erratic in their action against anaerobes (Bushby, 1973) although their combined effect may be somewhat better (Phillips and Warren, 1974). Rifampicin includes many anaerobes in its wide spectrum, but resistance emerges readily as with other organisms.

Nitrofurantoin displays rapid bactericidal activity against *B. fragilis* and *C. perfringens* (Ralph, 1978) but this is unlikely to have any therapeutic significance since use of this drug is restricted to the treatment of urinary tract infection.

Resistance to antimicrobial agents among anaerobes

Most large surveys of resistance in bacteria have been confined to aerobic organisms (Kresken and Wiedemann, 1986; O'Brien et al., 1986) and much less is known about resistance trends in anaerobes. The evidence that exists suggests that the prevalence of resistance to the commonly used anti-anaerobic agents remains at an encouragingly low level (Phillips and Tally, 1981; Finegold and Wexler 1988; Cuchural et al., 1988b; Cormick et al., 1990).

Over the years, resistance to tetracyclines has certainly increased substantially (Tally and Malamy, 1984) but clindamycin and chloramphenicol continue to exhibit reliable activity against most strains, although resistance to clindamycin has been increasingly reported in the USA (Rosenblatt, 1986). Resistance to metronidazole among anaerobes remains negligible, although the practice of using the drug in the presumptive identification of anaerobes may lead to an underestimate of resistance (Sprott and Kearns, 1988).

The development of resistance to metronidazole appears to be rare, even under the influence of mutagenic procedures in the laboratory (Britz and Wilkinson, 1979; Sindar et al., 1982). Moreover, laboratory derived resistant mutants display evidence of metabolic damage, including a much reduced growth rate, and this is likely to render them less virulent.

The mechanism of resistance to clindamycin has not been definitively elucidated, but is likely to be associated with changes in the ribosomal binding site of the drug (Tally et al., 1981). Chloramphenicol resistance appears to be due either to the production of chloramphenicol acetyltransferase, an enzyme which renders chloramphenicol inactive by acetylation of the 3-hydroxyl group (Britz, 1981), or to nitroreductase activity directed against the p-nitro group (Tally and Malamy, 1984).

Although most anaerobes are currently susceptible to β-lactam antibiotics the situation is changing and there is no guarantee that resistance to these agents will not continue to increase. Resistance is usually (but not always) mediated by β-lactamases. Such enzymes have been described in fusobacteria and clostridia as well as in many species of *Bacteroides* (Nord, 1986) but have not so far been found in anaerobic gram-positive cocci.

The constitutive β-lactamase activity possessed by nearly all strains that belong to the *B. fragilis* group exhibits cephalosporinase, rather than penicillinase, activity and consequently these strains are more resistant to cephalosporins than to penicillins. Nevertheless, the activity of penicillins is reduced to the extent of rendering them unreliable for therapeutic use and, contrary to popular belief, there is little difference between penicillins (including the acylureido compounds, azlocillin, mezlocillin and piperacillin) in this respect (Sutter and Finegold, 1976; Elwell and Wise, 1982).

In addition to constitutive β-lactamase activity, *Bacteroides* strains may elaborate enzymes which rapidly inactivate penicillins and even 'β-lactamase stable' compounds like cefoxitin, latamoxef and imipenem (Sato et al., 1982; Cuchural et al., 1983; Yotsuji et al., 1983; Eley and Greenwood, 1986a,b). Such strains are relatively uncommon at present: of 78 clinical isolates of the *B. fragilis* group examined in a study in Nottingham, 13 (17 per cent) were judged to produce elevated levels of β-lactamase, but only three of these strains inactivated cefoxitin or latamoxef. Two of these three strains rapidly inactivated imipenem and were not susceptible to β-lactamase inhibitors (Eley and Greenwood, 1986b).

Whether resistance to antimicrobial agents among anaerobes will remain at its present, relatively low, level or will continue to increase is hard to predict. However, since resistance to a number of these agents (but not metronidazole) can be mediated by transferable plasmids (Malamy and Tally, 1981) there is clearly no cause for complacency.

Sensitivity testing of anaerobes

Testing the susceptibility of anaerobes to antimicrobial agents poses methodological problems that are not encountered with aerobes. Anaerobic bacteria may require complex media for growth (many organisms require blood for optimal growth) and they also require an incubation environment that includes CO_2 (Watt, 1973). Although many anaerobes will grow well on solid media in 24 or 48 h, some species require 5–7 days for visible surface growth. As Phillips and Warren (1978) pointed out 'Conditions that are best for bacterial growth may be unsuitable for sensitivity testing'. Methods for susceptibility testing of anaerobes can be divided into five groups:
  disc susceptibility tests;
  broth disc tests;
  agar dilution tests;
  breakpoint methods;
  microsystems.
Many differences in methodology have been described within each group by different centres. This has led to a lack of standardization and to differing results being reported from different centres.

Disc tests

Sutter *et al.* (1972) first stressed the need for standardized methods for anaerobes, but there is still no general agreement on what these should be. Phillips and Warren (1978) described four different methods of seeding plates for disc tests:

- application of the inoculum taken from one plate to the surface of the test plates by a swab;
- preparation of 'pour-plates';
- flood-seeding the plate;
- suspension of organisms in nutrient broth to provide an inoculum that, when applied to the surface with a swab, produces semiconfluent growth.

In each of these methods the antibiotic discs are applied to the surface of the plates and the zones of inhibition read after incubation for 24 or 48 h in anaerobic conditions. In Edinburgh, Watt and colleagues (unpublished method) use an adaptation of the Stokes' method (Stokes and Ridgway, 1980) in which a suitable inoculum of the test organism is applied to the inner aspect of the plate and a suitable control organism (see page 423) applied concentrically in the outer surface. The discs are placed at the junction of the two inocula and the zone of inhibition of the test strain is compared with that of the control.

Broth-disc methods

The lack of standardization of disc diffusion methods has led to the development of 'broth-disc' methods, in which a disc containing a chosen amount of the antibiotic is placed in a tube of a suitable liquid medium (e.g., thioglycollate medium) and the tube is then seeded with the test strain. Growth of the organism indicates resistance to the antibiotic; no growth (with good growth in the control tube) is taken to indicate that the organism is sensitive. However, there is little standardization of medium composition, nor of the amount of antibiotic in the discs, so that variation from centre to centre is inevitable. One method commonly used in the USA involves aerobically incubated thioglycollate broth (Kurxynski *et al.*, 1976). The results of broth-disc methods appear to correlate well with those of agar dilution methods (Zabransky *et al.*, 1986).

Agar dilution methods

Because of the variability of other methods, some form of agar dilution technique is usually used as a reference standard for those other methods. Several variations on the theme have been described; in the USA many investigators use the NCCLS reference method (Rosenblatt, 1986) in which a turbidimetrically standardized inoculum is seeded on to Wilkins–Chalgren agar with a multiple inoculator. This reference method has the disadvantage that the agar, used without added blood, will not support good surface growth of demanding anaerobes. Investigators in the UK often use an agar (usually Diagnostic Sensitivity Test agar) with 5–10 per cent horse blood (Phillips and Warren, 1978). However, various other media have been used by other investigators.

Break-point methods

Little attention has so far been paid to the use of break-point susceptibility testing methods for anaerobes. Hoffler and Pulverer (1984) described a break-point method for anaerobes. They used an abbreviated form of the NCCLS agar dilution method in which the strains were tested for susceptibility to two or three concentrations of each antibiotic in agar. The concentrations were achieved with tablets of the antibiotics concerned (Adatabs, Mast Laboratories, Liverpool) in Wilkins–Chalgren agar. The concentrations used were chosen by the authors and so far there are no agreed break-point standards.

Microsystems

Several companies have produced microsystems, usually plastic trays with wells on to which varying amounts of test antibiotics have been dried. The wells are seeded with a suitable dilution of the test strain, each well receiving the same inoculum so that a range of final concentrations of the antibiotics is achieved. The plates are sealed and incubated anaerobically. Such methods give reasonable results for non-demanding anaerobes (Polomski *et al.*, 1983) but growth of fastidious organisms may be inconsistent in the small volumes of liquid media used.

Variables affecting sensitivity testing of anaerobes

Susceptibility testing of anaerobes can be affected by many factors. The most important of these are discussed below.

Medium composition

A wide variety of media have been used for susceptibility testing of anaerobes, especially in the USA. The NCCLS reference agar dilution method specifies Wilkins–Chalgren agar, but this medium has given variable results in collaborative studies (see Rosenblatt, 1986 for a discussion of this point); other media used have included Schaedler agar and Brucella agar. In the UK most investigators use Diagnostic Sensitivity Test agar. In a study of a range of solid media, Watt and Brown (1985) found that the surface growth of test anaerobes varied considerably from agar to agar—this may explain the variable results obtained in different centres. There is similar variation in the types of broth media used (Rosenblatt, 1986), these have included Wilkins–Chalgren, Schaedler, thioglycollate and brain–heart infusion broths.

Inoculum

The results of sensitivity tests for anaerobes are affected by inoculum size, especially in the case of β-lactam antibiotics, so that control of inoculum size is important. Inocula for sensitivity testing can be prepared in various ways, involving either seeding of broth with colonies from a blood agar plate and overnight incubation, or direct testing with isolated colonies that are suspended in a diluent and used immediately. In either case the suspensions are diluted to a turbidimetric standard, usually a 0.5 McFarland standard. Swenson and Thornsberry (1984) studied several methods of inoculum preparation, measuring the final inoculum for each method by means of viable counts. They showed that different organisms grew at different rates in different media and that a given turbidimetric standard would yield different viable counts for different organisms. They recommended preparation of inocula directly from solid media, with a suitable dilution to give inoculum sizes of about 10^5 cfu. (This is a convenient standard but sometimes it may not mimic the larger number of organisms found in infected lesions.) One alternative is to perform a range of viable counts for a range of species and, from these, to decide on an appropriate dilution of 24 or 48 h broth cultures to give standardized inocula of about 10^5 cfu.

Medium pH and incubation environment

The studies of Ingham et al. (1970) showed that the presence of CO_2 in the incubation environment for anaerobes led to a fall in the surface pH of blood agar plates after incubation. However, Watt and Brown (1975) showed that a larger, variable fall in surface pH was produced by surface growth of the test organism. Subsequent studies (Watt and Brown, 1985) confirmed these initial findings, showing that B. fragilis produced dramatic falls of 2 pH units or more. In a survey of different culture media it was clear that the fall in pH varied between different test media and different test strains, the fall being less in the case of clostridia and virtually nil in the case of anaerobic cocci. Efforts to control the pH by the use of buffers or by the design of a new medium were unsuccessful; in the latter case a medium on which the pH remained constant gave very poor surface growth.

It is clear that, although the gaseous environment is important, changes in surface pH due to bacterial growth may be a larger, more important uncontrolled variable in the testing of anaerobes, especially in tests against pH sensitive antibiotics such as erythromycin.

Presence or absence of blood

Many organisms give optimal surface growth in the presence of blood (usually 5–10 per cent horse blood). The inclusion of blood introduces additional variables. The binding of antibiotics to serum proteins, the presence of antibiotic antagonists and the possibility of antibacterial substances in the blood must all be borne in mind, but the variability introduced by the presence of blood has to be balanced against the variability of surface growth (especially of demanding strains) on media without blood. Some laboratories use whole blood while others use lysed blood. We have found that saponin-lysed blood increases the susceptibility of test strains to antibiotics such as erythromycin (Watt and

Brown, unpublished observations)—lysis by 'freezing and thawing' may be more reliable.

Jar variation

It has been established that when replicate inocula of demanding organisms on blood agar media are incubated in a series of anaerobic jars, the recovery of such organisms on solid media varies from jar to jar, even when all of the jars are functioning correctly, are well maintained and when careful anaerobic procedures are used (Watt et al., 1973). Such 'jar variation' has been observed in the case of sensitivity testing of anaerobes, notably in the case of erythromycin and clindamycin (Watt and Brown, unpublished observations). The sizes of the inhibition zones are constant within a given jar, but vary between jars. Such variation has been little studied but may be an important variable in susceptibility testing. It can be prevented if test plates are incubated in an anaerobic incubator or anaerobic cabinet.

Control organisms

There are no agreed control organisms for susceptibility testing of anaerobes in the UK. In the USA a list of control and reference strains has been published (Rosenblatt, 1986), this list includes gram-negative and gram-positive anaerobes. In the UK, many investigators have used a strain of *B. fragilis* (often strain NCTC 9343) as a control strain for all anaerobic sensitivity tests, but clearly it is important to control 'like with like', both in terms of the type of organism and in the speed of growth (i.e., one should not control tests of a *Fusobacterium* sp. with a *B. fragilis* strain. A minimum collection of control strains would include the following:

B. fragilis;
B. melaninogenicus or *B. asaccharolyticus*;
C. perfringens;
Pstr. anaerobius.

Relationship of susceptibility testing to therapy

The question is often asked: 'Why test the sensitivity of anaerobes to antimicrobials?' It is said that the susceptibility patterns are predictable, that clinicians pay little attention to them and that, in any case, the variation in methodologies is such as to make the validity of any results doubtful. These arguments can be countered as follows:
1. Although no clinically important resistance to metronidazole has been demonstrated in obligate anaerobes, susceptibility patterns for other antimicrobials are not predictable. Also, resistance to clindamycin is increasing (Rosenblatt, 1986). It is important to monitor changes in susceptibility patterns;
2. Clinicians are increasingly aware of the importance of anaerobes in human infections and take account of them in their choice of antibiotic therapy. However, many clinicians are unaware that some antibiotics are inactive against anaerobes and will not provide 'anti-anaerobe' cover in mixed infections;
3. There is good evidence that the results of *in vitro* susceptibility tests correlate well with clinical outcome, both in man (Gorbach and Bartlett, 1974) and in experimental models of anaerobic infections (Onderdonk et al., 1976).

However, there is still considerable debate about the general approach to sensitivity testing of anaerobic bacteria in the routine diagnostic laboratory. With most aerobic bacteria it is generally accepted that all significant isolates should have their sensitivity to an appropriate range of antimicrobials determined as soon as possible; sometimes the initial specimen (e.g., pus or urine) is used as the inoculum for a 'direct' sensitivity test (Stokes and Ridgway, 1980). This is necessary because the sensitivity of many aerobes to the commonly used antimicrobials is not predictable and resistance is an ever changing problem for antimicrobial therapy. It is also feasible, because simple, reliable and well controlled methods are available for testing aerobes. Moreover, most aerobes grow sufficiently quickly under laboratory conditions for results to be available the next day. However, the constraints on testing anaerobes, outlined above, may demand a different approach and the question should be 'need all anaerobic isolates be tested—and, if so, when?'

As has been shown above, the sensitivity patterns of the commonly isolated anaerobic bacteria are often much more predictable than those of the common aerobes. For practical purposes, anaerobes can be assumed to be sensitive to the 5-nitroimidazoles, and particularly to metronidazole. The clostridia, anaerobic cocci and non-*B. fragilis* group of bacteroides are generally sensitive to benzylpenicillin, and resistance to chloramphenicol and clindamycin is rare. Therefore, the need for immediate sensitivity tests is less and therapy can be started with confidence once the initial isolations have been made, or even on a strong suspicion of an anaerobic infection.

Furthermore, immediate tests for all anaerobic isolates are less feasible than with aerobes. The problems associated with sensitivity tests on anaerobes have been described above. If a disc sensitivity test method based on the Stokes' method, or if the broth-disc method is used, individual tests can be set up as isolates become available, but the whole test system needs careful control, as we have shown. However, break-point or agar-dilution methods are designed for testing batches of similar organisms and, with anaerobes, there are fewer isolates to test each day and they are likely to have more widely different requirements for media, growth conditions and length of incubation. There are rarely sufficient similar isolates in a day to provide reasonable batches of the different groups of anaerobic bacteria to test.

Even if immediate tests are set up on all significant isolates the results will rarely affect therapy. Many isolates will need more than 48 h for isolation and sensitivity testing to be completed, and by the time results are available the empirical, but usually reliable, therapy will be well established.

It would, perhaps, be foolhardy to deduce from this that sensitivity tests on anaerobic isolates need never be done, because sensitivity patterns in bacteria do change (O'Brien et al., 1986) and the laboratory needs to be able to monitor this. However, it could be suggested, especially in laboratories that do not have large numbers of anaerobes to test each day, that isolates should be held, collected and then tested in batches. The frequency of batches would depend upon the number of isolates to be tested. This approach would allow greater standardization and control of the test methods to achieve greater reliability and also enable the general pattern of sensitivity to be monitored.

Treatment of infections due to anaerobes

The decision to initiate treatment directed at anaerobes is based upon one of two premises:

1. The involvement of a particular anaerobe (e.g., *C. perfringens* or *C. tetani*) may have been proved by bacteriological investigation or strongly suspected on clinical grounds;
2. The infection is at a site where the involvement of anaerobic bacteria is considered likely.

In the first instance treatment is chosen specifically for the organism, whereas in the second case a range of possible pathogens must be covered; this range will depend upon the site of infection.

Treatment of specific anaerobic infections

Clostridial infections

In most clostridial infections treatment is geared to neutralization of exotoxin with wound toilet where appropriate (e.g., in gas gangrene); antibiotics are used to prevent further toxin production rather than to treat the established infection. In most clostridial diseases penicillin is the drug of choice, usually benzylpenicillin given by injection. However, a study comparing penicillin with metronidazole in the treatment of moderately severe tetanus showed that patients in the metronidazole group had a significantly better clinical outcome (Ahmadysah and Salim, 1985). Hyperbaric oxygen therapy is discussed elsewhere (Chapter 18).

In the case of diarrhoeal disease due to *C. difficile*, oral administration of vancomycin is the preferred therapy, with metronidazole as an alternative.

Infections due to non-sporing anaerobes

In infections due to a single species most clinicians in the UK favour the use of metronidazole given orally, by suppository or systemically, depending on the condition of the patient. The drug has very good tissue penetration and is rapidly bactericidal. Resistance to metronidazole is not a clinical problem, unlike the situation with clindamycin in the USA, where its widespread use as the drug of choice for anaerobic infections (metronidazole was not licensed for use in anaerobic infections in the USA until the mid-1980s) has led to an increasing problem of resistance among *Bacteroides* spp.

However, most infections with non-sporing anaerobes are polymicrobial and often require a combination of surgery (sometimes surgery alone is enough) with antibiotic therapy directed against both the aerobic and anaerobic components of the infection. Again, metronidazole is the drug of choice in the UK against the anaerobes, but has to be supplemented with an appropriate 'anti-aerobe' drug such as one of the cephalosporins or an aminoglycoside, as it has no activity against aerobes. Attention has turned in recent years to the possible use of 'broad-spectrum' compounds that could be used as a single agent in polymicrobial infections. The cephamycins (notably cefoxitin and cefotetan), imipenem and clavulanate-potentiated amoxycillin (Augmentin) all have good broad-spectrum activity *in vitro* but there have been few, if any, comparative trials of their use in the treatment, as opposed to prophylaxis, of anaerobic infections.

Treatment based upon site of infections

For consideration of treatment, anaerobic infections may be divided into four groups: abdominal; genitourinary; head, neck and lungs; soft-tissue.

Abdominal infections

Most of these infections are polymicrobial and the most common anaerobic species isolated is *B. fragilis*. Metronidazole is the drug of choice in this situation, usually combined with an aminoglycoside or cephalosporin for activity against the enterobacteria. None of the β-lactam agents is reliably effective against the *B. fragilis* group.

Genitourinary infections

Deep infections of the female genital tract (uterine sepsis, tubo-ovarian and pelvic abscesses) must be treated in the same way as abdominal infections. They are usually polymicrobial and although the vaginal anaerobes are mainly bacteroides of the melaninogenicus–oralis group or are anaerobic cocci, many of which are sensitive to penicillin and other β-lactam agents, about half of the *Bacteroides* isolates are *B. fragilis*. Vaginal infection (vaginosis) responds well to treatment with various agents (metronidazole, macrolide antibiotics, clavulanate-potentiated amoxycillin) that break the apparently synergic relationship between anaerobes, *G. vaginalis* and *Mobiluncus* spp.

Superficial anaerobic infections of the genitalia in men or women usually respond to local cleansing and systemic treatment with metronidazole or other antibiotics may promote healing in the more severe cases of ulceration and cellulitis.

Head, neck and lungs

Penicillin was used effectively for many years as the principal treatment for acute ulcerative gingivitis (Vincent's infection) although metronidazole is now probably the first choice for this and other types of periodontal disease. The antimicrobials may be given systemically, but much interest now centres on the use of small strips of acrylic, impregnated with drug, that are inserted into the deep pockets and provide a sustained release of the drug just where it is needed (Addy, 1988).

The other group of infections in this area that demand prompt and effective therapy are brain abscesses (and the chronic otitis media or sinusitis from which many originate) and lung abscesses. The mixed flora of these abscesses is derived from the upper respiratory tract or mouth and most of the anaerobes that are found in almost all of these abscesses are bacteroides of the melaninogenicus–oralis group and anaerobic cocci. However, an increasing proportion of these bacteroides are penicillin resistant and *B. fragilis* is also a fairly common component of the mixtures. Therefore, metronidazole is usually included in the combination of drugs used. Chloramphenicol is also useful, particularly in brain abscesses, because of its excellent penetration into CSF and brain tissue and because it is effective against almost all the anaerobes found in these infections (Ingham *et al.*, 1977; and see Chapter 16).

The rare, but often severe, form of tonsillitis (with or without bacteraemia) caused by *Fusobacterium necrophorum* responds to treatment with most antimicrobials active against anaerobes. *F. necrophorum* is very sensitive to β-lactam agents and the drugs of choice are metronidazole or benzylpenicillin.

Soft-tissue infections

The main aspect of treatment of most of these infections is adequate drainage and debridement. This has been the mainstay of the management of pressure sores (decubitus ulcers), varicose ulcers and peripheral gangrenous lesions. However, anaerobic bacteria are found in the depths of most of these lesions where continuing tissue damage occurs and there is some evidence that systemic treatment with metronidazole, or other anti-anaerobic agents, may promote faster healing of these lesions (Sapico et al., 1986).

Prophylactic use of antimicrobial agents

The rational approach to the use of antimicrobial drugs is generally in the treatment of infection that is bacteriologically proven, or at least strongly suspected on clinical grounds. The use of antimicrobials in the prevention of infection has proved useful in only a small number of well defined circumstances. Two of these are in the field of anaerobe infections:

> the prevention of postoperative wound infections after intestinal or gynaecological surgery;
> the prevention of clostridial infections.

The realization of the role of anaerobes, particularly *B. fragilis*, in postoperative infections after intestinal surgery (Chapter 13) brought with it a rational and successful new approach to antimicrobial prophylaxis in abdominal surgery. Antibiotic prophylaxis in abdominal surgery had been widely tried by the mid-1970s but it had been directed principally against the enterobacteria and had used agents that had little or no activity against anaerobes; it had been depressingly unsuccessful. Some agents had been used systemically others, such as neomycin, had been given by mouth in an attempt at preoperative gut sterilization. Mechanical cleansing of the bowel remains important but the emphasis in antimicrobial prophylaxis is on achieving high tissue levels of the agent at the time of surgery, i.e., at the time of contamination, and to use an agent effective against the principal pathogens, i.e., the anaerobes. Clindamycin was widely and successfully used in the mid-1970s but it carried an unacceptably high incidence of bowel disturbance including, at its most severe, pseudomembranous colitis (Lusk et al., 1977). Since then metronidazole has become established as the key agent for this type of prophylaxis. In a series of studies Willis and his co-workers (Willis et al., 1976, 1977 and Chapter 13) showed that metronidazole, given alone at the start of an abdominal operation and continued for 24 h or at most 48 h afterwards, reduced the incidence of postoperative sepsis to <5 per cent in appendicectomies, colectomies and similar operations. They found that the inclusion of an agent active against aerobes was unnecessary. In a parallel study (Willis et al., 1975) they obtained the same results with the same regimen in gynaecological surgery.

Although the importance of anti-anaerobic prophylaxis has been widely accepted many workers prefer to include an agent active against aerobes in their prophylactic regimens. This is usually a cephalosporin or an aminoglycoside. The introduction of new cephalosporins and other β-lactam agents with some activity against anaerobes, as well as against aerobes, led to the suggestion that these drugs might be used as single agents. Some studies with these agents have given results similar to those obtained with metronidazole alone, but not to any significant advantage (see Chapter 13).

This approach to prophylaxis has been a significant contribution to safer and more successful surgery.

Prevention of clostridial infections

Clostridial myonecrosis (gas gangrene) and tetanus result from growth and toxin production by *C. perfringens* or *C. tetani* in the depths of a dirty wound. Both clostridia are very sensitive to benzylpenicillin and this drug is widely used to prevent these infections. An injection of benzylpenicillin combined with a long acting penicillin is often given to patients with wounds that are at high risk of clostridial infection, so that if spores germinate the vegetative clostridia are immediately killed by the penicillin. Operation on the upper leg, e.g., amputations and limb operations, are also at risk of clostridial infection and benzylpenicillin or metronidazole is again used prophylactically from the start of the operation and for 48 h afterwards. For a more detailed account of tetanus prophylaxis, see Chapter 23.

References

Addy M. (1988). The use and delivery of antimicrobial agents in the management of periodontal disease. In *Anaerobes today*, pp. 187–193. Edited by Hardie J M, Borriello S P. John Wiley & Sons, Chichester.

Ahmadsyah I, Salim A. (1985). Treatment of tetanus: an open study to compare the efficacy of procaine penicillin and metronidazole. *British Medical Journal* **291**: 648–650.

Appelbaum P C, Chatterton S A. (1978). Susceptibility of

anaerobic bacteria to ten antimicrobial agents. *Antimicrobial Agents and Chemotherapy* 14: 371–376.

Britz M L. (1981). Resistance to chloramphenicol and metronidazole in anaerobic bacteria. *Journal of Antimicrobial Chemotherapy* 8 Suppl D: 49–57.

Britz M L, Wilkinson R G. (1979). Isolation and properties of metronidazole-resistant mutants of *Bacteroides fragilis*. *Antimicrobial Agents and Chemotherapy* 16: 19–27.

Bryan L E, Kwan S. (1981). Mechanisms of aminoglycoside resistance of anaerobic bacteria and facultative bacteria grown anaerobically. *Journal of Antimicrobial Chemotherapy* 8 Suppl D: 1–8.

Burdon D W. (1984). Treatment of pseudomembranous colitis and antibiotic-associated diarrhoea. *Journal of Antimicrobial Chemotherapy* 14 Suppl D: 103–109.

Bushby S R M. (1973). Trimethoprim-sulphamethoxazole: *in vitro* microbiological aspects. *Journal of Infectious Diseases* 128 Suppl: S442–S462.

Chow A W, Patten V, Guze L B. (1975). Comparative susceptibility of anaerobic bacteria to minocycline, doxycycline and tetracycline. *Antimicrobial Agents and Chemotherapy* 7: 46–49.

Cormick N A *et al*. (1990). The centimicrobial susceptibility patterns of the *Bacteroides fragilis* group in the United States, 1987. *Journal of Antimicrobial Chemotherapy* 25: 1011–1019.

Cuchural G J, Hurlbut S, Malamy M H, Tally F P. (1988a). Permeability to β-lactams in *Bacteroides fragilis*. *Journal of Antimicrobial Chemotherapy* 22: 785–790.

Cuchural G J, Tally F P, Jacobus N V, Marsh P K, Mayhew J W. (1983). Cefoxitin inactivation by *Bacteroides fragilis*. *Antimicrobial Agents and Chemotherapy* 24: 936–940.

Cuchural G J *et al*. (1988b). Susceptibility of the *Bacteroides fragilis* group in the United States: analysis by site of isolation. *Antimicrobial Agents and Chemotherapy* 32: 717–722.

Easmon C S F, Ison C A, Kaye C M, Timewell R M, Dawson S G. (1982). Pharmacokinetics of metronidazole and its principal metabolites and their activity against *Gardnerella vaginalis*. *British Journal of Venereal Diseases* 58: 246–249.

Edson R S, Rosenblatt J E, Lee D T, McVey E A. (1982). Recent experience with antimicrobial susceptibility of anaerobic bacteria. Increasing resistance to penicillin. *Mayo Clinic Proceedings* 57: 737–741.

Edwards D I. (1979). Mechanism of antimicrobial action of metronidazole. *Journal of Antimicrobial Chemotherapy* 5: 499–502.

Eley A, Greenwood D. (1984). Variations in susceptibility to latamoxef (moxalactam) and cefoxitin within the *Bacteroides fragilis* group. *Journal of Antimicrobial Chemotherapy* 13: 245–255.

Eley A, Greenwood D. (1986a). Beta-lactamases of type culture strains of the *Bacteroides fragilis* group and of strains that hydrolyse cefoxitin, latamoxef and imipenem. *Journal of Medical Microbiology* 21: 49–57.

Eley A, Greenwood D. (1986b). Characterization of β-lactamases in clinical isolates of *Bacteroides*. *Journal of Antimicrobial Chemotherapy* 18: 325–333.

Elwell A, Wise R. (1982). The β-lactamase stability of acylureidopenicillins. *Journal of Antimicrobial Chemotherapy* 10: 560–561.

Fekety R, Silva J, Buggy B, Deery H G. (1984). Treatment of antibiotic-associated colitis with vancomycin. *Journal of Antimicrobial Chemotherapy* 14 Suppl D: 97–102.

Felmingham D *et al*. (1985). Comparative *in-vitro* studies with 4-quinolone antimicrobials. *Drugs under Experimental and Clinical Research* 11: 317–329.

Finegold S M. (1984). Susceptibility of anaerobic bacteria to thiamphenicol. *Sexually Transmitted Diseases* 11 Suppl 4: 430–431.

Finegold S M, Wexler H M. (1988). Therapeutic implications of bacteriologic findings in mixed aerobic-anaerobic infections. *Antimicrobial agents and Chemotherapy* 32: 611–616.

Glupczynski Y, Labbe M, Crokaert F, Yourassowsky E. (1984). *In vitro* activity of teicoplanin and vancomycin against anaerobes. *European Journal of Clinical Microbiology* 3: 50–51.

Goldstein E J C, Citron D M. (1985). Comparative activity of the quinolones against anaerobic bacteria isolated at community hospitals. *Antimicrobial Agents and Chemotherapy* 27: 657–659.

Goldstein E J C, Citron D M. (1986). Comparative *in vitro* activities of amoxycillin-clavulanic acid and imipenem against anaerobic bacteria isolated from community hospitals. *Antimicrobial Agents and Chemotherapy* 29: 158–160.

Gorbach S L, Bartlett J G. (1974). Anaerobic infections. *New England Journal of Medicine* 390: 1177–1184.

Greenwood D, Palfreyman J. (1987). Comparative activity of LY146032 against anaerobic cocci. *European Journal of Clinical Microbiology* 6: 682–684.

Greenwood D, Palfreyman J, Eley A, Clarry T. (1988). Activity of teicoplanin against gram-negative anaerobes. *Journal of Antimicrobial Chemotherapy* 21: 500–501.

Gurthwith M J, Rabin H R, Love K and the Cooperative Antibiotic Diarrhea Study Group. (1977). Diarrhea associated with clindamycin and ampicillin therapy: preliminary results of a cooperative study. *Journal of Infectious Diseases* 135 Suppl: S104–S110.

Harvey K J, Miles H, Hurse A, Carson M. (1981). In-vitro activity of erythromycin against anaerobic organisms. *Medical Journal of Australia* 1: 474–475.

Hoffler V, Pulverer G. (1984). Antibiotic sensitivity testing of anaerobic bacteria by the breakpoint method *Chemotherapy* 30: 392–397.

Ingham H R, Selkon J B, Codd A A, Hale J H. (1970). The effect of carbon dioxide on the sensitivity of *Bacteroides fragilis* to certain antibiotics *in vitro*. *Journal of Clinical Pathology* 23: 259–261.

Ingham H R, Selkon J B, Roxby C M. (1977). Bacteriological study of otogenic brain abscesses: chemotherapeutic role of metronidazole. *British Medical Journal* **2**: 991–993.

Jenkins S G, Birk R J, Zabransky R J. (1982). Differences in susceptibilities of species of the *Bacteroides fragilis* group to several β-lactam antibiotics: indole production as an indicator of resistance. *Antimicrobial Agents and Chemotherapy* **22**: 628–634.

Jokipii L, Jokipii A M M. (1988). Comparative evaluation of the 2-methyl-5-nitroimidazole compounds, dimetridazole, metronidazole, secnidazole, ornidazole, tinidazole, carnidazole and panidazole against *Bacteroides fragilis* and other bacteria of the *Bacteroides fragilis* group. *Antimicrobial Agents and Chemotherapy* **28**: 561–564.

Jones B M, Geary I, Alawattegama A B, Kinghorn G R, Duerden B I. (1985). *In-vitro* and *in-vivo* activity of metronidazole against *Gardnerella vaginalis*, *Bacteroides* spp. and *Mobiluncus* spp. in bacterial vaginosis. *Journal of Antimicrobial Chemotherapy* **16**: 189–197.

Kresken M, Wiedemann B. (1986). Development of resistance in the past decade in central Europe. *Journal of Antimicrobial Chemotherapy* **18** Suppl C: 235–242.

Kurxynski T A, Yrios J W, Helstad A G, Field C R. (1976). Aerobically incubated thioglycollate broth disk method for antibiotic susceptibility testing of anaerobes. *Antimicrobial Agents and Chemotherapy* **10**: 727–732.

Lamothe F, Auger F, Lacroix J-M. (1984). Effect of clavulanic acid on the activities of ten β-lactam agents against members of the *Bacteroides fragilis* group. *Antimicrobial Agents and Chemotherapy* **25**: 662–665.

Lusk R H *et al.* (1977). Gastrointestinal side effects of clindamycin and ampicillin therapy. *Journal of Infectious Diseases* **135** Suppl: S111–S119.

Malamy M H, Tally F P. (1981). Mechanisms of drug-resistance transfer in *Bacteroides fragilis*. *Journal of Antimicrobial Chemotherapy* **8** Suppl D: 59–75.

Marrie T J, Haldane E V, Swantee C A, Kerr E A. (1981). Susceptibility of anaerobic bacteria to nine antimicrobial agents and demonstration of decreased susceptibility of *Clostridium perfringens* to penicillin. *Antimicrobial Agents and Chemotherapy* **19**: 51–55.

Martin D A, Sanders C V, Marier R L. (1982). *N*-formimidoyl thienamycin (MK0787): *in vitro* activity against anaerobic bacteria. *Antimicrobial Agents and Chemotherapy* **21**: 168–169.

Muto Y, Bando K, Watanabe K, Katoh N, Ueno K. (1989). Macrolide accumulation by *Bacteroides fragilis* ATCC 25285. *Antimicrobial Agents and Chemotherapy* **33**: 242–244.

Nord C E. (1986). Mechanisms of β-lactam resistance in anaerobic bacteria. *Reviews of Infectious Diseases* **8** Suppl 5: S543–S548.

Nord C E, Olsson-Liljequist B 1981 Resistance to β-lactam antibiotics in *Bacteroides* species. *Journal of Antimicrobial Chemotherapy* **8** Suppl D: 33–42.

O'Brien T F and the International Survey of Antibiotic Resistance Group. (1986). Resistance to antibiotics at medical centres in different parts of the world. *Journal of Antimicrobial Chemotherapy* **18** Suppl C: 243–253.

Onderdonk A B, Bartlett J G, Louie T J, Sullivan-Seigler N M, Gorbach S L. (1976). Microbial synergy in experimental intra-abdominal abscess. *Infection and Immunity* **13**: 22–26.

Phillips I, King A, Shannon K. (1988). *In vitro* properties of the quinolones. In *The quinolones*, pp. 83–117. Edited by V T Andriole. Academic Press, London.

Phillips I, Tally F P. (eds) (1981). Resistance in anaerobic bacteria. *Journal of Antimicrobial Chemotherapy* **8** Suppl D: 132 pp.

Phillips I, Warren C. (1974). Susceptibility of *Bacteroides fragilis* to trimethoprim and sulphamethoxazole. *Lancet* **i**: 827–829.

Phillips I, Warren C. (1978). Anaerobic bacteria. In *Laboratory methods in antimicrobial chemotherapy*. Edited by Reeves D S *et al*. Churchill Livingstone, Edinburgh.

Polomski J C, Bauer S H, McClatchey K D. (1983). Micromedia Systems anaerobe panel versus broth disk method in anaerobe antimicrobial testing. *Journal of Clinical Microbiology* **17**: 949–952.

Ralph E D. (1978). The bactericidal activity of nitrofurantoin and metronidazole against anaerobic bacteria. *Journal of Antimicrobial Chemotherapy* **4**: 177–184.

Reeves D S. (1986). The effect of quinolone antibacterials on the gastrointestinal flora compared with that of other antimicrobials. *Journal of Antimicrobial Chemotherapy* **18** Suppl D: 89–102.

Retsema J *et al*. (1987). Spectrum and mode of action of azithromycin (CP-62,993), a new 15-membered-ring macrolide with improved potency against gram-negative organisms. *Antimicrobial Agents and Chemotherapy* **31**: 1939–1947.

Reynolds A V, Hamilton-Miller J M T, Brumfitt W. (1975). A comparison of the *in vitro* activity of metronidazole, tinidazole and nimorazole against gram-negative anaerobic bacilli. *Journal of Clinical Pathology* **28**: 775–778.

Robbins M, Marais R, Felmingham D, Ridgway G L. (1987). The *in-vitro* activity of doxycycline and minocycline against anaerobic bacteria. *Journal of Antimicrobial Chemotherapy* **20**: 379–382.

Rolfe R D, Finegold S M. (1981). Comparative *in vitro* activity of new beta-lactam antibiotics against anaerobic bacteria. *Antimicrobial Agents and Chemotherapy* **20**: 600–609.

Rosenblatt J E. (1986). Antimicrobial susceptibility testing of anaerobes. In *Antibiotics in laboratory medicine*, 2nd edn. pp. 168–174. Edited by Lorian V. Williams & Wilkins, Baltimore.

Sapico F L *et al*. (1986). Quantitative microbiology of pressure sores in different stages of healing. *Diagnostic Microbiology and Infectious Diseases* **5**: 31–38.

Sato K, Matsuura Y, Inoue M, Mitsuhashi S. (1982). Properties of a new penicillinase type produced by

Bacteroides fragilis. Antimicrobial Agents and Chemotherapy **22**: 579–584.

Shinn D L S. (1962). Metronidazole in acute ulcerative gingivitis. *Lancet* **1**: 1191.

Sindar P, Britz M L, Wilkinson R G. (1982). Isolation and properties of metronidazole-resistant mutants of *Clostridium perfringens*. *Journal of Medical Microbiology* **15**: 503–509.

Sprott M S, Kearns A M. (1988). Metronidazole-resistant *Bacteroides melaninogenicus*. *Journal of Antimicrobial Chemotherapy* **22**: 951–952.

Steinkraus G E, McCarthy L R. (1979). *In vitro* activity of sodium fusidate against anaerobic bacteria. *Antimicrobial Agents and Chemotherapy* **16**: 120–122.

Stirling J, Goodwin S. (1977). Susceptibility of *Bacteroides fragilis* to fusidic acid. *Journal of Antimicrobial Chemotherapy* **3**: 522–523.

Stokes E J, Ridgway G I. (1980). Clinical bacteriology, 5th edn. p. 215. Edward Arnold, London.

Sutter V L. (1983). Frequency of occurrence and antimicrobial susceptibility of bacterial isolates from the intestinal and female genital tracts. *Reviews of Infectious Diseases* **5** Suppl 1: S84–S89.

Sutter V L, Finegold S M. (1976). Susceptibility of anaerobic bacteria to 23 antimicrobial agents. *Antimicrobial Agents and Chemotherapy* **10**: 736–752.

Sutter V L, Kwork Y-Y, Finegold S M. (1972). Standardized antimicrobial disc susceptibility testing of anaerobic bacteria. *Applied Microbiology* **23**: 268–275.

Swenson J M, Thornsberry C. (1984). Preparing inoculum for susceptibility testing of anaerobes. *Journal of Clinical Microbiology* **19**: 321–325.

Tally F P, Malamy M H. (1984). Antimicrobial resistance and resistance transfer in anaerobic bacteria: a review. *Scandinavian Journal of Gastroenterology* **19** Suppl 91: 21–30.

Tally F P, Sosa A, Jacobus N V, Malamy M H. (1981). Clindamycin resistance in *Bacteroides fragilis*. *Journal of Antimicrobial Chemotherapy* **8** Suppl D: 43–48.

Vanhoof R, Vanderlinden M P, Dierickx R, Lauwers S, Yourassowsky E, Butzler J P. (1978). Susceptibility of *Campylobacter fetus* subsp. *jejuni* to twenty-nine antimicrobial agents. *Antimicrobial Agents and Chemotherapy* **14**: 553–556.

Watt B. (1973). The significance of carbon dioxide on the growth of obligate anaerobes on solid media. *Journal of Medical Microbiology* **6**: 307–313.

Watt B, Hoare M V, Collee J G. (1973). Some variables affecting the recovery of anaerobic bacteria: a quantitative study. *Journal of General Microbiology* **77**: 447–454.

Watt B, Brown F V. (1975). Sensitivity testing of anaerobes on solid media. *Journal of Antimicrobial Chemotherapy* **1**: 440–442.

Watt B, Brown F V. (1985). Effect of the growth of anaerobic bacteria on the surface pH of solid media. *Journal of Clinical Pathology* **38**: 565–569.

Welling P G, Munro A M. (1972). The pharmacokinetics of metronidazole and tinidazole in man. *Arzneimittel-Forschung* **22**: 2128–2132.

Wexler H M, Harris B, Carter W T, Finegold S M. (1985). *In vitro* efficacy of sulbactam combined with ampicillin against anaerobic bacteria. *Antimicrobial Agents and Chemotherapy* **27**: 876–878.

Willis A T *et al.* (1975). An evaluation of metronidazole in the prophylaxis and treatment of anaerobic infections in surgical patients. *Journal of Antimicrobial Chemotherapy* **1**: 393–401.

Willis A T *et al.* (1976). Metronidazole in the prevention and treatment of bacteroides infections after appendicectomy. *British Medical Journal* **1**: 318–321.

Willis A T *et al.* (1977). Metronidazole in prevention and treatment of bacteroides infections in elective colon surgery. *British Medical Journal* **1**: 607–610.

Willis A T, Jones P H, Reilly S. (1981). Management of anaerobic infections: prevention and treatment. Research Studies Press, Letchworth.

Wise R. (1989). The development of macrolides and related compounds. *Journal of Antimicrobial Chemotherapy* **23**: 299–300.

Wise R, Andrews J M, Bedford K A. (1980). Clavulanic acid and CP-45, 899: comparison of their *in vitro* activity in combination with penicillins. *Journal of Antimicrobial Chemotherapy* **6**: 197–206.

Yotsuji A, Minami S, Inoue M, Mitsuhashi S. (1983). Properties of novel β-lactamase produced by *Bacteroides fragilis*. *Antimicrobial Agents and Chemotherapy* **24**: 925–929.

Zabransky R J, Birk R J, Kurxynski T A, Tookey K L. (1986). Predicting the susceptibility of anaerobes to cefoperazone, cefotaxime and cefoxitin with the thioglycollate broth disk procedure. *Journal of Clinical Microbiology* **24**: 181–185.

Index

Abdomen
  microflora 162–9, 199–201
  sepsis/infection 197–223, 312
    causes 209, 312
    clinical considerations 208–10
    pathogenesis 201–8
    predisposing factors 201–6
    prophylaxis 213–6
    treatment 210–3, 425
  specimen isolation from sites in 181
  wall, postoperative gas gangrene of 303
Abortion, septic (uterine sepsis), 234–5, 304, 306, 310
Abscesses
  abdominal 207, 209
  antimicrobial therapy 207, 279–80
    inefficacy 207
  brain 276–80, 425
  genitourinary 237
    pelvic 237
  lung 281–2
  oral/dental 250–5
  parapharyngeal 270–1
  peritonsillar 269–70
  retropharyngeal 270–1
  skin 289
Acetic acid, colonization resistance and 346
Acetyl CoA metabolism 27
Achlorhydria, gastic 403, 404
Acid end products of carbohydrate metabolism 193–4
Actinomyces and *Arachnia* spp. 138, 141, 143
Acne 289
Acquired immunodeficiency syndrome, intestinal spirochaetosis and 126–7
Actinobacillus spp.
  actinomycetemcomitans 256–7, 258–9
  oral and dental infection 256–7, 258–9
Actinobacteriales, classification 7, 8
Actinomyces spp. 135–50
  bordeovuli 142
  bovis 142
  bowellii 142
  classification 136–9

  culture 135, 139–40
  denticolens/denticola 139, 142
  eriksoni 171
  as gingival microflora 171
  identification 141–4
  isolation 139–41
  israelii 136, 138, 142, 144, 171, 248, 260, 261
  meyerii 142
  naeslundii 136, 138–9, 142, 171
  nature 133–6
  odontolyticus 139, 142, 171
  oral and dental infection 248, 256, 260–261
  pyogenes 142
  slackii 142
  uterine infection 236
  viscosus 136, 138–9, 142, 145–6, 171
Actinomycetaceae, classification 7
Actinomycetales, classification 7
Actinomycosis 260–1
Adenomas, colorectal 408
Adenosine triphosphate *see* ATP
Adhesins and adherence as virulence determinants 75, 201–2
  in abdominal infections 201–2
Aerobic bacteria
  anaerobic and, mixed infections 203–4
  phagocytosis, anaerobes inhibiting 203–4
Aerobic *Clostridium* spp. 88
Aesculin agar 192, 194
Agar dilution methods of antibiotic sensitivity testing 421
Agar media
  for *Clostridium* spp. 92
  pitting 191
Age, skin microflora and 288
AIDS, intestinal spirochaetosis and 126–7
Alimentary tract *see* Gastrointestinal tract
Aliphatic nitro compounds, reduction 31
Amine(s)
  secondary, production 398
  volatile, in bacterial vaginosis, tests for 232
Amino acid metabolites 396–8, 411

Aminoglycosides 418
  abdominal sepsis with use of 205
  in media 187
Amnionitis 233
Amoxycillin treatment for oral and dental infection 254–5
Anaerobic bag 183
Anaerobic chambers *see* Chambers
Anaerobic jars *see* Jars
Anaerobiospirillum spp. and diarrhoea 357
Angina, Ludwig's 260
Antagonism, bacterial 159, 289, 344–5, 346–7, *see also* Colonization resistance
Antibacterial substances, gastrointestinal tract secreting 345
Antibodies
  monoclonal, to *C.perfringens* 367–8
  staining with fluorescent 183
Antimicrobials/antibiotics 415–29, *see also specific agents*
  diarrhoea associated with use of *see* Diarrhoea
  poor/unreliable 416
  resistance to 419–20
    conjugal transfer 47–9
    plasmids conferring 42–4
    treponemal 121
  sensitivity/susceptibility testing to 420–4
Antimicrobials/antibiotics, treatment with 423, 424–6
  for abdominal infection/sepsis 210, 211–13, 214–15
    prophylactic 214–15
  abdominal infection/sepsis enhanced by 205, 206
  for abscesses *see* Abscesses
  in bacteraemia 332–3, 335–6
    prophylactic 332–3
  bacterial synergy and 204
  for botulism 391
  broad-spectrum vs narrow spectrum types 211–2
  combination 212–3
  for gas gangrene 316, 316–7
    prophylactic 316
  normal flora and colonization resistance adversely affected by 205, 344–5, 348
  for oral and dental infection 254–5, 257
  prophylactic (in general) 426
  for superficial infections 295
  susceptibility testing relating to 423–4
  for tetanus 380, 382–3
    prophylactic 382–3
  for treponematoses 120–1
Antitoxins
  botulism 390–1
  tetanus 375–7, 379–80
Appendicitis, acute, bacterial synergy and 202
Arachnia spp. 135–50

  classification 136–9
  identification 141–4
  isolation 139–41
  nature 133–6
  propionica 142
Archaebacteria, classification 8
Aromatic hydrocarbons, polycyclic 396, 409
Aromatic nitro compounds, reduction 31
Aromatic ring, cleavage 30
Aspiration (accidental fluid inhalation), in pulmonary infection 280–1
Aspiration (fluid withdrawal), in brain abscess surgery 279
Asporulales, classification 7
AT content and AT rich regions, clostridial 55
ATP
  determination, in bacteraemia 335
  generation 26–28

Babies *see* Infants
Bacilli
  Gram-negative
    in genitourinary infections 226–7
    as gingival microflora 170–1
  Gram-positive
    in genitourinary infections 226
    as gingival microflora 171
    identification 193
Bacillus subtilis
  plasmid-mediated transfer from 50–1
  transformation 50
Bacteraemia 207–8, 311–2, 324–42
  Bacteroides spp. 208, 325, 326–7, 332, 336
  Bifidobacteria spp. 330
  Capnocytophaga spp. 330
  clinical features 333
  Clostridium spp. 311–2, 327–9, 332
  Eubacterium spp. 330
  Fusobacterium spp. 325, 329–30
  incidence 324–30
  isolation and identification techniques 333–5
  mixed 330–2
  Peptococcaceae 329
  prophylaxis 332–3
  treatment 335–6
Bacteria
  antagonism/interference 159, 289, 344–5, 346–7, *see also* Colonization resistance as determinants
    in anaerobic infection predisposition 201–5
    in gas gangrene 305–6
  synergy 202–3, 204
Bacteriaceae, classification 7
Bacteriales, classification 7, 8

Bacteriocinogenic plasmids 40–1, 345
Bacteriophages 44
  colonization resistance and 345
Bacteroidaceae, classification 8
Bacteroides spp. 62–84, 326–7, 356–7
  abdominal infection 203–4
  antibiotic resistance 42–3, 47–8
  asaccharolyticus 78, 225
  in aspiration pneumonia 282
  in bacteraemia 208, 325–7, 332, 336
  bacteriocin production 40
  classification 62–3, 66–71
    revision 67, 69
  cloning 51, 52–3
  cloning vectors 49, 51
  in colonization resistance 346
  forsythus 258
  fragilis 47, 51, 69, 77–8, 171, 192, 200–4, 226, 282, 326, 332, 336, 346, 356–7
  gene transfer 44, 46–9, 51
  genetics 38–40, 42–3, 46–9, 51
  genitourinary infection 226–7, 230–2, 238–41
  as genitourinary microflora 172, 176, 225
  as gingival microflora 171
  gingivalis 77–8, 258–9
  historical background 2
  identification 64, 68, 73–4, 192
  intermedius 70, 122, 171, 258–9
  in intestinal infection 356–7
  as intestinal microflora 164–5, 168–9, 199–201
  isolation 71–3
  levii 70
  macacae 70
  melaninogenicus 69–70, 171, 192, 227, 239, 270
  multiacidus 70
  oral and dental infection 250–2, 258–9
  oralis 69–70, 227, 239
  pathogenicity and virulence 74–5, 77–8
  plasmids 39–40, 42–3
  pneumosintes 71
  splanchnicus 69
  superficial infection 291
  ureolyticus 192, 227, 240–1
  as vaginal microflora 174–5
  vulgatus 359
Bacteroides bile aesculin 194
Bag, anaerobic 183
Balanitis 238
Balanoposthitis 238
Benzylpenicillin 416
  clostridial infection prophylaxis 426
  minimum inhibitory concentrations 417
Bicozamycin agar 189
Bifidobacterium spp. 151–61, 330

  adolescentis 159
  bacteraemia 330
  bifidum 152
  breve 159
  classification 153
  ecology 157–60
  identification 154–7
  as intestinal microflora 165, 169, 199
  isolation 153–4
  nature 151–3
  pseudocatenulatum 159
Bile acids 399–400, 409–10
  faecal 409
  metabolites 399–400, 409–10
Bile salts, deconjugation/degradation 400
Biliary tract 405–6
  cancer, incidence 401
  microflora 201
Biochemical tests/characteristics
  Actinomyces and *Arachnia* spp. 141–2
  Clostridium spp. 87–8, 94–5
  Gram-negative non-sporing rods 64–5
Bioconversion 30–2
Birth, infections after giving 233–4, 304, 306, 310
Bisexual men, intestinal spirochaetosis in 126–7
Bladder
  gall-, cancer 406
  urinary, cancer see Urinary bladder cancer
Blood
  for antibiotic sensitivity tests, presence or absence of 422
  cultures
    in bacteraemia 328–9, 333–5
    Clostridia spp. 314–5, 328–9
    in gas gangrene 314–5
    new techniques 334–5
Borrelia spp. 113
  burgdorferi 113
Botulism 384–92
  action taken in outbreak of 390
  antitoxin 390–1
  laboratory identification 390
  management 389–90
    prophylaxis 391
  toxins 387–8
  wound 301, 391
Bowel see Intestine
Brachyspira spp. 115
Brain
  abscess 276–80
  infections 274, 276–80, 425
Break-point method of antibiotic sensitivity testing 421
Broth-disc method of antibiotic sensitivity testing 421

Broth enrichment *see* Enrichment cultures
Buruli ulcers, tropical and, differentiation 294
Butyric acid, colonization resistance and 346

Cabinets, anaerobic *see* Chambers
Campylobacter (Helicobacter) pylori 402–3
Cancer/malignant disease
    anaerobes in the aetiology of 395–414
    Clostridia spp. and 311–2, 327
    colorectal, antimicrobial management in 215
    incidence 401
Cancrum oris 240, 257, 295–6
Capnocytophaga spp., in bacteraemia 330
Capsule
    polysaccharides, of *C.perfringens*, serological typing 364–5
    as virulence determinant 75, 77
Carbohydrates
    acid end products *see* Acid end products
    cell wall, *Actinomyces* spp. 137–8, 143
Carbon dioxide, in anaerobic atmosphere, role 186
Carcinogen(s), production 395–400
Carcinogenesis *see* Cancer
Caries, dental 248–50
Catalase 20–1
Catalysts in anaerobic jars 185
Cavernous sinos thrombosis 275–6
Cavitation of wound, infection determined by 305
Cefoxitin
    in bacteraemia, prophylactic 336
    minimum inhibitory concentrations 417
    resistance, conjugal transfer 48
Cell
    composition/ultrastructure
        Clostridium spp. 89
        spirochaete 109–10, 118
    morphology *see* Morphology
    wall
        Actinomyces spp. 137–8, 143
        Arachnia spp. 143
        Bifidobacterium spp. 156–7
Cellulitis, clostridial 302–3, 307–8
Cellulose-degrading enzymes, *Clostridium* spp. 53–4
Central nervous system
    infection 276–80, 425
    specimen isolation from 181
    tetanus toxin effects on 376–7
Centrifugation, lysis 335
Cephalosporin 416
    for abdominal infection 210–11
Cephamycins 416
Chambers/cabinets, anaerobic, growth in 10, 185–6
    Clostridium spp. 90–1
    spirochaetes 116

Chemotherapy, antimicrobial *see* Antimicrobials
Childbirth, infections following 233–4, 306, 310
Children *see* Infants/neonates
Chloramphenicol 418
    in bacteraemia, prophylactic 336
    minimum inhibitory concentrations 417
    resistance, plasmids conferring 43
Cholecystectomy, *C.perfringens* bacteraemia following 311
Cholesteatoma 272, 273
Cholesterol metabolites 398–9
Chromatographic techniques of identification 183, 193–4
    Actinomyces and *Arachnia* spp. 143
    Clostridia spp. 315
Cladistic classification 5
Classification 4–15
    Actinomycetes 136–9
    Arachnia spp. 136–9
    Bacteroides *see Bacteroides* spp.
    Bifidobacteria 153
    Clostridium spp. 85–90, 386
    cocci 13, 100–4
    Fusobacterium spp. 62–6
    principles 4–6
    spirochaetes 111–5, 128–9
    techniques 5–6
Clindamycin 416–8
    in abdominal surgery, prophylactic 214
    in bacteraemia, prophylactic 336
    minimum inhibitory concentrations 417
    resistance
        conjugal transfer 48
        plasmids conferring 42–3, 43
Clinical material *see* Specimens
Cloning, gene 51–5
    vectors for 49–51
Clostridia, nuclear dehydrogenating 409–10
Clostridiaceae, classification 7
Clostridiales
    bacteriophages 44
    classification 7
Clostridium spp. 85–99, 299–323, 327–9, 351–6, 362–94
    absonum 95
    acetobutylicum 45, 50–1, 53
    aerobically growing 88
    antibiotic resistance 43–4, 48
    in bacteraemia 311–2, 327–9, 332
    bacteriocin production 40
    beijerinckii 346
    Bifidobacteria spp. and, interaction 159
    botulinum 2, 41, 89, 93, 95, 301, 384–92
        group I 385

group II 385
group III 385
group IV 385
type A 385–6, 391
type B 385–6, 391
type C 385–6
type D 385–6
type E 385–6
type G 385–6
butyricum 41, 358
carnis 96
in cellulitis 302–3, 307, 308
chauvoei 96
classification 85–90, 386
cloning 51–5
cloning vectors 49–50
colonizing, resistance to 343, 346–8
difficile 94, 96, 200, 226, 343, 346–8, 351–3
    non-toxigenic strain 348
fallax 96
fermentation 28
gas gangrene and *see* Gangrene
gene transfer 44–6, 48–50
genetics 38–46, 48–50
genitourinary infection 226, 233
haemolyticum 95
histolyticum 96, 308–9
historical background 1, 2
identification 93–5, 193, 313–5, 364–8, 374–5, 390
in intestinal infection 351–8, 362–71
as intestinal microflora 166–7, 200–1
isolation 90–3, 313–5, 364, 374–5
limosum 95
malenominatum 96
management of infection 315–7, 424
    prophylactic 426
novyi 93, 306, 308–9
    type A 95
    type B 95
pasteurianum 54
pathogenic 86–9, 95, 96, 299–323, 327–9, 351–6, 362–94
perfringens 2, 40–1, 46, 93, 95, 201, 226, 233, 235, 304, 306–7, 309, 311–3, 327, 332, 353–4, 357–8, 362–71
    food poisoning 362–71
    non-food poisoning 353–4
    type A 362–71
    type C 354–5, 356, 365
    type D 365
plasmids 39–42
ramosum 96
septicum 96, 309, 311–2, 328, 355–6
sordellii 95
spiroforme 96
sporogenes 28
tetani 2, 41–2, 93, 96, 373–84
    culture/identification 374–5
    occurrence 373–4
tetanomorphum 28
thermohydrosulfuricum 46
toxigenic 41–2, 88–9, 93–4, 306–7, 315, 352, 352
welchii 311
wound infection *see* Wound infection
Cocci 100–7
    classification 13, 100–4
    ecology 104–5
    Gram-negative
        classification 13
        as gingival microflora 171
        as intestinal microflora 165–6
    Gram-positive
        classification 13, 102–3
        genitourinary infection 227
        as gingival microflora 170
        as intestinal microflora 166
        as urethral microflora 225
        as vaginal microflora 173–4
    identification, provisional 191, 193
    isolation 101
    morphology/metabolism/culture 101
    nature 100–1
Colitis 351–61, *see also* Enterocolitis
    pseudomembranous 351
    ulcerative 358–9
Collapse, clostridial infection 310
Colon *see also* Intestines
    cancer *see* Colorectal cancer
    mechanical pre-operation 213
    microflora 200–201
    mucin in, *Bifidobacteria* spp. degrading 160
Colonization resistance (bacterial antagonism/resistance) 167, 343–52
    Bifidobacterium spp. in 159, 160
    to *Clostridium* spp. 343, 346–8, 351–2
    mechanisms 343–8
        direct 343–4
        indirect 344–5
    superficial sites 289
Colonopathy, tropical 127
Colony
    morphology *see* Morphology
    pigmentation 190–1
Colorectal cancer 406–10
    aetiology 408–10
    antimicrobial management in 215
    histopathology 408
    incidence 401

Communities of anaerobes, metabolic relationships 32–3
Computed tomography, brain abscesses 277–8
Conjugation 46–9
   cloning vectors introduced via 50–1
   intergeneric 48–9
   intragenic 47–8
Cooked meat broth 190
   for *Clostridium* spp. 91, 313–4
Cranial infections 274–80
p-Cresol production 396–7
Cristispira spp. 113
Crohn's disease 358–9
Cryptic plasmids 39–40
Cultural factors in colorectal cancer 407–8
Culture, anaerobic 9–10, 184–93
   Actinomyces spp. 135, 139–40
   Arachnia spp. 139–40
   in bacteraemia 333–5
   Bifidobacterium spp. 153–4
   Clostridium spp. 90–3, 313–5, 364–5, 374–5
   cocci 101
   enrichment *see* Enrichment cultures
   spirochaetes 114–6, 118–9
   Treponema spp. 114–8
Culture, tissue, for enterotoxin detection 367
Cycasin 396
Cysteine hydrochloride in media 187

Decarboxylases, as Na^+ pumps 29
Decubitus ulcers 290–1
Defence mechanisms, abdominal infections with compromised 205
4-Dehydrogenation 409
7-Dehydroxylation 409
Dental disease/infection 245–68, *see also* Periodontal disease
   pathogenesis 248
   source of organisms in 247–8
   spirochaetes in 121–4, 257, 258
Dental plaque, microflora in 170–1
Deoxycholic acid, colorectal cancer and 409
6-Deoxytalose, *Actinomyces* spp. 137–8
Dermis, anatomy 287–8
Dermogenital syndrome, necrotizing (Fournier's gangrene) 239–40, 297
Desulfotomaculum spp. as intestinal microflora 167
Diabetic ulcers 292–3
Diagnosis of anaerobic bacteria and their infection *see* Identification
Diarrhoea 351–61
   antibiotic-associated 351
      prevention by *Saccharomyces boulardii* 347–8
   C.perfringens 353–4, 363, 367

Diet *see* Nutrition
Digestive tract *see* Gastrointestinal tract
Dimethylamine production 398
Discs, for identification/diagnosis/antibiotic sensitivity testing 190, 194, 421
Disease *see* Infections
Disposable gas generator 9, 184–5
   Clostridium spp. grown in 91
DNA
   base ratios
      Bifidobacterium spp. 157
      Clostridium spp. 54–5, 89
      Treponema spp. 115
   homologies
      Actinomyces and *Arachnia* spp. 143
      Bifidobacterium spp. 157
      spirochaetal 111
   repair, in *E.coli*, *rec*-A-dependent 21
   transcription, *Clostridium* spp. 54–5
Dressings, specimen isolation from 182
Dysentery, swine 127

Ear, nose and throat infections 268–86
   antimicrobial therapy 425
Ecology (habitat etc.)
   Actinomyces spp. 134
   Arachnia spp. 134
   Bifidobacterium spp. 157–60
   coccal 104–5
   spirochaetal 110–1
Egg yolk media, *Clostridium* spp. on 92
   reactions produced on 92, 95–6
Eh *see* Redox potential
Electron acceptors 27
Electron carriers 27
Electron donors 27, 30
ELISA for enterotoxin 367
Empyema
   epidural 274–5
   subdural 274–5
   thoracic 282
Endocarditis 332–3
   treatment 336
Endometritis 235–6
Endosporaceae, classification 7
Endotoxin *see* Lipopolysaccharide
Energy conservation, free 25–9
Enrichment cultures
   for *Bifidobacterium* spp. 154, 189
   for *Clostridium* spp. 91–2, 314
   for spirochaetes 116–8
Enteritis necroticans 354–5
Enterocolitis
   neonatal necrotizing 357–8

neutropenic 355–6
Enterotoxin
  B.fragilis 356–7
  C.perfringens
    food poisoning 363, 365–8
    non-food poisoning 353–4
Enzyme(s) *see also specific (types of) enzymes*
  Bifidobacterium spp. 156
  exo- 25
  intestinal microflora 168
  oxygen toxicity-protecting 20
  oxygen-labile 19
  preformed, identification of anaerobes with 193
  as virulence determinants 75–6, 78
Enzyme-linked immunosorbent assay for enterotoxin 367
Epidermis, anatomy 287–8
Epididymitis 241
Epidural empyema 274–5
Epithelial barriers, abdominal infections with disrupted 205
Erythema test, skin, for enterotoxin 367
Erythromycin 418
  minimum inhibitory concentrations 417
  resistance, plasmids conferring 42
Escherichia coli
  B.fragilis and, mixed infection 203–4
  Bifidobacterium spp. and, interaction 159
  colonizing, resistance to 346
Ethionine 396
Eubacteria
  non-sporing, classification 7
  sporeforming, classification 7
Eubacterium spp.
  bacteraemia 330
  as gingival microflora 170
  as intestinal microflora 165–6, 169
Evacuation-replacement jars 9, 184
Exoenzymes 25
Extrachromosomal genetic elements 39–44, *see also specific types of elements*

FAB-Tween cultures *see* Fastidious anaerobe broth-Tween 80 cultures
Faeces, intestinal microflora isolated from 163–7, 200
Fallopian tubes, infection 236–7
Fascitis, anaerobic necrotizing 208–9
Fastidious anaerobic broth-Tween 80 cultures 190
  phase contrast microscopy 192
Fatty acids
  chromatography 194
  colonization resistance and 346
  polar and neutral 194
  spirochaetal/treponemal 119

  volatile and non-volatile 194
Females, genitourinary tract in
  abdominal infections and the 207
  infection 227–37
  microflora 172–6
  specimen isolation from 181
Fermentation 25–9
  Actinomyces and *Arachnia* spp. 142
  Bifidobacterium spp. 152, 155–6
  branched 27–8
  intestinal microflora 168
  linear 27
Ferredoxin
  gene 54–5
  role 29, 31
Filtration, lysis 335
Firmibacteria, classification 8
Firmicutes, classification 8
Flagella, spirochaete 109
Flame ionization detectors 193
Flora, micro- *see* Microflora
Fluorescence under ultraviolet light 183, 192
Fluorescent antibody staining 183
Food poisoning *see also* Nutrition/diet
  C.botulinum 389
  C.perfringens type A 362–71
Foreign bodies predisposing to surgical wound infection 206
Fournier's gangrene 239–40, 297
Free energy conservation 25–9
Free radical toxicity 19–22, 32
  protection against 20–2
Fructose-6-phosphate phosphoketolase, *Bifidobacterium* spp. 155–6
Fusidic acid 418
Fusobacterium spp. 62–84, 329–30
  bacteraemia 325, 329–30
  classification 62–6
  fusiformis 28
  genitourinary infection 227, 234
  as gingival microflora 171
  identification 73–4, 193
  as intestinal microflora 165
  isolation 71–3
  necrophorum 227, 234, 269, 329–30
  nucleatum 76–7
  oropharyngeal infection 269
  pathogenicity and virulence 74–7

Gallbladder cancer 406
Ganglioside receptors 377, 387
Gangrene 299–323
  Fournier's 239–40, 297
  gas (clostridial myonecrosis) 299–323

Gangrene – *cont'd*.
   classic 301–2
      clinical considerations 307–12
      endogenous 312
      genesis 304–7
      laboratory investigations 312–5
      management 315–7
      metastatic 311–2
      postabortal 235, 304, 306, 310
      postinjectional 303–4, 306
      postoperative 303, 309–11
      postpartum 233–4, 304, 306, 310
      prophylaxis 315–16, 426
      spontaneous/non-traumatic 311
   superficial 290–7
   synergic/synergistic 239–40, 296–7
      progressive (Meleney's gangrene) 104–5, 208–9, 239, 296–7
Gangrenous stomatitis (cancrum oris, noma) 240, 257, 295–6
Gardnerella vaginalis in genitourinary infection 226, 230–2
Gas capture system with blood cultures 335
Gas gangrene *see* Gangrene
Gas-liquid chromatography *see* Chromatographic techniques
Gas-producing envelopes *see* Disposable gas generator
Gastric cancer/microflora/pH etc. *see* Stomach
Gastritis, chronic atrophic 452–4
Gastrointestinal tract *see also specific portions*
   antibacterial substances secreted by 345
   function, intestinal flora and their role in 344
GC content and GC rich regions
   Bifidobacterium spp. 157
   Clostridium spp. 54
   Treponema spp. 115
Gender, skin microflora and 288
Genes
   cloning *see* Cloning
   transfer 44–51
Genetics of anaerobes 38–61
Genital lesions/ulcers 229, 238–9, *see also* Necrotizing dermogenital syndrome
   in men 238–9
   treponemes and 128
   in women 229, 239
Genitalia, external, infections, 228–9, 312
Genitourinary tract *see also* Urinary tract
   abdominal infections and the 207
   female *see* Females
   infections 224–44, 425
   male *see* Males
   microflora 171–6, 225

   specimen isolation from 181
Gingiva, microflora 170–1
Gingivitis
   acute ulcerative necrotizing 122, 257
   experimental 256
Glucose fermentation 26–7
β-Glucosidase 396
β-Glucuronidase 396, 406
Glucuronide hydrolysis 396, 406
Glutathione, oxygen tolerance and the role of 21
Glutathione dehydrogenase, oxygen tolerance and the role of 21
Glycoside hydrolysis, products of 396
Gracilicutes, classification 8
Gram-negative (anaerobes)
   bacilli *see* Bacilli
   cocci *see* Cocci
   identification, provisional 190–2
   non-sporing
      as intestinal microflora 165
      rods *see* Rods
   oral 246
Gram-positive (anaerobes)
   bacilli *see* Bacilli
   cocci *see* Cocci
   identification, provisional 190–2
   non-sporing
      as intestinal microflora 165
      rods *see* Rods
   oral 246
Gram stain 11, 183, 195
   classification based on 11–3
      Clostridium spp. 88
   method 195
Growth factors as virulence determinants 75
Gut *see* Intestine
Gynaecological surgery, infection following 237

Haber-Weiss reaction 19
Habit, anaerobic 11
Habitat *see* Ecology
Haemolysis (and haemolysins) 191
   Clostridium spp. 92
Haemorrhagic jejunitis, necrotizing 354–5
Head
   antimicrobial therapy 425
   specimen isolation from 181
Helical morphology, bacteria with 109
Helicobacter pylori 402–3
History
   of anaerobe science 1–2
   of classification of anaerobes 6
HIV and syphilis, coinfection 121
Holding jars 186

Homosexual men, intestinal spirochaetosis in 126–7
Host determinants
   abdominal infection 205–6
   gas gangrene 305–6
Human immunodeficiency virus and syphilis, coinfection 121
Humidity, skin microflora and 288
Hungate (roll-tube) technique 115–6, 186
Hydrocarbons, polycyclic aromatic 396, 409
Hydrogen (H_2) gas
   for growth in anaerobic jars 9, 184–5
   release 29, see also Proton-motive force
   transfer, in anaerobic communities 32
Hydrogen-consuming partner organism (hydrogenotroph) 32
Hydrogen ions (H^+) in redox reactions 22
Hydrogen-producing fermentative anaerobe (hydrogenogen) 32
Hydrogen peroxide
   formation in biological systems 19
   toxicity 19, 21
Hydrogenase 30
Hydrolytic enzymes
   intestinal microflora 168
   as virulence determinants 75, 78
Hydroxyl free radical (HO·), toxicity 19
Hyperbaric oxygen, for gas gangrene 316
Hypochlorhydria, gastric 405

Identification/diagnosis, laboratory 10–14, 31, 73–4, 180–196, see also Isolation
   Actinomyces spp. 141–4
   Arachnia spp. 141–4
   in bacteraemia 333–5
   Bacteroides spp. 64, 68, 73–4, 192
   Bifidobacterium spp. 154–7
   Clostridium spp. 93–5, 193, 313–5, 364–8, 374–5, 390
   cocci 103–4
   commercial kits 193
   Fusobacterium spp. 73–4, 193
   in oral and dental infections 253–4
   paper discs for 190, 194, 421
   spirochaetes/treponemes 118–20, 129
Imipenem, minimum inhibitory concentrations 417
Immunization, for tetanus
   active 382–3
   passive 380, 382–3
Immunological assays, frq enterotoxin 367–8
Immunological response
   Actinomyces spp. 145–6
   to intestinal pathogens 345
Incubation 186–7, 334
   blood cultures in bacteraemia 334
   length of 186–7
Infants/neonates
   botulism 391
   intestinal microflora 199
   necrotizing enterocolitis 357–8
   tenanus 383–4
Infections (and disease) 180–429, see also Sepsis and specific infections/diseases
   abdominal see Abdomen
   Actinomyces and *Arachnia* spp. 144–6
   antimicrobial therapy 424–6
   Bacteroides spp., pathogenesis 74–5, 77–8
   brain 274, 276–80, 425
   central nervous system 276–80, 425
   coccal 104–5
      diagnosis 104
      treatment 105
   dental/periodontal see Dental disease; Periodontal disease
   diagnosis see Identification/diagnosis
   ear, nose and throat see Ear, nose and throat
   Fusobacterium spp., pathogenesis 74–7
   genitourinary tract 224–44, 425
   intestinal 351–71
   mixed 203–4, 239–40, 248, 292–3, 301, 330–2
   oral see Oral disease
   respiratory 268–86, 425
   spirochaetal 120–8
   superficial (skin etc.) 287–98
   treponemal 120–1
   wound (surgical) see Wound
Inflammatory bowel disease 358–9
Injection, gas gangrene following 303–4, 306
Inoculum, for antibiotic sensitivity tests 422
Interference (antagonism), bacterial 159, 289, 344–7, see also Colonization resistance
Intestines 162–9, 343–50, see also specific portions
   cancer 406–10, see also Colorectal cancer
   colonization resistance 159–60, 167, 343–50
   infections 351–71
   inflammatory disease 358–9
   metaplasia 402
   microflora in 157–60, 162–9, 343–7
      anaerobic, protective effects 345–7
      geographical variations 168–9
      in infants 199
      metabolism 168
   preoperative preparation, infection incidence related to 213–5
   spirochaetosis 124–8
Intracranial infections 274–80
Intrauterine contraceptive devices, infections with 235–6
Ischaemic ulcers 291–2

Isolation 181–90, *see also* Identification
   Actinomyces spp. 139–41
   Arachnia spp. 139–41
   in bacteraemia 333–5
   Bacteroides spp. 71–3
   Bifidobacterium spp. 153–4
   Clostridium spp. 90–3, 313–5, 364–5, 374–5
   cocci 101
   Fusobacterium spp. 71–3
   media for *see* Media
   procedures 189–90
   sites on body suitable for 181
   spirochaetes 115–8
Isozymes, *Bifidobacterium* spp. 156

Jars
   anaerobic 9, 184–5
      antibiotic sensitivity testing and 423
      for *Clostridium* spp. 90–1
      failure, investigation/checking for 185
      maintenance 185
      for spirochaetes 116
      types 9, 184–5
   holding 186
Jaundice in bacteraemia 326
Jaws, osteomyelitis 261

Kanamycin agar 188
Kanamycin-vancomycin agar 188
Kopeloff method of Gram's staining 195

Laboratory diagnosis *see* Identification/diagnosis
β-Lactam agents 416
   abdominal infection and 202–3
   protection from 202–3
   resistance to 420
β-Lactamases
   abdominal infection and 202–3
   tonsillitis and 270
Lactic acid metabolism 26–7
   Lactobacillus acidophilus 173
Lactobacillus spp.
   acidophilus 173
   Bifidobacterium spp. and, differentiation 152–3
   in colonization resistance 347–8
   as gingival microflora 170
   as vaginal microflora 172–3
Large bowel cancer 406–10, *see also* Colorectal cancer
Latamoxef, minimum inhibitory concentrations 417
Lavage, peritoneal 211
Lecithinase reaction for *Clostridium* spp. 92
Leptospira spp.
   biflexa 112

   illini 112
   interrogans 112
Leptospiraceae, classification 111–2
Limb surgery, gas gangrene following 303
Lincomycin 416–8
Lipase reaction for *Clostridium* spp. 92
Lipids
   skin, skin microflora and 288
   spirochaetal/treponemal 119
Lipopolysaccharide (LPS, endotoxin)
   in abdominal sepsis 202
   as virulence determinant 75, 77
Lithocholic acid, colorectal cancer and 409
Ludwig's angina 260
Luminal contents in gastric atrophic gastritis 403–4
Lung infection 280–2, 425
Lysis centrifugation 335
Lysis filtration 335

Macrolides 418
Males, genitourinary tract in
   infections 237–41
   microflora 172, 176, 225
Maxillary sinus infection 261, 275
Media, culture 9–10, 187–9
   Actinomyces spp. 135, 139–40
   for antibiotic sensitivity tests 190, 194, 422
   Arachnia spp. 139–40
   in bacteraemia 333–4
   Bifidobacterium spp. 153–4
   Clostridium spp. 91–2, 364–5, 374–5
   cocci 101
   pre-reduced anaerobically sterilized 186
   selective 187–9, 194–5
   spirochaetes 116–7
   supplements 187
Meleney's (progressive synergistic) gangrene 104–5, 208–9, 239, 296–7
Men *see* Males
Mendosicutes, classification 8
Meningitis 274
Menstrual cycle, vaginal microflora variations with 175–6
Metabolism 16–37, *see also specific areas of metabolism*
   Actinomyces spp. 134–6
   Arachnia spp. 134–6
   cancer and 395–414
   cocci 101
   intestinal microflora 168
   Lactobacillus acidophilus 173
   spirochaete 118–9
Metabolites as virulence determinants 75
Metaplasia, intestinal 402

Methanobacteriaceae, classification 8
Methanobacteriales, classification 8
Methanococcaceae, classification 8
Methanococcales, classification 8
Methanomicrobiaceae, classification 8
Methanomicrobiales, classification 8
Methanosarcinaceae, classification 8
Methionine metabolites 396
Metronidazole 419–20
   for abdominal infection 210–11
   for bacteraemia/endocarditis 332–3, 336
      prophylactic 332–3
   for botulism 391
   for brain abscesses 280
   for gas gangrene prophylaxis 316
   for genitourinary infection 232
   minimum inhibitory concentrations 417
   for oral and dental infection 254
   reduction 31–2
   resistance to 420
   for tetanus 380
Micrococcaceae, classification 7
Micrococcales, classification 7–8
Microflora
   metabolism
      cancer and 395–411
      intestinal 168
   normal 162–79, 199–201, 225, 245–7, 288–9, 343–4
Microsystems for antibiotic sensitivity testing 421
Microtubules, spirochaete 109
Mobiluncus spp., genitourinary infection 227, 230–2
Monoclonal antibodies, to *C.perfringens* 367–8
Morphology
   C.botulinum 385
   cell 11–12
      Actinomyces spp. 133–4
      Arachnia spp. 133–4
      Clostridium spp. 87–8, 94–5
      cocci 101
      helical, bacteria with 109
      spiral, bacteria with 109
      spirochaete 109–10, 118
   colony 190–1
      Actinomyces spp. 133–4
      Arachnia spp. 133–4
      C.botulinum 385
Motility, spirochaete 109
Mouth *see* Oral cavity
Mucin, colonic, *Bifidobacterium* spp. degrading 160
Multiple sclerosis, treponemes and 128
Murein
   Actinomyces spp. 137
   Arachnia spp. 137
   Bifidobacterium spp. 156–7

Muscle *see also* Neuromuscular junction
   clostridial wound infection related to changes in 309
   relaxants, in tetanus management 379–80
Mutagens, production 395–400
Myonecrosis, clostridial *see* Gangrene
Myositis, streptococcal, clostridial cellulitis and myonecrosis compared with 308

NADH-ferredoxin reductase 29
Nalidixic acid-containing agar 188
Neck
   infection *see* Ear, nose and throat infection
   specimen isolation from 181
Necrobacillosis 269, 329–30
Necrosis
   myo-, clostridial *see* Gangrene
   skin/superficial sites 287–98
Necrotizing dermogenital syndrome (Fournier's gangrene) 239–40, 297
Necrotizing enterocolitis, neonatal 357–8
Necrotizing fasciitis, anaerobic 208–9
Necrotizing gingivitis, acute ulcerative 122, 257
Necrotizing haemorrhagic jejunitis 354–5
Neisseriaceae, classification 7
Neomycin agar 187–8
Neomycin-vancomycin agar 188
Neonates *see* Infants
Neoplastic disease, malignant *see* Cancer/malignant disease
Nervous system
   central *see* Central nervous system
   peripheral, botulinum toxin effects on 387–8
Neuromuscular junction, botulinum toxin effects at 387–8
Neuropathy, diabetic 292
Neurotoxins, systemic disease caused by 372–94
Neurotransmitters, tetanus toxin and 376
Neutropenic enterocolitis 355–6
Nitrate reduction 31
Nitrite, inhibition by/sensitivity to 31
Nitro compounds, aliphatic and aryl, reduction 31
Nitro free radical 32
Nitrofurantoin 419
Nitrogen metabolism, products of 396–7
Nitrogenase 30
Nitroimidazoles 419, *see also* Metronidazole
   reduction 31–2
   treatment with
      for abdominal infection 210
      for oral and dental infection 254
N-Nitroso compounds, production 397–8, 404, 406, 411
Noma 240, 257, 295–6

Nose *see* Ear, nose and throat infections
Nucleic acids *see also* DNA; RNA
   Clostridium spp. 89–90
   spirochaetal 111, 119–20
Nutrition/diet
   bacterial antagonism and 345
   colorectal cancer and 407–8
   intestinal flora and, relationship between 163

Oedema, wound 308
Oral cavity
   Actinomyces spp. 145
   microflora 145, 169–71
Oral disease/infection 121–4, 245–67
   pathogenesis 248
   source of organisms in 247–8
   spirochaetes in 121–4
Oropharyngeal infections 269–71
Osteomyelitis of jaws 261
Otitis media 271–3
   acute 272
   chronic suppurative 272
   secretory 272
   therapy 272–3
Oxidation-reduction *see* Redox
Oxygen *see also* Superoxide anion
   hyperbaric, for gas gangrene 316
   metabolism in absence of 29–32
   sensitivity/toxicity 17–22
     causes 18–9
     protection against 20–2
Oxygen tension, oral microflora and 246–7

Pain development in wound area 308
pAMβ 1 49, 50–51
Parametritis 236
Paranasal air sinuses, infection 273–4
   sequelae 274–5
Parapharyngeal abscesses 270–1
Paraplegia, bladder cancer risk 410
Parvobacteriaceae, classification 7
Pathogenicity *see* Infections
Patristic classification 5
pBF4 42–3
pBFTM10 42–3, 48
   conjugal transfer 48
pBR7 50
pCB101 50
pCL2 41–2
pCP1 42
pCW2/pCW3 33
pDP1/pDP1.1 51
pE5-2 51
Pelvic infections in non-pregnant women 235–7

Penicillin
   for bacteraemia prophylaxis 332–3
   for oral and dental infection 254
   resistance to 420
   for tetanus 380, 382–3
     prophylactic 382–3
   for treponematoses 120
Peptic ulcers, gastric cancer and 404
Peptococcaceae 329
   in bacteraemia 329
   classification 8, 102
Peptococcus spp.
   in bacteraemia 329
   identification, provisional 101, 191
   morphology/metabolism/culture 101
   Peptostreptococcus spp. and, differences between
     102–3
   as vaginal microflora 173–4
Peptostreptococcus spp.
   in bacteraemia 329
   identification, provisional 101, 191
   morphology/metabolism/culture 101
   Peptococcus spp. and, differences between 102–3
   as vaginal microflora 173–4
Perimandibular space infections 260
Perineum, infections 228–9, 312
Periodontal disease 254–60, *see also* Dental disease
   Actinomyces spp. in 145–6
   indicators 259–60
   pathogenic mechanisms 260
   spirochaetes in 121–4, 257–8
   types 255
Periodontitis 256–9
   adult 255, 257–9
   juvenile 255–7
Peripheral nervous system, botulinum toxin effects
   on 387–8
Peritoneal abscesses, intra- 209
Peritoneal lavage 211
Peritonsillar abscess 269–70
Peroxidases 21
pFD173/pFD176 49
pGD10 48
pH
   gastric 403
   medium, for antibiotic sensitivity tests 422
Phages *see* Bacteriophages
Phagocytosis of aerobic bacteria, anaerobes inhibiting
   203–4
Pharyngeal infections 269–71
pHB101 44
Phenetic classification 5
Phenols, urinary volatile, production 396–7, 408
Phosphorylation

photo-, anoxygenic 2
substrate level (SLP) 25–7
Photophosphorylation, anoxygenic 2
Physiology
Actinomyces spp. 134–6
Arachnia spp. 134–6
pIB143 49
pIB191 49
Pig-Bel 354–5
Pigmentation, colony 190–1
Pilin genes, *Bacteroides* spp. 52–3
Pillotinas 112
pIP136 48
pIP401/pIP402 43
Piperidine production 398
Pitting of agar 191
pJU plasmid series 49–50
Plaque, dental, microflora in 170–1
Plasmids 39–44, 46–7
antibiotic resistance 42–4
bacteriocinogenic 40–1, 345
cryptic 39–40
toxigenic 41–2
Plectridiaceae, classification 7
Plectridiales, classification 7
Pleuropulmonary infection 280–2, 425
Pneumonia
aspiration 281
chronic destructive 281
Poisoning, food *see* Food poisoning
Polycyclic aromatic hydrocarbons 396, 409
Polysaccharides
C.perfringens, serological typing 364–5
degradation (saccharolysis) 25
by intestinal microflora 168
Porphyrins, *Bacteroides* spp. 69–70
Port-a-Cul tube 182
Postoperative infection *see* Surgery
Postpartum infections 233–4, 304, 306, 310
Pregnancy
genitourinary infections related to 232–5
tetanus prevention in, neonatal 383
vaginal microflora variations with 176
Preoperative preparations, infection incidence
related to 213–5
Presynaptic block, tetanus toxin and 376
Prokaryotae, classification 8
Promoters of carcinogenesis, production 395–400
Propionibacterium spp.
acnes 169, 289, 332
avidum 169
granulosum 169
as skin flora 169, 289
skin infection and 289

Prostatitis 241
Proteins
Actinomyces spp. 138
Bifidobacterium spp. 156
fermentation by intestinal microflora 168
spirochaetal/treponemal 119
synthesis, inhibitors 416–8, *see also* Translation
Protohaem, *Bacteroides* spp. 69–70
Proton-motive force ($\Delta\mu H^+$), generation 28–9
Protoporphyrin, *Bacteroides* spp. 69–70
pSS-2 51
Puerperal infections 233–4, 304, 306, 310
Pulmonary infection 280–2, 425
Purification of bacteria *see* Isolation
Pus
specimen isolation from 182
transport 182
Pustules 289
Pyometra 236
Pyrrolidine production 398
Pyruvate metabolism 26–8

Quinolones 418–9

R-factors 42–4
R751 43
Radiology, gas gangrene 312–3, *see also* Computed tomography
Radiometric methods in bacteraemia 335
Rec-A-dependent DNA repair system of *E.coli* 21
Receptors
botulinum toxin and 387–8
ganglioside 377–8
tetanus toxin and 377
Rectal cancer *see* Colorectal cancer
Redox balance, maintenance 27
Redox potential (reduction-oxidation potential, Eh) 18, 22–4
bacterial synergy and 203
host factors reducing 205
indicators/measurement/values 23–4
of media, reduced, for *Clostridia* spp. 91
oral microflora and 247
wounded tissue, infection related to 305–6
Reproductive cycle, vaginal microflora variations with 175–6
Resazurin 24, 186
Respiration, anaerobic 25
Respiratory infections 268–86, 425
Retropharyngeal abscesses 270–1
Reversed passive latex agglutination for enterotoxin detection 367
R-factors 42–4

Ribosomal RNA sequence comparisons/homologies
  Clostridium spp. 90
  spirochaetal 111, 120
Rifampicin agar 189
Ristellaceae, classification 7
RNA
  sequence comparisons
    Clostridium spp. 90
    spirochaetal 111, 120
  translation, *Clostridium* spp. 55, *see also* Proteins
Robertson's cooked meat broth for *Clostridium* spp. 91
Rods
  gram-negative non-sporing
    biochemical/chemical characteristics used in differentiating 64–5
    classification 12, 64–5
    identification 72
  gram-positive non-sporing
    classification 13–4
    identification 192
    as intestinal microflora 165–6
  spore-forming
    identification 192
    as intestinal microflora 166–7
  systematics, techniques used to study 74
Root canal infections 249–50

Saccharolysis *see* Polysaccharides
Saccharomyces boulardii, in antibiotic-associated diarrhoea prevention 347–8
Saliva, microflora 169–70
Salpingitis 236–7
Schistosomiasis, bladder cancer risk 410
Schizomycetes 6
  anaerobic members 7
Sclerosis, multiple, treponemes and 128
Scotobacteria, classification 8
Scrotum, gas gangrene 312
Secretions, skin microflora and 288
Selective media 187–9, 194–5
Sepsis *see also* Infection
  abdominal *see* Abdomen
  uterine 234–5, 304, 306, 310–11
Septicaemia, clostridial 327–9
Serological tests/typing
  Actinomyces and *Arachnia* spp. 144–5
  Clostridium spp. 88–9, 94, 364–5, 385
  spirochaetes 118, 120
Sex, skin microflora and 288
Sf1Ep cells, *T.pallidum* and, coculture 118
Shigella flexneri, colonization resistance to 346
Shuttle vectors for cloning 49–51
Sinus infection (sinusitis)
  intracranial venous 275–6
  maxillary 261, 275
  paranasal air *see* Paranasal air sinuses
Skin 287–98, *see also* Superficial sites
  anatomy 287–8
  infections (ulceration and necrosis) 287–98
  microflora 169, 288
Skin erythema test for enterotoxin 367
Small bowel, microflora in 199–200
Sodium ion pumps, decarboxylases as 29
Soft tissue infection, treatment 425–6
Specimens
  examination, direct 183
  identification *see* Identification
  isolation *see* Isolation
  in oral and dental infections, handling 253–4
  transport 182–3
  types 181–2
Spherophoraceae, classification 7
Spinal cord, presynaptic block of inhibitory pathway in, tetanus toxin and 376
Spiral morphology, bacteria with 109
Spirallales, classification 7
Spirochaeta spp. 112–3
  plicatalis 112
Spirochaetaceae, classification 8, 111–2
Spirochaetales
  classification 8, 111–5
  disease caused by 120–8
Spirochaetes 108–32
  classification 111–5, 128–9
  culture 114–6, 118–9
  ecology 110–11
  identification 118–20, 129
  as intestinal microflora 167
  isolation 115–18
  morphology and ultrastructure 109–10, 118
  in oral and dental disease 121–4, 257–8
Spirochaetoses 120–8
Spore(s)
  C.botulinum 384–5, 391
  C.tetani 373
  classification based on detection/non-detection 11–12
Sporeforming anaerobes
  classification 7
  as intestinal microflora 166–7
Sporovibrionales, classification 7
Sporulales, classification 7
Sporulation and enterotoxin production, in *C.perfringens* 365
Steroid metabolites 398–400
Stomach 401–5, *see also* Gastritis
  cancer, incidence 401
  juice analysis 403–4

microflora 199
pH 403
Stomatitis, gangrenous (cancrum oris, noma) 240, 257, 295–6
Streptococcaceae, classification 8
Streptococcus spp.
   milleri 251–2
   mutans 248
   myositis due to, clostridial cellulitis and myonecrosis compared with 308
   oral and dental infection 248, 251–3
   oropharyngeal infection 269–70
   pyogenes 269–70
Subcutaneous tissue, anatomy 287–8
Subdural empyema 274–5
Succinate metabolism 27–8
Sulphonamides 419
Superficial sites 287–98, *see also* Skin
   specimen isolation from 181
   ulceration and necrosis 287–90
Superoxide anion ($O_2^{\cdot -}$)
   dismutation/transformation 20
   toxicity 19
Superoxide dismutase 20–1
   abdominal infection and 202
Supplements, media 187
Surgery
   for abdominal infection 210–11
   abdominal infection following 206–7, 212–6
      incidence and severity 215–6
      prophylaxis 213–6
   for brain abscesses 277–9
   gas gangrene following 303, 309–11
   gas gangrene prophylaxis in 315–6
   gas gangrene treatment in 316–7
   genitourinary infection following 237
   preparation before, infection incidence related to 213–5
   technique, infection incidence related to 213
   wound in *see* Wound
Swabs
   specimen isolation from 182
   transport 182–3
Synergic gangrene *see* Gangrene
Synergy, bacterial 202–4
Syphilis 121, 224
   HIV and, coinfection 121
Systemic antimicrobial prophylaxis 214–5
Systemic intoxication/toxigenic disease, clostridial 309, 372–94

Taxonomy, components of 4, *see also* Classification
Teicoplanin 419
Temperatures, growth, for *Clostridium* spp. 92

Tenericutes, classification 8
Terminosporaceae, classification 7
Tetanospasmin 375–6
Tetanus 373–4
   antitoxin 375–7, 379–80
   clinical diagnosis 379
   cryptogenic 378
   epidemiology 378–9
   local *vs* general 376–7
   prophylaxis 381–3
   toxin 375–7
   treatment 379–80
Tetracycline 418, 420
   resistance 42–3, 47–8
      conjugal transfer 47–8
      plasmids conferring 42–3
Tetracycline resistance-ERL 47–8
Thoracic empyema 282
Thoracic sites, specimen isolation from 181
Throat *see* Ear, nose and throat infections
Thrombosis
   cavernous sinus 275–6
   lateral sinus 276
Tissue
   soft, infection, treatment 425–6
   specimen isolation from 182
Tissue culture, for enterotoxin detection 367
Titanium (III) citrate 24
Tn4400 43
Tn43451 43
Tongue, microflora 169
Tonsillar infections 269–70
Tonsillitis 270, 425
Toxigenic plasmids 41–2
Toxin(s), *Clostridium* spp. 41–2, 88–9, 93–4, 306–7, 315, 352, 372–95, *see also* Antitoxins; Enterotoxin; Lipopolysaccharide
Transcription, clostridial 54–5
Transformation 45–6
   cloning vectors introduced via 49–50
Translation of RNA, clostridial 54, *see also* Proteins
Transmitters, neuro-, tetanus toxin and 376
Transport, specimens 182–3
Transposons, antibiotic resistance and 43
Trauma of wounding, infection determined by 305
Trench mouth (acute ulcerative necrotizing gingivitis) 122, 257
Treponema spp. 113
   characterization/typing 118–20
   culture/isolation 114–8
   denticola 123–4
   disease 120–1
   hyodysenteriae 115, 119, 127–8
   innocens 115, 119, 124–5, 128

Treponema spp. –cont'd.
   pallidum 113–14, 117–21, 224
   pallidum 'complex' 113–4
   vincentii 119, 124
Treponematoses 120–1
Trichomonas vaginalis and trichomoniasis 229–30
Trimethoprim 419
Tropical colonopathy 127
Tropical ulcer 293–5
   treponemes and 128
Tryptophan metabolites 397–8, 411
Tubes (fallopian), infection 236–7
Tubo-ovarian abscesses 237
Tumours, malignant *see* Cancer/malignant disease
Tween 80
   agar containing 188
   broth cultures containing *see* Fastidious anaerobe broth-Tween 80 cultures
Tyrosine metabolites 396–7
Tyzzer's disease 128

Ulcer(s)
   genital *see* Genital lesions/ulcers
   peptic, gastric cancer and 404
   skin/superficial sites 287–98
   tropical *see* Tropical ulcer
Ulcerative colitis 358–9
Ulcerative gingivitis, acute necrotizing 122, 257
Ultrastructure *see* Cell
Ultraviolet light, fluorescence under 183, 192
Urethra
   female, microflora 172, 176
   male
     infection 240–1
     microflora 172, 176
   strictures, bladder cancer risk 410
Urethritis 240–1
Urinary bladder cancer 410–11
   incidence 401
Urinary tract infections 228, *see also* Genitourinary tract
   urinary bladder cancer and 410–11
Urinary volatile phenols, production 396–7, 408
Uterus
   gas gangrene 235, 304, 306, 310, 317
     treatment 317
     true 310
   sepsis 234–5, 304, 306, 310–11, 317

Vagina
   infections 229–32
   microflora 172–6, 225
     quantitative and qualitative approaches to 175
   specimen isolation from 181
Vaginosis, bacterial 230–2
Vancomycin 419
   agar containing 188–9
Vectors for gene cloning 49–51
Veillonella spp.
   antibiotic susceptibility 105
   morphology/metabolism/culture 101
Veillonellaceae, classification 8
Veillonellales, classification 8
Vero cell assay for enterotoxin 367
Vibrionaceae, classification 7
Vincent's disease (acute ulcerative necrotizing gingivitis) 122, 257
Virulence
   Bacteroides spp. 74–5, 77–8
   Fusobacterium spp. 74–7
   spirochaetal (oral) 123–4
Visceral abscesses 209

Women *see* Females
Wound infection 299–323, 377–8, 391–2
   accidental, with *Clostridium* spp. 299–323, 377–8, 381–3, 391–2
   surgical 208
     with *Clostridium* spp. 377–8, 381–3
     foreign bodies predisposing to 206
Wounding, trauma of, infection determined by 305